HYDROLOGY
&
HYDRAULIC
SYSTEMS

RAM S. GUPTA, Ph.D., P.E.
Roger Williams University, Bristol, RI
Delta Engineers, Inc., Bristol, RI

WAVELAND

PRESS, INC.

Prospect Heights, Illinois

For information about this book, write or call:

    Waveland Press, Inc.
    P.O. Box 400
    Prospect Heights, Illinois 60070
    (847) 634-0081

CONTENTS

Contents

Contents

PREFACE

A course dealing with hydrologic principles and their application constitutes an essential part of the civil engineering curriculum. However, there are no standard contents adopted for this course by various universities. Universities formulate the course outline by selecting topics from the disciplines of hydrology, hydraulics, and water resources management in a proportion that depends on the emphasis placed on theory or design and on the number of courses on this subject included in the curriculum. *Hydrology and Hydraulic Systems* offers a wide selection and complete coverage of topics to fit a high degree of variability of course contents of various universities. It is a suitable textbook for at least two undergraduate- or graduate-level courses: a basic required course in hydrology and a course in hydrologic systems or structures offered by many universities separately. Supplemented by the instructor's notes, the book can be used for many other courses in the field of surface and ground water hydrology, and water resources engineering. For a one-semester course, the appropriate topics may be selected from throughout the book. Because of its features as listed below, the book will serve as a very useful reference for practicing engineers as well.

Hydrology and Hydraulic Systems provides a comprehensive understanding of all the elements involved in the development of water resources. The subject matter that follows a sequence in which resource development activities progress is tied together throughout the text. A fully up-to-date theoretical treatment has been presented covering the most recent research advances in the field of hydrology. A strong emphasis has been placed on practical applications and design. Over 300 illustrations and 150 tables have been provided along with the keys to the steps involved in the tabular computations. About 200 examples have been solved fully to facilitate understanding of the theory. In addition, over 400 exercise problems have been included for student assignment.

I wish to pay a tribute to the memory of the late Mrs. Patricia Hogan, who word-processed the entire manuscript very efficiently. I also express my appreciation to Apurv Gupta of Brown University, who helped edit some of the chapters, and to Sukirti Gupta, of the University of Rhode Island, for assisting in proofreading the manuscript and preparation of the solutions manual. My sincere thanks to Mrs. Carol Reineke, Secretary, Engineering, for extending a supporting hand from time to time in the assembly of the manuscript. I would also like to acknowledge the valuable assistance toward finalizing the text provided by the reviewers, Dr. M. L. Thatcher of Cooper Union, New York City; Dr. Raymond Wright of the University of Rhode Island, Kingston; Dr. Joseph Elmer of Water and Related Resources, Hampton, Connecticut; Professor Khalid Al-Hamdouni of Roger Williams College, Bristol, Rhode Island, and others.

DEVELOPMENT OF WATER RESOURCES

1.1 ELEMENTS OF DEVELOPMENT

Water resources are developed either to use or to control ground and surface water flows. A developmental activity for the use of water comprises studying the availability and demand of water and contemplating a project that can meet the expected needs from available supplies by means of engineering works and nonstructural measures. Water control activities are concerned with regulating the rate of water supply in order to improve conditions within an area. The purpose of the facilities provided by a project is to vary the quantity and quality, time of supply, and place of use of the resources in accordance with need. Developing a project constitutes arranging structural components, their layouts, and designs along with nonstructural measures, if any, to serve the intended purpose.

This book aims to provide a basic understanding of all the quantitative elements that are involved in the formulation of a water development project. The process of development can be explained by the following sequence of questions. The book is devoted to answering these questions.

1. How much water is needed?   For a water use project, it is essential to know the current requirements as well as to predict the requirements for the duration of the plan period for all of the anticipated uses. This is dealt with in Chapter 2.

2. How much water is available?   This question reflects the prime objective of the study of hydrology—a science related to the occurrence and distribution of natural water on the earth. Reliable estimates of fresh surface water and groundwater entering a basin and their spatial and temporal distributions are the most difficult of the problems to answer. In this book we study comprehensively all aspects of hydrology. This includes the fundamentals of natural water circulation through the hydrologic cycle and discussion of the elements of the cycle (Chapter 3), the theory and assessment of groundwater supplies (Chapters 4 and 5), the field measurements of streamflows (Chapter 6), and the assessment of surface water supplies with their distribution (Chapters 7 and 8). Estimation of peak flows (Chapters 8 and 12) is specifically important from the point of view of the control of water.

3. How are the requirements satisfied by the supplies? The formulation of a project is the answer to this question. The magnitude and distribution of supplies (Chapter 7) are compared with the variation in demands (Chapter 2). Under the condition when the supplies always exceed the demands, a run-of-river project (direct withdrawal from the source) is indicated. However, a storage of adequate size (Chapter 9) is needed in the case of seasonal or overall water deficiencies. Dams and control structures (Chapter 9) are designed to store as well as regulate water flows. Other facilities depend on the purpose to be served by the project. For example, a water supply project needs pumps and conveyance (channel) and distribution (pipe) systems; an irrigation project requires conveyance (canal) and drainage systems; a hydropower project includes channels and penstocks (conduits); a flood control project may require embankments, channel improvements, drainage works, and nonstructural measures; and a navigation project involves improvements to existing channels as well as construction of artificial channels. The basic hydraulic structures listed above under the categories of conveyance system or channels (Chapter 10), distribution system or pipes (Chapter 11), pumps (Chapter 11), and drainage system (Chapter 12) are included in the book. In addition, there are certain specialized facilities such as treatment plants for water supply, headworks and outlets for irrigation, powerhouses for hydropower, detention basins for flood control, and shiplocks for navigation. These structures are not covered in the book.

A development project may refer to a single unit, a regional development, or a basinwide development. It may be for a single purpose or may involve multipurpose uses.

4. How is the used-up water disposed of? The used-up waters or excess stormflows from urban areas, irrigated lands, highways, and airports are disposed of through the drainage system. Chapter 12 deals with this.

In addition to the questions above there are other questions that relate to the quality of water, such as: What type of water is needed? What kind of water is available? What treatment is required? What are the quality and effects of the wastewater drained into a river system? These are equally important questions in water resources development; however, the qualitative aspect is a distinct matter that is not included in the scope of this book.

It should be recognized that the planning and design studies provide only the conceptual framework of a water-development project. The complete process of development includes the construction, operation, and management of the project as well.

1.2 OBJECTIVES OF DEVELOPMENT

The objectives depend on the level of planning. Although the national objective can be as broad as "improving the quality of life of the people," a project component is designed with the narrow objective of serving the maximum usefulness. For water resources projects having federal involvement, there has been a continuous shift in the objectives. The Flood Control Act of 1936 established that the "benefits to whomsoever they may accrue" should be in excess of the estimated cost. The U.S. Inter-Agency Committee on Water Resources (1950), while recognizing the ultimate aim "to satisfy human needs and desires," iterated the economic criteria of benefits and costs for judging a project. In the late 1960s, a significant shift occurred in the national policy in the wake of the environmental movement.

1.2.1 The Concept of Multiobjective Planning

By an act of the Congress, the Water Resources Council was created in 1965 to establish the principles, standards, and procedures for formulation and evaluation of federal and federally assisted water and related land resources projects and to coordinate developmental activities in this field in the country. The principles provide the broad framework for the planning activities, whereas the standards provide for uniformity and consistency in comparing, measuring, and judging alternative plans (U.S. WRC, 1979). The procedures are developed within the framework of the principles and standards. The WRC suggested the following four objectives as an outcome of the recommendations of a 1969 task force.

1. To enhance national economic development by increasing the output of goods and services.
2. To enhance the quality of the environment by improving the quality of the natural and cultural resources and the ecological system.
3. To enhance social well-being through developmental activities.
4. To enhance regional development through increase in region's income, employment, economic base, and other components of the regional objective.

The first edition of the "Principles and Standards" of the WRC in 1973 retained only the first two of these objectives [i.e., national economic development (NED) and enhancement of environment quality (EQ)]. Four accounts were established to display completely all relevant beneficial and adverse effects of each alternative project proposal. The effects of a plan proposal under each account were reflected by the differences between the projected conditions "with the proposed plan" and projected conditions "without the plan." The four accounts were (1) the national economic development (NED) account to indicate the effects of a plan on the national economy, (2) the environmental quality (EQ) account to show the environmental impacts in quantitative and qualitative terms, (3) the regional economic development (RED) account to bring out the regional effects on income distribution and employment opportunity, and (4) the other social effects (OSE) account to demonstrate urban and community impacts and effects on life, health, and safety of people. From among the alternative projects, a project was selected based on an evaluation of the trade-offs between the objectives of NED and EQ and considering, where appropriate, the effects of the projects on regional economic development (RED) and other social effects (OSE) (U.S. WRC, 1973).

Over the period of time, some aspects of the "Principles and Standards" were considered to be an impediment to the development of water resources. In 1983, the new "Principles and Guidelines" (P&G) have been enacted to replace the previous "Principles and Standards" (P&S). These guidelines establish standards and procedures for use by federal agencies in formulating and evaluating alternative plans according to the stated principles. The guidelines do not create any procedural rights for private parties and departures from the guidelines are permitted in federal individual studies, if justified. The single objective retained in the new guidelines is "to contribute to the national economic development (NED) consistent with protecting the nation's environment pursuant to national environmental statutes" (U.S. WRC, 1982).

There continue to be four accounts, as stated earlier. However, the NED account is the only required account for a project. Other information having a material bearing on the decision-making process is included in the other accounts (EQ, RED, and OSE) or in some other appropriate format.

From among alternative projects, a project with the greatest net economic benefits is selected as the NED plan. Other proposals are also considered which reduce net NED benefits in order to contribute to environmental quality, regional development, and social benefits or to address further other federal, state, local, and international concerns not fully accommodated by the NED plan. Normally, the NED plan is accepted for implementation. However, another project is recommended when the benefits from other effects outweigh the corresponding NED losses.

1.3 PURPOSES OF DEVELOPMENT

Water resources are developed to fulfill one or more of the following purposes or functions. Subcategorization of the purpose and/or means by which the purpose is served are also indicated under each purpose.

1. Municipal and industrial water supply
   a. Surface water development
   b. Groundwater development
   c. Water desalination
2. Irrigation
   a. Crop growth
   b. Land reclamation
   c. Salinity control
3. Hydropower generation
   a. Run-of-river and storage projects
   b. Hydropower from pumped storage
   c. Tidal power
4. Flood control
   a. Reservoirs and detention basins
   b. Embankments or dikes

     c. Flood diversion
     d. River channel improvements
     e. Watershed management
     f. Flood zoning
     g. Flood forecasting and warnings
5. Water transport
     a. Navigation
     b. Port and harbor
6. Soil conservation
     a. Erosion control
     b. Sedimentation control
     c. Watershed management
7. Pollution control
     a. Prevention of saltwater intrusion
     b. Wastewater dilution
8. Scenic preservation
     a. Fish and wildlife
     b. Recreation
     c. Public health and sanitation

1.4 THE PROCESS OF PROJECT FORMULATION

As stated earlier, the development process comprises formulation, construction, and operation of a project. The project formulation is a very important step in this exercise, which covers the planning and design of structural or nonstructural facilities that constitute a project. The following six major steps are involved in the project formulation. Proceeding from one step to the other might suggest the necessity of new data or some revisions to be made in previous computations; this involves moving back and forth as well as iterating steps. As such, it is a dynamic process, as shown in Figure 1.1.

1. *Problem identification:* determination of the need for a project in the context of use/control of water, market demand for the product, and political incentive.

2. *Solution identification:* data collection, inventory of all resources, analysis of data, and developing solutions.

3. *Definition of alternatives:* formulation of alternative systems of structural and nonstructural measures in the light of constraints on resources.

4. *Evaluation of alternative projects:* performing economic, financial, environmental, and social analyses.

5. *Comparison of alternative projects:* displaying all relevant beneficial and adverse effects of each proposal under the NED account; indicating other significant effects either under three other accounts or in another appropriate format.

6. *Selection of a project:* picking up a project based on the NED objective of the optimum net economic benefit consistent with protecting the nation's environment.

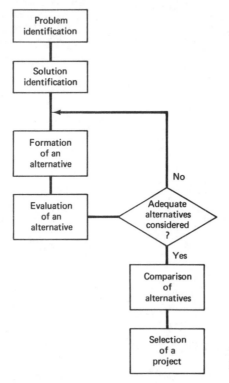

Figure 1.1 Project formulation process.

1.4.1 Developing a Multipurpose Project

Step 3 in the preceding section is the key to the formulation process wherein project elements are assembled together to assume the shape of a project proposal. The formulation of a multipurpose project starts from the nucleus of a single-purpose project. A nucleus representing the only purpose of a single-purpose project or the purpose of highest priority for a multipurpose project is selected. Elements essential to the needs of the nucleus are included and designed of a size proper for optimum development of the resource. To this nucleus project, an increment that may be a purpose or a portion of any aspect of potential development is added along with the associated project elements. The process is repeated to add increments corresponding to all other purposes. The combination that yields the highest increase in net benefits becomes the initial project. This incremental analysis may be performed by the systems approach.

A sufficient number of alternative plans are prepared by considering relevant alternative means of meeting the needs of each of the purposes identified. The alternative plans should be significantly differentiated from each other. Each alternative plan should include mitigation measures determined to be appropriate for protection of the environment. In the formulation of alternative plans an effort is made to include only those increments that provide net beneficial NED effects. However, incremental measures that do not provide net NED benefits may be included if they are cost-effective measures for addressing other federal, state, local, and international concerns not fully covered by the NED objective (U.S. WRC, 1982).

1.5 STAGES IN PROJECT FORMULATION

1.5.1 Studies for a Single Project

Generally, the project formulation process is carried out in three stages. Each successive level of study is more intense based on the results of the previous level. A project is dropped if not considered promising enough at any stage. The sequence of the study is as follows.

1. Reconnaissance or pre-feasibility study. This consists of a preliminary study based on the existing conditions to identify whether a feasible project is expected and if so, what further investigations are warranted. Office studies are made using the summarized data, previous reports, and maps. Office studies are supplemented by a minimum level of field work with respect to topographic surveys, streamflow measurements, and soil and geologic investigations to update the data and to confirm the estimates of the office studies. Preliminary benefit-cost and financial analyses and environmental and social evaluations of the various identified solutions are made. If the results are favorable, the study on more promising alternatives is carried forward to the next stage.

2. Feasibility study. The level of this study is in sufficient detail to make a decision on whether or not to implement a project. This is based on a detailed study of all data, field investigations, preparation of engineering designs, and analyses of economic and financial returns. A feasibility report includes the following:

1. Description of project features for a number of alternatives in sufficient detail
2. Estimates of the cost of project alternatives
3. Environmental impact analysis of alternatives
4. Economic and financial analyses of alternatives
5. Discussion of selection criteria and final recommendation for the project
6. Analysis of construction methods
7. Construction scheduling
8. Financial projections of costs and revenues and sources of financing
9. Institutional and other requirements

3. Final Design of the Project. This study leads to the construction of a project. For the selected alternative, the designs of each component and drawings and specifications are prepared in sufficient detail to obtain bids from the contractors. Often, a considerable period (even a decade or more) has elapsed since the feasibility study had been completed. Revisions are made to bring the project up to date, including adjustments in the financial aspects to secure funding for the project.

1.5.2 Stages in Regional Studies

The regional or basin studies, which involve a large number of projects, are categorized on the basis of the emphasis on the regional needs and problems and the extent of focus on individual projects. The U.S. Water Resources Council (1973) has proposed the following three-stage classification for regional planning.

1. Level A plan or framework studies. These studies are of a general nature to assess the extent of problems and needs of a region. Components of the individual projects are not identified at this level of study. These contain recommendations for plans and programs that can be dropped from further consideration and the additional studies that are needed for other programs. Such studies are suitable for broad-based long-term planning.

2. Level B plan or regional/river basin plan. These studies are made at the reconnaissance level and somewhat beyond to resolve complex long-range problems identified by level A studies. The studies identify the feasible and worthwhile projects to be included in a regional or river basin system. These studies consider the effects of alternative projects in multipurpose and multiunit settings and their possible trade-offs. The features of the projects included in the plan are specified. These studies are addressed to the middle-term needs — 15 to 25 years.

3. Level C plan or implementation studies. Out of the projects included in the basin (level B) plan, the feasibility studies of the projects that are to be implemented in the short term (10 to 15 years) are completed at this stage. These studies are undertaken by a single federal, state, or local authority for the purpose of authorization or development of plan implementation.

1.6 PRINCIPLES OF PROJECT ANALYSIS

One of the important problems of project formulation is to choose among alternative proposals. The basis of this selection is that of financial and economic analyses with trade-offs made in the light of environmental, regional, and social goals. In addition to economic and financial analyses, other analyses for water development projects include (1) the sensitivity and risk analyses, and (2) the allocation of cost among several functions of a multipurpose project. The principle features of the financial and economic analyses are described below.

1.6.1 Financial Analysis

The financial analysis provides a complete picture of the monetary condition of a project. The yearly expenditure during the construction of a project is compared with the corresponding yearly revenue from the project. If the revenues do not fully cover the expenses, either subsidies are needed or the deficits have to be covered from borrowed funds. In the latter case, repayment of loans with interest should be included under the expenses. The length of time for capital recovery, or the break-even point when the cumulative net revenue equals the cumulative cost of the construction and operation of a project, is determined from an annual analysis of the expense and income from the project. The analysis is continued beyond the break-even point to ascertain the financial return offered by the project. Even for a public-sector project, it is desirable that the break-even point be shorter than the useful life in order for a project not to become a public liability. Appropriate interest rates and inflation in prices are included in the financial analysis.

1.6.2 Economic Analysis

An economic analysis comprises comparing the economic costs and the economic benefits of the project. Generally, the comparison is made on an annual basis. The annual costs include annual payments for amortization of the total capital cost over the useful life of the project and annual operation and maintenance costs. In addition, there may be some adverse effects of a project, such as the pollution from industrial waste of a water supply project, that are included as the external costs (negative externalities) to make up the total economic cost.

Benefits are not the same as revenues or incomes from a project since the actual benefits are measured by the value of contribution to the national economic development (NED) objective. In addition, there may be certain other beneficial effects, such as facilities of recreation offered by a water supply reservoir, which are accounted for as the external benefits (positive externalities). Benefits are measured for the increment in output of goods and services and for the increase in its value over the output available without the project. The common measurement standards of the value of benefits in different kinds of development are (1) the willingness of users to pay in a water supply development, (2) the change in the net income of farmers in an irrigation development, (3) the cost of a most likely alternative in a hydroelectric development and for a navigation project, (4) the reduction in damages by a flood control project, and (5) an administratively assigned value (value of person-days spent at the site) for a recreation project. Usually, the economic analysis is made in terms of prices prevailing at the time of project planning. The analysis to compare project alternatives may be performed in several ways as follows:

1. Determine the net benefits, which is the economic benefits (B) minus the economic costs (C).
2. Determine the benefit-cost (B/C) ratio, which is the ratio of the economic benefits to the economic costs.
3. Determine the rate of return on investment, which is (annual benefits − annual cost)/(investment cost).
4. Determine the internal rate of return, which is the rate of discount applied to annual costs and benefits so that the present worth of all costs equals the present worth of all benefits.

Of these, the benefit-cost ratio is the most common indicator adopted in the economic analysis. A typical benefit-cost analysis for a hydropower project is given in Table 1.1.

1.7 LEVEL OF PLANNING

1.7.1 Single-Unit Planning

Planning for a specific project is the lowest level of planning. A specific project for any purpose should not be planned in isolation because there may be competitive demands for the same water for other purposes, or there may be conflicts with other uses.

TABLE 1.1 BENEFIT-COST ANALYSIS OF 400,000-KILOWATT CONVENTIONAL ISOLATED HYDROELECTRIC PLANT[a]

Project Data	
Dependable capacity	
$400,000 \times 0.90$	360,000 kW
Annual average energy	
$400,000 \times 0.55 \times 8760$	1,927,200,000 kWh
Annual Costs	
Private sponsorship	
$400,000 \times \$160.08$	$64,032,000
Government sponsorship	
$400,000 \times \$76.68$	$30,672,000
Annual Benefits	
Private alternative:	
Capacity: $360,000 \times 127.37$[b]	$45,853,200
Energy: $1,927,200,000 \times 0.01211$[c]	23,338,392
Total annual benefits	$69,191,592
Government alternative:	
Capacity: $360,000 \times 70.50$[b]	$25,412,400
Energy: $1,927,200 \times 0.01211$[c]	23,338,392
Total annual benefits	$48,750,792

Benefit-Cost Results

Private hydro compared with private thermal

Benefit-cost ratio $\dfrac{69,191,592}{64,032,000} = 1.08$

Annual net benefits $69,191,592 - 64,032,000 = \$5,159,592$

Govt. hydro compared with govt. thermal

Benefit-cost ratio $\dfrac{48,750,792}{30,672,000} = 1.59$

Annual net benefits $48,750,792 - 30,672,000 = \$18,078,792$

Govt. hydro compared with private thermal

Benefit-cost ratio $\dfrac{69,191,592}{30,672,000} = 2.26$

Annual net benefits $69,191,592 - 30,672,000 = \$38,519,592$

---

[a]Plant factor 0.55; dependable capacity 90% of rated capacity. *Source:* Goodman (1984).

[b]Cost of an alternative coal-fired thermal power plant.

[c]Fuel cost.

To realize fully the advantage of multiple use, reconcile competitive uses through choice of the best combination of uses, coordinate mutual responsibilities of different agencies and levels of government, river basins are usually the most appropriate geographical units for planning (U.S. President's Water Resources Council, 1962). Within the framework of the river basin plan, an individual project might be considered for single-purpose or multipurpose uses. The basis of formulating a project — single or multipurpose — is described in Section 1.4.1.

1.7.2 Multiunit Planning

The second level of planning is for a region within the river basin. Regional plans include a number of single- or multipurpose project units. Each project has a different characteristic with respect to its features and water usage. A project affects the function of other projects within the region and downstream from it. Adjustments in individual projects with respect to purposes, scope of each purpose and size of the facilities, and sometimes a rearrangement of the projects with respect to each other are made to maintain the limitation on resources, priorities of water usage, and optimization of the overall net benefits. For this purpose, several combinations of project elements are considered. Since a large number of combinations are involved with their interactions, the preferred method of analysis is the simulation study. In a simulation study, the project characteristics and operation rules are specified. The flows are distributed among various projects in accordance with these rules. The results are evaluated in terms of physical outputs and economic returns (benefits and costs).

By successive simulation exercises with different project characteristics and operating rules, an optimization is achieved. The foregoing technique considers the effects of only the hydrology and economic benefits. Separate studies are needed to consider the environmental quality and other noneconomic aspects.

Basinwide planning consists of regions within the basin. It involves great conceptual and computational complexity. The direct optimization for the entire basin is not practical. The plan is developed in steps through subbasin (regional) optimization studies. Sometimes, the studies extend beyond the basin boundary to a group of closely related river basins.

1.8 DYNAMIC PLANNING

Preparation of a master plan, which is in fact a comprehensive river basin plan, is an essential part of a basin development. Since this is done at level B study, the individual projects with their features are identified in the plan. There has been a tendency to treat the master plan as a decisive instrument and a project included in it as of unquestionable standing. On the other hand, planning is a dynamic process. The objectives, procedures, techniques, and needs all the change with time. The expanded data base, state of the development, results of the constructed projects, and unforeseen growth provide the basis for improvements in the plan. A master plan cannot be accepted in a single time framework. As a dynamic process, the prognosis should be reviewed after a short period, for example, five years, and extended for another five years. The master plan should be revised to reflect current trends. In the plan, a varying degree of revisions might be indicated, comprising abandoning, reducing scope, expanding, and staging of projects. The changing conditions make it desirable for the plan to be flexible.

1.9 AUGMENTATION OF SUPPLIES

There are many ways by which supplies available for development through the natural cycle can be augmented. These are suitable in different context. A summary of the measures is presented below.

1.9.1 Conjunctive Use of Surface and Ground Water

This envisages the use of surface water supplies for the intended purpose — directly from a stream system or from a reservoir impoundment — during times of excess to adequate precipitation and the use of groundwater sources in adjoining aquifers during dry periods. Supplies from the groundwater are made either directly to the demand center or into the stream to augment the flows. In the latter case, the pumping of groundwater can be accomplished in a short period at a fast rate if a surface reservoir exists. At the end of the dry period, the groundwater supply is depleted with use. Since the stream-aquifer system is hydraulically interconnected, at the onset of the wet season, recharge takes place from the stream system by induced infiltration. If necessary, this is supplemented by an artificial recharge so that the aquifer is full again at the end of the wet season for use in the next dry cycle.

1.9.2 Conservation of Water

Conservation relates to cutting down on the use of water by means of improved works, reduced wastes, changed use pattern, and imposed restrictions. It has been estimated that a reduction in the range 20 to 50% can be achieved in the use of irrigation water with increased efficiency, improved water scheduling, reduced runoff, and water-conserving irrigation methods. This can result in savings of 30,000 to 80,000 mgd of water. A reduction of 20% or more is possible in municipal requirements by installing water-saving devices which will bring in a savings of over 5000 mgd. As the opportunities for new development projects are becoming restricted, conservation policies are gaining favor with state and local governments. However, conservation measures cannot be considered as the sole means of meeting the needs of a project; they can be treated as supplemental measures together with the traditional structural measures.

1.9.3 Improved Management Practice or Nonstructural Measures

While water conservation is concerned primarily with effecting a change in the use pattern of water, management practices relate to modifications in the operating rules and the regulatory policy in order to enhance the supply from the existing projects and to better manage the natural distribution of water. To solve a water problem, it is not always necessary to adopt structural means. Nonstructural measures such as basin management, flow diversion, and planning usage that conform to natural water distribution and revision in the operation practice of the existing facilities are effective in the solution of some problems. Nonstructural measures in some cases offer a complete alternative to a traditional structural measure. In other cases, they may be combined with smaller traditional structural facilities to produce the solution.

In the light of the environmental movement it is recommended that a primarily nonstructural plan, or plan incorporating a combination of nonstructural measures, be included as one alternative whenever structural alternatives are considered.

1.9.4 Interbasin Transfer of Water

It is the aim of a water development project to supply water from a source to a remotely established demand center. Interbasin transfers extend this idea further to envision transporting water from a source basin containing excess water (exporting basin) to an adjoining drainage area (importing basin) through a ditch, canal, or pipeline, and pumps. Many projects exist on these lines; for example, the New York

City water supply system, the California State Water Project, the Los Angeles Aqueduct, and the Colorado River Aqueduct. Some very ambitious projects have been planned to interlink two river systems. A major objection to basin transfer projects has been raised on environmental grounds since they give rise to ecological problems. In the future, the economic factors might dominate to trade off some environmental impacts.

1.9.5 Wastewater Reuse

The practice of reusing wastewater as a supplemental source is becoming widespread in arid and semiarid regions. Treatment of wastewater for reuse is known as water reclamation. Municipal wastewater is reclaimed for nonpotable purposes, such as irrigation, groundwater recharge, industrial use, pasture for dairy animals, and recreational impoundments. Industrial wastewater without any treatment is recirculated for cooling purposes and, with some treatment, is used for plant process makeup. Wastewater from irrigation is reused to a limited extent for irrigation. However, when it is highly contaminated with salts from the soil, it is not practical for farmers to treat it for desalination. Where a separate storm sewer system exists, it is economically competitive to treat this water to render it suitable for reuse, ranging from high-quality application, to routine industrial supplies, to less rigorous lawn irrigation, and lawn protection and for recreation ponds.

1.9.6 Desalination of Water

Saline water is defined to contain over 1000 ppm (parts per million) of total dissolved solids (TDS). Brackish water contains 1000 to 35,000 ppm TDS. Seawater contains an average of 35,000 TDS. Over 35,000 ppm TDS, it is considered a brine solution.

The quantity of seawater is enormous. Also, technology exists for the conversion of saline water to fresh water. Still, the conversion is not undertaken on a large scale, for the reasons of costs and environment impact. At present the cost of conversion is about $1 per 1000 gallons. Environmental problems arise from the disposal of a large quantity of saline-water conversion wastes. The saline-water conversion process has a potential for municipal water supply since its priority is high and the requirements comparatively low.

1.9.7 Weather Modification

Weather modification is carried out by dispersal of materials into clouds to increase precipitation. Knowledge of rainmaking is, however, limited to certain types of clouds under certain conditions. As such, the process has only a limited application to localized cases at present. There is a need for expansion of technology on the processes to be altered under different conditions, the methods for such alterations, and prediction of the results. Also, legal and social issues of enormous size are involved because the process interferes with a natural phenomenon. These issues remain to be resolved.

DEMAND OF WATER

The need of a water development activity arises from the demand for water for some purpose. After the demand for a purpose is established, the availability of water resources in the vicinity of the demand center is assessed through application of the principles of hydrology. The reconciliation of the demand with the available resources in an optimal manner is the objective of water resources planning. Where the resources are restricted compared to the demand, as for irrigation in some regions, the problem is approached from a consideration of how much demand can be satisfied with the available water resources. There may be conflicting demands when more than one purpose is involved. These have to be resolved by establishing a priority ranking among water uses. The location, type, and components of a project as well as its functional characteristics are dependent on the purpose and magnitude of demands. Thus the project cost is a function of the demand. But the demand for water is affected to some extent by the cost of providing water via the project. Therefore, demand is not a static problem that can be finally determined at one time. The tentative estimates made initially are reviewed at a later stage in the planning process. The project estimates are also revised accordingly. The procedures for making tentative estimates of demand for each major purpose of development are discussed in subsequent sections.

2.1 DEMAND FOR WATER SUPPLY

Definitions of relevant terms follow.
Withdrawal uses are diversion of surface water or groundwater from its source of supply, such as irrigation and water supply. *Nonwithdrawal uses* are on-site uses such as navigation and recreation.

*Consumptive uses** are that portion of withdrawn quantity which is no longer available for further use because it is used up by crops, human beings, industrial plant processes, evapotranspiration, and so on.

Water supply requirements usually have the highest priority among the developmental uses, and water of good quality is needed. Although the total quantity of withdrawal in big cities may be relatively large, the consumptive water use is small since 80 to 90% of the total intake is returned to the river system (of course, its quality is degraded). Water requirements of a city can be divided into three broad categories:

1. Municipal requirements
2. Large industrial requirements
3. Waste dilution requirements

The order of magnitude of three kinds of requirements is indicated in Figure 2.1 for a typical city of a population of about 150,000. If the sewage and industrial waste

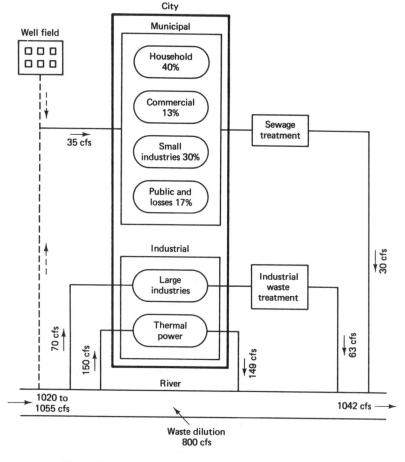

Figure 2.1 Water requirements of a city of 150,000 population.

*In the context of agriculture, the consumptive use requirement of a crop is defined as the amount of water needed for crop growth.

are discharged after a treatment, the waste dilution requirements can be reduced substantially as discussed subsequently. The total water supply requirement in a river system for a number of cities is not equal to the sum of the requirements of the individual city because the dilution requirement is a nonwithdrawal use that is available to all cities on the same river, and the consumptive requirement is only a small fraction of the total. Thus, if all cities are situated on the same river with sufficient distance in between for purification of the discharged wastes, the total requirements will be only slightly more than the largest city requirements.

2.2 MUNICIPAL REQUIREMENTS

This includes (1) such domestic uses as drinking, cooking, washing, sprinkling, and air conditioning, (2) public uses such as in public buildings and for firefighting, (3) commercial use in shopping centers, hotels, and laundries, (4) small industrial use by industries not having a separate system, and (5) losses in the distribution system. The municipal requirements are highly variable, depending on such factors as size of city, characteristics of the population, nature and size of commercial and industrial establishments, climatic conditions, and cost of supply.

Municipal requirements are estimated by the following simple relation:

$$\text{required quantity} = \begin{pmatrix} \text{population at} \\ \text{the end of} \\ \text{design period} \end{pmatrix} \times \begin{pmatrix} \text{per} \\ \text{capita} \\ \text{usage} \end{pmatrix} \quad [L^3] \quad (2.1)$$

The two parameters for assessing the municipal requirements in eq. (2.1)—population estimates and per capita water usage—are discussed in the following sections.

2.3 POPULATION FORECASTING

Like any other natural phenomenon, the prediction of future population is quite complex. Any sophisticated model has several implicit and explicit assumptions. The success of forecast lies in the judgment of the forecaster on the reliability of these assumptions. Most methods pertain to trend analysis wherein future changes in population are expected to follow the pattern of the past. However, dynamic human growth involves continuous deviations from past trends, which are difficult to assess. Long-term projection methods consider the probable shifts in the trends. There are four broad categories of population forecasting techniques: (1) graphical, (2) mathematical, (3) ratio and correlation, and (4) component methods.

Two types of population predictions required are (1) short-term estimates for 1 to 10 years, and (2) long-term forecasting for 10 to 50 years or more. The choice of methods to use for these two types of estimates is different, as described below.

2.4 SHORT-TERM ESTIMATES

Certain techniques from the categories of graphical and mathematical methods are used for short-term estimates. These methods are essentially trend analyses in graphic or mathematical form. The mathematical approach assumes three forms of population growth, shown by three segments of Figure 2.2. These are referred to as geometric growth, arithmetic growth, and declining rate of growth. Each segment has a separate relation. The historic population data of the study area may be plotted on a regular graph. According to the shape of the plot, the relationship of that segment should be used for population projection. The short-term methods are used for either intercensal estimates for any year between two censuses or postcensal estimates from the last census until the next census.

2.4.1 Graphical Extension Method

This consists of plotting the population of past census years against time, sketching a curve that fits the data, and extending this curve into the future to obtain the projected population. Since it is convenient and more accurate to project a straight line, it is aimed to get a straight-line fit to the past data by making a semilog or logarithmic plot, as necessary. The forecast may vary widely depending on whether the last two known points are joined and extended or other points are joined and extended.

2.4.2 Arithmetic Growth Method

This method considers that the same population increase takes place in a given period. Mathematically,

$$\frac{dP}{dt} = K_a$$

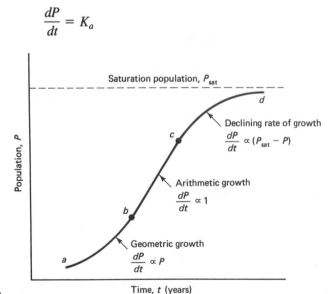

Figure 2.2 Population growth curve.

where

$$P = \text{population}$$
$$t = \text{time, years}$$
$$K_a = \text{uniform growth-rate constant}$$

By integrating the equation above, we obtain

$$P_t = P_0 + K_a t \qquad (2.2)$$

where

$$P_t = \text{projected population } t \text{ years after } P_0$$
$$P_0 = \text{present population}$$
$$t = \text{period of projection}$$

and

$$K_a = \frac{P_2 - P_1}{\Delta t} \qquad [\text{T}^{-1}] \qquad (2.3)$$

where P_1 and P_2 are recorded populations at some Δt interval apart.

Example 2.1

The population of a city has been recorded in 1970 and 1985 as 100,000 and 110,000, respectively. Estimate the 1995 population, assuming arithmetic growth.

Solution From eq. (2.3),

$$K_a = \frac{110,000 - 100,000}{15} = 667$$

From eq. (2.2),

$$P_{1995} = 110,000 + 667(10) = 116,670 \text{ persons}$$

2.4.3 Geometric Growth Method

This considers that the increase in population takes place at a constant percent of the current population. Mathematically,

$$\frac{dP}{dt} = K_p P$$

By integrating we obtain

$$\ln P_t = \ln P_0 + K_p t \qquad (2.4)$$

where

$$K_p = \frac{\ln P_2 - \ln P_1}{\Delta t} \qquad [\text{T}^{-1}] \qquad (2.5)$$

Example 2.2

In Example 2.1, estimate the population, assuming geometric growth.

Solution   From eq. (2.5),

$$K_p = \frac{\ln 110{,}000 - \ln 100{,}000}{15} = 0.0064$$

From eq. (2.4),

$$\ln P_{1995} = \ln 110{,}000 + 0.0064(10) = 11.67$$
$$P_{1995} = 117{,}000 \text{ persons}$$

2.4.4 Declining Growth Rate Method

This assumes that the city has a saturation population and the rate of growth becomes less as the population approaches the saturation level. In other words, the rate of increase is a function of the population deficit $(P_{sat} - P)$, that is,

$$\frac{dP}{dt} = K_D(P_{sat} - P)$$

Upon integration, we have

$$P_t = P_0 + (P_{sat} - P_0)(1 - e^{-K_D t}) \tag{2.6}$$

Rearranging eq. (2.6) gives

$$K_D = -\frac{1}{\Delta t}\ln\left(\frac{P_{sat} - P_2}{P_{sat} - P_1}\right) \qquad [\text{T}^{-1}] \tag{2.7}$$

Example 2.3

In Example 2.1, if the saturation population of the city is 200,000, estimate the 1995 population. Assume a declining rate of growth.

Solution   From eq. (2.7),

$$K_D = -\frac{1}{15}\ln\left(\frac{200{,}000 - 110{,}000}{200{,}000 - 100{,}000}\right) = 0.007$$

From eq. (2.6),

$$P_{1995} = 110{,}000 + (200{,}000 - 110{,}000)[1 - e^{-(0.007)(10)}]$$
$$= 116{,}085 \text{ persons}$$

Example 2.4

A city has a present population of 200,000, which is estimated to increase geometrically to 220,000 in the next 15 years. The existing treatment plant capacity is 51 mgd. The rate of input to the treatment plant is 165 gallons per person per day. For how long will the treatment plant be adequate?

Solution

1. From the known population data:

$$\ln 220{,}000 = \ln 200{,}000 + K_p(15)$$

or $K_p = 0.00635$.

2. Population that can be served by the plant:

$$\frac{51.0(10^6)}{165} = 309{,}090 \text{ persons}$$

3. Time to reach the design population:

$$\ln 309090 = \ln 200000 + 0.00635(t)$$

or $t = 68.55$ years.

2.5 LONG-TERM FORECASTING

Long-term predictions are made by techniques from all four categories. The entire past record of historic population data is used in long-term predictions. The mathematical curve-fitting approach is popular because it is relatively easy to apply. McJunkin (1964) indicates, however, that the component and ratio methods offer greater reliability than the traditional graphical–mathematical methods.

2.5.1 Graphic Comparison Method

Several larger cities in the vicinity are selected whose earlier growth exhibited characteristics similar to those of the study area. The population–time curves for these cities and for the study area are plotted as shown in Figure 2.3. From point O' corresponding to the last known population for study area A a horizontal line is drawn intersecting the other curves at O. At O', lines parallel to OB, OC, and OD are drawn as $O'B'$, $O'C'$, and $O'D'$, respectively. These lines establish a range of future growth within which $O'A'$ is extended. This method has a shortcoming since it is not certain that the future growth of the study area will be similar to the past growth of the other areas.

2.5.2 Mathematical Logistic Curve Method

This method is suitable for the study of large population centers such as large cities, states, or nations. On the basis of the study of the growth curve of Figure 2.2, certain mathematical equations of an empirical curve conforming to this shape (S-shape) were proposed. One of the best known functions is the logistic curve in the form

$$P_t = \frac{P_{\text{sat}}}{1 + ae^{bt}} \qquad (2.8)$$

where

P_t = population at any time t from an assumed origin

P_{sat} = saturation population

a, b = constants

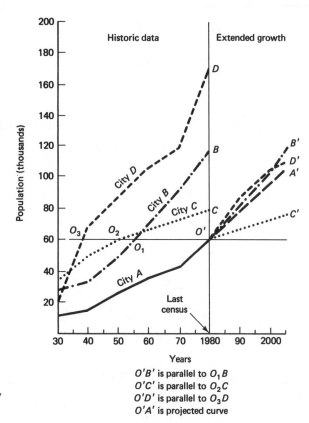

O'B' is parallel to $O_1 B$
O'C' is parallel to $O_2 C$
O'D' is parallel to $O_3 D$
O'A' is projected curve

Figure 2.3 Graphical projection by comparison.

The constants are determined by selecting three populations from the record: one in the beginning, P_0, one in the middle, P_1, and one near the end of the record, P_2, associated with the years T_0, T_1, and T_2 such that the number of years (interval) between T_0 and T_1 is the same as that between T_1 and T_2. This interval between T_0 and T_1 is designated as n. The constants are given by

$$P_{sat} = \frac{2P_0 P_1 P_2 - P_1^2 (P_0 + P_2)}{P_0 P_2 - P_1^2} \tag{2.9}$$

$$a = \frac{P_{sat} - P_0}{P_0} \tag{2.10}$$

$$b = \frac{1}{n} \ln \left[\frac{P_0 (P_{sat} - P_1)}{P_1 (P_{sat} - P_0)} \right] \tag{2.11}$$

In eq. (2.8), time t is counted from the year T_0. From eq. (2.9), P_{sat} must be positive and must exceed the latest known population. A test of the validity of logistic growth is that the population data plot as a straight line on specially scaled (logistic) graph paper. (The graph of $\log [(P_{sat} - P)/P]$ versus t is a straight line.)

Example 2.5

In two 20-year periods, a city grew from 45,000 to 258,000 to 438,000. Estimate (a) the saturation population, (b) the equation of the logistic curve, and (c) the population 40 years after the last period.

Solution

$$P_0 = 45,000$$
$$P_1 = 258,000$$
$$P_2 = 438,000$$
$$n = 20$$

(a) From eq. (2.9) (expressing numbers in thousand),

$$P_{sat} = \frac{2(45)(258)(438) - (258)^2(45 + 438)}{(45)(438) - (258)^2} = 469,000$$

(b) From eq. (2.10),

$$a = \frac{469 - 45}{45} = 9.42$$

From eq. (2.11),

$$b = \frac{1}{20} \ln\left[\frac{45(469 - 258)}{258(469 - 45)}\right] = -0.122$$

The equation of the logistic curve:

$$P_t = \frac{469,000}{1 + 9.42e^{-0.122t}} \tag{2.12}$$

(c) The time from the beginning, $t = 40 + 40 = 80$ years; thus

$$P = \frac{469,000}{1 + 9.42e^{-0.122(80)}} = 468,740 \text{ persons}$$

2.5.3 Ratio and Correlation Methods

A city or smaller area is a part of a region, state, nation, or larger area. There are many factors and influences affecting population growth occur throughout the region. Thus the growth of the smaller area has some relation to the growth of the larger area. Because a careful projection of the future population of the nation and states (larger area) is made by the authoritative organizations, these may be used to forecast the growth of the smaller area.

In the simplest technique, a constant ratio obtained from the most recent data is used as follows:

$$K_r = \frac{P_i}{P_i'} \tag{2.13}$$
$$P_t = K_r P_t'$$

where

$$P_i = \text{population of study area at last census}$$
$$P_i' = \text{population of larger area at last census}$$
$$P_t = \text{future population for study area}$$
$$P_t' = \text{estimated future population of larger area}$$
$$K_r = \text{constant}$$

In a refined technique using the variation in ratios, the ratios of the smaller area to the larger area population are calculated for a series of census years. By using any of the graphical or mathematical methods of short-term estimates (Section 2.4), the trend line of the ratios is projected. The projected ratio in the year of interest is applied to an estimate of the study area population.

In another statistical method, the correlation technique described in Chapter 7 is applied. Between the two series of census data relating to the study area and larger area, a relation is established through the regression analysis. For example, a simple regression equation may be of the following form:

$$P_f = aP_f' + b \qquad (2.14)$$

where

$$P_f = \text{population of the study area}$$
$$P_f' = \text{population of the region (larger area)}$$
$$a, b = \text{regression constants}$$

The future population is projected from eq. (2.14).

2.5.4 Component Methods

A population change can occur in only three ways: (1) by birth, (2) by death, and (3) through migration. These components of population change can be linked by the balance equation:

$$P_t = P_0 + B - D \pm M \qquad (2.15)$$

where

$$P_t = \text{forecast population at the end of time } t$$
$$P_0 = \text{existing population}$$
$$B = \text{number of births during time } t$$
$$D = \text{number of deaths during time } t$$
$$M = \text{net number of migrants during time } t \text{ (positive}$$
$$\text{value indicates moving into the study area)}$$

Because migration affects the births and deaths in an area, the estimates of net migration are made before estimating the natural change due to births and deaths. The migratory trends may be estimated by applying eq. (2.15) backward to the past census data on population, births, and deaths during a selected period. The school attendance method, comparing the actual children enrollment to the children from birth records, is also used for estimating migration.

For determining the natural change due to births and deaths, the simplest procedure is to multiply the existing population by the expected birth and death rates, that is,

$$B = K_1 P_0 \Delta t \qquad (2.16)$$

$$D = K_2 P_0 \Delta t \qquad (2.17)$$

where

$$K_1, K_2 = \text{birth and death rate, respectively}$$

$$\Delta t = \text{forecast period}$$

Better estimates of natural change (births and deaths) are made by the cohort-survival technique, which makes projections to each subcomponent (factor) related to the natural change.

2.6 PER CAPITA WATER USAGE

Per capita use is normally expressed as the average daily rate, which is the mean annual usage of water averaged for a day in terms of gallons (or liters) per capita per day (gpcd). The seasonal, monthly, daily, and hourly variations in the rate are given in percentages of the average. Which of these should be used for the design capacity depends on the component of the water supply system. The layouts of two water supply systems — one for direct pumping from a river or from a wellfield and one for an impounding reservoir — are shown in Figure 2.4. The period of design for which the population projection is to be made and the design capacity criteria of different component structures of the systems are indicated in Table 2.1.

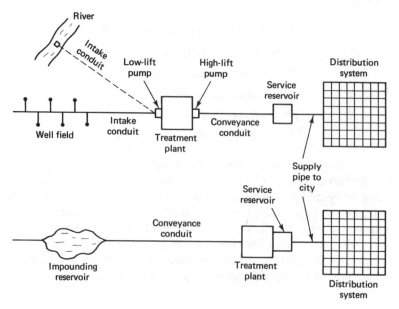

Figure 2.4 Layout of typical water supply systems. [from Fair, Geyer, and Okun (1966)]

Structure	Design Period[a] (years)	Required Capacity
1. Source of supply		
a. River	Indefinite	Maximum daily (requirements)
b. Wellfield	10–25	Maximum daily
c. Reservoir	25–50	Average annual demand
2. Conveyance		
a. Intake conduit	25–50	Maximum daily
b. Conduit to treatment plant	25–50	Maximum daily
3. Pumps		
a. Low-lift	10	Maximum daily plus one reserve unit
b. High-lift	10	Maximum hourly plus one reserve unit
4. Treatment plant	10–15	Maximum daily
5. Service reservoir	20–25	Working storage (from hourly demand and average pumping) plus fire demand plus emergency storage
6. Distribution		
a. Supply pipe or conduit	25–50	Greater of (1) maximum daily plus fire demand or (2) maximum hourly requirement
b. Distribution grid	Full development	

[a]"Design period" does not necessarily indicate the life of the structure. A design period takes into account other factors, such as subsequent ease of extension, rate of population growth and shifts in community, and industrial/commercial developments.

2.6.1 Average Daily per Capita per Day Usage

For household use, the per capita requirements of water range between 20 and 90 gallons per day, with a reasonable average of 60 gallons per day. Municipal water use should, however, also include commercial use, small industrial use, public use, and losses in the system. A typical distribution for an average city is given in Table 2.2.

TABLE 2.2 TYPICAL AVERAGE USAGE

Use	Average Use (gpcd)[a]	Percent of Total
Household	60	40
Commercial	20	13
Industrial	45	30
Public	15	10
Loss	10	7
Total	150	100

[a]Liter = gallons × 3.8.

The U.S. Water Resources Council (1968) has projected average usage in the year 2000 at 175 gpcd. When the industrial requirements become relatively important in a city, they should be considered separately.

2.6.2 Variations in Usage

The values in Table 2.2 refer to the daily average of the long-term (many years) usage. The consumption changes with the seasons, varies from day to day in the week, and fluctuates from hour to hour in a day. Knowledge of these variations is important for the design of project components, as indicated in Table 2.1. There are two common trends: (1) the smaller the city, the more variable is the demand; and (2) the shorter the period of flow, the wider is the variation from the average (i.e., the hourly peak flow is much higher than the daily peak). Typical variation in a city water supply in a day is shown in Figure 2.5. The variations are commonly indicated in terms of the percentage of the long-term average value. There are no fixed ratios; each city has its own trend. However, in the absence of data, the following formula of R. O. Goodrich is very convenient for estimating the maximum usage from 2 hours (2/24 day) to a year (365 days) for small cities.

$$p = \frac{180}{t^{0.1}} \quad \text{[unbalanced]} \tag{2.18}$$

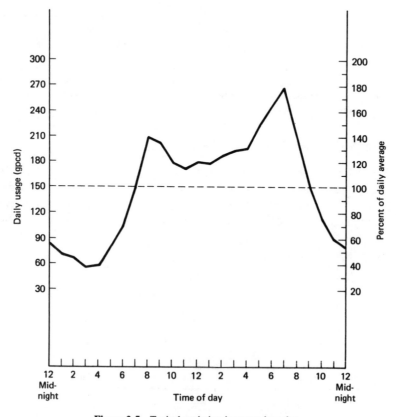

Figure 2.5 Typical variation in usage in a day.

where

$$p = \text{percentage of annual average daily usage}$$

$$t = \text{time, days}$$

From eq. (2.18), the maximum daily use is 180% of the (long-term) average daily usage, and the maximum monthly use is 128%. Larger cities may have smaller peaks.

The maximum hourly consumption in any day is likely to be 150% of the usage for that day (Steel and McGhee, 1979). For a distribution system, the fire demands have also to be added. It is unlikely that water will be drawn at the maximum hourly rate while a serious fire is raging. Hence the capacity is based on the maximum daily usage plus fire demand or maximum hourly usage, whichever is greater.

For pump design, the information on minimum flow rate which is considered to be 25 to 50% of the average daily flow is also important.

Example 2.6

For a city having an average daily water usage of 150 gpcd from the municipal supply, determine the maximum hourly requirement (excluding the fire demands).

Solution

1. Maximum daily usage = 180% of average daily

$$= \frac{180}{100}(150) = 270 \text{ gpcd}$$

2. For a maximum day,

maximum hourly usage = 150% of daily use

$$= \frac{150}{100}(270) = 405 \text{ gpcd}$$

2.6.3 Fire Demand

Although the total quantity of water used annually for firefighting purposes (which is included under the category of "public use") is very small, whenever demand rises, the rate of withdrawal is high. The service reservoir and distribution system should make provisions for the fire demands in their capacities.

Two empirical formulas have been suggested by insurance-related organizations for estimating fire demands. The American Insurance Association has suggested the following relation based on population:

$$Q = 1020\sqrt{P}\,(1 - 0.01\sqrt{P}) \qquad \text{[unbalanced]} \qquad (2.19)$$

where

$$Q = \text{required fire flow, gpm}$$

$$P = \text{population, thousands}$$

On the basis of construction type, floor area, and occupancy, the Insurance Services Office proposed

$$Q = 18C\sqrt{A} \qquad \text{[unbalanced]} \qquad (2.20)$$

where

Q = required fire flow, gpm

A = total floor area excluding the basement, ft^2

C = coefficient: 1.5 for wood frame construction, 1.0 for ordinary construction, 0.8 for noncombustible construction, and 0.6 for fire-resistant construction

The following limitations apply to eq. (2.20):

1. Flow should not exceed 6000 gpm for a single story, 8000 gpm for a single building, or 12,000 gpm for a single fire.
2. Flow should not be less than 500 gpm.

The duration for which the fire flow has to be maintained is given in Table 2.3. If the time periods indicated cannot be maintained, the community insurance rates are adjusted upward by the insurance companies.

TABLE 2.3 DURATION FOR FIRE FLOW[a]

Required Fire Flow (gpm)	Duration (hr)
<1000	4
1000–1250	5
1250–1500	6
1500–1750	7
1750–2000	8
2000–2250	9
>2250	10

[a]*Source:* Steel and McGhee (1979)

Example 2.7

A city with a population of 20,000 has an average daily usage of 150 gpcd. Determine the fire demand and the design capacity for different components of a water supply project. The working service storage is 1.5 mgd.

Solution

1. From Example 2.6,

$$\text{maximum daily usage} = 270 \text{ gpcd}$$
$$\text{maximum hourly usage} = 405 \text{ gpcd}$$

2. Average daily draft $= \dfrac{150(20,000)}{1,000,000} = 3$ mgd.

3. Maximum daily draft $= \dfrac{270(20,000)}{1,000,000} = 5.4$ mgd.

4. Maximum hourly draft $= \dfrac{405(20,000)}{1,000,000} = 8.1$ mgd.

5. Fire flow [from eq. (2.19)]:

$$Q = 1020\sqrt{20}\,(1 - 0.01\sqrt{20}) = 4358 \text{ gpm} \quad \text{or} \quad 6.27 \text{ mgd}$$

6. Maximum daily + fire flow = 5.4 + 6.27 = 11.67 mgd.

7. Pumps: Assume that the required flow is handled by three units and that one reserve unit is installed.

    a. Low-lift pumps $= \frac{4}{3}$ (maximum daily) $= \frac{4}{3}(5.4) = 7.2$ mgd

    b. High-Lift pumps $= \frac{4}{3}$ (maximum hourly) $= \frac{4}{3}(8.1) = 10.8$ mgd

8. Service reservoir:
    a. Fire flow duration (from Table 2.3) $= 10$ hr

    b. Total quantity of fire flow in a day $\frac{10}{24}(6.27) = 2.61$ mgd.

    c. Working storage (given) $= 1.5$ mgd.
    d. Emergency storage $=$ (days)(average daily draft) $= 3(3) = 9$ mgd.
    e. Service storage $= 2.61 + 1.5 + 9 = 13.11$ mgd.

9. Design capacities:

Structure	Basis	Capacity (mgd)
River flow	Maximum daily	5.4
Intake conduit, conduit to treatment	Maximum daily	5.4
Low-lift pumps	Maximum daily plus reserve	7.2
High-lift pumps	Maximum hourly plus reserve	10.8
Treatment plant	Maximum daily	5.4
Service storage	Working storage plus fire plus emergency	13.11
Distribution system	Maximum daily plus fire or maximum hourly	11.67

2.7 INDUSTRIAL REQUIREMENTS

Table 2.2 included a typical industrial water component from a public water supply system. This is not adequate for a community with large water-using industries. The large industries are usually served by separate supplies. About 5% of the industries use 80% of the industrial water demands; 70% of the industrial plants use as little as 2% of the demands. Thermal power stations are the heaviest water users, with a requirement of 600 gpcd or 80 gal/kWh on the average. Other major water-using industries are steel, paper, and beverages. The average industrial requirement for the entire country (excluding thermal power stations) is 250 to 300 gpcd. The requirements of individual industries per unit of the production are indicated in Table 2.4.

More than 90% of all the water used for industries, including thermal power stations, is for cooling purposes. Many measures can be taken to reduce the cooling requirements, such as constructing an artificial pond, recirculating the cooling water or using poor-quality water from a different source. The figures listed are applicable

TABLE 2.4 REQUIREMENTS OF MAJOR INDUSTRIES

Industry	Average Water Use
Thermal power	80 gal/kWh
Steel	35,000 gal/ton
Paper	50,000 gal/ton
Woolens	140,000 gal/ton
Coke	3,600 gal/ton
Oil refining	770 gal/barrel

when sufficient water is available. These can be substantially reduced — in the case of some industries by one-tenth — when water is scarce.

2.8 WASTE DILUTION REQUIREMENTS

In earlier times it was a common practice to dump raw municipal and industrial wastes into the same river that served as the source of supply, thus relying on the self-purification properties of the stream. As long as the streamflow is at least 40 times that of the wasteflow and there is a sufficiently long reach of river to the next city, both nuisance and unsafe conditions can be avoided. But with the rapid growth of cities and industrial activities and with increased use of water, the dumping of raw wastes into rivers is no longer permitted. The problem now is to what extent the municipal and industrial wastes* should be treated before discharge into the river. The amount of streamflow required for sufficient natural treatment of municipal and industrial wasteflow is a function of the pollutant characteristics of the waste- and streamflow properties with regard to oxygen content, dissolved minerals, water temperature, and length of the available downstream reach. This relationship is depicted in many models, the most common being the oxygen sag curve. For average conditions it has been found that the raw (fully untreated) waste from municipal and industrial sources, excluding thermal power plants, requires a ratio of streamflow to wasteflow of 40, and thoroughly treated waste requires a ratio of 2, with a linear variation in between as shown in Figure 2.6 (Kuiper, 1965).

If water supply is being planned from a reservoir project, there are three annual cost components to be considered: (1) cost of storage to provide the municipal (and industrial) requirements, (2) cost of storage to produce the required quantity for waste dilution, and (3) cost of waste treatment. Items (2) and (3) act opposite to each other, that is, when the degree of treatment is increased, the cost of treatment rises but the cost of waste dilution storage decreases, and vice versa. The various degrees of treatment and the annual costs associated with them are considered until the lowest cost is found that indicates the most economic treatment of the city waste.

Example 2.8

A city had a total withdrawal (excluding in-stream dilution requirements) of 140 mgd in 1987 distributed as follows: municipal usage, 30 mgd; manufacturing industries, 35 mgd;

---

*Industrial wastes have to be considered even if the industrial supplies are developed from a different source than the municipal supplies.

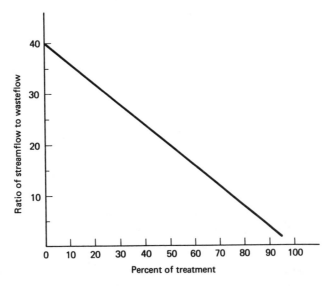

Figure 2.6 Requirements for waste dilution.

and thermal power, 75 mgd. The city had a population of 200,000, which is expected to rise to 220,000 in the year 2000. An industrial expansion of 20% and a thermal power increase of 80 MW is expected in the city by 2000. Estimate the total withdrawal in the year 2000. If the waste is to be discharged after 80% of the treatment, determine the total requirements of water.

Solution

1. Existing per capita municipal usage $= \dfrac{30(10^6)}{200,000} = 150$ gpcd.

   In the year 2000:

2. Municipal requirements $= \dfrac{(220,000)\,(150)}{1,000,000} = 33$ mgd.

3. Manufacturing industry requirements $= \dfrac{120}{100}(35) = 42$ mgd.

4. Thermal power requirements:
   a. Assuming a plant capacity factor of 0.6, the additional energy produced per day is

$$(80 \text{ MW}) \frac{(24 \text{ hr})}{(1 \text{ day})}(0.6)\frac{(1000 \text{ kW})}{(1 \text{ MW})} = 1.15 \times 10^6 \text{ kWh}$$

   b. From Table 2.4, water usage $= 80$ gal/kWh.
   c. Additional water required $= \dfrac{80(1.15 \times 10^6)}{10^6} = 92$ mgd.
   d. Total thermal power requirements $= 75 + 92 = 167$ mgd.

5. Total for municipal and manufacturing usage $= 33 + 42 = 75$ mgd.

6. Waste dilution requirements:
   a. From Figure 2.5, for 80% treatment the streamflow/wasteflow ratio $= 9$.
   b. Dilution requirement $= 9(75) = 675$ mgd.

7. City water supply requirements are summarized as follows:

Sector	mgd
Municipal requirements	33
Industrial requirements	209
Waste dilution requirements	675

2.9 DEMAND FOR IRRIGATION

The demand of water for irrigation depends on several factors, including the method of irrigation, type of crop to be grown, condition of soil, and prevailing climate. The following terms are relevant in the computation.

1. *Consumptive use or crop requirement:* the amount of water needed for crop growth.

2. *Irrigation requirement:* consumptive use *minus* effective rainfall available for plant growth. To this quantity the following items, whichever is applicable, are included: (a) irrigation applied prior to crop growth should be added; (b) water required for leaching should be added; (c) miscellaneous requirements of germination, frost protection, plant cooling, and so on, should be added; and (d) decrease in soil moisture should be subtracted.

3. *Farm delivery requirement:* irrigation requirement *plus* farm losses due to evaporation, deep percolation, surface waste, and nonproductive consumption. The losses are measured by the on-farm irrigation efficiency, which is the percent of farm-delivered water that remains in the root zone and is available for crop growth.

4. *Gross water requirement:* farm delivery requirement *plus* the seepage losses in the canal system from the headworks to the farm unit *plus* the waste of water due to poor operation.

The gross water requirement is indicated in terms of the depth of water over the irrigable area. The basic quantity of interest is the consumptive use of the crops, from which the effective precipitation is subtracted and various losses are added to establish the gross irrigation demand.

A considerable quantity of the water applied to farmland returns to the river system. This includes surface runoff during irrigation, wasted water, canal seepage, and deep percolation. This may be from 30 to 60% of the gross requirement. About one-half of this reaches the river through groundwater flow. The remainder reaches the river as surface runoff during the irrigation season and thus becomes available for use to downstream projects.

Irrigation projects must include subsurface drainage facilities, as a measure against waterlogging and salinity. Drainage is described in Chapter 12.

2.9.1 Consumptive Use of Crops

The consumptive use and evapotranspiration from a cropped area are considered synonymous. Numerous procedures have been devised to estimate the consumptive use. A method commonly applied to the cropped area is described here. Some other methods are discussed in Section 3.8. The data from actual farm experience or experimental basins, wherever available, are given preference over the computational procedures because of the empirical constants involved in the latter.

2.9.2 Blaney–Criddle Method

For the conditions in the arid western regions of the United States, Blaney and Criddle (1945) proposed a relation that determined consumptive use as the multiplication of the mean monthly temperatures, monthly percent of annual daytime hours, and a coefficient for individual crops that varied monthly and seasonally. The coefficients presented originally were the seasonal values for the entire growing season of crops. Subsequently, monthly coefficient values were suggested (Blaney and Criddle, 1962). However, these coefficients did not include the effects of humidity, wind movement, and other climatological factors. The modified Blaney–Criddle method (U.S. SCS, 1964) split the coefficient in two parts to consider these factors indirectly. The modified formula is

$$U = \sum K_t K_c t_m \frac{p}{100} \qquad \text{[unbalanced]} \qquad (2.21)$$

where

U = consumptive use for any specified growth period, in.

K_t = climatic coefficient related to mean monthly temperatures

K_c = growth stage coefficient

t_m = mean monthly temperature, °F

p = monthly percentage of annual daytime hours (Table 2.5)

The values of K_t are based on the formula

$$K_t = 0.0173 t_m - 0.314 \qquad \text{[unbalanced]} \qquad (2.22)$$

For $t_m < 36$ °F, use $K_t = 0.30$.

The monthly values of K_c are obtained from Table 2.6(a) for a perennial crop with a year-round growing season. For other seasonal crops, Table 2.6(b) is used, based on the percentage of the growing season covered by the month in question. The monthly consumptive amounts are summed over the growing season to obtain the seasonal consumptive use.

Example 2.9

For the growing season of sugar beets at Limberly, Idaho, located at latitude 43° N, the long-term mean monthly air temperatures are given in column 2 of Table 2.7. The crop is planted April 10 and harvested October 15. Estimate the seasonal consumptive use of water.

TABLE 2.5 MONTHLY PERCENTAGE OF DAYTIME HOURS OF THE YEAR

Month	Latitude (degrees north of equator)													
	24	26	28	30	32	34	36	38	40	42	44	46	48	50
Jan.	7.58	7.49	7.40	7.30	7.20	7.10	6.99	6.87	6.76	6.62	6.49	6.33	6.17	5.98
Feb.	7.17	7.12	7.07	7.03	6.97	6.91	6.86	6.79	6.73	6.65	6.58	6.50	6.42	6.32
Mar.	8.40	8.40	8.39	8.38	8.37	8.36	8.35	8.34	8.33	8.31	8.30	8.29	8.27	8.25
Apr.	8.60	8.64	8.68	8.72	8.75	8.80	8.85	8.90	8.95	9.00	9.05	9.12	9.18	9.25
May	9.30	9.38	9.46	9.53	9.63	9.72	9.81	9.92	10.02	10.14	10.26	10.39	10.53	10.69
June	9.20	9.30	9.38	9.49	9.60	9.70	9.83	9.95	10.08	10.21	10.38	10.54	10.71	10.93
July	9.41	9.49	9.58	9.67	9.77	9.88	9.99	10.10	10.22	10.35	10.49	10.64	10.80	10.99
Aug.	9.05	9.10	9.16	9.22	9.28	9.33	9.40	9.47	9.54	9.62	9.70	9.79	9.89	10.00
Sept.	8.31	8.31	8.32	8.34	8.34	8.36	8.36	8.38	8.38	8.40	8.41	8.42	8.44	8.44
Oct.	8.09	8.06	8.02	7.99	7.93	7.90	7.85	7.80	7.75	7.70	7.63	7.58	7.51	7.43
Nov.	7.43	7.36	7.27	7.19	7.11	7.02	6.92	6.82	6.72	6.62	6.49	6.36	6.22	6.07
Dec.	7.46	7.35	7.27	7.14	7.05	6.92	6.79	6.66	6.52	6.38	6.22	6.04	5.86	5.65
Annual	100.00	100.00	100.00	100.00	100.00	100.00	100.00	100.00	100.00	100.00	100.00	100.00	100.00	100.00

TABLE 2.6 CROP-GROWTH-STAGE COEFFICIENT K_c (MODIFIED BLANEY–CRIDDLE METHOD)

(a) PERENNIAL CROPS (NORTHERN HEMISPHERE)

Crop	Average k_c values by months												
	Jan.	Feb.	Mar.	Apr.	May	June	July	Aug.	Sept.	Oct.	Nov.	Dec.
Alfalfa	0.63	0.74	0.86	0.99	1.09	1.13	1.11	1.06	0.99	0.90	0.78	0.65
Grass pasture	0.48	0.58	0.74	0.85	0.90	0.92	0.92	0.91	0.87	0.79	0.67	0.55
Grapes	0.20	0.23	0.32	0.49	0.70	0.80	0.81	0.76	0.66	0.50	0.35	0.25
Citrus orchards	0.64	0.66	0.68	0.70	0.71	0.72	0.72	0.71	0.70	0.68	0.66	0.64
Deciduous, with cover	0.63	0.74	0.86	0.98	1.09	1.13	1.12	1.06	0.99	0.90	0.78	0.65
Deciduous, no cover	0.17	0.25	0.39	0.63	0.87	0.96	0.95	0.82	0.53	0.30	0.20	0.16
Avocados	0.27	0.42	0.58	0.71	0.78	0.81	0.78	0.71	0.63	0.54	0.43	0.36
Walnuts	0.10	0.14	0.23	0.43	0.68	0.92	0.98	0.88	0.69	0.49	0.31	0.15

(b) ANNUAL CROPS

Crop	K_c values at listed % of growing season										
	0	10	20	30	40	50	60	70	80	90	100
Field corn (grain)	0.44	0.49	0.58	0.71	0.93	1.05	1.08	1.06	1.01	0.93	0.85
Field corn (silage)	0.44	0.48	0.55	0.65	0.80	0.97	1.06	1.08	1.06	1.02	0.96
Grain sorghum	0.30	0.38	0.60	0.83	1.01	1.07	0.99	0.88	0.76	0.65	0.56
Winter wheat[a]	1.46	1.44	1.42	1.39	1.35	1.30	1.23	1.15	1.03	0.86	0.78
Spring grains	0.29	0.45	0.67	0.89	1.09	1.28	1.31	1.17	0.90	0.55	0.20
Cotton	0.20	0.25	0.33	0.50	0.79	0.97	1.02	0.95	0.81	0.65	0.29
Dry beans	0.50	0.59	0.71	0.87	1.02	1.10	1.12	1.06	0.94	0.81	0.67
Sugar beets	0.45	0.50	0.61	0.79	0.95	1.10	1.20	1.25	1.21	1.13	1.04
Potatoes	0.33	0.40	0.51	0.72	0.98	1.17	1.31	1.37	1.36	1.31	1.23
Tomatoes	0.45	0.45	0.47	0.56	0.75	0.95	1.03	0.99	0.90	0.80	0.70
Melons and cantaloupe	0.44	0.48	0.56	0.65	0.76	0.81	0.81	0.78	0.75	0.71	0.67
Small vegetables	0.29	0.40	0.57	0.69	0.77	0.81	0.82	0.79	0.72	0.58	0.38

[a]Data given only for springtime season of 70 days prior to harvest (after last frost). K_c increases from 0.50 at seeding to 1.46 during period with average temperature above 32 °F.

Source: Davis and Sorensen (1969).

Solution

TABLE 2.7 CONSUMPTIVE USE COMPUTATION

(1)	(2)	(3)	(4)	(5)	(6)	(7)	(8)
Period	Mean Monthly Temp. (°F)	Number of Days	Midperiod Percent of Total Season[a]	K_c from Table 2.6(b)	K_t^{b}	Percent p from Table 2.5	U^c (in.)
Apr. 11–30	44.5	20	5	0.48	0.46	9.00	0.88
May	55.2	31	19	0.60	0.64	10.14	2.15
June	60.6	30	35	0.87	0.73	10.21	3.93
July	69.5	31	51	1.10	0.89	10.35	7.04
Aug.	68.6	31	68	1.24	0.87	9.62	7.11
Sept.	57.9	30	84	1.18	0.69	8.40	3.96
Oct. 1–15	47.9	15	96	1.08	0.51	7.70	2.03
Total		188					

$_a\dfrac{\text{Number of days up to middle of the period}}{\text{Total days in growing season}}$.

$^{b}K_t = 0.0173(\text{col. 2}) - 0.314.$

$^{c}U = \dfrac{(\text{col. 2})(\text{col. 5})(\text{col. 6})(\text{col. 7})}{100}.$

Seasonal consumptive use:

$$U = \frac{20}{30}(0.88) + 2.15 + 3.93 + 7.04 + 7.11 + 3.96 + \frac{15}{31}(2.03)$$

$$= 25.75 \text{ in.}$$

2.9.3 Effective Rainfall

The portion of the rainfall during the growing season that is utilized in meeting the requirements of crops is termed the "effective rainfall." The remainder is lost through surface runoff and deep percolation. In humid areas, this may provide a major portion of the requirements, whereas in arid areas it may constitute only a small part. The necessity of irrigation in humid regions may arise due to unbalanced distribution of the rainfall.

Effective rainfall is influenced by many factors relating to the (1) soil moisture, (2) cropping pattern, (3) application of irrigation, and (4) rainfall characteristics. Based on the study of extensive data, the Soil Conservation Service (1964) suggested the relationship shown in Table 2.8. The limitation on use is given at the bottom of the table.

Whereas the crop consumptive use requirements vary from year to year by a small margin, the variations in rainfall are large. As such, the frequency analysis of effective rainfall is made as follows:

1. For the region under consideration, available data on monthly rainfall are collected.
2. Using Table 2.8, the effective rainfall figures for each month of record are determined.

TABLE 2.8 AVERAGE MONTHLY EFFECTIVE RAINFALL RELATED TO MEAN MONTHLY RAINFALL AND AVERAGE MONTHLY CONSUMPTIVE USE

Monthly mean rainfall (in.)	Average monthly consumptive use, U (in.)									
	1.0	2.0	3.0	4.0	5.0	6.0	7.0	8.0	9.0	10.0
	Average monthly effective rainfall (in.)[a]									
0.5	0.20	0.25	0.30	0.30	0.30	0.35	0.40	0.45	0.50	0.50
1.0	0.55	0.60	0.65	0.70	0.70	0.75	0.80	0.85	0.95	1.00
2.0	1.00	1.25	1.35	1.55	1.55	1.55	1.60	1.70	1.85	2.00
3.0	1.00	1.85	1.95	2.10	2.20	2.30	2.40	2.55	2.70	2.90
4.0	1.00	2.00	2.55	2.70	2.90	2.95	3.15	3.30	3.50	3.80
5.0	1.00	2.00	3.00	3.25	3.50	3.60	3.85	4.05	4.30	4.60
6.0	1.00	2.00	3.00	3.80	4.10	4.25	4.50	4.80	5.10	5.40
7.0	1.00	2.00	3.00	4.00	4.60	4.80	5.05	5.40	5.70	6.05
8.0	1.00	2.00	3.00	4.00	5.00	5.30	5.60	5.90	6.20	
9.0	1.00	2.00	3.00	4.00	5.00	5.75	6.05	6.35		

[a]Based on 3-in. net depth of application. For other net depths of application, multiply by the factors shown below.

Net depth of application	0.75	1.0	1.5	2.0	2.5	3.0	4.0	5.0	6.0	7.0
Factor	0.72	0.77	0.86	0.93	0.97	1.00	1.02	1.04	1.06	1.07

Note: Average monthly effective rainfall cannot exceed average monthly rainfall or average monthly consumptive use. When the application of the factors above results in a value of effective rainfall exceeding either, this value must be reduced to a value equal to the lesser of the two.

Source: U.S. SCS (1964).

3. For each year on record, the total effective precipitation for all the months of the growing season is determined.

4. From the resultant figures — one for each year on record — a frequency curve is prepared by the method of Section 8.6.

5. If an irrigation supply is desired which will be adequate 90% of the years (9 of 10 years), the effective rainfall corresponding to the 90% value of the frequency curve is observed.

6. The total effective rainfall is distributed over the months of the growing season in the ratio indicated by the 10 driest years on record.

2.9.4 Farm Losses

The losses that take place from the water delivered to the farm are measured by the on-farm efficiency. Thus

$$\text{on-farm efficiency} = \frac{\text{water utilized in crop growth}}{\text{water delivered to farm}}$$

The principal factors that affect efficiency include (1) the method of applying the water, (2) the texture and condition of the soil, (3) the slope of the land, (4) the preparation of the land (i.e., ditched or bordered), (4) the rate of irrigation flow in relation to the farm size, and (5) the management by the irrigator. The farm losses take place due to (1) the deep percolation beyond the root zone of crops, (2) the surface wastes from the fields, and (3) on-farm distribution losses and nonproductive consumption.

For deep percolation, a minimum allowance of about 20% of the applied water is made. This ensures adequate leaching if the applied water does not contain more than 1300 ppm of dissolved salts and the drained water can be accepted with 6000 ppm of dissolved salts. It may be necessary to pass additional water for leaching if the irrigation water applied is more saline or if the drained water has to have a lower salinity or if the soil requires reclamation. Refer to Hansen et al. (1980) for the leaching requirements.

Surface waste or runoff is inherent in most irrigation systems. This quantity is, however, recovered and becomes available for use within a project or elsewhere in the basin downstream. The surface wastes usually amounts to 6 to 10% of the quantity delivered to the fields. Distribution losses vary from a very small quantity in pipelines to about 15% in unlined ditches.

On-farm efficiency, which is a product of the efficiency of each of the items above, can be achieved in the range of 40 to 70% for a properly designed efficient irrigation system. The higher value in the range above is associated with the sprinkler system.

2.9.5 Conveyance Losses and Waste

Conveyance losses and waste relate to the water lost between the point of diversion from a stream or reservoir to the points of delivery to farms and is measured by the off-farm efficiency. The losses comprise the evaporation through the canal system, the water seeped through the canal system, and the operational wastes discharged into drains or streams.

Evaporation losses from the canal water surface are not too large unless the canal is very shallow and wide. Usually, these are less than 1% of the canal flow.

Seepage losses from canals depend on (1) the permeability of the soil, (2) the wetted surface of the canal, and (3) the difference in level of water in the canal and the adjacent groundwater table. The average unit seepage rates for the western United States are similar to those cited by Hart (Worstell, 1975); see Table 2.9.

TABLE 2.9 AVERAGE SEEPAGE LOSS FROM CANALS IN SOUTHERN IDAHO

Type of Soil	Seepage Loss (ft/day)
1. Medium clay and loam	0.5–1.5
2. Impervious clay	0.5
3. Medium soils	1.0
4. Somewhat pervious soils	1.5–2.0
5. Gravel	2.5–5.0

Source: Hart (1963).

From the lined canals, the loss rates are between about 0.1 and 1.0 ft/day (Worstell, 1975). To determine the total quantity of seepage from a canal system, the

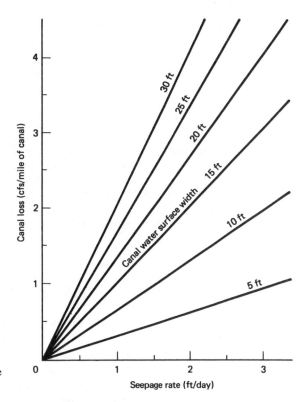

Figure 2.7 Chart to estimate the seepage losses from a canal (from Worstell, 1975).

data required are (1) the predominant soil texture to ascertain the average seepage rates, (2) the widths, and (3) the lengths of the canals. The chart developed by Worstell (1975), illustrated in Figure 2.7, can be used to estimate the seepage loss in cfs per mile for different canal widths. Broadly, the seepage losses range from 15 to 45% of diverted flow on unlined canals and from 5 to 15% on lined canals.

Operational wastes are unavoidable. These result from the inability to release into the canal system the quantity to match exactly all the requirements, operation of the canals at high levels to reduce siltation, unexpected rainfalls, and breaches in the system. These losses range from 5 to 30% of diversions on projects with ample supplies and from 1 to 10% with limited supplies.

Off-farm efficiency, comprising the foregoing items of conveyance losses, ranges between 50 and 90%. In cases where water originates on the farm itself, such as from a well, the off-farm efficiency is 100%. The average irrigation efficiency for the entire United States is indicated in Table 2.10.

TABLE 2.10 AVERAGE EFFICIENCY IN THE UNITED STATES

Year	Trend Efficiency (%)			High Efficiency (%)		
	On-farm	Off-farm	System	On-farm	Off-farm	System
1975	53	78	41			
1985	56	80	45	59	82	48
2000	59	83	49	66	88	58

Source: U.S. SCS (1976).

System efficiency is the overall efficiency obtained by multiplying the on-farm efficiency by the off-farm efficiency. Trend efficiency reflects irrigation/water improvements following the present trend in upgrading of systems. High efficiency considers an accelerated program of improving irrigation systems and water management using the best practical technology available.

Example 2.10

An irrigation project serves an area of 50,000 acres. The cropping pattern* is: alfalfa, 30%; wheat, 50%; rice, 30%; cotton, 20%. The monthly consumptive use values for these crops, which are calculated by the procedure of Section 2.9.2, are given in Table 2.11. The monthly effective rainfall values, which are calculated by the procedure of Section 2.9.3, are also given in the table. The irrigation water applied prior to crop growth and the soil moisture withdrawal in certain cases are also indicated in the table. On-farm efficiency is 60% and off-farm efficiency is 90%. Determine the monthly and total irrigation demands.

Solution   Refer to Table 2.11.

2.10 DEMAND OF WATER FOR HYDROPOWER

Water power involves the nonconsumptive use of water. This feature makes the water utilization distinct in two respects: (1) hydropower generation can readily be integrated with other development objects, and (2) all resources (streamflows) available at a site are evaluated from the consideration of power-producing potential. With proper planning, a very high percent of the total available streamflow in a river basin may be used for hydropower through a series of power plants. The problem pertains to locating the potential hydropower sites that are within a reasonable transmission distance of the power market under consideration. Since hydro energy is the product of the available head and the available flow (times a certain constant), the sites having a good combination of head and flow are investigated.

From a consideration of the head, rapids, falls, and dam locations offer good hydropower potential. Whereas the head at a site is practically constant, the available flows are highly variable. The study of maximum flows is important from the viewpoint of the design or installed capacity of the power plant; the average flows are important from the consideration of the energy output and minimum flows are required to predict the dependable plant capacity. Since the entire quantity available at a site (except the flood flows) is utilized in power production, the study of water demands for hydropower amounts to the collection of streamflow data and their analysis. Usually, the analysis relates to the preparation of the flow–duration curve discussed in Section 7.28.2, which indicates the magnitude of discharge against the percentage of time that discharge is exceeded at a site.

There are two types of hydropower plants: (1) a run-of-river plant uses direct streamflows, and its energy output is subjected to the instantaneous flow of the river; and (2) a storage plant with a reservoir is able to produce increased dependable en-

---

*The cropping pattern is defined as the percent of the total irrigable area devoted to each crop during each of the two principal growing seasons of a year. Each area used for crops in both seasons will be counted twice. The perennial crops using water in all 12 months will also be counted twice. Complete utilization of the land in both seasons will sum up to 200%.

TABLE 2.11 COMPUTATION OF MONTHLY IRRIGATION DEMANDS

Item	Jan.	Feb.	Mar.	Apr.	May	June	July	Aug.	Sept.	Oct.	Nov.	Dec.	Total
R (in.)	0.8	1.0	1.9	2.0	1.4	1.9	3.9	2.3	1.8	1.0	0.8	0.7	19.5
1. Alfalfa (30% of irrigable area)													
U (in.)	1.40	1.71	2.0	2.33	4.14	6.06	7.94	6.65	3.77	3.00	2.10	1.50	42.6
PP (in.) (+)													0
SM (in.) (−)													0
IR, gross (in.)	1.40	1.71	2.0	2.33	4.14	6.06	7.94	6.65	3.77	3.00	2.10	1.50	42.6
IR, net (in.)	0.6	0.71	0.10	0.33	2.74	4.16	4.04	4.35	1.97	2.00	1.30	0.80	23.1
IR, eff. (in.)	0.18	0.21	0.03	0.10	0.82	1.25	1.21	1.31	0.59	0.60	0.39	0.24	6.93
2. Wheat (50%)													
U (in.)	2.11	3.43	6.29	5.90					2.10	1.30	1.75	1.50	22.28
PP (in.) (+)									2.10				2.10
SM (in.) (−)				2.20									2.20
IR, gross (in.)	2.11	3.43	6.29	3.70					2.10	1.30	1.75	1.50	22.18
IR, net (in.)	1.31	2.43	4.39	1.70					0.3	0.30	0.95	0.80	12.18
IR, eff. (in.)	0.66	1.21	2.20	0.85					0.15	0.15	0.48	0.40	6.10

TABLE 2.11 (contd.)

Item	Jan.	Feb.	Mar.	Apr.	May	June	July	Aug.	Sept.	Oct.	Nov.	Dec.	Total
3. Rice (30%)													
U (in.)					6.80	10.42	12.53	9.50					39.25
PP (in.) (+)				9.00									9.00
SM (in.) (−)								2.50					2.50
IR, gross (in.)				9.00	6.80	10.42	12.53	7.00					45.75
IR, net (in.)				7.00	5.40	8.52	8.63	4.70					34.25
IR, eff. (in.)				2.10	1.62	2.56	2.59	1.41					10.28
4. Cotton (20%)													
U (in.)			1.34	2.82	2.96	5.36	6.86	4.14	1.14				24.62
PP (in.)		1.00											1.00
SM (in.)								1.5	0.5				2.0
IR, gross (in.)		1.00	1.34	2.82	2.96	5.36	6.86	2.64	0.64				23.62
IR, net (in.)		0	0	0.82	1.56	3.46	2.96	0.34	0				9.14
IR, eff. (in.)		0	0	0.16	0.31	0.69	0.59	0.07	0				1.82

TABLE 2.11 (contd.)

Item	Jan.	Feb.	Mar.	Apr.	May	June	July	Aug.	Sept.	Oct.	Nov.	Dec.	Total
Total, IR (eff.) (in.)	0.84	1.42	2.23	3.21	2.75	4.50	4.39	2.79	0.74	0.75	0.87	0.64	25.13
IR for area (acre-ft × 10³)	3.50	5.92	9.29	13.38	11.46	18.75	18.29	11.62	3.08	3.13	3.62	2.67	104.71
Farm delivery @ 60% efficiency (acre-ft × 10³)	5.83	9.87	15.48	22.30	19.10	31.25	30.48	19.37	5.13	5.21	6.03	4.45	174.52
Gross requirements (acre-ft × 10³)	6.48	10.97	17.20	24.78	21.22	34.72	33.87	21.52	5.70	5.79	6.70	4.94	193.90

Abbreviations:

R = effective rainfall
U = consumptive use
PP = irrigation applied prior to crop growth.
SM = soil moisture withdrawal
IR, gross = gross irrigation required = U + PP − SM
IR, net = net irrigation required = IR(gross) − R
IR, effective = IR(net) × percent irrigable area
IR for area = Farm area × Total IR (eff.)

$$\text{Farm delivery} = \frac{\text{IR for area}}{\text{farm efficiency}}$$

$$\text{Gross requirement} = \frac{\text{farm delivery}}{\text{off-farm conveyance efficiency}}$$

ergy on the basis of the controlled water release. If the reservoir serves only to smooth out the weekly fluctuations in streamflows, the plant is said to have a pondage capacity. On the other hand, a reservoir that serves to store water from the wet season to the dry season is said to have a storage capacity.

2.10.1 Power and Energy Production from Available Streamflows

Lowering a water quantity of Q ft^3/sec over h ft will release energy at a rate of $(62.4 \, Qh)$ ft-lb/sec. Converting this to kW units and including the efficiency term, an equation for power (rate of energy) can be given by

$$P = \frac{Qhe}{11.8} \quad [\text{FLT}^{-1}] \tag{2.23}$$

where

P = plant capacity, kW

Q = discharge through the turbines, cfs

h = net head on the turbines, ft

e = combined efficiency for turbines and generators

Flow–duration curves developed from long-term monthly streamflow records offer a convenient tool in plant capacity design. The procedure for preparation of a duration curve is described in Section 7.28.2. A typical curve is shown in Figure 2.8.

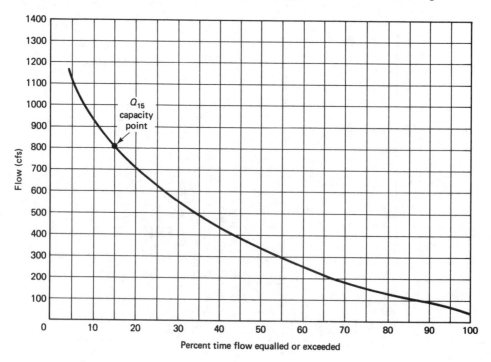

Figure 2.8 Flow–duration curve.

In eq. (2.23), with an average value of head, the efficiency and head are practically constant for a plant. Thus the power is directly proportional to the flow. In other words, the curve in Figure 2.8 indicates the power production with a suitable modification of the vertical scale. The design or installed capacity of a plant is based on the maximum flow, which is usually taken to be Q_{15} (i.e., flow exceeded 15% of the time). Floodflows above this magnitude are allowed to overflow without producing power. Thus the installed capacity is given by (taking $e = 0.84$)

$$P_{\text{instal}} = \frac{Q_{15}h}{14} \qquad [\text{FLT}^{-1}] \qquad (2.24)$$

where

$$P_{\text{instal}} = \text{installed capacity, kW}$$
$$Q_{15} = \text{discharge with 15\% exceedence, cfs}$$
$$h = \text{net head, ft}$$

If the time scale (abscissa) in Fig. 2.8 is expressed in terms of hours in a year, the area under the curve will provide the annual energy production. Mathematically,

$$E = \frac{Q_{\text{av}}h}{14}(8760) \qquad [\text{FL}] \qquad (2.25)$$

where

$$E = \text{annual energy, kWh.}$$
$$Q_{\text{av}} = \text{average discharge, cfs}$$
$$8760 = \text{number of hours in a year}$$

Q_{av} is the average discharge under the curve in Figure 2.8 taking Q_{15} as the highest magnitude of discharge, similar to eq. (7.59).

A plant capacity factor is the ratio of the average power production to the installed capacity. This is practically equal to the ratio Q_{av}/Q_{15}, assuming that the head and the efficiency are essentially constant. By reservoir storage, both Q_{av} and Q_{min} are improved, and thus the annual energy production and the dependable (firm) power are enhanced. The plant capacity factor also increases, resulting in a more efficient use of a plant. A plant capacity factor of 0.6 is common for storage-type power plants.

Energy computations assume that an adequate number and adequate sizes of turbine units have been installed to utilize the minimum available flow. If only one turbine unit is provided, its operative range is generally from 30 to 110% of the turbine design flow, which means that the turbine will be inoperative during the times the flow is less than 30% of the design value. Thus the energy production will be for a shorter period in a year and the total annual generation will, accordingly, be less.

Similarly, depending on the turbine type, there is an operating limitation on the head. Usually, a turbine can operate in the range 60 to 120% of the design head. It is considered that the available head is fairly constant or that an average value of head is used in energy computations by eq. (2.25) when there are small fluctuations, which is the case with run-of-river projects and projects with remote location of

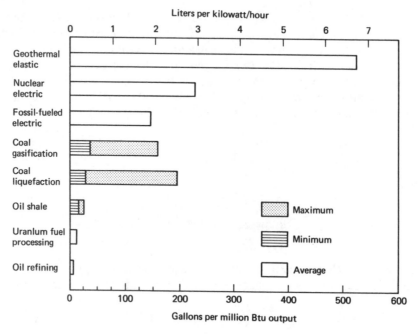

Figure 2.9 Water consumption in refining and conversion processes [from United States Geological Survey (USGS) Circular No. 703, 1974].

power plants. If variations in head are substantial, a sequential analysis is made wherein the energy calculations are made in steps at different intervals.

2.10.2 Demands for Other Energy Sources

In addition to the requirements in thermal power stations for cooling purposes as discussed in the context of industrial requirements and for hydropower generation described in the previous section, water is needed in other sources of energy production as well. It is needed for the processes related to the extraction of energy sources, such as the mining and refining of coal, uranium, and oil. It is also needed for the energy conversion processes of heat energy to mechanical energy to electrical energy. Water demands for extraction and conversion processes of various fuel energy sources are indicated in Figure 2.9. Per unit energy production, the consumption of water for cooling purpose in thermal power plants is 40 to 150 times greater than for the other sources of energy listed in Figure 2.9.

Example 2.11

At the Rimmon Pond site on the Naugatuck River near Seymourtown, Connecticut, in the Housatonic basin (drainage area 300 mi²), the flow–duration data from the monthly flow records are as given in Figure 2.8. The average head is 30 ft. Assess the site for its hydropower potential.

1. From Figure 2.8, Q_{15} = 810 cfs.
2. From eq. (2.24),

$$P_{instal} = \frac{810(30)}{14} = 1736 \text{ kW}$$

3. $Q_{av} = 0.175Q_{15} + 0.075Q_{20} + 0.10(Q_{30} + Q_{40} + Q_{50}$
$+ Q_{60} + Q_{70} + Q_{80} + Q_{90}) + 0.05\tilde{Q}_{100}$
$= 0.175(810) + 0.075(705) + 0.10(550 + 430 + 340 + 260 + 180$
$+ 130 + 90) + 0.05(40)$
$= 395$ cfs.

4. From eq. (2.25),

$$E = \frac{395(30)(8760)}{14} = 7.41 \times 10^6 \text{ kWh}$$

5. Plant capacity factor $= \dfrac{Q_{av}}{Q_{15}} = \dfrac{395}{810} = 0.49.$

2.11 DEMAND FOR NAVIGATION

There are three different methods to provide navigable waterways: (1) river regulation, (2) lock-and-dam, and (3) artificial canalization. In the first method, a river
channel is improved by means of river training works and dredging. In some sections of the river channel, the natural depth of water is often not sufficient to maintain navigability, which requires release of water from upstream reservoirs. This
demand from the reservoirs is likely to be on the order of several thousand cubic feet
per second successively for several months. Thus huge reservoir capacities of several
million acre-feet are needed for navigation purpose. One of the shortcomings of this
method is that the water deficiencies are usually in the lower reaches of a river, while
the reservoir sites are in its upper part. This results in many technical, operational,
and legal difficulties in maintaining the navigable flow in downstream reaches.

In the second method, the depth of water for low streamflow is increased behind
a series of dams through a succession of backwater curves. At each dam, a shiplock
is provided to negotiate the difference in water levels upstream and downstream of
the dam. The water demand relates to the (1) evaporation losses from the reservoir
pools, (2) water requirements for locking operations, and (3) leakages at shiplocks.

Each locking operation requires the release of water in the downstream direction
equivalent to the volume of the lock between the upstream and downstream levels.
This might involve a flow of over 1000 acre-ft/day (500 cfs). The water for lockages
is not accumulative since the water displaced by one lock can subsequently be used
by the next lock downstream. Compared to the locking requirements, the evaporation and leakages are insignificant.

The third method provides for an artificially constructed new channel with a
number of shiplocks. This method is adopted either to connect two different river
systems or in situations where the other two methods are not suitable. As regards the
water demand, a flow of several hundred cubic feet per second has to be maintained
through the channel. This is supplied from a stream with a natural dependable flow,
or by a reservoir. In addition, the requirements of evaporation, lockage operation,
and leakage as discussed for the second method, have to be provided for. If an unlined channel is constructed, the seepage losses have to be included also.

PROBLEMS

2.1. A community had a population of 12,000 in 1960, which is increased to 20,000 in 1985. The saturation population is 80,000. Estimate the 1995 population by (a) arithmetic growth, (b) constant percent increase, and (c) decreasing rate of increase.

2.2. Using the following census figures, estimate the population for 1990 by (a) the graphical method, and (b) the most appropriate mathematical method.

Year	Population (thousands)
1950	35.8
1960	38.2
1970	40.7
1980	43.3

2.3. From the following census data, estimate the 1965 and 1988 population by (a) the graphical method, and (b) the most appropriate mathematical method.

Year	Population
1950	25,000
1960	28,190
1970	31,780
1980	35,830

2.4. A water supply reservoir has a capacity of 25×10^3 acre-ft. It is serving a city having a present population of 60,000, which is expected to increase to 100,000 in 20 years. For how many years will the reservoir be adequate to supply the city? Assume arithmetic growth of the population and an average daily draft from the reservoir of 160 gallons per person. Neglect the losses.

2.5. A community has an estimated population 20 years hence of 15,000. The water treatment plant of the community has a capacity of 7.0 mgd, which is adequate for the next 35 years, with an input rate to the plant of 175 gallons per person per day. If the community is growing at a geometric rate, what is the present population?

2.6. A community has a population of 28,000. It is estimated that 20 years hence its population will be 38,000, and the saturation population is expected to be 80,000. The total water consumption at present has been estimated to be 4.0 million gallons per day. The existing treatment plant has a capacity of 9.2 million gallons per day. Determine in how many years the consumption will reach its design capacity if the community has a declining growth rate.

2.7. The continental United States registered the following populations. Determine (a) the saturation population, (b) the equation of the logistic curve, and (c) the projected population in the year 2000.

Year	Population (millions)
1820	9.6
1900	76.0
1980	225.1

2.8. A city has the following census data. Fit a logistic curve to the data and determine **(a)** the saturation population, and **(b)** the population in the year 2000.

Year	Population (thousands)	Year	Population (thousands)
1860	15.0	1930	67.8
1870	20.0	1940	78.0
1880	25.0	1950	83.4
1890	32.0	1960	91.8
1900	40.0	1970	96.6
1910	47.5	1980	103.2
1920	58.8		

2.9. A community located within the city of Problem 2.8 has the following census data. Estimate the population for 1980 by **(a)** the constant ratio method, and **(b)** the changing ratio method using the graphical extension.

Year	Population (thousands)	Year	Population (thousands)
1900	6.5	1940	16.0
1910	8.4	1950	17.4
1920	11.2	1960	19.5
1930	13.5		

2.10. Estimate the 1980 and 2000 population of the community in Problem 2.9 by the simple regression analysis.

2.11. Average daily usage of water in a city is 175 gallons per capita per day (gpcd), which excludes the fire demands. Determine **(a)** the maximum monthly usage in gpcd, **(b)** the maximum weekly usage in gpcd, and **(c)** the maximum daily usage in gpcd.

2.12. For Problem 2.11, determine the maximum hourly requirement for water.

2.13. The future population of a community has been estimated to be 100,000. Determine the rate of fire demand and its duration.

2.14. The fire demand of a community is dictated by a six-story building of ordinary construction having a floor area of 20,000 ft^2. Determine the daily requirement for firefighting purpose.

2.15. Determine the fire flow for a four-story wood frame building having a floor area of 1000 m^2 which is connected with a six-story building of noncombustible construction that has a floor area of 990 m^2.

2.16. If the population of the community in Problem 2.14 is 50,000 and the average daily usage of municipal supply is 175 gpcd, determine the design flow for the following:
(a) Groundwater source development
(b) Conduit to the treatment plant
(c) Water treatment plant
(d) Pumping plant
(e) Distribution system
(f) Service reservoir if the working storage is 2.5 mgd

2.17. In 1986 the water requirements of a city of 500,000 population were as follows:

Municipal:	87 mgd
Industries:	
Manufacturing	115 mgd
Thermal power	210 mgd
Waste dilution:	6.59 bgd

In the year 2000 it is expected that the population will increase by 10%, industries by 15%, and the thermal power by 100 MW. Determine the requirements by each sector assuming the same level of waste treatment as at present. Assume a plant capacity factor of 0.6.

2.18. At Boise, Idaho, latitude 43°54′N, the long-term mean monthly temperatures are as follows:

Month	Temp. (°F)	Month	Temp. (°F)
Jan.	27.9	July	72.5
Feb.	33.6	Aug.	71.0
Mar.	41.4	Sept.	61.2
Apr.	49.1	Oct.	50.1
May	56.1	Nov.	39.7
June	64.5	Dec.	30.4

Compute the seasonal consumptive use of water for an alfalfa crop having a growing season of April 1 to September 15.

2.19. For the Boise, Idaho, climate in Problem 2.18, compute the seasonal consumptive water use for potatoes. The growing season is May 10 to September 15.

2.20. For the Boise, Idaho, climate in Problem 2.18, compute the seasonal consumptive water use for grain sorghum. The growing season is June 5 to November 2.

2.21. An irrigation project serves an area of 100,000 acres. The cropping pattern is: wheat, 40%; potatoes, 30%; grain sorghum, 35%; and citrus, 25%. The monthly consumptive use and the effective irrigation for these crops are given below. The irrigation water applied prior to crop growth and the soil moisture withdrawals for certain months are also indicated. The on-farm irrigation efficiency is 65% and the off-farm conveyance efficiency is 85%. Determine the monthly diversions and total demand for irrigation.

Item	Jan.	Feb.	Mar.	Apr.	May	June	July	Aug.	Sept.	Oct.	Nov.	Dec.
R (in.)	0.8	0.9	1.5	1.9	1.7	2.5	4.5	3.2	2.1	1.4	0.6	0.5
Wheat												
U (in.)	2.1	3.2	5.95	5.70						1.5	1.75	1.40
PP (in.)									2.2			
SM (in.)				1.5								
Potatoes												
U (in.)					1.52	3.65	8.58	8.53	4.95			
PP (in.)				3.00								
SM (in.)								0.50	2.0			
(continued)												

Item	Jan.	Feb.	Mar.	Apr.	May	June	July	Aug.	Sept.	Oct.	Nov.	Dec.
Grain sorghum												
U (in.)						2.04	5.36	6.59	3.43	1.39	0.49	
PP (in.)					1.0							
SM (in.)												
Citrus												
U (in.)	1.35	1.75	2.75	4.1	4.4	5.0	6.8	7.5	5.9	4.9	2.5	1.6
PP (in.)												
SM (in.)												

Abbreviations:

R = effective rainfall
U = consumptive use
PP = irrigation applied prior to crop growth
SM = soil moisture withdrawal

2.22. Flow–duration data for the Housatonic River near New Milford Town, Connecticut, are indicated below. The average head at the site is 25 ft. Assess the site with respect to **(a)** potential capacity, **(b)** annual energy generation, and **(c)** plant capacity factor.

Flow (cfs)	450	930	1180	1370	1680	1950	2180	2360
Percent of time flow exceeded	100	90	80	70	50	30	20	15

2.23. At the Harrisville, New York, site on the West Oswegatchie River, the flow–duration data are as given below. The average head is 35 ft. Assess the site for **(a)** potential capacity, **(b)** annual energy generation, and **(c)** plant capacity factor.

Flow (cfs)	1000	600	400	300	200	120	40	20
Percent of time flow exceeded	9	12	15	19	28	47	85	100

2.24. In Problem 2.23, if the plant capacity factor is increased to 0.65 by the storage capacity, determine the percent increase in the annual energy generation.

CHAPTER
THREE

AVAILABILITY OF WATER

3.1 ASSESSMENT OF WATER QUANTITY

Design of water resources project is essentially an exercise in matching the demand with the supply of water. Two obvious sources of supply are surface water and groundwater. The techniques of measurement and issues related to quantitative assessment are the basic elements of hydrology that are covered in Chapters 4 through 8. In this chapter we provide a summary of the fundamental processes that contribute to the formation of surface and groundwater flows and a discussion of the key parameters of the process. An understanding of these parameters facilitates hydrologic analyses and planning.

3.2 HYDROLOGIC CYCLE

Both surface and groundwater flows originate from precipitation, which includes all forms of moisture falling on the ground from clouds, including rain, snow, dew, hail, and sleet. Precipitation at any place is distributed as follows:

1. A portion known as the interception is retained on buildings, trees, shrubs, and plants. This is eventually evaporated. The remaining quantity is known as the effective precipitation.

2. Some of the effective precipitation is evaporated back into the atmosphere directly.

3. Another portion is infiltrated into the ground. A part of the infiltration in the root zone is consumed by plants and trees and ultimately transpired into the atmosphere.

4. The water that percolates deeper into the ground constitutes the groundwater flow. It may ultimately appear as the baseflow in streams.

5. If the precipitation exceeds the combined evaporation and infiltration, puddles known as depression storage are formed. Evaporation takes place from these puddles.

6. After the puddles are filled, the water begins flowing over the surface to join a stream channel. With reference to the precipitation this is termed the net precipitation or precipitation excess. From a consideration of the surface water flow, this is known as direct runoff. Some evaporation takes place from the stream surface.

7. Runoff cannot occur unless a layer of water is formed in the path of motion of water. The water in this layer is known as detention storage. Evaporation takes place from this storage as well. When precipitation ceases, the water in detention storage eventually joins the stream channel.

8. The destination of all streams is open bodies of water, such as oceans, seas, and lakes, which are subject to extensive evaporation.

9. The evaporation from all the sources above, together with the transpiration, carries the moisture into the atmosphere. This results in the formation of clouds that contribute to the precipitation when steps 1 through 9 repeat.

This chain process, driven principally by energy from the sun, is known as the hydrologic cycle. The complete cycle is global in nature. Subcycles with smaller boundary limits also exist.

3.3 WATER BALANCE EQUATION

In quantitative terms the hydrologic cycle can be represented by a closed equation which represents the principle of conservation of mass, often referred to in hydraulics as the continuity equation. Many forms of this expression, called the water balance equation, are possible by subdividing, consolidating, or eliminating some of the terms, depending on the purpose of computation. The water balance can be expressed (1) for a short interval or for a long duration; (2) for a natural drainage basin or an artificially separated boundary or with respect to water bodies such as lakes, reservoirs, and groundwater basins; and (3) for the phase above the ground surface, that below the surface, or the entire phase. Three applications of the water balance equation are common; (1) a water balance equation for large basin areas, (2) a water balance equation for water bodies, and (3) a water balance equation for direct runoff. In the first two cases the entire phase above and below the ground surface is considered in the equation in terms of the streamflows. In its general form, the equation may be represented by

$$P + Q_{SI} + Q_{GI} - E - Q_{SO} - Q_{GO} - \Delta s - n = 0$$
$$[L^3 \quad \text{or} \quad L^3 T^{-1}] \quad (3.1)$$

where

$$P = \text{precipitation}$$

$$Q_{SI}, Q_{GI} = \text{surface and groundwater inflow into the boundary from outside}$$

$$E = \text{evaporation (including transpiration)}$$

$$Q_{SO}, Q_{GO} = \text{surface and groundwater outflow from the boundary}$$

$$\Delta s = \text{change of storage volume within the boundary}$$

$$n = \text{discrepancy term}$$

Since all water balance components are subject to errors of measurement and estimation, a discrepancy term has been included. The components of the eq. (3.1) are expressed as a volume of water or in the form of flow rates or as a mean depth over the basin. The last form is convenient for the balance equation of direct runoff.

3.3.1 Balance Equation for Large River Basins for Long Duration

In large river basins, the water balance equation is used for the quantitative evaluation of basin resources and for substantiation of projects for their intended use and proposed modifications. The study of mean water balances is usually performed on a long-duration basis (for an annual cycle). Over a long period, positive and negative water storage variations tend to balance, and the change in storage, Δs, may be disregarded. The groundwater exchange in large basins with neighboring basins is ignored (i.e., $Q_{GI} - Q_{GO} = 0$). There is no surface water inflow into a basin with a distinct watershed divide (i.e., $Q_{SI} = 0$). Ignoring the discrepancy term, eq. (3.1) reduces to

$$P - E - Q = 0 \quad [\text{L}^3 \quad \text{or} \quad \text{L}] \tag{3.2}$$

where Q is the discharge volume from the basin into the river.

3.3.2 Balance Equation for Water Bodies for Short Duration

The water balance equation for reservoirs, lakes, streams, and groundwater reservoirs is used to predict the consequences of the prevailing hydrologic conditions on the structure. The equation is relevant for day-to-day operation of the structure. The short time period is involved in these studies and the term Δs must be considered. Evaporation during this short period can, however, be neglected and inflow and outflow terms can be combined. The equation can be represented by

$$Q_i - Q_o - \frac{\Delta s}{\Delta t} = 0 \quad [\text{L}^3\text{T}^{-1}] \tag{3.3}$$

where

$$Q_i, Q_o = \text{average inflow and outflow rate, respectively}$$

$$\frac{\Delta s}{\Delta t} = \text{change of storage during } \Delta t$$

Example 3.1

At a particular time the storage in a river reach is 55.3 acre-ft. At that instant, the inflow into the reach is 375 cfs and the outflow is 563 cfs. After 2 hours, the inflow and outflow are 600 cfs and 675 cfs, respectively. Determine:

(a) The change of storage during 2 hours.
(b) The storage volume after 2 hours.

Solution

(a) **1.** Average inflow rate $= \dfrac{375 + 600}{2} = 487.5$ cfs.

2. Average outflow rate $= \dfrac{563 + 675}{2} = 619$ cfs.

3. From eq. (3.3),

$$487.5 - 619 - \frac{\Delta s}{\Delta t} = 0 \quad \text{or} \quad \frac{\Delta s}{\Delta t} = -131.5 \text{ cfs}$$

4. $\Delta s = -131.5(2) = -263$ cfs-hr

$$= \left(-263 \frac{\text{ft}^3}{\text{sec}} \text{hr}\right)\left(\frac{60 \times 60 \text{ sec}}{1 \text{ hr}}\right)\left(\frac{1 \text{ acre-ft}}{43,560 \text{ ft}^3}\right)$$

$$= -21.73 \text{ acre-ft.}$$

(b) $S_2 = S_1 + \Delta s = 55.30 - 21.73 = 33.57$ acre-ft.

3.3.3 Balance Equation for Direct Runoff in a Basin during a Storm

For determining the runoff from a storm, the water balance over the ground surface is considered. This is a short-time-interval equation of the form

$$P - L_i - E - R - I - S_D = 0 \qquad [\text{L}] \tag{3.4}$$

where

$L_i =$ interception

$R =$ direct surface runoff or net precipitation

$I =$ infiltration

$S_D =$ depression storage

The storm evaporation during the short period is small and can be neglected. Also ignoring the interception and depression storage in comparison with the infiltration (in a more exact determination, these terms are estimated separately), eq. (3.4) reduces to

$$R = P - I \qquad [\text{L}] \tag{3.5}$$

The application of eq. (3.5) is discussed in Section 3.9.

3.3.4 Water Balance Equation for Direct Runoff in a Basin for Longer than Storm Duration

The long duration in this balance equation means a period longer than the storm duration for which the component of evapotranspiration cannot be neglected. The values of the water balance components are averaged for this period. This can be a daily, weekly, monthly, or yearly duration. Models of this type have been developed by Thornthwaite and Mather (1955), Palmer (1965), and Haan (1972), in which the input of water from precipitation has been equated to the outflow of water by evapotranspiration, infiltration, and runoff. Conceptually, these models consider that moisture is either added to or subtracted from the soil, depending on whether precipitation for a period is greater than or less than the potential evapotranspiration.

When precipitation is less than the potential evapotranspiration, actual evapotranspiration in these models is treated as a function of the soil moisture content. This results in the loss of soil moisture and an increased moisture deficit.*

When precipitation for a period exceeds the potential evapotranspiration, moisture is added to the soil until it attains its capacity. Any excess water contributes to runoff.

A model presented by Thomas (1981), known as the *abcd* model, places an upper limit on the sum of evapotranspiration and soil moisture storage together rather than only on the soil moisture storage to its capacity. This provides a value of actual evapotranspiration less than the potential evapotranspiration and can simulate a decrease in soil moisture storage even when precipitation is in excess of potential evapotranspiration.

The Thomas model defines two state variables. One, known as the available water, is the sum of the precipitation to the end of a period i and the soil moisture storage to the end of the previous period $(i - 1)$; that is,

$$W_i = P_i + S_{i-1} \quad [\text{L}^3 \quad \text{or} \quad \text{L}] \tag{3.6}$$

The other state variable, Y_i, is the sum of actual evapotranspiration and soil moisture storage at the end of period i; that is,

$$Y_i = E_i + S_i \quad [\text{L}^3 \quad \text{or} \quad \text{L}] \tag{3.7}$$

The following nonlinear relation has been suggested by Thomas between the two state variables.

$$Y_i = \frac{W_i + b}{2a} - \left[\left(\frac{W_i + b}{2a} \right)^2 - \frac{W_i b}{a} \right]^{0.5} \quad [\text{L}^3 \quad \text{or} \quad \text{L}] \tag{3.8}$$

where a and b are the model parameters. Parameter a, according to Thomas, reflects "the propensity of runoff to occur before the soil is fully saturated." Its value of less than 1 results in runoff for $W_i < b$. Parameter b is an upper limit on the sum of evapotranspiration and soil moisture storage. Equation (3.8) assures that $Y_i < W_i$.

In order to allocate Y_i of eq. (3.7) between evapotranspiration and soil moisture storage at the end of the period, it is assumed that the rate of loss of soil moisture

---

*Moisture deficit is defined as the difference between the soil moisture capacity and the soil moisture storage at any time.

due to evapotranspiration is proportional to the soil moisture storage and potential evapotranspiration (PE), which leads to the relation

$$S_i = Y_i e^{-PE_i/b} \qquad [\text{L}^3 \text{ or } \text{L}] \qquad (3.9)$$

The difference $W_i - Y_i$ represents the sum of direct runoff $(DR)_i$ and infiltration contributing to groundwater recharge $(GR)_i$, since a part of the infiltrated water results in a change of soil moisture storage $(S_i - S_{i-1})$. The allocation between the direct runoff and groundwater recharge is suggested as follows:

$$(GR)_i = c(W_i - Y_i) \qquad [\text{L}^3 \text{ or } \text{L}] \qquad (3.10)$$

$$(DR)_i = (1 - c)(W_i - Y_i) \qquad [\text{L}^3 \text{ or } \text{L}] \qquad (3.11)$$

where c is a model parameter that is related to the fraction of mean runoff that comes from groundwater.

If G_i denotes the groundwater storage at the end of period i, then

$$G_i = \frac{(GR)_i + G_{i-1}}{d + 1} \qquad [\text{L}^3 \text{ or } \text{L}] \qquad (3.12)$$

The groundwater discharge is given by

$$(QG)_i = dG_i \qquad [\text{L}^3 \text{ or } \text{L}] \qquad (3.13)$$

where d is a model parameter for the fraction of groundwater storage discharged.

The streamflow at the end of period i is equal to $(DR)_i + (QG)_i$. Thus the model is applied to determine the averaged streamflow.

The values of parameters a, b, c, and d are obtained by calibrating the model from the known data for the water balance components. Alley (1984) estimated the following mean monthly values from the study of 10 sites in New Jersey each having a record of 50 years: $a = 0.992$, $b = 30$, $c = 0.16$, and $d = 0.26$. Runoff estimates are very sensitive to parameter a.

The application of the model requires initial estimates of soil moisture storage, S_0, and groundwater storage, G_0. Thomas suggests the use of optimized values of S_0 and G_0 from the basin study. Alley (1984) suggests to assume some initial values of S_0 and G_0 and to simulate data for some period (for a year in monthly data) prior to the beginning of the period of interest. The potential evapotranspiration for use in eq. (3.9) is computed by the Thornthwaite method, described in Section 3.8.2.

Example 3.2

The average monthly precipitation data recorded at the Whippany River Basin (drainage area 29.4 mi^2) at Morristown, New Jersey, during 1985 are given below. The monthly computed potential evapotranspiration values for the basin are also indicated. The model parameters are: $a = 0.98$, $b = 25$, $c = 0.10$, and $d = 0.35$. Initial soil moisture storage and groundwater storage are ascertained to be 8.0 and 2.0 in., respectively. For each month, determine (a) moisture storage, (b) direct runoff, (c) groundwater recharge, (d) groundwater storage, (e) groundwater discharge, and (f) streamflow.

Month:	J	F	M	A	M	J	J	A	S	O	N	D
Precipitation (in.)	1.0	1.8	0	0.60	0	0.1	8.78	9.80	4.80	0.74	0.40	6.88
Potential evapotranspiration (in.)	1.75	1.78	2.1	2.2	3.0	3.1	3.5	3.6	2.7	1.9	1.75	1.75

Solution

1. The computations are made for successive periods: monthly for this problem.

2. For January 1985:
   From eq. (3.6),

   $$W_1 = P_1 + S_0 = 1.0 + 8.0 = 9.0.$$

   From eq. (3.8),

   $$Y_1 = \frac{9.0 + 25.0}{2(0.98)} - \left\{\left[\frac{9.0 + 25.0}{2(0.98)}\right]^2 - \frac{(9.0)(25.0)}{(0.98)}\right\}^{0.5} = 8.90$$

   From eq. (3.9), soil moisture storage,

   $$S_1 = 8.90e^{-1.75/25.0} = 8.3 \text{ in.}$$

   From eq. (3.10), groundwater recharge,

   $$(GR)_1 = 0.10(9.0 - 8.9) = 0.01 \text{ in.}$$

   From eq. (3.11), direct runoff,

   $$(DR)_1 = (1 - 0.10)(9.0 - 8.9) = 0.09 \text{ in.}$$

   or $(0.09 \text{ in.})(29.4 \text{ mi}^2)\left(\frac{1 \text{ ft}}{12 \text{ in.}}\right)\left(\frac{5280^2 \text{ ft}^2}{1 \text{ mi}^2}\right) = 6.15 \times 10^6 \text{ ft}^3$

   or $(6.15 \times 10^6)\left(\frac{1 \text{ month}}{31 \times 24 \times 60 \times 60 \text{ sec}}\right) = 2.30 \text{ cfs}$

   From eq. (3.12), groundwater storage,

   $$G_1 = \frac{0.01 + 2.0}{1 + 0.35} = 1.49 \text{ in.}$$

   or $G_1 = (\text{depth})(\text{drainage area})$

   $$= (1.49 \text{ in.})(29.4 \text{ mi}^2)\left(\frac{1 \text{ ft}}{12 \text{ in.}}\right)\left(\frac{5280^2 \text{ ft}^2}{1 \text{ mi}^2}\right)$$

   $$= 101.8 \times 10^6 \text{ ft}^3$$

   From eq. (3.13), groundwater discharge,

   $$(QG)_1 = G_1 d = (101.8 \times 10^6 \text{ ft}^3)\left(\frac{0.35}{\text{month}}\right)\left(\frac{1 \text{ month}}{31 \times 24 \times 60 \times 60 \text{ sec}}\right)$$

   $$= 13.3 \text{ cfs}$$

   $$\text{streamflow} = (DR)_1 + (QG)_1 = 2.30 + 13.30 = 15.60 \text{ cfs}$$

3. Similar computations are performed for the month of February with starting values of S and G as 8.3 and 1.49, respectively, and so on.

3.4 ERRORS OF COMPUTATION OF WATER BALANCE COMPONENTS

Evaluation of water balance components always involves errors due to measurement and interpretation (Winter, 1981). Precipitation and streamflow are the only components of the balance equation that are observed extensively from the network of stations. The data for evaporation are observed on a limited scale and for infiltration from the experimental basins. The water storage variations are obtained from water-level and soil moisture observations and by snow surveys. Also, empirical formulas are used for the computation of evaporation, infiltration, and water storage. Winter (1981) discussed various types of errors involved in measurement and computation of components of the water balance equation. The time frame is very important since the long-term averages have smaller errors than the short-term values. The errors associated with annual and monthly estimates of water balance components are indicated in Table 3.1 based on commonly used methodologies.

TABLE 3.1 ERRORS IN HYDROLOGIC COMPONENTS BY COMMONLY USED METHODOLOGIES

	Percent Error	
	Annual Estimate	Monthly Estimate
1. Precipitation		
Gage observation	2	2
Gage placement (height)	5	5
No windshield		20
Areal averaging	10	15
Gage density	13	20
2. Streamflow		
Current-meter measurement	5	5
Stage–discharge relationship	20	30
Channel bias	5	5
Regionalization of discharge	70	
3. Evaporation		
Energy budget	10	
Class A pan	10	10
Pan to lake coefficient	,15	50
Areal averaging	15	15

Source: Based on Winter (1981).

The water balance equation usually does not balance out; accordingly, a residual term for the discrepancy has been included in eq. (3.1). When a component is estimated through an empirical formula, the error of imperfection of the formula is added into the residual term. If a component is evaluated indirectly by substituting other known terms in the water balance equation, this component includes the errors of computation of other variables. Winter (1981) computed the groundwater inflow of lake in three different geometric and climatic settings as the balance term of the water budget equation. The errors involved in the other three components of the equation (i.e., precipitation, evaporation, and streamflow) propagate as the sum of variances and covariances. Thus the variance of the error of the residual term is equal to the summation of variances of error of each of the other components of the equation, plus twice the covariances of the error of components with each other. Covariance terms relate to interrelationships of the measurement error of the components, not to the interrelationships of the components themselves. As indicated by Winter, overall error is the standard deviation of the error of the residual. For a worst possible estimate of the total error, the error due to each component is considered additive (of the same sign). Due to these errors, the groundwater inflow has been overestimated by 60% in one case and about 400% in two cases on an annual basis. Thus the errors of measurement and estimation have a significant impact on water balance calculations. This is particularly serious when a component is computed as the residual quantity. To minimize the error it is desirable to measure or compute all components using the best independent methods.

3.5 PRECIPITATION

Precipitation, largely in the form of rain and snow, is the source of moisture coming to the earth. It is a key source parameter in the water balance equation. The accuracy of measurement and computation of precipitation determines to a considerable extent the reliability of all water balance computations (Sokolov and Chapman, 1974). The rainfall and snowfall at any place are measured by gages of self-recording or manual observation type. These gages record the depth of rainfall or snowfall at any place within a given time. Snow measurements are made by standard rain gages equipped with shields to reduce the effect of wind. Snow boards and stakes are also used. The direct method of measuring snowfall is, however, not entirely satisfactory. It is supplemented by snow surveying. For this purpose, snow courses are established. Each course comprises a series of sampling points from where the samples of snowpack are taken by core-cutting equipment on a regular basis. In addition to the ground survey, aerial snow surveying is performed in remote places. Snow surveying provides information on snow depth variation, water equivalent, density, and snow quality. Each gage catches the precipitation falling within a circle 8 in. in diameter and hence indicates only a point measurement. To obtain the precipitation for an area on a representative basis, a number of gages are needed. A network of over 15,000 gages (stations) exist throughout the country, about 80% of which are operated by the National Weather Service and the remainder by other government and private agencies.

The measured precipitation data are subject to errors due to (1) water displacement by the dipstick, (2) amount for wetting of the gage collector, (3) evaporation during precipitation and manual readings, (4) height of the gage above the ground, and (5) the effect of wind.

The density and arrangement of the network and the method of analysis influence the estimate of areal distribution of rainfall from point data. Numerous papers have been published on precipitation measurement errors. Studies indicate that wind is the major cause of error in precipitation gage measurements (Larson and Peck, 1974). The errors increase with wind speed and are much greater for snowfall than for rainfall. A properly selected and well-protected site can reduce wind errors considerably. Gage shields can further reduce catch deficiency* for snow, although they

*Catch deficiency $= 1 - \dfrac{\text{gage catch}}{\text{true catch}}$.

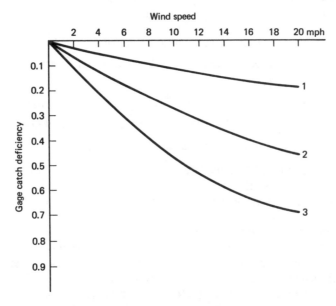

Figure 3.1 Gage catch deficiency versus wind speed: 1, rain gage; 2, snow gage with a shield; 3, snow gage without a shield (from Larson and Peck, 1974).

have very little effect on rain measurement. However, shields are not effective at wind speeds above 20 mph. Larson and Peck have summarized the catch deficiency versus wind speed based on the studies at two gages of the National Weather Service, as shown in Figure 3.1. At 20 mph, a catch deficiency of 70% can be experienced in snow measurement by an unshielded gage. A shield can reduce the error to about 50%. For rainfall, a deficiency of about 20% is expected with or without the shield at 20 mph. The curves of Figure 3.1 can be used to apply corrections to observed precipitation data.

3.5.1 Analysis of Point Precipitation Data

The point observations from a precipitation gage are subject to two regular problems. A gage site (station) may have a short break in the record because of instrument failure or absence of the observer. It is often necessary to estimate the missing record. Another problem is that the recording conditions at a gage site may have changed significantly some time during the period of record, due to relocation or upgrading of a station in the same vicinity, difference in observational procedure, or any other reason. The problem is resolved in both cases by comparison with neighboring gage sites.

Estimating missing data. The precipitation value missing at a site can be estimated from concurrent observations at three or more neighboring stations that are located as close to and evenly spaced from the missing-data station as possible, known as index stations. The normal-ratio method is used, according to which

$$\frac{P_x}{N_x} = \frac{1}{n}\left(\frac{P_1}{N_1} + \frac{P_2}{N_2} + \frac{P_3}{N_3} + \cdots + \frac{P_n}{N_n}\right) \qquad \text{[dimensionless]} \qquad (3.14)$$

where

$$P_x = \text{missing precipitation value for station X}$$

$$P_1, P_2, P_3, \ldots, P_n = \text{precipitation values at the neighboring station for the concurrent period}$$

$$N_x = \text{normal long-term, usually annual, precipitation at station X}$$

$$N_1, N_2, \ldots, N_n = \text{normal long-term precipitation for neighboring stations}$$

$$n = \text{number of index (neighboring) stations}$$

Equation (3.14) can be also applied for estimating the missing storm depth at a site by treating $P_1, P_2, \ldots$ as related to a particular storm. Also, $N_1, N_2, \ldots$ can be taken as the average values for a particular month for all years on record for index stations.

Checking consistency of data: double-mass analysis. The double-mass analysis is a consistency check used to detect whether the data at a site have been subjected to significant change in magnitude due to external factors such as tampering of the instrument, change in the recording conditions, or shift in observation practices. The change due to meteorological factors will equally affect all stations involved in the test and thus will not cause a lack of consistency created by the external effects. The analysis also provides a means of adjusting the inconsistent data. In the analysis, a plot is made of accumulated annual or seasonal precipitation values at the site in question (being checked for consistency) against the concurrent accumulated values of several surrounding stations. More conveniently, the mean of the surrounding stations is used in the accumulation and the plot, as shown in Figure 3.2. If the data are consistent, the plot will be a straight line. On the other hand, the inconsistent data will exhibit a change in slope or break at a point where inconsistency has occurred. This is shown by point V in Figure 3.2 in year 1961. If the slope of the line UV is a and of the line VW is b, the adjustment of the inconsistent data is made by the ratio of the slopes of two line segments. Two ways of adjustment are possible.

1. The data are adjusted to reflect the conditions that existed prior to the indicated break. This is done by multiplying each recent precipitation value after break-point V of station X (being tested) by the ratio a/b.
2. The data are adjusted to reflect recent conditions following the break. This is achieved by multiplying each value of the precipitation before the breakpoint by the ratio of b/a.

An adjustment of the second type is usually made. More than one break (change in the slope) in the data are observed in certain cases. Sometimes an apparent change in slope is noticed because of a natural variation in the data which is not to be associated with changes in gage location, gage environment, or observation procedure. In the case of doubt, a test of hypothesis should be performed by the Fisher distribution (Section 8.4.2) on the two sets of data (before and after the apparent break) to check whether the data are homogeneous and the break is purely by chance. Searcy and Hardison of U.S. Geological Survey (1960) recommend that if fewer than 10 stations are grouped together to check the consistency of a station, the record of each

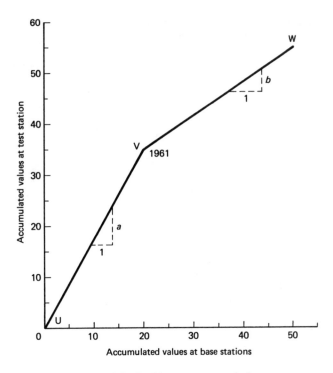

Figure 3.2 Double-mass-curve analysis.

station should be tested by double-mass analysis for consistency by plotting it against the group of all other stations, and those records that are inconsistent should be eliminated from the group.

The double-mass curve technique should seldom be used in mountainous areas. It is also not suitable for adjusting daily or storm precipitation. When the records of the stations have different starting dates, the mass curve can be formed by accumulating the data from recent to the past, in the reverse order to that indicated earlier.

Example 3.3

The annual records of five precipitation stations are given in Table 3.2. Check the consistency of station A. Adjust the record if it is inconsistent.

Solution:

1. The mean of a group of stations (B, C, D, and E) is computed in column 7.
2. The accumulated values for station A and the group of stations are given in columns 8 and 9.
3. Column 8 is plotted against column 9 in Figure 3.3. The breakpoint is observed at 1974.
4. The ratio of recent to past slope $= 1.06/0.78 = 1.36$.
5. The data prior to the breakpoint (1970–1973) are corrected by the factor 1.36, as indicated in Table 3.3

TABLE 3.2 ANNUAL PRECIPITATION FOR DOUBLE-MASS ANALYSIS

(1)	(2)	(3)	(4)	(5)	(6)	(7)	(8)	(9)
	Annual Precipitation for Station (in.)					Mean of Stations	Cumulated Precipitation for	Cumulated Precipitation for Mean of
Year	A	B	C	D	E	B, C, D, E	Station A	B, C, D, E
1970	26.28	29.89	24.55	36.56	31.80	30.70	26.28	30.70
1971	22.46	24.70	32.79	30.82	31.66	29.99	48.74	60.69
1972	26.81	33.60	32.35	38.61	33.61	34.54	75.55	95.23
1973	23.66	31.94	25.99	27.71	33.11	29.69	99.21	124.92
1974	19.00	29.06	29.38	36.10	25.24	29.95	118.21	154.87
1975	46.71	29.29	49.88	42.62	44.43	41.56	164.92	196.43
1976	36.99	30.89	38.28	32.06	38.49	34.93	201.91	231.36
1977	24.27	21.51	26.19	23.66	31.88	25.81	226.18	257.17
1978	37.42	25.95	28.90	33.34	36.32	31.13	263.60	288.30
1979	30.45	33.25	24.58	38.50	35.91	33.06	294.05	321.36
1980	34.26	25.06	28.32	31.58	26.11	27.77	328.31	349.13
1981	30.34	35.31	31.33	35.29	36.70	34.66	358.65	383.79
1982	40.53	40.50	34.62	31.15	36.84	35.78	399.18	419.57
1983	37.48	32.87	39.88	33.26	39.81	36.46	436.66	456.03
1984	40.42	31.21	38.29	39.73	37.81	36.76	477.08	492.79
1985	27.50	27.56	25.72	25.54	29.66	27.12	504.58	519.91

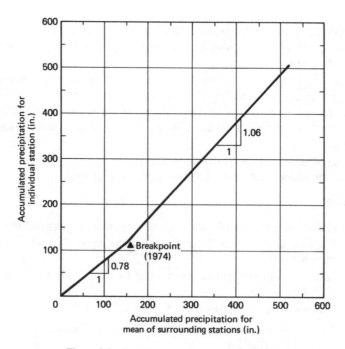

Figure 3.3 Double-mass curve for Example 3.3.

TABLE 3.3 ADJUSTED PRECIPITATION OF STATION A

Year	Recorded Precipitation (in.)	Adjusted Precipitation (in.)
1970	26.28	35.74
1971	22.46	30.55
1972	26.81	36.46
1973	23.66	32.18
1974–1985		Same as recorded

3.5.2 Conversion of Point Precipitation to Areal Precipitation

The representative precipitation over a defined area is required in engineering applications, whereas the gaged observation pertains to the point precipitation. The areal precipitation is computed from the record of a group of rain gages within the area by the following methods.

1. Arithmetic or station average method
2. Weighted average method
   a. Thiessen polygon method
   b. Isohyetal method

Arithmetic average method. This simple method consists of computing the arithmetic average of the values of the precipitation for all stations within the area. This method assigns equal weight to all stations irrespective of their relative spacings and other factors. The outside stations in proximity are included.

Thiessen polygon method. In this method, the weight is assigned to each station in proportion to its representative area defined by a polygon. These polygons are formed as follows:

1. The stations are plotted on a map of the area drawn to a scale (Figure 3.4).
2. The adjoining stations are connected by the dashed lines.
3. Perpendicular bisectors are constructed on each of these dashed lines, as shown by the solid lines in Figure 3.4.
4. These bisectors form polygons around each station. Each polygon is representative of the effective area for the station within the polygon. For stations close to the boundary, the boundary lines form the closing limit of the polygons.
5. The area of each polygon is determined* and then multiplied by the rainfall value for the station within the polygon.
6. The sum of item 5 divided by the total drainage area provides the weighted average precipitation.

*This is done by using a planimeter or drawing the figure to a scale on the graph paper and counting the division squares covered by the polygon, duly multiplied by the map scale.

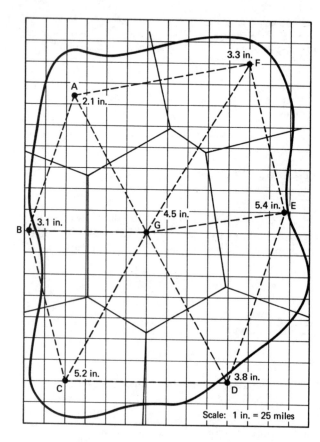

Figure 3.4 Thiessen polygon method of average areal precipitation.

Example 3.4

The rain gages located in and around a drainage area are shown in Figure 3.4, along with the rainfall recorded at these stations due to a storm. Determine the average precipitation for the drainage area by (a) the arithmetic average method, and (b) the Thiessen polygon method.

Solution

(a) Arithmetic average = $\dfrac{2.1 + 3.1 + 5.2 + 3.8 + 5.4 + 3.3 + 4.5}{7}$

$\bar{p} = 3.91$ in.

(b) Thiessen polygon method: Refer to Table 3.4.

Isohyetal method. This is the most accurate of the three methods and provides a means of considering the orographic (mountains) effect. The procedure is as follows:

1. The stations and rainfall values are plotted on a map to a suitable scale.
2. The contours of equal precipitation (isohyets) are drawn as shown in Figure 3.5. The accuracy depends on the construction of the isohyets and their intervals.

Availability of Water   Chap. 3

TABLE 3.4 AVERAGE PRECIPITATION COMPUTATION
BY THIESSEN POLYGON METHOD

(1) Observed Precipitation (in.)	(2) Area of Polygon (mi^2)	(3) Precipitation × area (col. 1 × col. 2)
2.1	735	1,543.5
3.1	475	1,472.5
5.2	640	3,328.0
3.8	620	2,356.0
5.4	740	3,996.0
3.3	685	2,260.5
4.5	1,210	5,445.0
Total	5,105	20,401.5

$$\text{Average } \overline{p} = \frac{20,401.5}{5,105} = 4.0$$

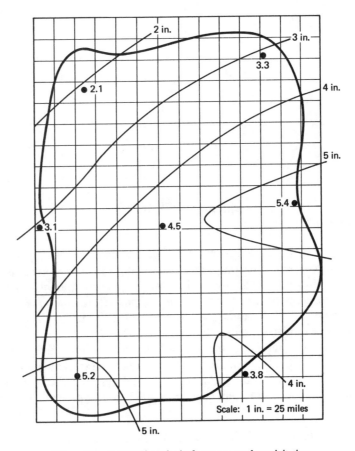

Figure 3.5 Isohyetal method of average areal precipitation.

3. The area between successive isohyets is computed and multiplied by the numerical average of the two contour (isohyets) values.

4. The sum of item 3 divided by the drainage area provides the weighted average precipitation.

Example 3.5

Solve Example 3.4 by the isohyetal method.

Solution Refer to Table 3.5.

TABLE 3.5 AVERAGE PRECIPITATION COMPUTATION BY ISOHYETAL METHOD

(1) Isohyet (in.)	(2) Area Covered by the Isohyet (mi²)	(3) Area between Two Isohyets (mi²)	(4) Average of Two Isohyets (in.)	(5) Precipitation × Area (col. 3 × col. 4)
<2	0			
		125	1.8 (est.)	225
2	125			
		820	2.5	2,050
3	945			
		1,100	3.5	3,850
4	2,045			
		2,555	4.5	11,498
5	4,600			
		500	5.2	2,600
>5	5,100			
Total		5,100		20,223

$$\text{Average } \bar{p} = \frac{20{,}223}{5{,}100} = 3.96 \text{ in.}$$

3.5.3 Intensity–Duration–Frequency (IDF) Analysis of Point Precipitation

The point precipitation data of various storms are analyzed in an IDF study. Since the precipitation data serve the purpose of estimating the streamflows in many instances, not only the total quantity of precipitation but its rate, known as the intensity, and the duration are important in a consideration of peak-flow study. A point or gaged observation can be considered to be representative of a 10-mi² drainage area. Hence the studies of point-rainfall extremes are extensively used in the design of the small-area drainage systems comprising storm sewers, drains, culerts, and so on. The application of intensity, duration, and frequency data in the rational method is discussed in Chapter 12. The intensity–duration–frequency analysis can be carried out only where the data from a recording rain gage are available. The procedure of analysis is as follows:

1. A specific duration of rainfall, such as 5 min, is selected.

2. From the record of the rain gage, which indicates the accumulated amount of precipitation with respect to time, the maximum rainfall of this duration in

each year is noted. This is the maximum incremental precipitation (difference between accumulated precipitation values) for the selected (5 min) duration obtained from the gage record. For a partial duration series, all values in the record that exceed a level given by the excessive precipitation for the selected duration are noted. The excessive precipitation as defined by the National Weather Service is precipitation that falls at a rate equaling or exceeding that indicated by the following formula:

$$p = \frac{t + 20}{100} \quad \text{[unbalanced]}$$

where

p = precipitation, in.

t = precipitation duration, min

3. The precipitation values are arranged in descending order and the return period for each value is obtained using the formula $T = n + 1/m$, where m is the rank of the data and n is the total number of years of data in the record (Section 8.7). For partial duration series (Section 8.4.1), the adjustment in the precipitation values is made by applying the following empirical multiplication factors:

Return Period (years)	Conversion Factor
2	0.88
5	0.96
10	0.99
>10	1.0

4. Similar analyses are carried out for other selected durations $(10, 15, 20, \ldots$ min), as shown in Example 3.6.

5. For each frequency level computed, the precipitation amounts (depths) are plotted for different durations. These are the depth–duration–frequency curves. The precipitation depths can be converted to intensities by $i = 60p/t$. For instance, a precipitation of 0.5 in. of 30 min duration has an intensity of 1 in./hr. These values are plotted as the intensity–duration–frequency curve on arithmetic (ordinary) graph paper as in Figure 3.6, or on log-log paper. Interpolation between the return periods can be done from the curves of the lower and higher return periods.

Example 3.6

For the precipitation data arranged for different durations in Table 3.6, prepare intensity–duration–frequency curves for 20-year and 10-year frequencies.

Solution

1. For each duration, the precipitation depths are arranged in descending order. The highest value has been assigned a rank of 1 and the lowest a rank of 22. The return periods are obtained in column 8 of Table 3.6.

2. The depths of different duration corresponding to 20-year frequency are interpolated from the plots for 23- and 11.5-year frequencies and converted to intensities in Table 3.7. Similar calculations are done for 10-year frequency. These are plotted in Figure 3.6.

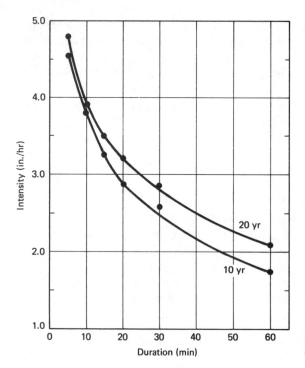

Intensity (in./hr)

Duration (min)

Figure 3.6 Intensity–duration–frequency curve.

TABLE 3.6 FREQUENCY ANALYSIS OF DIFFERENT DURATION OF PRECIPITATION DEPTHS

(1)	(2)	(3)	(4)	(5)	(6)	(7)	(8)
			Precipitation (in.) of Duration of:				Return Period
Rank m	5 min	10 min	15 min	20 min	30 min	60 min	$T = n + 1/m$
1	0.40	0.66	0.89	1.07	1.48	2.15	23
2	0.38	0.63	0.83	0.97	1.29	1.92	11.5 ←20 yr
3	0.37	0.62	0.79	0.91	1.26	1.48	7.7 ←10 yr
4	0.36	0.60	0.76	0.86	1.06	0.91	5.8
5	0.35	0.60	0.73	0.86	0.83		4.6
6	0.33	0.58	0.72	0.77	0.82		3.8
7	0.33	0.50	0.72	0.77	0.78		3.3
8	0.31	0.50	0.63	0.70	0.78		2.9
9	0.30	0.49	0.57	0.67	0.67		2.6
10	0.28	0.44	0.56	0.62	0.66		2.3
⋮							
22	0.13	0.23	0.32	0.40	0.40	0.43	1.05

The National Weather Service has prepared a series of intensity–duration–frequency maps for the United States for several combinations of return period and duration of precipitation. These generalized maps are used in the absence of recording gage data.

TABLE 3.7 VALUES FOR 20-YEAR AND 10-YEAR PRECIPITATION INTENSITIES OF DIFFERENT DURATION

Return Period (years)	Intensity (in./hr) of Duration of:					
	5 min	10 min	15 min	20 min	30 min	60 min
20	4.74	3.9	3.50	3.13	2.86	2.09
10	4.51	3.78	3.24	2.85	2.58	1.75

The intensity–duration–frequency curves are also prepared using empirical relations of the type

$$i = \frac{A}{t + B} \qquad \text{[unbalanced]} \qquad (3.15)$$

where

i = intensity, in./hr

t = duration, min

A, B = constants that depend on the return period and climatic factors

The constants for the different parts of the country as shown in Figure 3.7 are given in Table 3.8.

3.5.4 Depth–Area–Duration (DAD) Analysis of a Storm

DAD is an areal precipitation analysis of a single storm. The analysis is performed to determine the maximum amounts of precipitation of various durations over areas of various sizes. The procedure is, as such, applied to a storm that produces an excessive depth of precipitation. In a study of the probable maximum precipitation (PMP), which is defined as a rational upper limit of precipitation of a given duration

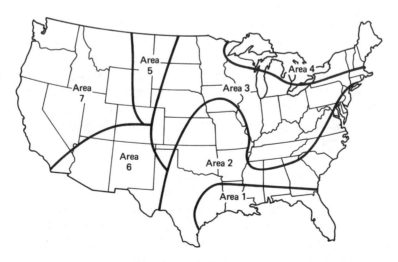

Figure 3.7 Map of similar rainfall characteristics (from Steel and McGhee, 1979).

TABLE 3.8 INTENSITY–DURATION CONSTANTS FOR VARIOUS REGIONS[a]

Frequency (years)		Area 1	Area 2	Area 3	Area 4	Area 5	Area 6	Area 7
2	A =	206	139.75	102	70	70	68	31.9
	B =	30	21	17	13	16	14	11
5	A =	246.9	190.2	131.1	96.9	81.1	74.8	48
	B =	29	25	19	16	13	12	12
10	A =	300	229.9	170	111	111	122	59.8
	B =	36	29	23	16	17	23	13
25	A =	326.8	259.8	229.9	170	129.9	155.1	66.9
	B =	33	32	30	27	17	26	10
50	A =	315	350	250	187	187	159.8	65
	B =	28	38	27	24	25	21	8
100	A =	366.9	374.8	290	220	240.2	209.8	77.2
	B =	33	36	31	28	29	26	10

[a]Conversion of units by the author.
Source: Steel and McGhee (1979).

over a particular basin, several severe storms are analyzed and the maximum values for various durations are selected for each size of area. The generalized depth–area–duration values of the probable maximum precipitation for the eastern United States, as prepared by the National Weather Service, are illustrated in Table 8.14.

The information from a recording gage is needed for this study. The procedure comprises of first determining the depth–area relation for the total depth of a storm and then breaking down the overall depth relating to each area among different durations. The points of equal duration on the depth and area graph produce curves of the depth–area–duration. The steps are explained below.

1. Resulting from a storm, prepare the accumulated precipitation or mass curves for each station in the basin, which, in fact, are the records from the rain gages.
2. From the total amounts of precipitation from the storm at various stations, prepare an isohyetal map. The simpler storms present a single isohyetal pattern. The complex storms that are produced by two or more closely spaced bursts of rainfall have closed isohyetal pattern divided into zones.
3. The isohyets are assumed to be the boundaries of individual areas. Determine the average depth of precipitation for the areas enclosed by successive isohyets. This provides the total storm depth and area relation.
4. Consider the smallest isohyet. Within an area enclosed by this isohyet, there will be a certain number of gage stations. Determine the weight of each station by drawing the Thiessen polygons for these stations.
5. For each of the above stations, using the mass curve from step 1 determine the incremental (difference of) precipitation values for various durations. Multiply these values by the respective weight of each station.

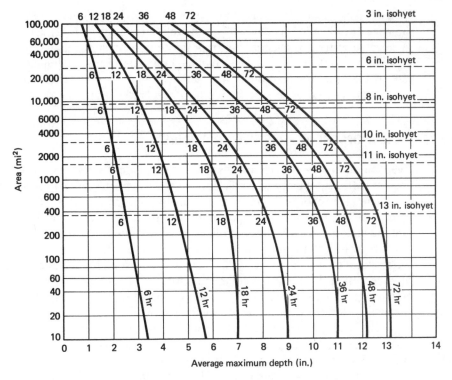

Figure 3.8 Depth–area–duration curve.

6. For all stations within the area, sum up the values of step 5 for different durations separately, with the last duration equal to the storm period.

7. The ratio of the average depth of precipitation from step 3, corresponding to the area within the smallest isohyet, to the value from step 6 for the total storm period is the factor by which all station values of different durations in step 6 are multiplied to derive the adjusted values.

8. Steps 4 through 7 are repeated for successive areas (isohyets). The values are plotted on semilog paper and the lines through the similar durations are drawn as shown in Figure 3.8.

3.6 EVAPORATION AND TRANSPIRATION

Precipitation that does not ultimately become available as surface or subsurface runoff is termed water loss. It consists of evaporation, which is the amount of water vaporized into the atmosphere from free water surface and land areas, and transpiration, which is the water absorbed by plants and crops and eventually discharged into the atmosphere. As discussed in Section 3.2, evaporation and transpiration take place in each stage of the hydrologic cycle. They form a major segment of the hydrologic cycle since about 70% of the precipitation in the United States is returned back to the atmosphere as evaporation and transpiration. From open water bodies such as lakes, reservoirs, seas, and oceans, the loss is by direct evaporation. From a drainage basin the loss is due to (1) evaporation from the soil, (2) evaporation of the intercepted water, (3) evaporation from the depression storage, and (4) transpiration of water by plants and trees. This combined total loss from the drainage basin is also termed the evapotranspiration. The methodologies for estimation are grouped according to the type of surface from which evaporation/transpiration occurs.

3.7 EVAPORATION FROM FREE-WATER BODIES

Evaporation is a two-phase process. The first phase occurs when water molecules acquire sufficient energy to break through the water surface and escape into the atmosphere. This energy is provided principally by solar radiation. The second phase consists of transporting the vapor molecules from the vicinity of the water surface into the atmosphere. This is controlled by the difference between the vapor pressure of the body of water and that of the air for molecular diffusion and by the wind speed for evaporation due to convection. There are at least eight factors on which the rate of evaporation depends: (1) solar radiation, (2) air temperature, (3) atmospheric pressure, (4) relative humidity, (5) water temperature, (6) wind speed, (7) quality of water, and (8) geometry of the evaporating surface.

The methods of estimating evaporation comprise (1) balance methods, such as water budget and energy budget; (2) comparative methods, such as pan evaporation and atmometers; and (3) aerodynamic methods, such as eddy correlation, gradient, and mass transfer. These are discussed in detail in the *National Handbook of Recommended Methods for Water Data Acquisition* (U.S. Government, 1977).* Three of the widely used methods, one from each category, are described below.

3.7.1 Evaporation Using Pans

The most common method of finding evaporation from a free-water body is by means of an evaporation pan. The Standard National Weather Service Class A pan is widely used. This, built of unpainted galvanized iron, is 4 ft in diameter by 10 in. in depth, and is mounted on a wooden frame 12 in. above the ground, to circulate the air beneath the pan. It is filled to a depth of 8 in. The water surface level is measured daily by a hook gage in a stilling well. The evaporation is computed as the difference in the observed levels adjusted for any precipitation during observation intervals. It has been observed that evaporation occurs more rapidly from a pan than from larger water bodies. A coefficient is accordingly applied to pan observations to derive the equivalent lake or reservoir evaporation.

$$E_L = KE_p \quad [L] \tag{3.16}$$

where

$$E_L = \text{evaporation from a water body}$$
$$E_p = \text{evaporation from the pan}$$
$$K = \text{pan coefficient}$$

*The chapter on evaporation was prepared in 1982.

It is normal to compute the pan coefficient on an annual basis but monthly coefficients have also been used. Gray (1973) and Linsley et al. (1982) have summarized the coefficient values, which range from 0.6 to 0.8 with an average of 0.7.

Kohler et al. (1955) proposed a formula to be used before applying the pan coefficient, to adjust the pan observation for heat exchange through the pan and the effect of wind (advected energy). The *National Handbook* suggests another correction for splashout and blowout whenever precipitation is greater than 8 mm. Kohler and Parmele (1967) indicate that the corrected pan data represent the "free water evaporation" that will occur in a very shallow water body. For a natural water body, it is necessary to consider significant heat storage and energy advected by water coming in or going out. An equation for this adjustment has been suggested by Kohler and Parmele (1967).

A refinement to eq. (3.16), which is suitable for estimating monthly or even daily evaporation losses, considers the saturation vapor pressures of the lake (water body) and the pan.

$$E_L = K' \frac{e_{sL} - e_Z}{e_{sp} - e_Z} E_p \quad [\text{L}] \tag{3.17}$$

where

e_{sL} = saturation vapor pressure for maximum temperature just below the lake surface

e_{sp} = saturation vapor pressure for maximum temperature in evaporation pan

e_Z = mean vapor pressure at a height Z above the lake surface

K' = a coefficient, mainly a function of the type of pan, equal to 1.5 for U.S. Class A pan at $Z = 4$ m (Webb, 1966)

Daily computed values of E_L are summed up to get the monthly evaporation. For direct monthly computation, mean monthly values of e_{sL}, e_{sp}, and e_Z are used to compute E_L. Saturation vapor pressure is given in Appendixes C-1 and C-2.

The pan-to-lake coefficient is a major cause of error in the pan method. Errors in the range of 10 to 15% of annual estimates and up to 50% for monthly estimates have been reported. The method is, as such, preferred for long-duration estimates.

Example 3.7

Find the daily evaporation from a lake for a day on which the following data were observed:

1. Air temperature (mean) = 85°F
2. Lake temperature
   Maximum   = 68°F
   Minimum   = 63°F
3. Pan temperature
   Maximum   = 83°F
   Minimum   = 78°F
4. Relative humidity   = 25%
5. Wind speed   = 10 mph
6. Pan evaporation   = 0.35 in.

Compute the results by (a) the simple relation, and (b) the refined formula.

Solution

(a) From eq. (3.16),

$$E_L = KE_p = 0.7(0.35) = 0.245 \text{ in.}$$

(b) Saturation vapor pressure at maximum temperature for lake, (from Appendix C-1), $e_{sL} = 0.692$ in. Hg.

Saturation vapor pressure at maximum temperature for pan $e_{sp} = 1.141$ in. Hg. Saturation vapor pressure at mean air temperature = 1.214 in. Hg. Moisture in air = 25%. Vapor pressure of the air,

$$e_z = \left(\frac{25}{100}\right)(1.214) = 0.30 \text{ in. Hg}$$

From eq. (3.17),

$$E_L = (1.5)\left(\frac{0.692 - 0.30}{1.141 - 0.30}\right)(0.35) = 0.244 \text{ in.}$$

3.7.2 Evaporation by Mass-Transfer Method

In 1802, John Dalton suggested a diffusion equation for evaporation, known as Dalton's law, according to which the evaporation is proportional to the difference between saturated vapor pressure at water temperature and the vapor pressure due to the moisture in the air.

$$E = f(u)(e_s - e_a) \qquad \text{[unbalanced]} \tag{3.18}$$

where

$$f(u) = \text{wind function}$$
$$e_s = \text{saturation vapor pressure}$$
$$e_a = \text{vapor pressure of air}$$

The wind function has the following form:

$$f(u) = a + NW^n \tag{3.19}$$

where

$$a, n = \text{constants}$$
$$N = \text{mass-transfer coefficient}$$
$$W = \text{wind speed}$$

Many relations of the form of eq. (3.19) have been proposed. Generally, n is assumed to be unity for lakes and a is assumed to be zero. The mass-transfer coefficient, N, accounts for many variables. It is considered constant for a specific lake.

The mass-transfer coefficient, N, is commonly determined by calibration against the energy budget evaporation. If a plot is made of the product $W(e_s - e_a)$ against the independent estimate of evaporation, the slope of the line is N.

From the study of 20 reservoirs from 1 to 30,000 acres, Harbeck (1962) derived $N = 0.105A_s^{-0.05}$;* thus

$$E_{\text{day}} = 0.105A_s^{-0.05}W(e_s - e_a) \qquad \text{[unbalanced]} \qquad (3.20)$$

where

E_{day} = evaporation per day, in.

A_s = surface area of lake, acres

e_s = saturation vapor pressure at the water surface temperature, in. Hg

e_a = vapor pressure of the air at height a above the water surface, in. Hg

W = wind speed, mph

This is a widely used method in which the devices needed are (1) an anemometer to measure wind speed which is placed at 2 m (6.6 ft) above the water surface, (2) a hygrothermograph or psychrometer for humidity, and (3) thermometers for air and water temperatures.

Example 3.8

Solve Example 3.7 by the mass-transfer method. $N = 0.07$.

Solution

1. Mean lake temperature = 65.5°F.
2. Saturation vapor pressure at 65.5°F, $e_s = 0.634$ in. Hg.
3. Saturation vapor pressure at mean air temperature = 1.214 in. Hg.
4. Relative humidity = 25%.
5. Vapor pressure of air, $e_a = (0.25)(1.214) = 0.304$ in. Hg.

From eq. (3.20),

$$E_{\text{day}} = (0.07)(10)(0.634 - 0.304)$$
$$= 0.23 \text{ in.}$$

3.7.3 Evaporation by Energy Budget Method

This method, which is based on the conservation of heat energy, is rigorously correct in theory. It is most accurate in application for period of a week or longer. However, the instruments required are very expensive, and observation and processing of the data are time consuming. Hence the method is not very common in use.

In the method, the incoming and outgoing radiant energy, both long and short wave, are accounted for in addition to changes in energy of the body of water. The difference in energy contributes to evaporation, the amount of which is computed from the known latent heat of evaporation of water. Approximately 600 times more energy is required to evaporate water than to increase its temperature 1°C.

Evaporation is estimated by

$$E = \frac{Q_s - Q_r + Q_a - Q_{ar} - Q_{br} + Q_v - Q_x}{L(1 + B)} \qquad \text{[L]} \qquad (3.21)$$

*Unit conversion by the author.

where

Q_s = incoming solar radiation

Q_r = reflected solar radiation

Q_a = incoming long-wave radiation

Q_{ar} = reflected long-wave radiation

Q_{br} = long-wave radiation from the water

Q_v = net energy advected into the lake

Q_x = increase in stored energy of water

L = latent heat of evaporation in energy per unit depth

B = Bowen ratio = Q_h/Q_e

Q_h = energy exchanged from water to atmosphere as sensible heat

Q_e = energy lost by water by evaporation

Measurements to be made in this method include (1) incoming short-wave and long-wave radiation, (2) air and water temperature and dew point to compute the Bowen ratio, and (3) temperature surveys of the lake to measure changes in heat stored.

The Hydrologic Research Laboratory of the National Weather Service prefers to apply the energy method using an insulated pan. Combining the energy budget and the mass-transfer equation, Kohler and Parmele (1967) derived a relation that is used to determine the "free water" evaporation using the observed energy data from the pan. For evaporation from natural lakes, an adjustment is made for stored energy and advection effect, as discussed earlier.

3.8 EVAPOTRANSPIRATION FROM DRAINAGE BASIN

Evapotranspiration considers evaporation from natural surfaces whether the water source is in the soil, in plants, or in a combination of both. With respect to the cropped area, the consumptive use* denotes the total evaporation from an area plus the water used by plant tissues, thus having the same meaning as evapotranspiration. The determination of evaporation and transpiration as separate elements for a drainage basin is unreliable. Moreover, their separate evaluation is not required for most studies.

Evapotranspiration is one of the most popular subjects of research in the field of hydrology and irrigation. Numerous procedures have been developed to estimate evapotranspiration. These fall in the categories of (1) water balance methods, such as evapotranspirometers, hydraulic budget on field plots, and soil moisture depletion; (2) energy balance method; (3) mass-transfer methods, such as windspeed function, eddy flux, and use of enclosures; (4) a combination of energy and mass-transfer methods, such as the Penman method, (5) prediction methods, such as the empirical equations and the indices applied to pan-evaporation data; and (6) methods for specific crops. These have been described in the *National Handbook of Recommended Methods for Water Data Acquisition* (U.S. Government, 1977).

*In different words, as defined in Chapter 2, this denotes the amount of water required to support the optimum growth of a particular crop under field conditions.

In the context of evapotranspiration, Thornthwaite in 1948 introduced the term "potential evapotranspiration" to define the evapotranspiration that will occur when the soil contains an adequate moisture supply at all times (i.e., when moisture is not a limiting factor in evapotranspiration). The prediction methods estimate potential evapotranspiration. Most other methods apply to estimation of actual evapotranspiration when the moisture supply is limited, as also to potential evapotranspiration under the condition of sufficient water at all times.

3.8.1 Evapotranspirometers

A properly constructed and installed evapotranspirometer provides the most accurate estimates of evapotranspiration and is a reliable means of calibrating other methods. It is an instrument consisting of a block of soil with some planted vegetation enclosed in a container. If there is a provision for drainage of the soil water, it is referred to as a lysimeter. Evapotranspiration is ascertained by maintaining a water budget for the container, that is, accounting for the water applied, water drained off the bottom, and change in the moisture content of the soil in the lysimeter. However, these are rare and expensive and relate to a particular place, soil type, and vegetation.

3.8.2 Empirical Relation of Thornthwaite

During the past three decades, a number of empirical equations based on climatological data have been developed for potential evapotranspiration. All of these equations cannot be applied for all purposes. It is necessary to identify the character of the evaporation surface to which the equation applies, such as cropped area, natural watershed, or open water body. Also, the various equations are not universally applicable to all regions. For instance, the Thornthwaite equation was derived under humid conditions. Its utility for semiarid and arid areas is limited. A summary of various empirical equations is presented in the *National Handbook of Recommended Methods for Water Data Acquisition* (U.S. Government, 1977). A commonly used method proposed by Thornthwaite is discussed here.

The potential evapotranspiration is considered as a function of temperature and sunshine hours. For a simplified case of 30 days in each month and 12 hours of sunshine each day, the equation has the form

$$\text{ET}_{\text{month}} = 1.62 \left(\frac{10 T_m}{I} \right)^a \quad \text{[unbalanced]} \qquad (3.22)$$

with

$$a = 675 \times 10^{-9} I^3 - 771 \times 10^{-7} I^2 + 179 \times 10^{-4} I + 492 \times 10^{-3} \qquad (3.23a)$$

$$I = \sum_{m=1}^{12} \left(\frac{T_m}{5} \right)^{1.514} \qquad (3.23b)$$

where

$$\text{ET}_{\text{month}} = \text{monthly potential evapotranspiration, cm}$$
$$T_m = \text{mean monthly temperature, °C}$$
$$I = \text{annual heat index, excluding negative temperature}$$

TABLE 3.9 ADJUSTMENT FACTOR FOR POSSIBLE HOURS OF SUNSHINE IN VARIOUS MONTHS

Northern Latitude	Month											
	J	F	M	A	M	J	J	A	S	O	N	D
0°	1.04	0.94	1.04	1.01	1.04	1.01	1.04	1.04	1.01	1.04	1.01	1.04
10°	1.00	0.91	1.03	1.03	1.08	1.06	1.08	1.07	1.02	1.02	0.98	0.99
20°	0.95	0.90	1.03	1.05	1.13	1.11	1.14	1.11	1.02	1.00	0.93	0.94
30°	0.90	0.87	1.03	1.08	1.18	1.17	1.20	1.14	1.03	0.98	0.89	0.88
35°	0.87	0.85	1.03	1.09	1.21	1.21	1.23	1.16	1.03	0.97	0.86	0.85
40°	0.84	0.83	1.03	1.11	1.24	1.25	1.27	1.18	1.04	0.96	0.83	0.81
45°	0.80	0.81	1.02	1.13	1.28	1.29	1.31	1.21	1.04	0.94	0.79	0.75
50°	0.74	0.78	1.02	1.15	1.33	1.36	1.37	1.25	1.06	0.92	0.76	0.70

Source: Grey (1973).

Equation (3.22) is a simplification of the basic equation for 12 hours of daily sunshine per 30-day month. The factors to adjust for the expected hours of sunshine in actual number of days of each month are given in Table 3.9. The method is not applied for periods shorter than the monthly estimates.

Example 3.9

Estimate the monthly potential evapotranspiration at Boston. The mean monthly temperatures are shown in column 2 of Table 3.10. If the growing season for a crop is from May 15 to September 15, determine the seasonal consumptive use.

Solution   Refer to Table 3.10. Latitude of Boston $\approx 42°N$.

TABLE 3.10 COMPUTATION OF POTENTIAL EVAPOTRANSPIRATION

(1) Month	(2) Mean Monthly Temperature, T_m (°F)	(3) T_m (°C)	(4) $i = \left(\dfrac{T_m}{5}\right)^{1.514}$	(5) ET (cm), Unadjusted	(6) Daylight Factor (Table 3.9)	(7) ET (cm), Adjusted
Jan.	−1.5	−18.6	—*	—	—	—
Feb.	5.2	−14.9	—	—	—	—
Mar.	30.2	−1.0	—	—	—	—
Apr.	40.2	4.6	0.88	2.01	1.11	2.23
May	58.1	14.5	5.01	7.06	1.26	8.90
June	75.5	24.2	10.88	12.36	1.26	15.57
July	70.3	21.3	8.97	10.75	1.28	13.76
Aug.	67.5	19.7	7.97	9.87	1.18	11.65
Sept.	51.0	10.6	3.12	5.01	1.04	5.21
Oct.	40.2	4.6	0.88	2.01	0.96	1.93
Nov.	31.2	−0.4	—	—	—	—
Dec.	15.2	−9.3	—	—	—	—

*Exclude negative temperatures.

$$I = \sum = 37.71$$

$$a = 675 \times 10^{-9}(37.71)^3 - 771 \times 10^{-7}(37.71)^2 + 179 \times 10^{-4}(37.71) + 492$$
$$\times 10^{-3}$$
$$= 0.0362 - 0.110 + 0.675 + 0.492$$
$$= 1.093$$

Seasonal potential evapotranspiration

$$= \frac{16}{31}(8.9) + 15.57 + 13.76 + 11.65 + \frac{15}{30}(5.21)$$

$$= 48.18 \text{ cm} \quad \text{or} \quad 19 \text{ in.}$$

3.8.3 Combination Method of Penman

Since energy is required to sustain evaporation and a mechanism is required to remove the vapor, Penman, in 1948, combined the energy balance approach and the vapor transfer process to compute the evaporation, E_o. Multiplying the evaporation values by empirical constants, evapotranspiration, E_t, is estimated. The relationships are as follows:

$$H = R_A(1 - r)(0.18 + 0.55n/N) - \sigma T_a^4(0.56 - 0.092\sqrt{e_d})[(0.10 + 0.9n/N)] \qquad [\text{LT}^{-1}] \qquad (3.24)$$

$$E_o = 0.35(e_a - e_d)(1 + 0.0098u_2) \qquad [\text{LT}^{-1}] \qquad (3.25)$$

$$E_t = \frac{\Delta H + 0.27E_0}{\Delta + 0.27} \qquad [\text{LT}^{-1}] \qquad (3.26)$$

where

H = daily heat budget at surface, mm of water/day

R_A = mean monthly extraterrestrial radiation, mm of water/day (Table 3.11)

r = reflection coefficient of surface

n = actual duration of bright sunshine

N = maximum possible duration of bright sunshine

T_a = mean daily air temperature (absolute)

σ = Boltzmann constant (2.01×10^{-9} mm/day)

σT_a^4 = variable given in Table 3.12, mm/day

e_a = saturation vapor pressure at mean air temperature, mm of Hg

e_d = actual vapor pressure in air ($e_a \times$ relative humidity)

u_2 = mean wind speed at 2 m above the ground, mi/day [speed u_h measured at height, h can be corrected to 2 m by $u_2 = u_h(\log 6.6 / \log h)$]

Δ = $\Delta e_a/\Delta T_a$ (Figure 3.9)

E_o = evaporation, mm/day

E_t = potential evapotranspiration, mm/day

TABLE 3.11 MIDMONTHLY INTENSITY OF SOLAR RADIATION (R_A) ON A HORIZONTAL SURFACE (MM OF WATER EVAPORATED PER DAY)[a]

	Southern Hemisphere										Northern Hemisphere								
	90°	80°	70°	60°	50°	40°	30°	20°	10°	0°	10°	20°	30°	40°	50°	60°	70°	80°	90°
Jan.	17.6	17.3	16.5	16.6	17.1	17.3	17.3	16.8	15.8	14.5	12.8	10.8	8.5	6.0	3.6	1.3	—	—	—
Feb.	10.7	10.5	11.2	12.7	14.1	15.2	15.8	16.0	15.7	15.0	13.9	12.3	10.5	8.3	5.9	3.5	1.1	—	—
Mar.	1.9	3.6	6.1	8.4	10.5	12.2	13.6	14.6	15.1	15.2	14.8	13.9	12.7	11.0	9.1	6.8	4.3	1.8	—
Apr.	—	—	1.9	4.3	6.6	8.8	10.8	12.5	13.8	14.7	15.2	15.2	14.8	13.9	12.7	11.1	9.1	7.8	7.9
May	—	—	0.1	1.9	4.1	6.4	8.7	10.7	12.4	13.9	15.0	15.7	16.0	15.9	15.4	14.6	13.6	14.6	14.9
June	—	—	—	0.8	2.8	5.1	7.4	9.6	11.6	13.4	14.8	15.8	16.5	16.7	16.7	16.5	17.0	17.8	18.1
July	—	—	—	1.2	3.3	5.6	7.8	10.0	11.9	13.5	14.8	15.7	16.2	16.3	16.1	15.7	15.8	16.5	16.8
Aug.	—	—	0.8	2.9	5.2	7.5	9.6	11.5	13.0	14.2	15.0	15.3	15.3	14.8	13.9	12.7	11.4	10.6	11.2
Sept.	—	1.3	3.8	6.2	8.5	10.5	12.1	13.5	14.4	14.9	14.9	14.4	13.5	12.2	10.5	8.5	6.8	4.0	2.6
Oct.	—	7.1	8.8	10.7	12.5	13.8	14.8	15.3	15.3	15.0	14.1	12.9	11.3	9.3	7.1	4.7	2.4	0.2	—
Nov.	15.3	15.0	14.5	15.2	16.0	16.5	16.7	16.4	15.7	14.6	13.1	11.2	9.1	6.7	4.3	1.9	0.1	—	—
Dec.	19.3	18.9	18.1	17.5	17.8	17.8	17.6	16.9	15.8	14.3	12.4	10.3	7.9	5.5	3.0	0.9	—	—	—

[a]Computed from *Manual of Meteorology* by Napier Shaw, Vol. II, *Comparative Meteorology*, 2nd ed., Cambridge University Press, 1936, pp. 4 and 5. Values from the table by Shaw multiplied by 0.86 and divided by 59, give the radiation in mm of water per day.
Source: Criddle (1958).

TABLE 3.12   VALUES OF σT_a^4 IN THE PENMAN EQUATION[a]

Absolute Temperature	σT_a^4 (mm water/day)	Temperature (°F)	σT_a^4 (mm water/day)
270	10.73	35	11.48
275	11.51	40	11.96
280	12.40	45	12.45
285	13.20	50	12.94
290	14.26	55	13.45
295	15.30	60	13.96
300	16.34	65	14.52
305	17.46	70	15.10
310	18.60	75	15.65
315	19.85	80	16.25
320	21.15	85	16.85
325	22.50	90	17.46
		95	18.10
		100	18.80

[a]The heat of vaporization was assumed to be constant at 590 gal/g of water.
Source: Criddle (1958).

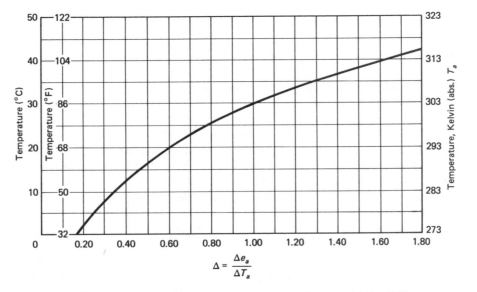

Figure 3.9   Temperature versus Δ in the Penman equation (from Criddle, 1958).

Example 3.10

In Example 3.9, estimate the monthly potential evapotranspiration for June. The average relative humidity is 50%. The wind speed is 130 mi/day. Assume that $n/N = 70\%$ and $r = 25\%$.

Solution   From Table 3.11, $R_A = 16.7$.

$$R_A(1 - r)(0.18 + 0.55n/N) = 16.7(1 - 0.25)[0.18 + 0.55(0.70)]$$
$$= 7.08 \text{ mm/day}.$$

Sec. 3.8   Evapotranspiration from Drainage Basin   **83**

From Table 3.12, $\sigma T_a^4 = 15.7$ mm/day.
From Appendix C-2, $e_a = 22$ mm Hg.

$$e_d = e_a \times \text{R.H.} = 22(0.5) = 11.$$

$$\sigma T_a^4[0.56 - 0.092(\sqrt{e_d})] = 15.7(0.56 - 0.092\sqrt{11}) = 4.00 \text{ mm/day}.$$

$$(0.1 + 0.9n/N) = 0.10 + 0.9(0.7) = 0.73.$$

$$H = 7.08 - 4(0.73) = 4.16 \text{ mm/day}.$$

$$E_o = 0.35(22 - 11)[1 + 0.0098(130)]$$
$$= 8.75 \text{ mm/day}.$$

From Figure 3.9, $\Delta = 0.75$.

$$E_t = \frac{(0.75)(4.16) + (0.27)(8.75)}{0.75 + 0.27} = 5.38 \text{ mm/day}$$

or   6.35 in./month

3.8.4 Blaney–Criddle Method for Specific Crops

There is another category of methods that are suitable for the cropped area. These methods provide the results for a specific crop. Crop coefficients that reflect the water-using characteristics of each crop are incorporated in these methods. The Blaney–Criddle method, developed for the conditions in the arid western regions of the United States, is described in Section 2.9.2.

3.8.5 Relation between Actual and Potential Evapotranspiration

Moisture deficiency affects evapotranspiration. The available soil moisture ranges from field capacity, the maximum amount of water held in the soil after the excess of gravitational water has drained away, to permanent wilting point, the moisture content of the soil when the leaves of plants become permanently wilted. There are various theories on evapotranspiration as a function of soil moisture content. According to Veihmeyer and Hendrickson (1955; cited in Alley, 1984), evapotranspiration may proceed at the potential rate until soil moisture approaches the permanent wilting point. On the other hand, Thornthwaite and Mather (1955) consider that the ratio of actual to potential evapotranspiration is a linear function of the ratio of available soil moisture to moisture capacity. An in-between approach adopted by Penman (1949; cited in Alley, 1984) postulates that evapotranspiration occurs at the potential rate until the soil moisture deficit exceeds the root constant; then it proceeds at a slower rate. Palmer's model (1965), referred to in Section 3.3.4, is based on this concept. Thomas (1981) considers that evapotranspiration is a nonlinear function of available water, which is the precipitation plus the soil moisture storage. The upper limit is at potential evapotranspiration. The treatment depends on the type of model considered. For various crop surfaces, coefficients are applied to potential evapotranspiration to account for partial plant cover during growth, crop characteristics, and effects of maturity of crops. A relatively new approach to estimate evapotranspiration by crops at various stages of growth involves the area of leaves per unit of land area known as the leaf area index (LAI) (Hanks, 1974).

3.9 DIRECT RUNOFF FROM RAINFALL OR NET STORM RAIN

Information on net storm rain is necessary in hydrograph analysis, discussed in Chapter 7. As indicated by the water balance equation (3.5), the net storm rain or direct runoff contributing to immediate streamflow is assessed by subtracting the infiltration from the total rainfall. A simple model, a homogeneous soil column with a uniform initial water content, is considered. There are three distinct cases of infiltration.

1. When a rainfall intensity, i, is less than the saturated hydraulic conductivity, K_s,* all the rainfall infiltrates, as shown by line I in Figure 3.10.
2. The effect of the rainfall rate, which is greater than the saturated conductivity $(i > K_s)$, is shown by curve II. Initially, water infiltrates at the application rate. After a time t_p, the capacity of soil to infiltrate water falls below the rainfall rate. Surface ponding begins, which results in depression storage and runoff.
3. For a rainfall intensity that exceeds the capacity of soil to infiltrate water from the beginning, water is always ponded on the surface. In this case, the rate of infiltration is controlled only by the soil-related factors. This rate, shown by curve III in Figure 3.10, is called the infiltration capacity of a given soil, f_p.

The infiltration capacity, f_p, decreases with time, due primarily to reduction in the hydraulic gradient between the surface and the wetting front.[†] It approaches a constant rate, f_c, which is considered to be equal to the apparent saturated hydraulic conductivity, K_s.

After the surface ponding (beyond time t_p for case 2 and from the beginning for case 3), for a continuous uniform rain of intensity i, the surface runoff hydrograph has a shape indicated by q in Figure 3.11. The difference between rainfall and runoff appears as the curve marked $(i - q)$. The curve f_p relates to the infiltration rate. The difference between the dotted $(i - q)$ curve and the f_p curve signifies interception and other minor losses (storages) at the beginning. After the surface storage is filled in, the two curves coincide (i.e., direct runoff results from subtraction of the infiltration out of the rainfall). If there is knowledge of the minor losses,[‡] these are deducted from the first part of the precipitation after ponding. Ordinarily, these are ignored because they are relatively minor and cannot be assessed reliably. The basic problem thus relates to determination of the infiltration loss rate under different conditions. This is known as the infiltration approach to surface runoff assessment, as against the direct rainfall–runoff correlation (Section 7.15) and multivariate runoff relation (Linsley et al., 1982, pp. 244–249).

---

*Natural soils are usually not completely saturated even below the water table, due to air entrapment during the wetting process. The hydraulic conductivity, K_s, is taken to be the residual air saturation conductivity and is sometimes referred to as the apparent saturated conductivity. For a definition of hydraulic conductivity, refer to Chapter 4.

[†]This is the limit of water penetration into the soil. The front separates the wet soil from the dry soil.

[‡]The minor losses are considered in several ways, depending on the available information: (1) only interception is excluded from the precipitation; (2) surface retention is excluded, comprising interception, depression storage, and evaporation during the storm; or (3) initial storm loss is subtracted, which is the interception and only a small fraction of the depression storage. Other depressions are considered as a part of the drainage.

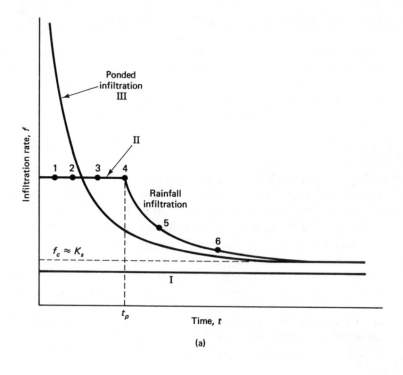

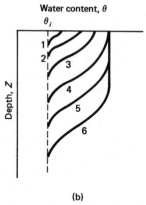

Figure 3.10 Infiltration behavior under different rainfall conditions (from Skaggs and Khaleel, 1982).

There are four approaches to determining the net storm rainfall using the infiltration concept. Two of these, the infiltration capacity curve and the nonlinear loss rate function, are detailed methods that consider the time-varying infiltration rates. In the simplified index approach the average rate of infiltration for the period of storm is used. The SCS method uses the time-averaged parameters and indirectly considers the infiltration rate through the soil characteristics.

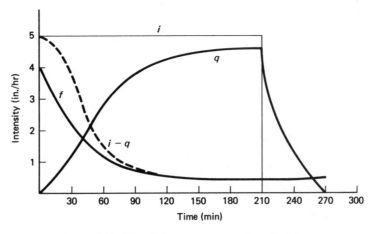

Figure 3.11 Water balance components of overland flow.

3.9.1 Infiltration Capacity Curve Approach

Green and Ampt proposed in 1911 a relation for infiltration capacity based on Darcy's law of soil water movement. Extensive research on the theory of infiltration was carried out during the 1930s and mid-1940s. Kostiakov and Horton suggested empirical relations for the infiltration capacity that became quite popular because of simplicity. Subsequent empirical equations were formulated by Philip in 1957 and Holton in 1961.

For unsaturated soil, the equation for flux (volume of water moving per unit area per unit time) is given by Darcy's law, in which the hydraulic conductivity is a function of water content. When combined with the equation of conservation of mass, this relation yields the following:

$$\frac{\partial \theta}{\partial t} = \frac{\partial}{\partial z}\left(K\frac{dh}{d\theta}\frac{\partial \theta}{\partial z}\right) - \frac{\partial K}{\partial z} \qquad [\text{T}^{-1}] \qquad (3.27)$$

where

θ = water content of soil

K = hydraulic conductivity

h = pressure head on soil medium

z = distance measured positively downward from the surface

Equation (3.27), known as the Richards equation, is the governing equation of infiltration through saturated and unsaturated soil. Exact analytical solutions to the Richards equation are limited to a few cases. Numerical solutions have been developed for various initial and boundary conditions of interest.

The elaborate procedures of the numerical method have been of limited value in practice because of computational cost, time, and soil properties data requirements. On the other hand, the simple equation of Green and Ampt has been a focus of renewed interest by many researchers. This, together with other approximate procedures, are described in sections 3.9.2 through 3.9.4.

3.9.2 Empirical Infiltration Models

Empirical models are of two types. In the Kostiakov, Horton, and Philip models, the infiltration capacity is expressed as a function of time that decreases rapidly with time during the early part of the infiltration process if rainfall intensity is greater than the infiltration capacity. However, if the rainfall intensity is less than the capacity, the decay in the curve is less in the ratio of actual infiltration to potential (capacity) infiltration. This necessitates an adjustment of the capacity curve. The Holton model removes the problem of capacity curve adjustment for the rainfall application rate by relating the infiltration capacity to the soil moisture deficiency. The moisture deficiency (available storage) is reduced with time due to infiltrated water, and so is the infiltration capacity. The Green–Ampt model is also based on the storage concept.

Horton equation. Horton (1939) presented a three-parameter equation expressed as

$$f_p = (f_0 - f_c)e^{-kt} + fc \qquad [\text{LT}^{-1}] \qquad (3.28)$$

where

f_0 = initial infiltration capacity, in./hr

f_c = final constant infiltration capacity (equal to apparent saturated conductivity), in./hr

k = factor representing the rate of decrease in the capacity, 1/time

The parameters f_0 and k have no physical basis; that is, they cannot be determined from soil water properties and must be ascertained from experimental data.

Holton equation. For agriculture watersheds, Holton et al. in the Agriculture Research Service of the U.S. Department of Agriculture developed infiltration models during the mid-1960s and 1970s. The modified equation used in the USDAHL-70 Watershed Model has the form

$$f_p = \text{GI} \cdot aS^{1.4} + f_c \qquad (3.29)$$

where

f_p = infiltration capacity, in./hr

GI = growth index of crop, percent of maturity

a = index of surface connected porosity

S = available storage in the surface layer, in.

f_c = constant rate of infiltration after long wetting, in./hr

The Agriculture Research Service has developed the experimental GI curves for several crops (see, e.g., Holton et al., 1975).

Index a is a function of surface conditions and the density of plant roots. Estimates of a are given in Table 3.13.

The values for f_c are based on the hydrologic soil groups, as categorized in the *SCS National Engineering Handbook* and explained in Section 3.7. Estimates of f_c are given in Table 3.14.

TABLE 3.13 ESTIMATES OF INDEX *a* IN THE HOLTON EQUATION

	Basal Area Rating[a]	
Land Use or Cover	Poor Condition	Good Condition
Fallow[b]	0.10	0.30
Row crops	0.10	0.20
Small grains	0.20	0.30
Hay (legumes)	0.20	0.40
Hay (sod)	0.40	0.60
Pasture (bunch grass)	0.20	0.40
Temporary pasture (sod)	0.20	0.60
Permanent pasture (sod)	0.80	1.00
Woods and forests	0.80	1.00

[a]Adjustments needed for "weeds" and "grazing."

[b]For fallow land only, poor condition means "after row crop" and good condition means "after sod."

Source: Skaggs and Khaleel (1982).

TABLE 3.14 ESTIMATES OF FINAL INFILTRATION RATE

Hydrologic Soil Grade	f_c (in./hr)
A	0.45–0.30
B	0.30–0.15
C	0.15–0.05
D	0.05–0

Source: Skaggs and Khaleel (1982).

Available storage, S, is computed by $S = (\theta_s - \theta)d$, where, θ_s is the water content at saturation that equals porosity (in fact, it is the water content at residual air saturation), θ is the water content at any instant, and d is the surface-layer depth. For control depth, d, using the depth of the plow layer or the depth to the first impeding layer has been suggested by Holton and Creitz. However, Huggins and Monke (1966) consider that determining the depth is uncertain since it is highly dependent on surface condition and practices of preparing the seedbed.

The procedure of application of the Holton model is as follows: (1) first measure or estimate the initial moisture content, θ_i; (2) compute the initial available storage by $S_0 = (\theta_s - \theta_i)d$; (3) determine the initial infiltration capacity f_p from eq. (3.29); (4) determine S after a period of time Δt by $S = S_0 - F + f_c \Delta t + ET\Delta t$, where F is the minimum of $f_p \Delta t$ and $i \Delta t$ (the available storage is reduced by the infiltration water but partly recover due to drainage from the surface layer at the rate of f_c and by evapotranspiration, ET, through plants); (5) determine f_p after period Δt by eq. (3.29); and (6) repeat the process.

3.9.3 Approximate Infiltration Model of Green–Ampt

The Green–Ampt model (1911) has received considerable renewed research attention recently and has found favor in field applications because of the fact that (1) it is a simple model, (2) it has a theoretical base on Darcy's law (it is not strictly empirical), (3) its parameters have physical significance that can be computed from soil properties, and (4) it has been used with good results for profiles that become dense with depth, for profiles where hydraulic conductivity increases with depth, for soils with partially sealed surfaces, and for soils having nonuniform initial water contents. The model is developed below.

Consider a column of homogeneous soil of unlimited depth with an initial uniform water content θ_i. It is assumed that a ponding depth H is maintained over the surface from time 0 on and a sharply defined wetting front is formed as shown in Figure 3.12. The length of the wet zone increases as infiltration progresses.

The application of Darcy's law results in the following form of the Green–Ampt equation:

$$f_p = K_s(H + S_f + L)/L \qquad [\mathrm{LT^{-1}}] \qquad (3.30)$$

where

$$K_s = \text{saturated hydraulic conductivity}$$
$$H = \text{ponding depth}$$
$$S_f = \text{suction (capillary) head at the wetting front}$$
$$L = \text{depth to the wetting front}$$

If the total (cumulative) infiltration is expressed as $F = (\theta_s - \theta_i)L$ or ML and the ponding depth is very shallow, $H \approx 0$, then eq. (3.30) can be written as

$$f_p = K_s + \frac{K_s M S_f}{F} \qquad [\mathrm{LT^{-1}}] \qquad (3.31)$$

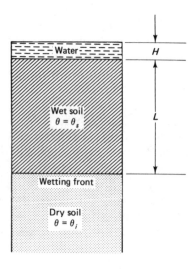

Figure 3.12 Simplified wetting front in the Green–Ampt model.

Availability of Water   Chap. 3

where $M = (\theta_s - \theta_i)$ is the initial soil water deficit.

Since $f_p = dF/dt$, the integration of eq. (3.31) with the condition $F = 0$ at $t = 0$ provides a cumulative infiltration as

$$K_s t = F - S_f M \ln\left(1 + \frac{F}{MS_f}\right) \qquad [\text{L}] \qquad (3.32)$$

Seytoux and Khanji (1974) indicated that the form of eqs. (3.30) through (3.32) remain the same when simultaneous movement of both water and air take place. The terms on the right-hand side would, however, have to be divided by a viscous resistance correction factor, ranging from 1.1 to 1.7.

The effective suction at the wetting front, S_f, has been a subject of further research (refer to Skaggs and Khaleel, 1982). Many suction-related terms have been 'd to represent it. Mein and Larson (1973) suggested the average suction at the wetting front, S_{av}, for S_f and used the ratio of unsaturated hydraulic conductivity to saturated conductivity as a weighting factor to define it. Many investigators found the application of S_{av} satisfactory.

Equations (3.30) through (3.32) are for the case when a ponding exists ($i \geq f_p$) from the beginning. If $i < f_p$,* the surface ponding effect will not take place until time t_p. Under this condition, for a steady rainfall, the actual infiltration rate, f, can be summarized as follows:

1. For $t < t_p$,

$$f = i \qquad [\text{LT}^{-1}] \qquad (3.33)$$

2. For $t = t_p$,

$$f = f_p = i \qquad [\text{LT}^{-1}] \qquad (3.34a)$$

The cumulative infiltration at the time of surface ponding, F_p, can be obtained from eq. (3.31) after substituting S_{av} for S_f.

$$F_p = \frac{S_{av} M}{i/K_s - 1} \qquad [\text{L}] \qquad (3.34b)$$

$$t_p = \frac{F_p}{i} \qquad [\text{T}] \qquad (3.34c)$$

3. For $t > t_p$, as given in eq. (3.31),

$$f = f_p = K_s + \frac{K_s S_{av} M}{F} \qquad [\text{LT}^{-1}] \qquad (3.35a)$$

For cumulative infiltration, Mein and Larson suggested an equation analogous to (3.32):

$$K_s(t - t_p + t_p') = F - MS_{av} \ln\left(1 + \frac{F}{MS_{av}}\right) \qquad [\text{L}] \qquad (3.35b)$$

where t_p' is the equivalent time to infiltrate F_p under the condition of surface ponding from the beginning as obtained from eq. (3.32) after substituting S_{av} for S_f.

*It is assumed that $i > K_s$. If not, surface ponding will not occur at all as discussed in Example 3.12.

For unsteady rainfall, the Green–Ampt model provides good results if the rainfall variations are not excessive and the rainfall contributes to an extension of the wetted profile. However, if there are relatively long periods of low or zero rainfall, the model predictions are less accurate, due to redistribution of the soil water. For rainfall after a long dry period, a new soil water distribution should be considered.

Net rainfall or runoff is computed from the following equation of the water balance at the surface, neglecting evaporation:

$$RO = i\Delta t - \Delta F - \Delta S \qquad [L] \qquad (3.36)$$

where

RO = net rainfall during time Δt

ΔF = difference during Δt in cumulated infiltration F, computed by eq. (3.35b)

ΔS = change in surface storage during Δt

3.9.4 Determination of Parameters in Green–Ampt Model

As stated earlier, an advantage of the Green–Ampt model as compared to the empirical models is that its parameters can be ascertained from the physical properties of soil. The saturated volumetric water content, θ_s, is measured by the porosity of soil, although it is somewhat less due to entrapped air even below the water table. Similarly, the value of K_s is less than the saturated hydraulic conductivity, K_0. Bouwer (1966) described an air-entry parameter to measure K_s. In the absence of a field-measured value, he suggested that K_s be estimated as $K_s \approx 0.5 K_0$. For five soils ranging in texture from sand to light clay, Mein and Larson (1973) reported the values of K_s and porosity as reproduced in Table 3.15.

Mein and Larson (1973) also provided the relations of capillary suction (S) versus relative conductivity (K/K_s) for the five soils as given in Figure 3.13. The parameter of average capillary suction is defined as

$$S_{av} = \int_0^1 S\, dK_r \qquad [L] \qquad (3.37)$$

where K_r = relative hydraulic conductivity = K/K_s. Thus S_{av} is the area under the curve of a particular soil in Figure 3.13.

Usually, it proves advantageous to determine the parameters of the model from field measurements by fitting measured infiltration data into the equation (Skaggs and Khaleel, 1982).

TABLE 3.15 SOIL PROPERTIES OF SELECTED SAMPLES

Soil	K_s (mm/hr)	Porosity
Plainfield sand (disturbed sample)	123.8	0.477
Columbia sandy loam (disturbed sample)	50.0	0.518
Guelph loam (air dried, sieved)	13.2	0.523
Ida silt loam (undisturbed sample)	1.1	0.530
Yolo light clay (disturbed sample)	0.4	0.499

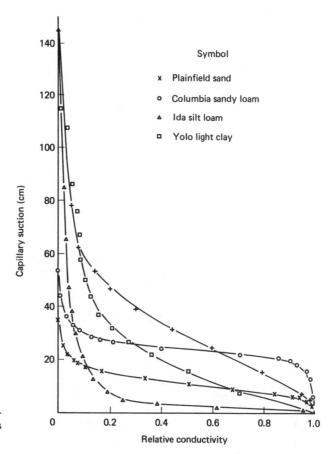

Figure 3.13 Capillary suction versus relative conductivity relation for the selected soils (from Mein and Larson, 1973).

Example 3.11

Rainfall at a constant intensity of 6 mm/hr falls on a homogeneous soil which has an initial uniform moisture content of 0.23. The soil property data obtained are $K_s = 1.24$ mm/hr and θ_s (porosity) = 0.48. The value of S_{av} estimated from Figure 3.13 is 150 mm. Determine the net rainfall. Assume no interception and depressional storage.

Solution

    A. Time to surface ponding, t_p:
      1. $M = \theta_s - \theta_i = 0.48 - 0.23 = 0.25$.
      2. From eq. (3.34b),

$$F_p = \frac{150(0.25)}{6/1.24 - 1} = 9.8 \text{ mm}$$

      3. From eq. (3.34c), $t_p = 9.8/6 = 1.63$ hr.
    B. Infiltration after the ponding:
      4. First to determine t_p' from eq. (3.32) as if ponding is from the beginning

$$(1.24)t_p' = (9.8) - 150(0.25) \ln\left[1 + \frac{9.8}{0.25(150)}\right]$$

$$t_p' = 0.88 \text{ hr}$$

Sec. 3.9   Direct Runoff from Rainfall or Net Storm Rain   **93**

5. From eq. (3.35a),

$$f = 1.24 + \frac{1.24(150)\,(0.25)}{F}$$

or

$$f = 1.24 + \frac{46.5}{F} \qquad (a)$$

6. From eq. (3.35b),

$$1.24(t - 1.63 + 0.88) = F - 0.25(150)\ln\left[1 + \frac{F}{0.25(150)}\right]$$

or

$$t = 0.75 + 0.81F - 30.24\ln(1 + 0.0267F) \qquad (b)$$

A graph of F versus t for eq. (b) is plotted in Figure 3.14.

7. At different time levels, the value of F is noted from Figure 3.14 as given in Table 3.16. Using eq. (3.36), the net rainfall for successive intervals is computed in the table. Actual infiltration, f, for various values of F can be computed from eq. (a), although it is not required to calculate the net rainfall.

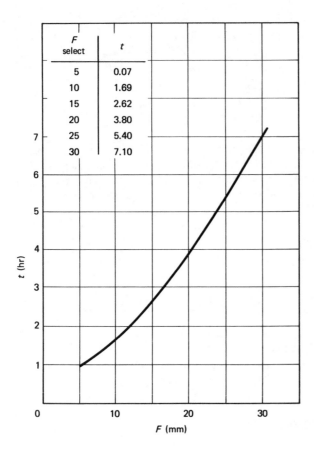

F select	t
5	0.07
10	1.69
15	2.62
20	3.80
25	5.40
30	7.10

Figure 3.14 Plot for graphical solution of infiltration equation in Example 3.11.

Availability of Water   Chap. 3

TABLE 3.16 INFILTRATION AND NET RAINFALL COMPUTATIONS WITH THE GREEN–AMPT MODEL[a]

(1) Time t (hr)	(2) F (mm)	(3) Δt (hr)	(4) ΔF (mm)	(5) RO (i.e., $i\Delta t - \Delta F$) (mm)
1.63	9.8 (F_p)			
		0.37	2.2	0.02
2.0	12.0			
		1.00	4.6	1.4
3.0	16.7			
		1.00	3.8	2.2
4.0	20.5			

[a]Columns 3 and 4 indicate the difference in successive values (increment) of columns 1 and 2, respectively.

Example 3.12

A storm pattern for a watershed is as follows:

t (min)	Intensity (in./hr)
0–10	0.5
10–20	2.0
20–30	6.5
30–40	5.0
40–50	0.9
50–60	2.0
60–70	3.0

The soil texture is sandy with a saturated moisture content (porosity) of 0.50, a saturated hydraulic conductivity of 1.0 in./hr, and an average capillary suction of 6 in. The initial moisture content is 0.3. Determine the net rainfall for successive 10-min periods. Assume a depression storage of 0.5 in.

Solution

A. First rainfall period (0–10 min)

1. Since $i < K_s$, there is no ponding and the entire rain infiltrates.
2. Cumulated infiltration, $F_1 = 0.5(10/60) = 0.08$ in.
3. The values are listed in Table 3.17 to determine the net rain.

B. Second rainfall period (10–20 min)

4. $F_p = \dfrac{6(0.2)}{2/1 - 1} = 1.2$ in.

5. $\Delta F_p = 1.2 - 0.08 = 1.12$ in.

6. $t_p = \dfrac{1.12}{2} = 0.56$ hr or 33.6 min > 10 min. Hence there is no ponding in the second period.

7. Infiltration during second period, $\Delta F_2 = 2(10/60) = 0.33$ in.
8. Cumulative infiltration to end of the period, $F_2 = 0.33 + 0.08 = 0.41$ in.
9. Infiltration capacity, f_p [from eq. (3.31)]

$$f_p = 1 + \frac{(1)(0.2)(6)}{0.41} = 3.93 \text{ in./hr}$$

TABLE 3.17 COMPUTATIONS FOR UNSTEADY RAINFALL WITH THE GREEN–AMPT MODEL

(1) Time (min)	(2) F (in.)	(3) Δt (hr)	(4) ΔF^a (in.)	(5) i (in./hr)	(6) $i\Delta t$ (in.)	(7) ΔS^b (in.)	(8) $\Sigma\Delta S$ (in.)	(9) RO^c (in.)
0	0							
		0.167	0.08	0.5	0.08	0	0	0
10	0.08							
		0.167	0.33	2.0	0.33	0	0	0
20	0.41							
		0.167	0.49	6.5	1.09	0.5	0.5	0.10
30	0.90							
		0.167	0.40	5.0	0.84	0	0.5	0.44
40	1.30							
		0.167	0.30	0.9	0.15	−0.15	0.35	0
50	1.60							
		0.167	0.25	2.0	0.33	0.08	0.43	0
60	1.85							
		0.167	0.25	3.0	0.50	0.07	0.50	0.18
70	2.10							

[a]Successive difference of column 2.
[b]For the condition of RO = 0, $\Delta S = i \cdot \Delta t - \Delta F$ which is subject to a maximum value of (depression storage capacity − ΔS of previous step).
[c]$RO = i\Delta t - \Delta F - \Delta S$.

C. Third rainfall period (20–30 min)

10. Rainfall rate increases to 6.5 in./hr, but f_p is 3.93 in./hr, so the surface ponding occurs at the outset of this period (i.e., t_p = 20 min or 0.33 hr).

11. From eq. (3.32), writing t_p' for t in the equation;

$$1 \cdot t_p' = 0.41 - 6(0.2) \ln\left[1 + \frac{0.41}{0.2(6)}\right]$$

$$t_p' = 0.06 \text{ hr}$$

12. From eq. (3.35b),

$$1(t - 0.33 + 0.06) = F - 0.2(6) \ln\left[1 + \frac{F}{0.2(6)}\right]$$

or

$$t = 0.27 + F - 1.2 \ln(1 + 0.83F)$$

A plot of F versus t for this equation is given in Figure 3.15.

13. At the end of the period, when t = 30 min or 0.5 hr, F = 0.90 in. from the equation above.

14. Ponding will accumulate up to depression storage capacity of 0.5 in., then runoff will commence to be computed by eq. (3.36) as explained in Table 3.17.

D. Fourth rainfall period (30–40 min)

15. Surface ponding continues from the third period; hence the equation of step 12 holds good.

16. At the end of the period, when t = 40 min or 0.67 hr, F = 1.3 in., from the equation of step 12. No further storage is available, runoff is computed by eq. (3.36) in Table 3.17.

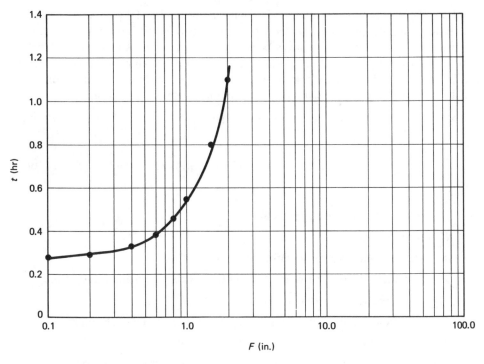

Figure 3.15 Graphical solution of cumulative infiltration equation in Example 3.12.

E. Fifth rainfall period (40–50 min)

   17. Infiltration capacity at the beginning of the period from eq. (3.35a):

$$f_p = (1) + \frac{(1)(6)(0.2)}{1.3} = 1.92 \text{ in./hr}$$

   18. Since the rainfall rate of 0.9 in./hr is less than f_p but the depression storage is full at 0.5 in. in the previous period (see Table 3.17), the infiltration at full capacity will continue until the storage is exhausted.

   19. If the infiltration continues at full capacity (i.e., the ponding effect continues throughout the period), the cumulative infiltration at the end of the period can be computed from the equation of step 12 plotted in Figure 3.15. For $t = 50$ min or 0.83 hr, $F = 1.60$ in.

   20. Depression storage will first meet infiltration without any runoff, thus

$$O = 0.9(0.167) - (1.60 - 1.30) - \Delta S$$

   or $\Delta S = -0.15$ (i.e., storage reduces by 0.15 in. to 0.35 in.).

F. Sixth and seventh rainfall periods

   21. Computation continues similar to the fourth period by the application of the equation of step 12 and as shown in Table 3.17.

3.9.5 HEC's Nonlinear Loss-Rate Function Approach

The Hydrologic Engineering Center (HEC) of the U.S. Army Corps of Engineers has used the term "loss" for the precipitation not available to direct runoff and has indicated that the rate of loss is related nonlinearly to the rainfall intensity and a loss-rate function that decreases with increased ground wetness. The HEC studies indicated that a fairly definite quantity of water loss by interception and infiltration is required to satisfy initial moisture deficiencies before runoff can occur. An allowance for this initial loss or initial abstraction is made according to various antecedent soil moisture conditions. After the initial abstraction, the loss takes place at the following rate, which does not exceed the amount of precipitation for any time interval:

$$f = Kp^E \quad [LT^{-1}] \tag{3.38a}$$

with

$$K = K_0 C^{-0.1L} \quad [\text{dimensionless}] \tag{3.38b}$$

where

f = loss rate, in. or mm per hour

K = loss rate function

p = rainfall intensity, in. or mm per hour

E = exponent ranging between 0.3 and 0.9, with a frequent value of 0.7

K_0 = loss coefficient at start of a storm

C = coefficient controlling the rate of decrease of the loss-rate function

L = accumulated loss during the storm, in. or mm

A typical plot for the loss-rate function is shown in Figure 3.16. The loss-rate coefficients are determined from the rainfall and runoff data. The HEC has developed a loss-rate optimization program for this purpose (HEC, 1973).

Example 3.13

Determine the net rainfall for successive periods for the storm of Example 3.12 by the loss-rate function approach. Assume that Figure 3.16 applies for the loss-rate function. Use $E = 0.7$.

Solution

1. Initial accumulated loss from Figure 3.16 is 1.0 in.
2. Assuming a uniform distribution of the rain within 10-min observation periods, the rainfall of first 26 min will be abstracted in the initial loss of 1.0 in., as follows:

0–10 min	0.08 in.
10–20 min	0.33 in.
20–26 min	0.59 in.

3. The direct runoff will appear after 26 min. The computation is shown in Table 3.18.

3.9.6 Net Storm Rain (Runoff) by the SCS Approach

By studying the infiltration behavior of different types of soils, the Soil Conservation Service has developed a method of computing the direct runoff resulting from a rainfall storm (U.S. SCS, 1972). The factors affecting infiltration are: hydrologic soil group, type of land cover, hydrologic condition and antecedent (prestorm) moisture condition, and cropping practice in the case of cultivated agriculture land. Each of

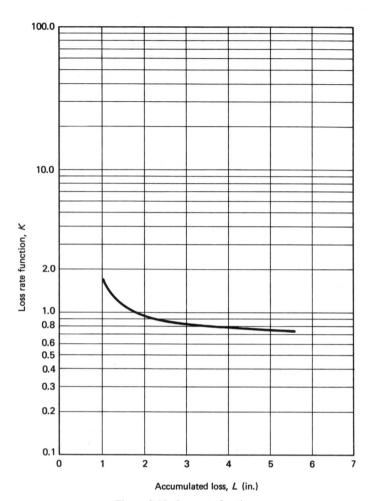

Figure 3.16 Loss-rate function.

TABLE 3.18 COMPUTATION OF NET RAINFALL BY THE LOSS-RATE FUNCTION METHOD

(1) Time (min)	(2) Rainfall Intensity (in./hr)	(3) Rainfall during Period (in.)	(4) K Value (Fig. 3.16)	(5) Loss Rate, $f = Kp^{0.7}$ (in./hr)	(6) Loss during Period (in.)	(7) Accumulated Loss (in.)	(8) Net Rainfall (in.) (col. 3 − col. 6)
0–10	0.50	0.08					
10–20	2.00	0.33				1.00 Initial loss	
20–26	6.50	0.59					
26–30	6.50	0.49	1.70[a]	6.30[b]	0.42[c]	1.42	0.07
30–40	5.0	0.83	1.17	3.61	0.60	2.02	0.23
40–50	0.9	0.15	0.93	0.86	0.14	2.16	0.01
50–60	2.0	0.33	0.91	1.48	0.25	2.41	0.08
60–70	3.0	0.50	0.88	1.90	0.32	2.73	0.18

[a]The value corresponding to accumulated loss in the preceding step.
[b]For example, $f = (1.7)(6.5)^{0.7} = 6.3$.
[c](Column 5)(duration in hours): i.e., $(6.3)(4/60) = 0.42$.

Sec. 3.9   Direct Runoff from Rainfall or Net Storm Rain

these factors is subdivided into many classes. Hydrologically, soils are assigned four groups on the basis of intake of water on bare soil when thoroughly wetted, as shown below. With urbanization the soil profile is disturbed considerably. The group classification can be based on the texture of disturbed soil.

Group	Minimum Infiltration Rate (in./hr)	Texture[a]
A	0.3–0.45	Sand, loamy sand, or sandy loam
B	0.15–0.30	Silt loam or loam
C	0.05–0.15	Sandy clay loam
D	0–0.05	Clay loam, silty clay loam, sandy clay, silty clay, or clay

[a]Reproduced from U.S. Soil Conservation Service (1986).

Type of land cover, such as bare soil, vegetation, impervious surface, and so on, establishes runoff production potential. Important cover types for urban areas, cultivated agriculture lands, other agriculture lands, and arid rangelands are given in Table 3.19. Cultivated agricultural lands are further subdivided by treatment or cropping practice, such as straight row, contoured, and contoured and terraced. The hydrologic conditions reflect the level of land management. Hydrologically poor conditions represent a state of land use that will provide higher runoff as compared to the good condition. The antecedent moisture condition (AMC) is the index of the soil condition with respect to runoff potential before a storm event. It has three categories:

Category	Condition
I	Dry soil but not to the wilting point
II	Average conditions
III	Saturated soils; heavy rainfall or light rainfall with low temperatures have occurred in the last 5 days

The SCS has evolved a system of curve numbers. A distinct curve number (CN) is assigned on the basis of the combination of each of factors above. Table 3.19 gives curve numbers (CN) for antecedent moisture condition II. Table 3.20 provides conversion of CN to other conditions. For an area with many different subareas, the composite CN is determined by adding the product of CN and respective area and dividing by the total area.

The SCS runoff equation is

$$Q = \frac{(P - 0.2S)^2}{(P + 0.8S)} \quad [L] \tag{3.39}$$

TABLE 3.19 CURVE NUMBERS FOR ANTECEDENT MOISTURE CONDITION II

Use	Cover Type	Treatment	Hydrologic Condition	Hydrologic soil group			
				A	B	C	D
Urban	Fully developed						
	Open space (lawns, parks)		Poor (cover <50%)	68	79	86	89
			Fair	49	69	79	84
			Good (grass cover >75%)	39	61	74	80
	Impervious areas (paved parking, roofs, driveways, paved roads)			98	98	98	98
	Dirt roads			72	82	87	89
	Urban districts						
	Commercial and business			89	92	94	95
	Industrial			81	88	91	93
	Developing areas			77	86	91	94
Cultivated agriculture lands	Fallow	Bare soil	...	77	86	91	94
	Row crops	Straight row	Poor	72	81	88	91
		Straight row	Good	67	78	85	89
		Contoured	Poor	70	79	84	88
		Contoured	Good	65	75	82	86
		Contoured and terraced	Poor	66	74	80	82
		Contoured and terraced	Good	62	71	78	81
	Small grain	Straight row	Poor	65	76	84	88
		Straight row	Good	63	75	83	87
		Contoured	Poor	63	74	82	85
		Contoured	Good	61	73	81	84
		Contoured and terraced	Poor	61	72	79	82
		Contoured and terraced	Good	59	70	78	81
	Close-seeded legumes or rotation meadow	Straight row	Poor	66	77	85	89
		Straight row	Good	58	72	81	85
		Contoured	Poor	64	75	83	85
		Contoured	Good	55	69	78	83
		Contoured and terraced	Poor	63	73	80	83
		Contoured and terraced	Good	51	67	76	80

TABLE 3.19 (contd.)

Use	Cover Type	Treatment	Hydrologic Condition	Hydrologic soil group			
				A	B	C	D
Agriculture lands	Pasture or range		Poor	68	79	86	89
			Fair	49	69	79	84
			Good	39	61	74	80
	Meadow		Poor	30	58	71	78
	Woods		Poor	45	66	77	83
			Fair	36	60	73	79
			Good	30	55	70	77
	Farmsteads (building, lanes, driveways)		. . .	59	74	82	86
Arid and semiarid rangelands	Herbaceous (mixture of grass, weeds, and low-growing brush)		Poor (<30% ground cover)		80	87	93
			Fair		71	81	89
			Good (>70% cover)		62	74	85
	Oak–aspen (mountain brush mixture)		Poor		66	74	79
			Fair		48	57	63
			Good		30	41	48
	Pinyon–juniper		Poor		75	85	89
			Good		41	61	71
	Sagebrush with grass understory		Poor		67	80	85
			Good		35	47	55
	Desert shrub		Poor	63	77	85	88
			Fair	55	72	81	86
			Good	49	68	79	84

Source: Condensed from U.S. Soil Conservation Service (1986).

TABLE 3.20 CROSS-LINKING OF
CURVE NUMBERS FOR VARIOUS
ANTECEDENT MOISTURE CONDITIONS

Curve Number for Condition II	Corresponding Curve Number for Condition:	
	I	III
100	100	100
95	87	99
90	78	98
85	70	97
80	63	94
75	57	91
65	45	83
60	40	79
55	35	75
50	31	70
45	27	65
40	23	60
35	19	55
30	15	50
25	12	45
20	9	39
15	7	33
10	4	26
5	2	17
0	0	0

Source: U.S. Soil Conservation Service (1972).

where

Q = accumulated runoff, in. or mm depth over the drainage area

P = accumulated rainfall depth, in. or mm

S = potential maximum retention* of water by the soil, in. or mm

The potential maximum retention, S, is related to the curve number, CN, by the following relation:

$$CN = \frac{1000}{10 + S} \quad \text{[unbalanced]} \quad (3.40)$$

Thus, once a curve number is ascertained from Tables 3.19 and 3.20 for the known conditions, the direct runoff can be computed from eqs. (3.40) and (3.39). The TR-55 (SCS, 1986) contains a graph and a table that solve eq. (3.39) directly. The tabular solution is reproduced in Table 3.21 for a certain range of CNs and rainfall values.

*This is mostly the infiltration. The term is distinct from the surface retention, which does not include the infiltration.

TABLE 3.21 RUNOFF DEPTH FOR SELECTED CNs AND RAINFALL AMOUNTS[a]

Rainfall	| Runoff Depth (in.) for Curve Number of:												
	40	45	50	55	60	65	70	75	80	85	90	95	98
1.0	0.00	0.00	0.00	0.00	0.00	0.00	0.00	0.03	0.08	0.17	0.32	0.56	0.79
1.2	0.00	0.00	0.00	0.00	0.00	0.00	0.03	0.07	0.15	0.27	0.46	0.74	0.99
1.4	0.00	0.00	0.00	0.00	0.00	0.02	0.06	0.13	0.24	0.39	0.61	0.92	1.18
1.6	0.00	0.00	0.00	0.00	0.01	0.05	0.11	0.20	0.34	0.52	0.76	1.11	1.38
1.8	0.00	0.00	0.00	0.00	0.03	0.09	0.17	0.29	0.44	0.65	0.93	1.29	1.58
2.0	0.00	0.00	0.00	0.02	0.06	0.14	0.24	0.38	0.56	0.80	1.09	1.48	1.77
2.5	0.00	0.00	0.02	0.08	0.17	0.30	0.46	0.65	0.89	1.18	1.53	1.96	2.27
3.0	0.00	0.02	0.09	0.19	0.33	0.51	0.71	0.96	1.25	1.59	1.98	2.45	2.77
3.5	0.02	0.08	0.20	0.35	0.53	0.75	1.01	1.30	1.64	2.02	2.45	2.94	3.27
4.0	0.06	0.18	0.33	0.53	0.76	1.03	1.33	1.67	2.04	2.46	2.92	3.43	3.77
4.5	0.14	0.30	0.50	0.74	1.02	1.33	1.67	2.05	2.46	2.91	3.40	3.92	4.26
5.0	0.24	0.44	0.69	0.98	1.30	1.65	2.04	2.45	2.89	3.37	3.88	4.42	4.76
6.0	0.50	0.80	1.14	1.52	1.92	2.35	2.81	3.28	3.78	4.30	4.85	5.41	5.76
7.0	0.84	1.24	1.68	2.12	2.60	3.10	3.62	4.15	4.69	5.25	5.82	6.41	6.76
8.0	1.25	1.74	2.25	2.78	3.33	3.89	4.46	5.04	5.63	6.21	6.81	7.40	7.76
9.0	1.71	2.29	2.88	3.49	4.10	4.72	5.33	5.95	6.57	7.18	7.79	8.40	8.76
10.0	2.23	2.89	3.56	4.23	4.90	5.56	6.22	6.88	7.52	8.16	8.78	9.40	9.76
11.0	2.78	3.52	4.26	5.00	5.72	6.43	7.13	7.81	8.48	9.13	9.77	10.39	10.76
12.0	3.38	4.19	5.00	5.79	6.56	7.32	8.05	8.76	9.45	10.11	10.76	11.39	11.76
13.0	4.00	4.89	5.76	6.61	7.42	8.21	8.98	9.71	10.42	11.10	11.76	12.39	12.76
14.0	4.65	5.62	6.55	7.44	8.30	9.12	9.91	10.67	11.39	12.08	12.75	13.39	13.76
15.0	5.33	6.36	7.35	8.29	9.19	10.04	10.85	11.63	12.37	13.07	13.74	14.39	14.76

[a]Interpolate the values shown to obtain runoff depths for CNs or rainfall amounts not shown.
Source: U.S. Soil Conservation Service (1986).

To use the method for sequential rainfall, the intensity is converted to the rainfall depth for each period of sequence and accumulated to the end of each period. From the accumulated rainfall to the end of successive rain periods, the accumulated direct runoff or net rainfall is derived for each time period using the SCS relation. The accumulated direct runoff is then converted to the increments of the runoff.

Example 3.14

Determine the direct runoff (net storm rain) for successive 10-min periods of the storm of Example 3.12. The soil in the basin belongs to hydrologic group B. The basin is mostly wooded in good hydrologic condition. The saturated soil condition (condition III) exists in the basin.

Solution

1. For given hydrologic characteristics and for the moisture condition II, CN, from Table 3.19 = 55.
2. Corresponding CN for condition III, from Table 3.20 = 75.
3. Computations for accumulated rain and runoff are given in Table 3.22.

TABLE 3.22 COMPUTATION OF RUNOFF BY THE SCS METHOD

(1) Time (min)	(2) Rainfall Intensity (in./hr)	(3) Amount of Rain[a] (in.)	(4) Accumulated Rainfall (in.)	(5) Accumulated Direct Runoff (Table 3.21) (in.)	(6) Runoff Increments[b] (in.)
0–10	0.5	0.08	0.08	0	0
10–20	2.0	0.33	0.41	0	0
20–30	6.5	1.08	1.49	0.16	0.16
30–40	5.0	0.83	2.32	0.55	0.39
40–50	0.9	0.15	2.47	0.65	0.10
50–60	2.0	0.33	2.80	0.84	0.19
60–70	3.0	0.50	3.30	1.17	0.33

[a]For example, (0.5 in./hr $\times$ 10 min)(1 hr/60 min).
[b]Difference between successive values, $0.55 - 0.16 = 0.39$.

3.9.7 Infiltration—Index Approach

The index approach is the simplest procedure to estimate the total volume of storm runoff. The object of this method is to obtain a coefficient that may be applied to an entire rain period, or to an entire storm if it is made up of several rain periods, to arrive at an estimate of the direct runoff (Cook, 1946). Three types of indices are common: (1) the ϕ index, which represents a level (horizontal line) of intensity that divides the rainfall intensity diagram in such a manner that the depth of rain above the index line is equivalent to surface runoff depth over the basin, as illustrated in Figure 3.17; (2) the f_{av} index, which indicates the average rate of infiltration during a period in which the rainfall intensity is equal to or more than the infiltration capacity, f_p; and (3) the W index, which is a mean of f_{av} when it varies across a watershed.

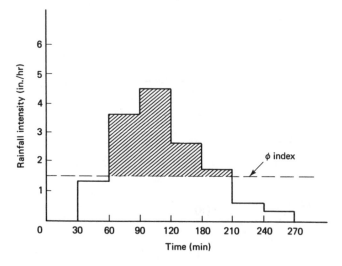

Figure 3.17 Representation of ϕ index.

The value of W for a rain occurring after the watershed is wetted and the infiltration capacity is reduced to the minimum is known as the W_{min} index.

The ϕ index is the simplest of these indices. For its determination, the rainfall due to a storm is measured and the amount of runoff is obtained from the corresponding direct hydrograph. The difference is the ϕ index. The value of ϕ increases with the increase of rain intensity up to a certain level and then approaches to a constant number. It is also affected by the rainfall pattern (Cook, 1946). Whenever possible, the ϕ index should be applied to a similar storm from which it is derived. However, it must be appreciated that the use of indices does not constitute a rational application of the infiltration theory. Rainfall intensities less than ϕ-index are not considered in determining ϕ. Trial and error is involved.

Example 3.15

The rainfall intensities during each 30 min of a 150-min storm over a 500-acre basin are 4.5, 3, 2, 3.5, and 2 in./hr, respectively. The direct runoff from the basin is 105 acre-ft. Determine the ϕ index for the basin.

Solution

1. Total rainfall $= \left(4.5\dfrac{\text{in.}}{\text{hr}}\right)(30 \text{ min})\left(\dfrac{1 \text{ hr}}{60 \text{ min}}\right)$

$$+ 3\left(\dfrac{30}{60}\right) + 2\left(\dfrac{30}{60}\right) + 3.5\left(\dfrac{30}{60}\right) + 2\left(\dfrac{30}{60}\right)$$

$$= 7.5 \text{ in. or } 0.625 \text{ ft.}$$

2. Rainfall volume $= (500)(0.625) = 312.5$ acre-ft.

3. Runoff volume $= 105$ acre-ft (given).

4. Volume under ϕ index $= 312.5 - 105 = 207.5$ acre-ft.

5. Infiltration depth $= \dfrac{207.5}{500} = 0.415$ ft or 5 in.

6. ϕ index $= (5 \text{ in.})\left(\dfrac{1}{150 \text{ min}}\right)\left(\dfrac{60 \text{ min}}{1 \text{ hr}}\right)$

$$= 1.98 \text{ in./hr.}$$

Example 3.16

Using the ϕ index of 2 in./hr, determine the net rainfall from the storm of Example 3.12.

Solution

1. Runoff intensities = rainfall intensities − ϕ index.

2. Hence the runoff intensities for successive 10-min periods are 0, 0, 4.5, 3.0, 0, 0, and 1.0, respectively.

3. Total runoff $= \left(\dfrac{4.5}{60}\right)(10) + \left(\dfrac{3.0}{60}\right)(10) + \left(\dfrac{1.0}{60}\right)(10)$

$$= 1.42 \text{ in.}$$

3.10 DIRECT RUNOFF FROM SNOWMELT

Snowfall is the second major form of precipitation after rainfall. In mountainous regions it is the primary form of precipitation. It has a particular relevance in mountainous and high plains basins, where it accumulates throughout the season and produces substantial runoff in spring melt. Whether a precipitation will fall as rain or snow depends on surface temperature, for snowfall occurs wherever the temperatures are below 34 to 36°F.

The direct measurement of snowfall is obtained through the use of rain gages equipped with shields for better catch or by using snow boards and snow stakes. However, measurement at points (gage sites) over large areas does not always serve very useful purposes, due to drift and blowing effects. Snowpack surveys are accordingly performed to ascertain the areal extent of the snow-covered area, the water equivalent, and the depth of the snowpack at selected points in the basin before and during melting of the snow. Where the snowpack varies considerably, the water equivalent should be measured at various elevations and exposures.

3.10.1 Snowmelt Process

To a hydrologist, it is the runoff produced from the snowpack that is of direct concern in floods and water supply studies. The snowmelt is essentially a process of conversion of ice into water by the heat energy. The sources of heat energy for snowmelt are all natural, comprising; (1) absorbed solar radiation; (2) net long-wave terrestrial and atmospheric radiation; (3) condensation from air, releasing latent heat of vaporization to the snowpack; (4) convection heat transfer by wind; (5) heat content of rain water; and (6) conduction of heat from ground.

A snowpack accumulated for a prolonged cold period has a temperature well below the freezing point. Mild-weather conditions cause melting of the snowpack surface. This initial meltwater moves slightly into the snowpack and again refreezes, through contact with the colder underlying snow. In the process the temperature of the snowpack rises slightly, due to the release of heat of fusion by the meltwater. When the warm conditions persist, the temperature of the entire snowpack rises to 32°F. Melt from the surface flows down through the pack. Although it does not refreeze, it is still retained by the snowpack because of the water-holding capacity or liquid-water deficiency. Once the holding capacity is completely satisfied, the snowpack is considered to be "ripe." The total water stored within the snowpack has been elaborately discussed by the U.S. Army Corps of Engineers (1965). Any additional heat energy input to the snowpack causes the melt to reach the snow–soil interface. This is similar to the rainfall reaching the ground. Infiltration and other losses take place from the snowmelt, as in the case of rainfall, and the melt excess appears as direct runoff.

Snowmelt quantities cannot be measured directly. A number of equations have been developed to estimate snowmelt. These are presented with examples of their application by the U.S. Army Corps of Engineers (1960, 1965). Two common procedures are described here. From the estimated snowmelt, deductions should be made for infiltration and other losses to compute melt excess.

3.10.2 Temperature Index or Degree-Day Method

The simplest approach for computing snowmelt is based on degree-days, or simply the air temperature, which is the single most important factor in the melting of snow. The relation is given by

$$M = C(T - T_B) \quad [\text{LT}^{-1}] \tag{3.41}$$

where

M = snowmelt, in./day

C = melt rate coefficient or degree-day factor, ranging from 0.015 to 0.2; a mean value of 0.08 is used in the western United States

T = mean daily air temperature or maximum daily air temperature, °F

T_B = base temperature above which snowmelt is assumed to occur, typically 32°F for mean daily or 40°F for maximum daily value

3.10.3 Generalized Equation of the Corps of Engineers

The U.S. Army Corps of Engineers conducted extensive study of the snowmelt for regions in the western United States. The theory developed has been based on the theoretical process of heat transfer, as well as the empirical approach of experimental observations. The set of equations are summarized below.

A. Snowmelt During a Rainstorm
  1. For open or partly forested (0–60% cover) basin

$$M = 0.09 + (0.029 + 0.0084KW + 0.007R)(T - 32)$$
$$[\text{unbalanced}] \tag{3.42}$$

  2. For heavily forested (60–100% cover) basin

$$M = 0.05 + (0.074 + 0.007R)(T - 32) \tag{3.43}$$

B. Snowmelt During No-Rain Period
  1. For partly forested (10–60% cover) basin

$$M = K'(1 - F)(0.004S)(1 - A) + K(0.0084W)(0.22T' + 0.78T_D')$$
$$+ F(0.029T') \tag{3.44}$$

  2. For heavily forested (≥ 80% cover) basin

$$M = 0.074(0.53T' + 0.47T_D') \tag{3.45}$$

where

M = snowmelt, in./day

T = air temperature at 10-ft height, °F

T' = difference between air temperature at 10-ft height and at snow surface, °F

T_D' = difference between dew point temperature at 10-ft height and at the snow surface, °F

R = rainfall, in./day

F = estimated basin forest cover, fraction

W = wind speed at 50 ft above the snow, mph

S = solar radiation at the snow surface per day, langleys/day

A = albedo or reflectivity of snow determined by .7 to $.75/D^{0.2}$, should be above 0.4

D = days since last snowfall

K = basin condensation − convection melt factor; range from 1.0 for an open basin to 0.3 for densely forested basin

K' = basin short-wave radiation melt factor; usually 0.9 to 1.1

In mountainous regions, the temperature drops on the average about 3°F per 1000 ft. Therefore, for application of these equations, a number of zones that are 1000 ft or its multiple in height are selected. The melt in each zone by the appropriate equation is determined for each day. The melt of each zone is multiplied by the area of that zone. The sum of all zones is then divided by the total basin area to obtain the basin-mean snowmelt.

Example 3.17

The basin characteristics and initial conditions (by snow survey at the beginning of the melt) for a mountainous region are given in Table 3.23 and the meteorologic data are given in Table 3.24. Determine the snowmelt for each day of the record. Assume that $K = 0.6$, $K' = 1.0$, and forest cover $F = 0.5$. Consider that the last snowfall occurred on January 31.

TABLE 3.23 INITIAL CONDITIONS

Characteristics	Zone 1	Zone 2
Elevation (ft)	3000–4000	4000–5000
Area (mi^2)	315	730
Snowpack (in.)	18	13

TABLE 3.24 METEOROLOGICAL DATA

Date	Rainfall (in.)	Air Temperature at 3000 ft (°F)	Dew Point at 3000 ft (°F)	Wind Speed (mph)	Solar Radiation (langleys)
Feb. 3, 1986	0.05	40	31	10.5	400
Feb. 4, 1986	0	48	29.2	11.0	567
Feb. 5, 1986	0	42	28.5	8.2	397

Solution

A. Considering the lapse rate of 3°F per 1000 ft, the air temperature and dew point temperature at various elevations are calculated in Table 3.25. The zone temperatures are average of two respective elevation temperatures.

TABLE 3.25 AIR AND DEW POINT TEMPERATURES AT VARIOUS ELEVATIONS

Date	Air Temperature (°F)			Dew Point Temperature (°F)		
	At 3000 ft	At 4000 ft	At 5000 ft	At 3000 ft	At 4000 ft	At 5000 ft
Feb. 3, 1986	40	37	34	31	29	26
Feb. 4, 1986	48	45	42	29.2	26.2	23.2
Feb. 5, 1986	42	39	36	28.5	25.5	22.5

B. Rainfall occurs on the first day (Feb. 3, 1986). The total melt is determined using eq. (3.42).
  1. Zone 1

$$M = 0.09 + [0.29 + 0.0084(0.6)(10.5) + 0.007(0.05)](38.5 - 32)$$
$$= 0.625 \text{ in./day}$$

  2. Zone 2

$$M = 0.09 + [0.29 + 0.0084(0.6)(10.5) + 0.007(0.5)](35.5 - 32)$$
$$= 0.378 \text{ in./day}$$

  3. The weighted basin average snowmelt for the first day:

$$M = \frac{0.625(315) + 0.378(730)}{315 + 730} = 0.452 \text{ in./day}$$

  4. The status of the snowpack at the end of first day is as follows:

	Zone 1	Zone 2
(a) Snowpack at beginning (in.)	18	13
(b) Snowmelt on first day (in.)	0.625	0.378
(c) Snowfall during the day (in.)	0	0
(d) Snowpack at the end of first day (in.)	17.375	12.622

C. For the second day (Feb. 4, 1986), there is no rain. Equation (3.44) is used to compute the snowmelt.
  1. Zone 1

$$D = 4 \qquad A = \frac{0.7}{4^{0.2}} = 0.53$$

$$T' = 46.5 - 32 = 14.5$$
$$T'_D = 27.7 - 32 = -4.3$$
$$M = 1(1 - 0.5)(0.004 \times 567)(1 - 0.53) + 0.6(0.0084 \times 11.0)$$
$$\times [0.22(14.5) + 0.78(-4.3)] + 0.5(0.029 \times 14.5)$$
$$= 0.73 \text{ in./day}$$

  2. Zone 2

$$M = 1(1 - 0.5)(0.004 \times 567)(1 - 0.53) + 0.6(0.0084 \times 11)$$
$$\times [0.22(11.5) + 0.78(-7.3)] + 0.5(0.29 \times 11.5)$$
$$= 0.52 \text{ in./day}$$

3. Weighted average snowmelt for the basin $= \dfrac{0.73(315) + 0.52(730)}{315 + 730}$

$$= 0.58 \text{ in./day.}$$

D. The melt can be similarly computed for the third day.

PROBLEMS

3.1. It is estimated that 60% of the annual precipitation in a basin of drainage area 20,000 acre is evaporated. If the average annual river flow at the outlet of the basin has been observed to be 2.5 cfs, determine the annual precipitation in the basin.

3.2. In a river reach the rate of inflow at any time is 350 cfs and the rate of outflow is 285 cfs. After 90 min, the inflow and outflow are 250 cfs and 200 cfs, respectively, and the storage of 10.8 acre-ft has been observed. Determine the change in storage in 90 min and the initial storage volume.

3.3. The evaporation losses from a reservoir of the constant surface area of 500 acres are 150 acre-ft per day. If the outflow from the reservoir is 50 cfs, determine the change in the water level of the reservoir in a day without inflow.

3.4. In Problem 3.3, a precipitation of 3 in. falls on the reservoir surface in a day. What is the change in the reservoir depth?

3.5. The average monthly rainfall recorded at the Metapoisset Basin (drainage area 30.5 mi^2) near New Bedford, Massachusetts, during 1986 and the computed potential evapotranspiration are given below. The initial moisture storage and groundwater storage are assessed to be 7.5 in. and 2.5 in., respectively. The fitted parameters of the Thomas *abcd* model are $a = 0.96$, $b = 20$, $c = 0.15$, and $d = 0.11$. Determine the monthly values of the (**a**) moisture storage, (**b**) direct runoff, (**c**) groundwater recharge, (**d**) groundwater storage, (**e**) groundwater discharge, and (**f**) streamflow.

Month:	J	F	M	A	M	J	J	A	S	O	N	D
Rainfall (in.)	0	1.7	2.66	1.24	0.36	2.12	4.44	7.48	4.66	2.22	0.6	0.24
Evapotranspiration (in.)	1.05	1.48	1.95	2.10	2.80	2.80	3.10	3.20	2.80	1.90	1.50	1.30

3.6. During a storm, gage station A was inoperative but the surrounding stations, B, C, and D, recorded rainfall of 5.2, 4.5, and 6.1 in., respectively. The average annual precipitation for stations A, B, C, and D are 48, 53, 59, and 67 in., respectively. Estimate the storm precipitation for station A.

3.7. The following rainfall record exists for five stations in a basin for the month of August. Estimate the missing rainfall for August 1985 for station A.

Station	All Years' Average for August (in.)	Measured in August 1985 (in.)
A	4.5	?
B	3.0	3.5
C	2.0	1.9
D	2.5	2.7
E	4.8	5.0

3.8. The precipitation data from October to April for six stations are given below. Check the consistency of data of station A and make an adjustment to reflect the recent conditions.

	Precipitation for October to April (in.)					
Year	Station A	Station B	Station C	Station D	Station E	Station F
1960	13.75	16.31	14.04	14.15	15.16	15.13
1961	20.22	20.79	21.85	21.06	17.16	22.23
1962	18.74	21.89	18.30	21.93	18.08	20.61
1963	20.27	20.26	17.13	19.04	22.27	22.29
1964	15.17	20.18	19.41	17.23	19.42	16.69
1965	17.13	14.36	17.37	15.58	13.78	18.84
1966	15.22	19.75	21.18	13.52	18.29	16.76
1967	18.71	19.98	18.34	15.90	14.27	20.58
1968	12.14	13.57	13.01	14.40	15.83	13.35
1969	18.50	21.17	17.64	21.05	16.99	20.35
1970	23.36	24.43	23.44	27.43	16.11	25.69
1971	9.5	13.88	19.86	16.16	15.98	10.45
1972	11.83	18.21	15.24	14.30	17.56	13.02
1973	13.40	18.48	21.23	17.65	18.48	14.74
1974	11.23	17.41	16.95	18.03	13.58	12.36
1975	13.14	17.49	20.11	13.05	16.44	14.45

3.9. The annual precipitation at station X and the mean annual precipitation at 12 surrounding stations are given below.
(a) Determine the consistency of the data of station X.
(b) In what year did the breakpoint (change in regime) occur?
(c) What is the unadjusted mean annual flow at station X?
(d) What is the adjusted mean annual flow according to recent regime conditions?
(e) Plot the double-mass curve of the adjusted data.

	Annual Precipitation (mm)			Annual Precipitation (mm)	
Year	Station X	12-Station Mean	Year	Station X	12-Station Mean
			1968	400	325
1954	517		1969	396	442
1955	264		1970	385	364
1956	462		1971	308	299
1957	297	338	1972	319	429
1958	275	299	1973	352	430
1959	385	390	1974	429	455
1960	319	338	1975	275	338
1961	396	340	1976	330	377
1962	407	339	1977	253	364
1963	385	364	1978	407	442
1964	638	520	1979	374	429
1965	451	338	1980	330	455
1966	374	312	1981	308	338
1967	220	286	1982	297	325

3.10. The rainfall recorded at the network of stations due to a storm over a trapezoidal watershed is shown in Fig. P3.10. Determine the average areal rainfall by the **(a)** average method, and **(b)** Thiessen polygon method.

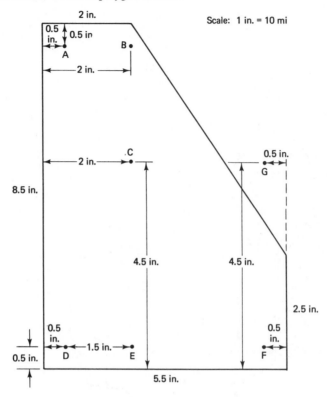

Station	Precipitation (in.)
A	2.8
B	3.5
C	4.2
D	5.1
E	6.0
F	4.5
G	3.9

Figure P3.10

3.11. Solve Problem 3.10 by the isohyetal method.

3.12. The annual precipitation (in mm) observed on a network of stations is shown in Fig. P3.12. Trace the drainage basin along with the location and values of the rainfall to compute the average areal rainfall by the **(a)** arithmetic average, and **(b)** Thiessen polygon method.

3.13. Solve Problem 3.12 by the isohyetal method.

3.14. From the precipitation data of various durations given in Example 3.6, prepare the intensity–duration–frequency curves for 5-year and 3-year return periods.

3.15. Construct the depth–area–duration curves of maximum probable precipitation for a basin located in New York City, using Figure 8.8 and Table 8.14.

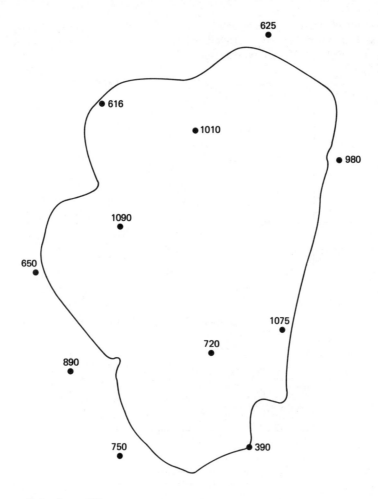

625

● 616

●1010

● 980

1090
●

650
●

1075
●

720
●

890
●

750
●

● 390

Scale: 1 cm = 16 km

Figure P3.12

3.16. In pan measurements close to a reservoir, the following data were observed on a certain day. Determine the daily evaporation by **(a)** the simple formula assuming that $K = 0.7$, and **(b)** the revised formula, assuming that $K' = 1.5$.

1. Pan evaporation   = 0.60 in.
2. Mean air temperature   = 74°F
3. Maximum reservoir temperature = 65°F
4. Minimum reservoir temperature = 59°F
5. Maximum pan temperature   = 75°F
6. Minimum pan temperature   = 65°F
7. Relative humidity   = 47%

3.17. The temperature and relative humidity averaged for the month of October have the following values. During the month, the evaporation measured from the pan was 7.5 in.

Determine the monthly evaporation from the lake by (a) the simple formula, assuming that $K = 0.75$, and (b) the revised formula, assuming that $K' = 1.5$.

1. Monthly average mean air temperature   = 62°F
2. Monthly average maximum lake temperature = 55°F
3. Monthly average maximum pan temperature = 61°F
4. Monthly average relative humidity   = 55%

3.18. Solve Problem 3.16 by the mass-transfer method of evaporation if the wind speed is 40 mph and the reservoir area is 5000 acres.

3.19. The following data are recorded at a meteorological station. Determine the average pan coefficient for the simple pan equation. (*Hint:* Compute E_{day} by the mass-transfer method and substitute in the pan equation to determine the daily coefficient.) $(N = 0.05)$

Date	Mean Water Temperature (°F)	Mean Air Temperature (°F)	Relative Humidity (%)	Wind Speed (mph)	Pan Evaporation (in.)
June 8, 1986	62	78	45	18	0.164
June 15, 1986	63	87	20	10	0.227
June 21, 1986	59	74	47	40	0.300
June 27, 1986	74	90	35	30	0.761

3.20. Estimate the monthly potential evapotranspiration for a place located at a latitude of 45°N. The mean monthly temperatures are given below. If the crop season is from April 20 to September 5, determine the seasonal consumption for the crop.

Month	Mean Temp. (°F)	Month	Mean Temp. (°F)
Jan.	−2.5	July	72.5
Feb.	7	Aug.	70.3
Mar.	39.1	Sept.	50.1
Apr.	45.5	Oct.	45.0
May	52.8	Nov.	38.2
June	68.0	Dec.	18.5

3.21. For Problem 3.20, estimate the potential evapotranspiration by the Penman method for the month of June. The relative humidity is 40% and the wind speed 15 ft from the surface is 135 mi/day. Assume that $n/N = 65\%$ and $r = 28\%$.

3.22. The standard f_p curve can be given by the equation

$$f_p = 1.2 + (9 - 1.2)e^{-0.076t}$$

where f_p is in./hr and t is in min. Plot the
(a) standard infiltration capacity curve,
(b) mass (cumulated) infiltration curve,
(c) storm infiltration capacity curve if the rainfall intensity is 1.5 in./hr during first 40 min. and 6 in./hr thereafter.

3.23. Rainfall occurs at a constant rate of 10.5 mm/hr on a sandy loam having an initial moisture content of 0.20. Determine the net rainfall rate with the Green–Ampt model, assuming no interception. At saturation, the moisture content is 0.52 and the hydraulic conductivity is 2×10^{-4} mm/sec. The average capillary suction is 250 mm.

3.24. Rainfall rates for successive 20-min periods of a 140-min storm are 1.5, 1.5, 6.0, 4.0, 1.0, 0.8, and 3.2 in./hr, totaling 6.0 in. Determine the net rain for successive periods by the Green–Ampt model. Assume a saturated hydraulic conductivity of 0.8 in./hr, a porosity of 0.62, an initial moisture content of 0.28, and an average capillary suction of 7.2 in. The depression storage is 0.5 in.

3.25. Following are the rates of rainfall in in./hr for successive 30-min periods from a storm of 180 min: 0.5, 0, 0, 4.5, 5.0, and 3.0. Determine the net rain in each successive period with the Green–Ampt model. The soil is sandy with a porosity of 0.59 and the saturated conductivity of 3.3×10^{-4} in./sec. The initial moisture content is 0.15 and the average capillary suction is 6.5 in. Neglect the depression storage.

3.26. From the rainfall sequence of Problem 3.24, determine the net rain in successive 20-min periods by the loss-rate function approach. Assume an initial loss of 1.5 in. The following values may be used for the loss-rate function. Use $E = 0.7$.

Accumulated Loss (in.)	K
1.5	1.50
2.0	0.95
3.0	0.72
4.0	0.65
5.0	0.55

3.27. For the storm of Problem 3.25, determine the successive net rainfall by the loss-rate function approach. Use Figure 3.16 for the loss-rate function. Initial loss = 1 in., $E = 0.7$.

3.28. From the rainfall sequence of Problem 3.24, determine the net rainfall during successive 20-min periods by the SCS method. The soil in the basin belongs to group A. It is an irrigated land with contoured cropping pattern in good hydrologic condition. The soil is in average condition before the storm (moisture condition II).

3.29. From the storm of Problem 3.25, determine the net rainfall during successive 30-min periods by the SCS method. The watershed is a part of the hard-surfaced urban drainage system. The hydrologic soil group is C and the soil is in a dry condition (condition I).

3.30. The rainfall intensity for each 20-min period are given below for a storm. The direct runoff from the basin is 2.5 in. Determine **(a)** the total storm rain, and **(b)** the ϕ index for the basin.

20-minute period	1	2	3	4	5
Rain intensity (in./hr)	2.5	5.0	6.5	3.5	1.0

3.31. Determine the volume of direct surface runoff in acre-ft that will result from the following storm. The basin area is 1000 acres. The ϕ index is 0.5 in./hr.

30-minute period	1	2	3	4
Rain intensity (in./hr)	0.9	0.05	0.60	0.5

3.32. The direct surface runoff volume computed from the hydrograph of a 5-mi^2 basin area is 13040 cfs-hr. The hydrograph was produced by a 5-hr storm of uniform intensity of 2.05 in./hr. Determine **(a)** the ϕ index, and **(b)** the net storm rain.

3.33. For a plain area, the following climatic data are observed on a certain rain-free day in a region covered with snow. Determine the snowmelt rate by (a) the degree-day method, $C = 0.08$; and (b) the energy-budget method, $K' = 1.0$, $K = 0.5$, forest cover = 0.45. The snowfall did not occur for the last 4 days from the date of measurement.

$$\text{Air temperature (°F)} = 53$$

$$\text{Dew point (°F)} = 28$$

$$\text{Average wind (mph)} = 11.0$$

$$\text{Solar radiation (langleys/day)} = 560$$

3.34. For a mountainous region, the following snow survey and climatic data are observed on three consecutive days. Determine the mean snowmelt rates for each day. The last snowfall occurred on February 27, 1986. Assume that $K = 0.5$, $K' = 1.0$, and $F = 0.55$.

Initial Conditions

	Zone 1	Zone 2	Zone 3
Elevation (ft)	2000–3000	3000–4000	4000–5000
Area (mi²)	250	730	400
Snowpack (in.)	9	11	15

Meteorological Conditions

Date	Rainfall (in.)	Air Temp. (°F) at 2000 ft	Dew Point (°F) at 2000 ft	Wind Speed (mph)	Solar Radiation (langleys/ day)
Mar. 1	0	45	29	8.8	450
Mar. 2	0.1	40	28.5	6.5	560
Mar. 3	0	50	31	10.2	470

THEORY OF GROUNDWATER FLOW

4.1 SCOPE

The terms "subsurface water" and "groundwater" have been given different meanings by different research workers all over the world. These terms have been used in a broader sense to include all waters below the surface of the earth in liquid, solid, or vapor forms, appearing as physically or chemically bound waters, as free water in the zones of aeration and saturation, and in a supercritical state in the zone of dense fluids extending below the zone of saturation having a pressure greater than 218 atm and temperature higher than 374°C. These terms have also been used to refer only to the "free water" that can move through rock and soil, comprising water in capillary fringe, gravitational water infiltered through the zone of aeration, and moving groundwater of the zone of saturation. Further, these terms have been used to mean water in the zone of saturation only. The use of these terms in the United States, however, almost stabilized when in 1923 Meinzer defined subsurface water to designate all waters that occur below the earth surface, and groundwater to mean the water in the zone of saturation. The *International Glossary of Hydrology,* prepared by WMO and UNESCO (1974), adopted the same meanings for these terms. Meinzer's concept of subsurface water as all variety of water in the interior of the earth is very broad, something very similar to the present definition of subsurface hydrosphere. In the common sense, subsurface flow is meant to indicate water moving in the zones of aeration and saturation and the deep percolation.

Hydrogeology covers the study of subsurface water in all its phases: origin, manner of deposition, laws of motion, distribution, physical and chemical properties, interrelationship with atmospheric and surface waters, effects of human activities, economic values, and so on. On the other hand, civil engineers are more concerned about the movement and distribution of groundwater and its application, which is the subject matter of this chapter. Most laws governing the movement of water below the earth surface have been established only in recent years. Many other terms are also of recent origin and are still developing or being defined in newer senses.

4.2 CLASSIFICATION OF SUBSURFACE WATER

Hydrogeologists have classified subsurface water based on the fundamental ideas of geological structures containing such water. However, the classification based on the manner in which water is deposited is widely accepted by both hydrogeologists and engineers. This scheme takes into consideration the fact that the physical, geographical, geological, and thermodynamic conditions are responsible factors in the deposition of water in the interior of the earth. In this classification, zonal divisions of subsurface water are made. Nineteenth-century scientists noted that there existed a law of zonation of natural phenomena such as climatic zonation, soil zonation, and vertical zonation of the material of the globe. All natural water supply is distributed in three zones: atmospheric, surface, and subsurface waters.

The zonation was traced in subsurface water as well. From 1910 onward, a number of classification schemes based on the manner in which water is deposited were proposed by Soviet, German, French, American, and other West European scientists. A very original classification was suggested by Meinzer (1923) which is very widely accepted today. This classification, shown in Table 4.1, established two broader divisions: interstitial (rock voids) water and internal (deep-lying) water. Interstitial water is subdivided into suspended (vadose) water in the zone of aeration and groundwater in the zone of saturation. Suspended water has three further subdivisions: soil water zone, intermediate zone, and capillary zone. The water in the zone of saturation was divided by Meinzer into free water and pressure water.

The French scientist Schoeller, in 1962, distinguished between the following zones beneath the surface: (1) the evaporation zone, (2) the infiltration zone, (3) the capillary fringe, and (4) the zone of groundwater accumulation. In the last zone, free surface and pressure surface are recognized.

In Lange's classification of 1969, often used by hydrogeologists in the USSR, three basic groups of water are recognized: soil water, subsurface water,* and interstratal water.

Davis and DeWiest (1966) of the United States suggested certain minor changes in Meinzer's classification. The original classification and suggested changes are shown schematically in Figure 4.1. Davis and DeWiest combined the collecting rock properties, thus describing groundwater as (1) water of igneous and metamorphic rocks, (2) water of hard sedimentary rocks, (3) water of unconsolidated sediments, and (4) water of regions of extreme climatic conditions.

Pinneker (1983) considered that present classifications are concerned only with the distribution of water pertaining to landmasses. Groundwater below the oceans and seas is not covered. Also, the deep-lying water in the zone of saturation which is pressured by the action of geostatic pressure or other internal forces is not identified in these classifications, although artesian water pressured by hydrostatic pressure has

---

*In the USSR the term "subsurface water" is commonly used in the sense of groundwater.

TABLE 4.1 GROUNDWATER CLASSIFICATION SCHEME

Zone of fissured rocks			Water in voids in rocks (pore, fissure, cavern water)
Zone of aeration	Soil water zone	Soil water	
	Intermediate zone	Intermediate zone water	
	Capillary zone	Capillary water	
Zone of saturation		Groundwater (phreatic water)	
Zone of rock flowage		Deep water	

Source: After Meinzer (1923).

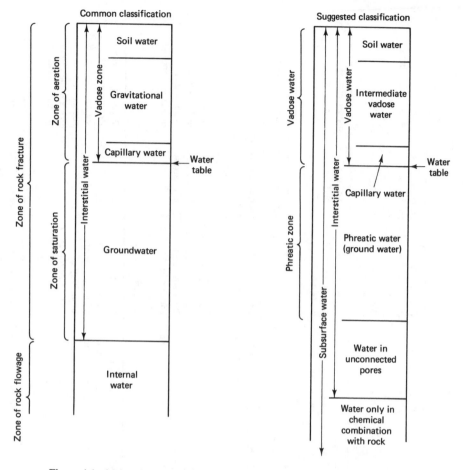

Figure 4.1 Meinzer's classification and modification by Davis and DeWiest (1966).

been recognized. Including the process in which groundwater deposits are formed and retaining the concept of Meinzer, in 1979 he suggested the classification indicated in Table 4.2. This has the following scheme:

- *Groups:* depending on the position of groundwater in the earth's crust
- *Sections:* according to the degree of saturation of rock formation with water
- *Types:* on the basis of hydraulic features
- *Classes:* basic varieties of groundwater according to the way in which they are formed
- *Subclasses:* on the basis of water-collecting properties of rocks
- *Special conditions:* specific nature of surroundings

4.3 WATER-BEARING FORMATIONS

Formations that can yield significant quantities of water are known as aquifers. This characteristic is imparted to the formations by interconnected openings or pores

TABLE 4.2 GROUNDWATER CLASSIFICATION ACCORDING TO THE MANNER IN WHICH IT HAS BEEN FORMED

Group	Section	Type	Class	Subclass		Special conditions	
				Water in strata of porous rocks (pore and stratal water)	Water in fissured cavernous rocks (fissure and vein-fissure water)	Water in permafrost regions	Water in volcanically active regions
Continental groundwater	Groundwater of the zone of aeration	Suspended water	Perched water (in the broad sense)	Salt water and infiltrating water, perched water	Salt water and infiltrating water, perched water	Active layer	Upper part of lava cover
	Groundwater of the zone of saturation on continents	Mainly nonpressure water	Groundwater	Aquifer nearest to the surface on stable impermeable layer	Upper parts of the zone of intensive fissuring and karst massif	Suprapermafrost / Interpermafrost and intrapermafrost	Lower part of lava cover
		Pressure water	Artesian water	Industrial water under hydrostatic pressure	Buried fissured zone under hydrostatic pressure	Subpermafrost	Water of hydrothermal systems under hydrostatic pressure
			Deep-lying	Sedimentary layers, which are subjected to the action of geostatic pressure and endogenic forces	Water of deep-lying faults within the sphere of activity of endogenic forces	Absent	Water of volcanic structures and hot spring systems, connected with a rising stream from the magma chamber
Groundwater below seas and oceans	Groundwater of the submarine zone of saturation	Mainly pressure water	Water connected with the land mass	Shelf and marine deposits	Karsted rock of the shelf and fault zones	Subpermafrost shelf of the northern seas	Submarine volcanic structures and marine hot spring systems
			Water not connected with the land mass	Water of deep basins	Trenches and midoceanic rifts	Absent	Absent

Source: Pinneker (1983).

122

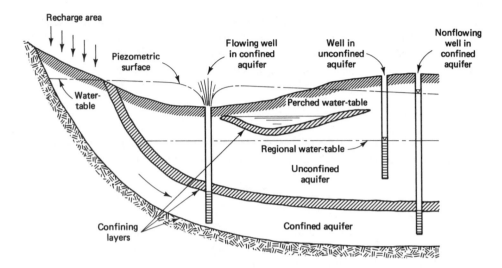

Figure 4.2 Types of aquifers.

through which water can move. Alluvial deposits are thus the best form of aquifers: probably 90% of all developed aquifers are in such formations. Such aquifers often have the advantage of direct replenishment by seepage from streams. Rock formations of a volcanic nature, limestone, and sandstone possess cracks, fissures, cavities, faults, caverns, and joints through which they yield water. The quality of such rocks as an aquifer depends on the extent of such openings; sometimes they form highly permeable aquifers. Generally, metamorphic and igneous rocks are in solid forms and serve as poor aquifers. Similarly, clays, although having a high level of porosity, prove to be poor aquifers because their pores are too small and isolated.

Types of aquifers are shown schematically in Figure 4.2. Mainly there are two types: unconfined aquifers and confined aquifers. In unconfined aquifers, the upper surface of the groundwater body is exposed to atmospheric pressure or a "water table" exists. Confined aquifers, also known as pressure or artesian aquifers, occur where groundwater is under greater-than-atmospheric pressure due to an overlaying confined layer of a relatively impermeable medium.

A special case of unconfined aquifers involves perched aquifers, where a stratum of relatively impermeable material exists above the main body of groundwater. The downward-percolating water is intercepted by this stratum and a groundwater body of limited areal extent is thus formed. The zone of aeration is present between the bottom of the perching bed and the main water table.

A special case of confined aquifers is the leaky aquifer, also known as semi-confined aquifer. Such an aquifer is either overlain or underlain by a semipervious layer through which water leaks in or out of the confined aquifer.

4.4 PARAMETERS OF GROUNDWATER STORAGE

Two aspects important in the study of groundwater are movement of water underground to streams and wells and underground storage in which an aquifer serves as a storage reservoir. The volume of water taken or released from storage with the changes in the water levels is reflected in the parameters of specific yield or specific retention for water-table aquifers and by specific storage or storage coefficient for confined aquifers. The fundamental parameter in groundwater phenomenon, however, is porosity; this provides the designation of porous medium to the soil.

4.4.1 Porosity

An element of soil, separated in three phases, is shown schematically in Figure 4.3.
Porosity is defined as the ratio of the volume of voids to the total volume, or

$$\eta = \frac{V_v}{V_t} \quad \text{[dimensionless]} \tag{4.1}$$

There is another term, void ratio, which is commonly used in soil mechanics to provide an indication of voids or pores in the soil. It is defined as the ratio of the volume of voids to the volume of solids in a soil sample, or

$$e = \frac{V_v}{V_s} \quad \text{[dimensionless]} \tag{4.2}$$

This term, however, is rarely used in groundwater flow.
Porosity and void ratio are interrelated by the expression

$$e = \frac{\eta}{1 - \eta} \quad \text{[dimensionless]} \tag{4.3}$$

Bulk (dry) density of soil is the mass of soil solids per unit gross volume of soil, and the density of soil particles (grains) is equal to the mass of soil solids per unit volume of soil solids. For the same mass of soil solids,

$$\rho_d \propto \frac{1}{V_t} \quad \text{and} \quad \rho_s \propto \frac{1}{V_s}$$

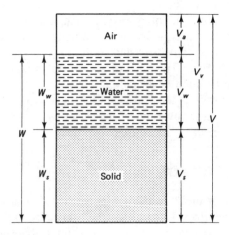

Figure 4.3 Three phases in a soil element.

where

$$\rho_d = \text{dry (bulk) density}$$
$$V_t = \text{total soil volume}$$
$$\rho_s = \text{grain density}$$
$$V_s = \text{dry soil volume}$$

From these relations and eq. (4.1), the following relation emerges:

$$\eta = 1 - \frac{\rho_d}{\rho_s} \quad \text{[dimensionless]} \tag{4.4}$$

Porosity is a measure of the water bearing capacity of a formation. However, it is not just the total magnitude of porosity that is important from the consideration of water extraction and transmission but the size of voids and the extent to which they are interconnected since pores may be open (interconnected) or closed (isolated). For instance a clay formation may have a very high porosity but it is a poor medium as an aquifer. Specific yield or effective porosity and specific retention, as discussed below, are important from this consideration.

Example 4.1

A sample of sandy soil is collected from an aquifer. The sampler with a volume of 50 cm³ is filled with the soil. When the soil is poured into a graduated cylinder, it displaces 30.5 cm³ of water. What is the porosity and the void ratio of the sand.

Solution The volume of water displaced is equal to the volume of soil particles (solids); thus

$$V_s = 30.5 \quad V_t = 50$$

Hence

$$V_v = V_t - V_s = 50 - 30.5 = 19.5$$

From eq. (4.1),

$$\eta = \frac{V_v}{V_t} = \frac{19.5}{50.0} = 0.39 \quad \text{or} \quad 39\%$$

From eq. (4.3),

$$e = \frac{\eta}{1 - \eta} = \frac{0.39}{1 - 0.39} = 0.64$$

Example 4.2

A soil sample occupies 0.132 ft³. When dried, it weighs 15.8 lb. If the specific gravity of soil solids is 2.65, calculate (a) the bulk density of the soil, and (b) the porosity of the soil.

Solution

$$\text{Dry unit weight} = \frac{15.8}{0.132} = 119.7 \text{ lb/ft}^3$$

$$\text{Dry (bulk) density, } \rho_d = \frac{119.7}{g} = \frac{119.7}{32.2} = 3.72$$

$$\text{Unit weight of soil grains} = G_s \gamma_w = (2.65)(62.4)$$
$$= 165.36 \text{ lb/ft}^3$$

$$\text{Density of soil grains,} \quad \rho_s = \frac{165.36}{g} = \frac{165.36}{32.2} = 5.14$$

Hence from Eq. (4.4),

$$\eta = 1 - \frac{3.72}{5.14} = 0.276 \quad \text{or} \quad 27.6\%$$

[Instead of dry weight, the natural (wet) weight of 18 lb/ft^3 with a moisture content, ω of 14% could have been given in the problem. In such a case,

$$W_s = \frac{W_t}{1 + \omega/100} = \frac{18}{1 + 14/100} = 15.8 \text{ lb}$$

Other steps are the same as above.]

4.4.2 Specific Retention (of Water-Table Aquifer)

Lowering of the water table is accompanied by draining of the water from the pore spaces of an aquifer and its replacement with air. This process occurs because the pressure of water inside the pores becomes less than the surrounding air pressure. However, a part of the water is retained within the pores, due to forces of adhesion (attraction between pore walls and adjacent water molecules) and cohesion (attraction between molecules of water), which are stronger than the pressure difference between the air pressure and the water pressure. The difference of air pressure and water pressure is known as capillary pressure, P_c. The volume of water thus retained against the force of gravity, compared to the total volume of rock (soil), is termed the specific retention. It is also known as the field capacity or water-holding capacity. This is a measure of the water-retaining capacity of the porous medium. Specific retention is thus dependent on both pore characteristics and factors affecting the surface tension, such as temperature, viscosity, mineral composition of water, and so on.

As stated above, the amount of water drained from the saturated soil is a function of capillary pressure. A characteristic curve has been shown in Figure 4.4. As P_c increases, the volumetric-moisture content* decreases. At a large value of P_c, the volumetric-moisture content tends toward a constant value because of adhesion and cohesion (explained above) and $\Delta\omega/\Delta P_c$ approaches zero. The volumetric-moisture content at this state is equal to the specific retention.

A simple device consisting of a porous plate, capillary tube, and leveling bottles is used to measure volumetric-moisture content and capillary pressure head on a saturated sample. The data are plotted as in Figure 4.4 to obtain the specific retention of the representative sample.

---

*Moisture content, ω is a weight parameter. Here this term is used to indicate the quantity of water inside the pores in terms of volume. The following relation holds:

$$\text{volumetric moisture content} = \frac{\text{weight moisture content} \times \text{bulk density of soil}}{\text{density of water}}$$

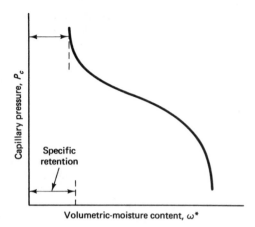

Figure 4.4 Water retention curve.

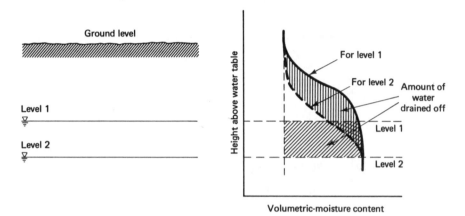

Figure 4.5 Water drained with lowering of water table.

  The porewater pressure at any depth h below the water table is equal to γh, like hydrostatic pressure, or simply h in terms of water head. Thus pressure above the water table, with reference to the water table as a datum, will be negative and equal to the height of the point from the water table. This negative pressure is simply the capillary pressure, P_c. If we follow the relationship of Figure 4.4 between capillary pressure and water content, the same curve indicates moisture content (in volumetric terms) of the soil at various heights above the water table. Consider the water table at level 1; the moisture distribution curve will be as shown in Figure 4.5. Suppose that the water table drops down to level 2. When equilibrium is achieved, the moisture distribution curve will be similar to level 1 but will be displaced to level 2, as shown in Figure 4.5. The shaded area between the two curves or at the base between two water-table lines represents the amount of water drained from the soil with the reduction of water table from level 1 to level 2.

Example 4.3

  The following capillary pressure head (negative pressure head) and moisture content data were obtained from a porous plate test on a coarse-textured soil. If the water table was initially located 300 cm below the surface and subsequently receded to 350 cm below the surface, calculate (a) the volume of water removed per unit area, and (b) the specific retention for the soil.

P_c, cm	0	20	50	80	100	120	150	170	200	230	250	280	300
w, %	44	44	43	40.5	38	35	30	27	22	18	15.5	14	14

Solution

(a) **1.** The capillary head versus moisture content diagram had been drawn for initial water table and final water table conditions in Figure 4.6.
2. The volume of water removed per unit soil area is represented by the area between the two curves.
3. The area between the curves = 1500.
4. The volume of water removed = 1500/100* = 15 cm³ per cm² area.

*Since the moisture content is plotted in percent.

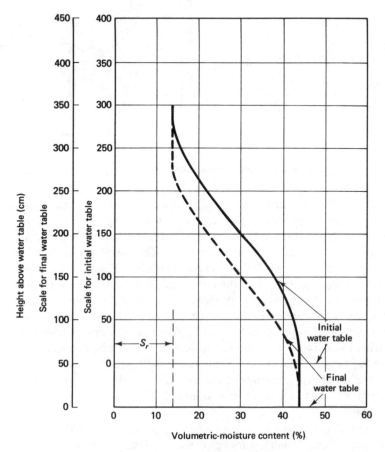

Figure 4.6 Amount of water drained with lowering of level in Example 4.3.

Theory of Groundwater Flow   Chap. 4

5. The total volume of soil per unit surface area between two water tables =
$1 \times 50 = 50 \text{ cm}^3$.

6. The water removed per unit (1 cm^3) volume of soil $= 15/50 = 0.30$.

(b) The specific retention $= 14\%$ (from the figure).

4.4.3 Specific Yield (of Water-Table Aquifer)

This term, also known as effective porosity, is defined as the volume of water released from storage in a water-table aquifer per unit horizontal area of aquifer with unit decline of water table. This concept was introduced in parallel with the storage coefficient in the confined aquifer, although the release in confined aquifers is not derived from drainage but as a result of aquifer compressibility and changes in water density. According to this definition,

$$S_y = \frac{1}{A} \frac{dv}{dh} \quad \text{[dimensionless]} \tag{4.5}$$

where

$$S_y = \text{specific yield}$$
$$A = \text{area of soil formation}$$
$$dv = \text{volume of water drained}$$
$$dh = \text{change in water table}$$

Since some water remains in the soil, the sum of specific yield and specific retention is equal to the porosity, or

$$S_y = \eta - S_r \quad \text{[dimensionless]} \tag{4.6}$$

where

$$S_y = \text{specific yield}$$
$$S_r = \text{specific retention}$$

In addition to the capillary head–moisture content technique discussed in the preceding section, there are other procedures to determine specific yield, including the well pumping tests discussed subsequently.

Table 4.3 indicates the representative values of specific yield for various types of soils and rocks. The specific yield of most aquifer formations ranges from about 0.1 to about 0.3 and averages 0.2.

Example 4.4

For Example 4.3, determine the porosity and specific yield of the soil.

Solution

Porosity = volumetric moisture content at saturation or zero capillary head

$$\eta = 44\% \quad \text{(from Figure 4.6)}$$

From eq. (4.6),

TABLE 4.3 REPRESENTATIVE VALUES
OF SPECIFIC YIELD FOR SOILS AND ROCKS

Material	Specific Yield (%)
Gravel, coarse	23
Gravel, medium	24
Gravel, fine	25
Sand, coarse	27
Sand, medium	28
Sand, fine	23
Silt	8
Clay	3
Sandstone, fine-grained	21
Sandstone, medium-grained	27
Limestone	14
Dune sand	38
Loess	18
Peat	44
Schist	26
Siltstone	12
Till, predominantly silt	6
Till, predominantly sand	16
Till, predominantly gravel	16
Tuff	21

Source: Todd, 1980.

$$S_y = 44 - 14 = 30\%$$

(Note that the water drained out in Example 4.3 is equal to the specific yield.)

Example 4.5

The water table drops 5 ft over an area of 3.5 acres. If the soil has a specific yield of 4%, how much water has drained from the area?

Solution

1. The area of 3.5 acres = $3.5 \times 43,560 = 152,460$ ft^2.
2. The total volume of soil drained off = $5 \times 152,460 = 762,300$ ft^3.
3. The volume of water = $S_y \times$ total volume of soil drained off (by definition)

   $$= 0.04 \times 762,300$$
   $$= 30,492 \text{ ft}^3.$$

In water-table aquifers, some quantity is derived from compression of the aquifer and change of density of the water. Thus the storage coefficient for water-table aquifers is the total specific yield and fraction attributable to compressibility. The latter is, however, negligible compared to gravity drainage, specific yield provides an indication of aquifer release. The term "storage coefficient" is generally used in relation to confined aquifers.

4.4.4 Storage Coefficient and Specific Storage (for Confined Aquifer)

The storage coefficient is the volume of water that is released or taken into storage by a confined aquifer per unit area of the aquifer per unit decline or rise in pressure head. This is also called the storativity. This definition is similar to that of the specific yield for a water-table aquifer. However, confined aquifers remain saturated at all times and, as such, water release is not derived from drainage of the voids by gravity as in the case of unconfined aquifers. In confined aquifers, the release or addition of water is attained due to the change in pore pressure.

In an equilibrium condition, the forces due to the weight of the formations overlying the aquifer and all other loads from the top are balanced by the skeleton and water within the pores of the aquifer. Due to the pumping of a well, the water pressure inside the pores is reduced. This results in a slight compaction of the skeleton of the aquifer and expansion of the water permitted by its elasticity. A certain amount of water is thus released from storage. The reverse process takes place in response to recharge.

Jacob made the first attempt in 1940 to introduce an analytical expression for the storage coefficient. For an elastic confined aquifer, he defined

$$S = \eta \gamma_w b \left(\frac{1}{E_w} + \frac{1}{nE_s} \right) \quad \text{[dimensionless]} \tag{4.7a}$$

or

$$S = \rho g b (\alpha + \eta \beta) \tag{4.7b}$$

where

S = storage coefficient

b = thickness of aquifer

E_w = bulk modulus of elasticity of water (3×10^5 psi at ordinary temperatures)

E_s = bulk modulus of elasticity of soil solids

α = aquifer compressibility ($1/E_s$)

β = water compressibility ($1/E_w$)

η = porosity

The first term of the expression relates to the expansibility of water and the second term to the compressibility of the aquifer.

The expression above includes a parameter denoting the aquifer thickness, which may not be constant for an aquifer. Another term, known as specific storage, was introduced by deleting b (considering $b = 1$) in the expression above. The specific storage is a constant property of an aquifer and, as such, is a more fundamental parameter. It is defined as the volume of water released from storage per unit decline in pressure head within the unit volume of an aquifer; or

$$S_s = \frac{S}{b} \quad [L^{-1}] \tag{4.8}$$

where S_s represents specific storage.

DeWiest (1966) criticized the foregoing derivation of Jacob, who had considered deformation of the aquifer (one side of the volume element was considered deformable). For an undeformed unit volume of aquifer, DeWiest derived a groundwater flow equation similar to that of Jacob, retaining the storage coefficient as that of Jacob's in eq. (4.7), but in the process redefined the specific storage which was not connected to the aquifer thickness by a simple relation of eq. (4.8). According to DeWiest,

$$S_s = \gamma_w[(1 - \eta)\alpha + \eta\beta] \qquad [L^{-1}] \qquad (4.9)$$

Cooper (1966) made a further refinement considering flow rate relative to moving grains of the aquifer medium and the flow rate across the fixed boundaries of the control volume. For a very small grain velocity, his form reduces to Jacob's formulation. In contrast to the specific yield, the storage coefficient of the confined aquifer is much smaller, ranging from about 10^{-5} to 10^{-3}.

Example 4.6

A confined aquifer of 40 m thickness has a porosity of 0.3. Determine the storage coefficient and specific storage according to the Jacob and DeWiest analysis. $\alpha = 1.5 \times 10^{-9}$ cm^2/dyn, $\beta = 5 \times 10^{-10}$ cm^2/dyn.

Solution

1. Storage coefficient [from eq. (4.7)]:

$$S = \gamma_w b(\alpha + \eta\beta)$$

$$= (980) \frac{\text{dyn}}{\text{cm}^3} (40 \times 100) \text{ cm} (1.5 \times 10^{-9} + 0.3 \times 5 \times 10^{-10}) \frac{\text{cm}^2}{\text{dyn}}$$

$$= 6.47 \times 10^{-3}$$

2. Specific storage:
   (a) According to Jacob [from eq. (4.8)],

$$S_s = \frac{S}{b} = \frac{6.47 \times 10^{-3}}{40 \times 100} = 1.62 \times 10^{-6} \text{ per cm}$$

   (b) According to DeWiest [from eq. (4.9)],

$$S_s = \gamma_w[(1 - \eta)\alpha + \eta\beta]$$
$$= (1 \times 980)[(1 - 0.3)1.5 \times 10^{-9} + (0.3)5 \times 10^{-10}]$$
$$= 1.18 \times 10^{-6} \text{ per cm}$$

Example 4.7

The storage coefficient determined from a pumping test in an aquifer is 4×10^{-4} at a location where the aquifer depth is 100 ft. If the average volume of the aquifer per square foot is 80 ft^3, how much water will be released by the aquifer with a drop in head of 70 ft?

Solution

$$S_s = \frac{4 \times 10^{-4}}{100} = 4 \times 10^{-6} \text{ per foot per unit volume}$$

The amount of water released is

$$(4 \times 10^{-6})(80)(70) = 0.022 \text{ ft}^3 \text{ per ft}^2 \text{ of area}$$

Note that if the storage coefficient is used directly, the volume of water released is

$$(4 \times 10^{-4})(1 \text{ ft}^2)(70) = 0.028 \text{ ft}^3 \text{ per ft}^2$$

This is incorrect because the average thickness of the aquifer is 80 ft, whereas the storage coefficient is based on a 100-ft depth. To use the storage coefficient, the following procedure has to be followed:

$$S_s = 4 \times 10^{-6} \text{ per foot}$$

$$S \text{ for aquifer of 80 ft} = (4 \times 10^{-6})(80) = 3.2 \times 10^{-4}$$

The amount of water released is

$$(3.2 \times 10^{-4})(1)(70) = 0.022 \text{ ft}^3 \text{ per ft}^2$$

4.5 PARAMETERS OF GROUNDWATER MOVEMENT

Besides serving as an underground storage reservoir, an aquifer acts as a conduit through which water is transmitted from a higher level to a lower level of energy. The difference in energies at various locations is caused by a continuous process of infiltrations and extraction of water underground. A basic parameter connected with water movement through a porous medium is the coefficient of permeability or hydraulic conductivity, described below.

4.5.1 Hydraulic Conductivity

This is a coefficient of proportionality between the rate of flow and the energy gradient causing that flow. As such, it combines the properties of a porous medium and the fluid flowing through it. The relevant fluid properties are viscosity, μ, and specific weight, γ. The medium properties comprise porosity, grain-size distribution, and shape of grains. A term used to communicate the effectiveness of the porous medium alone as a transmitting medium is the intrinsic (specific) permeability k, which has a dimension of L^2. When fluid properties are combined, too, the term is called the coefficient of permeability or hydraulic conductivity, expressed as

$$K = k\frac{\gamma}{\mu} \quad [\text{LT}^{-1}] \tag{4.10}$$

where

$$K = \text{hydraulic conductivity}$$
$$k = \text{intrinsic permeability}$$
$$\gamma = \text{specific weight of fluid}$$
$$\mu = \text{dynamic viscosity of fluid}$$

The relationship of flow and energy gradient, known as Darcy's law, in which the hydraulic conductivity appears as a constant of proportionality, itself is used to define the term. Thus

$$q = Ki$$

or

$$K = \frac{q}{i} \quad [LT^{-1}] \tag{4.11}$$

where

q = specific discharge or discharge per unit cross-sectional area

i = hydraulic gradient = $\Delta h/l$

Δh = change in head over length, l

From the relation (4.11) a medium is said to have a hydraulic conductivity of 1 (unit length per unit time) if it transmits a unit discharge through a cross section of unit area under a hydraulic gradient of unit change in head through unit length of flow.

Similarly, for purposes of the definition of the intrinsic permeability, eq. (4.10) is substituted in eq. (4.11), and k thus becomes

$$k = \frac{q\nu}{gi} \quad [L^2] \tag{4.12}$$

where ν = kinematic viscosity = μ/ρ. Accordingly, a medium is said to have an intrinsic permeability of 1 (unit of length squared) if it transmits a unit discharge of fluid of unit kinematic viscosity through a cross section of unit area under a unit potential gradient. The units of cm^2, ft^2, and darcy are used for intrinsic permeability. Their equivalence is shown in Table 4.4.

The units used for hydraulic conductivity are ft/day, m/day, and gallons per day/ft^2. The last unit, also known as meinzer, has been adopted by the U.S. Geological Survey. For laboratory measurement, a water temperature of 60°F is considered standard, whereas the actual temperature in the field is used to measure the field coefficient of permeability. The equivalence of these terms is also indicated in Table 4.4.

The hydraulic conductivity varies from aquifer to aquifer, from liquid to liquid, from location to location, from direction to direction, and from temperature to temperature. When K is the same in all places (space), it is a homogeneous medium. When it varies in space, the medium is said to be heterogeneous. Even in a homogeneous medium, K can vary with the direction of flow when it is known as an an-

TABLE 4.4 EQUIVALENCE OF INTRINSIC PERMEABILITY HYDRAULIC CONDUCTIVITY, AND TRANSMISSIVITY

	Intrinsic Permeability	
Darcy	cm^2	ft^2
1	0.987×10^{-8}	1.062×10^{-11}
	Hydraulic Conductivity	
Meinzer or gpd/ft^2	ft/day	m/day
1	0.134	0.041
	Transmissivity	
gpd/ft	ft^2/day	m^2/day
1	0.134	0.0124

TABLE 4.5 REPRESENTATIVE VALUES OF HYDRAULIC CONDUCTIVITY FOR SOILS AND ROCKS

Material	Hydraulic Conductivity (m/day)
Gravel, coarse	150
Gravel, medium	270
Gravel, fine	450
Sand, coarse	45
Sand, medium	12
Sand, fine	2.5
Silt	0.08
Clay	0.0002
Sandstone, fine-grained	0.2
Sandstone, medium-grained	3.1
Limestone	0.94
Dolomite	0.001
Dune sand	20
Loess	0.08
Peat	5.7
Schist	0.2
Slate	0.00008
Till, predominantly sand	0.49
Till, predominantly gravel	30
Tuff	0.2
Basalt	0.01
Gabbro, weathered	0.2
Granite, weathered	1.4

Source: Todd, 1980.

isotropic medium. In large instances, however, an aquifer can be considered to be homogeneous and isotropic.

Values of hydraulic conductivity can be obtained from empirical formulas, from laboratory measurements, and from field tests. Field tests to derive this coefficient are described subsequently. The representative values for various aquifer mediums are given in Table 4.5.

Example 4.8

Determine the hydraulic conductivity of a medium of which intrinsic permeability is 1 Darcy and through which water flows at 60°F.

Solution At 60°F, $\rho = 0.999$ g/cm^3 and $\mu = 1.12$ cP or 1.12×10^{-2} P or g/cm $\cdot$ sec.

From eq. (4.10), $K = k\dfrac{\gamma}{\mu} = k\dfrac{\rho g}{\mu}$

$$K = (1\ \text{Darcy})\left(0.999\ \frac{\text{g}}{\text{cm}^3}\right)\left(980\ \frac{\text{cm}}{\text{sec}^2}\right)\left(\frac{1}{1.12 \times 10^{-2}}\ \frac{\text{cm} \cdot \text{sec}}{\text{g}}\right)$$
$$\cdot \left(\frac{0.987 \times 10^{-8}}{1}\ \frac{\text{cm}^2}{\text{Darcy}}\right)$$

$$\uparrow$$
from Table 4.4

$$= 862.8 \times 10^{-6}\ \text{cm/sec}$$

Conversion to Meinzer:

$$K = \left(862.8 \times 10^{-6} \frac{cm}{sec}\right)\left(\frac{1\ m}{100\ cm}\right)\left(\frac{24 \times 60 \times 60\ sec}{1\ day}\right)\left(\frac{1\ Meinzer}{0.041\ m/day}\right)$$

$$= 18.2\ Meinzer$$

Thus 1 Darcy = 18.2 Meinzer for water at 60°F.

Example 4.9

At station A the water-table elevation is 650 ft above sea level, and at B, which is 1000 ft apart from A, the elevation is 632 ft. The average velocity of flow is observed to be 0.1 ft/day. Determine the coefficient of permeability in the Meinzer unit.

Solution From eq. (4.11),

$$K = \frac{q}{i}$$

where

$$q = \text{specific discharge} = \text{velocity} = 0.1\ \text{ft/day}$$
$$i = \text{hydraulic gradient} = 650 - 632/1000 = 0.018$$
$$\text{thus,}\ K = 0.1/0.018 = 5.56\ \text{ft/day}$$

Conversion to Meinzer:

$$K = \left(5.56 \frac{ft}{day}\right)\left(\frac{1\ Meinzer}{0.134\ ft/day}\right)$$

$$= 41.46\ Meinzer$$

4.5.2 Transmissivity

Transmissivity determines the ability of an aquifer to transmit water through its entire thickness. In an aquifer of uniform thickness,

$$T = \overline{K}b \qquad [\text{L}^2\text{T}^{-1}] \tag{4.13}$$

where

$$T = \text{transmissivity}$$
$$\overline{K} = \text{average hydraulic conductivity}$$
$$b = \text{thickness of aquifer}$$

When the hydraulic conductivity is a continuous function of depth,

$$\overline{K} = \frac{1}{b}\int_0^b K_z\,dz \qquad [\text{LT}^{-1}] \tag{4.14}$$

When a medium is stratified, two conditions can exist: the direction of flow is either parallel to the stratifications or normal to it. When flow direction is parallel to the stratifications, as shown in Figure 4.7, the average value of hydraulic conductivity can be given by

$$\overline{K} = \frac{1}{b}(K_1b_1 + K_2b_2 + K_3b_3 + \cdots + K_nb_n) \qquad [\text{LT}^{-1}] \tag{4.15}$$

Theory of Groundwater Flow   Chap. 4

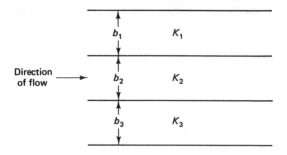

Figure 4.7 Flow parallel to stratifications.

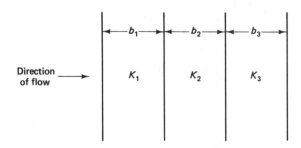

Figure 4.8 Flow normal to stratifications.

For flow perpendicular to stratifications, as shown in Figure 4.8,

$$\overline{K} = \frac{b}{b_1/K_1 + b_2/K_2 + b_3/K_3 + \cdots + b_n/K_n} \quad [LT^{-1}] \quad (4.16)$$

Field tests to determine the transmissivity of a medium are described subsequently.

Example 4.10

The soil under a dam consists of four layers as follows:

Layer	Hydraulic Conductivity (cm/hr)	Depth (m)
1	5	4.8
2	2	8.0
3	0.6	18.0
4	1.0	3.0

(a) What is the average vertical conductivity of the soil?

(b) What is the transmissivity of the soil when the water table is at the ground surface?

Solution From eq. (4.16),

$$\overline{K} = \frac{4.8 + 8.0 + 18.0 + 3.0}{4.8/5 + 8.0/2 + 18.0/0.6 + 3.0/1} = 0.89 \text{ cm/hr}$$

$$= \left(0.89 \frac{\text{cm}}{\text{hr}}\right)\left(\frac{1 \text{ m}}{100 \text{ cm}}\right)\left(\frac{24 \text{ hr}}{1 \text{ day}}\right)$$

$$= 0.214 \text{ m/day}$$

From eq. (4.13),

$$T = (0.214)(33.8) = 7.23 \text{ m}^2/\text{day}$$

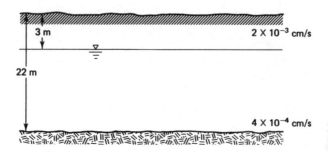

3 m

2 × 10⁻³ cm/s

22 m

4 × 10⁻⁴ cm/s

Figure 4.9 Stratum with uniformly varying hydraulic conductivity, in Example 4.11.

Example 4.11

In a soil stratum, the hydraulic conductivity at the surface is 2×10^{-3} cm/sec. It uniformly reduces to 4×10^{-4} cm/sec at a depth of 22 m, as shown in Figure 4.9. If the water table is 3 m below the surface, determine the transmissivity of the stratum.

Solution The hydraulic conductivity at a height x can be expressed (bottom as datum) as

$$K = 4 \times 10^{-4} + \left(\frac{2 \times 10^{-3} - 4 \times 10^{-4}}{22}\right) x$$

or

$$K = 4 \times 10^{-4} + 0.727 \times 10^{-4} x$$

From eq. (4.14),

$$\overline{K} = \frac{1}{19} \int_0^{19} (4 \times 10^{-4} + 0.727 \times 10^{-4} x) \, dx$$

$$= \frac{1}{19} \left\{ 4 \times 10^{-4} [x]_0^{19} + 0.727 \times 10^{-4} \left[\frac{x^2}{2}\right]_0^{19} \right\} = 10.9 \times 10^{-4} \text{ cm/sec}$$

or

$$\overline{K} = \left(10.9 \times 10^{-4} \frac{\text{cm}}{\text{sec}}\right) \left(\frac{1 \text{ m}}{100 \text{ cm}}\right) \left(\frac{24 \times 60 \times 60 \text{ sec}}{1 \text{ day}}\right) = 0.942 \text{ m/day}$$

From eq. (4.13),

$$T = (0.942)(19) = 17.89 \text{ m}^2/\text{day}$$

4.5.3 Leakance, Retardation Coefficient, and Leakage Factor (for Leaky Aquifer)

Hantush (1964) introduced leakance or coefficient of leakage as a term characteristic of the semipervious confining layer through which water leaks out from an aquifer. Defined, as follows, it is a measure of the ability of the confining layer to transmit vertical leakage:

$$L_e = \frac{K'}{b'} \qquad [\text{T}^{-1}] \tag{4.17}$$

where

L_e = leakance

K' = coefficient of permeability of semipervious layer of thickness b'

Other factors, introduced by Hantush to indicate areal distribution of leakage and used for the solution of the equation of the flow through leaky aquifer, were retardation coefficient and leakage factor, defined as

$$a = \frac{K}{K'/b'} \quad \text{[L]} \tag{4.18a}$$

and

$$B = \sqrt{\frac{Kb}{K'/b'}} \quad \text{[L]} \tag{4.18b}$$

where K is the coefficient of permeability of aquifer of thickness b.

Example 4.12

The banks (and bottom) of a stream consist of silty clay of hydraulic conductivity 0.008 m/day having an average depth of 150 cm. The underlying aquifer of fine sand has an average thickness of 20 m. Determine the (a) coefficient of leakage, (b) retardation coefficient, and (c) leakage factor. Hydraulic conductivity of fine sand = 2.5 m/day.

Solution From eq. (4.17),

$$L_e = \frac{0.008}{150/100} = 0.0053 \text{ per day}$$

From eq. (4.18a),

$$a = \frac{2.5}{0.0053} = 471.7 \text{ m}$$

From eq. (4.18b),

$$B = \sqrt{\frac{2.5 \times 20}{0.0053}} = 97.1 \text{ m}$$

4.6 BASIC EQUATION OF GROUNDWATER FLOW: DARCY'S LAW

The fundamental law of groundwater movement was discovered by Henri Darcy in 1856. He ran an experiment on a vertical pipe filled with sand under conditions simulated by Figure 4.10.

He concluded that the flow rate Q was proportional to the cross-sectional area A, inversely proportional to the length L of the sand-filter flow path and proportion to head drop $(h_1 - h_2)$. This provided the famous Darcy equation

$$Q = \frac{KA(h_1 - h_2)}{L} \quad [L^3T^{-1}] \tag{4.19}$$

where K is the hydraulic conductivity, which represented the constant of proportionality.

The ratio $h_1 - h_2/L$ is known as the hydraulic gradient. Defining specific discharge, q, or discharge velocity, v, as discharge per unit cross-sectional area, (4.19) becomes

$$q = v = \frac{-K\Delta h}{L} \quad [LT^{-1}] \tag{4.20}$$

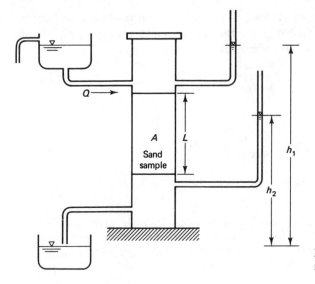

Figure 4.10 Simulation of Darcy's experiment.

where

$$q = \text{specific discharge}$$

$$v = \text{Darcy velocity or discharge velocity}$$

$$\Delta h = \text{drop of head in length } L \text{ (negative sign indicates flow in the direction of decreasing head)}$$

4.6.1 Darcy Velocity and Seepage Velocity

In eq. (4.20), v, known as the Darcy velocity, is a fictitious velocity since it assumes that flow occurs through the entire cross section of the material, whereas the flow is actually limited to the pores space only. If v_v is the seepage velocity and A_v is the area of voids, then, from the continuity equation,

$$Q = Av = A_v v_v$$

or

$$v_v = v\frac{A}{A_v}$$

Multiplying both sides by the length of the medium,

$$v_v = v\frac{AL}{A_v L} = v\frac{V}{V_v}$$

By definition, $\eta = V_v/V$; thus

$$v_v = \frac{v}{\eta} \qquad [LT^{-1}] \qquad (4.21)$$

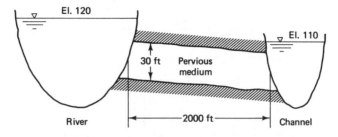

Figure 4.11 Model of river and channel in Example 4.13.

where

$$v = \text{Darcy velocity}$$
$$v_v = \text{seepage or interstitial velocity}$$
$$\eta = \text{porosity}$$

Example 4.13

A channel runs almost parallel to a river as shown in Figure 4.11. The water level in the river at an elevation of 120 ft and in the channel at an elevation of 110 ft. The river and channel are 2000 ft apart and a pervious formation of average 30 ft thickness and hydraulic conductivity of 0.25 ft/hr joins them together. Determine the rate of seepage flow from the river to the channel.

Solution Consider a 1-ft length of river (and channel) perpendicular to the paper. From eq. (4.19),

$$Q = \frac{KA(h_1 - h_2)}{L}$$

area of cross section of the aquifer normal to flow,

$$A = (30 \times 1) = 30 \text{ ft}^2$$
$$K = \left(0.25 \frac{\text{ft}}{\text{hr}}\right)\left(\frac{24 \text{ hr}}{1 \text{ day}}\right) = 6 \text{ ft/day}$$
$$Q = \frac{6(30)(120 - 110)}{2000}$$
$$= 0.9 \text{ ft}^3/\text{day/ft length}$$

Example 4.14

A semi-impervious (aquitard) separates an overlying water-table aquifer from an underlying confined aquifer as shown in Figure 4.12. Determine the rate of flow, if any, taking place between the aquifer.

Solution

1. Since the water table is above the piezometric surface and a semi-impervious (leaky) layer exists, flow will take place from the water-table aquifer to the confined aquifer.

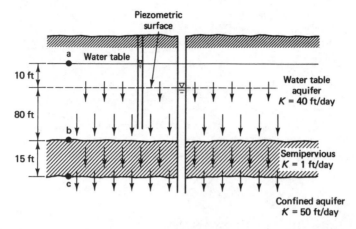

Piezometric
surface

a   Water table

10 ft

Water table
aquifer
$K = 40$ ft/day

80 ft

b

15 ft

Semipervious
$K = 1$ ft/day

c

Confined aquifer
$K = 50$ ft/day

Figure 4.12   Vertical downward flow through semipervious layer, Example 4.14.

2. Head loss will take place when water moves through the water-table aquifer, which is not known.

3. Assume that the head at point b shown in the figure is h_b and consider the unit horizontal area through which flow takes place.

4. Between points a and b, from eq. (4.20),

$$q = \frac{40(90 - h_b)}{90} \qquad (1)$$

5. Between points b and c, from eq. (4.20),

$$q = (1)\frac{(h_b + 15) - (80 + 15)}{15} \qquad (2)$$

6. Solving eqs. (1) and (2) provides

$$h_b = 88.70 \text{ ft}$$

from eq. (1), $q = 0.58$ ft³/day per square foot

Example 4.15

A confined aquifer has a source of recharge as shown in Figure 4.13. The hydraulic conductivity of the aquifer is 50 m/day and its porosity is 0.2. The piezometric head in two wells 1000 m apart is 55 m and 50 m, respectively, from a common datum. The average thickness of the aquifer is 30 m and the average width is 5 km. (a) Determine the rate of flow through the aquifer. (b) Determine the time of travel from the head of the aquifer to a point 4 km downstream (assume no dispersion or diffusion).

Solution

1. Area of cross section of flow $= 30 \times 5 \times 1000 = 15 \times 10^4$ m²

2. Hydraulic gradient $= \dfrac{55 - 50}{1000} = 5 \times 10^{-3}$

3. Rate of flow, from eq. (4.19):

$$Q = (50 \text{ m/day})(15 \times 10^4 \text{ m}^2)(5 \times 10^{-3}) = 37{,}500 \text{ m}^3/\text{day}$$

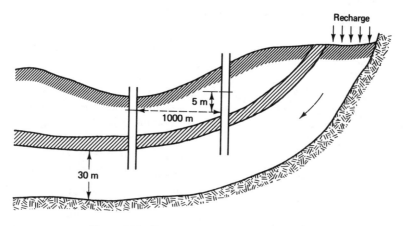

Figure 4.13 Travel time in uniform-sized aquifer.

4. Darcy velocity:

$$v = \frac{Q}{A} = \left(37,500\ \frac{m^3}{day}\right)\left(\frac{1}{15 \times 10^4\ m^2}\right) = 0.25\ m/day$$

5. Seepage velocity, from eq. (4.21):

$$v_v = \frac{v}{\eta} = \frac{0.25}{0.2} = 1.25\ m/day$$

6. Time to travel 4 km downstream:

$$t = \frac{4 \times 1000\ m}{1.25\ m/day} = 3200\ days \quad or \quad 8.77\ years$$

This shows that water moves very slowly underground.

4.6.2 Validity of Darcy's Law

1. In laminar flow, velocity bears a linear relationship to the hydraulic gradient. Since Darcy's law states that the discharge velocity is proportional to the first power of the hydraulic gradient, it is valid only within a laminar flow condition. As for pipes, Reynolds number is used to distinguish laminar flow from turbulent flow. For a porous medium, Reynolds number is expressed as

$$\mathrm{Re} = \frac{\rho v d_{10}}{\mu} \quad [\text{dimensionless}] \tag{4.22}$$

where

v = Darcy velocity

d_{10} = effective grain size (i.e., 10% of materials are finer than size indicated)

For $\mathrm{Re} < 1$, laminar flow occurs, as in most cases; for $\mathrm{Re} \geq 1$ but < 10, there is no serious departure from laminar flow; and for $\mathrm{Re} > 10$, there is turbulent flow, as in the immediate vicinity of pumped wells.

2. The Darcy's law is not valid where water flows through extremely fine-grained materials (e.g., colloidal clays).

3. The Darcy's law is not valid where the medium is not fully saturated.

Example 4.16

A 0.3-m well has a 25-m-long screen that covers the entire depth of an aquifer. The aquifer medium has an effective grain size of 1.5 mm. The well is pumped at a rate of 0.2 m³/s. Assess the validity of Darcy's law near the well.

Solution

1. Area through which flow into well takes place $= (\pi \times d)$ (thickness of aquifer)

$$= (\pi \times 0.3)(25) = 23.55 \text{ m}^2.$$

2. Darcy velocity $= \dfrac{Q}{A} = \dfrac{0.2}{23.55} = 8.5 \times 10^{-3}$ m/s or 0.85 cm/s.

3. Assume that $\rho = 1$ g/cm³ and $\mu = 0.01$ P or g/cm · sec.

4. From eq. (4.22),

$$\text{Re} = \frac{(1)(0.85)(0.15)}{0.01} = 12.75 > 10$$

5. Since Re > 10, flow is turbulent and Darcy's law is not applicable.

4.6.3 Generalization of Darcy's Law

Above, Darcy's law has been presented for one-dimensional flow and in the form in which it was empirically proposed by Darcy. However, at any point of a fluid in three-dimensional flow, there are three velocity components, a pressure component, and a density component. In groundwater flow, density is commonly considered constant (water is assumed to be incompressible unless specifically stated to the contrary, as in storage coefficient computation for a artesian aquifer). When water flows through an inclined medium as shown in Figure 4.14, its pressure (piezometric) head, which is a scalar quantity, is written as

$$h = Z + \frac{p}{\gamma} \quad [\text{L}] \tag{4.23}$$

There is a kinetic energy term also which can be neglected in considering the loss of head due to flow.

The velocity (specific discharge) component, however, is a vectorial quantity and can be expressed in three directions by Darcy's equation (4.20):

$$v_x = -K\frac{\partial h}{\partial x}$$

$$v_y = -K\frac{\partial h}{\partial y} \quad [\text{LT}^{-1}] \tag{4.24}$$

$$v_z = -K\frac{\partial h}{\partial z}$$

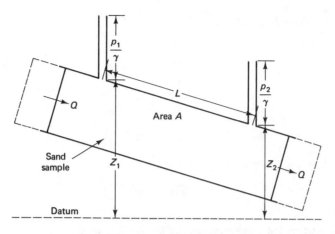

Figure 4.14 Flow through inclined medium.

If $\mathbf{i}$, $\mathbf{j}$, $\mathbf{k}$ represent standard unit vectors in the x, y, and z directions, respectively, the velocity (specific discharge) components in the three coordinate directions will be $\mathbf{i}v_x$, $\mathbf{j}v_y$, and $\mathbf{k}v_z$. The resultant velocity (specific discharge) vector will be given by

$$\mathbf{v} = \mathbf{i}v_x + \mathbf{j}v_y + \mathbf{k}v_z$$

Treating K as constant and substituting eq. (4.24), we have

$$\mathbf{v} = -K\left\{\mathbf{i}\frac{\partial h}{\partial x} + \mathbf{j}\frac{\partial h}{\partial y} + \mathbf{k}\frac{\partial h}{\partial z}\right\}$$

or

$$\mathbf{v} = -K\nabla h \qquad [\mathrm{LT^{-1}}] \qquad (4.25)$$

where ∇h denotes the head-gradient vector.

Equation (4.25) is a generalized form of Darcy's law expressed in vectorial notation. This equation is for isotropic soil in which hydraulic conductivity K is constant in all directions. For anisotropic aquifers, in which a different hydraulic conductivity is assigned to each of the coordinate directions, eqs. (4.24) and (4.25) will have the following forms:

$$v_x = -K_x\frac{\partial h}{\partial x} \qquad v_y = -K_y'\frac{\partial h}{\partial y} \qquad v_z = -K_z\frac{\partial h}{\partial z}$$

and

$$\mathbf{v} = \left\{-\mathbf{i}K_x\frac{\partial h}{\partial x} - \mathbf{j}K_y\frac{\partial h}{\partial y} - \mathbf{k}K_z\frac{\partial h}{\partial z}\right\} \qquad [\mathrm{LT^{-1}}] \qquad (4.26)$$

4.6.4 Velocity Potential

For the case when hydraulic conductivity K is constant, the velocity potential ϕ is defined as a scalar quantity having the following relation:

$$\phi = Kh \qquad [\mathrm{L^2T^{-1}}] \qquad (4.27)$$

In terms of the velocity potential

$$v_x = -\frac{\partial \phi}{\partial x}$$

$$v_y = -\frac{\partial \phi}{\partial y} \quad [LT^{-1}]$$

(4.28)

$$v_z = -\frac{\partial \phi}{\partial z}$$

and Darcy's law, eq. (4.25), takes the form:

$$\mathbf{v} = -\nabla\phi \quad [LT^{-1}]$$

(4.29)

Example 4.17

A homogeneous but anisotropic aquifer has the following values of hydraulic conductivity and head gradient:

	K (cm/sec)	Gradient h
x direction	0.03	0.22
y direction	0.035	0
z direction	0.002	−0.98

(a) Calculate the Darcy velocity vector.
(b) Plot the vectors $K\ \nabla h$.
(c) Compute the magnitude of the Darcy velocity.

Solution

(a) 1. Positive coordinate directions are set up as shown in Figure 4.15(a).

2. $\mathbf{i}K_x\dfrac{\partial h}{\partial x} = \mathbf{i}(0.03)(0.22) = 6.6 \times 10^{-3}\mathbf{i}$

$\mathbf{k}K_z\dfrac{\partial h}{\partial z} = \mathbf{k}(0.002)(-0.98) = -2.0 \times 10^{-3}\mathbf{k}$.

3. $\mathbf{v} = -\{6.6 \times 10^{-3}\mathbf{i} - 2.0 \times 10^{-3}\mathbf{k}\}$.

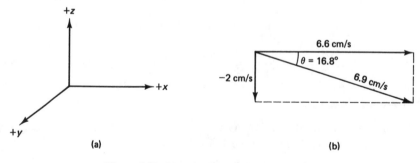

(a) (b)

Figure 4.15 Plot of velocity-vector components.

(b) 1. Suppose that standard unit vectors $\mathbf{i}, \mathbf{j}, \mathbf{k}$ are represented by 10^3 in./cm per second in the respective directions.

2. $6.6 \times 10^{-3}\mathbf{i} = (6.6 \times 10^{-3})(10^3) = 6.6$ in.

$-2.0 \times 10^{-3}\mathbf{k} = (-2.0 \times 10^{-3})(10^3) = -2.0$ in.

These are plotted in Figure 4.15(b).

(c) 3. Resultant magnitude, $v = 6.9$ cm/sec:

$$\tan\theta = \frac{2}{6.6} = 0.303$$

$$\theta = 16.8° \text{ from horizontal}$$

4.7 GENERAL EQUATION OF GROUNDWATER FLOW

There are two types of groundwater flow. In a steady or equilibrium state of flow, the water-table or piezometric surface is stabilized in a position and does not change with time as flow takes place, which means that inflow of water matches outflow of water. In an unsteady or nonequilibrium state, the water-table or piezometric head varies with time; thus water is either added or withdrawn from the groundwater storage during the flow. The fundamental equation of groundwater flow is Darcy's law. For steady state it can be used, as in the previous item, to determine the specific discharge (and discharge) if the head gradient is known or to compute the gradient for a given specific discharge. Darcy's law by itself, however, does not provide all the necessary conditions to solve groundwater flow problems in general. It gives three relations among four variables: three velocity (specific discharge) vectors and the head. For a general equation of flow, a fourth relation is provided by the equation of continuity or conservation of mass.

The groundwater flow equation contains three components: Darcy's law, the continuity equation, and the storage component. Conceptually, the flow equation is developed as follows. The continuity equation is written in terms of mass rate of inflow and outflow and accumulation of matter within an elemental volume situated in the field of flow. The terms of the mass rate are substituted by Darcy's law in the form of the piezometric (water-table) head, and the term relating to matter accumulation is expressed through the storage coefficient. This equation, usually in terms of piezometric head, is recognized as the flow equation.

The form of the equation differs for confined and unconfined aquifers because of different expressions for Darcy's law in two cases. In the case of confined flow, the area of flow is constant. For unconfined flow, however, area is a function of head (saturated depth). Also, the storage coefficient has different meanings in two cases, although this does not affect the form of the equation. The final equation for confined aquifer is a linear diffusion type of equation and for an unconfined aquifer it is the nonlinear Boussinesq equation. Derivations are presented in Sections 4.7.1 and 4.7.2 separately for two cases.

4.7.1 Equation for Confined Aquifers

I. Darcy's law:
From eq. (4.24), three velocity (specific discharge) vectors:

$$v_x = -K_x \frac{\partial h}{\partial x}$$

$$v_y = -K_y \frac{\partial h}{\partial y}$$

$$v_z = -K_z \frac{\partial h}{\partial z}$$

II. Continuity equation:
1. Mass discharge is equal to the water density times the volume discharge. The mass balance equation is written since the water density is considered variable in the storage coefficient. Mass discharge (flux) through face 1 into the element of Figure 4.16 in the x direction is

$$(\rho Q_x)_1 \quad \text{or} \quad (\rho v_x)_1 \, \Delta y \, \Delta z$$

2. Mass discharge through face 2 out of the element is

$$(\rho Q_x)_2 \quad \text{or} \quad (\rho v_x)_2 \, \Delta y \, \Delta z$$

3. If ρv_x is considered as a continuous function, the Taylor series may be used to expand $(\rho v_x)_2$ in terms of $(\rho v_x)_1$ as follows:

$$(\rho v_x)_2 = (\rho v_x)_1 + \frac{\partial (\rho v_x)}{\partial x} \Delta x + \frac{1}{2} \frac{\partial^2 (\rho v_x)}{\partial x^2} (\Delta x)^2 + \cdots$$

Taking the first two terms (neglecting the others) yields

$$(\rho v_x)_2 = (\rho v_x)_1 + \frac{\partial (\rho v_x)}{\partial x} \Delta x$$

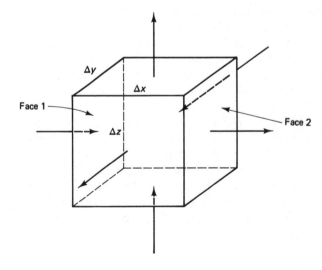

Figure 4.16 Elemental volume in the field of flow.

4. Net influx in the x direction

Outflow rate $-$ inflow rate

$$= \left[(\rho v_x)_1 + \frac{\partial (\rho v_x)}{\partial x} \Delta x\right] \Delta y \, \Delta z - (\rho v_x)_1 \Delta y \, \Delta z$$

$$= \frac{\partial (\rho v_x)}{\partial x} \Delta x \, \Delta y \, \Delta z$$

Similar terms could be written for the y and z directions.

5. The total net flux (outflow $-$ inflow) will be the summation of fluxes in the x, y, z directions and will be equal to the rate of change of mass within the element:

$$\frac{\partial (\rho v_x)}{\partial x} \Delta x \, \Delta y \, \Delta z + \frac{\partial (\rho v_y)}{\partial y} \Delta x \, \Delta y \, \Delta z + \frac{\partial (\rho v_z)}{\partial z} \Delta x \, \Delta y \, \Delta z = -\frac{\partial M}{\partial t}$$

The negative sign is to make the net flux positive when the mass is depleted. The equation above can be rewritten as

$$\frac{\partial (\rho v_x)}{\partial x} + \frac{\partial (\rho v_y)}{\partial y} + \frac{\partial (\rho v_z)}{\partial z} + \frac{1}{\Delta x \, \Delta y \, \Delta z} \frac{\partial M}{\partial t} = 0$$

This is the continuity equation.

III. Rate of change of mass:

6. Mass accumulated within the element

$$M = \rho \eta \Delta x \, \Delta y \, \Delta z$$

7. Considering that compression and expansion occur in the z direction only and Δx and Δy are constant:

$$\frac{\partial M}{\partial t} = \left[\eta \, \Delta z \frac{\partial \rho}{\partial t} + \rho \, \Delta z \frac{\partial \eta}{\partial t} + \rho \eta \frac{\partial (\Delta z)}{\partial t}\right] \Delta x \, \Delta y$$

8. The first term relates to the compression of water and the other two with compression of material. The expression above reduces to (Marino and Luthin, 1982, pp. 146–147)

$$\frac{\partial M}{\partial t} = (\alpha + \beta \eta)\rho \, \Delta x \Delta y \, \Delta z \frac{\partial p}{\partial t}$$

where

$$\alpha = 1/E_s$$
$$\beta = 1/E_w$$
$$E_s = \text{bulk modulus of elasticity of aquifer solids}$$
$$E_w = \text{bulk modulus of elasticity of water}$$
$$\eta = \text{porosity}$$
$$p = \text{pressure}, \, p = \gamma h$$

In terms of h (since $p = \gamma h$),

$$\frac{\partial M}{\partial t} = (\alpha + \eta\beta)\rho\,\Delta x\,\Delta y\,\Delta z\,\gamma\frac{\partial h}{\partial t}$$

or

$$\frac{\partial M}{\partial t} = \rho S_s\,\Delta x\,\Delta y\,\Delta z\,\frac{\partial h}{\partial t} \qquad \text{since } S_s = \gamma(\alpha + n\beta)$$

IV. Manipulation of continuity equation
    9. Since

$$\frac{\partial\,(\rho v_x)}{\partial x} = \rho\frac{\partial v_x}{\partial x} + v_x\frac{\partial\rho}{\partial x}$$

10. If the second term on the right side relating to change of water density is dropped in comparison to the first term,

$$\frac{\partial\,(\rho v_x)}{\partial x} = \rho\frac{\partial v_x}{\partial x}$$

11. Substituting v_x by Darcy's law from eq. (4.24) as shown in item I:

$$\frac{\partial\,(\rho v_x)}{\partial x} = \rho\frac{\partial v_x}{\partial x} = -\rho\frac{\partial}{\partial x}\left(K_x\frac{\partial h}{\partial x}\right)$$

Similar terms could be written for the y and z directions.

12. Substituting these and the rate of mass term of step 8 into the continuity equation of step 5:

$$\frac{\partial}{\partial x}\left(K_x\frac{\partial h}{\partial x}\right) + \frac{\partial}{\partial y}\left(K_y\frac{\partial h}{\partial y}\right) + \frac{\partial}{\partial z}\left(K_z\frac{\partial h}{\partial z}\right) = S_s\frac{\partial h}{\partial t} \qquad [\mathrm{T}^{-1}] \qquad (4.30)$$

13. Equation (4.30) is for *nonhomogeneous, anisotropic* confined aquifers. For *homogeneous, anisotropic* cases, when the hydraulic conductivity will be the same in space, eq. (4.30) will become

$$K_x\frac{\partial^2 h}{\partial x^2} + K_y\frac{\partial^2 h}{\partial y^2} + K_z\frac{\partial^2 h}{\partial z^2} = S_s\frac{\partial h}{\partial t} \qquad [\mathrm{T}^{-1}] \qquad (4.31)$$

    This is a linear second-order partial differential equation for unsteady-state flow in confined aquifer. Equations of similar form appear in the flow of heat and electricity. The derivation above is based on the concept of Jacob (1950). DeWiest (1965) followed a different approach and in the process redefined the specific storage, S_s, term, as discussed in Section 4.4.4.

4.7.2 Equation for Unconfined Aquifers

Development here is on the same lines as in the case of a confined aquifer. Consider an elemental volume cutting through the entire saturated thickness, as shown in Figure 4.17.

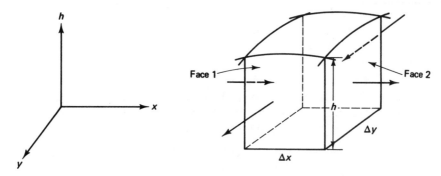

Figure 4.17 Elemental volume through the entire depth of an unconfined aquifer.

I. Darcy's laws: Assuming Dupuit assumption of horizontal flow, we have

$$Q_x = -K_x \frac{\partial h}{\partial x} h \, \Delta y$$

$$Q_y = -K_y \frac{\partial h}{\partial y} h \, \Delta x$$

There is no vertical flow according to the assumption above.

II. Equation of Continuity:
1. It can be written in volume discharge form here since compressibility of water is not involved (neglected).
2. In the x direction, net flux:

$$\text{outflow} - \text{inflow} = Q_{x_1} + \frac{\partial (Q_x)}{\partial x} \Delta x - Q_{x_1}$$

$$= \frac{\partial (Q_x)}{\partial x} \Delta x$$

3. In the y direction $= \dfrac{\partial (Q_y)}{\partial y} \Delta y$.

4. The total net flux is equal to the change in stored water volume.

$$\frac{\partial (Q_x)}{\partial x} \Delta x + \frac{\partial (Q_y)}{\partial y} \Delta y = -\frac{\partial V_w}{\partial t}$$

III. Rate of change of stored water volume:

$$V_w = S_y \, dh \, \Delta x \, \Delta y \qquad \text{(by definition)}$$

$$\frac{\partial V_w}{\partial t} = S_y \frac{\partial h}{\partial t} \Delta x \, \Delta y$$

IV. Manipulation of continuity equation: Substitute in eq. of step II(4), the values of Q from Darcy's law from Step I and V_w from step III.

$$-\frac{\partial}{\partial x}\left(K_x h \frac{\partial h}{\partial x}\right) \Delta x \, \Delta y - \frac{\partial}{\partial y}\left(K_y h \frac{\partial h}{\partial y}\right) \Delta x \, \Delta y = -Sy \frac{\partial h}{\partial t} \Delta x \, \Delta y$$

or

$$\frac{\partial}{\partial x}\left(K_x h \frac{\partial h}{\partial x}\right) + \frac{\partial}{\partial y}\left(K_y h \frac{\partial h}{\partial y}\right) = S_y \frac{\partial h}{\partial t} \qquad [\mathrm{LT}^{-1}] \qquad (4.32)$$

This is the nonlinear Boussinesq equation. Linearization of this could be achieved if the change in water table is small compared to the water-table depth h. In that case, the average aquifer thickness, b, could be substituted for h. For a homogeneous case, then, eq. (4.32) will become

$$K_x \frac{\partial^2 h}{\partial x^2} + K_y \frac{\partial^2 h}{\partial y^2} = \frac{S_y}{b} \frac{\partial h}{\partial t} \qquad [\mathrm{T}^{-1}] \qquad (4.33)$$

Thus eq. (4.33) for an unconfined aquifer becomes exactly like eq. (4.31) in two dimensions for confined aquifer. Only storage terms will have different meanings.

4.7.3 An Overview of Groundwater Flow Equation

Incorporating a source or sink term in eq. (4.30), a most general form of the equation of saturated flow through a porous medium is

$$\frac{\partial}{\partial x}\left(K_x \frac{\partial h}{\partial x}\right) + \frac{\partial}{\partial y}\left(K_y \frac{\partial h}{\partial x}\right) + \frac{\partial}{\partial z}\left(K_z \frac{\partial h}{\partial z}\right) \pm W(x, y, z, t) = S_s \frac{\partial h}{\partial t} \qquad [\mathrm{T}^{-1}]$$

$$(4.34)$$

where $W(x, y, z, t)$ is the source or skin term of discharge per unit volume representing recharge or discharge (well) point. Dropping out the source/sink term, for a homogeneous, anisotropic medium, above equation becomes eq. (4.31), as reproduced again

$$K_x \frac{\partial^2 h}{\partial x^2} + K_y \frac{\partial^2 h}{\partial y^2} + K_z \frac{\partial^2 h}{\partial z^2} = S_s \frac{\partial h}{\partial t} \qquad [\mathrm{T}^{-1}] \qquad (4.31)$$

For a homogeneous, isotropic medium, this reduces to

$$\frac{\partial^2 h}{\partial x^2} + \frac{\partial^2 h}{\partial y^2} + \frac{\partial^2 h}{\partial z^2} = \frac{S_s}{K} \frac{\partial h}{\partial t} \qquad [\mathrm{L}^{-1}] \qquad (4.35)$$

Equations (4.34), (4.31) and (4.35) are also expressed in terms of transmissivities instead of hydraulic conductivities. In such a case, the specific storage term is replaced by the storage coefficient or specific yield and the source/sink term is substituted by $bW(x, y, z, t)$.

Equation (4.35) is recognized as the "equation of heat conduction," "equation of flow of electricity," "diffusion equation," and as the "nonequilibrium equation of groundwater flow."

In steady flow, the pressure distribution does not change with time. Accordingly, eq. (4.35) becomes

$$\frac{\partial^2 h}{\partial x^2} + \frac{\partial^2 h}{\partial y^2} + \frac{\partial^2 h}{\partial z^2} = 0 \qquad [\mathrm{L}^{-1}] \qquad (4.36a)$$

or

$$\nabla^2 h = 0 \qquad (4.36b)$$

This is the Laplace equation which appears in mathematics and many branches of physics: for steady-state conduction of heat and electricity, for steady diffusion, and in the elastic membrane theory. An equation identical in form to the Laplace equation for saturated flow was developed independently by Jules Dupuit (in 1863), P. Forchheimer (in 1886), and Charles Slichter (in 1899).

The equation of flow in an unconfined aquifer is the nonlinear Boussinesq equation, which is extremely difficult to solve. With certain permissible assumptions, it is linearized to the form of eq. (4.34). All discussion above, as such, holds good for an unconfined aquifer as well, with specific yield being substituted for the storage coefficient.

PROBLEMS

4.1. What volume of solid material is present in 1 ft^3 of sandstone if the porosity of the sandstone is 0.35?

4.2. A soil has a bulk density of 1.7 g/cm^3. The specific gravity of soil solids is 2.6. What is the porosity and the void ratio of the soil?

4.3. A wet soil weighs 300 g and occupies a volume of 200 cm^3. If it contains 15% moisture by weight:
   (a) What is its bulk density?
   (b) What is its porosity for specific gravity of soil grains of 2.65?
   (c) What is its void ratio?

4.4. In Example 4.3, if the water table was initially 170 cm below the surface and finally went down to 220 cm, how much water was drained from the unit area of the soil?

4.5. In Example 4.3, the water table was initially at 280 cm. If it is lowered by 20 cm, determine the quantity of water released.

4.6 A 100-g dry soil sample is tested for capillary head moisture distribution. The negative pressure head changes and the incremental amount of water released from the sample are indicated below. The bulk density of the soil is 1.4, and at saturation it contained 38% moisture by weight.

Negative Pressure Head (cm)	Water Released (cm^3)
0	0.0
10	0.2
20	0.3
30	0.6
40	0.9
50	1.3
60	1.7
70	1.4
80	0.8
90	0.4
100	0.2

(a) Calculate the volumetric moisture content at each pressure head (ratio of volume of water retained to volume of soil).

(b) Plot a moisture distribution curve.

(c) If the water table drops from 100 cm to 150 cm below the surface, determine the amount of water drained from the soil.

4.7. In Problem 4.6:

    (a) What is the porosity of the soil?

    (b) What is the specific retention of the soil?

    (c) What is the specific yield of the soil?

4.8. In a clayey formation, the water table drops by 0.6 m over an area of 8 ha. How much volume of water is drained off if the specific yield is 23%?

4.9. An unconfined aquifer system covers an area of 20 million square meters. The water table is 22 m below the land surface. When 50 million m^3 of water is added to the aquifer, the water table rises by 10 m. What is the specific yield of the aquifer?

4.10. The coefficient of storage of an artesian aquifer is 3×10^{-4}. If the average thickness of the aquifer is 250 ft and its porosity is 0.35, estimate the fraction of coefficient of storage resulting from the expansion of water and that from the compressibility of the aquifer. ($E_w = 3 \times 10^5$ psi.)

4.11. From a pumping test on a confined formation, the storage coefficient is found to be 3×10^{-3} for a location having a depth of 50 m. If the average depth of an area of 3.2 km^2 is 35 m, estimate the volume of water contributed by the area when the pressure head is dropped by 10 m.

4.12. A confined aquifer has an average thickness of 150 ft and a porosity of 0.35. If the compressibility of water is 3×10^{-6} in.^{2}/lb and the compressibility of material is 2.5×10^{-6} in.^{2}/lb, what is the storage coefficient of the aquifer?

4.13. For the aquifer in Problem 4.12 the compressibility of the material is not known but the storage coefficient is 0.0003. Determine the compressibility of the material, and also the storage coefficient components contributed by compressibility of water and material, respectively.

4.14. The hydraulic conductivity of a soil at 50°F is 0.015 ft/sec. What is its intrinsic permeability?

4.15. A tracer element was introduced into an aquifer at an upstream location and from its appearance at a downstream location, the average flow velocity was found to be 0.5 in./day. The slope of the piezometric surface was 1 ft/mi. Determine the hydraulic conductivity of the aquifer.

4.16. A layered soil is shown in Fig. P4.16. Estimate the transmissivity for the formation.

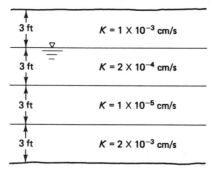

3 ft	$K = 1 \times 10^{-3}$ cm/s
3 ft	$K = 2 \times 10^{-4}$ cm/s
3 ft	$K = 1 \times 10^{-5}$ cm/s
3 ft	$K = 2 \times 10^{-3}$ cm/s

Figure P4.16

4.17. For Fig. P4.17, determine the coefficient of leakage and leakage factor.

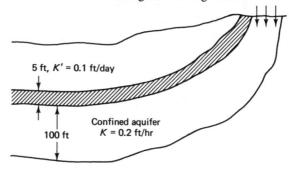

5 ft, $K' = 0.1$ ft/day

Confined aquifer
$K = 0.2$ ft/hr

100 ft

Figure P4.17

4.18. A confined aquifer slopes gradually from 12 m to 8 m thickness. The slope of piezometric surface is 0.25 m per kilometer. If the hydraulic conductivity of the aquifer is 25 m/day, how much water flows through the aquifer from its width of 2.5 km?

4.19. Flow in a valley takes place as shown in Fig. P4.19. The formation in the valley has a hydraulic conductivity of 400 ft/day and a porosity of 0.25. The difference in the water levels in two wells shown is 1 ft. Between the observation wells, the average depth is 100 ft.

(a) Determine the rate of flow per mile width of the aquifer.

(b) How long will it take the groundwater to travel from head of the valley to the stream bank?

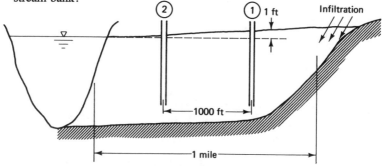

2

1 1 ft

Infiltration

1000 ft

1 mile

Figure P4.19

4.20. A porous medium is oriented at an angle of 40° with the horizontal plane. If the hydraulic grade line is parallel to the orientation of the medium and the specific discharge is 5 m³/day per square meter, determine the hydraulic conductivity of the medium.

4.21. Two observation wells have been constructed in the formation shown in Fig. P4.21. If the flow rate is 0.01 m³/hr per unit width of the formation, determine K_2.

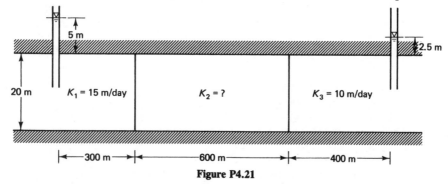

5 m

2.5 m

20 m

$K_1 = 15$ m/day

$K_2 = ?$

$K_3 = 10$ m/day

|← 300 m →|← 600 m →|← 400 m →|

Figure P4.21

4.22. A well of 6 in. radius is drilled through a confined aquifer of 100 ft thickness. The aquifer consists of uniform sand of average grain size of 0.3 in. The well is pumped at a rate of 100 ft^3/hr. Determine whether Darcy's flow conditions exist near the well.

4.23. A homogeneous, isotropic aquifer has a hydraulic conductivity of 0.02 ft/s. Its head gradients in the x, y, and z directions are 0, 0.19, and 0.87, respectively.
   (a) Determine the velocity vector components
   (b) plot the velocity vectors for the aquifer
   (c) find the magnitude and direction of the velocity for the aquifer.

CHAPTER
FIVE

APPLICATIONS OF
GROUNDWATER FLOW

5.1 GRAPHICAL SOLUTION OF STEADY-STATE FLOW EQUATION: TECHNIQUE OF FLOW NET

As discussed in Section 4.7.3, the Laplace equation represents the steady-state groundwater flow through a homogeneous, isotropic medium. The geometric solution of the Laplace equation in two dimensions consists of two families of curves intersecting each other orthogonally. The lines of equal hydraulic head, called equipotentials, and the lines describing flow paths of water particles through the aquifer media, called streamlines, represent such a family of curves. Since water moves in the direction of steepest hydraulic gradient, streamlines are perpendicular to equipotential lines. Together they form a flow net. A flow net thus represents a solution of the steady-state flow equation for homogeneous, isotropic soil.

5.1.1 Flow Net in Isotropic Soil

The equation for flow nets originates from Darcy's law. However, where Darcy's law cannot be applied directly due to the irregularity of flow zones or directions of flow and difficulty in defining of the boundaries, the method of flow net can be used.

Consider the portion of a flow net drawn in Figure 5.1(a). The flow taking place through a flow channel between equipotential lines 1 and 2 as shown in Figure 5.1(b) per unit width (perpendicular to the paper) is (from Darcy's law)

$$\Delta q = K(d_m \times 1)\left(\frac{\Delta h_1}{dl}\right) \qquad (a)$$

157

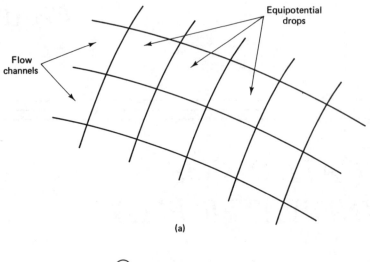

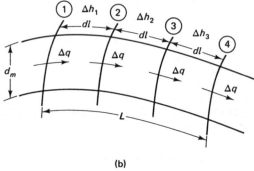

(b)

Figure 5.1 (a) Portion of a flow net; (b) Flow through a single flow channel.

The flow through equipotential lines 2 and 3 is

$$\Delta q = K(d_m \times 1)\left(\frac{\Delta h_2}{dl}\right) \tag{b}$$

Since the same flow continues, eq. (a) = eq. (b), or

$$\Delta h_1 = \Delta h_2 \tag{c}$$

that is, the head drop is the same in each potential drop (two successive equipotential lines). If there are n_d such drops, then

$$\Delta h = \frac{h}{n_d} \tag{d}$$

where h is the total head loss between the first and last equipotential lines. Substituting (d) in (a) yields

$$\Delta q = K\frac{dm}{dl}\frac{h}{n_d} \tag{e}$$

Applications of Groundwater Flow   Chap. 5

Equation (e) is for one flow channel. If there are n_f such channels, the total flow per unit width

$$q = \frac{n_f}{n_d} K \frac{dm}{dl} h \tag{f}$$

If the flow net is drawn as squares, so that $dm \approx dl$, eq. (f) becomes

$$q = \frac{n_f}{n_d} Kh \qquad [\mathrm{L^2 T^{-1}}] \tag{5.1}$$

where

q = rate of flow or seepage per unit width

n_f = number of flow channels

n_d = number of equipotential drops

h = total head loss in flow system

K = hydraulic conductivity

Since each potential line indicates the available head (total head minus loss of head), the uplift pressure at the base of the structure can be given by

$$u = \left(\frac{n}{n_d} h + Z \right) \gamma_w \qquad [\mathrm{FL^{-2}}] \tag{5.2}$$

where

u = uplift water pressure

n = number of the equipotential line counting last line on the downstream as zero

n_d = number of potential drops

h = total head loss in flow system

Z = depth of the base below the datum (if the base is above the datum, Z is negative)

5.1.2 Procedure to Draw a Flow Net

Drawing a flow net is a trial procedure. Certain rules to be observed are:

(a) Equipotential lines cross flow lines at a right angle.
(b) Shapes formed by equipotential and flow lines should be as close to square as possible.
(c) Impermeable boundaries are flow lines.
(d) The soil-water interface at upstream and downstream of a structure is an equipotential line.
(e) The seepage surface is a flow line.

The method of sketching the flow net is as follows:

1. Draw to a convenient scale the cross section of the structure, water elevations, and soil profiles.
2. Establish boundary conditions and draw flow lines and equipotential lines for the boundaries.
3. Sketch in intermediate flow lines and equipotential lines by smooth curves adhering to right-angle intersections and square figures. Where the direction of flow is in a straight line, flow lines are an equal distance apart and parallel.
4. Continue sketching until an inconsistency develops. Each inconsistency will indicate changes to be made in flow lines and equipotential lines. Successive trials will result in a reasonably consistent flow net.
5. It is for the student to decide the number of flow lines to be drawn. Three to five flow lines are usually sufficient. Depending on the number of flow lines selected, the number of equipotential lines will be automatically fixed because of the requirement for the square figures of the net.

Example 5.1

A dam is constructed on a permeable stratum underlain by an impermeable rock as shown. A row of sheet pile is installed at the upstream face. If the permeable soil has a hydraulic conductivity of 150 ft/day, determine (a) the rate of flow from upstream to downstream, and (b) the uplift pressure acting in the bottom of the dam.

Solution   The flow net is drawn as shown in Figure 5.2.

$$n_f = 5 \qquad n_d = 17$$

(a) From eq. (5.1),

$$q = \frac{n_f}{n_d} Kh$$

$$= \frac{5}{17}(150)(35) = 1554 \text{ ft}^3/\text{day per foot} \quad \text{or} \quad 0.018 \text{ cfs per foot}$$

(b) From eq. (5.2),

$$u = \left(\frac{n}{n_d}h + Z\right)\gamma_w$$

$$= \left[\frac{n}{17}(35) + 2\right]62.4$$

$$= (2.06n + 2)(62.4)$$

Position:	A	B	C	D	E	F	G	H	I	J
Distance from front toe (ft)	0	3	22	37.5	50	62.5	75	86	94	100
n		16.5	9	8	7	6	5	4	3	2	1.2
u (ksf)		2.25	1.28	1.15	1.02	0.90	0.77	0.64	0.51	0.38	0.28

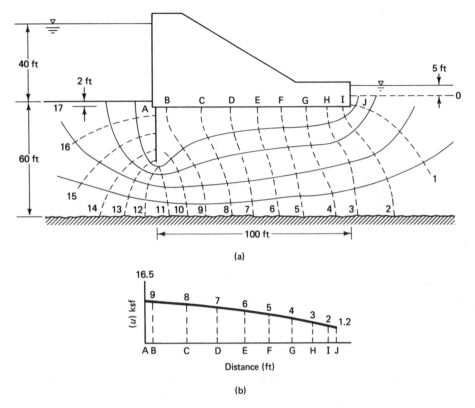

Figure 5.2 (a) Flow net for a dam with sheet pile; (b) Distribution of uplift pressure.

5.1.3 Flow Net in Anisotropic Soil

In anisotropic soil when K_x is not equal to K_z, the groundwater flow equation is not of the Laplace form and the flow net is not the solution to the equation. However, by transposing $x' = \sqrt{K_z/K_x}\,x$, the flow equation can be converted to Laplace form in terms of x' and z. Thus the flow net solution for an anisotropic soil can be obtained by distorting the horizontal scale for the system.

The procedure is as follows:

1. Transform all horizontal dimensions into notional dimensions using

$$x' = x\sqrt{K_z/K_x} \quad [\text{L}] \tag{5.3}$$

where

$$x = \text{natural horizontal dimensions}$$
$$x' = \text{notional horizontal dimensions}$$

2. Draw the cross section of the structure in natural vertical dimensions and notional (distorted) horizontal dimensions, to a convenient scale.
3. Sketch the flow net as usual.

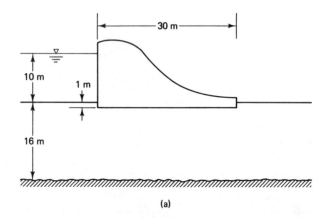

(a)

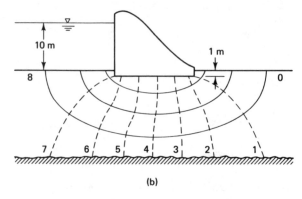

(b)

Figure 5.3 (a) Profile of a dam on an isotropic stratum; (b) Transformed section and
flow net for a dam.

4. Determine the rate of flow using

$$q = \frac{n_f}{n_d}\sqrt{K_x K_z}\, h \qquad [L^2 T^{-1}] \qquad (5.4)$$

Example 5.2

A concrete dam is found on a permeable stratum having $K_x = 30$ m/day and $K_z = 6$ m/day, as shown in Figure 5.3(a). Determine (a) the rate of flow below the dam, and (b) the distribution of the uplift pressure.

Solution Factor for converting horizontal dimensions $= \sqrt{\dfrac{K_z}{K_x}} = \sqrt{\dfrac{6}{30}} = 0.45.$

Length of dam $= 30 \times 0.45 = 13.5$ m.

The profile of the structure with distorted x scale is shown in Figure 5.3(b).
The flow net has also been sketched in the Figure 5.3(b).

(a) $n_f = 4$, $n_d = 8$. From eq. (5.4),

$$q = \frac{n_f}{n_d}\sqrt{K_x K_z}\, h$$

$$= \frac{4}{8}\sqrt{(30\,(6)}\,(10) = 67 \text{ m}^3/\text{day per meter width}$$

(b) From eq. (5.2),

$$u = \left(\frac{n}{n_d} h + Z\right)\gamma_w$$

$$= \left[\frac{n}{8}(10) + 1\right](9.81)$$

$$= (1.25n + 1)(9.81)$$

Position:	A	B	C	D	E	F	G
Distance on distorted model (m)	0	0.65	3.5	6.75	9.65	12.2	13.5
Natural distance (m)	0	1.45	7.8	15	21.4	27	30
n		7	6	5	4	3	2	1
u (kN/m²)	95.65	83.4	71.1	58.9	46.6	34.3	22.1

5.2 ANALYTICAL SOLUTION OF THE STEADY-STATE FLOW EQUATION

Groundwater flow equations are summarized in Section 4.7.3. When an appropriate equation is solved to satisfy the initial and boundary conditions of a given flow system, the distribution of head throughout the system is obtained. These equations are, however, difficult to solve for many flow problems since the theory of partial differential equations is very limited. The steady-state flow is represented by the Laplace equation, for which a general solution can be obtained from the theory of partial differential equations. For actual flow problems, however, the solutions should meet certain boundary conditions or specifications regarding the distribution and variation of head at one or more boundaries or the rate of flow across them. Solutions to some problems have been obtained either by adopting some approximate differential equations that are easier to solve *or* by idealizing the conditions for the system. A few cases are analyzed below. The medium has been considered homogeneous and isotropic in all cases.

5.2.1 Confined Groundwater Flow between Two Water Bodies

Figure 5.4 shows the model for one-dimensional flow [eq. (4.36a)] in one-dimensional form:

$$\frac{\partial^2 h}{\partial x^2} = 0 \qquad \text{(a)}$$

General solution of this is

$$h = Ax + B \qquad \text{(b)}$$

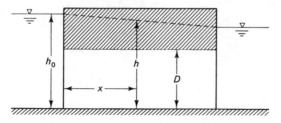

Figure 5.4 Confined flow between two water bodies.

Substituting the boundary condition in (b): At $x = 0$, $h = h_0$:

$$h_0 = A(0) + B \quad \text{or} \quad B = h_0 \tag{c}$$

Differentiating (b) yields

$$\frac{dh}{dx} = A \tag{d}$$

Darcy's law:

$$q = -KD\frac{dh}{dx} \tag{e}$$

Substituting (d) in (e), we obtain

$$A = -q/KD \tag{f}$$

Substituting (c) and (f) in (b) gives us

$$q = \frac{h_0 - h}{x}KD \qquad [L^2T^{-1}] \tag{5.5}$$

Example 5.3

A channel runs parallel to a river. The water level in the river is at an elevation of 120 ft and in the channel at an elevation of 110 ft. The river and channel are 2000 ft apart. A confined aquifer of 30 ft thickness joins them. The hydraulic conductivity is 0.25 ft/hr. Determine the rate of seepage from the river to the channel.

Solution From eq. (5.5),

$$q = \frac{h_0 - h}{x}KD$$

$$= \left(\frac{120 - 110}{2000}\right)(0.25)(30)$$

$$= 0.0375 \text{ ft}^3/\text{hr} \quad \text{or} \quad 0.9 \text{ ft}^3/\text{day}$$

5.2.2 Unconfined Flow between Two Water Bodies

Figure 5.5 shows a one-dimensional model for this condition. Equation (4.32) in one-dimensional steady-state form:

$$Kh\frac{\partial^2 h}{\partial x^2} = 0 \tag{a}$$

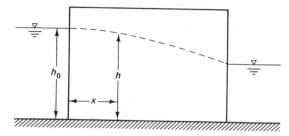

Figure 5.5 Unconfined flow between two water bodies.

or

$$\frac{\partial^2 h}{\partial x^2} = 0 \qquad \text{(b)}$$

Equation (b) is similar to the differential equation of the previous case; hence following similar steps,

$$q = \frac{h_0 - h}{x} KD \qquad [L^2T^{-1}] \qquad (5.6)$$

where D is the average thickness of the aquifer.

Example 5.4

Consider the cross section shown in Figure 5.6. Determine the rate of flow.

Solution

$$\text{Average } \overline{K} = \frac{b}{b_1/K_1 + b_2/K_2} = \frac{1800}{1000/20 + 800/30} = 23.48 \text{ m/day}$$

From eq. (5.6),

$$q = (23.48) \left(\frac{50 - 45}{1800} \right) \left(\frac{50 + 45}{2} \right)$$

$$= 3.1 \text{ m}^3/\text{day per meter}$$

Alternative Solution

1. Suppose that the head is h at the interface of two soil media.

2. Per meter width of the first medium, from eq. (5.6),

$$q = \left(\frac{50 - h}{1000} \right) (20) \left(\frac{50 + h}{2} \right)$$

$$= \left(\frac{1}{100} \right) (50^2 - h^2) \qquad \text{(a)}$$

3. Per meter width of the second medium, from eq. (5.6):

$$q = \left(\frac{h - 45}{800} \right) (30) \left(\frac{h + 45}{2} \right)$$

$$= \left(\frac{3}{160} \right) (h^2 - 45^2) \qquad \text{(b)}$$

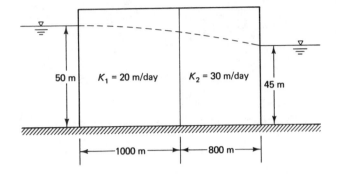

Figure 5.6 Unconfined nonhomogeneous aquifer between two water bodies, Example 5.4.

4. Equating eqs. (a) and (b) gives us

$$\frac{1}{100}(50^2 - h^2) = \frac{3}{160}(h^2 - 45^2)$$

$$h = 46.8 \text{ m}$$

5. Substituting in eq. (a) yields

$$q = 3.1 \text{ m}^3/\text{day per meter}$$

5.2.3 Confined Flow to a Well

Flow toward a well in a homogeneous and isotropic aquifer is radially symmetric. When the well screen, perforated pipe, or open well bore extends through the entire thickness of an aquifer, it is known as a fully penetrating well, as shown in Figure 5.7. Usually, a gravel pack is provided around the well screen. In such cases the well radius is considered from the center of the well to the outside of the gravel pack. Loss of head in the well and gravel pack is known as well loss; this quantity is very small.

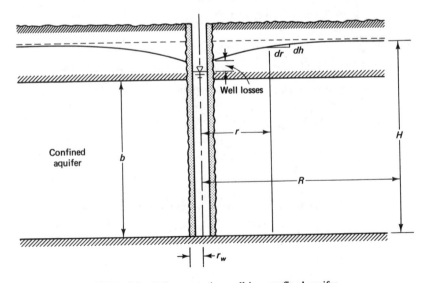

Figure 5.7 Fully penetrating well in a confined aquifer.

Applications of Groundwater Flow   Chap. 5

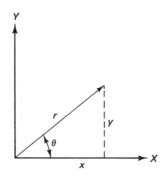

Figure 5.8 Polar coordinates.

Equation (4.36a) refers to the steady-state flow in a homogeneous, isotropic aquifer, which is reproduced below in two dimensions:

$$\frac{\partial^2 h}{\partial x^2} + \frac{\partial^2 h}{\partial y^2} = 0 \tag{a}$$

Polar coordinates are convenient for the problems concerning radial flow, as indicated in Figure 5.8.

$$x = r \cos \theta$$
$$y = r \sin \theta \tag{b}$$
$$r = (x^2 + y^2)^{1/2} \qquad \theta = \tan^{-1} \frac{y}{x}$$

Using the chain rule of differentiation gives

$$\frac{\partial^2 h}{\partial x^2} = \frac{x^2}{r^2} \frac{\partial^2 h}{\partial r^2} + \frac{y^2}{r^3} \frac{\partial h}{\partial r} + \frac{y^2}{r^4} \frac{\partial^2 h}{\partial \theta^2} - \frac{2xy}{r^4} \frac{\partial h}{\partial \theta}$$
$$\frac{\partial^2 h}{\partial y^2} = \frac{y^2}{r^2} \frac{\partial^2 h}{\partial r^2} + \frac{x^2}{r^3} \frac{\partial h}{\partial r} + \frac{x^2}{r^4} \frac{\partial^2 h}{\partial \theta^2} + \frac{2xy}{r^4} \frac{\partial h}{\partial \theta} \tag{c}$$

Adding the two equations above yields

$$\frac{\partial^2 h}{\partial r^2} + \frac{1}{r} \frac{\partial h}{\partial r} + \frac{1}{r^2} \frac{\partial^2 h}{\partial \theta^2} = 0 \tag{d}$$

When flow is directed toward or originates from the origin of the coordinate system, it is independent of θ, and above eq. (d) reduces to the following ordinary differential equation, since the head is a function of radial coordinate only:*

$$\frac{d^2 h}{dr^2} + \frac{1}{r} \frac{dh}{dr} = 0 \qquad [\text{L}^{-1}] \tag{5.7}$$

This is the groundwater flow equation in polar coordinates.
Equation (5.7) can be written in the following form:

$$\frac{1}{r} \frac{d}{dr} \left(r \frac{dh}{dr} \right) = 0 \tag{e}$$

*For direct derivation of this equation, refer to Lohman (1972) or any other groundwater text.

Sec. 5.2   Analytical Solution of the Steady-State Flow Equation   **167**

Integrating (e) yields

$$r\frac{dh}{dr} = C_1 = \text{constant} \qquad \text{(f)}$$

The constant C_1 has to be evaluated from the boundary condition. Radial flow from the periphery of circle of radius r and aquifer thickness b is (from Darcy's law)

$$Q = 2\pi r b K \frac{dh}{dr}$$

or

$$r\frac{dh}{dr} = \frac{Q}{2\pi bK} \qquad \text{(g)}$$

Substituting in (f) gives us

$$C_1 = \frac{Q}{2\pi bK}$$

Hence

$$r\frac{dh}{dr} = \frac{Q}{2\pi bK}$$

or

$$dh = \frac{Q}{2\pi bK}\frac{dr}{r} \qquad \text{(h)}$$

Integrating (h) again, we have

$$h = \frac{Q}{2\pi bK}\ln r + C_2 \qquad \text{(i)}$$

for the initial boundary condition if at a radial distance R from the well, the head is H (as at the boundary on an island or at the end of the cone of depression).
Substituting in (i) gives*

$$h = \frac{Q}{2\pi bK}\ln \frac{r}{R} + H \qquad [L] \qquad (5.8a)$$

*Equation (5.8b) can be derived easily by direct application of Darcy's law. Referring to Fig. 5.7 and applying the Darcy's law at a distance r, we have

$$Q = 2\pi r b K \frac{dh}{dr}$$

$$\int_h^H dh = \frac{Q}{2\pi bK}\int_r^R \frac{dr}{r}$$

$$H - h = \frac{Q}{2\pi bK}\ln \frac{R}{r}$$

the same as eq. (5.8).

Applications of Groundwater Flow   Chap. 5

or

$$H - h = \frac{Q}{2\pi bK} \ln \frac{R}{r} \quad [L] \tag{5.8b}$$

where

H = piezometric head at radial distance R

h = piezometric head at any distance r

Q = discharge from the well

b = thickness of confined aquifer

bK = transmissivity of aquifer

This form of the equation is known as the Theim equation. Field tests for determination of the coefficient of permeability are made under steady-state conditions wherein water levels in test and observation wells are stabilized after a long period of pumping. Equation (5.8b) is used to determine the coefficient of permeability. For a known value of K, this formula is used to compute the discharge. Instead of pumping, the test is sometimes performed by adding water into the well. Equation (5.8b) holds good for that case also.

Example 5.5

A well is pumped from a confined aquifer at a rate of 0.08 m³/s for a long time. In two observation wells located 50 m and 10 m away from the well, the difference in elevation has been observed as 1.5 m. What is the transmissivity of the aquifer?

Solution From eq. (5.8b),

$$T = bK = \frac{Q}{2\pi(H-h)} \ln \frac{R}{r}$$

$$= \frac{0.08}{2\pi(1.5)} \ln \frac{50}{10}$$

$$= 0.0137 \text{ m}^2/\text{s}$$

Example 5.6

An aquifer of 20 m average thickness is overlain by an impermeable layer of 30 m thickness. A test well of 0.5 m diameter and two observation wells at a distance of 10 m and 60 m from the test well are drilled through the aquifer. After pumping at a rate of 0.1 m³/s for a long time, the following drawdowns are stabilized in these wells: first observation well, 4 m; second observation well, 3 m. Determine the hydraulic conductivity and the drawdown in the test well.

Solution Arrangement is shown in Figure 5.9. Apply eq. (5.8b) between the two observation wells:

$$(H-3) - (H-4) = \frac{0.1}{2\pi(20)K} \ln \frac{60}{10}$$

$$K = \frac{0.1}{2\pi(20)(1)} \ln \frac{60}{10}$$

$$= 1.43 \times 10^{-3} \text{ m/s}$$

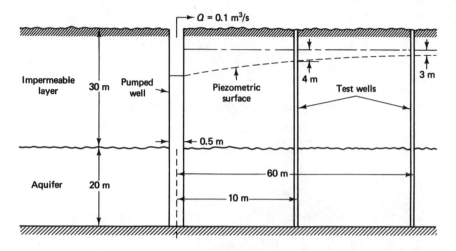

Figure 5.9 Testing of a confined aquifer, Example 5.6.

Apply eq. (5.8b) between the test well and the first observation well.

$$s - 4 = \frac{0.1}{2\pi(20)\,(1.43 \times 10^{-3})} \ln \frac{10}{0.25}$$

$$= 2.05$$

$$s = 6.05 \text{ m}$$

5.2.4 Unconfined Flow to a Well

Analysis of unconfined aquifer is made on the assumptions of Dupuit that consider (1) the flow to be horizontal, and (2) the velocity of flow to be proportional to the tangent of the hydraulic gradient instead of its sine. An essential difference between confined flow and unconfined flow is that in unconfined flow as the water table slopes, the saturated thickness changes and the area of cross section of flow varies as shown in Figure 5.10.

Equation (5.7) represents the steady-state groundwater flow equation for an unconfined aquifer as well. As previously, integration of eq. (5.7) provides

$$r\frac{dh}{dr} = C_1 \qquad (a)$$

Substituting similar boundary condition as for the confined case, that is, radial flow at the periphery of circle of radius r having saturated thickness, h,

$$Q = 2\pi r h K \frac{dh}{dr} \qquad (b)$$

(Note that saturated thickness here is h, not the aquifer thickness b as for the confined aquifer.) Substituting in eq. (a) yields

$$C_1 = \frac{Q}{2\pi h K}$$

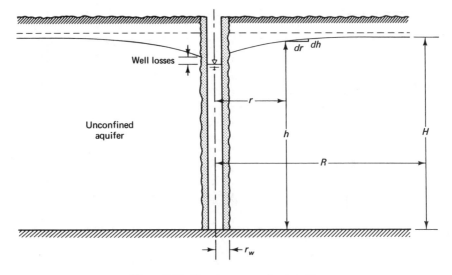

Figure 5.10 Flow in an unconfined aquifer.

Hence

$$r \frac{dh}{dr} = \frac{Q}{2\pi hK}$$

or

$$h\,dh = \frac{Q}{2\pi K} \frac{dr}{r} \qquad (c)$$

Integrating eq. (c) gives

$$h^2 = \frac{Q}{\pi K} \ln r + C_2 \qquad (d)$$

For boundary conditions in eq. (d), if at a radial distance R, the head is H, then

$$h^2 = \frac{Q}{\pi K} \ln \frac{r}{R} + H^2 \qquad [\text{L}^2] \qquad (5.9a)$$

or*

$$H^2 - h^2 = \frac{Q}{\pi K} \ln \frac{R}{r} \qquad [\text{L}^2] \qquad (5.9b)$$

*For a direct derivation, apply Darcy's law at a distance r (see Fig. 5.10):

$$Q = 2\pi rhK \frac{dh}{dr}$$

$$\int_h^H h\,dh = \frac{Q}{2\pi K} \int_r^R \frac{dr}{r}$$

$$H^2 - h^2 = \frac{Q}{\pi K} \ln \frac{R}{r}$$

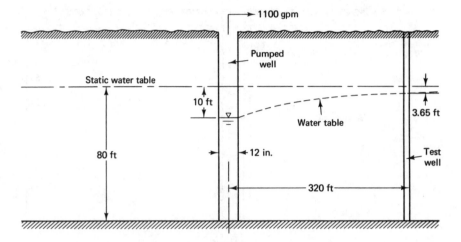

Figure 5.11 Test well in an unconfined aquifer.

Equation (5.9b) is used in a similar manner as eq. (5.8b) for a confined aquifer, that is, to assess the hydraulic coefficient by performing field tests or to compute the steady-state discharge.

Example 5.7

A fully penetrating 12-in.-diameter well has its bottom 80 ft below the static water table. After 24 h of pumping at 1100 gpm, the water level in the test well stabilizes to 10 ft below the static water table. A drawdown of 3.65 ft is noticed in an observation well 320 ft away from the test well. Determine the hydraulic conductivity of the aquifer.

Solution Refer to Figure 5.11.

$$Q = 1100 \text{ gpm} = 2.45 \text{ cfs}$$

From eq. (5.9b),

$$K = \frac{Q}{\pi(H^2 - h^2)} \ln \frac{R}{r}$$

$$= \frac{2.45}{\pi(76.35^2 - 70.0^2)} \ln \frac{320}{0.5}$$

$$= 0.0054 \text{ ft/s} \quad \text{or} \quad 469 \text{ ft/day}$$

5.2.5 Travel Time of Groundwater

Refer to Figure 5.7 or 5.10. Assume that a water particle takes time dt to move a distance dr:

$$\text{velocity of water movement,} \quad v_r = \frac{dr}{dt} \tag{a}$$

According to Darcy's law,

$$\text{Darcy velocity,} \qquad v = -K\frac{dh}{dr}$$

$$\text{seepage velocity,} \qquad v_s = \frac{v}{\eta} = -\frac{K}{\eta}\frac{dh}{dr} \qquad (b)$$

Equating eqs. (a) and (b) yields

$$dt = -\frac{\eta}{K}\frac{dr}{dh}dr \qquad (c)$$

For flow toward a well, the radial discharge from the periphery of a circle of radius r and depth D is found by

$$Q = 2\pi r D K \frac{dh}{dr}$$

or

$$\frac{1}{K}\frac{dr}{dh} = \frac{2\pi D r}{Q} \qquad (d)$$

Substituting eq. (d) in eq. (c) and integrating gives us

$$\int_0^t dt = \int_R^r -\frac{2\pi D r}{Q}\eta\, dr$$

$$t = \frac{\pi D \eta}{Q}(R^2 - r^2) \qquad [\text{T}] \qquad (5.10)$$

where

$t = $ time of travel from R to r

$r = $ any radial distance

$R = $ radial distance at the boundary from where the time of travel to be computed

$D = $ thickness of the confined aquifer, b, or average saturated thickness between radial distances R and r

$\eta = $ porosity

Example 5.8

In Example 5.6, determine the time of travel of groundwater from the observation well at a distance 60 m to the pumped well if the porosity of the aquifer is 0.3.

Solution   From eq. (5.10),

$$t = \frac{\pi D \eta}{Q}(R^2 - r^2)$$

$$= \frac{\pi(20)(0.3)}{0.1}(60^2 - 0.25^2)$$

$$= 678 \times 10^3 \text{ sec} \quad \text{or} \quad 7.8 \text{ days}$$

5.2.6 Flow to a Well in Semiconfined Aquifer: Theory of Leaky Aquifer

A leaky confined aquifer is underlain by an impervious bed and overlain by a semipervious layer. Above the semiconfining layer is a water-table aquifer, as shown in Figure 5.12. Initially, the artesian (piezometric) level in a semiconfined aquifer and the water table in an unconfined aquifer coincide. Pumping of the semiconfined aquifer lowers its piezometric level, thus creating a head difference between unconfined and confined aquifers, thereby inducing leakage through the semipervious layer. In the theory of leaky aquifers the flow is considered vertical through the semiconfining layer and radially in horizontal direction in the confined aquifer.

For the semiconfined aquifer, in the equation for groundwater flow a source term, in line with eq. (4.34), is added reflecting leakage from water-table aquifer into semiconfined aquifer. This source term is derived applying Darcy's law across a semipervious stratum:

$$q' = K' \frac{H_0 - h}{b'}$$

Thus the equation of a leaky aquifer in one dimension is

$$Kb\frac{d^2h}{dr^2} + Kb\frac{1}{r}\frac{dh}{dr} + K'\frac{H_0 - h}{b'} = 0 \qquad \text{(a)}$$

In terms of drawdown s, since $s = H_0 - h$,

$$-\frac{d^2s}{dr^2} - \frac{1}{r}\frac{ds}{dr} + \frac{K'}{Kbb'}s = 0 \qquad \text{(b)}$$

or

$$\frac{d^2s}{dr^2} + \frac{1}{r}\frac{ds}{dr} - \frac{s}{B^2} = 0 \qquad [\text{L}^{-1}] \qquad (5.11)$$

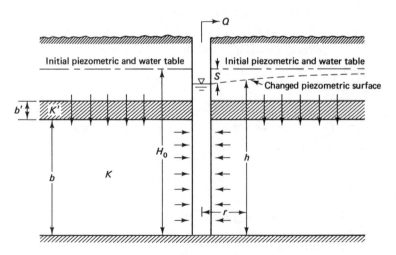

Figure 5.12 Flow in a semiconfined aquifer.

where B = leakage factor = $\sqrt{Kb/K'/b'}$, by definition. Equation (5.11) is a general equation of flow through a leaky aquifer. The general solution to this equation is

$$s = C_1 K_0\left(\frac{r}{B}\right) + C_2 \Gamma_0\left(\frac{r}{B}\right) \qquad [\text{L}] \qquad (5.12)$$

where

s = drawdown, and K'/b' = leakage coefficient

$\Gamma_0(r/B)$ = zero-order modified Bessel function of the first kind

$K_0(r/B)$ = zero-order modified Bessel function of the second kind

C_1, C_2 = constants for integration to be evaluated by boundary conditions

For an infinite aquifer, the following boundary conditions apply: at ∞, $s = 0$; and at $r = r_w$, $r(ds/dr) = -Q/2\pi T$. The final equation reduces to

$$s = \frac{Q}{2\pi T} K_0\left(\frac{r}{B}\right) \qquad [\text{L}] \qquad (5.13)$$

where T is the transmissivity of the aquifer. The values of $K_0(r/B)$ for various values of r/B are tabulated in Table 5.1. For $r/B < 0.05$, eq. (5.13) can be written as

$$s = \frac{Q}{2\pi T} \ln\left(1.123\frac{B}{r}\right) \qquad [\text{L}] \qquad (5.14)$$

Equations (5.13) and (5.14) apply to both confined and unconfined leaky aquifers.

Example 5.9

An infinite aquifer is underlain by an impervious stratum and overlain by a semipervious layer 5 ft thick having a coefficient of permeability of 1×10^{-8} ft/sec. The aquifer has an average thickness of 100 ft and a coefficient of permeability of 1×10^{-3} ft/sec. Groundwater is pumped from the aquifer at a rate of 0.15 ft^3/sec through a fully penetrating well of 12 in. diameter. Determine the drawdown (a) at 1000 ft from the well, and (b) at the face of the well.

Solution

$$T = bK = (100)(1 \times 10^{-3}) = 0.1$$

Leakage coefficient, $\quad \dfrac{K'}{b'} = \dfrac{1 \times 10^{-8}}{5} = 2 \times 10^{-9}$ per second

Leakage factor $B = \sqrt{\dfrac{T}{K'/b'}}$

$$= \sqrt{\dfrac{0.1}{2 \times 10^{-9}}} = 7071 \text{ ft}$$

TABLE 5.1 ZERO-ORDER MODIFIED BESSEL FUNCTION

x	$K_0(x)$	x	$K_0(x)$	x	$K_0(x)$
0.010	4.721	0.086	2.576	0.70	0.661
0.012	4.539	0.088	2.553	0.72	0.640
0.014	4.385	0.090	2.531	0.74	0.620
0.016	4.251	0.092	2.509	0.76	0.601
0.018	4.134	0.094	2.488	0.78	0.583
0.020	4.029	0.096	2.467	0.80	0.565
0.022	3.933	0.098	2.447	0.82	0.548
0.024	3.846	0.100	2.427	0.84	0.532
0.026	3.766	0.10	2.427	0.86	0.517
0.028	3.692	0.12	2.248	0.88	0.501
0.030	3.624	0.14	2.098	0.90	0.487
0.032	3.559	0.16	1.967	0.92	0.473
0.034	3.499	0.18	1.854	0.94	0.459
0.036	3.442	0.20	1.753	0.96	0.446
0.038	3.388	0.22	1.662	0.98	0.433
0.040	3.337	0.24	1.580	1.00	0.421
0.042	3.288	0.26	1.505	1.0	0.421
0.044	3.242	0.28	1.436	1.2	0.319
0.046	3.197	0.30	1.373	1.4	0.244
0.048	3.155	0.32	1.314	1.6	0.188
0.050	3.114	0.34	1.259	1.8	0.246
0.052	3.075	0.36	1.208	2.0	0.114
0.054	3.038	0.38	1.160	2.2	0.0893
0.056	3.002	0.40	1.115	2.4	0.0702
0.058	2.967	0.42	1.072	2.6	0.0554
0.060	2.933	0.44	1.032	2.8	0.0438
0.062	2.900	0.46	0.994	3.0	0.0347
0.064	2.869	0.48	0.958	3.2	0.0276
0.066	2.838	0.50	0.924	3.4	0.0220
0.068	2.809	0.52	0.892	3.6	0.0175
0.070	2.780	0.54	0.861	3.8	0.0140
0.072	2.752	0.56	0.832	4.0	0.0112
0.074	2.725	0.58	0.804	4.2	0.0089
0.076	2.698	0.60	0.778	4.4	0.0071
0.078	2.673	0.62	0.752	4.6	0.0057
0.080	2.648	0.64	0.728	4.8	0.0046
0.082	2.623	0.66	0.704	5.0	0.0037
0.084	2.599	0.68	0.682		

(a) At 1000 ft from the well,

$$\frac{r}{B} = \frac{1000}{7071} = 0.14$$

From Table 5.1, for $x = 0.14$, $K_0(x) = 2.098$. From eq. (5.13),

$$s = \frac{0.15}{2\pi(0.1)}(2.098)$$

$$= 0.5 \text{ ft}$$

(b) At the well face, since $r/B < 0.05$,

$$\frac{B}{r} = \frac{7071}{0.5} = 14{,}142$$

From eq. (5.14),

$$s = \frac{0.15}{2\pi(0.1)} \ln[(1.123)(14{,}142)]$$

$$= 2.31 \text{ ft}$$

5.2.7 Well Flow near Boundaries: Theory of Images

An inherent assumption in the equations of groundwater flow in Section 4.7 is that the aquifer is of areally infinite extent. All aquifers are, however, bounded either by recharge boundaries such as streams and lakes or by impermeable boundaries such as buried rocks or tight faults. When wells are located close to such boundaries, their influence is experienced by the wells, and the formulas based on an infinite aquifer become inapplicable. In order that such cases can be analyzed, it is necessary to make the aquifer appear to be of infinite extent. This is achieved through the theory of images. In this method, imaginary wells are introduced in such a manner that the conditions produced by the presence of the boundary(ies) are duplicated utilizing the concepts of infinite aquifer. Thus the equivalent hydraulic systems are created to which equations developed previously can be applied.

Well near a stream. Along a stream a constant head equivalent to the water level exists. The cone of depression of a well pumped near a stream should thus terminate at the water surface in the stream. This could be achieved by assuming that a recharging imaginary well is present on the other side of the stream an equal distance opposite the real discharging well. The water is injected into this imaginary well at the same rate as in the real well, so that the increase in head due to the cone of impression of the imaginary well and decrease in head due to the cone of depression of the real well exactly cancel each other along the line of the stream, as shown in Figure 5.13. The flow net for the system of Figure 5.13 is shown in Figure 5.14. As discussed previously, the graphical solution can be used to analyze such cases.

Analytically, if the well is located a distance from the stream, its imaginary recharge well will be as shown in Figure 5.15. Consider any point I having coordinates x and y:

$$r_1 = \sqrt{(a - x)^2 + y^2} \tag{a}$$

$$r_2 = \sqrt{(x + a)^2 + y^2} \tag{b}$$

For a real well, applying eq. (5.8b) between points O and I, we have

$$s_1 = \frac{Q}{2\pi b K} \ln \frac{a}{r_1} \tag{c}$$

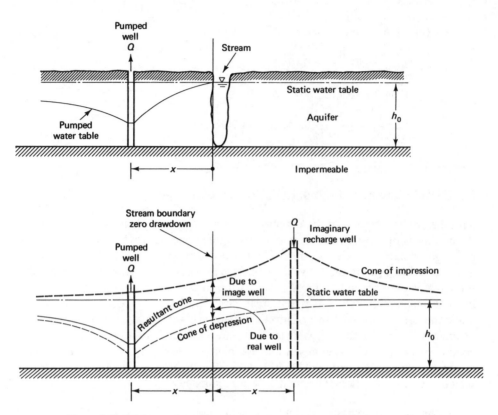

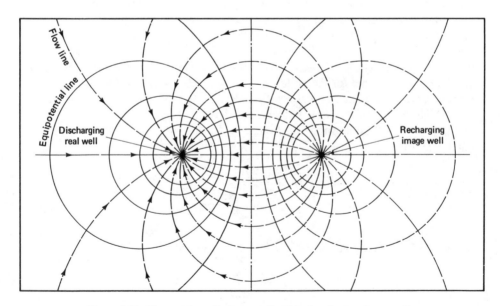

Figure 5.13 Well near a stream and its equivalent imaginary system in aquifer of infinite extent.

Figure 5.14 Flow net for a discharge well and its imaginary recharge well.

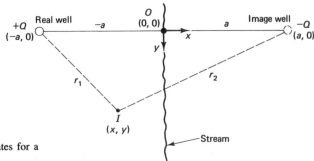

Figure 5.15 Setting up coordinates for a well near a stream with its image.

where s_1 is the drawdown between O and I. For the imaginary well, applying eq. (5.8b) between O and I, we have

$$s_2 = - \frac{Q}{2\pi bK} \ln \frac{a}{r_2} \qquad (d)$$

Adding (c) and (d) gives the total drawdown as

$$s = \frac{Q}{4\pi bK} \ln \frac{y^2 + (a + x)^2}{y^2 + (a - x)^2} \qquad [L] \qquad (5.15)$$

where

a = horizontal distance of the well from the stream

x, y = coordinates of the point where drawdown is desired (the head is h)

The origin of the coordinate system is at the intersection of the horizontal line from the well and the vertical stream axis.

The analytical procedure above can be used to derive a relation for an unconfined aquifer in which eq. (5.8b) will be used for the head difference as given in eqs. (c) and (d). The formula will be the same as eq. (5.15), where b will represent the average saturated thickness of the aquifer. The same procedure can be used to derive a relation for a leaky aquifer by using eq. (5.13) for relations (c) and (d).

Example 5.10

A 0.5-m well fully penetrates a 30-m-thick confined aquifer of hydraulic conductivity 20 m/day. Due to continuous pumping, if a drawdown of 1 m is registered in the well:

(a) What is the rate of pumping if the well is located 50 m from a stream?

(b) What would the rate of pumping be for the same drawdown if the well was located 5000 m from the stream?

Solution

(a) $a = 50$ m, $x = 50 - 0.25 = 49.75$, and $y = 0$. From eq. (5.15),

$$s = \frac{Q}{4\pi bK} \ln \frac{y^2 + (a + x)^2}{y^2 + (a - x)^2}$$

or

$$Q = \frac{4\pi b K s}{\ln \dfrac{y^2 + (a + x)^2}{y^2 + (a - x)^2}}$$

$$= \frac{4\pi (30)(20)(1)}{\ln[(50 + 49.75)^2/(50 - 49.75)^2]}$$

$$= 629 \text{ m}^3/\text{day}$$

(b) $Q = \dfrac{4\pi(30)(20)(1)}{\ln[(5000 + 4999.75)^2/(5000 - 4999.75)^2]}$

$$= 355.6 \text{ m}^3/\text{day}$$

The steeper gradient in the first case contributed to about 80% higher flow.

Well near an impermeable boundary. The desired condition here is that no flow take place across the boundary. If an imaginary discharge well is placed opposite the pumping well an equal distance away from the boundary and both wells pump at the same rate, they will offset each other at the boundary, as shown in Figure 5.16. The flow net for this case is shown in Figure 5.17.

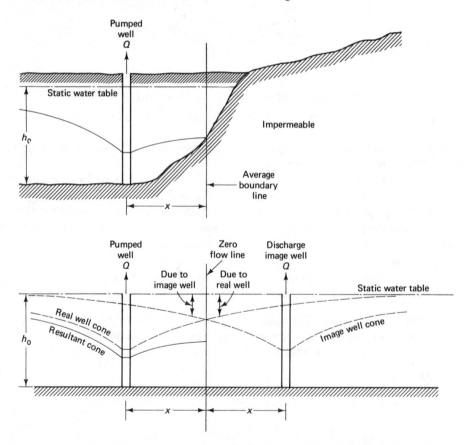

Figure 5.16 Well near an impermeable boundary and its equivalent system.

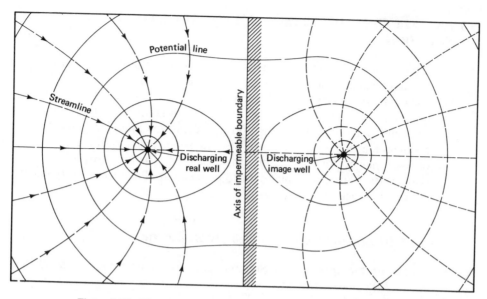

Figure 5.17 Flow net for a discharge well and its imaginary discharge well.

Refer to Figure 5.15. Since both wells are discharging in this case, the sign of eq. (d) is positive. Since there is no stream, a reference distance R is established where there is no drawdown effect (i.e., the head is H_0). This is either the radius of influence or the boundary of an island. Addition of eq. of (c) and (d) types:

$$s = \frac{Q}{2\pi bK} \ln \frac{R^2}{r_1 r_2} \quad [\mathrm{L}] \qquad (5.16)$$

where

$$r_1, r_2 \text{ are given by eqs. (a) and (b) of section 5.2.7}$$
$$R = \text{radius of influence or boundary of the island}$$

Example 5.11

In Example 5.10, if an impermeable boundary is located 50 m from the well having a radius of influence of 1.5 km, what is the rate of pumping?

Solution $r_1 = 0.25$ m and $r_2 = 99.75$ m. From eq. (5.16),

$$Q = \frac{2\pi K b s}{\ln(R^2/r_1 r_2)}$$

$$= \frac{2\pi (20)(30)(1)}{\ln[(1500)^2/(0.25)(99.75)]}$$

$$= 330 \text{ m}^3/\text{day}$$

Thus the flow in the well is less than half of the flow in Example 5.10 for the well near a stream.

5.3 ANALYTICAL SOLUTION OF UNSTEADY-STATE FLOW EQUATION

Equations (4.31) and (4.35) represent transient or unsteady-state flow conditions. Certain boundary value problems for simplified cases have been solved using these equations. An extensive application of eq. (4.35), in terms of polar coordinates, has, however, been made on problems relating to radial flow toward wells; this has practical significance. The unconfined flow, in the form of the linear Boussinesq equation, becomes identical to the confined flow equation. Thus, for the mathematical solution of two kinds of flow, no distinction is required. The unconfined flow, however, imposes certain limitations that have to be recognized and specifically handled, as discussed subsequently.

5.3.1 (Unsteady) Flow to a Well: Theis Equation

Following the approach of Section 5.2.3 for conversion into polar coordinates, eq. (4.35) can be written as

$$\frac{\partial^2 h}{\partial r^2} + \frac{1}{r}\frac{\partial h}{\partial r} = \frac{S}{T}\frac{\partial h}{\partial t} \qquad [L^{-1}] \tag{5.17a}$$

or, in terms of drawdown,

$$\frac{\partial^2 s}{\partial r^2} + \frac{1}{r}\frac{\partial s}{\partial r} = \frac{S}{T}\frac{\partial s}{\partial t} \qquad [L^{-1}] \tag{5.17b}$$

Equation (5.17b) has been solved by first converting it to base u defined by the Boltzmann variable (i.e., $u = r^2 S/4Tt$). This will reduce it to an ordinary differential equation of the type of eq. (5.7). Its double integration with constants evaluated by appropriate boundary conditions will yield the solution.

C. V. Theis, in 1935, was the first to obtain the solution for eq. (5.17b) based on an analogy between groundwater flow and heat conduction. He based his solution on the following assumptions and for initial and boundary conditions indicated below.

ASSUMPTIONS

1. The aquifer is homogeneous, isotropic, and of infinite extent (this is a built-in assumption of the groundwater flow equation).
2. The transmissivity of the aquifer is practically constant.
3. The water derived is entirely from storage and is released instantaneously with decline of head.
4. The well penetrates the entire thickness of the aquifer, and its diameter is very small compared to pumping rates, so that storage in well is negligible.

INITIAL AND BOUNDARY CONDITIONS

1. At time = 0, drawdown = 0, at any distance.
2. At time > 0, drawdown = 0, at infinite distance.

3. At the well face, r_w, $\partial s/\partial r = -Q/2\pi r_w T$ (according to Darcy's law, i.e., flow into the well is equal to its discharge).

The solution to eq. (5.17b) is

$$s = \frac{Q}{4\pi T} \int_u^\infty \frac{e^{-u}}{u}\, du \quad [L] \tag{5.18}$$

This is known as the Theis equation, and the exponential integral is referred to as the well function, $W(u)$; thus the set of equations is

$$s = \frac{Q}{4\pi T} W(u) \quad [L] \tag{5.19a}$$

when

$$u = \frac{r^2 S}{4Tt} \tag{5.19b}$$

$$W(u) = -0.5772 - \ln u + u - \frac{u^2}{2(2!)} + \frac{u^3}{3(3!)} - \cdots \tag{5.19c}$$

where

r = any distance from the center of pumping well

t = any time when drawdown is s

S = storage coefficient

T = transmissivity

The set of equations above is generally used to determine the hydraulic properties of transmissivity and storage coefficient of an aquifer. The required values of other variables to solve the equations above are obtained from aquifer testing in the field. The aquifer test comprises pumping a well at a constant rate for a period ranging from several hours to several days and measuring the change in water levels at fixed time intervals in observation wells located at different distances from the pumped well. The Theis equation is a significant contribution since it permits making tests in much less time in an unsteady state itself without waiting to achieve steady-state conditions.

Many procedures have been suggested to solve the equations above, including the one suggested by Theis himself. Earlier methods were related to procedures for solving the foregoing equations only. Subsequently, refinements were introduced by making modifications in the equations, particularly the well function $W(u)$, to reflect actual conditions of flow systems as they differ from the assumptions listed earlier.

5.3.2 Aquifer-Test Analysis

Aquifer-test analysis involves application of field data from an aquifer or pumping test to compute hydraulic properties of the aquifer. Equation (5.19a) has a general form

$$s = \frac{Q}{4\pi T} W(u, \alpha, \beta, \ldots) \quad [L] \tag{5.20}$$

where

$$u = r^2 S/4Tt$$

α, β = dimensionless factors to define particular aquifer-system conditions

In general procedure, a graph from the field data is prepared between s versus t/r^2 (or r^2/t). Instead of drawdown, the recovery data after pumping ceases could be used to prepare this curve. Standard curves are drawn between W and u for various controlled values of $\alpha, \beta, \ldots$, which are known as the type curves. By a curve-matching process between a type curve and a field data curve, as explained in the next section, eq. (5.20) is solved, to compute the values of T and S.

There can be many site conditions in a well–aquifer system, as listed below.

I. Areal extent of aquifer
   1. Aquifer of infinite extent
   2. Aquifer bound by an impermeable boundary
   3. Aquifer bound by a recharge boundary
II. Depth of well
   1. Fully penetrating well
   2. Partially penetrating well
III. Confined aquifer
   1. Nonleaky aquifer
   2. Leaky confining bed releasing water from the storage
   3. Leaky confining bed not yielding water from storage but transmitting water from overlying aquifer
   4. Leaky aquifer in which head in the overlying aquifer changes
IV. Unconfined Aquifer
   1. Aquifer in which significant dewatering (reduction in saturated thickness) occurs
   2. Aquifer in which vertical flow occurs near the well
   3. Aquifer with delayed yield (i.e., water from the storage does not release quickly)

The combination of a condition from one category to any other condition of another category can lead to numerous site conditions for which a special type curve or set of curves has to be developed. The contribution in this field has been significant; over 100 papers have been written dealing with different situations. Theis made a beginning by introducing the concept of the type curve. Jacob and Hantush advanced the theory to cover leaky aquifer problems and produced many type curves for such cases. Others who made contribution to leaky aquifers are Neuman and Witherspoon. Boulton did extensive research on unconfined aquifers and developed many tables and type curves for vertical flow and delayed yield. Neuman and Streltsova also made significant contributions in unconfined flow hydraulics.

This chapter deals with the basic analysis of confined and unconfined aquifers. Reed (1980) has compiled the type-curve solutions for 11 conditions of flow in confined aquifers. Marino and Luthin (1982) provided detailed coverage to unconfined aquifers.

The selection of a proper type curve or a set of curves is imperative for the data analysis. A type curve that fits the site conditions should be used. Stallman (1971) found that an error of many orders of magnitude could be committed by improper use of type curves.

5.3.3 Analysis of Confined Aquifer

This confined aquifer analysis is applicable to an impermeable (nonleaky) aquifer of infinite extent. Besides the type-curve method originally developed by Theis, there are two other methods of analysis. All of these methods are described below.

Type-curve method. From eq. (5.19a),

$$s = \frac{Q}{4\pi T} W(u) \tag{5.19a}$$

or

$$\log s = \left[\log \frac{Q}{4\pi T} \right] + \log W(u) \tag{a}$$

$$u = \frac{r^2 S}{4Tt} \quad \text{or} \quad \frac{t}{r^2} = \frac{S}{4T} \frac{1}{u} \tag{5.19b}$$

or

$$\log \frac{t}{r^2} = \left[\log \frac{S}{4T} \right] + \log \frac{1}{u} \tag{b}$$

For a constant Q, the bracketed parts of equations (a) and (b) are constant. Thus if a constant equal to $\log (Q/4\pi T)$ is added to $\log W(u)$, $\log s$ is obtained. Similarly, when $\log (S/4T)$ is added to $\log(1/u)$, the result is $\log (t/r^2)$. In other words, a graph between $\log W(u)$ and $\log 1/u$ is similar to a graph between $\log s$ and $\log t/r^2$. It is offset by constant amounts, as shown in Figure 5.18.

The procedure is summarized below.

1. Prepare a plot on log-log paper of $W(u)$ (on vertical coordinates) and $1/u$ on (horizontal coordinates). This is known as the type curve. For various values of u, $W(u)$ can be calculated from eq. (5.19c) or more conveniently from Table 5.2. Three type curves are shown in Figure 5.19; curve A covers the range of $1/u$ from 10^{-1} to 10^2, curve B from 10^2 to 10^5, and curve C from 10^5 to 10^8.

2. From given pumping test data, prepare a plot, on transparent log-log paper, of drawdown, s versus t/r^2.* The length of each cycle of this log-log paper should be same as used for the type curve of step 1. This is known as the data curve.

---

*There are many other ways in which the type curve and the data curve are plotted. Two of these are: (1) plot $W(u)$ and u for the type curve and s and r^2/t for the data curve; and (2) plot $W(u)$ and $1/u$ for the type curve and s and t for the data curve. The plotting of $1/u$ and t/r^2 is, however, preferred because it eliminates the necessity for computing $1/t$ for various values of s.

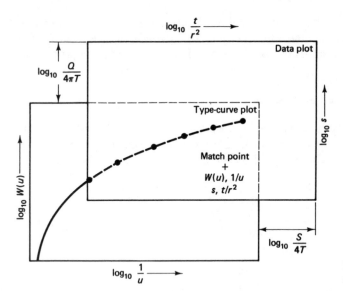

Figure 5.18 Relation of $W(u)$ versus $1/u$ and s versus t/r^2.

TABLE 5.2   VALUES OF WELL FUNCTION $W(u)$ FOR VALUES OF $1/u$

$1/u$	$1/u \times 10^{-1}$	1	10	10^2	10^3	10^4	10^5
1.0	0.00000[a]	0.21938	1.82292	4.03793	6.33154	8.63322	10.93572
1.2	0.00003	0.29255	1.98932	4.21859	6.51369	8.81553	11.11804
1.5	0.00017	0.39841	2.19641	4.44007	6.73667	9.03866	11.34118
2.0	0.00115	0.55977	2.46790	4.72610	7.02419	9.32632	11.62886
2.5	0.00378	0.70238	2.68126	4.94824	7.24723	9.54945	11.85201
3.0	0.00857	0.82889	2.85704	5.12990	7.42949	9.73177	12.03433
3.5	0.01566	0.94208	3.00650	5.28357	7.58359	9.88592	12.18847
4.0	0.02491	1.04428	3.13651	5.41675	7.71708	10.01944	12.32201
5.0	0.04890	1.22265	3.35471	5.63939	7.94018	10.24258	12.54515
6.0	0.07833	1.37451	3.53372	5.82138	8.12247	10.42490	12.72747
7.0	0.11131	1.50661	3.68551	5.97529	8.27659	10.57905	12.88162
8.0	0.14641	1.62342	3.81727	6.10865	8.41011	10.71258	13.01515
9.0	0.18266	1.72811	3.93367	6.22629	8.52787	10.83036	13.13294

$1/u$	$1/u \times 10^7$	10^8	10^9	10^{10}	10^{11}	10^{12}	10^{13}
1.0	15.54087	17.84344	20.14604	22.44862	24.75121	27.05379	29.35638
1.2	15.72320	18.02577	20.32835	22.63094	24.93353	27.23611	29.53870
1.5	15.94634	18.24892	20.55150	22.85408	25.15668	27.45926	29.76184
2.0	16.23401	18.53659	20.83919	23.14177	25.44435	27.74693	30.04953
2.5	16.45715	18.75974	21.06233	23.36491	25.66750	27.97008	30.27267
3.0	16.63948	18.94206	21.24464	23.54723	25.84982	28.15240	30.45499
3.5	16.79362	19.09621	21.39880	23.70139	26.00397	28.30655	30.60915
4.0	16.92715	19.22975	21.53233	23.83492	26.13750	28.44008	30.74268
5.0	17.15030	19.45288	21.75548	24.05806	26.36064	28.66322	30.96582
6.0	17.33263	19.63521	21.93779	24.24039	26.54297	28.84555	31.14813
7.0	17.48677	19.78937	22.09195	24.39453	26.69711	28.99969	31.30229
8.0	17.62030	19.92290	22.22548	24.52806	26.83064	29.13324	31.43582
9.0	17.73808	20.04068	22.34326	24.64584	26.94843	29.25102	31.55360

[a]Value shown as 0.00000 is nonzero but less than 0.000005.
Source: Reed (1980).

Data for this curve are obtained from a pumping test in which discharge is kept constant. Drawdowns can be observed in an observation well at any distance r for different time intervals; that is, r is constant and time varies. Thus the drawdown-time analysis is made or alternatively, drawdowns can be observed at the same time in wells located at different distances, thus involving drawdown-distance analysis. In both cases t/r^2 is computed and plotted against s in the form of a data curve.

3. The data plot is superimposed (placed) over the type-curve plot. The data curve plot is moved up or down, right or left, keeping its x and y axes parallel to the type-curve axes, until the data curve overlaps over a certain portion of the type curve.

4. Any arbitrary point is selected on the overlapping part of two sheets (plots). This point need not be on the curves itself. It is often convenient to select a point on the type curve whose coordinates are a multiple of 10. Record $W(u)$ and $1/u$ coordinates, and the corresponding s and t/r^2 coordinates, of this matching point.

5. Transmissivity is computed from eq. (5.19a), rearranged as

$$T = \frac{Q}{4\pi s} W(u) \quad [L^2 T^{-1}] \tag{5.21a}$$

and the storage coefficient from eq. (5.19b), rearranged as

$$S = 4T \frac{t}{r^2} u \quad \text{[dimensionless]} \tag{5.21b}$$

In the type curve, the maximum variation in $W(u)$ takes place in the range of $1/u$ of 10^{-1} to 10^2 (type curve A, Figure 5.19) when from almost vertical the curve becomes almost horizontal. Gradually, the curve becomes more flat (horizontal). In many instances, the data curve might match curve A. The data curve should be visually compared with Figure 5.19 to decide which of the type curves might be appropriate for a comparison.

Example 5.12

A confined aquifer is pumped at a rate of 1.11 ft³/sec. In an observation well a distance of 200 ft from the well, the following drawdown data were observed. Determine the transmissivity and storage coefficient of the aquifer.

Time since pumping started (min)	1	1.5	2.0	2.5	3.0	4.0	5.0	8.0
Observed drawdown (ft)	0.66	0.87	0.99	1.11	1.21	1.36	1.49	1.75

Time (min)	10.0	14.0	18.0	24.0	30.0	40.0	50.0	60.0
Drawdown (ft)	1.86	2.08	2.20	2.36	2.49	2.65	2.78	2.88

Time (min)	80.0	100.0	120.0	150.0	180.0	210.0	240.0
Drawdown (ft)	3.04	3.16	3.28	3.42	3.51	3.61	3.67

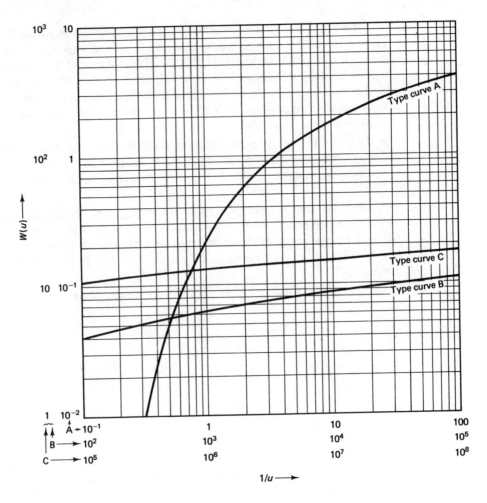

Figure 5.19 Type curves.

Solution

1. From the pumping test data;

$\frac{t}{r^2}$	2.5×10^{-5}	3.75×10^{-5}	5×10^{-5}	6.25×10^{-5}	7.5×10^{-5}	1×10^{-4}	1.25×10^{-4}	2×10^{-4}
s	0.66	0.87	0.99	1.11	1.21	1.36	1.49	1.75

$\frac{t}{r^2}$	2.5×10^{-4}	3.5×10^{-4}	4.5×10^{-4}	6.0×10^{-4}	7.5×10^{-4}	1×10^{-3}	1.25×10^{-3}	1.5×10^{-3}
s	1.86	2.08	2.20	2.36	2.49	2.65	2.78	2.88

$\frac{t}{r^2}$	2×10^{-3}	2.5×10^{-3}	3×10^{-3}	3.75×10^{-3}	4.5×10^{-3}	5.3×10^{-3}	6×10^{-3}
s	3.04	3.16	3.28	3.42	3.51	3.61	3.67

2. These data (s versus t/r^2) are plotted on Figure 5.20. The curve matches with the type curve A. The match point corresponding to $W(u) = 1$ and $1/u = 10$ on the data sheet is $s = 0.55$ and $t/r^2 = 0.5 \times 10^{-5}$.

3. Substituting in eq. (5.21a) yields

$$T = \frac{Q}{4\pi s} W(u)$$

or

$$T = \frac{1.11}{4\pi(0.55)} (1) = 0.161 \text{ ft}^2/\text{sec} \quad \text{or} \quad 13,880 \text{ ft}^2/\text{day}$$

4. From eq. (5.21b),

$$S = \frac{t}{r^2} 4Tu$$

$$= \left(5 \times 10^{-5} \frac{\text{min}}{\text{ft}^2}\right)\left(4 \times 0.161 \frac{\text{ft}^2}{\text{sec}}\right)\left(\frac{1}{10}\right)\left(\frac{60 \text{ sec}}{1 \text{ min}}\right)$$

$$= 1.93 \times 10^{-4}$$

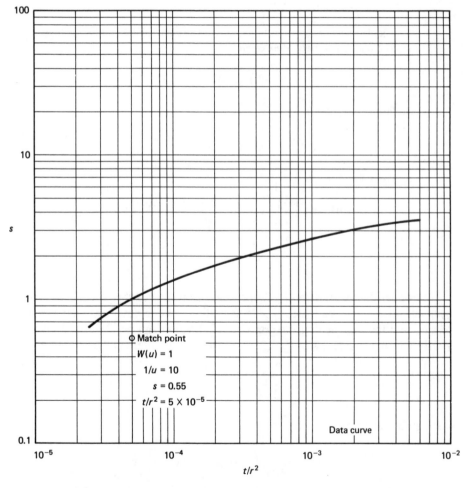

Figure 5.20 Data plot for Example 5.12 on drawdown–time analysis by Theis method.

Example 5.13

In a confined aquifer test, the following drawdown data were measured at the observation wells 30 hr after pumping began. The well is pumped at 3000 m³/day. Determine the transmissivity and the storage coefficient.

Observation	Distance from Pumped Well (m)	Drawdown (m)
1	150	1.44
2	300	1.17
3	600	0.9

Solution

1. From the data above:

t/r^2 (min/m²)	8×10^{-2}	2×10^{-2}	5×10^{-3}
s (m)	1.44	1.17	0.9

2. The data plot is shown in Figure 5.21. The curve matches with the type curve B. The match point coordinates are:

$$W(u) = 5 \qquad \frac{1}{u} = 2.8 \times 10^3 \qquad s = 1 \qquad \frac{t}{r^2} = 1 \times 10^{-1}$$

3. From eq. (5.21a),

$$T = \frac{3000}{4\pi(1)}(5) = 1194 \text{ m}^2/\text{day}$$

4. From eq. (5.21b),

$$S = 4\left(1194 \frac{\text{m}^2}{\text{day}}\right)\left(1 \times 10^{-1} \frac{\text{min}}{\text{m}^2}\right)\left(\frac{1}{2.8 \times 10^3}\right)\left(\frac{1 \text{ day}}{24 \times 60 \text{ min}}\right)$$

$$= 1.2 \times 10^{-4}$$

Cooper–Jacob method. Cooper and Jacob showed that when $u \ (= r^2S/4Tt)$ becomes sufficiently small, steady-state conditions tend to develop in the cone of depression and the exponential integral or well function can be closely approximated by only the first two terms of eq. (5.19c). Thus eq. (5.19a) can be written

$$s = \frac{Q}{4\pi T}(-0.5772 - \ln u) \tag{a}$$

This may be rewritten

$$s = \frac{Q}{4\pi T}\left(\ln 0.562 - \ln \frac{r^2S}{4Tt}\right) \tag{b}$$

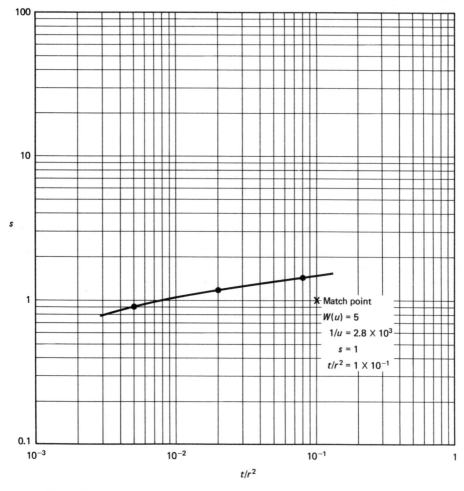

Figure 5.21 Data plot for Example 5.13 on drawdown–distance analysis by Theis method.

or

$$s = \frac{Q}{4\pi T} \ln \frac{2.25Tt}{r^2 S}$$

or

$$s = \frac{2.3Q}{4\pi T} \log \frac{2.25Tt}{r^2 S} \qquad [\text{L}] \qquad (5.22)$$

On semilog paper, eq. (5.22) represents a straight line with a slope of $2.3Q/4\pi T$. It is plotted in three different ways.

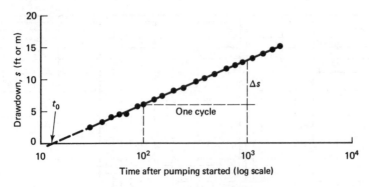

Figure 5.22 Plot of s versus log time.

Drawdown–Time Analysis. The drawdown measurements are made in an observation well at various times; distance r is a constant. A plot is made between drawdown, s (ordinary scale on vertical coordinate), and time, t (log scale on horizontal coordinate), as shown in Figure 5.22. Slope of the line

$$m = \frac{2.3Q}{4\pi T} \tag{c}$$

or

$$\frac{\Delta s}{\log(t_2/t_1)} = \frac{2.3Q}{4\pi T} \tag{d}$$

If a change in the drawdown, Δs, is considered for one log cycle, then $\log t_2/t_1 = 1$ and eq. (d) reduces to

$$\Delta s = \frac{2.3Q}{4\pi T} \tag{e}$$

or

$$T = \frac{2.3Q}{4\pi \Delta s} \qquad [L^2 T^{-1}] \tag{5.23}$$

Where the straight-line intersects the x-axis, the drawdown is zero and the time is t_0. Substituting these values in eq. (5.22), we have

$$0 = \frac{2.3Q}{4\pi T} \log \frac{2.25Tt_0}{r^2S} \tag{f}$$

or

$$0 = \log \frac{2.25Tt_0}{r^2S}$$

or

$$1 = \frac{2.25Tt_0}{r^2S}$$

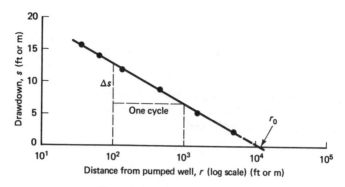

Figure 5.23 Plot of s versus log r.

or

$$S = \frac{2.25Tt_0}{r^2} \quad \text{[dimensionless]} \tag{5.24}$$

Drawdown–Distance Analysis. When drawdown measurements are made at a given time in various wells, time is a constant. A plot is made between s and distance as shown in Figure 5.23. From a similar consideration as in drawdown–time analysis above

$$T = \frac{2.3Q}{2\pi\Delta s} \quad [\text{L}^2\text{T}^{-1}] \tag{5.25}$$

and

$$S = \frac{2.25Tt}{r_0^2} \quad \text{[dimensionless]} \tag{5.26}$$

where r_0 is the intercept at the x-axis.

Measurements in Many Wells at Various Times for Either Drawdown–Time or Drawdown–Distance Analysis. A general procedure is to plot drawdown versus log t/r^2 as shown in Figure 5.24. In this case,

$$T = \frac{2.3Q}{4\pi\Delta s} \quad [\text{L}^2\text{T}^{-1}] \tag{5.23}$$

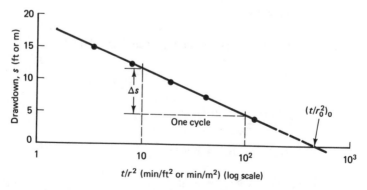

Figure 5.24 Plot of s versus t/r^2.

Sec. 5.3 Analytical Solution of the Unsteady-State Flow Equation

and

$$S = 2.25T\left(\frac{t}{r^2}\right)_0 \qquad \text{[dimensionless]} \qquad (5.27)$$

where $(t/r^2)_0$ is the intercept of straight-line on the x-axis.

It should be recognized that the derivations by Cooper–Jacob above are based on the assumption of u being small, that is,

$$u \le 0.05$$

or

$$\frac{r^2S}{4Tt} \le 0.05 \qquad (g)$$

or

$$t \ge \frac{5r^2S}{T} \qquad \text{[T]} \qquad (5.28)$$

The data points will begin to fall on a straight line after time, t, is sufficiently long to satisfy eq. (5.28).

Example 5.14

Solve Example 5.12 by the Cooper–Jacob method.

Solution

1. Drawdown versus time data are plotted in Figure 5.25.
2. From eq. (5.23), for one cycle, $\Delta s = 1.26$ m; thus

$$T = \frac{2.3(1.11)}{4\pi(1.26)} = 0.161 \text{ ft}^2/\text{sec} \quad \text{or} \quad 13{,}880 \text{ ft}^2/\text{day}$$

3. From eq. (5.24),

$$S = 2.25\left(13{,}880 \; \frac{\text{ft}^2}{\text{day}}\right)\left(\frac{1}{200^2 \text{ ft}^2}\right)(0.35 \text{ min})\left(\frac{1 \text{ day}}{24 \times 60 \text{ min}}\right)$$

$$= 1.90 \times 10^{-4}$$

4. From eq. (5.28), the Cooper–Jacob formulation is valid when

$$t \ge \frac{5r^2S}{T}$$

$$t \ge \frac{5(200)^2(1.90 \times 10^{-4})}{13{,}880}$$

$$\ge 0.0027 \text{ day} \quad \text{or} \quad 4 \text{ minutes}$$

Measurements after 4 min are valid.

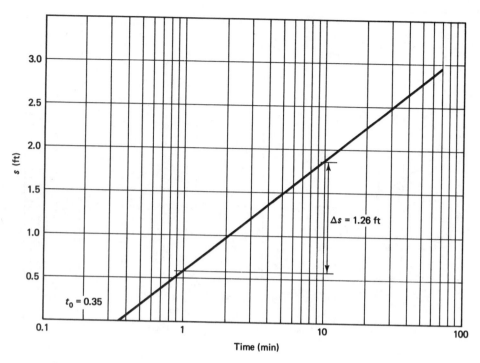

Figure 5.25 Plot of s versus log t for Example 5.14 on drawdown–time analysis by Cooper–Jacob method.

Example 5.15

Solve Example 5.13 by the Cooper–Jacob Method.

Solution

1. Drawdown versus distance data are plotted in Figure 5.26.
2. From eq. (5.25), for one cycle, $\Delta s = 0.9$ m; thus

$$T = \frac{2.3(3000)}{2\pi(0.9)} = 1220 \text{ m}^2/\text{day}$$

3. From eq. (5.26),

$$S = 2.25\left(1220 \frac{\text{m}^2}{\text{day}}\right)(30 \text{ hr})\left(\frac{1}{6000^2 \text{ m}^2}\right)\left(\frac{1 \text{ day}}{24 \text{ hr}}\right)$$

$$= 1.0 \times 10^{-4}$$

4. To check whether the Cooper–Jacob formulation is valid to the farthest well, from eq. (5.28),

$$t \geq \frac{5(600)^2(1 \times 10^{-4})}{1220}$$

$$\geq 0.148 \text{ day} \quad \text{or} \quad 3.5 \text{ hr}$$

Since the measurement was made after 30 hr, it is valid.

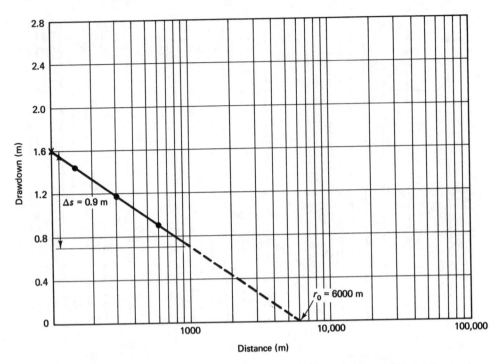

Figure 5.26 Plot of s versus log r for Example 5.15 on drawdown–distance analysis by Cooper–Jacob method.

Chow method.   Chow (1952) combined the approach of Theis and Cooper–Jacob in suggesting a solution for impermeable (nonleaky) aquifer. He indicated

$$\frac{W(u)e^{u}}{2.3} = \frac{s}{\Delta s/\log(t_2/t_1)} \qquad \text{[dimensionless]} \qquad (5.29)$$

where

$$W(u) = \text{well function of well parameter } u$$

$$s = \text{drawdown in an observation well at any time } t$$

$$\Delta s/\log(t_2/t_1) = \text{slope of semilog plot of } s \text{ versus log } t$$

The left-hand term of eq. (5.29) was referred to as $F(u)$ and a plot of $F(u)$, $W(u)$, and u was prepared by Chow as shown in Figure 5.27. Thus

$$F_u = \frac{s}{\Delta s/\log(t_2/t_1)} \qquad \text{[dimensionless]} \qquad (5.30)$$

For one log cycle on time scale, $\log(t_2/t_1) = 1$ and

$$F_u = \frac{s}{\Delta s} \qquad \text{[dimensionless]} \qquad (5.31)$$

Applications of Groundwater Flow   Chap. 5

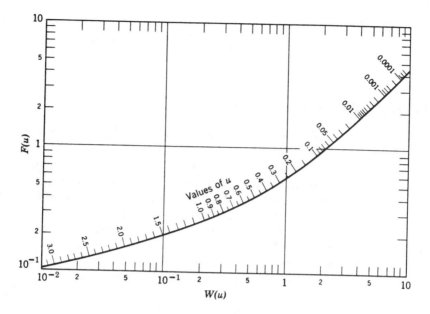

Figure 5.27 Relation among $F(u)$, $W(u)$, and u (after Chow, 1952).

The procedure is summarized as follows:

1. Plot drawdown–time data on a semilog plot in the same manner as for the Cooper–Jacob method.
2. On this plot, for any arbitrarily selected point, note the coordinates s and t. Also determine the drawdown difference, Δs, per log cycle of time from this plot. The ratio of $s/\Delta s$ will be equal to F_u as per eq. (5.31).
3. For F_u from step 2 read values of $W(u)$ and u from Figure 5.27 [for $F(u) > 2$, $W(u) = 2.3F_u$, and u is obtained from Table 5.2].
4. Since $W(u)$, u, s, and t are now known from steps 2 and 3, follow the Theis method [i.e., substitute in eqs. (5.21a) and (5.21b) to determine T and S].

Example 5.16

Solve Example 5.12 by the Chow method.

Solution

1. Drawdown–time data are plotted on semilog paper in Figure 5.25.
2. From this plot, for any selected point A, $s = 1.5$ ft, $t = 5.1$ min. For one log time cycle, $\Delta s = 1.26$; thus

$$F_u = \frac{s}{\Delta s} = \frac{1.5}{1.26} = 1.19$$

3. From Figure 5.27, $u = 0.04$, $W(u) = 2.68$.
4. From eq. (5.21a),

$$T = \frac{(1.11)(2.68)}{4\pi(1.5)} = 0.158 \text{ ft}^2/\text{sec} \quad \text{or} \quad 13,640 \text{ ft}^2/\text{day}$$

Sec. 5.3 Analytical Solution of the Unsteady-State Flow Equation **197**

5. From eq. (5.21b),

$$S = 4\left(0.158 \frac{\text{ft}^2}{\text{sec}}\right)(5.1 \text{ min})\left(\frac{1}{200^2 \text{ ft}^2}\right)(0.04)\left(\frac{60 \text{ sec}}{1 \text{ min}}\right)$$

$$= 1.93 \times 10^{-4}$$

5.3.4 Analysis of Unconfined Aquifer

Although the equation of groundwater flow through unconfined aquifer reduces to the same form as for the confined flow, the solution developed in Section 5.3.3 is not entirely applicable because of (i) dewatering of the aquifer, (ii) vertical flow near the well, and (iii) delayed yield due to gravity drainage as mentioned earlier.

If the drawdown is small compared to the depth of the aquifer, the effect of dewatering and vertical flow can be neglected. Also, if pumping continues long enough, the effect of delayed yield becomes negligible. In such situations, the approach of confined aquifer can be applied to water-table aquifers as well.

According to Hantush, near a well (in the region less than 0.2 times the depths of the aquifer), the vertical effect is not significant after time period:

$$t > 5b\frac{S_y}{K_z} \quad [\text{T}] \tag{5.32}$$

where

$$t = \text{time period}$$
$$b = \text{aquifer thickness}$$
$$S_y = \text{specific yield}$$
$$K_z = \text{vertical hydraulic conductivity}$$

According to Stallman, the delayed yield is pronounced for the period

$$t = 10S_y\frac{S}{K_z} \quad [\text{T}] \tag{5.33}$$

For a pumping test of a duration shorter than computed by eqs. (5.32) and (5.33), the type curves developed by Boulton and Neuman involving vertical flow and delayed yield should be utilized. However, if the pumping test can be extended to surpass the time requirements evident from eqs. (5.32) and (5.33), the approach of confined aquifer can be followed. As for the dewatering (lowering of the saturated thickness), the methods of a confined aquifer can be applied if the drawdown is less than 25% of the initial depth of saturation. The observed (measured) values of drawdown are corrected by the equation

$$s' = s - \frac{s^2}{2b} \quad [\text{L}] \tag{5.34}$$

where

$$s' = \text{corrected drawdown}$$
$$s = \text{measured drawdown}$$
$$b = \text{thickness of aquifer}$$

The value of the storage coefficient obtained using the method of a confined aquifer is adjusted again as follows:

$$S_y = \frac{(b - \bar{s})S_y'}{b} \qquad \text{[dimensionless]} \qquad (5.35)$$

where

S_y = adjusted specific yield

S_y' = computed specified yield

$\bar{s}$ = drawdown at the end of pumping at the geometric mean radius of all observation wells

Example 5.17

A well fully penetrating an unconfined aquifer of saturated thickness 50 ft is pumped at a rate of 0.8 ft³/sec. The drawdowns as measured in an observation well 30 ft from the pumped well are shown below. Determine the aquifer properties.

Time (min)	20	50	70	110	200	400	800	1200	1700	2000
Drawdown, s (ft)	1.52	1.78	2.0	2.39	2.88	3.42	4.07	4.39	4.72	4.91

Solution

(a) Applying corrections to drawdowns by eq. (5.34):

t (min)	20	50	70	110	200	400	800	1200	1700	2000
s' (ft)	1.5	1.75	1.96	2.33	2.80	3.30	3.90	4.20	4.50	4.67

(b) By the Cooper–Jacob method:
1. A plot of drawdown and log time is shown in Figure 5.28.
2. For one log time cycle, $\Delta s = 1.85$ ft; $t_0 = 5.7$ min.
3. From eq. (5.23),

$$T = \frac{2.3(0.8)}{4\pi(1.85)} = 0.079 \ \text{ft}^2/\text{sec} \quad \text{or} \quad 6842 \ \text{ft}^2/\text{day}$$

4. From eq. (5.24),

$$S_y' = 2.25\left(6842 \ \frac{\text{ft}^2}{\text{day}}\right)(5.7 \ \text{min})\left(\frac{1}{30^2 \ \text{ft}^2}\right)\left(\frac{1 \ \text{day}}{24 \times 60 \ \text{min}}\right)$$
$$= 6.8 \times 10^{-2}$$

(c) Adjusted specific yield, from eq. (5.35):

$$S_y = \left(\frac{50 - 4.67}{50}\right)(6.8 \times 10^{-2}) = 6.16 \times 10^{-2}$$

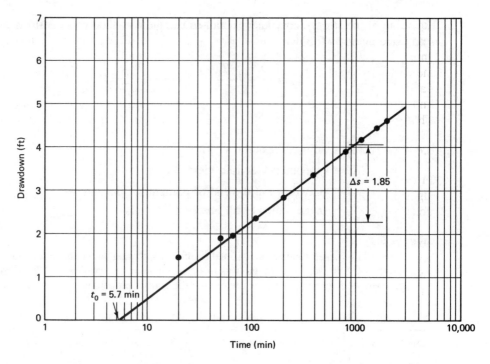

Figure 5.28 Adjusted drawdown versus log time for Example 5.17 on unconfined aquifer.

(d) To check whether the pumping was for a long enough period to neglect vertical flow and delayed yield effects:

1. From eq. (5.32),

$$t > \frac{5(50)(6.16 \times 10^{-2})}{6842/50}$$

$$> 0.11 \text{ day} \quad \text{or} \quad 160 \text{ min}$$

The data over 160 minutes are free from a vertical flow effect.

2. From eq. (5.33),

$$t = \frac{10(6.16 \times 10^{-2})(4.67)}{6842/50}$$

$$= 0.021 \text{ day} \quad \text{or} \quad 30 \text{ min}$$

Thus the delayed yield effect can also be neglected. (The question above could have been solved by the type-curve method or by the Chow method as well.)

5.4 NUMERICAL SOLUTION TO FLOW EQUATION

The graphic and analytical solutions, as presented previously, are essentially restricted to simpler flow problems involving simple boundary conditions, mostly in homogeneous, isotropic media. This is due to the inherent problem of obtaining a solution of the governing groundwater flow equation. As it is, no solution exists for the general equation (4.34). Simplifications are introduced to obtain the solutions. Many times, real flow problems cannot be adequately represented. Faced with this situation, scientists and engineers, in dealing with complicated problems, have dispensed with the analytical solution and have attempted to get the desired information out of partial differential equation directly. There are three common techniques of such nature, known as finite-difference method, similarity solution, and finite-element method. The first has a common application in groundwater and is described here briefly.

5.4.1 Finite-Difference Method

In this method a differential equation (of continuous nature) is replaced by a set of algebraic equations called finite difference at discrete points. These algebraic equations are then solved by methods of matrices.

   To understand the mechanism of this conversion, consider a plot of head versus horizontal distance as shown in Figure 5.29. The x axis is divided into intervals of Δx. The differential can be approximated with respect to a subsequent interval known as forward difference or over the preceding interval called backward difference, as shown in the Figure 5.29. Similarly, the central-difference approximation will be

$$\left(\frac{\partial h}{\partial x}\right)_n = \frac{h_{n+1} - h_{n-1}}{2\Delta x} \qquad \text{[dimensionless]} \qquad (5.36)$$

The approximation for the second derivative at n interval will be

$$\left(\frac{\partial^2 h}{\partial x^2}\right)_n = \frac{h_{n+1} - 2h_n + h_{n-1}}{(\Delta x)^2} \qquad [L^{-1}] \qquad (5.37)$$

   The flow equation (4.34), in two dimensions and expressed in terms of transmissivity, has the form

$$\frac{\partial}{\partial x}\left(T_x\frac{\partial h}{\partial x}\right) + \frac{\partial}{\partial y}\left(T_y\frac{\partial h}{\partial y}\right) - q = S\frac{\partial h}{\partial t} \qquad [LT^{-1}] \qquad (5.38)$$

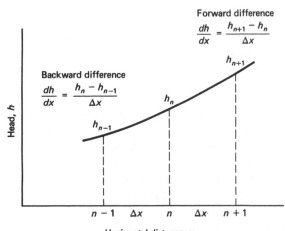

Forward difference
$$\frac{dh}{dx} = \frac{h_{n+1} - h_n}{\Delta x}$$

Backward difference
$$\frac{dh}{dx} = \frac{h_n - h_{n-1}}{\Delta x}$$

Figure 5.29 Finite difference of a derivative.

Horizontal distance, x

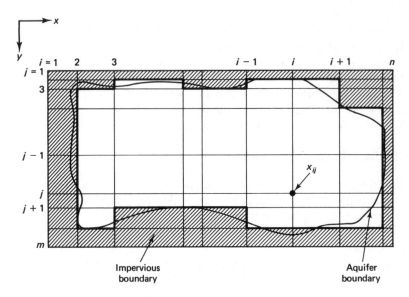

Figure 5.30 Grid overlay on an aquifer system.

where q is the groundwater withdrawal rate per unit area; for recharge, q will have a negative sign. For homogeneous, isotropic soil, eq. (5.38) reduces to

$$\frac{\partial^2 h}{\partial x^2} + \frac{\partial^2 h}{\partial y^2} = \frac{S}{T}\frac{\partial h}{\partial t} + \frac{q}{T} \qquad [\text{L}^{-1}] \qquad (5.39)$$

As a first step in obtaining a finite-difference solution, a grid is superimposed on the flow region of interest. Each intersection is called a "node," and as shown in Figure 5.30, indicates coordinates x, y for space dimensions at a particular time level, k (i.e., x_i, y_j, t_k). Thus $h_{i,j,k}$ denotes the head at space dimension x_i, y_j at a time level t_k.

When the space derivatives are approximated at current time level k using an equation of the form (5.37) and the time derivative is approximated with a forward difference, it is known as the explicit approximation (method) and contains only one unknown relating to head at advanced level $h_{i,j,k+1}$. When the space derivatives are replaced at an advanced time level $(k + 1)$ and the time derivative is produced by a backward approximation at k (current) level, it is known as implicit approximation and contains five unknown values of head of $(k + 1)$ level and only one known value for the current time level. In an alternative procedure, one space derivative is replaced at the advanced time level $(k + 1)$ and the next one is expressed at the current time level. In this way the unknowns are reduced to three, and thus results in a computationally desirable form of tridiagonal matrix. This is known as the alternating direction implicit (ADI) method. This method, in this way or with certain alterations, is preferred in finite-difference approximations.

5.4.2 Alternating Direction Implicit Method

Suppose that the first space derivative $\partial^2 h/\partial x^2$ is replaced at the advanced time level $(k + 1)$ and the second at the current time level (k). Using difference approximations, we have

$$\frac{h_{i-1,j,k+1} - 2h_{i,j,k+1} + h_{i+1,j,k+1}}{\Delta x^2} + \frac{h_{i,j-1,k} - 2h_{i,j,k} + h_{i,j+1,k}}{\Delta y^2} =$$

$$\frac{S}{T} \frac{h_{i,j,k+1} - h_{i,j,k}}{\Delta t} + \frac{q_{i,j,k}}{T} \qquad [L^{-1}] \qquad (5.40)$$

There are three unknowns for the $(k + 1)$ time level. Other values at the current time level are known. Collecting the terms together and multiplying both sides by Δx^2 yields

$$h_{i-1,j,k+1} + B_i h_{i,j,k+1} + h_{i+1,j,k+1} = D_i \qquad [L] \qquad (5.41)$$

where

$$B_i = -\left(2 + \frac{S}{T} \frac{\Delta x^2}{\Delta t}\right)$$

$$D_i = -h_{i,j-1,k} \frac{\Delta x^2}{\Delta y^2} - h_{i,j,k}\left(\frac{S}{T} \frac{\Delta x^2}{\Delta t} - \frac{2\Delta x^2}{\Delta y^2}\right) - h_{i,j+1,k} \frac{\Delta x^2}{\Delta y^2} + \frac{q_{i,j,k}}{T} \Delta x^2$$

Thus B_i and D_i are constants based on known values at the current time level. If eq. (5.41) is written for all nodes along the x-axis with unknown head for that node and two adjacent nodes, there will be n equations for a point j on the y-axis. The values of head at two outer nodes are known from boundary conditions. There will be $(n - 2)$ simultaneous equations to be solved belonging to a tridiagonal coefficient matrix. These equations are for a horizontal line of the grid. Thus, by solving $(n - 2)$ equations, it is possible to solve one line at a time parallel to the x-axis (i.e., marching in the direction of Y).

At the next time interval, $\partial^2 h/\partial y^2$ will be expressed at time level $(k + 2)$, while $\partial^2 h/\partial x^2$ retained at the current level of $(k + 1)$; thus

$$\frac{h_{i-1,j,k+1} - 2h_{i,j,k+1} + h_{i+1,j,k+1}}{\Delta x^2} + \frac{h_{i,j-1,k+2} - 2h_{i,j,k+2} + h_{i,j+1,k+2}}{\Delta y^2} =$$

$$\frac{S}{T} \frac{h_{i,j,k+2} - h_{i,j,k+1}}{\Delta t} + \frac{q_{i,j,k+1}}{T} \qquad (5.42)$$

$$[L^{-1}]$$

There are three unknowns for the $(k + 2)$ time level. Arranging terms yields

$$h_{i,j-1,k+2} + B_j h_{i,j,k+2} + h_{i,j+1,k+2} = D_j \qquad [L] \qquad (5.43)$$

where

$$B_j = -\left(2 + \frac{S}{T} \frac{\Delta y^2}{\Delta t}\right)$$

$$D_j = -h_{i-1,j,k+1} \frac{\Delta y^2}{\Delta x^2} - h_{i,j,k+1}\left(\frac{S}{T} \frac{\Delta y^2}{\Delta t} - \frac{2\Delta y^2}{\Delta x^2}\right) - h_{i+1,j,k+1} \frac{\Delta y^2}{\Delta x^2} + \frac{q_{i,j,k+1}}{T} \Delta y^2$$

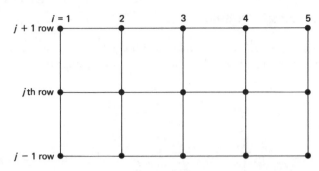

Figure 5.31 Grid of a five-node row and three-node column.

Writing eq. (5.43) for all nodes will provide m equations for each value of i on the x-axis (i.e., these belong to a vertical line on the grid). One line at a time can be evaluated by solving $(m - 2)$ equations for the interior nodes. For a complete cycle covering eqs. (5.41) and (5.43), two advanced time levels would be required for two-dimensional problems and three advanced levels for three-dimensional problems.

To demonstrate the application, consider a small grid of a five-node row as shown in Figure 5.31. Heads at $i = 1$ and $i = 5$ are known from boundary conditions. Writing eq. (5.41) for node 2 (dropping the time subscript) gives

$$h_{1,j} + B_2 h_{2,j} + h_{3,j} = D_2 \tag{a}$$

Similarly for nodes 3 and 4, respectively:

$$h_{2,j} + B_3 h_{3,j} + h_{4,j} = D_3 \tag{b}$$

and

$$h_{3,j} + B_4 h_{4,j} + h_{5,j} = D_4 \tag{c}$$

In matrix form, the equations above are

$$\begin{bmatrix} B_2 & 1 & 0 \\ 1 & B_3 & 1 \\ 0 & 1 & B_4 \end{bmatrix} \begin{bmatrix} h_{2,j} \\ h_{3,j} \\ h_{4,j} \end{bmatrix} = \begin{bmatrix} D_2 \\ D_3 \\ D_4 \end{bmatrix} \tag{d}$$

Equation (d) is a tridiagonal coefficient matrix with nonzero values on the main and adjacent diagonals only. This can be conveniently converted into an upper triangular matrix to solve for heads. One row at a time, heads can be solved for various values of j. For the next time step, equations can be formed similarly for any column.

5.4.3 Iterative and Strongly Implicit Methods

In a revised technique known as the iterative alternate direction implicit (IADI) method, both space derivatives and time derivatives are replaced by advanced time-level approximations as in the implicit method. However, while solving these equations, one space derivative (in fact, its approximation) is solved one iteration level below compared to the other, on an alternative basis.

In the two-dimensional and three-dimensional model of the USGS (Trescott, 1976; Trescott et al., 1976), a relatively new technique developed by Stone in 1968 has been used in which approximations of both spaces and time are made at an advanced time level. Known as the "strongly implicit procedure" or the "approximate factorization technique," it is much faster than the alternate direction implicit procedure. In the strongly implicit procedure, the approximate equations are not in tridiagonal form but are reduced to such form through a matrix modification procedure.

5.4.4 Initial and Boundary Conditions

The condition in which an aquifer system exists at the beginning of its operation with respect to a certain dynamic characteristic is required to be defined for the system to be evaluated. Usually, distribution of piezometric heads or water tables are given at initial time $t = 0$.

Boundary conditions are a means to describe the real system appropriately. By proper control of parameters, they are incorporated in finite-difference formulations.

Rivers and lakes are recharge boundaries having constant heads. At these nodes, head may be kept unchanged at initial levels. Some computer models set a very high value for the storage coefficients at such nodes.

Impermeable rocks form no flow boundaries. These can be treated by considering no hydraulic gradient or the same head as in the preceding node for all such nodes along the boundary. Some models assign zero transmissivities outside the boundary.

If only a part of the flow region is to be considered, it may be important to account for a specified flux into the region. The flux boundary can be treated by assigning recharge or discharge wells to appropriate nodes.

The methods above provide an approximation to the differential equation. When the grid space and time steps are not selected properly, the difference between the approximate and exact solution grows as t increases. The approximate solution is said to be unstable in such cases. The solution is stable in the one-dimensional case when

$$t < \frac{1}{2} \frac{S(\Delta x)^2}{T} \quad [T] \tag{5.44}$$

Example 5.18

Two fully penetrating wells are located in a confined aquifer system as shown in Figure 5.32. Well 1 pumps at a rate of 0.8 cfs and well 2 pumps at a rate of 0.5 cfs. The initial piezometric head is horizontal, in level with the stream. Determine the distribution of head after 1 day of pumping. The aquifer is bound by a stream on one side and impermeable boundaries on two sides. It is extended semi-infinitely on the remaining side, but a width of 270 ft is represented on the flow system under consideration.

Solution   The grid layout is shown in Figure 5.33. $T = 40 \times 50 = 2000$ ft²/day.
Initial conditions: All $h = 60$ ft at $t = 0$.
Boundary conditions

$h_{1,1} = h_{1,2} = h_{1,3} = h_{1,4} = 60$ ft (recharge boundary)

$\left. \begin{array}{l} h_{5,1} = h_{4,1} \\ h_{5,2} = h_{4,2} \\ h_{4,3} = h_{3,3} \\ h_{4,4} = h_{3,4} \end{array} \right\}$ impermeable boundary

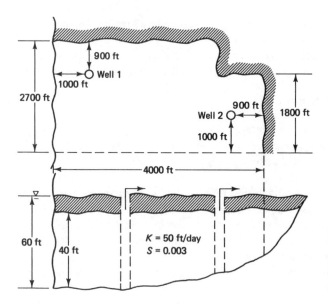

Figure 5.32 (a) Plan of an aquifer system (not to scale); (b) section of a confined aquifer (vertical scale exaggerated).

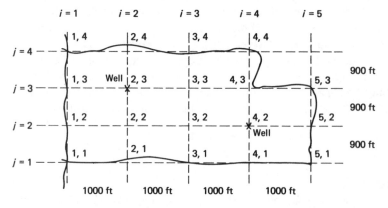

Figure 5.33 Finite-difference grid with grid coordinates.

and

$$\left.\begin{aligned} h_{2,4} &= h_{2,3} \\ h_{3,4} &= h_{3,3} \\ h_{4,4} &= h_{4,3} \end{aligned}\right\} \text{impermeable boundary}$$

The maximum time step to satisfy the stability requirement:

$$\Delta t < \frac{1}{2}\left[\frac{(0.003)(900)^2}{2000}\right] = 0.61 \text{ day}$$

A time step Δt of 0.5 day is selected.

I. First time step, $\Delta t = 0.5$. Apply eq. (5.41), for one row at a time. In the first time step, heads in row 1 ($j = 1$) will not be affected since there is no pumping on this row. For row 2, $j = 2$.

Matrix coefficients: From initial conditions, storage coefficient, and transmissivity,

$$B_2 = -\left(2 + \frac{S}{T}\frac{\Delta x^2}{\Delta t}\right) = -\left[2 + \frac{0.003(1000)^2}{2000(0.5)}\right]$$

$$= -5$$

Similarly, $B_3 = B_4 = -5$.

$$D_2 = -60\frac{(1000)^2}{(900)^2} - 60\left[\frac{0.003(1000)^2}{2000(0.5)} - \frac{2(1000)^2}{(900)^2}\right] - 60\frac{(1000)^2}{(900)^2}$$

$$= -180$$

$$D_3 = -180$$

Since

$$q = \frac{0.5}{900(1000)} = 0.555 \times 10^{-6} \text{ ft/sec} \quad \text{or} \quad 0.048 \text{ ft/day}$$

hence

$$D_4 = -180 + \frac{0.048(1000)^2}{2000}$$

$$= -156$$

For node 2,

$$h_{1,2,0.5} + B_2 h_{2,2,0.5} + h_{3,2,0.5} = D_2$$

For node 3,

$$h_{2,2,0.5} + B_3 h_{3,2,0.5} + h_{4,2,0.5} = D_3$$

For node 4,

$$h_{3,2,0.5} + B_4 h_{4,2,0.5} + h_{5,2,0.5} = D_4$$

Substituting B and D, initial and boundary conditions, gives

$$60 - 5h_{2,2,0.5} + h_{3,2,0.5} = -180$$

$$h_{2,2,0.5} - 5h_{3,2,0.5} + h_{4,2,0.5} = -180$$

$$h_{3,2,0.5} - 5h_{4,2,0.5} + h_{4,2,0.5} = -156$$

or

$$\begin{bmatrix} -5 & 1 & 0 \\ 1 & -5 & 1 \\ 0 & 1 & -4 \end{bmatrix}\begin{bmatrix} h_{2,2,0.5} \\ h_{3,2,0.5} \\ h_{4,2,0.5} \end{bmatrix} = \begin{bmatrix} -240 \\ -180 \\ -156 \end{bmatrix}$$

Solving by Gaussian elimination

$$h_{2,2,0.5} = 59.7$$

$$h_{3,2,0.5} = 58.7$$

$$h_{4,2,0.5} = 53.7$$

Row 3 is solved by a similar procedure.
Row 4 is an impermeable boundary, hence will have same heads as in row 3.
Head distribution obtained after the first time step is shown in Figure 5.34.

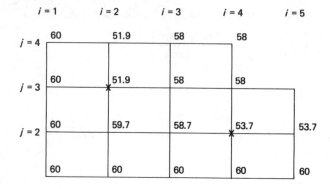

	$i=1$	$i=2$	$i=3$	$i=4$	$i=5$

$j=4$ 60   51.9   58   58

$j=3$ 60   51.9   58   58

$j=2$ 60   59.7   58.7   53.7   53.7

60   60   60   60   60

Figure 5.34 Heads at the end of $\Delta t = 0.5$ day.

II. Second time step, $\Delta t = 0.5$

The head matrix in Figure 5.34 forms the initial heads for the second time step. Computations now are made column by column. For column 2 ($i = 2$),

$$B_1 = -\left[2 + \frac{0.003(900)^2}{2000(0.5)}\right] = -4.43$$

Similarly, $B_2 = B_3 = -4.43$:

$$D_1 = -60\frac{(900)^2}{(1000)^2} - 60\left[\frac{0.003(900)^2}{2000(0.5)} - \frac{2(900)^2}{(1000)^2}\right] - 60\frac{(900)^2}{(1000)^2}$$

$$= -145.8$$

$$D_2 = -60\frac{(900)^2}{(1000)^2} - 59.7\left[\frac{0.003(900)^2}{2000(0.5)} - \frac{2(900)^2}{(1000)^2}\right] - 58.7\frac{(900)^2}{(1000)^2}$$

$$= -144.5$$

Since

$$q = \frac{0.8}{900(1000)} = 0.889 \times 10^{-6} \text{ ft/sec} \quad \text{or} \quad 0.077 \text{ ft/day}$$

$$D_3 = -60\frac{(900)^2}{(1000)^2} - 51.9\left[\frac{0.003(900)^2}{2000(0.5)} - \frac{2(900)^2}{(1000)^2}\right] - 58\frac{(900)^2}{(1000)^2}$$

$$+ \frac{0.077(900)^2}{2000}$$

$$= -106.5$$

For node 1,*

$$h_{2,0,1} + B_1 h_{2,1,1} + h_{2,2,1} = D_1$$

For node 2,

$$h_{2,1,1} + B_2 h_{2,2,1} + h_{2,3,1} = D_2$$

*Since the aquifer extends beyond the grid boundary in this direction, the head outside the boundary is included here and assigned a value of the initial head of 60 ft.

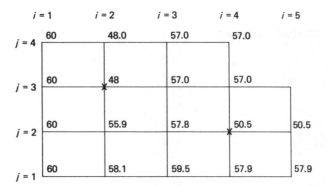

Figure 5.35 Head distribution at the end of second time step, (i.e., after 1 day).

For node 3,

$$h_{2,2,1} + B_3 h_{2,3,1} + h_{2,4,1} = D_3$$

Substituting B and D, initial and boundary conditions, yields

$$60 - 4.43h_{2,1,1} + h_{2,2,1} = -145.8$$
$$h_{2,1,1} - 4.43h_{2,2,1} + h_{2,3,1} = -144.5$$
$$h_{2,2,1} - 4.43h_{2,3,1} + h_{2,3,1} = -106.5$$

or

$$\begin{bmatrix} -4.43 & 1 & 0 \\ 1 & -4.43 & 1 \\ 0 & 1 & -3.43 \end{bmatrix} \begin{bmatrix} h_{2,1,1} \\ h_{2,2,1} \\ h_{2,3,1} \end{bmatrix} = \begin{bmatrix} -205.8 \\ -144.5 \\ -106.5 \end{bmatrix}$$

Solving gives

$$h_{2,1,1} = 59.3$$
$$h_{2,2,1} = 56.8$$
$$h_{2,3,1} = 47.6$$

Other columns can be solved similarly. At the end of the second time step, $\Delta t = 0.5$ (i.e., after $t = 1$ day) the head distribution is as shown in Figure 5.35.

5.5 DESIGN CRITERIA

5.5.1 Well Loss

The total drawdown in a pumping well consists of two components. One is the drawdown at the face of the well shown as s_a in Figure 5.36 and the other is the drawdown that occurs as water moves through the well screen and inside the well to the pump intake, indicated as s_w in the figure. The first is known as the formation loss and the latter as the well loss.

The well loss is given by

$$s_w = CQ^n \quad [L] \tag{5.45}$$

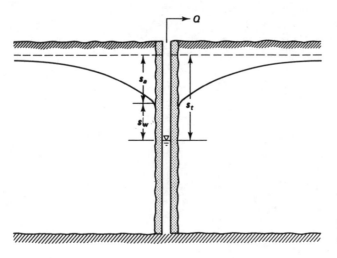

Figure 5.36 Formation loss and well loss in a pumped well.

where

$$C = \text{constant}, <0.5 \text{ min}^2/\text{m}^5, \text{ for a properly designed well}$$
$$n = \text{power of the discharge, generally 2}$$

Thus the total drawdown for steady state is

$$s_t = \frac{2.3Q}{2\pi T} \log \frac{r_0}{r_w} + CQ^n \quad [\text{L}] \tag{5.46}$$

For unsteady state,

$$s_t = \frac{2.3Q}{4\pi T} \log \frac{2.25Tt}{r_w^2 S} + CQ^n \quad [\text{L}] \tag{5.47}$$

5.5.2 Well Efficiency

Well efficiency is related to the well loss as follows:

$$E = \frac{s_a}{s_t} \times 100 \quad [\text{dimensionless}] \tag{5.48a}$$

or

$$E = \left(1 - \frac{s_w}{s_t}\right) \times 100 \quad [\text{dimensionless}] \tag{5.48b}$$

If the well losses are zero, the efficiency is 100%.

5.5.3 Specific Capacity

Specific capacity is defined as the well yield per unit drawdown. Thus

$$\text{specific capacity} = \frac{Q}{s_t} \quad [\text{L}^2\text{T}^{-1}] \tag{5.49}$$

Depending on steady-state or unsteady-state conditions, s_t will be determined by eq. (5.46) or (5.47), respectively. Specific capacity decreases with pumping rate and time. A reduction of up to 40% in the specific capacity has been observed in 1 year in wells deriving water entirely from storage.

Equation (5.49), combined with eq. (5.47) for the value of s_t, provides a means to determine the transmissivity for a known value of the specific capacity based on pumping a well of known diameter for a given period of time, usually 12 hr.

Example 5.19

A fully penetrating 12-in. well in a confined aquifer is pumped at a rate of 400 gpm for 12 hr when a drawdown of 20 ft is recorded in the well. The transmissivity and storage coefficient for the aquifer are 5000 ft^2/day and 6×10^{-3}, respectively. Determine the (a) well losses, (b) efficiency of the well, and (c) specific capacity of the well.

Solution

$$s_a = \frac{2.3Q}{4\pi T} \log \frac{2.25Tt}{r_w^2 S}$$

$$= \frac{2.3}{4\pi} \left(400 \; \frac{\text{gal}}{\text{min}} \right) \left(\frac{1}{5000} \; \frac{\text{day}}{\text{ft}^2} \right) \left(\frac{1 \; \text{ft}^3}{7.48 \; \text{gal}} \right) \left(\frac{60 \times 24 \; \text{min}}{1 \; \text{day}} \right)$$

$$\times \log \left[\frac{2.25}{(0.5^2 \; \text{ft}^2)(6 \times 10^{-3})} \right] \left(5000 \; \frac{\text{ft}^2}{\text{day}} \right) (12 \; \text{hr}) \left(\frac{1 \; \text{day}}{24 \; \text{hr}} \right)$$

$$= 2.82 \log(37.5 \times 10^5)$$

$$= 18.54 \; \text{ft}$$

(a) Well loss, $s_w = 20.00 - 18.54 = 1.46$ ft.

(b) Efficiency, $E = \dfrac{s_a}{s_t} \times 100 = \dfrac{18.54}{20} \times 100 = 92.7\%.$

(c) Specific capacity $= \dfrac{Q}{s_t} = \left(400 \; \dfrac{\text{gal}}{\text{min}} \right) \left(\dfrac{1}{20 \; \text{ft}} \right) \left(\dfrac{1 \; \text{ft}^3}{7.48 \; \text{gal}} \right)$

$$= 2.67 \; \text{ft}^2/\text{min}.$$

5.5.4 Well Field Design

Well field design comprises determining the location of wells, deciding the number of wells required, and fixing the spacing of these wells. Hydrogeologic characteristics of the aquifer and pumping pattern affect the location of the wells. Two common criteria are the wells should be located (1) closer to the recharging boundaries and parallel to such boundaries, and (2) perpendicular to the impermeable boundary, as far away as possible.

To determine the number of wells that will be needed, the total quantity required to be withdrawn from a well field is divided by the potential yield of a well. The potential yield of a well is obtained in either of the following two ways:

1. The specific capacity of a well, when known, is multiplied by a fraction (one-half to two-thirds) of the available drawdown. The available drawdown is equal to the difference between the static water level and the lowest pumping level

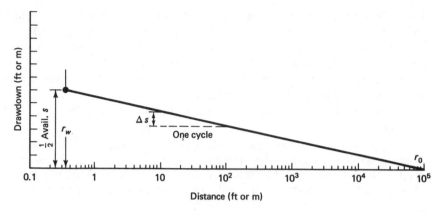

Figure 5.37 Drawdown–distance analysis for well yield.

that can be imposed on the well, which is normally a meter above the top of the screen.

2. From the Jacob equation reproduced below, the value of r_0 is obtained based on the transmissivity and storage coefficient of the flow region and considering pumping period of 1 year (365 days).

$$r_0^2 = \frac{2.25Tt}{S} \quad [L^2] \tag{5.50}$$

where r_0 is the distance from pumping well to zero drawdown.

A semilog plot is prepared by drawing a line connecting point r_0 to a point representing one-half of the available drawdown at the proposed radius of the well, as shown in Figure 5.37.

From this figure Δs, corresponding to one log cycle of the distance, is read and used in the following Jacob formula to derive the estimated pumping rate:

$$Q = 2.7T\Delta s \quad [L^3T^{-1}] \tag{5.51}$$

Only 50% of the available drawdown has been used to determine the pumping rate; the other 50% is estimated to be lost in well losses, well interferences, boundary effects, and so on. This fraction of the available drawdown can be revised based on the well performance.

From spacing consideration, the farther apart the wells are, the less their mutual interference, but the greater the cost of interconnecting pipelines and power equipment. Theis gave a formula for spacing from economic consideration by equating the added cost of pumping due to mutual interference against the capitalized cost of these installations.

From a consideration of well interference alone when wells affect each other's drawdown, the following procedure will indicate the required spacings:

1. If three wells numbered 1, 2, and 3 are located in a straight line, well 2 in the center will be affected by both wells 1 and 3.

2. In the well yield analysis above, one-half of the available drawdown has been allocated for well losses, well interferences, and boundary effects. The boundary

effects are neglected (or determined separately). The well losses are determined from eq. (5.45). The remaining drawdown can be assigned to well interference.
3. Since two wells interfere, this value of the drawdown is divided by 2. Plot the position of this drawdown (effect of each well) on Figure 5.37 and read the distance, which should be the spacing of wells 1 and 3 from well 2.

Example 5.20

It is proposed to withdraw a quantity of 2 million gallons of water per day from an aquifer having a transmissivity of 8000 ft^2/day and a storage coefficient of 6×10^{-4}. The maximum permissible drawdown in the aquifer is 40 ft. Wells of 12 in. diameter are to be installed. Determine the number of wells required.

Solution   From eq. (5.50),

$$r_0^2 = 2.25 \left(8000 \, \frac{\text{ft}^2}{\text{day}} \right) \left(\frac{1}{6 \times 10^{-4}} \right) (365 \text{ days})$$

$$r_0 = 1.05 \times 10^5 \text{ ft}$$

Available drawdown = 40 ft.
One-half of available drawdown = 20 ft.
A line connecting one-half available drawdown to r_0 is shown in Figure 5.38. Δs for one distance cycle = 3.8. From eq. (5.51),

$$Q = 2.7 T \Delta s$$

$$= 2.7(8000)(3.8) = 82,080 \text{ ft}^3/\text{day} \quad \text{or} \quad 0.95 \text{ cfs} \quad \text{or} \quad 427 \text{ gal/min}$$

$$\text{number of wells} = \frac{\text{withdrawal from well field}}{\text{yield of each well}}$$

$$= \frac{2 \times 10^6}{427 \times 24 \times 60} = 3.25$$

Thus four wells will be required, to withdraw the desired quantity.

Example 5.21

For Example 5.20 determine the spacing among the wells if the well efficiency is 80% and interference is produced only by two outer wells surrounding an interior well.

Solution   From eq. (5.48a),

$$E = \frac{s_a}{s_t} \times 100$$

$$s_t = \frac{s_a}{E} \times 100 = \frac{20}{80} \times 100 = 25 \text{ ft}$$

Drawdown assigned to well interference = $40 - 25 = 15$.
Drawdown interference per well = $15/2 = 7.5$.
A point corresponding to 7.5 ft drawdown is marked on Figure 5.38. The corresponding distance = 1.3×10^3 ft. Thus the well spacing = 1300 ft.

5.5.5 Yield of an Aquifer

The water that can be withdrawn from an aquifer depends on the rate of recharge of (inflow into) the aquifer. In the past the term "safe yield" was used to indicate the annually extractable quantity, which was limited to the average annual recharge of the

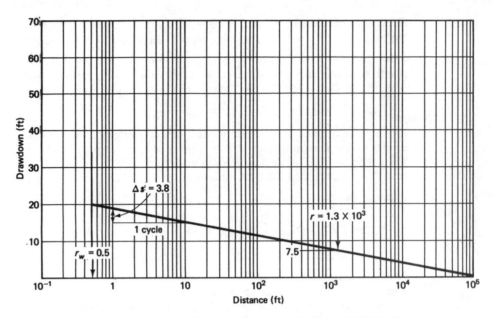

Figure 5.38 Plot of the line connecting r_0 to one-half of available drawdown.

basin (aquifer). Perennial yield has now replaced the concept of safe yield. According to this, the quantity of groundwater available is equal to the recharge capability of the basin, with a limit placed by the storage volume available in the underground basin. Yield is also controlled by economic, legal, and political constraints. Thus the yield of the aquifer has to be adjusted to the recharge potential of the basin. However, the more water available for recharge, the greater the expected yield.

5.6 RECHARGE OF GROUNDWATER

Recharge is the replenishment of groundwater storage from surface supplies. Precipitation is a natural source of recharge of a basin. There is also incidental recharge that occurs as a result of human activities unrelated to a recharge project. Artificial recharge of a basin is carried out by a variety of methods when it is necessary to augment groundwater supplies. As stated in the preceding section, the basin recharge governs the perennial yield of an aquifer.

5.6.1 Methods of Artificial Recharge

There are five common techniques of artificial recharge. A number of methods are included in the first technique, which is the most widely practiced.

1. Water spreading
   a. Basin method
   b. Stream-channel method
   c. Ditch method
   d. Flooding method

2. Recharge pits
3. Recharge wells
4. Induced recharge
5. Wastewater disposal

Basin method. Water diverted from or pumped out of a stream is filled into a single or a series of basins. It percolates through the soil to replenish the groundwater supply. Basins are formed by construction of dikes or by excavation. Multiple basins are favored because of the continuity of operation. Studies have indicated that the recharge rates (percolation) are directly proportional to the head difference between surface water level and water table. Hence recharging through narrow strips that are spaced apart proves almost as effective as spreading over an entire area. Studies have also indicated that recharge (infiltration) capacity decreases with time. This has been attributed to the clogging of soil pores by fine silt or clay particles settled from water as well as to the microbial growth.

Stream-channel method. In alluvial streams water infiltrates underground through bed and banks. To improve recharge potential, certain measures designed to enhance both area and time of infiltration are adopted. These include widening and leveling the stream channel, and construction of temporary check dams and dikes by river-bottom material itself. In streams having reservoirs, the releases could be regulated to match the absorptive capacity of downstream channels and thus enhance infiltration.

Ditch method. Similar to the concept of stream channels, a series of flat-bottomed shallow ditches are constructed spaced close to each other, in which water is diverted from a stream. The width of the ditches ranges from 1 to 5 ft, and their gradient is sufficient to prevent the deposition of fine suspended material at the bottom of ditches, thus clogging the voids.

Flooding method. This is similar to the basin method, but basins are not constructed by dikes or excavation. On a relatively flat natural terrain, water is spread out to infiltrate through the soil. To prevent erosion and avoid disturbing the soil, water is distributed with a low velocity in a thin sheet. An embankment is provided surrounding the flooding area if a natural control boundary does not exist. This is the simplest and least expensive method of recharge.

Recharge pits. If a permeable formation exists at a shallow depth underlying a relatively hard stratum, excavation of a pit serves as a very effective method of recharge. Sometimes the sides and bottom of these pits are covered by a material of higher permeability such as gravel. This gravel layer is replaced every few years as it gets clogged with fine particles. Fine material usually settles to the bottom, leaving the sides unclogged for a relatively longer period.

Recharge wells. These are similar to the pumping wells. Their operation is, however, inverse, wherein water is admitted into the well from the surface. The cone of recharge is formed, which is a mirror image of the cone of depression. The formulas of pumping wells are also applicable to recharge wells. It should, however, be recognized that the rate of recharge is not equal to the rate of pumping for a cone of recharge that is equivalent to a cone of depression. This is due to the fact that the coefficient of permeability is lower in recharge because of the clogging of the aquifer material by fine particles in recharged water and due to the growth of bacteria on the well screen. The well recharge has a limited scope. It is desirable for deep, confined aquifers.

Induced recharge. In a hydraulically connected stream-aquifer system, when a well near a stream is pumped, the lowered water level inside the well creates a steeper gradient from the stream to the well, thus inducing higher discharge toward the well. This is known as induced infiltration. In many pumping schemes, the well field is operated in such a manner that replenishment of the aquifer by induced infiltration is attained when the river is carrying excess flow.

Wastewater disposal. Using wastewater after its secondary treatment for the purpose of recharge serves two objectives: it improves the quality by removal of physical, chemical, and biological impurities, and it replenishes the underground water supply. Methods such as the spreading technique, recharge pits, and recharge wells can be used to recharge through wastewater. The selection of a method depends on availability of land, intended use of wastewater, and soil conditions.

5.7 SALINE WATER INTRUSION

Mixing of saline water is a most common type of pollution problem of groundwater. This occurs due to invasion of seawater in coastal aquifers, seepage of saline wastes from the surface, and upward movement of saline waters of geologic origin in other aquifers. The first category has been recognized very widely.

5.7.1 Freshwater and Saltwater Interface

Figure 5.39 shows a cross section of a coastal aquifer. If at any point on the interface the pressure from the top of the fresh water balances the pressure of saline water from bottom, then

$$\rho_s g Z = \rho_f g (Z + h)$$

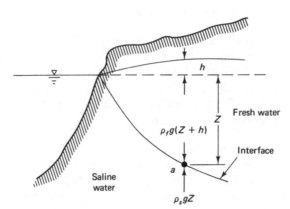

Figure 5.39 Fresh and saline water balance in an unconfined coastal aquifer.

or

$$Z = \frac{\rho_f}{\rho_s - \rho_f} h \quad [L] \qquad (5.52)$$

where

ρ_s = density of saline water

ρ_f = density of fresh water

Z = depth of the interface at any point

h = water table or piezometric head above seawater level at any point

Equation (5.52) is recognized as the Ghyben–Herberg relation. The movement of water has been ignored in this relation. More exact solutions for the interface have been obtained from potential flow theory by Glover and Charmonman.

For typical seawater, $\rho_s = 1.025$ and $\rho_f = 1.0$; then

$$Z = 40h \qquad (a)$$

It had been observed in the past by many investigators that salt water occurred at a depth below sea level of 40 times the height of the fresh water as given by eq. (a). If h is not positive (i.e., the piezometric or water-table head is not above sea level, sea-water will advance directly inland.

5.7.2 Upconing of Saline Water

In a situation where a saline water layer underlies a freshwater zone and a well penetrating only the freshwater portion is pumped, a local rise of the interface of saline and fresh water occurs as shown in Figure 5.40. This phenomenon is known as upconing.

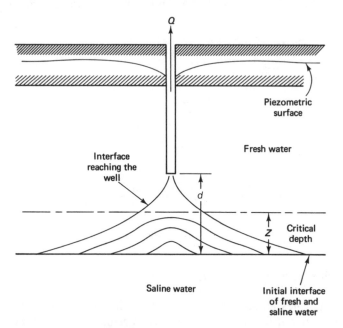

Figure 5.40 Upconing of salt water under a pumping well.

The rise of upconing under steady-state conditions is given by

$$Z = \frac{Q\rho_f}{2\pi \, dK(\rho_s - \rho_f)} \qquad [\text{L}] \qquad (5.53)$$

where

K = hydraulic conductivity
d = depth of the initial freshwater–saline water interface below the bottom of the well

When this rise becomes critical (i.e., $Z/d = 0.3$ to 0.5) salt water reaches the well, contaminating the supply. Thus the maximum discharge to keep the rise below the critical limit can be given by substituting $Z = 0.5d$ in eq. (5.53):

$$Q_{max} = \pi \, d^2 K \frac{\rho_s - \rho_f}{\rho_f} \qquad [\text{L}^3\text{T}^{-1}] \qquad (5.54)$$

In a real situation, a zone of brackish water occurs between salt water and fresh water. Even with a low rate of pumping, some saline water reaches the well. However, the upconing effect can be minimized by increasing d (i.e., separating wells from the saltwater layer as far as possible) and by decreasing Q, the flow rate.

Example 5.22

In a deep aquifer, fresh water extends to a depth of 120 ft, below which there is a deposit of salt water of specific weight 64 lb/ft^3. A water well is drilled through the freshwater zone to a depth of 80 ft. Determine the maximum rate at which this well can be pumped without drawing saline water from the well if the aquifer permeability is 1000 ft/day.

Solution: From eq. (5.54),

$$Q_{max} = \pi \, d^2 K \frac{\rho_s - \rho_f}{\rho_f}$$

or

$$Q_{max} = \pi d^2 K \frac{\gamma_s - \gamma_f}{\gamma_f}$$

$$= \pi(40^2 \text{ ft}^2)\left(1000 \, \frac{\text{ft}}{\text{day}}\right)\left(\frac{64.0 - 62.4}{62.4}\right)$$

$$= 128{,}800 \text{ ft}^3/\text{day} \quad \text{or} \quad 1.49 \text{ cfs}$$

5.7.3 Protection against Intrusion

Before it contaminates an aquifer system, it is desirable to control the intrusion of salinity because once it develops it is not easy to remove. It might require years to

TABLE 5.3 CONTROLLING SALTWATER INTRUSION
OF VARIOUS CATEGORIES

Source or Cause of Intrusion	Control Methods
Seawater in coastal aquifer	Modification of pumping pattern
	Artificial recharge
	Extraction barrier
	Injection barrier
	Subsurface barrier
Upconing	Modification of pumping pattern
	Saline scavenger wells
Oil field brine	Elimination of surface disposal
	Injection wells
	Plugging of abondoned wells
Defective well casings	Plugging of faulty wells
Surface infiltration	Elimination of source
Saline water zones in freshwater aquifers	Relocation and redesign of wells

Source: Todd (1980).

restore normal conditions. Many control methods have been suggested for the various category of problems summarized in Table 5.3 (Todd, 1980).

For a coastal region, these methods include (1) manipulation of well field location and operation to create a seaward hydraulic gradient; (2) artificial recharge that will maintain a low profile of freshwater–saltwater interface based on hydraulic equilibrium, as discussed in Section 5.7.1; (3) constructing a line of wells adjacent and parallel to the sea that is continuously pumped to create a trough in which mostly salt water flows from the sea and some fresh water within the basin flows seaward toward the trough; (4) reverse of method 3, in which fresh water is injected through a line of well, thus creating a pressure ridge through which fresh water flows both seaward and toward land; and (5) construction of an impermeable core throughout the depth of the aquifer, creating a watertight barrier against intrusion. Some of these methods may not be economically justified.

For upconing, methods comprise (1) separating of wells from the saltwater layer as far as possible; (2) pumping wells at lower rates, as discussed in Section 5.7.2; and (3) providing scavenger wells that pump saline water from below the fresh water. In the case of surface infiltration, the solution lies in preventing such occurrences.

PROBLEMS

5.1. Figure P5.1 shows the cross section of a line of sheet piling driven to a depth of 7 m into a homogeneous sandy soil of thickness 15 m. If the soil has a permeability of 25 m/day, determine the (a) quantity of seepage under the pile, and (b) pore (uplift) pressure at A, B, and C.

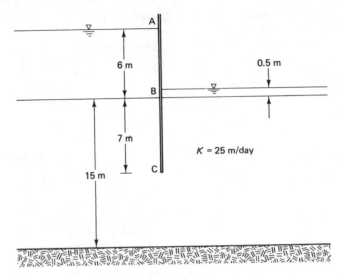

Figure P5.1

5.2. The concrete dam shown in Fig. P5.2 founded on isotropic soil of permeability 30 m/day. Determine **(a)** the rate of flow under the foundation, and **(b)** the uplift pressure on the foundation.

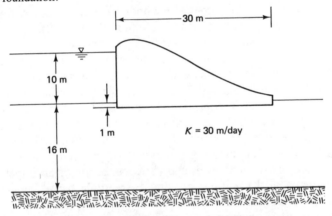

Figure P5.2

5.3. For the concrete dam with a line of pile at the downstream toe, as shown in Fig. P5.3, determine **(a)** the rate of seepage under the dam, and **(b)** the uplift pressure at the base.

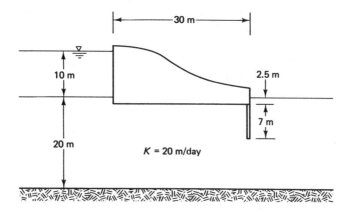

Figure P5.3

5.4. A masonry dam having a sheetpiling cutoff at the upstream end is located on a pervious stratum having a horizontal hydraulic conductivity of 0.01 ft/min and a vertical hydraulic conductivity of 0.005 ft/min (Fig. P5.4). Determine the **(a)** rate of seepage, and **(b)** uplift force acting at points A and B on the base.

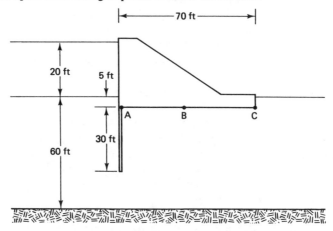

Figure P5.4

5.5. Two streams are separated by a confined aquifer of average thickness 40 ft and hydraulic conductivity 0.03 ft/sec. The water level in a stream at a higher level is 60 ft. A piezometer located 1000 ft from this stream has a level of 59.2 ft. Find the rate of flow from one stream to another.

5.6. In an unconfined aquifer having $K = 30$ m/day, two observation wells, 2 km apart, have water levels of 50 and 40 m, respectively, above the base of the aquifer. Determine the rate of flow.

5.7. A confined aquifer underlies an unconfined aquifer as shown in Fig. P5.7. Determine the flow rate from one stream to another.

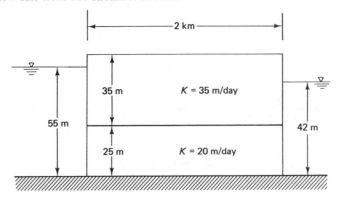

Figure P5.7

5.8. A well of 0.4 m diameter fully penetrates a 25-m-thick confined aquifer of coefficient of permeability of 12 m/day. The well is located in the center of a circular island of radius 1 km. The water level at the boundary of the island is 80 m. At what rate should the well be pumped so that the water level in the well remains 60 m above the bottom?

5.9. A 12-in. well is drilled through a confined aquifer of average thickenss 80 ft. Two observation wells located 50 and 120 ft away register a difference in drawdown of 10 ft. The steady-state pumping rate is 1000 gpm. Determine the transmissivity.

5.10. A 0.3-m-diameter well fully penetrates an unconfined aquifer. The water table is at a height of 70 m from the bottom and the coefficient of permeability is 20 m/day. The well is pumped so that water level in the well remains steady 20 m below the original water table. The pumping has no effect on the water table at a distance of 1 km. Determine the rate of flow of the well.

5.11. A fully penetrating unconfined well of 12 in. diameter is pumped at a rate of 1 ft^3/sec. The coefficient of permeability is 750 gal/day per square foot. The drawdown in an observation well located 200 ft away from the pumping well is 10 ft below its original depth of 150 ft. Find the water level in the well.

5.12. For Problem 5.8, find the time of travel of groundwater from the boundary of the island to the well. The aquifer has a porosity of 0.25.

5.13. For Problem 5.9, determine the travel time of water from one observation well to another if the porosity is 0.2.

5.14. For Problem 5.11, determine the time of flow from the observation well to the pumping well for the aquifer of porosity 0.2.

5.15. An areally infinite aquifer with a transmissivity of 25×10^{-3} m^2/s is overlain by a semipervious stratum of a leakage coefficient of 5×10^{-9} per second. A well of 0.5 m diameter is drilled through the aquifer. When the well is pumped at a rate of 0.05 m^3/s, determine the drawdown at a distance of 500 m from the well and just outside the well.

5.16. A well of 0.4 m diameter fully penetrates a leaky confined aquifer having a hydraulic conductivity of 20 m/day and a thickness of 30 m. The leakage coefficient of the confining layer is 0.05 per day. A drawdown of 1 m is recorded at a distance of 100 m from the well. What is the rate of pumping of the well?

5.17. A system is shown below, in Fig. P5.17. The well is 0.5 m in diameter and fully penetrates a confined aquifer of 50 m thickness and a hydraulic conductivity of 20 m/day. If the pumping rate is 5 m³/min, determine the drawdown **(a)** at point X and **(b)** at the well face.

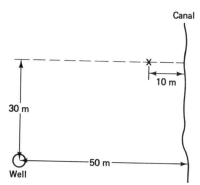

Figure P5.17

5.18. An unconfined aquifer with $K = 30$ m/day is bounded by a stream on one side as shown in Fig. P5.18. At a distance of 80 m from the stream, a well of 0.4 m diameter is pumped at a rate of 3000 m³/day. What is the drawdown in the well?

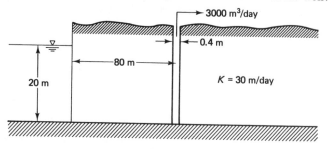

Figure P5.18

5.19. In Problem 5.17, if an impermeable boundary exists in place of the canal, what is the drawdown at point X? The well has a radius of influence of 2 km.

5.20. For the system shown in Fig. P5.20, determine the drawdown for the pumping rate of 4000 ft³/day.

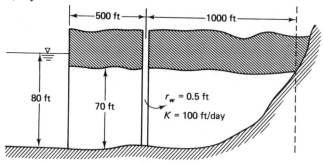

Figure P5.20

Chap. 5   Problems

5.21. From a well fully penetrating a confined aquifer, the following drawdowns were observed from an observation well 200 ft from the well being pumped at 500 gpm. Determine the value of transmissivity and storage coefficient by the type-curve method.

Time since pumping (min)	0	1	1.5	2.0	2.5	3	4	5		8
Drawdown (ft)	0	0.66	0.87	0.99	1.11	1.21	1.36	1.49		1.75

Time	10	14	18	24	30	40	50	60	100	150
Drawdown	1.86	2.08	2.20	2.36	2.49	2.65	2.78	2.88	3.16	3.42

Time	180	210	240
Drawdown	3.51	3.61	3.67

5.22. From a pumping test on a confined aquifer, the following data were noted on an observation well 50 ft from a well pumped at 300 gpm. Find the transmissivity and storage coefficient by the type-curve method.

Time (min)	30	50	70	90	120	150	200	400	600	900
Drawdown (ft)	6.5	9.0	11	12.4	14.1	15.4	17.0	21.2	23.6	26.0

5.23. From a 2-hour test on a confined aquifer at 400 gpm, the following drawdowns were measured on a number of wells. Determine the transmissivity and storage coefficient by the type-curve method.

Well number:	1	2	3	4	5	6	7
Distance (ft)	29	35	44	60	85	100	163
Drawdown (ft)	14.9	13.8	12.7	11.7	10.1	9.6	7

5.24. During a test on a confined aquifer of 50 m thickness the following data were noted. The coefficient of permeability was determined from the lab test to be 25 m/day. What were the rate of pumping and the storage coefficient of the aquifer by the type-curve method? $r = 61$ m.

Time (min)	1	3	5	10	18	30	50	100	150	210	240
Drawdown (m)	0.2	0.37	0.45	0.57	0.67	0.76	0.85	0.97	1.05	1.1	1.12

5.25. In a confined aquifer test, the following drawdowns were noted in various observation wells at a time 502 min. after pumping commenced at a rate of 300 m³/day.

Well number:	1	2	3	4	5	6
Distance (m)	4.6	10.4	20	34.5	65.8	91.8
Drawdown (m)	1.35	0.9	0.65	0.41	0.20	0.11

Determine the transmissivity and storage coefficient by the type-curve method.

.5.26. In an unsteady flow, the following data are given: $Q = 0.1$ cfs, $T = 650$ ft^2/day, $t = 35$ days, $r = 0.6$ ft, and $S = 6 \times 10^{-3}$. What is the drawdown?

5.27. Solve Problem 5.21 by the Cooper–Jacob method. Check whether this method is valid for this test.

5.28. Solve Problem 5.24 by the Cooper–Jacob method. Check the validity of this approach.

5.29. A well fully penetrating an artesian aquifer was continuously pumped at a rate of 500 gpm. After 200 days, the following drawdowns were noted in three observation wells. Compute the transmissivity and storage coefficient of the aquifer by the Cooper-–Jacob method. Also check whether the method is applicable to all wells.

Distance from Pumped Well (ft)	Drawdown (ft)
100	8.5
1000	5.65
10000	2.84

5.30. Solve Problem 5.21 by the Chow method.

5.31. Solve Problem 5.24 by the Chow method.

5.32. From a pumping test on an unconfined aquifer of 40 ft thickness following drawdown, measurements were made on an observation well 60 ft from the pumped well. The rate of pumping is 1500 gal/min. Determine the transmissivity and specific yield of the aquifer. Is it justified to treat this similar to a confined aquifer?

Time (min)	30	60	90	120	300	500	1000	2400
Drawdown (ft)	8	10.6	12.72	13.47	15.83	18.1	23.0	29.0

5.33. In the confined aquifer system shown in Fig. P5.33, the pumping rates in well 1 and well 2 are 0.8 cfs and 1 cfs, respectively. The aquifer length extends 4000 ft and a width of 2700 ft has been included in the flow system being considered. The initial piezometric head is 60 ft everywhere. Applying the alternate direction implicit procedure, determine the head distribution after 1 day of pumping.

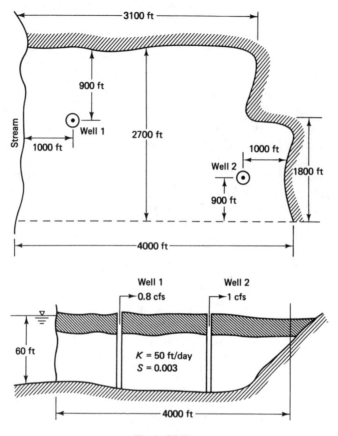

Figure P5.33

5.34. The water requirement from an aquifer is 3 million gallons per day. The aquifer has a transmissivity of 6000 ft²/day and a storage coefficient of 8×10^{-4}. The available drawdown is 50 ft. It is proposed to install wells of 8 in. diameter. Design the well field for well efficiency of 90%.

5.35. An 8-in. well pumps a confined aquifer at a rate of 300 gal/min. The transmissivity and storage coefficient for the aquifer are 200 ft²/hour and 0.0003, respectively. If the well has an efficiency of 70%, what is the total drawdown after 12 hr? Also, what is the specific capacity of the well?

5.36. In a water-table aquifer, the rate of pumping is 2000 m³/day. The aquifer has a specific yield of 0.02. The specific capacity after 12 hr of pumping a well of 0.3 m diameter is 150 m²/day. If the well has an efficiency of 70%, determine the transmissivity of the aquifer.

5.37. Determine the well yield in the aquifer above if the available drawdown is 30 m.

5.38. In a deep saline water aquifer, there is a zone of fresh water of 100 ft depth. A well is drilled through a depth of 80 ft to pump out the water at a rate of 0.5 cfs. The aquifer permeability is 100 ft/hr. Determine the upconing rise of the saline water under steady state. (Assume a saltwater unit weight of 64 lb/ft³.)

5.39. In Problem 5.38, will the pump draw the saline water? If not, what will be the maximum rate of pumping without pumping the saline water out of the well?

CHAPTER
SIX

MEASUREMENT OF SURFACE WATER FLOW

6.1 DETERMINATION OF STREAMFLOW

The quantity of water flowing in a stream, its distribution in space, and its variability with time are important information that are required to plan any surface water supply project or to design a hydraulic structure. The most direct and desirable method is to measure the quantity of flow per unit time, referred to as the *streamflow* or *discharge*. For this purpose, a stream gaging station is set up. Since long-term flow records are needed for the planning of a project because of the high variability of flow, a network of stream gaging stations is designed from which data are continuously collected for use at any time in the future. For measurements in small creeks and open channels, hydraulic instruments such as weirs, notches, and flumes are very convenient means for computing the discharge. In the time of floods, it is not always possible to make direct measurements due to such problems as inaccessibility of the site, damage to measuring structure, and short duration of peak. In such a situation, indirect methods are used by making measurements of certain data after the floods.

It is not possible to gage every site where flow data are desired. If time permits, a temporary gage is installed to collect direct information. However, when the project formulation has to proceed without delay, three alternatives are available. In order of preference, they are: extend information from nearby gaging sites, estimate streamflows from precipitation data, and use generalized information or the empirical approach.

The procedure to assess the streamflow can be summarized as follows:

  I. Measurement of streamflow
     A. Direct measurement or stream gaging
        1. Measurement by current meter method

2. Measurement by floats
   3. Tracer-dilution technique
   4. Ultrasonic method
   5. Electromagnetic method
   B. Measurement through hydraulic devices
   1. Weirs and notches
   2. Orifices
   3. Flumes
   C. Indirect measurement of peak flows
   1. Slope-area method
   2. Contracted–opening method
   3. Flow-over-structure methods

II. Estimation of streamflow
   A. Application of precipitation data
   1. Precipitation–runoff relation
   2. Hydrograph analyses
   3. Empirical formulas
   B. Extension of gage-sites data
   C. Generation of synthetic flows
   D. Use of generalized data, charts, tables, and empirical approach

In this chapter we discuss direct measurement of streamflow. Indirect methods are described in Chapter 8 and the estimation procedure of streamflow is described in Chapter 7.

6.2 STREAM GAGING

Stream gaging or hydrometry is a procedure of measuring the water stage (level) and discharge at a gaging station with an objective of obtaining a continuous record of stage and discharge at the station. For this purpose, installation of equipment is made at the stream site, which enables continuous or regular observation of water stage and frequent measurement of discharge and optional recording of any other hydrologic parameter, such as sediment load. A number of such stations in a basin form a hydrologic network that provides information on the water resources of the basin. The network of continuous-record stations, known as the basin network, is often augmented by an auxiliary network of partial-record stations: for example, to provide data on peak discharge only.

The systematic records of streamflows as published by the U.S. Geological Survey for each gaging site from year to year involve both measurement and computational steps as outlined below.

1. Selection of gaging station sites
2. Measurement of water level (stage) on a continuous or daily basis
3. Measurement of discharge from time to time
4. Establishing a relation between stage and discharge

5. Conversion of the measured daily stage into discharge using the relation of step 4.

6. Presentation and publication of measured and computed data

The measurement of discharge in step 3 is performed using a current meter, floats, tracer dilution, or the ultrasonic or electromagnetic method, although the current meter is in most common use.

6.3 MEASUREMENT OF STAGE

The stage, also known as gage height, is the height of the water surface in a stream above a fixed datum. The datum can be a recognized reference level, such as mean sea level, or an arbitrary level chosen for convenience. Two or three reference marks of known gage height are established on stable structures to maintain a permanent datum.

There are two broad categories of gages: nonrecording gages and recording gages. Nonrecording gages are manually observed at fixed hours. These comprise the staff gage, wire weight gage, float-tape gage, and crest-state gage. The simplest is the staff gage, which consists of porcelain-enameled iron sheet sections (gaged plates) of 4 in. or 0.1 m width and 3.4 ft or 1 m length graduated every 0.02 ft or 1 cm.

Automatic or self-recording gages provide a continuous record of stage. These have advantages over nonrecording gages. Their features are discussed below.

6.4 COMPONENTS OF AUTOMATIC GAGES

There are three components of automatic recording gages. These pertain to sensing of the water stage, method of its recording, and storage of the data. Various types of each of these components and their combination in a recording system (gage) are shown in Figure 6.1. A brief description follows.

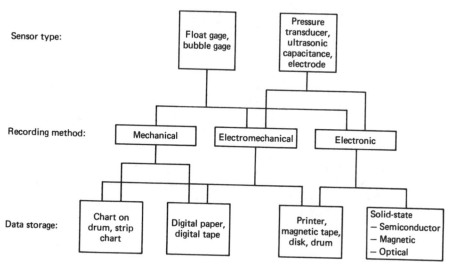

Figure 6.1 Components of self-recording gages.

6.4.1 Sensor Devices

Float sensor. This consists of a float attached to one end of a cable that passes over a pulley and is counterweighted at the other end. The float follows the rise and fall of the water level rotating the pulley. Through a system of other pulleys, this moves a pen up and down recording the stage.

The float is installed inside a stilling well, which protects the float and dampens the water surface fluctuations. Stilling wells can be made of bricks, concrete blocks, concrete, concrete pipe, or steel pipe. They are placed directly in the stream or on a bank in the vicinity of the stream. In the latter case, an intake pipe connects the stream to the well as shown in Figure 6.2. The bottom of the well is at least 1 ft below the minimum stage and the top is above the 100-year flood level. The dimension of the well is usually 4 ft in diameter or 4 ft by 4 ft in size. For intake the most common size is a 2-in.-diameter pipe.

Three errors associated with float gages are float-lag error, line-shift error, and counterweight submergence error. These are discussed by Rantz and others (1982).

Bubble-gage sensor. This is a pressure-actuated system in which an orifice at the end of a tubing is installed underneath the water surface at the location of the gage datum. The water level is directly proportional to the pressure experienced at the orifice. A gas, usually nitrogen, is passed through the tube to bubble freely into the stream through the orifice. The gas pressure is equal to the head on the orifice or the gage height. A servomanometer or a bellows system converts the pressure to the

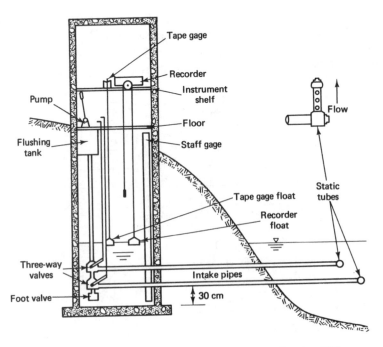

Figure 6.2 Stilling well for a float-type recorder (from Herschy, 1985).

Measurement of Surface Water Flow   **Chap. 6**

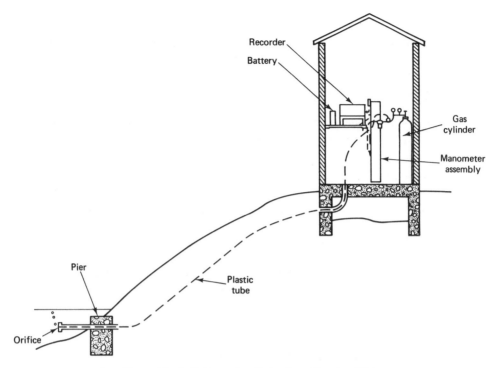

Figure 6.3 Bubble-gage installation (from Herschy, 1985).

shaft movement for stage recording. For this system the orifice is installed directly inside a stream; the tubing runs along the stream bank and the recorder can be installed away from the stream as shown in Figure 6.3.

Pressure transducers for stage sensing. These devices convert the water pressure into electrical signals that can be recorded at the gage site or at a remote location. A transducer has two components: a force-summing-up device that responds to pressure changes and an output device to convert force-summing-up device signals to electrical signals.

Capacitance, resistance, and ultrasonic sensors. These have not been widely used in stream gaging. The capacitance sensor operates by sensing the change in capacitance between an insulated wire electrode mounted vertically for the full depth of the stream and the surrounding water. The change in water level varies the capacitance in a controlled manner which is sensed electronically.

Since water is an electrical conductor, the resistance sensor operates by detecting the change in the electrical resistance of a vertically mounted sensing element as the water surface moves over it.

In an ultrasonic sensor, an acoustic pulse is directed toward the water surface by the sensor. This pulse is reflected from the surface back to the sensor. The time of travel from transmission to reception is measured and from the velocity of sound, the distance is computed.

6.4.2 Recording Mechanism

Mechanical device. The device consists of a time element and a water-level element. The time element, controlled by a clock, moves the chart or tape at a predetermined fixed rate. The water-level element, actuated by a float or bubble gage, moves the pen stylus or the punch block.

Electromechanical device. The time element is controlled by an electrically driven clock. Where an AC power source is available at the site, it can be used to power the device.

Solid-state device. Modern devices have been developed that are capable of accepting and storing data entirely electronically. The functions of such devices comprise:

1. Receiving a signal from a sensor
2. Converting that signal to compatibility with the recording format
3. Recording corresponding time information
4. Recording other site identification data
5. Transferring the data to solid-state memory

A recorder may be made as a single-site unit or operations of many recorders may be controlled by a central processing unit. These devices represent a significant advancement in stage recording technology.

6.4.3 Storing of Data

Graphic (analog) record on a drum. A chart is mounted on a drum which is installed inside a box either horizontally or vertically. With the rotation of the drum and the movement of the stylus pen a continuous trace of the water stage with respect to time is provided.

Graphic record on a strip chart. A roll of paper (chart) moves horizontally from one small drum to another at a preadjusted fixed rate. Over the chart, the stylus pen makes a continuous record of the gage height, similar to that on the drum chart.

Digital record on paper tape. The paper tape moves slowly vertically from one small drum to another. At preselected time intervals, the tape is punched for the gage height transmitted to the instrument by rotation of the input shaft, which drives two code disks. Electronic translators are used to read the punch tape on to a tape suitable for input into a digital computer.

Magnetic tape. Signals from transducers, capacitance, resistance, and ultrasonic sensors can be recorded in digital format on a magnetic tape. For magnetic tape records a replay system is required that is capable of reconstructing the data. It is often necessary to translate the recorded tape data into a computer tape format.

Computer storage of data. From a solid-state recording device, the stage records are transferred to the data memory, which may be either an integral part of the device or may be detachable for data extraction at a processing center. The data memory is capable of accepting very large numbers of records, which enables broad-based data gathering, including recording of air and water temperatures and other site information, as well as correction and checking of the data.

6.5 REMOTE TRANSMISSION OF STAGE DATA: TELEMETRING SYSTEM

The stream gaging stations are usually visited by the hydrologic field parties at regular intervals for the servicing of the stations and the conducting of some hydrologic measurements. The record of the stage data is collected at the time of such visits. However, when the stage information is needed at frequent intervals, or it is not practical to visit the gage site for a considerable time, the telemetring system is used for remote transmitting of the data. There are two types.

6.5.1 Continuous Transmission of Data

Position-motor system. The system, for a distance up to 15 mi, employs a pair of self-synchronizing motors. One motor, on the transmitter side, is actuated by a float or bubble-gage, and the other, on the receiving unit, follows the rotary motion of the first one, to which it is electrically connected by a transmission line.

Impulse system. This system, far longer than a position-motor system, operates over telephone lines. An impulse sender at the gaging site is actuated by a float or bubble-gage and sends electrical impulses over the line to the receiver.

6.5.2 Intermittent Transmission at Predetermined Intervals

Telemark system. The system codes the instantaneous stage and signals it over the telephone circuit or by radio. The distance is unlimited in this case. It consists of a positioning element which is actuated by a sensor and a signaling element which makes contact across the signaling drums positioned in correspondence with the stage.

Resistance system. The system, for a distance up to 40 mi, consists of two potentiometers and a microammeter null indicator. One potentiometer, at the gage site, is actuated by the sensor. The other potentiometer, at the observation site, is adjusted for a null balance and the gage height is read from a dial coupled to the potentiometer.

Satellite data-collection system. From stream gaging stations, stage data are transmitted using inexpensive battery-operated radios, known as data-collection platforms (DCPs). Data for transmission are obtained either directly from a digital recorder or through a memory device. Through an earth resources technology satellite (ERTS), these data are directed to the receiving centers.

6.6 THEORY OF DISCHARGE MEASUREMENT

Discharge or streamflow is the volume rate of flow of water in a stream expressed as cubic feet per second or cubic meters per second. It is a product of the area of cross section and the velocity of flow. A natural stream channel can have any irregular shape for which any standard formula cannot be used to compute the area. Similarly, there is no fixed velocity which varies in both width and depth in a stream section. Thus the discharge can be given by

$$Q = \int_A v \, dA \qquad [L^3 T^{-1}] \qquad (6.1)$$

Except for the tracer-dilution technique, in which the equation of the mass rate of flow is used, the other discharge methods are based on eq. (6.1). The current meter and float methods perform algebraic integration (summation). The stream section is divided into a number of subsections. The depth and velocity measurements are arranged to determine an average velocity for each subsection. The discharge is computed by

$$Q = \sum av \qquad [L^3 T^{-1}] \qquad (6.2)$$

where

a = individual subsection area

v = mean velocity of flow in the subsection

Measurements using ultrasonic and electromagnetic methods give the average velocity for the entire stream section. Then the discharge is obtained by multiplying the average velocity by the entire area of cross section.

Example 6.1

A profile of a stream cross section has been divided into five areas as shown in Figure 6.4. The area of each section and its mean velocity are indicated below. Determine the stream discharge at the cross section.

Section	Area (acres)	Mean Velocity (ft/sec)
I	2.5×10^{-3}	0.15
II	3.8×10^{-3}	0.22
III	6.2×10^{-3}	0.35
IV	8.0×10^{-3}	0.28
V	3.0×10^{-3}	0.19

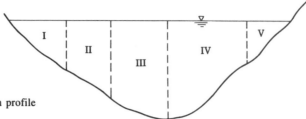

Figure 6.4 Stream cross-section profile for Example 6.1.

Solution

(1)	(2) Area, *a*		(3)	(4)	(5)
				Velocity, *v*	$Q = av$ (ft³/sec)
Section	Acres × 10⁻³	ft²		(ft/sec)	(col. 3 × col. 4)
I	2.5	108.9		0.15	16.34
II	3.8	165.5		0.22	36.41
III	6.2	270.0		0.35	94.50
IV	8.0	348.5		0.28	97.58
V	3.0	130.7		0.19	24.83
Total					269.66

6.7 MEASUREMENT BY CURRENT METER

The use of the current meter is a most common method of discharge measurement. The current meter consists of a cup- or propeller-type rotor. The number of revolutions of the rotor in a given period is directly proportional to the velocity of water. The relation between revolutions per second, n, and velocity of flow, v, is of a straight-line form ($v = a + bn$). Values of a and b are established from the calibration of the meter by the manufacturer and are known as the meter rating. For convenience the rating data are produced in a table form.

The current meter is divided into two broad categories of vertical-axis meter and horizontal-axis meter, depending on the direction of the rotor shaft. The vertical-axis rotors are mounted with cups or vanes that rotate with the current. The horizontal-axis rotors have a propeller-type attachment. In both cases, each revolution of the rotor completes a circuit through a battery connection that produces an audible click in a headphone or moves a digital counter. A stopwatch is used to measure the time over which revolutions are counted. Both types are available in a standard size and a miniature size for use in very small depths.

The vertical-axis cup meter, known as the Price current meter after its inventor, is most common in the United States. The main parts of the type AA Price current meter are shown in Figure 6.5(a). A complete assembly of suspension cable, headphone, battery unit, and sounding weight is shown in Figure 6.5(b).

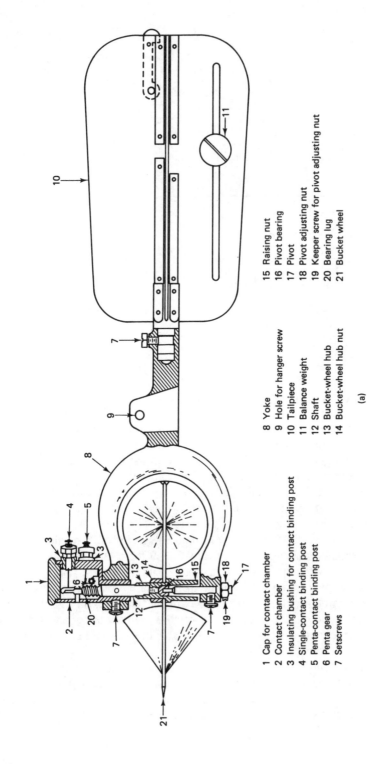

1 Cap for contact chamber
2 Contact chamber
3 Insulating bushing for contact binding post
4 Single-contact binding post
5 Penta-contact binding post
6 Penta gear
7 Setscrews

8 Yoke
9 Hole for hanger screw
10 Tailpiece
11 Balance weight
12 Shaft
13 Bucket-wheel hub
14 Bucket-wheel hub nut

15 Raising nut
16 Pivot bearing
17 Pivot
18 Pivot adjusting nut
19 Keeper screw for pivot adjusting nut
20 Bearing lug
21 Bucket wheel

(a)

Figure 6.5 (a) Main parts of type AA current meter (from Buchanan and Somers, 1969).

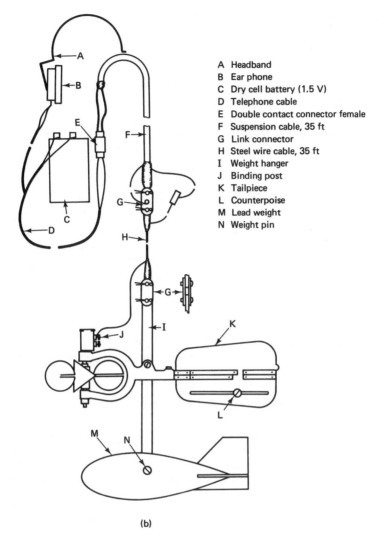

A	Headband
B	Ear phone
C	Dry cell battery (1.5 V)
D	Telephone cable
E	Double contact connector female
F	Suspension cable, 35 ft
G	Link connector
H	Steel wire cable, 35 ft
I	Weight hanger
J	Binding post
K	Tailpiece
L	Counterpoise
M	Lead weight
N	Weight pin

(b)

Figure 6.5 (b) assembly of a type AA current meter (courtesy of Geophysical Instrument and Supply Co.)

The U.S. Geological Survey has developed an optical current meter which is a stroboscopic device used to measure surface velocities at the time of floods without immersing the instrument.

6.7.1 Procedures of Current Meter Measurement

Measurements by current meter are classified as follows in terms of the procedure used to cross a stream during the measurement.

1. By wading
2. From a bridge

3. From a cableway
4. By boat
5. Over ice cover

In the wading procedure, measurements are made by entering the stream. The method is thus applicable to shallow depths up to 4 ft and velocities of less than 3 to 4 ft/sec.

Either a handline or a sounding reel supported by a bridge board or a portable crane is used to suspend the meter and the sounding weight. The size of the sounding weight should be greater than the maximum product of velocity and depth in the cross section.

Cableway measurement is superior to bridge measurement because there is no obstruction of the flow passage, but it involves more initial and operational expenses. The sounding reel carrying the meter and the weight is attached to the cable car.

In deep rivers, where no cableways or suitable bridges are available, the measurement is made by boat. A tag line is first stretched across the section. The tag line serves the dual purpose of holding the boat in position during the measurement and measuring of the width of the river.

For measurement under an ice cover, the most desirable section is just upstream from a riffle because the ice cover is thickest there. At least 20 holes are cut across the section using an ice drill. The effective depth of the water is the total depth minus the depth of the ice cover. A meter with vanes is preferred because the vanes do not become filled with slush ice.

6.7.2 Velocity Distribution in a Stream Section

The velocity in a stream section is not uniformly distributed, due to the presence of the free surface and friction along the stream wall. It varies both across the width and along the depth. Figure 6.6 indicates the general pattern of velocity distribution in a stream channel. The maximum velocity usually occurs below the free surface near

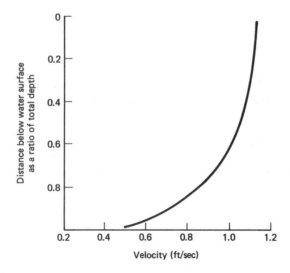

Figure 6.6 Typical vertical-velocity curve.

the center of the channel section. The velocity decreases toward the banks. Also, the closer to the banks, the deeper is the point of highest velocity in a vertical section. The factors that affect the velocity distribution are the shape of the section, the roughness of the channel, and the presence of bends. The surface wind has very little effect. A spiral type of motion has been observed in laboratory investigations. In natural rivers, the spiral motion is usually so weak that its effect is practically eliminated by the channel friction (Chow, 1959).

The problem of the horizontal variation of velocity is resolved by dividing the width of the river into a number of segments while performing the velocity measurements. The vertical variation has, however, to be considered at each segment. The vertical velocity distribution is based on the concept of the boundary layer* theory because analogies have been found between turbulent boundary-layer flow and turbulent pipe and channel flows. The turbulent channel flow can be visualized as a turbulent layer that has become as thick as the depth of flow.

6.7.3 Boundary Layer Theory

On a flat plate placed inside a fluid along the direction of flow, a laminar boundary layer develops which grows in the downstream direction and becomes a turbulent boundary layer. The turbulent boundary layer has three zones of (1) laminar sublayer, (2) zone of logarithmic velocity distribution, and (3) zone of velocity defect law, as illustrated in Figure 6.7. The logarithmic velocity distribution law has the widest application since it covers a major portion of the boundary layer, overlaps into the velocity defect law, and because flows broadly agree to this law. This law, derived from Prandtl's mixing-length theory (see Robertson and Crowe, 1980, pp. 331–333), has the following form for a flat plate:

$$\frac{v}{\sqrt{\tau_0/\rho}} = \frac{1}{K} \ln \frac{y\sqrt{\tau_0/\rho}}{\nu} + C \qquad [\text{dimensionless}] \qquad (6.3)$$

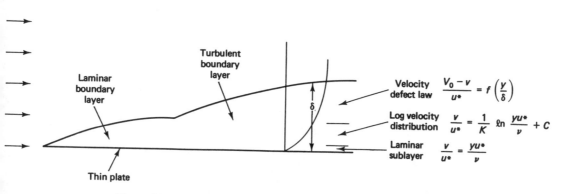

Figure 6.7 Velocity distribution in turbulent boundary layer on a flat plate.

*A boundary layer is a region next to the boundary of an object in which the fluid velocity is diminished because of the shear resistance created by the boundary.

where

$$v = \text{velocity at any distance } y \text{ from the plate}$$
$$\tau_0 = \text{shear stress at the plate (wall)}$$
$$\rho, \nu = \text{density and kinematic viscosity of fluid}$$
$$K = \text{von Kármán universal turbulent constant, } \approx 0.4$$
$$C = \text{a constant}$$

Nikuradse's experiments and numerous other tests on rough pipes confirmed the validity of the logarithmic velocity law.

For pipes, the von Kármán logarithmic velocity distribution has the following form, which is a direct derivation from eq. (6.3) after substitution of V_{max} at r_0 and eliminating constant C.

$$\frac{v - V_{max}}{\sqrt{\tau_0/\rho}} = \frac{1}{K} \ln \frac{y}{r_0} \qquad [\text{dimensionless}] \qquad (6.4)$$

where r_0 represents the radius of the pipe.

For open channels, by substituting, $\tau_0 = gds$ and $r_0 = d$ and manipulating eq. (6.4), Vanoni (1941) obtained the following relation:

$$v = \overline{V} + \frac{1}{K} \sqrt{gds} \left(1 + \ln \frac{y}{d} \right) \qquad [\text{LT}^{-1}] \qquad (6.5)$$

where

$$\overline{V} = \text{average velocity}$$
$$d = \text{depth of flow}$$
$$s = \text{slope of channel}$$

Analyses have shown that the power law equation of the following form provides a result similar to the logarithmic distribution law that conforms to the experimental data very closely in the boundary layer, as well as pipe and channel flows. Also, it is more convenient to apply. An extensive study by Dickinson (1967) led to the consensus that the distribution in streams fit the parabolic curve given by the power law.

$$v = V_0 \left(\frac{y}{a} \right)^{1/m} \qquad [\text{LT}^{-1}] \qquad (6.6)$$

where

$$v = \text{velocity at a distance } y \text{ from the bed}$$
$$V_0 = \text{a known velocity at a distance } a \text{ from the bed}$$
$$m = \text{a constant that varies from 6 to 10 depending on the Reynolds}$$
$$\text{number (Daily and Harleman, 1966); usually, } m = 7$$

Integrating eq. (6.6), for average velocity $\overline{V}$,

$$\overline{V} = \frac{1}{d} \int_0^d v \, dy \qquad (a)$$

or

$$\overline{V} = \frac{1}{d} \int_0^d V_0 \left(\frac{y}{a}\right)^{1/m} dy \qquad \text{(b)}$$

or

$$\overline{V} = \frac{1}{d} \frac{am}{m+1} V_0 \left[\left(\frac{y}{a}\right)^{1+1/m}\right]_0^d \qquad \text{(c)}$$

or

$$\overline{V} = \frac{m}{m+1} V_0 \left(\frac{d}{a}\right)^{1/m} \qquad [LT^{-1}] \qquad \text{(6.7)}$$

Suppose that the mean velocity occurs at a distance Z from the bottom.

Making $v = \overline{V}$ in eq. (6.6) by substituting $y = Z$ and equating to eq. (6.7), we obtain

$$V_0 \left(\frac{Z}{a}\right)^{1/m} = \frac{m}{m+1} V_0 \left(\frac{d}{a}\right)^{1/m} \qquad \text{(d)}$$

or

$$Z = \left(\frac{m}{m+1}\right)^m d \qquad [L] \qquad \text{(6.8)}$$

For values of m between 6 and 10, eq. (6.8) provides Z to be approximately equal to ·0.4d (i.e., the average velocity occurs at 0.6 depth below the surface).

If $V_{0.2}$ is the velocity at $y = 0.2d$ and $V_{0.8}$ is the velocity at $y = 0.8d$,

$$V_m = \frac{1}{2}\left(V_{0.2} + V_{0.8}\right) \qquad \text{(e)}$$

From eq. (6.6),

$$V_m = \frac{1}{2}\left[V_0 \left(\frac{0.2d}{a}\right)^{1/m} + V_0 \left(\frac{0.8d}{a}\right)^{1/m}\right] \qquad \text{(f)}$$

For $m = 7$,

$$V_m = 0.88 V_0 \left(\frac{d}{a}\right)^{1/7} \qquad [LT^{-1}] \qquad \text{(6.9)}$$

For $m = 7$, eq. (6.7) gives $\overline{V} = 0.88V_0(d/a)^{1/7}$, which is equal to V_m of eq. (6.9). Thus the mean of 0.2-depth and 0.8-depth velocities is equal to the average velocity.

6.7.4 Measurement of Velocity

The current meter or any other instrument measures velocity at a point, whereas a mean value of the velocity in a vertical is required to evaluate the discharge. The mean velocity in a vertical is obtained from the point velocity measurements by one of the following 10 methods:

1. Two-point method
2. Six-tenths-depth method

3. Vertical-velocity curve method
4. Integrated measurement method
5. Three-point method
6. Five-point method
7. Six-point method
8. Two-tenths-depth method
9. Subsurface-velocity method
10. Surface-velocity method

The first two methods are common. As proved in the preceding section, the velocity at 0.6 depth or mean of 0.2 and 0.8 depths is the average velocity by the logarithmic distribution and power laws. A field study by Savini and Bodhaine (1971) indicated that the average velocity determined by both one-point and two-point methods differed from 10-point measurement by 0.7%. The two-point method is slightly better, but it is not used where the depth is less than 2.5 ft. An indication as to whether the two-point method is adequate is derived from two conditions; the 0.2-depth velocity should be greater than the 0.8-depth velocity, and the 0.2-depth velocity should be less than twice the 0.8-depth velocity.

Although there is a striking similarity between observed velocity distribution and the logarithmic (and power) law, the actual distribution in an open channel is not strictly logarithmic. According to the logarithmic (and power) law, the maximum velocity should occur at the surface, which is not the actual case. In natural streams there is further deviation from the theoretical distribution. For this reason, methods 3 through 7 involving observations at a larger number of points are used. Methods 7 through 10 are employed in special circumstances.

In the vertical-velocity curve method, a number of velocity observations are made at points well distributed between the water surface and the streambed at each vertical. A plot is made between observed velocities and observation depths as a ratio of total depth, as shown in Figure 6.6. A graphic integration is carried out by measuring the area between the curve and the ordinate axis. The mean velocity is obtained from dividing the area by the length of the ordinate axis. The arithmetic integration (summation) is also a very convenient way to obtain the mean velocity. These procedures on the velocity curve method are demonstrated in Examples 6.2 through 6.5.

In the integrated measurement method the current meter is lowered to the bed and raised to the surface at a uniform rate. The measurement of velocity thus obtained represents the mean velocity for the section. The vertical-axis current meter is not used in this method.

The three-point method combines the two-point and six-tenths-depth methods. In the two-tenths depth method the velocity is observed at 0.2 of the depth below the surface and a coefficient is applied to the velocity observed. The USGS studies determined a coefficient of 0.87. In the subsurface velocity method observations are made at some arbitrary distance below the surface when it is not possible to obtain the depths with reliability at very high flow conditions. The coefficients are necessary to convert these to the mean velocity. To determine the coefficients, depths of measurement, as compared to total depth, are estimated after the stage has receded. In conditions of very high flow (i.e., floods), the surface velocity method is preferred over subsurface velocity if an optical current meter is available. A coefficient between 0.85 and 0.90 is used to compute the mean velocity. For smoother sections a value toward the upper limit of 0.9 is applied.

Example 6.2

The water velocity in a stream channel has a distribution across a vertical section given by $v = 2(4 - y)^{1/7}$, where v is the velocity in f/s and y is the distance from the water surface. The water depth at this section is 4 ft. (a) Determine the mean velocity across the vertical section. (b) Also, if the vertical section represents the average condition for the channel of rectangular shape of 50 ft width, determine the discharge in the channel.

Solution   The velocity profile is shown in Figure 6.6.

(a) $\overline{V} = \dfrac{1}{4} \displaystyle\int_0^4 v\, dy$

$\phantom{(a) \overline{V}} = \dfrac{1}{4} \displaystyle\int_0^4 2(4 - y)^{1/7}\, dy$

$\phantom{(a) \overline{V}} = -\dfrac{1}{4}\,(2) \left(\dfrac{7}{8}\right) [(4 - y)^{8/7}]_0^4$

$\phantom{(a) \overline{V}} = 2.13 \text{ ft/s}$

(b) $Q = A\overline{V} = (50)(4)(2.13)$

$ = 426.7 \text{ cfs}$

Example 6.3

The vertical-velocity distribution in a channel is given by $v = 2y^{1/2}$, where y is the distance from the bottom. Determine the mean velocity by the graphic integration.

Solution   The velocities for various depths are computed below and plotted in Figure 6.8.

Depth from Bottom (ft)	Depth/Total Depth	Velocity, $v = 2y^{1/2}$ (ft/sec)
1	0.25	2
2	0.50	2.83
3	0.75	3.46
4	1.0	4.0

x scale: 10 divisions = 1 ft/sec

y scale: 10 divisions = 0.2

100 squares* = 0.2 ft/sec

1 square = 0.002 ft/sec

Area covered by the curve = 1340 squares

or $\overline{V} = 1340 \times 0.002$

$= 2.68 \text{ ft/sec}$

Example 6.4

Solve Example 6.3 by the algebraic summation.

*A small square formed by one division on x and y scales.

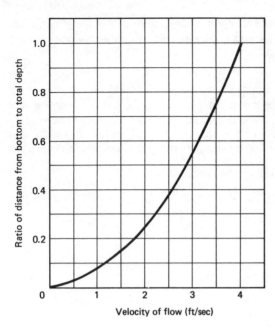

Figure 6.8 Plot of velocity versus depth for Example 6.3.

Solution Either from the velocity equation of Example 6.3 or from the plot in Fig. 6.8, the velocities corresponding to different values of depths are as follows:

1 Depth (ft)	2 Velocity (ft/sec)	3 Mean V (ft/sec)	4 Area of Curve, (ft²/s) (Depth in col. 1 × col. 3)
0	0		
0.5	1.41	0.71	0.36
1.0	2	1.70	0.85
1.5	2.45	2.22	1.11
2.0	2.83	2.64	1.32
2.5	3.16	3.0	1.50
3.0	3.46	3.31	1.66
3.5	3.74	3.60	1.80
4.0	4.0	3.87	1.94
Total			10.54

$$\overline{V} = \frac{1}{4}(10.54) = 2.63 \text{ ft/sec}$$

Example 6.5

The following point-velocity observations were made in a vertical section of a stream channel. Determine the mean velocity by various method and compare their results. The total depth of flow is 4 m.

Ratio of Observation to Total Depth	Velocity (m/s)
0.05	0.36
0.2	0.35
0.4	0.34
0.6	0.32
0.8	0.28
0.95	0.20

Solution

1. Vertical-velocity curve method:

Depth versus velocity data are plotted in Figure 6.9.

$$x \text{ scale:} \quad 10 \text{ divisions} = 0.04 \text{ m/s}$$
$$y \text{ scale:} \quad 10 \text{ divisions} = 0.2$$
$$100 \text{ squares} = 0.008 \text{ m/s}$$
$$1 \text{ square} = 8 \times 10^{-5} \text{ m/s}$$
$$\text{Area under the plot} = 1420 \text{ squares}$$
$$\text{or} \quad (1420)(8 \times 10^{-5}) = 0.114 \text{ m/s}$$
$$\text{Mean velocity} = \text{datum} + 0.114$$
$$\overline{V} = 0.2 + 0.114 = 0.314 \text{ m/s}$$

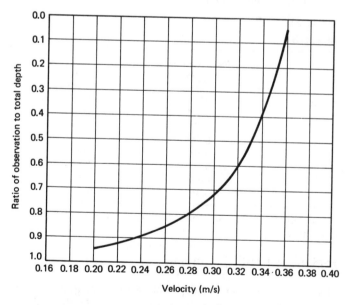

Figure 6.9 Vertical-velocity profile for Example 6.5.

2. Two-point method:

$$\overline{V} = 0.5(V_{0.2} + V_{0.8}) = 0.5(0.35 + 0.28) = 0.315 \text{ m/s}$$

3. Six-tenths depth method:

$$\overline{V} = V_{0.6} = 0.32 \text{ m/s}$$

4. Three-point method:

$$\overline{V} = 0.25(V_{0.2} + 2V_{0.6} + V_{0.8})$$
$$= 0.25[0.35 + 2(0.32) + 0.28] = 0.318 \text{ m/s}$$

5. 0.2-depth method:

$$V_{0.2} = 0.35$$
$$\overline{V} = (\text{coefficient})V_{0.2} = (0.87)(0.35) = 0.305 \text{ m/s}$$

6. Surface-velocity method:

$$\overline{V} = (\text{coefficient})V_{\text{surface}} = (0.85)(0.36) = 0.306 \text{ m/s}$$

7. Five-point method:

$$\overline{V} = 0.1(V_{\text{surf}} + 3V_{0.2} + 3V_{0.6} + 2V_{0.8} + V_{\text{bed}})$$
$$= 0.1[0.36 + 3(0.35) + 3(0.32) + 2(.28) + 0.20] = 0.313 \text{ m/s}$$

8. Six-point method:

$$\overline{V} = 0.1(V_{\text{surf}} + 2V_{0.2} + 2V_{0.4} + 2V_{0.6} + 2V_{0.8} + V_{\text{bed}})$$
$$= 0.1[0.36 + 2(0.35) + 2(0.34) + 2(0.32) + 2(0.28) + 0.20]$$
$$= 0.314 \text{ m/s}$$

SUMMARY OF RESULTS

Method	Mean Velocity (m/s)	Deviation	Error (%)
1. Vertical-velocity curve	0.314	0	0
2. Two-point	0.315	0.001	0.3
3. Six-tenths	0.320	0.006	1.9
4. Three-point	0.318	0.004	1.3
5. 0.2-depth	0.305	0.009	2.9
6. Surface-velocity	0.306	0.008	2.5
7. Five-point	0.313	0.001	0.3
8. Six-point	0.314	0	0

Thus, the vertical-velocity curve, six-point, five-point, and two-point methods provide better results in this case.

6.7.5 Measurement of Depth (Sounding)

Along with velocity, measurement of depth at vertical sections is required to compute discharge. The following four methods are used for depth measurement:

1. Wading rod
2. Sounding weight suspended by a hand line

3. Sounding weight suspended by a reel line
4. Sonic sounder

Wading rod.   This is a graduated steel rod of hexagonal or round shape, $\frac{1}{2}$ in. in size. The rod is placed in the stream so that the base plate rests on the streambed and the depth of water is read on the graduated rod. The current meter can be set at a desired position of 0.2, 0.6, or 0.8 depth.

Weight with a hand line.   When it is not possible to use a wading rod due to deep or swift water, a sounding weight is suspended below the current meter. The assembly is attached to a cable and is used from a bridge, boat, or cableway to perform measurements. The weights are streamlined to a bomb shape.

Weight with a reel line.   For high-water measurements requiring heavier weights, a sounding reel is used. It has a drum for winding the sounding cable, a crank and ratchet assembly for lowering, raising, and holding the current meter and weight assembly, and a depth indicator.

Air Correction for Depth.   The position that a sounding line will take is shown in Figure 6.10. The air correction is *de* and from trigonometry given by

$$de = \left(\frac{1 - \cos\theta}{\cos\theta}\right)ab \qquad [\text{L}] \tag{6.10}$$

The corrections from eq. (6.10) as a percent of vertical depth *ab* are given in Table 6.1 for various values of θ.

Wet-Line Correction for Depth.   Below the water surface, the tangent at any point of the cable is equal to the total horizontal force (of the current) divided by the total vertical force (of the sounding weight) at that point. This provides the value of the angle that the cable makes at any point below the water surface. These angles

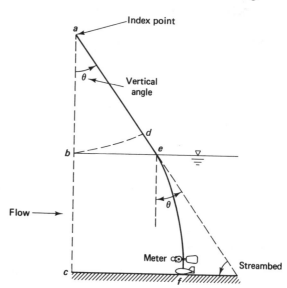

Figure 6.10   Deflection of current meter cable in deep, swift water.

TABLE 6.1 AIR CORRECTION

Vertical Angle (deg)	Correction (%)	Vertical Angle (deg)	Correction (%)
4	0.24	18	5.15
6	0.55	20	6.42
8	0.98	22	7.85
10	1.54	24	9.46
12	2.23	26	11.26
14	3.06	28	13.26
16	4.03	30	15.47

TABLE 6.2 WET-LINE CORRECTION

Vertical Angle (deg)	Correction (%)	Vertical Angle (deg)	Correction (%)
4	0.06	18	1.64
6	0.16	20	2.04
8	0.32	22	2.48
10	0.50	24	2.96
12	0.72	26	3.50
14	0.98	28	4.08
16	1.28	30	4.72

for incremental depths are computed and using the relation of eq. (6.10), corrections are computed. The summation of these provides total wet-line correction, which has been tabulated as a function of wet-line depth *ef* in Table 6.2 for various values of θ.

The following procedure is followed to apply this correction:

1. Depth, *aef*, is measured by the sounding line.
2. Measure the vertical distance, *ab*, by taking the reading when the weight is placed at the water surface. Determine the air correction from Table 6.1 for *ab*.
3. Wet-line depth, *ef* = *aef* − (*ab* + air correction). Determine wet-line correction from Table 6.2 for *ef*.
4. The corrected *bc* is computed to be

$$bc = aef - ab - (\text{air correction} + \text{wet-line correction}) \quad [\text{L}] \quad (6.11)$$

5. To position the current meter at 0.2 depth: For 0.2 depth, the wet-line curvature is neglected.

$$\begin{pmatrix} \text{vertical distance} \\ \text{to 0.2 depth} \end{pmatrix} = ab + 0.2bc + \begin{pmatrix} \text{distance from the} \\ \text{bottom of weight to} \\ \text{current meter} \end{pmatrix} \quad [\text{L}] \quad (6.12a)$$

$$(\text{corrected 0.2 depth}) = \begin{pmatrix} \text{vertical distance} \\ \text{to 0.2 depth} \end{pmatrix} + \begin{pmatrix} \text{air correction} \\ \text{for vertical} \\ \text{distance to} \\ \text{0.2 depth} \end{pmatrix} \quad [\text{L}] \quad (6.12b)$$

6. To position the meter at 0.6- or 0.8-depth:

$$\begin{pmatrix} \text{corrected} \\ \text{0.8 depth} \end{pmatrix} = aef - 0.2\left(bc + \begin{matrix} \text{wet-line} \\ \text{correction} \end{matrix} \right) + \begin{pmatrix} \text{distance from} \\ \text{weight to the} \\ \text{current meter} \end{pmatrix} \quad [L] \quad (6.13)$$

Equations (6.12) and (6.13) are used for placing the current meter at 0.2 and 0.8 depths in cable-suspended measurements even when air and wet-line corrections are not involved. In such cases the correction terms are treated as being equal to zero.

Example 6.6

In gaging a deep, swift stream through a cableway, the total depth of the sound line was found to be 25.2 ft. The depth from the guide pulley to the surface was measured to be 10.3 ft. The protractor measured the vertical angle of 24°. The weight hanger separates the current meter from the weight by 1 ft. Determine the

(a) true depth of the water,
(b) position for 0.2 depth, and
(c) 0.8 depth of the current meter.

Solution Refer to Figure 6.10.

(a) $aef = 25.2$ ft
$ab = 10.3$

From Table 6.1, for 24° and ab of 10.3,

$$\text{air correction} = \frac{9.46}{100}(10.3) = 0.974 \text{ ft}$$

wet-line depth, $ef = aef - ab - \text{air correction}$

$$= 25.2 - 10.3 - 0.974 = 13.93 \text{ ft}$$

From Table 6.2, for 24° and ef of 13.93,

$$\text{wet-line correction} = \frac{2.96}{100}(13.93) = 0.412 \text{ ft}$$

corrected depth, $bc = aef - ab - (\text{air correction} + \text{wet-line correction})$

$$= 25.2 - 10.3 - (0.974 + 0.412) = 13.51 \text{ ft}$$

(b) Position of 0.2 depth

$$\text{vertical distance} = 10.3 + 0.2(13.51) + 1 = 14.00$$

$$\text{air correction for 14-ft depth} = \frac{9.46}{100} \times 14.0 = 1.32$$

$$0.2 \text{ depth} = 14.00 + 1.32 = 15.32 \text{ ft}$$

Position of 0.8 depth:

$$0.8 \text{ depth} = 25.2 - 0.2(13.51 + 0.412) + 1 = 23.41 \text{ ft.}$$

Example 6.7

A stream is gaged using a hand line from a bridge. The total depth of the sound line from the rail of the bridge is measured to be 8.25 m. The depth up to the water surface is 4.4 m. If the distance from the center of the current meter to the bottom of the weight is 0.3 m, determine the position where the current meter is to be placed for 0.2 depth and 0.8 depth, respectively.

Solution With the hand line no air and wet-line corrections are involved. Depth of water = 8.25 − 4.4 = 3.85 m.

From eq. (6.12a),

$$0.2 \text{ depth} = ab + 0.2bc + \text{distance from meter to weight}$$

$$= 4.4 + 0.2(3.85) + 0.3 = 5.47 \text{ m}$$

From eq. (6.13)

$$0.8 \text{ depth} = aef - 0.2b + \text{distance of meter to weight}$$

$$= 8.25 - 0.2(3.85) + 0.3 = 7.78 \text{ m}$$

Sonic Sounder. Based on the principle of echo sounding, a sonic sounder provides a continuous strip-chart record of the depth of the stream. The portable sounder works on a 6- or 12-V storage battery. Its transducer releases pulses of ultrasonic energy at fixed intervals. The instrument measures the time taken by these pulses of energy to travel to the streambed, to be reflected, and to return to the transducer. With a known velocity of sound in water, the instrument computes and records the depth.

6.7.6 Computation of Discharge

Measurements of velocity and depth, made at a number of locations across a stream channel, are used to compute discharge by summing up the product of mean velocity and area of cross section of the segment between successive locations. Usually, between 20 and 30 verticals of equidistant or variable spacings are used to divide a stream width. These spacings should be arranged so that no segment contains more than 10% of the total flow. Depending on the procedure used to obtain the multiplication of velocity and area of various elements constituting the channel section, methods are known as midsection, mean-section, velocity-depth integration, and velocity-contour methods of discharge computation. The first two methods are arithmetic summation procedures and the last two are graphic methods. Midsection is a preferred method.

Midsection method. In this method it is assumed that the velocity at each vertical represents a mean velocity for a section that extends half the distance into the preceding and following segments, as shown in Figure 6.11.

$$\text{Area for subsection 3} = \frac{W_2 + W_3}{2} d_3 \tag{a}$$

$$\text{Discharge through subsection 3} = \overline{V}_3 \frac{W_2 + W_3}{2} d_3 \tag{b}$$

$$\text{Discharge through subsection } x = \overline{V}_x \frac{W_{x-1} + W_x}{2} dx \qquad [L^3T^{-1}] \tag{6.14}$$

When the cross section is such that there is a depth at the edge of the water as at the last vertical in Fig. 6.11, the velocity is estimated as a certain percentage (between 65 and 90%) of the adjacent vertical because it is not possible to measure ve-

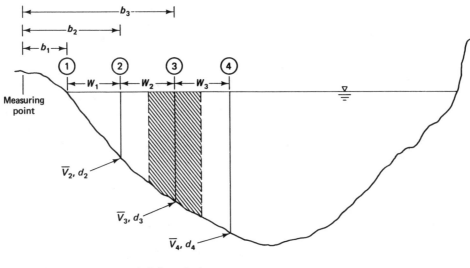

1, 2, 3, . . . Stations

$b_1, b_2, b_3, . . .$ Distance from the initial point to the station (observation verticals)

$d_1, d_2, d_3, . . .$ Depth of water at the observation verticals

$W_1, W_2, W_3, . . .$ Width between successive verticals

Figure 6.11 Subsection in the midsection method (enlarged subsection has been shown).

locity by the current meter. At the beginning section, in using eq. (6.14), W_0 will have no significance and should be dropped. Example 6.8 provides the data in the format as they are recorded in the field book while taking measurments.

Example 6.8

Compute the discharge by midsection method for the following measurement data. The current-meter rating is given by $v = 0.1 + 2.2N$, where v is velocity in ft/sec and N is the number of revolutions per second.

Distance from Initial Point (ft)	Depth (ft)	Observed Depth	Revolutions	Time (sec)
10	1			
12	3.5	0.2	35	50
		0.8	22	50
14	5.2	0.2	40	60
		0.8	30	55
17	6.3	0.2	45	60
		0.8	30	55
19	4.4	0.2	33	45
		0.8	30	50
21	2.2	0.6	22	50
23	0.8	0.6	10	45
25	0			

Solution The computations are shown in Table 6.3.

Width (col. 8) is the successive difference of col. 1.

Effective width (col. 9) is the average of preceding and following widths in col. 8.

Area (col. 10) = col. 2 × col. 9.

Discharge (col. 11) = col. 7 × col. 10.

Note that the conditions of a minimum 20 verticals and less than 10% flow in any subsection are violated to reduce computation. Also for the first vertical, having a depth of 1 ft, a velocity of 0.65 times the adjacent velocity has been taken. For the last vertical, this value is taken as zero since there is no water depth.

TABLE 6.3 COMPUTATION OF DISCHARGE BY MIDSECTION METHOD (EXAMPLE 6.8)

(1)	(2)	(3)	(4)	(5)	(6)	(7)	(8)	(9)	(10)	(11)
					velocity					
Distance from Initial Point	Depth (ft)	Observed Depth	Revolutions	Time (sec)	At points	Mean in Section	Width (ft)	Effective Width (ft)	Area (ft^2)	Discharge (ft^3/sec)
10	1					0.88[a]		1	1.0	0.88
							2			
12	3.5	0.2	35	50	1.64	1.36		2	7.0	9.52
		0.8	22	50	1.07					
							2			
14	5.2	0.2	40	60	1.57	1.44		2.5	13.0	18.72
		0.8	30	55	1.30					
							3			
17	6.3	0.2	45	60	1.75	1.53		2.5	15.75	24.10
		0.8	30	55	1.30					
							2			
19	4.4	0.2	33	45	1.71	1.57		2	8.8	13.82
		0.8	30	50	1.42					
							2			
21	2.2	0.6	22	50	1.07	1.07		2	4.4	4.71
							2			
23	0.8	0.6	10	45	0.59	0.59		2	1.6	0.94
							2			
25	0					0		1	0	0
Total									51.55	72.69

[a]0.65 × 1.36 of subsequent vertical = 0.88

Mean-section method. The segment area (subsection) extends from vertical to vertical as shown in Figure 6.12.

$$\text{Area for subsection 3–4} = \frac{d_3 + d_4}{2} W_3 \tag{c}$$

$$\text{Discharge through subsection 3–4} = \left(\frac{\overline{V}_3 + \overline{V}_4}{2}\right)\left(\frac{d_3 + d_4}{2}\right) W_3 \tag{d}$$

$$\text{Discharge through subsection } x \text{ and } x+1 = \left(\frac{\overline{V}_x + \overline{V}_{x+1}}{2}\right)\left(\frac{d_x + d_{x+1}}{2}\right) W_x \quad [L^3T^{-1}] \tag{6.15}$$

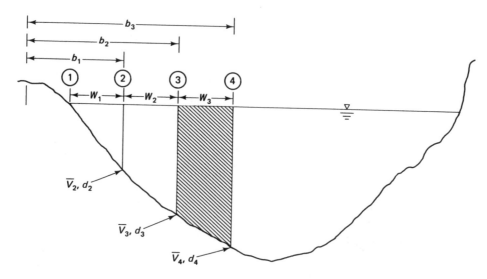

Figure 6.12 Subsection in the mean-section method.

Example 6.9

Solve Example 6.8 by the mean-section method.

Solution   Computations are arranged in Table 6.4.
Average velocity (col. 8) is the average of mean velocities of two verticals in col. 7.
Average depth (col. 9) is the average of depths of two verticals in col. 2.
Width (col. 10) is the successive difference of col. 1.
Area (col. 11) = col. 9 × col. 10.
Discharge = col. 8 × col. 11.

Velocity–depth integration method.   This is a graphic method in which velocity measurements in each vertical should preferably be performed at a number of depths to plot the vertical-velocity curve for each vertical. The procedure is as follows:

1. Draw the vertical-velocity curve for each vertical and determine the area under this curve that will represent (velocity × depth) value at each vertical. (If the mean velocity in a vertical has been determined by any other method, it can be multiplied by the vertical depth.)
2. Plot (velocity × depth) values at the location of respective verticals across the stream cross section as shown in Figure 6.13. Draw a smooth curve through these points. The area enclosed by this curve will provide the discharge.

Example 6.10

The velocity–depth values for various verticals of a stream cross section as obtained from vertical-velocity curve analyses are indicated in Figure 6.14. Determine the discharge of stream by the velocity–depth integration method.

Solution   The velocity–depth versus distance has been plotted in Figure 6.15.

Scale factor: *x*-scale: 10 divisions = 10 ft

 y-scale: 10 divisions = 2 ft^2/sec

TABLE 6.4 COMPUTATION OF DISCHARGE BY MEAN-SECTION METHOD (EXAMPLE 6.9)

(1) Distance from Initial Point (ft)	(2) Depth (ft)	(3) Observed Depth	(4) Revolutions	(5) Time (sec)	(6) Velocity At points	(7) Velocity Mean in Vertical	(8) Average Velocity for Subsection (ft/sec)	(9) Average Depth for Subsection (ft)	(10) Width (ft)	(11) Area (ft²)	(12) Discharge (ft³/sec)
10	1					0.88					
							1.12	2.25	2	4.5	5.04
12	3.5	0.2 0.8	35 22	50 50	1.64 1.07	1.36					
							1.40	4.35	2	8.7	12.18
14	5.2	0.2 0.8	40 30	60 55	1.57 1.30	1.44					
							1.49	5.75	3	17.25	25.70
17	6.3	0.2 0.8	45 30	60 55	1.75 1.30	1.53					
							1.55	5.35	2	10.70	16.59
19	4.4	0.2 0.8	33 30	45 50	1.71 1.42	1.57					
							1.32	3.3	2	6.60	8.71
21	2.2	0.6	22	50	1.07	1.07					
							0.83	1.5	2	3.00	2.49
23	0.8	0.6	10	45	0.59	0.59					
							0.30	0.40	2	0.80	0.24
25	0				0	0					
Total										51.55	70.95

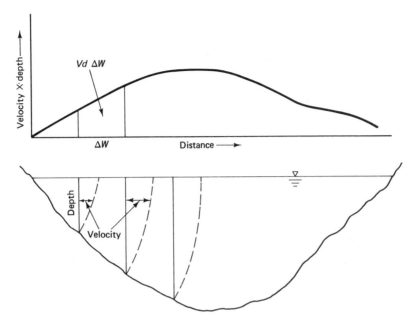

Figure 6.13 Graphic procedure of velocity–depth integration (subsections are enlarged).

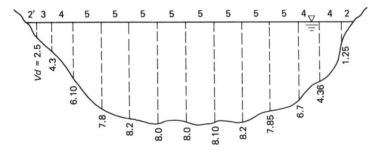

Figure 6.14 Results of vertical curve analysis for a stream cross section.

$$100 \text{ squares}^* = 20 \text{ ft}^3/\text{sec}$$
$$1 \text{ square} = 0.2 \text{ ft}^3/\text{sec}$$
$$\text{Area under the curve} = 1900 \text{ squares}$$
$$\text{Discharge} = 1900 \times 0.2 = 380 \text{ ft}^3/\text{sec}$$

Velocity-contour method. This is also a graphical method. Since velocity contours have to be drawn, velocity measurements at a number of points in each vertical are required for the application of this method. The procedure is as follows:

1. Draw the river cross section to a convenient scale. On each vertical, write the point velocity measurements. Connect points of equal velocity to draw the velocity contours (isovels) at intervals of 0.1 to 0.5 ft/sec (0.03 to 0.15 m/s), as shown in Figure 6.16(a).

*A small square formed by one division on x and y scales.

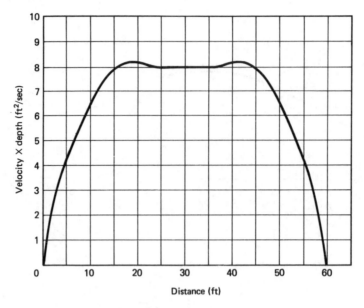

Figure 6.15 Plot of velocity times depth versus discharge for Example 6.10.

2. Starting from the highest value, determine the areas between successive velocity contours. Plot velocity against cumulated area up to that velocity contour, as shown in Figure 6.16(b). The area enclosed by the curve represents the total discharge.

Example 6.11

The velocity distribution (contours) in a river cross section has the pattern shown in Figure 6.16(a). The areas measured by a planimeter between successive contour lines are tabulated below. Determine the discharge at the site.

Contours (ft)	Area between Contours (ft^2)	Cumulated Area (ft^2)
>2.5	5.0	5.0
2.5–2.0	20.5	25.5
2 –1.5	25.2	50.7
1.5–1.0	16.3	67.0
1.0–0.5	10.8	77.8
0.5–0	4.2	82.0

Solution The lower limit of contour velocity and the corresponding cumulated area are plotted in Figure 6.17.

$$\text{Scale factor:} \quad x\text{-scale:} \quad 10 \text{ divisions} = 10 \text{ ft}^2$$

$$y\text{-scale:} \quad 10 \text{ divisions} = 0.5 \text{ ft/sec}$$

$$100 \text{ squares} = 5 \text{ ft}^3/\text{sec}$$

$$1 \text{ square} = 0.05 \text{ ft}^3/\text{sec}$$

$$\text{Area under the curve} = 2691 \text{ squares}$$

$$\text{Discharge} = 2691(0.05) = 134.6 \text{ ft}^3/\text{sec}$$

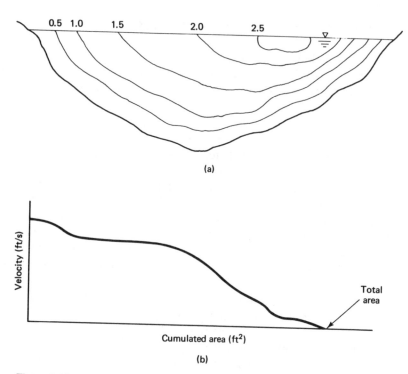

(a)

(b)

Figure 6.16 (a) Velocity distribution in a channel cross section; (b) plot of velocity versus area to compute discharge.

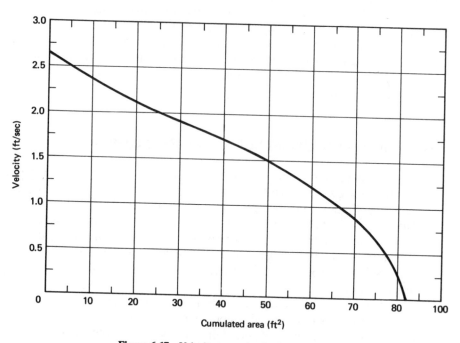

Figure 6.17 Velocity–area plot for Example 6.11.

6.8 MEASUREMENT OF DISCHARGE BY FLOATS

During high-flow conditions, when excessive velocities, depths, or floating drifts prohibit the use of a current meter, the float method is used to measure discharge. It is essentially observing the time required for a float to transverse a known length, so that

$$V = \frac{L}{t} \quad [LT^{-1}] \tag{6.16}$$

The three types of floats used are (1) surface floats, (2) subsurface floats with a submerged canister, and (3) rod floats.

Two cross sections are selected along a reach of a straight channel that are about four or five times the stream width apart. A number of floats are distributed uniformly across the stream width, and their position with reference to distance from the bank is posted. The floats are introduced a short distance upstream from the upper cross section so that they take up the speed of the current when they reach the upper cross section. They are introduced from a bridge or cableway, or tossed in from the bank. A boat is used for wide rivers.

The velocity of each float derived from eq. (6.16) is converted to the mean velocity in the vertical by a reduction coefficient, which is commonly 0.85 for surface float and 1.0 for subsurface and rod float.

At the time of measurement, water surface elevation is referenced to stakes along the bank at each cross section and also at one or more intermediate sites. At a later date these sections are surveyed to derive an average stream cross section to compute the area.

The computation of discharge is similar to that of current meter computation. The discharge in each subsection of the average cross section is obtained by multiplying the area of subsection by the mean velocity for that subsection.

When it is impractical to obtain proper float movement across the entire width or when floats tend to move toward the center of the flow, an unadjusted discharge is obtained based on the mean of surface velocity. A coefficient is applied to determine the stream discharge.

6.9 MEASUREMENT OF DISCHARGE BY TRACER DILUTION

This method is used when conditions are not favorable for current meter measurement. In rock-strewn shallow streams or rough channels carrying highly turbulent flow, the dilution technique provides an effective method of flow measurement. In this method, a tracer solution of known concentration, in known quantity, is injected into the stream to be diluted by the discharge of the stream. At a cross section downstream from the injection site, water is sampled to ascertain the concentration of the tracer solution. Applying the principle of conservation of mass, the discharge is determined directly.

There are two basic injection techniques, several sampling techniques, and a large number of tracers of chemical, fluorescent dye, and radioactive types. The injection techniques differ in terms of period of injection of the tracer solution into the stream. The two methods are described below.

6.9.1 Constant Injection Rate Method

A solution of concentration C_1 of a selected tracer is injected at a constant rate q at a beginning section. At a second section downstream, the concentration of the tracer is measured for a sufficient period of time at a number of points to ensure proper mixing and attaining of constant dilution. This has been shown in Figure 6.18. Assuming no concentration of the same tracer in natural flow:

$$\text{Mass rate at station 1} = qC_1$$

$$\text{Mass rate at station 2} = (Q + q)C_2$$

Equating two rates gives

$$qC_1 = (Q + q)C_2$$

or

$$Q = \frac{C_1 - C_2}{C_2}q$$

In general, C_1 is much greater than C_2; then

$$Q = \frac{C_1}{C_2}q \qquad [\text{L}^3\text{T}^{-1}] \tag{6.17}$$

where

q = rate of injection of tracer solution, ft^3/sec, m^3/s, or liters/s

Q = discharge of the stream, ft^3/sec, m^3/s, or liters/s

C_1 = concentration of tracer as injected, lb/ft^3, kg/m^3, or mg/liter

C_2 = diluted concentration at downstream, lb/ft^3, kg/m^3, or mg/liter

$\dfrac{C_1}{C_2} = N$ = dilution ratio

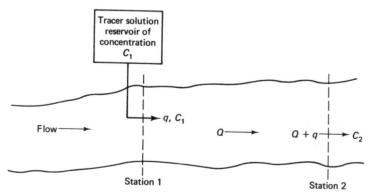

Figure 6.18 Constant rate injection of tracer solution.

Example 6.12

In tracer dilution measurement by the constant injection rate method, the following observations are made. Determine the streamflow.

1. Rate of injection of tracer = 2.5×10^{-4} ft^3/sec.
2. Concentration of injected solution = 1.8 g/liter.
3. Concentration at three sampling points evenly spaced.
   At point A, $C_2 = 5.42$ μg/liter (5.42×10^{-6} g/liter).
   At point B, $C_2 = 5.35$ μg/liter.
   At point C, $C_2 = 5.50$ μg/liter.

Solution Because the three points are spaced evenly, each is given equal weight in computing mean C_2:

$$C_2 = \frac{5.42 + 5.35 + 5.50}{3} = 5.42 \text{ μg/liter} \quad \text{or} \quad 5.42 \times 10^{-6} \text{ g/liter}$$

From eq. (6.17),

$$Q = \frac{C_1}{C_2} q$$

$$= \frac{1.8}{5.42 \times 10^{-6}} (2.5 \times 10^{-4}) = 83 \text{ ft}^3/\text{sec}$$

6.9.2 Sudden Injection Method

A volume V of a tracer solution of concentration C_1 is injected over a short period at the beginning of the section (station 1). At a second section downstream (station 2), the concentration of the tracer, C_i, is determined from time to time for a period sufficiently long enough to ensure that all of the tracer has passed through the second section.

$$\text{Mass of tracer at station 1} = VC_1$$

$$\text{Mass of tracer at station 2} = Q \int_0^\infty C_i \, dt$$

$$= Q \sum_{i=1}^N \frac{C_i(t_{i+1} - t_{i-1})}{2}$$

where N is the number of samples collected. Equating yields

$$Q = \frac{VC_1}{\displaystyle\sum_{i=1}^N C_i(t_{i+1} - t_{i-1})/2} \qquad [\text{L}^3\text{T}^{-1}] \qquad (6.18)$$

where

V = volume of injected solution

C_1 = concentration of injected tracer

C_i = measured tracer concentration d/s at a given time in sample i

i = sequence number of sample

N = total number of samples

t_i = time when sample C_i is obtained

Measurement of Surface Water Flow   Chap. 6

Example 6.13

Gaging by the sudden injection method of the tracer dilution technique provided the following data. Calculate the discharge.

1. Volume of tracer injected = 3.12 liters.
2. Concentration of injected tracer = 1.75 g/liter.
3. Concentrate of tracer in samples collected at various times:

Time (min)	Concentration (μg/liter)
0	0.0
2	5.0
3	12.1
4	11.5
5	9.5
6	5.2
•7	1.8
8	1.0
10	0.3
15	0.0

Solution

Sample	Concentration C_i (μg/liter)	t_i (min)	$(t_{i+1} - t_{i-1})/2$ (min)	$\dfrac{C_i(t_{i+1} - t_{i-1})}{2}$
1	0	0	1	0
2	5	2	1.5	7.5
3	12.1	3	1[a]	12.1
4	11.5	4	1	11.5
5	9.5	5	1	9.5
6	5.2	6	1	5.2
7	1.8	7	1	1.8
8	1.0	8	1.5	1.5
9	0.3	10	3.5	1.1
10	0	15		0
Total				50.2

[a]For example, for sample 3, $(t_4 - t_2)/2 = (4 - 2)/2 = 1$.

From eq. (6.18),

$$Q = \frac{VC_1}{\sum_{i=1}^{N} C_i(t_{i+1} - t_{i-1})/2}$$

$$= \frac{3.12(1.75)}{50.2 \times 10^{-6}} = 108{,}765 \text{ liters/min} \quad \text{or} \quad 1.81 \text{ m}^3/\text{s}$$

6.10 DISCHARGE MEASUREMENT BY THE ULTRASONIC (ACOUSTIC) METHOD

This method has been applied selectively to measure discharge of rivers, canals, penstocks, conduits, and tunnels. The method uses two transducers, two receivers, and a digital processor. The block diagram is shown in Figure 6.19. The transducers

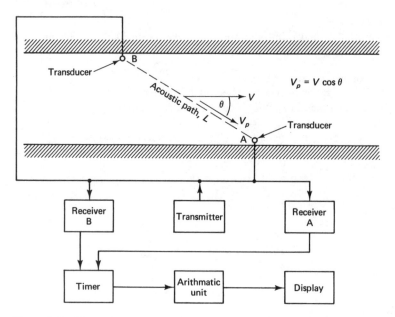

Figure 6.19 Ultrasonic method of discharge measurement (from Holmes, Whirlow and Wright, 1973).

are mounted on each bank in an oblique direction, as shown in the figure. Sound pulses sent by A are received by B, and in the opposite direction, pulses transmitted by B are received by A. Sound waves traveling downstream have a higher velocity than those traveling upstream, due to the stream velocity component parallel to the flight (acoustic) path. Since the stream velocity is much less than the sound velocity in water, the difference in upstream and downstream time is very small and needs to be recorded precisely.

The travel time downstream is $t_{BA} = L/(C + v_p)$ and that upstream is $t_{AB} = L/(C - v_p)$, C being the acoustic velocity in water. Since $v_p = v \cos \theta$, from the difference of time or difference of frequency (l/time) the following is derived, for the average velocity:

$$\overline{V} = \frac{L \, \Delta t}{2 t_{AB} t_{BA} \cos \theta} \qquad [LT^{-1}] \qquad (6.19)$$

When multiplied by the average depth of flow and channel width, this leads to the following discharge equations:

1. Travel-time difference method:

$$Q = \frac{L^2 \overline{d} \, \Delta t \, \tan \theta}{2 t_{AB} t_{BA}} \qquad [L^3 T^{-1}] \qquad (6.20a)$$

2. Frequency difference method:

$$Q = \frac{L^2}{2} \left(\frac{1}{t_{BA}} - \frac{1}{t_{AB}} \right) \overline{d} \, \tan \theta \qquad [L^3 T^{-1}] \qquad (6.20b)$$

where

$$L = \text{flight (path) length}$$
$$\theta = \text{angle of acoustic path to direction of flow}$$
$$\overline{d} = \text{average depth of flow along AB}$$
$$t_{AB} = \text{travel time from A to B}$$
$$t_{BA} = \text{travel time from B to A}$$
$$\Delta t = t_{AB} - t_{BA}$$

In a single-path system, measurement is made at one depth only with a pair of transducers set at 0.6 of the most frequently occurring depth. More common, however, is the multipath system, in which several pairs of transducers are installed at various water depths. The average velocity of each path when multiplied by spacing between paths (transducers), and the length of each, provides the total discharge. Holmes, et al. (1973) have presented the Gaussian quadrature method of computation of discharge with three or more paths of measurement of velocity.

6.11 DISCHARGE MEASUREMENT BY THE ELECTROMAGNETIC METHOD

This method is capable of measuring flow in weedy rivers and rivers with moving beds. However, it requires on-site calibration with a current meter. According to the theory of electromagnetic induction, the flowing water in a river cuts the vertical component of the earth's magnetic field and an electromotive force (emf) is induced in the water. This emf, sensed by electrodes at each bank, is directly proportional to the velocity of flow. The emf due to earth's magnetic field is, however, very feeble. To generate a measurable potential in the electrodes, a magnetic field is generated by means of coils buried in the river across its width.

Based on empirical tests, the following equation has been developed:

$$Q = K\left(\frac{E}{I}h\frac{r_w}{r_b}\right) \qquad [L^3T^{-1}] \qquad (6.21)$$

where

$$K = \text{a constant}$$
$$E = \text{voltage at electrodes, mV}$$
$$I = \text{coil current, A}$$
$$h = \text{depth of flow, m or ft}$$
$$r_w = \text{water resistivity, } \Omega \cdot \text{m or } \Omega\text{-ft}$$
$$r_b = \text{bed resistance, } \Omega$$

The value of the constant K is evaluated by calibration using current-meter measurements.

6.12 MEASUREMENTS THROUGH HYDRAULIC DEVICES

In relatively shallow rivers, small creeks, and open channels, certain devices can be installed to measure the discharge. These, comprising weirs, orifices, and flumes, may be permanently constructed structures across the river or may be portable devices. The measurement on these is based on the energy principle. Localized losses involved due to inertia and viscous effects are included in the form of a coefficient of discharge, which is preferably ascertained experimentally.

6.12.1 Weirs and Notches

A weir may be defined as a regular obstruction across a channel section over which flow takes place. It may be a vertical flat plate with a sharpened upper edge; then it is known as a sharp-crested weir or notch. It may have a solid broad section of concrete or other material; then it is known as a broad-crested weir. Weirs are classified according to their shapes [i.e., rectangular, triangular (V-notch), trapezoidal (Cippoletti), and parabolic]. The rectangular weir is the most popular. A rectangular section that spans the full width of the channel is known as a suppressed weir. If the width of the weir section is less than the width of the channel, it is a weir with end contraction. Where the downstream water level is lower than the crest, the weir is said to have a free discharge. If the downstream level is higher than the crest level, it is known as a drowned or submerged weir. All cases discussed in Sections 6.12.2 and 6.12.3 relate to free discharge. The submerged weir is discussed separately in Section 6.12.4.

6.12.2 Flow over Sharp-Crested Weir

The water flowing over a sharp-crested weir under free discharge conditions falls away from the downstream face of the weir. This forms a nappe, as shown in Figure 6.20. Air is trapped between the lower nappe surface and downstream face of the weir. Thus the underside of the jet or lower nappe is exposed to the atmospheric pressure. If means of restoring this air are not provided, the entrapped air will be carried away by the flowing water, creating a negative pressure. This can increase the discharge as much as 25% but can damage the structure. In a contracted weir, air is restored from the sides, and in a suppressed weir, through ventilation pipes.

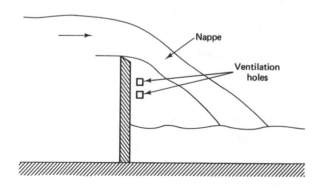

Figure 6.20 Flow over free-discharging sharp-crested weir.

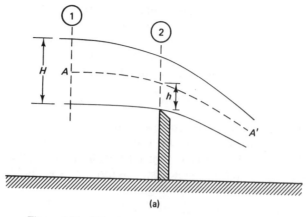

Figure 6.21   Thin plate weir: (a) free-discharging profile.

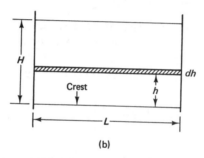

Figure 6.21   (b) weir section.

Rectangular sharp-crested suppressed weir.   Apply the energy equation at points 1 and 2 on the streamline AA' in Figure 6.21, and assume (after Weisbach) that the pressure at point 2 is atmospheric. Crest as a datum:

$$H + \frac{v_1^2}{2g} = h + \frac{v_2^2}{2g} \tag{a}$$

When the approach velocity, v_1, is neglected,

$$v_2 = \sqrt{2g(H - h)} \tag{b}$$

The flow through the elemental strip of area $L\,dh$ is

$$dQ = L\,dh\,\sqrt{2g(H - h)} \tag{c}$$

After introducing a discharge coefficient to account for the inertia and shear losses, the total discharge

$$Q = C_d\sqrt{2g}\,L\int_0^H \sqrt{H - h}\,dh \tag{d}$$

or

$$Q = \frac{2}{3}C_d\sqrt{2g}\,LH^{3/2} \qquad [\mathrm{L^3T^{-1}}] \tag{6.22}$$

The coefficient of discharge should preferably be determined experimentally. Empirical formulas by Hamilton–Smith, White, Rehbock, and many others have been proposed to determine the coefficient. When the head is not greater than one-third of the weir length, an approximate value of the discharge coefficient is taken as 0.6 to 0.62. This provides the following Francis formula in FPS units:

$$Q = 3.33LH^{3/2} \qquad [L^3T^{-1}] \tag{6.23}$$

Rectangular weir with end contractions. When the weir length is less than the width of the channel, it is known as the weir with end contractions. The effective length of weir in this case is less than the actual weir length, due to contraction of the flow jet caused by the sidewalls. The formula is given by

$$Q = \frac{2}{3}C_d\sqrt{2g}\,(L - 0.1nH)H^{3/2} \qquad [L^3T^{-1}] \tag{6.24}$$

where n is the number of end contractions; if both ends are contracted, $n = 2$.

Rectangular weir with velocity of approach. The effective head responsible for the discharge is the sum of water head and the head due to velocity of approach. If the latter is not very small, the limits of integration in eq. (d) should be modified between $v_0^2/2g$ and $(H + v_0^2/2g)$. The resulting equation is

$$Q = \frac{2}{3}C_d\sqrt{2g}(L - 0.1nH)\left[\left(H + \frac{v_0^2}{2g}\right)^{3/2} - \left(\frac{v_0^2}{2g}\right)^{3/2}\right] \qquad [L^3T^{-1}] \tag{6.25}$$

In eq. (6.25), v_0 cannot be found unless Q is known. Thus the equation is first solved by neglecting v_0. The approximate discharge thus determined is used to find v_0. A revised value of Q is determined using this v_0. The process is repeated until the final discharge is within 1% of the preceding discharge.

Example 6.14

An end-contracted weir of total length 286 ft and crest height 5 ft is used to discharge water from a tank without exceeding a head of 2.5 ft. The wasteweir carries piers that are 10 ft clear distance apart and 2 ft wide, to support a footway. Determine the discharge. $C_d = 0.6$.

Solution

1. Let N represent the number of weir sections.
2. Number of piers = $N - 1$.
3. Total length, $286 = 10N + 2(N - 1)$.
4. Thus, $N = 24$.
5. Number of end contractions = $2N = 2(24) = 48$.
6. From eq. (6.25), assuming that $v_0 = 0$,

$$Q = \frac{2}{3}(0.6)\sqrt{2(32.2)}\,[286 - 0.1(48)\,(2.5)][2.5]^{3/2}$$

$$= 3476.7 \text{ cfs.}$$

7. $v_0 = \dfrac{Q}{A} = \dfrac{3476.7}{(7.5)\,(286)} = 1.62 \text{ ft/sec.}$

8. $\dfrac{v_0^2}{2g} = \dfrac{(1.62)^2}{2(32.2)} = 0.041$ ft.

9. Revised $Q = \dfrac{2}{3}(0.6)\sqrt{2(32.2)}\,[286 - 0.1(48)(2.5)](2.541^{3/2} - 0.041^{3/2})$

$= 3555$ cfs.

Triangular (V-notch) weir. These are suitable for low discharges, because the head increases more rapidly on a triangular section. The area of elemental strip in Figure 6.22,

$$dA = b\,dh$$

or

$$dA = 2h\,\tan\frac{\theta}{2}\,dh \tag{a}$$

From the preceding section, at height h the velocity

$$v = \sqrt{2g(H - h)} \tag{b}$$

Discharge through the elemental area

$$dQ = C_d\!\left(2h\,\tan\frac{\theta}{2}\,dh\right)\sqrt{2g(H - h)} \tag{c}$$

Total discharge

$$Q = 2C_d\sqrt{2g}\,\tan\frac{\theta}{2}\int_0^H h\sqrt{H - h}\,dh \tag{d}$$

or

$$Q = \frac{8}{15}c_d\sqrt{2g}\,\tan\frac{\theta}{2}H^{5/2} \qquad [\mathrm{L^3T^{-1}}] \tag{6.26}$$

Ordinarily, V-notch weirs are not appreciably affected by the velocity of approach (U.S. Bureau of Reclamation, 1984).

Trapezoidal weir. The discharge is the sum of discharges over the rectangular section with end contractions and over the triangular section. A trapezoidal weir with a side slope of 1 horizontal to 4 vertical is known as a Cippoletti weir. The discharge through a Cippoletti weir is given by the Francis formula for a suppressed rectangular weir [eq. (6.23)], in which the coefficient is increased by about 1%:

$$Q = 3.367LH^{3/2} \qquad [\mathrm{L^3T^{-1}}] \tag{6.27}$$

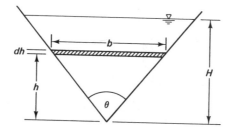

Figure 6.22 V-notch section.

6.12.3 Flow over Broad-Crested Weir

When the thickness of the crest of a weir is more than 0.47 times the head, it is classified as a broad-crested weir. As the stream of water flows over the broad crest, the head drops from H to h, due to the acceleration of water as a result of a sudden reduction of sectional area, as illustrated in Figure 6.23.

The acceleration raises the discharge, which attains a maximum value at $h = \frac{2}{3}H$ (precisely two-thirds of the energy head) when the flow is critical.

Applying the energy equation at points 1 and 2, we obtain

$$H = h + \frac{v^2}{2g} \tag{a}$$

or

$$v = \sqrt{2g(H - h)} \tag{b}$$

and

$$Q = C_d Lh\sqrt{2g(H - h)} \tag{c}$$

or

$$Q = C_d\sqrt{2g}\,L(\tfrac{2}{3}H)\sqrt{H - \tfrac{2}{3}H} \tag{d}$$

or

$$Q = 0.385 C_d\sqrt{2g}\,LH^{3/2} \qquad [\text{L}^3\text{T}^{-1}] \tag{6.28}$$

A large number of discharge coefficient values are encountered because a wide variety of broad-crested weir shapes can be included. For measuring flows, broad-crested weirs offer no advantages over sharp-crested weirs and thus the U.S. Bureau of Reclamation seldom uses them for measuring purposes (U.S. Bureau of Reclamation, 1984). The weir crest should be calibrated either by field tests on the actual structure or by model studies of it. The value of the discharge coefficient is generally between 0.85 and 1.1.

6.12.4 Flow over Submerged Weir

When the downstream water level exceeds the crest height, it influences the discharge over the weir. The submergence reduces the discharge through the weir. Herschel, Villemonte, and Marvis have suggested relations between free (unsubmerged)

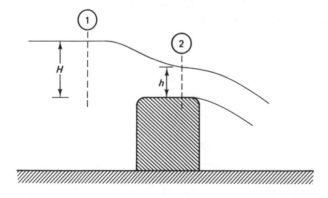

Figure 6.23   Flow over broad-crested section.

and submerged discharge over a weir as a function of upstream and downstream heads. The Villemonte relation for various types of weirs is represented by

$$\frac{Q_s}{Q} = \left[1 - \left(\frac{H_2}{H_1} \right)^n \right]^{0.385} \qquad \text{[dimensionless]} \qquad (6.29)$$

where

Q = free (unsubmerged) weir discharge

Q_s = submerged discharge

H_1 = upstream head

H_2 = downstream head

n = coefficient, $n = 1.44$ contracted rectangular,

$n = 1.50$ for suppressed rectangular, and

$n = 2.50$ for 90° (notch weir)

Example 6.15

A stream has a width of 30 m, depth of 3 m, and a mean velocity of 1.25 m/s. Find the height of a weir to be built on the stream floor to raise the water level by 1 m.

Solution

1. $Q = 30(3)(1.25) = 112.5$ m³/s.
2. Raised water level $= 3 + 1 = 4$ m.
3. Velocity of approach, $v_0 = \dfrac{112.5}{30(4)} = 0.94$ m/s.
4. Velocity head $= \dfrac{(0.94)^2}{2(9.81)} = 0.045$ m.

5. It is not known whether the weir has a free or a submerged discharge. First assuming a free discharge:

$$Q = (0.385)C_d\sqrt{2g}\,L\left[\left(H + \frac{v_0^2}{2g} \right)^{3/2} - \left(\frac{v_0^2}{2g} \right)^{3/2} \right]$$
$$112.5 = 0.385(0.95)(4.43)(30)\left[(H + 0.045)^{3/2} - (0.045)^{3/2} \right]$$
$$H = 1.71 \text{ m}$$

Thus the head required over the crest is 1.71 m. Since the total water depth is 4 m, the crest height has to be less than 3 m or it is a submerged weir.

6. For a sbmerged weir, let the height of crest be X meters. Hence $H_1 = 4 - X$, $H_2 = 3 - X$, $Q_s = 112.5$ m³/s.

$$Q = (0.385)(0.95)(4.43)(30)(4 - X)^{3/2}$$

neglecting approach velocity or

$$Q = 48.61(4 - X)^{3/2} \text{ m}^3/\text{s}$$

From eq. (6.29),

$$\frac{112.5}{48.61(4 - X)^{1.5}} = \left[1 - \left(\frac{3 - X}{4 - X} \right)^{1.5} \right]^{0.385}$$

or

$$\frac{8.86}{(4 - X)^{3.9}} = 1 - \left(\frac{3 - X}{4 - X}\right)^{1.5}$$

By trial and error, $X = 2$ m.

6.12.5 Orifices

An orifice is a hole or an opening in a barrier placed in a stream through which water discharges under pressure. The orifice is also made in the side or bottom of a tank or vessel or in a plate placed between the flanges of a pipeline to measure flow through these structures. The orifice is classified according to size (small and large), shape (circular, rectangular, triangular), and shape of the upstream edge (sharp edged or round cornered). It can be without or with a mouthpiece, which is a cylindrical extension of an orifice. An orifice may discharge free or may be submerged under a downstream level.

Flow through a small orifice.   When the area of an orifice is sufficiently small with respect to the size of the container, the velocity of flow can be considered constant throughout the orifice. For the orifice section shown in Figure 6.24, apply the Bernoulli's theorem at points 1 and 2 with the datum at the center of the orifice.

$$0 + \frac{v_1^2}{2g} + h = 0 + \frac{v_2^2}{2g} + 0 \qquad \text{(a)}$$

The approach velocity, v_1, is very small compared to v_2 and can be neglected. Hence

$$v_2 = \sqrt{2gh} \qquad \text{(b)}$$

The actual velocity is slightly less, due to the viscous shear effect between water and orifice edge. Hence, including a coefficient of velocity, we have

$$v_2 = C_v\sqrt{2gh} \qquad \text{(c)}$$

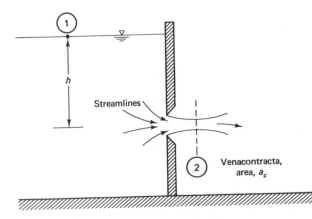

Figure 6.24 Stream jet through an orifice.

The size of the jet is narrowest at a distance of about one-half the orifice diameter. At the narrowest section, the *vena contracta*, the streamlines are parallel and perpendicular to the orifice. At the *vena contracta*, discharge

$$Q = a_c C_v \sqrt{2gh} \qquad \text{(d)}$$

In terms of the orifice area,

$$Q = C_c C_v A \sqrt{2gh} \qquad \text{(e)}$$

where C_c is the ratio of the area of jet at the *vena contracta* to the area of the orifice, known as the coefficient of contraction. The two coefficients are combined into a single coefficient of discharge, C_d, whose value ranges from 0.60 to 0.68 depending on size, shape, and water head on the orifice. Thus

$$Q = C_d A \sqrt{2gh} \qquad [\text{L}^3\text{T}^{-1}] \qquad (6.30)$$

In the case of a tank or vessel, if the water level is not kept constant by an inflow, the level will drop due to discharge from the orifice. The rate of flow through the orifice will vary with the change in head. Consider that at any instant the head over the orifice is h, and in time dt it falls by dh. If the volume of water leaving the tank is equated to the volume of flow through the orifice, then

$$-A_t \, dh = C_d A \sqrt{2gh} \, dt \qquad [\text{L}^3] \qquad (6.31)$$

By expressing the tank area, A_t, by a suitable formula for a specified shape and by integrating between two levels, the time needed to lower the water surface can be determined.

Flow through large orifice. When the head over the orifice is less than five times the size (diameter) of orifice, it is a large orifice for which eq. (6.30) is not true because the streamlines of the jet are not normal to the orifice plane and the velocity is not constant throughout the orifice. Instead, it acts like a weir under pressure, with the water level always above the top edge of the weir on the upstream side.

In the rectangular orifice under the low head as shown in Figure 6.25, the velocity of flow through an elemental strip at a depth of h from the free surface is $\sqrt{2gh}$, and the discharge is

$$dQ = (B \, dh)\sqrt{2gh} \qquad \text{(a)}$$

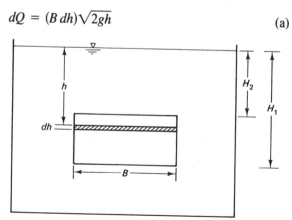

Figure 6.25 Large rectangular orifice.

For the total discharge, integrating between the limits of H_1 and H_2 and introducing a coefficient,

$$Q = \frac{2}{3} C_d \sqrt{2g} B (H_1^{3/2} - H_2^{3/2}) \qquad [\text{L}^3\text{T}^{-1}] \tag{6.32}$$

For a circular or any other shape of orifice, the area term in eq. (a) is expressed in terms of h and then the equation is integrated. If the velocity of approach cannot be neglected, the velocity head should be added in both H_1 and H_2 of eq. (6.32).

Example 6.16

In a stream of 5 ft width and 3 ft depth, a plate is placed that has a rectangular orifice 3 ft in length and 1.2 ft in height. The upper edge of the orifice is 9 in. below the water surface. Determine the discharge. $C_d = 0.6$.

Solution

1. Neglecting the velocity of approach.
2. $H_2 = 0.75$ ft, $H_1 = 0.75 + 1.2 = 1.95$ ft.
3. From eq. (6.32),

$$Q = \frac{2}{3} (0.6) \sqrt{2(32.2)} \, (3) \, (1.95^{3/2} - 0.75^{3/2})$$

$$= 19.96 \text{ cfs}$$

4. Velocity of approach $= \dfrac{19.96}{(5)\,(3)} = 1.33$ fps.

5. Velocity head $= \dfrac{(1.33)^2}{2(32.2)} = 0.03$ ft.

6. Including the velocity head, we have

$$Q = \frac{2}{3} (0.6) \sqrt{2(32.2)} \, (3) \, (1.98^{3/2} - 0.78^{3/2})$$

$$= 20.18 \text{ cfs}$$

6.12.6 Flumes

Flumes are devices in which the flow is locally accelerated by means of (1) a lateral contraction in the channel sides, or (2) combining the lateral contraction with a hump in the channel bed. The first type is known as a venturi flume. The equation for discharge through a flume is based on the energy principle (Bernoulli's theorem). Usually, flumes are designed to achieve the critical flow in the contracted (throat) section. Flumes have four advantages: (1) they can operate with small head loss; (2) they are insensitive to the velocity of approach; (3) they make good measurements without submergence as well as under submerged conditions; and (4) there is no related sediment deposition problem.

A venturi flume section is shown in Figure 6.26. Specific energy at the throat

$$H = h_2 + \frac{v_2^2}{2g} \tag{a}$$

or

$$v_2 = \sqrt{2g(H - h_2)} \tag{b}$$

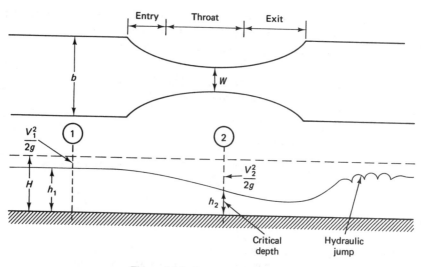

Figure 6.26 Venturi flume section.

For a rectangular throat section,

$$Q = (Wh_2)\sqrt{2g(H - h_2)} \qquad \text{(c)}$$

If the flow is critical at the throat, then $h_2 = \frac{2}{3}H$. Hence

$$Q = 0.385C_d\sqrt{2g}WH^{3/2} \qquad [L^3T^{-1}] \qquad (6.33)$$

(including the coefficient of discharge).

In eq. (6.33), H is the total head at the approach channel, including the velocity head. This formula is similar to the broad-crested weir equation (6.28). If a factor is incorporated for the velocity of approach, it will be seen that the discharge through a critical section flume is dependent on the head in the approach section and the width of the throat. A special group of flumes with standard designs, including standard throat dimensions, have been developed, known as the Parshall flumes. The relations for these have been developed by the U.S. Bureau of Reclamation through extensive calibration experiments. Equations for different flume sizes are summarized in Table 6.5.

TABLE 6.5 PARSHALL FLUME DISCHARGE RELATIONS

Throat width	Flow capacity (cfs)	Equation*	
3 in.	0.03–1.9	$Q = 0.992H_a^{1.547}$	(6.34)
6 in.	0.05–3.9	$Q = 2.06H_a^{1.58}$	(6.35)
9 in.	0.09–8.9	$Q = 3.07H_a^{1.53}$	(6.36)
1– 8 ft	Up to 140	$Q = 4WH_a^{1.522W^{0.026}}$	(6.37)
10–50 ft	Up to 2000	$Q = (3.687W + 2.5)H_a^{1.6}$	(6.38)

*H_a, water level in a well in the converging (approach) section (ft); W, Throat width (ft); Q, discharge (cfs).

Sec. 6.12   Measurements Through Hydraulic Devices

In the case of submergence, a flow-rate correction is obtained from the graphs (U.S. Bureau of Reclamation, 1984). This correction is subtracted from the value computed by the foregoing equations.

Example 6.17

In a 6-ft Parshall flume, the gage reading in the approach well is 2 ft. The submergence is 90%, for which the correction is 14 cfs. Determine the discharge.

Solution From eq. (6.37),

$$Q_0 = 4(6)[2^{(1.522)(6)^{0.026}}] = 72.48 \text{ cfs}$$

The correction of flow rate, $Q_c = 14$ cfs, and the corrected flow, $Q = 72.48 - 14 = 58.48$ cfs.

6.13 DISCHARGE RATING

The discharge rating depicts the relation between stage and discharge for a gaging station that is applied to records of stage of a stream to convert them into discharge. The rating for a site is established by performing periodic field measurements of discharge and stage. Measured discharge is plotted against concurrent stage to define a rating curve for the site. At least 10 to 12 points covering the range of low to high flows are needed to determine the stage and discharge relation, and periodic measurements are needed thereafter to check the validity of the relation. Certain physical characteristics at the gaging section or in the channel bed, known as station controls, stabilize the stage and discharge relation. When an appropriate control is missing at a gaging site, the rating or stage–discharge relation will change (shift) from time to time. When different types of controls become operative at different stages, different relations will hold from stage to stage, which is usually the case.

The discharge rating may be simple if there is a direct relation between stage and discharge. It may, however, be complex if any other parameter is also needed to define the stage–discharge relation. Usually, there are three types of ratings.

1. Simple stage–discharge, or two-parameter discharge relation
2. Slope–stage–discharge, or three-parameter discharge relation with slope
3. Velocity index–stage–discharge, or three-parameter discharge relation with velocity index

6.13.1 Controls for Stage–Discharge

As stated above, controls tend to make a stage–discharge curve stable. There are two types of controls: section control and channel control. Section controls comprising physical features at a particular section, such as riffle, ledge of rock, weir, or spillways, can be natural or man-made. Channel controls include all features, such as size, shape, slope, roughness, alignment, constriction, or expansion in a reach of channel downstream of the gage, that provide rigidity and stability to the bed and banks of the stream. If a control is effective for the entire range of low to high flows, it is known as a complete control. More commonly, however, control is partial. Section control is often effective only at low stages and is submerged by channel control at medium or high stages unless it is a high dam. Channel control is generally effective at high stages, but the reach of the channel acting as the control may lengthen with increasing stage, inducing new features that may affect the stage–discharge relation.

6.14 SIMPLE STAGE DISCHARGE RELATION

The rating curve or stage–discharge relation for each gaging site has its own features, based on the control characteristics for the station. A plot of a series of discharge measurements made at medium and high stages will indicate whether a simple stage–discharge relation applies, because in an unsteady flow situation of complex relation, plotted points will have a scattered pattern. For a simple stage–discharge relation, the curve has a parabolic form, given by

$$Q = A(h \pm a)^n \quad [L^3T^{-1}] \quad (6.39)$$

where

$$Q = \text{discharge}$$
$$h = \text{gage height}$$
$$a = \text{stage reading at zero flow (datum correction)}$$
$$A, n = \text{constants}$$

Traditionally, discharge measurements are plotted on the horizontal scale (abscissa) and the gage height on the vertical scale (ordinate).* A curve is fitted by eye to the plotted points. A plot on rectangular-coordinate paper is shown in Figure 6.27. Two different controls coming into effect at different stages result in producing a compound curve formed of two different parabolic curves.

6.14.1 Logarithmic Rating Curve

Taking logarithm of eq. (6.39) will transform it to a straight line as follows:

$$\log Q = n \log (h \pm a) + \log A \quad (6.40)$$

A plot of Q and $(h \pm a)$ on a log-log paper will produce a straight line. A straight-line plot is preferred because (1) it can be extended or extrapolated, (2) it can be described by a simple mathematical equation, and (3) by noting changes in the slope of the line, the ranges in state for which the individual controls are effective can be identified. However, often an additional plot of of the low-flow data on rectangular-coordinate paper is prepared so that the point of zero flow may be plotted. For this purpose, logarithmic rating-curve sheets have been designed with a rectangular-coordinate scale in one corner. Sometimes, when the stage–discharge equation changes too frequently with stage, the logarithmic method may not be suitable and a parabolic curve on rectangular-coordinate paper is used.

The log plot is between Q and $(h \pm a)$, not gage height. If the control is a section control of regular shape, the value of datum control, a, is the distance between zero gage height and the lowest point of the control. However, for a channel control

---

*This is unusual. In eq. (6.39), Q is a dependent variable and, as such, should normally be plotted on the y-scale. The arrangement on a rating curve is, however, reversed. The slope n is accordingly computed as the ratio of horizontal to vertical distance.

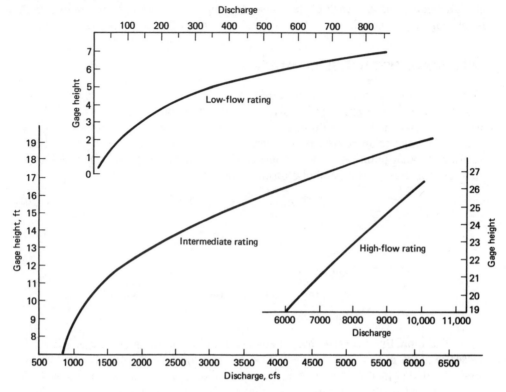

Figure 6.27 Simple stage-discharge relation.

or section control of irregular shape, the value of a is a mathematical constant, to maintain the concept of a logarithmic linear relation. It is thus necessary to ascertain the value of the datum control to prepare the plot.

The logarithmic rating relation is seldom one straight line throughout the entire range of stage for the gaging stations. It is usually necessary to fit two or more lines, each corresponding to the range over which a particular control is applicable.

6.14.2 Determination of the Stage of Zero Flow

When discharge (x-scale) and stage (y-scale) are plotted on log-log paper, the shape of the plot determines the type of equation, as follows:

Type of plot	Type of equation	Remark
Straight line	$Q = Ch^n$	Stage of zero flow coinciding with zero gage height
Concave up [Fig 6.28(a)]	$Q = C(h - a)^n$	Stage of zero flow above zero gage height
Concave down [Fig 6.28(b)]	$Q = C(h + a)^n$	Stage of zero flow below zero gage height

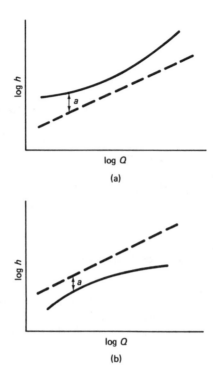

Figure 6.28 Type of curves and zero flow correction.

The following two methods are used to determine the stage of zero flow, a.

Trial-and-error procedure

1. Prepare a log-log plot of gage height and discharge, shown as curve *ef* in Figure 6.29.
2. If the plot is concave upward, add the trial value of a to the chosen scale (i.e., slide the scale downward) and make a plot, as shown by *gh*. (If it is concave downward, subtract the trial value of a.)
3. Go on subtracting (or adding) trial values of a until a value is found that results in a straight-line plot as *ij*.

Arithmetic Procedure

1. Select two points on the gage–discharge plot and read values Q_1, Q_2 and h_1, h_2. Compute Q_3 as follows:

$$Q_3 = \sqrt{Q_1 Q_2}$$

2. From the plot, read h_3 corresponding to Q_3.
3. According to the straight-line property on the log plot,

$$a = \frac{h_1 h_2 - h_3^2}{h_1 + h_2 - 2h_3} \quad [\text{L}] \tag{6.41}$$

Sec. 6.14   Simple Stage Discharge Relation   **277**

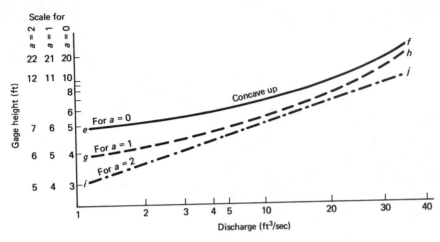

Figure 6.29 Trial-and-error procedure to determine the stage of zero flow.

Example 6.18

Discharge measurements and the corresponding stages observed at a stream gaging station are listed below. Determine the datum correction (stage of zero flow) by (a) the trial-and-error procedure, and (b) the arithmetic method.

Discharge (ft³/sec)	Stage (ft)	Discharge (ft³/sec)	Stage (ft)
0.40	0.85	12.23	3.40
0.61	0.98	18.31	4.27
1.26	1.30	19.84	4.62
1.61	1.46	25.10	5.26
3.77	2.02	35.24	6.46
6.78	2.62	45.75	7.44
8.98	2.94	58.12	8.67
9.73	3.06	72.02	10.09

Solution   The data are plotted on log-log graph paper in Figure 6.30. The curve has two segments. At point A, corresponding to a gage height of 3.06 ft, the high-stage control became operative.

There will be two stage–discharge relations, one below point A and the other for higher stages. This example solves the relation for the lower stage up to 3.06 ft. A similar procedure will apply for the other curve.

(a) Trial-and-error procedure
1. A log-log plot of discharge and stage for the lower range is reproduced in Figure 6.31, as shown by curve *ab*.
2. Since this curve is slightly concave upward, the trial value of $a = 0.1$ ft is added to the stage scale (*y*-scale shifted down by 0.1) and the data are plotted again as *cd*, which is a straight line. Hence $a = 0.1$.
(b) Arithmetic Procedure
1. Select $Q_1 = 0.40$ and $Q_2 = 8.98$ ft³/sec, corresponding $h_1 = 0.85$ and $h_2 = 2.94$ ft.

Measurement of Surface Water Flow   Chap. 6

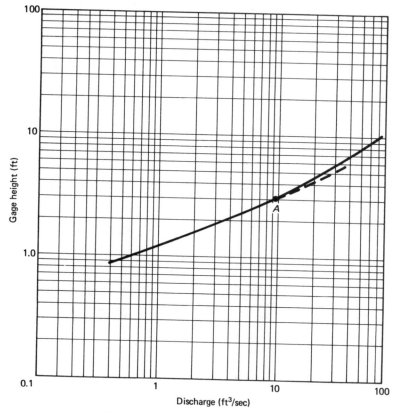

Figure 6.30 Log-log plot of stage and discharge data for Example 6.18.

2. $Q_3 = \sqrt{Q_1 Q_2} = \sqrt{0.4(8.98)} = 1.90 \text{ ft}^3/\text{sec}$
$h_3 = 1.56 \text{ ft}$ (from Fig. 6.31)

3. $a = \dfrac{h_1 h_2 - h_3^2}{h_1 + h_2 - 2h_3} = \dfrac{0.85(2.94) - (1.56)^2}{0.85 + 2.94 - 2(1.56)} = 0.098 \text{ ft}$

6.14.3 Equation of Stage–Discharge Curve

It is desirable to express the stage–discharge relation in mathematical form by determining the equation so that it can be used directly for discharge conversion or in preparation of a conversion (rating) table. Once a straight-line form of the stage and discharge has been obtained after ascertaining the value of the stage of zero flow, the equation of this line defining the stage–discharge relation can conveniently be determined graphically or by linear regression analysis. When the rating curve is composed of more than one straight-line segment, the equation for each is determined separately.

Graphic procedure to determine rating equation.   In log Q versus log $(h \pm a)$ plot of the equation log $Q = n$ log $(h \pm a) + $ log A [eq. (6.40)], n is the slope of the line. Since in a rating curve, the dependent variable Q is plotted on the x-axis, the slope is taken as the ratio of the horizontal distance to the vertical distance. Further, since both the x- and y-scales are logarithmic (fixed), the slope can

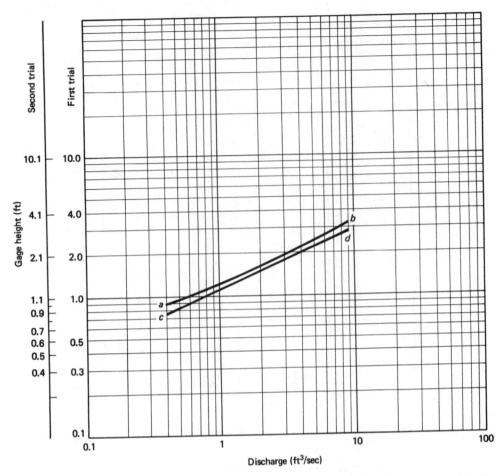

Figure 6.31 Trial and error procedure for determining the stage of zero flow for Example 6.18.

be determined by measuring horizontal and vertical projections of the line using a ruler/scale marked in inches or millimeters.

When $(h \pm a) = 1$, $\log (h \pm a) = 0$ and $Q = A$. Thus, the value of A is obtained by reading Q corresponding to $(h \pm a)$ equal to 1. If the scale does not contain $(h \pm a) = 1$, any value of Q and corresponding $(h \pm a)$ are read and substituted in the rating equation (6.40) to obtain A.

Linear regression analysis to determine rating equation. Regression analysis is the procedure to establish a curve that fits a given set of data. Simple regression involves two variables, one dependent and one independent, as opposed to multiple regression involving several independent variables. If the equation of the curve relates to a straight line, it is known as linear regression. The best-fitting curve through the set of data is obtained based on the principle of least squares, according to which the sum of the squares of deviations (differences) of the measured value of the dependent variable in the data set from the value estimated from the fitted curve should be minimal. Whether the derived equation represents the relationship ade-

Measurement of Surface Water Flow   Chap. 6

quately is indicated by the correlation coefficient. Its value of 1 represents a perfect relation, and 0 indicates no relation among variables.

For a straight-line equation,

$$y = mx + C \quad [L] \tag{6.42}$$

the least-squares line has properties given by

$$\sum y = CN + m \sum x \quad [L]$$

$$\sum xy = C \sum x + m \sum x^2 \quad [L^2] \tag{6.43}$$

In the equations above, known as normal equations for the least-squares line, N represents the total number of observations between variables x and y.

Equation (6.43), when solved, provides the following values for m and C:

$$C = \frac{\left(\sum y\right)\left(\sum x^2\right) - \left(\sum x\right)\left(\sum xy\right)}{N\left(\sum x^2\right) - \left(\sum x\right)^2} \quad [L] \tag{6.44a}$$

$$m = \frac{N\left(\sum xy\right) - \left(\sum x\right)\left(\sum y\right)}{N\left(\sum x^2\right) - \left(\sum x\right)^2} \quad [\text{dimensionless}] \tag{6.44b}$$

The standard deviation, standard error, and correlation coefficient are related as follows:

$$S_y^2 = \frac{\sum y^2 - \left(\sum y\right)^2/N}{N - 1} \quad [L^2] \tag{6.45a}$$

$$S_{yx}^2 = \frac{\sum y^2 - C \sum y - m \sum xy}{N - 2} \quad [L^2] \tag{6.45b}$$

$$r = \left(1 - \frac{S_{yx}^2}{S_y^2}\right)^{1/2} \quad [\text{dimensionless}] \tag{6.45c}$$

When using the equations above on the rating equation (6.40), the variable y is denoted by log Q and variable X by log $(h \pm a)$. Thus, first, the logarithmic values of Q and $(h \pm a)$ have to be determined to be used in eq. (6.44) to ascertain constant parameters C and n. The process has been explained in Table 6.6 (Example 6.19).

Example 6.19

Find the equation of the rating curve in Example 6.18 by (a) a graphic procedure, and (b) regression analysis.

Solution From Example 6.18, $a = 0.1$. Hence the equation of the line is log $Q =$ n log $(h - 0.1) + \log A$.

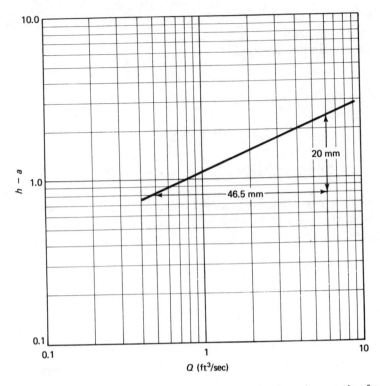

Figure 6.32 Plot of log Q and log($h \pm a$) to determine the rating equation for Example 6.19.

(a) Graphic procedure. A log-log plot of Q and $(h - 0.1)$ for the discharge (Q) and stage (h) data in Example 6.18 is given in Figure 6.32. From the graph

$$n = \text{slope} = \frac{\text{horizontal distance}}{\text{vertical distance}} = \frac{46.5}{20} = 2.325$$

For $(h - 0.1) = 1$, $Q = 0.8$; therefore, $A = Q = 0.8$ or log $A = -0.097$. Hence the equation of the curve

$$\log Q = -0.097 + 2.325 \log (h - 0.1)$$

or

$$\log Q = \log 0.8 + 2.325 \log (h - 0.1)$$

or

$$Q = 0.8(h - 0.1)^{2.325}$$

(b) Regressional analysis. Refer to Table 6.6.

TABLE 6.6 REGRESSION ANALYSIS FOR THE RATING EQUATION[a]

(1)	(2) Q from data	(3) h from data	(4)	(5)	(6)	(7)	(8)	(9)
No.			$h - 0.1$	$\log Q = y$	$\log (h - 0.1) = x$	x^2	y^2	xy
1	0.4	0.85	0.75	−0.40	−0.125	0.016	0.16	0.050
2	0.61	0.98	0.88	−0.215	−0.056	0.003	0.046	0.012
3	1.26	1.30	1.20	0.100	0.079	0.006	0.01	0.008
4	1.61	1.46	1.36	0.207	0.134	0.018	0.043	0.028
5	3.77	2.02	1.92	0.576	0.283	0.080	0.332	0.163
6	6.78	2.62	2.52	0.831	0.401	0.161	0.691	0.333
7	8.98	2.94	2.84	0.953	0.453	0.205	0.908	0.432
Σ				2.052	1.169	0.489	2.19	1.026

[a]$N = 7$; $\log Q$ (col. 5) = logarithm of col. 2; $\log (h - 0.1)$ (col. 6) = logarithm of col. 4; x^2 (col. 7) = square of col. 6; xy (col. 9) = col. 5 × col. 6, y^2 (col. 8) = square of col. 5.

From eq. (6.44a),

$$\log A \text{ or } C = \frac{\left(\Sigma y\right)\left(\Sigma x^2\right) - \left(\Sigma x\right)\left(\Sigma xy\right)}{N\left(\Sigma x^2\right) - \left(\Sigma x\right)^2}$$

$$= \frac{2.052(0.489) - 1.169(1.026)}{7(0.489) - (1.169)^2}$$

$$= -0.095$$

From eq. (6.44b),

$$m = \frac{N\left(\Sigma xy\right) - \left(\Sigma x\right)\left(\Sigma y\right)}{N\left(\Sigma x^2\right) - \left(\Sigma x\right)^2}$$

$$= \frac{7(1.026) - 1.169(2.052)}{7(0.489) - (1.169)^2}$$

$$= 2.325$$

Hence the equation is

$$\log Q = -0.095 + 2.325 \log (h - 0.1)$$

or

$$\log Q = \log 0.8 + 2.325 \log (h - 0.1)$$

or

$$Q = 0.8(h - 0.1)^{2.325}$$

Next, we check the adequacy of the relationship. From eq. (6.45a),

$$S_y^2 = \frac{\left(\Sigma y^2\right) - \left(\Sigma y\right)^2/N}{N - 1}$$

$$= \frac{2.19 - (2.052)^2/7}{6}$$

$$= 0.265$$

From eq. (645b),

$$S_{yx}^2 = \frac{\sum y^2 - C\sum y - m\sum xy}{N - 2}$$

$$= \frac{2.19 - (-0.095)(2.052) - 2.325(1.026)}{5}$$

$$= 0$$

From eq. (645c),

$$r = \left(1 - \frac{0}{0.265}\right)^{1/2} = 1.0$$

That is, there is a perfect correlation.

6.15 SLOPE–STAGE–DISCHARGE RELATION

When variable backwater* conditions exist in a stream due to channel constriction, artificial structures, downstream tributaries, or natural flood waves, the discharge is not merely a function of stage but is also affected by the slope of the water surface (energy gradient). The slope or fall is used as a third parameter in such cases. In addition to a main gage, known as a base gage, an auxiliary gage is installed downstream of the base gage and simultaneous gage readings are made along with the discharge measurements at both gages. The procedure described here to establish the rating is known as the unit fall method.

According to both Chezy and Manning, the discharge is directly proportional to the square root of the slope. Following this relation, we obtain.

$$\frac{Q}{Q_r} = \left(\frac{F}{F_r}\right)^n \qquad \text{[dimensionless]} \tag{6.46}$$

where

Q = measured discharge of a stream for a given base gage height

Q_r = discharge from the rating curve corresponding to the same base gage height

F = measured fall between base and auxiliary gages

F_r = fall between base and auxiliary gages corresponding to rating curve discharge, Q_r

n = close to 0.5, between 0.4 and 0.6

---

*A constant backwater, such as that caused by section control, does not affect a simple stage–discharge relation.

For $F_r = 1$,

$$Q_r = \frac{Q}{F^n} \qquad [L^3 T^{-1}] \tag{6.47}$$

First the rating curves are prepared as follows:

1. Adopting $n = 0.5$ in eq. (6.47), determine Q_r from the measured discharge values Q.
2. Plot Q_r against the corresponding base gage stage. Fit a curve to the plotted points.
3. Repeat steps 1 and 2 using $n = 0.4$ and subsequently, 0.45, 0.55, and 0.6.
4. From the five curves with different values of n, select the one that best fits the points plotted.
5. Compute ratios Q/Q_r for various measured values of discharge and make another plot between Q/Q_r and corresponding fall measurements, F.

Two curves from steps 4 and 5, as shown in Figure 6.33, are used together to convert the stage values to discharges. For an observed base gage height, determine

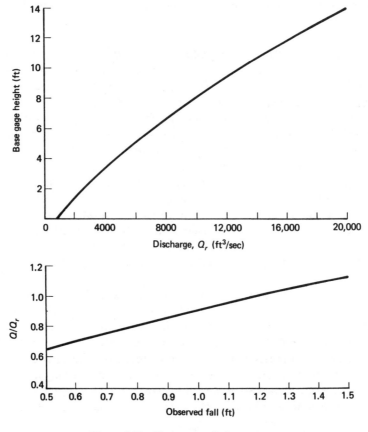

Figure 6.33 Slope–stage–discharge curves.

Q_r using the first curve. From the second curve, for the fall observed, read Q/Q_r, and substituting the determined value of Q_r, find Q.

6.16 CONVERSION OF STAGE RECORDS INTO DISCHARGE

For a water year, the daily record of discharge is computed from the record of stage using the discharge rating for the gaging station. The process, performed manually or by computer, involves the following steps.

1. Station analysis. As a preliminary to computing the discharge record, a study is performed of the data collected at each station. This includes (a) a review of the stage data and determination of the datum corrections, if any, to be applied to observed stages; (b) appraisal of the accuracy of the discharge measurements and computations; (c) analysis of the discharge rating to determine whether the last rating used is applicable for part or all of the year, or whether departures are more than 5% and thus require a new rating curve; and (d) preparation of tables from the rating curves on the standard rating-table form that lists discharge at intervals of 0.1 ft of stage or checking the range and validity of formulas when mathematical expressions are to be used.

2. Preparation of gage-height record. For a nonrecording gage, the first step is to check and reconcile observer's readings with hydrographer's readings made on field visits. The datum corrections, if any, are applied next. For gradually changing stages, the mean daily gage heights are computed as an average of two observed readings for each day. For rapid changes, a stage hydrograph is sketched through plotted points of gage heights using the graphic stage record from a nearby recording gage station. Peaks are determined from high-water marks or crest-stage readings.

For a recording station, the corrections that need to be made to a stage record are (a) time correction due to a slow or fast clock, which is prorated by straight-line interpolation and applied by changing the positions of the midnight lines for the affected days; (b) gage-height corrections, based on differences in readings of the recorder pen and inside staff gage at inspection times prorated with time, and also reversal errors that occur when the pen reverses direction on reaching the top and bottom of the chart; and (c) datum correction, if required.

3. Determination of daily mean gage height. This is usually a template which is placed over a 24-hour segment of the recorder chart and adjusted so that areas above and below a line on the template approximately balance each other. For the chart from a recording gage, this is the uncorrected gage height, which is corrected for gage-height correction and datum corrections referred above. From the sketched hydrograph for a nonrecording gage, corrected daily mean gage heights are obtained. When there is a large variation in stage during a day, the day is subdivided into smaller increments and the mean gage height for each time increment is determined.

4. Applying rating tables to the gage heights. A form containing columns for daily mean gage height and discharge for the 12-month period, and also a place to record monthly and annual summaries, is used. Discharges are determined by applying the appropriate rating tables to the gage heights.

6.17 PRESENTATION AND PUBLICATION OF DATA

A summary of daily mean discharges for a gaging site for a water year is prepared from computations of the discharge records. The USGS publishes an annual report that includes a sheet each for daily mean discharges of each gaging station. In addition, it contains a map showing the location of gaging stations, introductory text, reservoir records, tabulation of discharge records for partial record stations, and water quality data. Groundwater data are also included in the annual report.

6.18 FUTURE TRENDS IN STREAMFLOW MEASUREMENTS

Knowledge is growing very fast in the fields of electronics, acoustics, optics, and nuclear energy. Based on advances in these fields, many instruments capable of more accurate measurements are either under development or are in the experimental stage in the field of water resources. Water stage observation and discharge measurement are two basic procedures in stream gaging.

For water stage recording the digital recorders are gaining preference, to take advantage of computation and processing by the computer and automatic graph plotter. A level recording device based on an acoustic technique has been developed to record stage data in a form that is directly accessible to computer usage.

Use of a telemetry system, including a satellite data collection system, is limited today but is expected to grow in the future. The trend is toward multidata recording systems, where different kinds of data are recorded on the same plot, and also toward multistation data collection at a centralized platform for remote transmission to receiving centers.

Discharge measurement involves width, depth, and velocity elements. For depth measurements, the echo sounder, which is already in widespread use, is likely to be in even greater use. Electronic instruments that permit precise location of the position of observations are available now. Also, sophisticated instruments for measuring the bed profile with sufficient resolution have been developed (e.g., Hydrodist, Ralog, and Hi-fix). The current meter is still a popular method of discharge measurement. A number of new types of velocity meters are in the experimental stage. These include the acoustic velocity meter, doppler velocity meter, electromagnetic velocity meter, optical current meter, nuclear current meter, and the eddy shedding current meter.

In the dilution method, radio element generators, such as the indium-tin generator, open up the possibility for continuous discharge measurements. The ultrasonic method has proved successful, but the electromagnetic method has not made a significant impact, due to its complexity, unsatisfactory accuracy, and high cost.

In the search for new methods of discharge measurement, one that has already been established is the "moving boat" method, which uses a sonic counter to record the geometry of the cross section and a continuously operating current meter sensing the combined stream and boat velocities during a traverse, which is made without stopping. An installation that automatically measures discharge and produces a record by means of magnetic induction has been described by Gils (1970).

With regard to indirect measurement through structures, electronic measurement and recording of the head of water constitute notable advances.

Automated observation platforms for the acquisition of observations from a large number of stations and the use of aircraft and earth satellites for acquisition and relay of data to receiving centers have promising prospects.

PROBLEMS

6.1. A river cross section has been divided into 20 sections. The area of each section and the mean velocity observed are indicated below. Determine the rate of flow in the river.

Section:	1	2	3	4	5	6	7	8	9	10
Area (m²)	2.2	4.5	6.2	6.8	7.2	5.5	4.5	3.2	2.1	1.0
Velocity (m/min)	3.5	3.8	4.5	4.8	4.2	4.0	3.8	3.7	3.6	3.2

6.2. The equation for vertical velocity distribution (in ft/sec) in a stream was found to be $V = 1.9y^{1/7}$, where y is the distance from the bottom. If the depth of the flow is 3.5 ft, what is the mean velocity of flow by the direct integration method? What is the discharge per foot width?

6.3. Solve Problem 6.2 by the graphic integration method.

6.4. Solve Problem 6.2 by the arithmetic summation method.

6.5. The equation for the vertical velocity distribution with respect to the water surface is $V = 3.8(4 - y)^{1/2}$. For a depth of flow of 4 ft, what is the average velocity of flow?

6.6. The equation for a fully developed turbulent flow has the logarithmic form

$$V = 4.45 + 2.62 \log \frac{y}{d}$$

in which y is the distance from the bed. For the depth of flow, d, of 5 ft, determine the mean velocity of flow. (The equation applies a small distance above the bed, beyond the laminar sublayer.)

6.7. The following observations were recorded for point velocities in a 15-ft-deep vertical section of a stream. Determine the mean velocity in the section using different methods. Determine the percent error in each method compared to the velocity-curve method.

Observation Point Depth	Velocity (ft/sec)
0.04	2.35
0.2	2.30
0.4	2.20
0.6	2.00
0.8	1.75
0.9	1.50

6.8. In a gaging measurement of a deep swift stream from a bridge, the total depth of the sound line was measured to be 7.55 m. The depth from the guide pulley to the surface was 3.0 m. The angle of sound line from the vertical was 20°. The distance from the centerline of the current meter to the bottom of the weight was 0.3 m. Determine (a) the true depth of the water, and (b) the position of 0.2 depth and 0.8 depth.

6.9. In a hand-line measurement of a stream, the total depth from the pulley was measured to be 25.2 ft. The depth to the water surface from the pulley was 10.5 ft. The meter was suspended by a hanger having 1-ft distance from the centerline of the meter to the bottom of the weight. Determine the position of the hand line to place the current meter at an 0.6 depth.

6.10. A river cross-section profile is shown in Fig. P6.10. The depths measured at various verticals are indicated together with the point measurements of velocity in each vertical at 0.2 and 0.8 depths or 0.6 depth. Compute the discharge at the cross section by the midsection method.

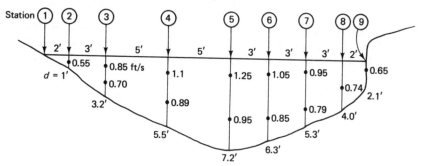

Figure P6.10

6.11. Field observations for discharge measurement at a site are recorded below. Determine the streamflow by the midsection method. The current-meter rating is $v = 0.03 + .25N$, where v is velocity in m/s and N is the number of revolutions per second.

Distance from Initial Point (m)	Depth (m)	Meter Position	Revolutions	Time (sec)
4	0		0	
6	0.4	0.6	45	59
8	0.85	0.6	52	61
10	1.58	0.2	58	62
		0.8	46	61
12	1.71	0.2	65	62
		0.8	51	63
14	1.17	0.2	51	62
		0.8	39	60
16	0.81	0.6	41	63
17	0.48	—	—	—

6.12. Solve Problem 6.10 by the mean-section method.

6.13. Solve Problem 6.11 by the mean-section method.

6.14. Solve Problem 6.10 by the velocity–depth integration method.

6.15. Velocity measurement data were plotted in the form of vertical-velocity curves for various verticals of a stream cross section. Analyses of these data yielded the velocity–distance values shown in Fig. P6.15. Determine the discharge of the stream.

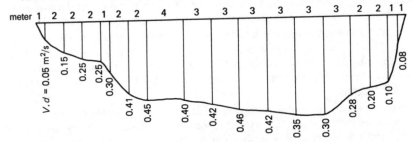

Figure P6.15

6.16. To determine discharge by the velocity–contour method, velocity measurements at a number of points on each vertical were made and isovels were drawn. The areas computed between successive isovels are listed below. Determine the streamflow.

Isovel (m/s)	Area between Isovels (m²)
>0.25	0.8
0.25–0.2	2.2
0.2 –0.15	3.42
0.15–0.10	4.16
0.10–0.05	1.62
0.05–0	1.7

6.17. In measuring discharge by the tracer-dilution technique using the constant injection rate method, a tracer of concentration 6.23 g/liter was injected at a rate of 8 ml/s. The samples collected at a downstream cross section from four points evenly distributed across the width registered concentrations of 0.048, 0.05, 0.045 and 0.052 mg/liter, respectively. Calculate the stream discharge.

6.18. Discharge measurement by the sudden tracer injection method provided the data listed below. Volume of dye injected = 10.85×10^{-3} m³. Concentration of dye injected = 3.55 g/liter. Samples were collected at various times. Determine the stream discharge.

Sample	Time (min)	Concentration (μg/liter)
1	0	0
2	2	0.6
3	4	4.5
4	6	10.2
5	7	10.8
6	9	8.0
7	12	3.2
8	15	1.0
9	20	0.0

6.19. Results from a sudden injection of tracer into a stream to measure streamflow by the dilution technique provided the following data: Weight (mass) of tracer = 1800 g.

Sampling interval = 30 sec. Number of samples = 25. Average concentration of 25 samples = 7.55 mg/liter. Calculate the discharge. [*Hint:* Mass injected = VC_1 = 1800; mass at downstream station = $N\bar{C} \Delta tQ = 25(7.55 \times 10^{-3})(30)Q$.]

6.20. A suppressed sharp-crested weir 15 ft long has a crest of 3 ft height. Determine the discharge for a head over the crest of 25 in. The width of the approach channel is 25 ft. $C_d = 0.65$.

6.21. Determine the length of a sharp-crested weir required to discharge 4000 cfs at a head of 2.25 ft. The weir is divided into sections by vertical posts of 2 ft width and 10 ft clear distance. The approach channel has a width of 400 ft and a depth of 3.5 ft. $C_d = .6$.

6.22. Determine the discharge through the stepped notch shown in Fig. P6.22. $C_d = 0.6$. (*Hint:* For example, the head variation, and hence the range of integration on the middle section, is from 0.75 ft to 1.75 ft.).

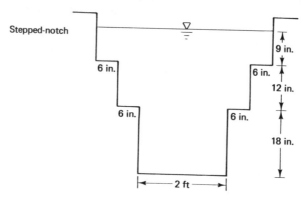

Figure P6.22 Stepped notch.

6.23. Water passes over a rectangular weir of 10 ft width at a depth of 1 ft. If the weir is replaced by an 80° V-notch, determine the depth of water over the notch. Neglect the end contractions. C_d for notch = 0.59, C_d for rectangular weir = 0.63.

6.24. Determine the side slopes of a trapezoidal weir of length L and head H so that the discharge through it (taking into consideration the end-contraction effect on rectangular section) is the same as that through a suppressed rectangular weir of length L and head H.

6.25. A channel of 10 ft width and 3 ft depth is installed with a sharp-crested weir of crest height 1.8 ft and length 8.5 ft. The flow depth over the weir is 1.2 ft. The rectangular weir is to be replaced by a Cippoletti weir. Determine the crest length for the new weir if other conditions remain unchanged. Correct for the velocity of approach. $C_d = 0.6$ for a rectangular weir.

6.26. Determine the discharge over a broad-crested weir of 100 ft length. The upstream water level over the crest is 2 ft and the crest has a height of 2.25 ft. The width of approach channel is 150 ft. $C_d = 0.95$.

6.27. A rectangular channel 4 m wide has a uniform depth of 2 m. If the channel discharge is 10 m³/s, determine the height of the weir to be built at the channel end for free discharge. $C_d = 0.95$.

6.28. A submerged weir in a pond is 10 ft long. The crest of the weir is 9 in. below the upstream level and 6 in. below the tailwater level. Determine the discharge. $C_d = 0.8$.

6.29. A stream is 200 ft wide and 10 ft deep. It has a mean velocity of flow of 4 ft/sec. If a weir of 8 ft height is installed, how much will be the rise of water upstream? $C_d = 1.0$.

6.30. A tank has a rectangular orifice of 3 ft length and 2 ft height. The top edge of the orifice is 8 ft below the water surface. Determine the discharge through the orifice, treating it as (a) a small orifice, and (b) a large orifice. $C_d = 0.62$.

6.31. Determine the diameter of the orifice located at the bottom of a tank if the water level in the tank is kept at 15 ft above the orifice and the discharge is 20 cfs. $C_d = 0.6$.

6.32. How much time will it take to empty the tank in Problem 6.31 if the area of cross section of the tank is 10 ft × 10 ft?

6.33. Determine the discharge through a 15-ft Parshall flume under a head of 3 ft discharging free.

6.34. Determine the discharge through a 4-ft Parshall flume if the approach head is 1.0 m and the submergence is 80% with a flow-rate correction of 5.9 cfs.

6.35. For the upper range of stages in Example 6.18 when high-water control became effective, determine the datum correction (stage of zero flow) by (a) the trial-and-error procedure, and (b) the arithmetic procedure. The discharge stage data are reproduced below.

Discharge (ft³/sec)	Stage (ft)	Discharge (ft³/sec)	Stage (ft)
9.73	3.06	35.24	6.46
12.23	3.40	45.75	7.44
18.31	4.27	58.12	8.67
19.84	4.62	72.02	10.09
25.10	5.26		

6.36. On Zor Creek at Gbanka, in Liberia (West Africa), the discharge measurements made during the period February 1973–January 1975 are recorded below along with corresponding gage heights. Determine the datum correction (stage of zero flow) by (a) the trial-and-error method, and (b) the arithmetic method.

Gage Height (cm)	Discharge (m³/s)	Gage Height (cm)	Discharge (m³/s)
30	0.02	80	2.35
40	0.16	100	4.4
50	0.40	120	7.5
70	1.55	150	13.5

6.37. The following rating table was developed from field measurements during 1975–1976 on the St. John River at Baila in Liberia (West Africa). Determine the stage of zero flow by (a) trial-and-error method, and (b) the arithmetic method.

Gage Height (ft)	Discharge (cfs)	Gage Height (ft)	Discharge (cfs)
1.0	70	7.0	9,500
2.0	430	8.0	13,500
3.0	1,200	9.0	17,500
4.0	2,350	10.0	22,500
5.0	4,250	11.0	27,500
6.0	6,500	11.5	30,000

6.38. Find the equation for the rating curve of Problem 6.35 by (a) the graphic procedure, and (b) regression analysis.

6.39. Determine the rating equation for the stage–discharge data in Problem 6.36 by **(a)** the graphic procedure, and **(b)** regression analysis.

6.40. Determine the rating equation for the stage–discharge data in Problem 6.37. ·

6.41. The following observations relate to discharge and stage at the base station and stage at an auxiliary gage 2000 m downstream. Develop the slope–stage–discharge relationship. What is the estimated discharge when the base and auxiliary stages are 9.0 m and 8.25 m, respectively?

Discharge (m³/s)	Stage at Base Gage (m)	Stage at Auxiliary Gage (m)
34	1.012	0.951
206	2.206	1.279
78	1.359	1.155
165	1.963	1.347
164	1.755	1.054
200	2.139	1.331
995	7.638	4.986
780	7.108	5.188
1415	10.558	7.678
445	4.026	2.429
760	6.105	3.923
580	4.907	2.990

ESTIMATION OF SURFACE WATER FLOW

Part A: Hydrograph Analysis

7.1 RUNOFF AND STREAMFLOW

The term "runoff" is used for water that is "on the run" or in a flowing state, in contrast to water held in storage or evaporated into the atmosphere. Since such flow conditions take place in various stages of a hydrological cycle, there are various types of runoff, as shown in Figure 7.1. In this figure, boxes indicate storage units. Surface runoff or overland runoff is that part of the runoff that travels over the surface of the ground to reach a stream channel and through the channel to the basin outlet. (To be precise, the surface runoff includes the precipitation directly falling over the channel reach, and the overland runoff excludes the channel precipitation.) Surface runoff appears relatively quickly as streamflow.

Subsurface runoff is that part of the runoff that travels under the ground to reach a stream channel and to the basin outlet. It consists of two parts. One part moves laterally through the upper soil horizons within the unsaturated zone or through the shallow perched saturated zone toward the stream channel. This is known as the subsurface storm flow, interflow, or throughflow. Another part infiltrates deeper to the saturated zone to form the groundwater flow. This flow discharges into the stream channel as the base runoff or baseflow. The groundwater flow into a stream is due to the infiltrated precipitation in the past at any time. It is, accordingly, referred to as delayed runoff. The interflow or subsurface storm flow has an intermediate travel time to the stream between the surface and base flows. When the response time is short, the interflow is sometimes considered as a contribution to the surface flow. Total

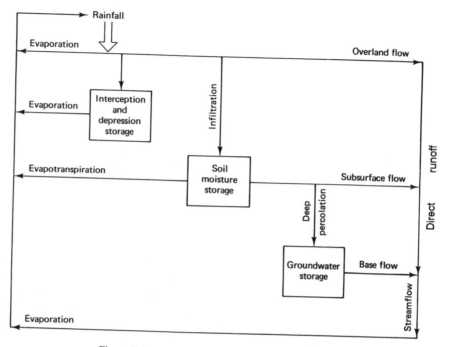

Figure 7.1 Forms of runoff in the hydrologic cycle.

runoff, comprising surface runoff and subsurface runoff at the downstream end of any reach of a stream channel, forms the streamflow. Thus the runoff is that part of the precipitation that eventually appears as streamflow. Excluding the baseflow contribution, the balance is the direct streamflow.

The runoff, particularly the overland runoff, is measured in terms of the depth of the water. The streamflow, on the other hand, is measured as a volume rate of flow in cubic feet per second or cubic meters per second. The volume is an equating term in two units of measurement; that is, when a runoff depth is multiplied by the contributing area, it provides the same volume as the direct streamflow multiplied by the time for which the flow has contributed the runoff. The baseflow, accounted for separately because of its different nature and area of contribution, is discussed subsequently.

A plot of streamflow (discharge) against time at any section of a stream channel is known as a hydrograph. The runoff due to a precipitation storm and the resulting streamflow hydrograph are, thus, directly related, as both indicate the same volume of water.

7.2 MECHANISM OF RUNOFF GENERATION

The baseflow of a stream is contributed by the groundwater discharge, shown as path 2 in Figure 7.2. The runoff process that leads to the direct streamflow is, however, not that straightforward. Despite considerable research in this field, the mechanism of runoff is not fully resolved. There are three widely accepted theories.

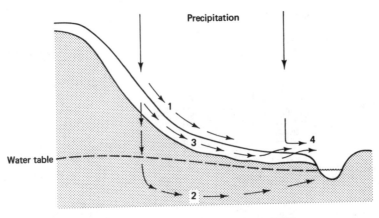

Figure 7.2 Paths of runoff (after Dunne, 1982).

The classic concept of Horton (1933) holds that any soil surface has a certain maximum rate of water absorbance, known as the infiltration capacity. This capacity is high at the onset of rainfall and then declines rapidly, to achieve a constant rate. If rainfall intensity at any time during a storm exceeds the infiltration capacity of the soil, water accumulates on the surface, fills small depressions, and runs downslope as overland flow. According to this theory, the major contribution to direct stream-flow is from the overland runoff, recognized as the Horton overland flow, shown as path 1 in Figure 7.2. Practically the entire basin area contributes to this overland flow. Horton's runoff concept serves as a base for the unit hydrograph technique (Section 7.8) and the infiltration curve technique (Section 3.7). Horton's theory has applicability in arid and semiarid landscapes and cultivated fields, paved areas, construction sites, and rural roads of humid regions that lack a dense vegetation cover and well-aggregated topsoil.

For forests and densely vegetated humid regions, Hewlett and co-workers (see Hewlett and Hibbert, 1967), Kirby and Chorley (1967), and others suggested the theory of subsurface storm flow (throughflow). According to this, a densely vege-tated humid region has the capacity to absorb all except the rarest, most intense storms. A major part of this absorbed water moves laterally through the shallow soil horizon in the zone of aeration, shown as path 3 in Figure 7.2. This process of trans-mission effectively contributes to the streamflow. The flow is confined to intergranu-lar pores, root holes, worm holes, and structural openings. It travels more slowly than Horton overland flow, but some of it arrives quickly enough to produce floods. Freeze (1972b) and many others hold the opinion that this is a viable mechanism but it can-not provide a very large contribution to the total quantity of direct streamflow.

The third type of storm runoff for humid regions is based on the concept of satu-ration overland flow (Musgrave and Holton, 1964). Rainfall causes a thin layer of soil on some parts of a basin to saturate upward from some restricting boundary to the ground surface, especially in zones of shallow wet or less permeable soil. Then the rainfall cannot infiltrate further in the saturated soil and runs over as the saturation overland flow. Some water moving through the top soil also appears as the return flow, shown as path 4 in Figure 7.2. Thus direct precipitation on the saturated soil with or without return flow contributes to the streamflow. This process occurs fre-

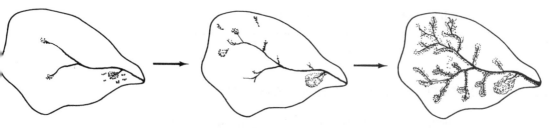

Figure 7.3 Expansion of source area.

quently on the footslopes of a hill, bottoms of valleys, swamps, and shallow soils. It expands outward from the stream channels as shown in Figure 7.3. Unlike Horton's concept of the entire area contributing to runoff, the flow at any moment is contributed by the saturated area, which is only a small percentage of the total basin area. This source area expands and shrinks. The source area increases at the beginning and decreases at the end of a rainstorm. Accordingly, the process is referred to as the variable source area concept (Hewlett and Hibbert, 1967) or the dynamic watershed concept (Tennessee Valley Authority; see Dunne, 1982). Studies on hillslope hydrology are based on this theory.

The general consensus is that in densely vegetated humid regions, streamflow is mostly generated by a combination of subsurface storm flow (throughflow) and saturation overflow. The relative contributions by each mechanism depend on soil and topographic conditions. Figure 7.4 summarizes the conditions affecting the runoff process.

In the literature, another descriptor, "partial source area," has been used for a runoff process (Betson, 1964; Dunne and Black, 1970). This concept is based on

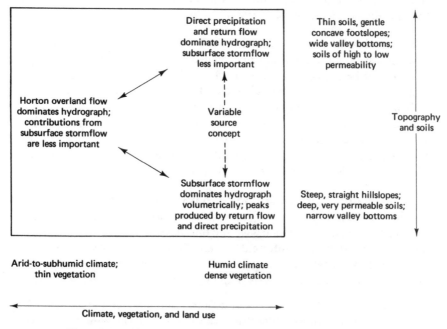

Figure 7.4 Conditions controlling runoff mechanism (after Dunne, 1982).

Horton's theory of predominantly overland flow. The contributing area is, however, considered only a portion of the basin, which is taken as a relatively fixed source area, as against the variable one in the saturation flow concept.

7.3 TECHNIQUES OF ESTIMATION OF STREAMFLOW

Measurements of streamflow from a hydrological stream-gaging network, as described in Chapter 6, are the main and best source of the surface water flow data. However, no national data collection program anywhere in the world collects sufficient data to satisfy all the design and decision-making needs in any one catchment (Fleming, 1975). The World Meteorological Organization recommended (1976) that when the data are inadequate, project activity should begin with installation of a hydrological gaging network. Project planning and design necessitate at least one decade of hydrological data, and it is often not feasible to postpone implementation of a project for this duration after setting up a network — hence arises the need for streamflow estimation. The data to be estimated relate to natural flow conditions comprising average annual flows and distribution of the flow during a year over days, months, or seasons, and peak discharges and minimum discharges for various durations.* We discuss natural streamflows in this chapter and describe extreme flow conditions in Chapter 8.

There are four approaches to the estimation of streamflows.

1. Hydrograph analysis.   Since streamflow represents precipitation returning to a stream, a comparison of a precipitation storm in a basin and the resulting streamflow hydrograph at the outlet of the basin provides a site-specific rainfall–runoff model to convert all precipitation storms to streamflow hydrographs. Based on this concept, the unit hydrograph method has been developed.

2. Correlation with meteorological data.   Hydrological and meteorological processes are both natural and interrelated. The precipitation and streamflow data collected can be considered as statistical samples derived from indefinite natural series of meteorological and hydrological events. The standard statistical techniques and probability theory, including cross-correlation or regression analysis and frequency analysis, can be applied to these data. In multiple linear[†] regression analysis, the drainage basin characteristics and other meteorological parameters are also included in the correlation beside the precipitation. Since the meteorological data series is often longer, it is used to extend the short hydrological data series.

3. Correlation with hydrological data at another site.   Many similarities have been noticed in the natural streamflow of various streams over a fairly large region. The data generated at one gage site thus have a transferable value for a neighboring site on the same or a different stream. This may involve a simple extrapolation or interpolation of information gathered at two stations on the same stream or cross-correlation by regression analysis or duration analysis between two different streams.

---

*Additional hydrological data include sediment transport and ice phenomena.

[†]In the case of nonlinear relationships, a linearization is usually possible by an appropriate transformation, such as a logarithmic transformation.

4. Sequential data generation. In the techniques described above, the data were reconstituted for the past period up to the present time by using either meteorological data or hydrological information from another site. When the data have to be extended into the future or when there are no parallel data to provide a relation, the synthetic data are produced based on a time series that includes a random component. Further, the estimation of a large number of values by the correlation technique produces a series that has a low variance. It is, therefore, necessary to use a technique that incorporates a random component. The data generated have statistical characteristics similar to the short-duration series used in the generation process. This involves a stochastic process.

Any of the techniques above is capable of estimating streamflows, but depending on the available data and need of their extension, a specific technique is used because of its suitability and convenience. There are five situations in which streamflows have to be estimated. The information that exists in each of these cases and applicable techniques are indicated in Table 7.1. The correlation technique, besides estimating streamflows in the situations when they are lacking, is also used to extend a record

TABLE 7.1 DATA SITUATION AND ESTIMATION TECHNIQUES

Case	Available Data	Technique
I. Gaged site		
1. Assessing all streamflow data from precipitation	Precipitation data for the site	Hydrograph analysis
2. Augmenting streamflow data	1. Short-term streamflow data and long-term precipitation data for the site	Rainfall–runoff relation
	2. Short-term streamflow data for the site and long-term streamflow data for another site	1. Correlation of stream-gaging stations 2. Comparison of flow duration curves
3. Estimating gaps in streamflow data	(Same as item 2)	
4. Generation of data	Short-term streamflow data	Synthetic flow generation
II. Ungaged site		
5. Assessing stream-flow data	1. Streamflow data at one or two neighboring sites on the same river	Interpolation or extrapolation
	2. Overall precipitation and other meteorological data	Hydrologic cycle model
	3. Overall precipitation and soil data	SCS method
	4. Drainage basin characteristics	Generalized regional relation
	5. Channel geometry	Generalized regional relation

by using another record, which is more than 25% longer, to improve the data because the error introduced by the correlation is usually less than the sampling error of the short-duration record. Also, extension and filling in of data are necessary for regional studies, in which every record should be adjusted to the same length.

7.4 HYDROLOGICAL PROCESSES IN STREAMFLOW ESTIMATION

The descriptive hydrology presents the theory of water distribution in a subjective manner, which by means of the quantitative hydrology is expressed in terms of numbers, either measured or calculated. The functional relationship between the numbers in a quantitative representation is termed the mathematical hydrology. In this context the process of representing a phenomenon mathematically is known as the mathematical model.

Under the mathematical hydrology, there are divisions of physical or deterministic hydrology, statistical hydrology, probabilistic hydrology, stochastic hydrology, and systems hydrology. The deterministic hydrology has two subdivisions of empirical and conceptual procedures.

The empirical method yields an output for a given set of inputs without giving consideration to the relationships of the parameters involved in a process being considered. It considers only the extremes or treats discrete time periods. The hydrograph analysis for inadequate data and filling in of the gaps and the use of generalized relations for the ungaged sites constitute the empirical approach.

In the conceptual approach, the various processes and their interrelationships are identified, although sometimes use of the empirical relationship is also made. The meteorological–runoff relationship is initially approached by the conceptual method.

The correlation and regression techniques for inadequate data and data with gaps are the statistical methods that establish functional relationships between the two sets of data. The relation is assessed in statistical terms by the standard deviation, correlation coefficient, and significance tests.

Probabilistic methods involve the concept of frequency or probability. Peak flow and minimum flow estimates, which ignore the sequence of events and treat the data as time independent, are based on the probabilistic approach.

The generation of a synthetic long-duration data series from a limited sample of data is the stochastic process. This treats the sequence of events as time dependent and incorporates a random component to represent the natural phenomenon.

Thus, all hydrological processes are made use of in the estimation of streamflows, except for the systems approach, which is essentially an optimization process used in the planning.

7.5 HYDROGRAPH ANALYSIS FOR ESTIMATION OF STREAMFLOW

As defined earlier, a hydrograph is a graphic representation of river discharge with respect to time. The strip chart from a water-level recorder provides a stage hydrograph, which is transformed by the application of a rating curve into a discharge hydrograph of a stream. A streamflow hydrograph is a result of the runoff processes, as discussed in Section 7.2, comprising overland flow, interflow, and baseflow that

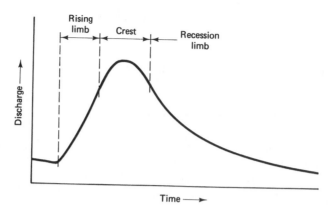

Figure 7.5 Simple storm hydrograph.

are generated by precipitation storms. A hydrograph resulting from a precipitation storm is known as a storm hydrograph. The streamflow hydrograph is thus a cumulative effect of storm hydrographs. The shape of a single storm hydrograph has a typical pattern, as shown in Figure 7.5.

There is a rising limb, the shape of which is characterized by the basin properties and duration, intensity, and uniformity of the rain. The crest segment includes the part of the hydrograph from the inflection point on the rising limb to an inflection point on the recession limb. It contains the peak flow rate. The peak represents the arrival of flow at the outlet from all parts of the basin. For short-duration rain that does not last until the entire area contributes, the peak represents the flow from that portion of the basin receiving the highest concentration of runoff. The end of the crest segment or inflection point marks the time when direct run from the overland flow (excess rainfall) into the stream outlet ceases. The recession limb thus indicates the storage contribution from detention storage (depth of water built up over the land surface), interflow, and groundwater flow. The recession curve is independent of the characteristics of the rainstorm. It can be considered as a rate of discharge resulting from the draining-off process. If there is no added inflow, the time variation of discharge due to the draining-off process, and hence the equation of the recession curve, can be given by

$$Q_t = Q_0 K^t \qquad [L^3 T^{-1}] \tag{7.1}$$

where

$$Q_t = \text{discharge } t \text{ time units after } Q_0$$
$$Q_0 = \text{initial discharge (at } t = 0)$$
$$K = \text{recession constant}$$

This equation, on a semilogarithm plot of log Q_t and time, represents a straight line. The plot is usually a curve which can be approximated by three straight lines of decreasing slope or increasing values of K, representing three different types of storage contribution to the recession: surface detention, interflow, and groundwater. The slope of each of the three lines provides a value of the recession constant for each type of storage contribution.

7.6 DIRECT RUNOFF HYDROGRAPH AND BASEFLOW HYDROGRAPH

It is convenient to consider the total flow to be divided into two parts: the storm, or direct, runoff* and the baseflow. A stream carries baseflow during most of the year when there are no storms in the basin. This comes from the groundwater. Groundwater accretion resulting from any storm is released over an extended period; thus a particular storm contributing to direct runoff is not directly concerned with the baseflow. The precipitation excess of a storm (i.e., net rainfall after all abstractions) constitutes the direct runoff. The arrival of direct runoff at the stream outlet is the starting point for the direct runoff hydrograph (DRH). As time elapses, progressively distant areas add to the outlet flow until a peak flow is attained. If the rainfall continues beyond this period and maintains a constant intensity for a long period of time, a state of equilibrium will be reached and the constant peak flow should continue. The condition of equilibrium is, however, seldom attained because even in extended rainfall, variations in intensity occur throughout its duration. After the peak, DRH begins to descend. A point of inflection comes when the overland flow to the outlet ceases and the storage contribution from surface detention and interflow begins. When the contribution pertains to baseflow only, it indicates the end of DRH.

The first step in the hydrograph analysis is to separate the baseflow and direct runoff hydrograph. When multiple storms occur it is sometimes necessary to separate the overlapping parts of consecutive direct runoff hydrographs.

7.7 HYDROGRAPH SEPARATION

There are two common approaches for separating the baseflow from the direct runoff. The first approach relates to the use of the recession curve equation (7.1), since the last part of the curve is for the baseflow. This approach makes it possible to separate interflow also, if desired. The second approach is of an arbitrary nature. There are many techniques under the second approach.

7.7.1 Separation by Recession Curve Approach

When the data from a stream-gaging station are available for rainless periods reflecting baseflow only, several time intervals of equal value are selected and considered as a unit time. Flow at the beginning of each interval is analogous to Q_0, and at the end of each interval is analogous to Q_1. The values of Q_0 and Q_1 are plotted on ordinary graph (grid) paper. A straight line is fitted graphically to these points. The slope of the line is $Q_1/Q_0 = K$ of eq. (7.1). Once the equation is known, beginning with Q at the starting point of the direct runoff hydrograph and with the selected time as the unit time, the groundwater recession curve is plotted.

Commonly, however, the separation has to be done without a dry period streamflow record. The hydrograph is plotted on semilogarithmic paper with log of discharge values on the y-axis and corresponding time on the x-axis. Since the logarithmic form of eq. (7.1) is a straight line, the tail part of the hydrograph extended back by a

---

*Interflow is included with the direct runoff. Sometimes it is treated separately and the hydrograph is separated into three components.

straight line under the hydrograph will approximate the groundwater flow. The residuals are plotted as the DRH.

Example 7.1

The daily streamflow data for Fall River, Massachusetts, at a site having a drainage area of 6500 km^2 are given in Table 7.2. Separate the baseflow from the direct runoff hydrograph (DRH) by the recession curve method. Determine the equivalent depth of the direct runoff.

TABLE 7.2 DAILY DISCHARGE OF FALL RIVER, MASS.

Time (days)	Flow (m^3/s)	Time (days)	Flow (m^3/s)
1	1,600	9	2,800
2	1,680	10	2,200
3	5,000	11	1,850
4	11,300	12	1,600
5	8,600	13	1,330
6	6,500	14	1,300
7	5,000	15	1,280
8	3,800		

Solution

1. The semilogarithm plot of log Q versus t is shown in Figure 7.6.

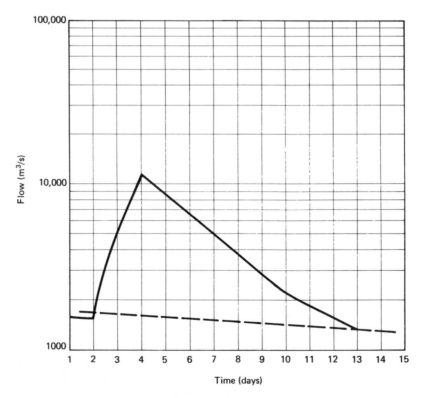

Figure 7.6 Baseflow separation by the recession curve approach.

2. The last straight-line segment of the straight line is extended backward as shown by the dashed line.
3. The ordinates of DRH (residual ordinates) are listed in Table 7.3 and the area under DRH is computed, which signifies the runoff volume.

TABLE 7.3 COMPUTATION OF DIRECT RUNOFF VOLUME

Time (days)	Direct Runoff (m^3/s)	Average Runoff (m^3/s)	Duration (days)	Runoff × Time $(m^3 \cdot day/s)$
1	0			
		0	1.1	0
2.1	0			
		1,700	0.9	1,530
3	3,400			
		6,550	1.0	6,550
4	9,700			
		8,360	1.0	8,360
5	7,020			
		5,995	1.0	5,995
6	4,970			
		4,235	1.0	4,235
7	3,500			
		2,910	1.0	2,910
8	2,320			
		1,835	1.0	1,835
9	1,350			
		1,070	1.0	1,070
10	790			
		620	1.0	620
11	450			
		345	1.0	345
12	240			
		120	1.0	120
13	0			
Total				33,570

$$\text{Volume of runoff} = \left(33,570 \, \frac{m^3}{s} \, day\right)\left(\frac{24 \times 60 \times 60 \text{ s}}{1 \text{ day}}\right)$$

$$= 2900.4 \times 10^6 \, m^3$$

$$\text{Runoff depth} = \frac{\text{runoff volume}}{\text{drainage area}} = \frac{2900.4 \times 10^6 \, m^3}{(6500 \text{ km}^2)(1 \times 10^6 \, m^2/km^2)}$$

$$= 0.446 \text{ m}$$

7.7.2 Separation by Arbitrary Approach

Since the method of Section 7.7.1 is approximate and there is no clear basis for making a precise distinction between direct and baseflow after they have been intermixed in a stream, many other approximate procedures are commonly used. The inaccuracies involved in the separation of the baseflow are usually not important.

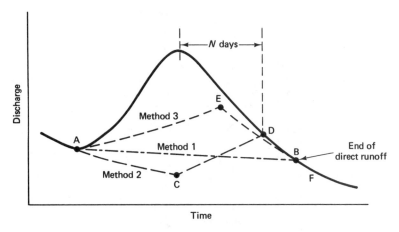

Figure 7.7 Methods of baseflow separation.

Method 1. Refer to Figure 7.7. The simplest method is to join the beginning of the direct runoff (point A) to the end of direct runoff (point B) by a straight line. If point B is not well defined, draw a horizontal line from point A.

Method 2. Extend the recession curve before the storm to point C beneath the peak. Connect point C to point D by a straight line. Point D on the hydrograph represents N days after the peak, given by the formula

$$N = aA^{0.2} \qquad \text{[unbalanced]} \tag{7.2}$$

where

$$N = \text{time, days}$$
$$A = \text{drainage area}$$
$$a = 0.8 \text{ when } A \text{ is in square kilometers}$$
$$\text{or } 1.0 \text{ when } A \text{ is in square miles}$$

The value of N is best determined by an inspection so that the time base is neither too long nor too short.

Method 3. Extend the recession curve backward to point E below the inflection point. Connect A to E by a straight line or an arbitrary shape. If a recession curve is fitted to the hydrograph, the point of departure F of the computed curve from the actual curve marks the end of direct runoff. Otherwise, the endpoint B is arbitrarily selected.

Example 7.2

For the data of Example 7.1, separate the baseflow by the methods of Section 7.7.2.

Solution

1. The hydrograph has been plotted on Figure 7.8.
2. *Method 1.* Join point A, the beginning of direct runoff, to point B, the end of direct runoff. Both points are selected by judgment.

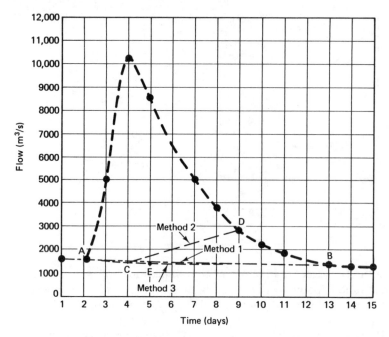

Figure 7.8 Baseflow separation for Example 7.2.

3. *Method 2.* Extend the recession curve before the storm up to point C below the peak. Join point C to point D, computed from eq. (7.2), as follows:

$$N = 0.8A^{0.2}$$
$$= 0.8(6500)^{0.2} = 4.6 \text{ days} \approx 5 \text{ days}$$

4. *Method 3.* Extend the recession curve backward to point E. Join point E to A.
5. The ordinates of DRH (residual ordinates) by three methods are indicated in Table 7.4.

TABLE 7.4 ORDINATES OF DRH BY DIFFERENT METHODS

Time (days)	Direct Runoff (m³/s)		
	Method 1	Method 2	Method 3
1	0	0	0
2	0	0	0
3	3480	3520	3500
4	9800	9900	9850
5	7150	6900	7200
6	5050	4550	5100
7	3550	2700	3600
8	2400	1250	2400
9	1420	0	1420
10	820	0	820
11	470	0	470
12	250	0	250
13	0	0	0

Estimation of Surface Water Flow   Chap. 7

7.8 CONCEPT OF THE UNIT HYDROGRAPH AND INSTANTANEOUS UNIT HYDROGRAPH

7.8.1 Time Parameters

Time base of hydrograph. There are three time parameters related to the direct runoff hydrograph and unit hydrograph. The time base of a hydrograph is considered to be the time from the beginning to the end of the direct or unit hydrograph, shown as T in Figure 7.9a.

Lag time. The lag time or basin lag, t_p, is a basic time parameter which is defined as the difference in time between the center of mass of net rainfall and center of mass of runoff (or peak rate of flow), as indicated in Figure 7.9a. Many other time intervals between rainfall and its hydrograph are also referred to as the time lag. Singh (1988) has listed various other definitions of the time lag.

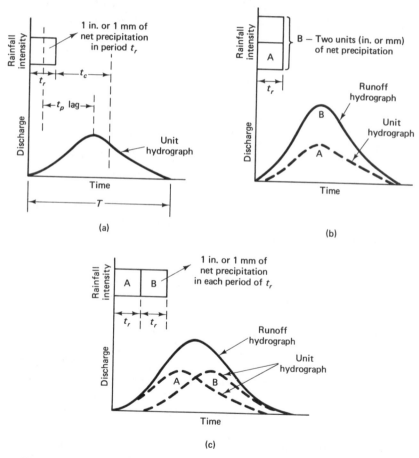

Figure 7.9 Principles of unit hydrograph: (a) unit hydrograph; (b) runoff hydrograph for two units of precipitation of tr duration; (c) runoff hydrograph from unit precipitation for two consecutive periods of tr duration.

Time of concentration. The time of concentration, t_c, is defined in two ways. In terms of physical characteristics of a watershed, which is more important in peak flow assessment, it is defined as the travel time of a water particle from the hydraulically most remote point in the basin to the outflow location. This is explained in Section 12.7.5. Based on rainfall and hydrograph characteristics, it is taken as the time from the end of the net rainfall to the point of inflection on the falling limb of the DHR or the unit hydrograph that signifies the end of direct rainfall inflow into the stream and the start of the detention storage contribution, as indicated in Figure 7.9a.

7.8.2 Unit Hydrograph

In 1932, L. K. Sherman introduced the concept of the unit hydrograph, which is a very important contribution that is used commonly to transform rainfall to streamflow. The unit hydrograph is defined as a hydrograph of direct runoff (excluding the baseflow) observed at the downstream limit of a basin due to one unit of rainfall excess (net precipitation) falling for a unit time t_r (Figure 7.9a). The unit of excess precipitation is taken as 1 in. or 1 mm. The unit of time for precipitation may be 1 day or less, but must be less than the time of concentration explained above. Sherman (1942) suggested as follows:

Basin Area (mi²)	Unit of Time (Duration) of Precipitation Excess (hr)
Over 1000	12 to 24 (preferably 12)
100–1000	6, 8, or 12
20	2
Small areas	One-third to one-fourth of the time of concentration

The effect of a small difference in storm (rainfall) duration is not significant; a tolerance of ±25% in duration is acceptable (Linsley et al., 1982).

The following characteristics of a hydrograph form the basis of the unit hydrograph concept:

1. A hydrograph reflects all of the combined physical characteristics of the drainage basin (shape, size, slope, soils) and that of the causative storm (pattern, intensity, duration).
2. Since the basin features do not change from storm to storm, hydrographs from storms of similar duration and pattern are considered to have a similar shape and time base. The theory of superposition applies and linearity of the relation is assumed. Thus, if 2 in. of net precipitation of a specified unit time occurs, the resulting hydrograph will have the same shape as the hydrograph from 1 in. of net precipitation of the same unit duration except that all of the ordinates will be twice as large, as shown in Figure 7.9b.

    Similarly, when 1 in. of net precipitation occurs in each of the two consecutive unit time durations, the resulting hydrograph will be the sum of two 1-in. hydrographs, with the second hydrograph beginning 1 time unit later, as shown in Figure 7.9c.

3. Variations in storm characteristics do, however, have a significant effect on the shape of hydrographs. This includes (a) rainfall duration, (b) intensity, and (c) areal distribution in the basin.

If the rainfall time duration is increased, the time base of the unit hydrograph will be lengthened. Since a unit hydrograph contains 1 unit of runoff by definition, the peak will be lowered for an increased duration. There is a technique, discussed later, to develop unit hydrographs for storms of longer duration from a unit hydrograph of short duration.

The effect of rainfall intensity on the hydrograph is related to basin size. On large basins, the changes in storm intensity have to last for a considerable time to cause a noticeable effect on the hydrograph, whereas in very small basins, short-duration heavy bursts of rain will produce marked peaks in the hydrographs.

As regards the areal pattern, precipitation concentrated in the lower part of the basin will produce a hydrograph of rapid rise, sharp peak, and rapid recession. On the other hand, the precipitation occurring in the upper part will result in slower rise, low peak, and slower recession. Thus, only one unit hydrograph is not sufficient for a basin. Theoretically, a separate unit hydrograph is necessary for each possible duration of rainfall, but a few unit hydrographs of short durations are adequate, since a tolerance of $\pm 25\%$ is permissible and techniques are available for adjustment of the duration. Also, the unit hydrograph concept is applicable to relatively small basins where the difference in areal distribution of the rainfall does not significantly affect the hydrograph. This concept should not be used for basins much larger than 2000 mi^2 (5000 km^2) unless reduced accuracy is acceptable (Linsley et al., 1982).

When only daily rainfall records are available (hydrograph unit duration is 24 hours), there is a lower limit of 1000 mi^2 for the area below which the unit hydrograph theory should not be applied (Gray, 1973).

According to Sherman (1942), the unit hydrograph method does not apply to runoffs originating from snow or ice. The Hydrologic Engineering Center of the U.S. Army Corps of Engineers, however, makes use of the unit hydrograph technique in snowmelt excess as well.

7.8.3 Distribution Graph

In 1935, M. Bernard suggested a dimensionless form of the unit hydrograph, in which each ordinate of the hydrograph is obtained by dividing the runoff volume at a particular time by the total runoff volume under the hydrograph, thus representing the relative fraction or percentage values at different times. This is known as the distribution graph. The area under the graph is 100% and the graph holds true for all storms of the same duration in the basin, regardless of their intensity.

7.8.4 Instantaneous Unit Hydrograph

In 1943, C. O. Clark used the concept of an instantaneous unit hydrograph in routing analysis. As the duration of net precipitation approaches zero, an instantaneous hydrograph (IUH) results. It is a hydrograph that results from 1 unit of net precipitation applied instantly (in an infinitesimally small time) over a basin. This is a fictitious situation used for the purpose of analysis. The IUH is indicative of the basin storage characteristics, since the rainfall-duration effects are eliminated. Further, since storms of different durations produce varying hydrograph shapes, the IUH, which is independent of duration, is unique for a basin. However, this is true in the linear theory on which the concept of unit hydrograph is based. Several studies since 1960 indicate that the IUH varies from storm to storm, and a nonlinearity exists in the hydrograph properties. But the simplified concept is still very useful. For determination of an IUH, the data from a basin on a storm and the corresponding hydrograph are required. The IUH can then be used to compute a unit hydrograph (UH).

7.9 DERIVATION OF UNIT HYDROGRAPH

The unit hydrograph has a very important application in the estimation of natural streamflows and peak flows from rainfall records.

There are two approaches to deriving a unit hydrograph. In the first approach, the unit hydrograph is derived directly from a storm hydrograph recorded in the basin. This is an inverse problem, since the basic use of the unit hydrograph is to construct storm or streamflow hydrographs. The second approach is to make use of the instantaneous unit hydrograph concept.

7.9.1 Derivation by the Inverse Procedure

It is necessary to have data on a rainfall storm and the corresponding runoff (streamflow). A hydrograph resulting from an isolated, intense, short-duration storm of nearly uniform distribution in space and time is most satisfactory. If such a well-defined single-peaked hydrograph is unavailable, the unit hydrograph is derived from a complex hydrograph.

The duration (unit time) of the unit hydrograph will be the same as the duration of the storm that had produced the storm hydrograph. It is, however, possible to adjust the duration of the unit hydrograph by the technique of superposition described subsequently.

The procedure to derive a unit hydrograph is as follows:

1. The hydrograph associated with a storm is plotted. The baseflow is separated by the technique of Section 7.7, thus obtaining the direct runoff hydrograph (DRH).
2. The area under the DRH that represents the volume of surface runoff is computed. This volume of runoff is converted to a depth, P_n, of the net storm over the basin by the equation

$$P_n = \frac{KV}{A} \quad [\text{L}] \quad (7.3)$$

where

P_n = runoff depth of the storm

K = conversion factor, as given in Table 7.5

V = volume under the hydrograph

A = drainage area of the basin

3. Each of the ordinates of the DRH is divided by P_n. The result is a unit hydrograph of duration equal to the duration of the storm.

TABLE 7.5 FACTOR TO CONVERT RUNOFF VOLUME TO DEPTH

Unit of Runoff Ordinate	Unit of Time Base	Unit of Volume	Unit of Area	Unit of Depth	K
cfs	day	cfs · day	mi²	in.	3.72×10^{-2}
cfs	hour	cfs · hr	mi²	in.	1.55×10^{-3}
m³/s	day	m³/s · day	km²	mm	86.4
m³/s	hour	m³/s · hr	km²	mm	3.6

Example 7.3

The hydrograph of Example 7.1 was produced by a storm of 12-hour duration considered to have uniform intensity over the basin. Determine the unit hydrograph.

Solution

1. The data are tabulated in Table 7.6. The baseflow separated by the recession curve technique in Example 7.1 is given in column 3 and the direct runoff in column 4 of the table. The runoff volume, from Example 7.1, is 33,570 m³/s · day.

2. From eq. (7.3),

$$P_n = \frac{86.4(33,570)}{6500} = 446 \text{ mm}$$

3. Each ordinate of the DRH (column 4) is divided by 446. The resultant unit hydrograph of 12-hour duration is given in column 5. This has been plotted in Figure 7.10.

TABLE 7.6 COMPUTATION OF UNIT HYDROGRAPH

(1) Time (days)	(2) Total Runoff (m³/s)	(3) Baseflow (m³/s)	(4) Direct Runoff (m³/s)	(5) Unit Hydrograph (m³/s) per mm
1	1,600	1,600	0	0
2.1	1,680	1,680	0	0
3	5,000	1,600	3,400	7.62
4	11,300	1,600	9,700	21.75
5	8,600	1,580	7,020	15.74
6	6,500	1,530	4,970	11.14
7	5,000	1,500	3,500	7.85
8	3,800	1,480	2,320	5.20
9	2,800	1,450	1,350	3.03
10	2,200	1,410	790	1.77
11	1,850	1,400	450	1.01
12	1,600	1,360	240	0.54
13	1,330	1,330	0	0

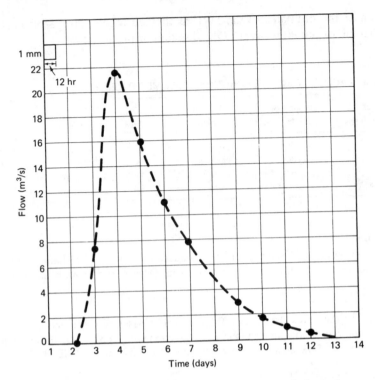

Figure 7.10 Derivation of unit hydrograph from a storm hydrograph.

7.9.2 Derivation by the IUH Technique

The unit hydrographs derived from storms of different durations by the inverse procedure will have different shapes. The instantaneous unit hydrograph eliminates the duration effect and defines a unique unit hydrograph for the basin (in the linear theory). Of the many procedures for preparing an IUH, the method due to Clark (lag and route technique), who was the first to use the IUH concept, has been described here.

There are two steps in the procedure: (1) preparation of an IUH from the field data of a storm and the corresponding runoff, and (2) conversion of the IUH to a unit hydrograph.

In developing the IUH, Clark (1943) conceived a fictitious linear reservoir located at the outlet of the stream such that the storage is proportional to the outflow [i.e., $S = KO$, where K is an attenuation constant, equivalent to the recession constant of eq. (7.1)]. Since the difference between inflow, I, and outflow, O, is the rate of change of storage,

$$\frac{I_1 + I_2}{2} - \frac{O_1 + O_2}{2} = \frac{\Delta S}{\Delta t} \tag{a}$$

or

$$\frac{I_1 + I_2}{2} - \frac{O_1 + O_2}{2} = \frac{K(O_2 - O_1)}{\Delta t} \quad \text{(since } S = KO) \tag{b}$$

or

$$O_2 = C_1 I_1 + C_2 O_1 \qquad \text{(treating } I_1 = I_2) \qquad \text{(c)}$$

In general terms,

$$O_i = C I_i + (1 - C)O_{i-1} \qquad [L^3 T^{-1}] \qquad (7.4)$$

where

$$O_i = \text{outflow at the end of period } i$$
$$I_i = \text{inflow at the end of period } i$$
$$C = \frac{2\Delta t}{2K + \Delta t} \qquad \text{[dimensionless]} \qquad (7.5)$$

Clark routed the inflow comprising the runoff over the basin divided into several areas (subbasins) and represented by a time–area curve. The outflow from eq. (7.4) thus provided the IUH.

The following input data are needed for application of the Clark method:

1. *Time of concentration, t_c.* This information is obtained from the runoff hydrograph of a storm. It is estimated as the time from the end of the net rainfall to the inflection point on the recession limb of the hydrograph. If field data are not available, t_c is estimated empirically.

2. *Attenuation constant, K.* This parameter accounts for the effect of storage in the channel on the hydrograph. It is also estimated from the runoff hydrograph by dividing the flow at the point of inflection of the direct runoff hydrograph by the rate of change of discharge at the same time (slope of the hydrograph at the inflection point), as shown in Figure 7.11.

3. *Time–area relation.* On the map of the basin, isochrones defining points with equal travel times to the outlet are marked, as shown in Figure 7.12a. The area between each pair of isochrones is measured. A curve of travel time versus area (between each pair or cumulated) is drawn, known as the time–area relation. From this curve, the values of areas $a_1, a_2, a_3, \ldots$ that are Δt apart are read, where Δt is the computational interval that must be equal to or less than the duration of the net rainfall. The areas are in the units of volume for per in. or

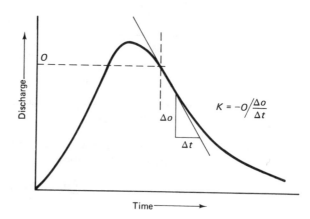

Figure 7.11   Determination of the attenuation constant.

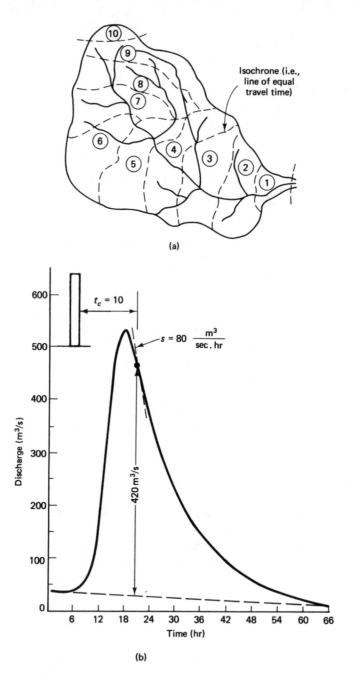

Figure 7.12 (a) Isochronal map of the St. Paul Basin; (b) 2-hr hydrograph of the St. Paul River at the Walker Bridge.

mm depth (in.–mi^2 or mm–km^2). These are converted to discharge units, representing the inflows, by the relation

$$I_i = \frac{F a_i}{\Delta t} \quad [\mathrm{L^3 T^{-1}}] \tag{7.6}$$

where

I_i = ordinate in discharge unit cfs or m^3/s at the end of period i of inflow (time–area curve)

a_i = ordinate of time–area relation at the end of period i

Δt = time period of computational interval in hours

F = conversion factor:

= 645 to convert in.–mi^2 to cfs

= 0.278 to convert mm–km^2 to m^3/s

Equation (7.6), together with (7.4) and (7.5), provides an IUH. The IUH is converted to a unit hydrograph of duration Δt by averaging two IUHs spaced an interval Δt apart as follows:

$$Q_i = 0.5(O_i + O_{i-1}) \qquad [L^3T^{-1}] \tag{7.7}$$

where Q_i is the ordinate of unit hydrograph at time i.

Example 7.4

The watershed of the St. Paul River at the Walker Bridge is shown in Figure 7.12a, on which isochrones have been marked. A hydrograph observed at the Walker Bridge site due to a storm of 2-hour duration is also shown in Figure 7.12b. Determine (a) the instantaneous unit hydrograph, and (b) the 2-hour duration unit hydrograph.

Solution

1. Time of concentration (from the hydrograph) = 10 hr.

2. Attenuation constant, K (from the hydrograph)

$$K = \frac{\text{ordinate of DRH at inflection}}{\text{slope of DRH at inflection}} = \frac{420}{80} = 5.25 \text{ hr}$$

3. Derivation of time–area curve (from isochronal diagram):

Map Area Number	Area (km^2)	Cumulated Area (km^2)	Travel Time (hr)	Travel Time (%)
1	10	10	1	10
2	25	35	2	20
3	50	85	3	30
4	55	140	4	40
5	70	210	5	50
6	65	275	6	60
7	80	355	7	70
8	100	455	8	80
9	110	565	9	90
10	30	595	10	100

4. Plot the percent time versus cumulated area as shown in Figure 7.13.

5. From Figure 7.13, areas are read at points that are one computational interval, Δt, apart (equal to the desired unit hydrograph duration) and recorded in column 3 of Table 7.7.

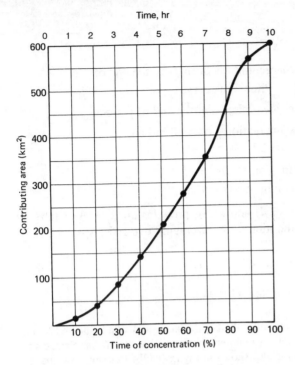

Time, hr

Contributing area (km²)

Time of concentration (%)

Figure 7.13 Time–area relation for St. Paul Basin.

TABLE 7.7 UNIT HYDROGRAPH COMPUTATION BY THE CLARK METHOD

(1)	(2)	(3)	(4)		(5)	(6)	(7)
			Inflow (from Fig. 7.13)				2-hr
No.	Time (hr)	Cumulated Area	Incremental Area, a_i		I (m³/s)	IUH, O_i (m³/s)	Unit Hydrograph Q_i (m³/s)
1	0	0	0		0	0	0
2	2	35	35		4.9	1.57	0.79
3	4	140	105		14.6	5.74	3.66
4	6	275	135		18.8	9.92	7.83
5	8	460	185		25.7	15.00	12.46
6	10	595	135		18.8	16.22	15.61
7	12		0		0	11.03	13.63
8	14					7.50	9.27
9	16					5.10	6.30
10	18					3.47	4.29
11	20					2.36	2.92
12	22					1.60	1.98
13	24					1.09	1.35
14	26					0.74	0.92
15	28					0.50	0.62
16	30					0.34	0.42
17	32					0.23	0.29
18	34					0.16	0.20

6. From eq. (7.6),

$$I_i = \frac{0.278}{2}a_i = 0.139a_i$$

By this formula, a_i is converted to I_i in column 5 of the table.

7. From eqs. (7.5) and (7.4),

$$C = \frac{2\Delta t}{2K + \Delta t} = \frac{2(2)}{2(5.25) + 2} = 0.32$$

$$O_i = CI_i + (1 - C)O_{i-1} = 0.32I_i + 0.68O_{i-1}$$

Route the inflow (column 5) to outflow (column 6) using the relation above.

8. From eq. (7.7),

$$Q_i = 0.5(O_{i-1} + O_i)$$

Thus average the ordinate of the IUH with another ordinate of the IUH one interval before to obtain the unit hydrograph in column 7.

7.10 CHANGING THE UNIT HYDROGRAPH DURATION

A unit hydrograph represents 1 inch (or 1 mm) of direct runoff from the rainfall of a specified period of time. Rainfall of different durations will produce different shapes of the unit hydrograph. The longer duration of the rainfall will lengthen the time base and lower the peak, and vice versa, since a unit hydrograph by definition contains 1 unit of direct runoff. There are two common techniques by which a unit hydrograph can be adjusted from one duration to another.

7.10.1 Lagging Method

This is restricted to cases when a duration has to be converted to a longer duration which is a multiple of the original duration. If a unit hydrograph of duration, t_r, is added to another identical unit hydrograph lagged by t_r, the resulting hydrograph represents the hydrograph of 2 units of storm occurring in $2t_r$ time, as shown in Figure 7.14. If the ordinates of this hydrograph are divided by 2, a unit hydrograph will result. In general terms,

$$\text{UH of } nt_r \text{ duration} = \frac{\substack{\text{sum of } n, \text{ UH of } t_r \text{ duration} \\ \text{each lagged by } t_r \text{ time}}}{n} \qquad [L^3T^{-1}] \qquad (7.8)$$

Example 7.5

The following unit hydrograph results from a 2-hour storm. Determine the hourly ordinates of a 6-hour unit hydrograph.

Time (hr)	0	1	2	3	4	5	6
Q (m³/s)	0	1.42	8.50	11.30	5.66	1.45	0

Solution   See Table 7.8.

$$t_r = 2 \text{ hr}$$

$$n = \frac{6\text{-hr duration}}{2\text{-hr duration}} = 3$$

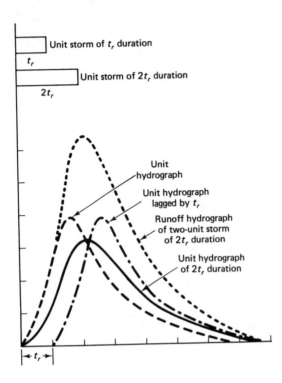

Unit storm of t_r duration

t_r

Unit storm of $2t_r$ duration

$2t_r$

Unit
hydrograph

Unit hydrograph
lagged by t_r

Runoff hydrograph
of two-unit storm
of $2t_r$ duration

Unit hydrograph
of $2t_r$ duration

$\leftarrow t_r \rightarrow$

Figure 7.14 Lagging procedure to convert unit hydrograph duration.

TABLE 7.8 CONVERSION OF UNIT HYDROGRAPH DURATION BY LAGGING

(1)	(2)	(3)	(4)	(5)	(6)	(7)
						6-hr Unit Hydrograph (m^3/s)
	2-hr Unit	Three 2-hr Hydrographs Each Lagged by 2 hr				
Time (hr)	Hydrograph (m^3/s)	$1 \times$ UH	$1 \times$ UH	$1 \times$ UH	Total	(col. 6/3.0)
0	0	0			0	0
1	1.42	1.42			1.42	0.47
2	8.50	8.50	0		8.50	2.83
3	11.30	11.30	1.42		12.72	4.24
4	5.66	5.66	8.50	0	14.16	4.72
5	1.45	1.45	11.30	1.42	14.17	4.72
6	0	0	5.66	8.50	14.16	4.72
7			1.45	11.30	12.75	4.25
8			0	5.66	5.66	1.89
9				1.45	1.45	0.48
10				0	0	0

7.10.2 S-Curve Method

By the summation-curve method, a unit hydrograph can be converted to any other duration of shorter or longer time period. The S-curve results when the unit rate of excess rainfall continues indefinitely. The S-curve is thus constructed by adding together a series of t_r-duration unit hydrographs each lagged the unit duration, t_r, with respect to the preceding one, as shown in Figure 7.15. The curve assumes the S-shape

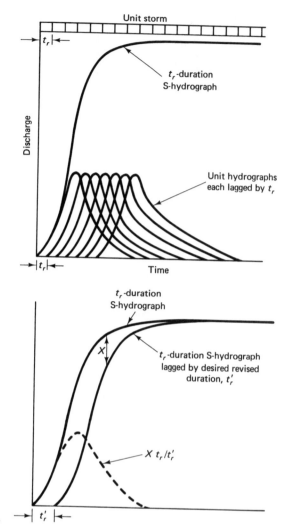

Figure 7.15 Illustration of the S-curve.

and its ordinates in its equilibrium condition acquire a constant outflow rate equivalent of the net rainfall. If T is the time base of the unit hydrograph, the summation of T/t_r unit hydrographs will produce the S-curve. It is not necessary, however, to add so many unit hydrographs to prepare a S-curve. A simplified procedure is illustrated in Table 7.9. In column 3, the values of column 4, corresponding to the previous t_r duration, have been repeated. Column 4, which represents the S-curve, is the sum of columns 2 and 3.

The next three columns of the table demonstrate the conversion process. In column 5, the t_r-duration S-curve (values of column 4) are lagged by the desired revised duration, t_r'. Column 6 indicates the difference of S-curve ordinates, that is, column 4 − column 5. The ordinates of the t_r'-duration hydrograph are computed by multiplying the S-curve differences by the ratio t_r/t_r'.

Example 7.6

Solve Example 7.5 by the S-curve method.

TABLE 7.9 COMPUTATION OF S-CURVE AND CONVERSION OF UNIT HYDROGRAPH DURATION

Computation of S-Curve from Unit Hydrograph				Conversion of t_r-Duration Unit Hydrograph to t_r'-Duration UH		
(1) Time	(2) t_r-Duration Unit Hydrograph	(3) S-Curve Additions	(4) t_r-Duration S-Curve (col. 2 + col. 3)	(5) t_r-Duration S-Curve Lagged by t_r'	(6) S-Curve Difference (col. 4 − col. 5)	(7) t_r'-Duration UH (col. 6 × t_r/t_r')
Time = duration of effective rainfall = t_r (↕ t_r)	a	0	a'	↕ t_r'		
	b	0	b'	a'		
	c	0	c'	b'		
	d	0	d'	c'		
	e	0	e'			
	f	a'	f'	d'		
	g	b'	g'	e'		
	h	c'	h'	f'		
	i	d'	i'	etc.		
	j	e'	j'			
	k	f'	k'			
	l	g'	l'			
	m	h'	m'			
	n	i'	n'			
	o	j'	o'			
	etc.					

TABLE 7.10 COMPUTATION OF 2-HOUR S-CURVE AND 6-HOUR UNIT HYDROGRAPH

(1)	(2)	(3)	(4)	(5)	(6)	(7)
				2-hr		6-hr Unit
	2-hr Unit			S-Curve		Hydrograph
Time	Hydrograph	S-Curve	2-hr	Lagged	S-Curve	(col. 6 × $\frac{2}{6}$)
(hr)	(m³/s)	Addition	S-Curve	by 6-hr	Difference	
0	0	+0	0		0	0
1	1.42	+0	1.42		1.42	0.47
2	8.50	+0	8.50		8.50	2.83
3	11:30	+1.42	12.72		12.72	4.24
4	5.66	+8.50	14.16		14.16	4.72
5	1.45	+12.72	14.17		14.17	4.72
6	0	14.16	14.16	0	14.16	4.72
7		14.17	14.17	1.42	12.75	4.25
8		14.16	14.16	8.50	5.66	1.89
9		14.17	14.17	12.72	1.45	0.48
10		14.16	14.16	14.16	0	0
11		14.17	14.17	14.17	0	0

7.11 FORMULATION OF SYNTHETIC UNIT HYDROGRAPH

For sites where hydrologic records are not available to derive a unit hydrograph, the unit graph is synthesized from the physical characteristics of the watershed. Several methods (Snyder, 1938; Commons, 1942; Williams, 1945; Mitchell, 1948; Taylor and Schwarz, 1952; U.S. SCS, 1957; Hickok et al., 1959; Bender and Roberson, 1961; Gray, 1961) have been developed, some to serve special purposes, such as flood estimation. Two of these common methods are described below.

7.11.1 Snyder's Method

The four parameters lag time, peak flow, time base, and standard duration of net rainfall for the unit hydrograph, have been related to the physical geometry of the basin by the following relations:

$$t_p = C_t (LL_c)^{0.3} \quad \text{[unbalanced]} \qquad (7.9)$$

$$Q_p = \frac{C_p A}{t_p} \quad \text{[unbalanced]} \qquad (7.10)$$

$$T = 3 + \frac{t_p}{8} \quad \text{[T]} \qquad (7.11)$$

$$t_D = \frac{t_p}{5.5} \quad \text{[T]} \qquad (7.12)$$

When the duration of net rainfall, t_r, is other than the standard duration, t_D, the following adjustments in lag time and peak discharge are made:

$$t_{pR} = t_p + 0.25(t_r - t_D) \qquad [\text{T}] \qquad (7.13)$$

$$Q_{pR} = Q_p \frac{t_p}{t_{pR}} \qquad [\text{L}^3\text{T}^{-1}] \qquad (7.14)$$

where

  t_D = standard duration of net rainfall, hour

  t_r = duration of net rainfall other than standard duration adopted in the study, hour

  t_p = lag time from midpoint of net rainfall duration, t_D, to peak of the unit hydrograph, hour

  t_{pR} = lag time from midpoint of duration, t_r, to the peak of the unit hydrograph, hour

  T = time base of unit hydrograph, days

  Q_p = peak flow for standard duration, t_D

  Q_{pR} = peak flow for duration, t_r

  L_C = stream mileage from the outlet to a point opposite the basin centroid

  L = stream mileage from the outlet to the upstream limits of the basin

  A = drainage area, mi^2 or km^2

  C_t = coefficient representing slope of the basin; varies from 1.8 to 2.2 for distance in miles or from 1.4 to 1.7 for distance in kilometers; Taylor and Schwarz stated that $C_t = 0.6/\sqrt{S}$, S being the basin slope

  C_p = coefficient indicating the storage capacity; varies from 360 to 440 for English units and from 0.15 to 0.19 for metric units

If the ungaged basin and the gaged basin are located in close proximity to each other within a region, the coefficients C_t and C_p are computed from the data of the gaged basin. The coefficients so obtained are used in the equations above to construct the unit hydrograph for the ungaged basin. Otherwise, generalized values are used for the coefficients.

A unit hydrograph is sketched, from the lag time, peak discharge, and time base computed from eqs. (7.9) through (7.14), to represent a unit runoff amount (area under the graph). Equation (7.11) usually gives long base length for small to medium basins. The following formulas of the Corps of Engineers give additional assistance in plotting time width, W_{50}, in hours, at the discharge point equal to 50% of the peak discharge, and the width, W_{75}, in hours, at the discharge point equal to 75% of the peak flow.

$$W_{50} = \frac{770A^{1.08}}{Q_{pR}^{1.08}} \qquad \text{(English units)} \qquad [\text{unbalanced}] \qquad (7.15a)$$

or

$$W_{50} = \frac{0.23A^{1.08}}{Q_{pR}^{1.08}} \quad \text{(metric units)} \quad \text{[unbalanced]} \quad (7.15\text{b})$$

and

$$W_{75} = \frac{440A^{1.08}}{Q_{pR}^{1.08}} \quad \text{(English units)} \quad \text{[unbalanced]} \quad (7.16\text{a})$$

or

$$W_{75} = \frac{0.13A^{1.08}}{Q_{pR}^{1.08}} \quad \text{(metric units)} \quad \text{[unbalanced]} \quad (7.16\text{b})$$

In eqs. (7.15a) and (7.16a), A is in mi^2 and Q in cfs, and in eqs. (7.15b) and (7.16b), A is in km^2 and Q in m^3/s.

As a rule of thumb, the widths W_{50} and W_{75} are proportioned each side of the unit hydrograph peak in the ratio 1:2, with the short side on the left of the synthetic unit hydrograph.

Example 7.7

For a basin of 500 km^2 having $L = 25$ km and $L_C = 10$ km, derive the 4-hour unit hydrograph. Assume that $C_t = 1.6$ and $C_p = 0.16$.

Solution $t_r = 4$ hr (given). From eq. (7.9),

$$t_p = 1.6(25 \times 10)^{0.3} = 8.38 \text{ hr}$$

From eq. (7.10),

$$Q_p = \frac{0.16(500)}{8.38} = 9.55 \text{ m}^3/\text{s}$$

From eq. (7.11),

$$T = 3 + \frac{8.38}{8} = 4.05 \text{ days or 97 hr}$$

From eq. (7.12),

$$t_D = \frac{8.38}{5.5} = 1.5 \text{ hr}$$

From eq. (7.13),

$$t_{pR} = 8.38 + 0.25(4 - 1.5) = 9 \text{ hr}$$

From eq. (7.14),

$$Q_{pR} = \frac{9.55(8.38)}{9.0} = 8.89 \text{ m}^3/\text{s}$$

Time from beginning to peak,

$$p_r = \frac{t_r}{2} + t_{pR} = 2 + 9 = 11 \text{ hr}$$

From eq. (7.15b),

$$W_{50} = \frac{0.23(500)^{1.08}}{(8.89)^{1.08}} = 18 \text{ hr}$$

From eq. (7.16b),

$$W_{75} = \frac{0.13(500)^{1.08}}{(8.89)^{1.08}} = 10 \text{ hr}$$

The unit hydrograph has been sketched in Figure 7.16.

7.11.2 Soil Conservation Service (SCS) Method

The SCS employs an average dimensionless hydrograph developed from an analysis of a large number of unit hydrographs from field data of various-sized basins in different geographic locations.

This dimensionless hydrograph has its ordinate values of discharge expressed as the dimensionless ratio with the peak discharge and its abscissa values of time interval as the dimensionless ratio with the period of rise (time from beginning to the peak flow). The ratios for the SCS dimensionless unit hydrograph are given in Table 7.11.

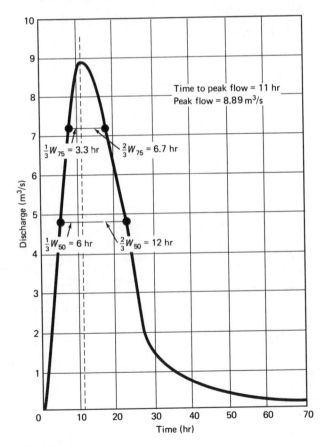

Figure 7.16 Synthetic unit hydrograph by Snyder's method.

Estimation of Surface Water Flow   Chap. 7

The unit hydrograph ordinates for different time periods can be obtained from Table 7.11. However, to use this table, the values of P_r and Q_p are required, which are computed as follows:

$$Q_p = \frac{484A}{P_r} \quad \text{(English units)} \quad \text{[unbalanced]} \quad (7.17a)$$

or

$$Q_p = \frac{0.208A}{P_r} \quad \text{(metric units)} \quad \text{[unbalanced]} \quad (7.17b)$$

TABLE 7.11 RATIOS FOR THE SCS DIMENSIONLESS UNIT HYDROGRAPH

Time Ratio, t/P_r	Hydrograph Discharge Ratio, (Q/Q_p)
0	0
0.1	0.015
0.2	0.075
0.3	0.16
0.4	0.28
0.5	0.43
0.6	0.60
0.7	0.77
0.8	0.89
0.9	0.97
1.0	1.00
1.1	0.98
1.2	0.92
1.3	0.84
1.4	0.75
1.5	0.66
1.6	0.56
1.8	0.42
2.0	0.32
2.2	0.24
2.4	0.18
2.6	0.13
2.8	0.098
3.0	0.075
3.5	0.036
4.0	0.018
4.5	0.009
5.0	0.004
infinity	0

Source: Gray (1973).

$$P_r = \frac{t_r}{2} + t_p \quad [T] \quad (7.18)$$

The time lag, t_p, is computed by eq. (7.9) or by a regional empirical relation, or by the SCS equation involving SCS curve number.

Example 7.8

Solve Example 7.7 by the SCS method.

Solution $t_p = 8.38$ hr, from eq. (7.9) computed in Example 7.7.
From eq. (7.18), $P_r = 4/2 + 8.38 = 10.38$ hr ≈ 10.5 hr.
From eq. (7.17b),

$$Q_p = \frac{0.208(500)}{10.5} = 9.90 \text{ m}^3/\text{s}$$

Using Table 7.11, the hydrograph ordinates are given in Table 7.12.

TABLE 7.12   SYNTHETIC UNIT HYDROGRAPH BY SCS METHOD

t/P_r	t (hr)	Q/Q_p (from Table 7.11)	Q (m³/s)
0	0	0	0
0.2	2.1	0.075	0.74
0.5	5.25	0.43	4.26
0.8	8.4	0.89	8.81
1.0	10.5	1.00	9.90
1.5	15.75	0.66	6.53
2.0	21.0	0.32	3.17
3.0	31.5	0.075	0.74
4.0	42.0	0.018	0.18
5.0	52.5	0.004	0.04

7.12 ESTIMATION OF STREAMFLOW FROM UNIT HYDROGRAPH

A direct application of the unit hydrograph is to synthesize the storm runoff (DRH) and the streamflow (by adding the baseflow) from a series of rainfall events of varying intensity. If the rainfall records are available on a daily basis, the resulting hydrograph is of daily streamflow. When the rainfall record belongs to a heavy storm, the hydrograph produced is for the flood flow, as discussed in Chapter 8.

Consider that a storm consists of a series of rainfall excesses $i_1, i_2, \ldots, i_n$, each of duration d. To formulate the storm hydrograph, a unit hydrograph of d duration (unit time) will be required for the basin. The ordinates of the d-duration unit hydrograph will be multiplied by i_1. Shifting the base by time d, the unit hydrograph ordinates will be multiplied by i_2, and so on, covering all excesses. Each of these hydrographs represents the DRH for individual rainfall excesses, and their sum is the DRH for the entire storm. The estimated baseflow added to this will provide the streamflow (storm) hydrograph.

Example 7.9

Given below is a hydrograph that resulted from an isolated 2-hour duration storm of 1.5 in. net rainfall. Determine the streamflow hydrograph from the storm sequence indicated. Assume that the losses amount to 60% of the precipitation.

Hydrograph data:

Time (hr)	1	2	3	4	5	6	7	8	9	10	11
Flow (cfs)	151	146	268	562	660	630	510	370	250	190	150

Storm sequence data:

Time units = 2 hr	Unit 1	Unit 2	Unit 3	Unit 4
Precipitation (in.)	1	2.75	4.5	2.25

Solution

1. The base flow is separated from the hydrograph by the technique of Section 7.7.2. From the DRH thus obtained, the unit hydrograph is derived using the procedure of Section 7.9.1. The values are shown in columns 1 and 2, respectively, of Table 7.13.

TABLE 7.13 STREAMFLOW HYDROGRAPH FROM UNIT HYDROGRAPH

(1)	(2)	(3)	(4)	(5)	(6)	(7)	(8)	(9)
				DRH Ordinates				Stream
	2-hr						Base-	Flow
Time	UH	Unit 1	Unit 2	Unit 3	Unit 4	Total	flow	Hydrograph
(hr)	(cfs)	$0.4 \times$ UH	$1.1 \times$ UH	$1.8 \times$ UH	$0.9 \times$ UH	(cfs)	(cfs)	(cfs)
1	0					0	151	151
2	0	0				0	146	146
3	81	32				32	146	178
4	277	111	0			111	147	258
5	342	137	89			226	147	373
6	321	128	305	0		433	148	581
7	241	96	376	146		618	148	766
8	147	59	353	499	0	911	149	1060
9	67	27	265	616	73	981	149	1130
10	27	11	162	578	249	1000	150	1150
11	0	0	74	434	308	816	150	966
12			30	265	289	584	150	734
13			0	121	217	338	150	488
14				49	132	181	150	331
15				0	60	60	150	210
16					24	24	150	174
17					0	0	150	150

2. The net storm (net rainfall) sequence is obtained, excluding 60% losses from the precipitation.
3. The ordinates of the UH are multiplied by the successive values of the net rainfall, each lagged by the effective duration as shown in columns 3, 4, 5, and 6.
4. The total of columns 3 through 6 results in the DRH. Adding baseflow to this provides the streamflow hydrograph.

Example 7.10

During the month of September 1985, the weighted average values of the daily rainfall recorded at various stations in the Housatonic basin above New Milford (CT) are as given below. A representative 24-hour duration unit hydrograph* for the basin is also indicated, along with the baseflow observed at the New Milford site. A study indicates that the infiltration and other losses constitute 85% of the precipitation. Compute the streamflow during September 1985.

*This is obtained from a 1-day segregated rainfall and the corresponding record of streamflow. C. W. Sherman indicated that the actual duration of a 1-day rainstorm is 10 to 13 hours.

Date	5	8	15	21
Rainfall (in.)	1.21	0.48	3.22	1.02

Time (days)	0	1	2	3	4	5
Unit hydrograph (cfs)	0	1000	2950	125,000	5600	1750
Baseflow (cfs)	120	120	120	125	130	135

Time (days)	6	7	8	9	10	11	12
Unit hydrograph (cfs)	1000	620	500	350	270	250	0
Baseflow (cfs)	140	125	120	120	120	120	120

Solution   Computations are shown in Table 7.14.

TABLE 7.14 ESTIMATION OF DAILY STREAMFLOW BY UNIT HYDROGRAPH

(1)	(2)	(3)	(4)	(5)	(6)	(7)	(8) DRH (cfs)	(9)	(10)	(11)	(12)
Date	Rainfall (in.)	Losses (in.)	Net Rainfall (in.)	UH (cfs)	Sept. 5	Sept. 8	Sept. 15	Sept. 21	Total	Baseflow (cfs)	Streamflow (cfs)
1				1,000					0	120	120
2				2,950					0	120	120
3				12,500					0	125	125
4				5,600					0	130	130
5	1.21	1.03	0.18	1,750	180				180	135	315
6				1,000	531				531	140	671
7				620	2,250				2,250	125	2,375
8	0.48	0.41	0.07	500	1,008	70			1,078	120	1,198
9				350	315	207			522	120	642
10				270	180	875			1,055	120	1,175
11				250	112	392			504	120	624
12				0	90	123			213	120	333
13					63	70			133	120	253
14					49	43			92	120	212
15	3.22	2.74	0.48		45	35	480		560	120	680
16					0	25	1,416		1,441	120	1,561
17						19	6,000		6,019	120	6,139
18						18	2,688		2,706	120	1,826
19						0	840		840	120	960
20							480		480	120	600
21	1.02	0.87	0.15				298	150	448	120	568
22							240	443	683	120	803
23							168	1,875	2,043	120	2,163
24							130	840	970	120	1,090
25							120	263	383	120	503
26							0	150	150	120	270
27								93	93	120	213
28								75	75	120	195
29								53	53	120	173
30								41	41	120	161

Part B: Correlation Technique

7.13 AUGMENTING SHORT STREAMFLOWS AND FILLING-IN MISSING DATA

The common techniques for extending data of short duration are (1) precipitation–runoff relation, (2) correlation of two sets of streamflow data, and (3) comparison of flow–duration curves. The same methods are applicable for filling in the gaps of a stream-gaging station record. The application of the multiple correlation technique by utilizing meteorological and drainage basin data has been made to develop the generalized relations for the ungaged sites, as described in Part D. Commonly, the technique of rainfall–runoff relation is used to estimate annual flows for missing years, and correlation of the stream-gaging station records is used to extend short-term monthly records.

7.14 STATIONARY AND HOMOGENEOUS CHECK OF DATA

For the standard statistical techniques to be applied, the long-duration data series should satisfy the conditions of stationariness and homogeneity. When a series is divided into several segments and a statistical parameter such as the mean value is used to characterize the data of each segment, the expected value of the statistical parameter is practically the same for each segment in a stationary series.

The temporal homogeneity check is performed to detect any sudden changes or inconsistency in the data at any time in the period of record. The double-mass curve analysis (Section 3.5.1) is often used for this purpose. The spatial homogeneity is checked to ensure that the data are representative of hydrologically and meteorologically similar areas. A test for areal homogeneity has been developed by Langbein, as described by Chow (1964, pp. 8–36).

7.15 PRECIPITATION–RUNOFF RELATION FOR ESTIMATION OF STREAMFLOW

The runoff in this context signifies the streamflow. The rainfall and runoff, being parts of the same hydrologic cycle, are intimately correlated. The runoff relates to the flow passing through a section in a stream and it reflects the cumulative effect of the precipitation falling anywhere in the area representing the catchment of that section. An areal average of the precipitation from the measured values at the rain gages within the catchment is determined by Thiessen's polygon or Isohyetal method (Section 3.5.2). The average monthly or annual precipitation is related to the corresponding short-duration record of monthly or annual runoff (streamflow). The variability of streamflow is not reflected in the relation. Further, the precipitation in the catchment area takes time to reach the point of flow, depending on the distance. Hence the runoff at a stream point may in part be a result of the precipitation that had occurred in the catchment in the past. The effective precipitation includes all carryover portions that contribute to a current level of runoff. The procedure of determining the effective precipitation is essentially trial and error.

In a monthly correlation study, the precipitation carryover effect extends to many months in the past. The procedure is as follows: For each individual month, the correlations are attempted by assuming a relation of the type

$$Q = b_0 + b_1 X_1 + b_2 X_2 \quad \text{[unbalanced]} \qquad (7.19)$$

where

$$Q = \text{mean monthly flow for the month in question}$$
$$X_1, X_2 = \text{values of monthly precipitation for the current month}$$
$$\text{and series of previous months in different combinations}$$
$$b_0, b_1, b_2 = \text{constants obtained from the multiple regression}$$

For example,

$$X_1 = P_i + P_{i-1} + P_{i-2} \quad \text{[L]} \qquad (7.20a)$$
$$X_2 = P_{i-3} + P_{i-4} + P_{i-5} \quad \text{[L]} \qquad (7.20b)$$

where

$$P_i = \text{average precipitation for the month in question}$$
$$P_{i-1}, P_{i-2}, \ldots = \text{precipitation in previous months}$$

The relationship yielding the best value of the correlation coefficient is adopted for an individual month. The separate relations are derived for each month. The trial nature of the problem, with many possibilities of combinations, makes it desirable to be solved by the computer. An element of complexity is introduced in case of a significant groundwater contribution to the streamflow when the precipitation effect may extend to years in the past.

The approach becomes less tedious in correlating annual series because the effective precipitation consists of the current and the previous year or, at most, a few additional years. The effective precipitation in an annual correlation study is ascertained by a trial-and-error procedure applying rank analysis. Two steps involved are (1) determination of effective precipitation by rank analysis, and (2) derivation of a precipitation–runoff equation by simple regression analysis.

7.15.1 Rank Analysis for Effective Precipitation

The effective precipitation is that portion of the current year's precipitation and the proportion of the preceding years' precipitation that furnishes the current year's streamflow:

$$P_e = aP_0 + bP_1 + cP_2 + \cdots \quad \text{[L]} \qquad (7.21)$$

where

$$P_e = \text{effective annual precipitation}$$
$$P_0 = \text{current year's precipitation}$$
$$P_1, P_2, \ldots = \text{previous years' precipitation}$$
$$a, b, c, \ldots = \text{coefficient that must add to unity}$$

The trial should start with the two years' precipitation, P_0 and P_1, and should be extended beyond if good correlation is not achieved with these two years. The procedure can be explained by an example.

Example 7.11

The Warren River in Warren, Rhode Island, has an annual streamflow record from 1959 to 1975 as shown in Table 7.15. The average annual precipitation computed from the stations in the drainage basin are also listed. Both the streamflow and precipitation data are checked by the double-mass technique and found consistent. Determine the effective precipitation for the basin.

TABLE 7.15 ANNUAL FLOW OF THE WARREN RIVER, RHODE ISLAND

Year:	1959	1960	1961	1962	1963	1964	1965	1966
Streamflow (cfs)	2908	3450	3450	1100	3840	1930	1782	2350
Precipitation (in.)	32.32	31.18	32.44	17.13	36.02	22.44	23.07	26.42

Year:	1967	1968	1969	1970	1971	1972	1973	1974	1975
Streamflow (cfs)	2480	2750	3150	2290	3700	1100	1050	2740	2050
Precipitation (in.)	26.82	28.58	30.59	25.23	34.29	17.99	19.60	29.13	23.93

Year:	1976	1977	1978	1979	1980	1981	1982	1983	1984	1985
Streamflow (cfs)	—	—	—	—	—	—	—	—	—	—
Precipitation (in.)	19.10	28.66	17.32	18.27	30.11	24.0	19.17	17.80	20.50	23.80

Solution

1. Refer to Table 7.16.
2. Annual runoff is arranged in column 2 and precipitation in column 4 of the table.
3. The yearly runoff is assigned a rank number beginning with the highest runoff as number 1, as shown in column 3. When two values are identical, they are both assigned the average of the two sequence numbers they would have if they were slightly different from each other, as for years 1960 and 1961, and 1962 and 1972.
4. The yearly precipitation has also been assigned a rank number in column 5. The difference in rank between the precipitation and the runoff is squared and noted in column 6. The sum of column 6 is obtained.
5. A formula for effective precipitation is assumed and values of effective precipitation are computed. In the first assumption, a relation of $P_e = 0.9P_0 + 0.1P_1$ is assumed. The computed values of effective precipitation are again assigned rank numbers and the difference of the rank has been squared as before. The sum of the squares of the difference in rank (i.e., 4.0) is compared with that for the previous case (i.e., 11.0). A lower value means that the trial relation is better than the previous case.

Estimation of Surface Water Flow   Chap. 7

TABLE 7.16 COMPUTATION OF EFFECTIVE PRECIPITATION

(1)	Runoff		Observed Precipitation			First Assumption			Second Assumption			Third Assumption		
	(2)	(3)	(4)	(5)	(6)	(7)	(8)	(9)	(10)	(11)	(12)	(13)	(14)	(15)
Year	Yearly (cfs)	Rank	Average Yearly (in.)	Rank	Col. $(5-3)^2$	$0.9P_0 + 0.1P_1$	Rank	Col. $(8-3)^2$	$0.8P_0 + 0.2P_1$	Rank	Col. $(11-3)^2$	$0.7P_0 + 0.3P_1$	Rank	Col. $(14-3)^2$
1959	2908		32.32											
1960	3450	3.5	31.18	4	0.25	31.29	4	0.25	31.40	4	0.25			
1961	3450	3.5	32.44	3	0.25	32.29	3	0.25	32.19	3	0.25			
1962	1100	14.5	17.13	16	2.25	18.66	16	2.25	20.18	15	0.25			
1963	3840	1	36.02	1	0	34.13	1	0	32.24	2	1.0			
1964	1930	12	22.44	13	1	23.8	12	0	25.15	11	1.0			
1965	1782	13	23.07	12	1	23.0	13	0	22.94	13	0.0			
1966	2350	9	26.42	9	0	26.09	9	0	25.76	10	1.0			
1967	2480	8	26.82	8	0	26.80	8	0	26.76	8	0.0			
1968	2750	6	28.58	7	1	28.41	6	0	28.24	6	0			
1969	3150	5	30.59	5	0	30.39	5	0	30.19	5	0			
1970	2290	10	25.23	10	0	25.77	10	0	26.30	9	1			
1971	3700	2	34.29	2	0	33.38	2	0	32.47	1	1			
1972	1100	14.5	17.99	15	0.25	19.62	14	0.25	21.25	14	0.25			
1973	1050	16	19.60	14	4	19.44	15	1	19.25	16	0			
1974	2740	7	29.13	6	1	28.18	7	0	27.22	7	0			
1975	2050	11	23.93	11	0	24.45	11	0	24.96	12	1			
Total					11.00			4.0			7.00			

A second assumption of $P_e = 0.8P_0 + 0.2P_1$ is made and the procedure above is repeated. Since the sum of the squares of the difference in rank of 7.0 is now greater than that of the first assumption, the relation of the first assumption will correlate best with the runoff. Different trial relations are assumed until a low point in the sum of the squares is reached. Beyond this point, the sums of the squares increase with increasing portion of the preceding year's precipitation.

For this example, the relation is

$$P_e = 0.9P_0 + 0.1P_1 \qquad [\text{L}] \tag{7.22}$$

7.15.2 Correlation of Effective Precipitation and Runoff by Regression Analysis

The equation of the relation between effective precipitation and runoff is a straight line of the form

$$Q = C + mP_e \qquad [\text{unbalanced}] \tag{7.23}$$

where

$$Q = \text{runoff (streamflow)}$$
$$C, m = \text{constants representing abstractions}$$
$$P_e = \text{effective precipitation}$$

The yearly values of effective precipitation computed in Section 7.15.1 are plotted against the annual runoff and a straight line is drawn to average the pattern of plotted points. In many instances it is difficult to draw a line through the shotgun pattern of the points. A line of least-squares fit is drawn by simple regression analysis as described in Chapter 6.

For a straight-line relation, $y = mx + C$, the following set of equations provide the values of constants C and m and other statistical parameters.

$$C = \frac{\left(\sum y\right)\left(\sum x^2\right) - \left(\sum x\right)\left(\sum xy\right)}{N\sum x^2 - \left(\sum x\right)^2} \qquad [\text{L}] \tag{6.44a}$$

$$m = \frac{N\sum xy - \left(\sum x\right)\left(\sum y\right)}{N\sum x^2 - \left(\sum x\right)^2} \qquad [\text{dimensionless}] \tag{6.44b}$$

Standard variance:

$$S_y^2 = \frac{\sum y^2 - \left(\sum y\right)^2 \big/ N}{N - 1} \qquad [\text{L}^2] \tag{6.45a}$$

Standard error:

$$S_{yx}^2 = \frac{\sum y^2 - C\sum y - m\sum xy}{N - 2} \qquad [\text{L}^2] \tag{6.45b}$$

Correlation coefficient:

$$r = \left(1 - \frac{S_{yx}^2}{S_y^2}\right)^{1/2} \quad \text{[dimensionless]} \qquad (6.45c)$$

where N is the number of pairs of data.

Example 7.12

For Example 7.11, (a) determine the precipitation–runoff relation, and (b) using this relation, extend the streamflow record through 1985.

Solution

(a) Precipitation–runoff relation

1. Refer to Table 7.16.
2. Runoff data are given in column 2 and the effective precipitation data in column 7 of the table.
3. These data are plotted in Figure 7.17, which indicates a straight-line pattern.

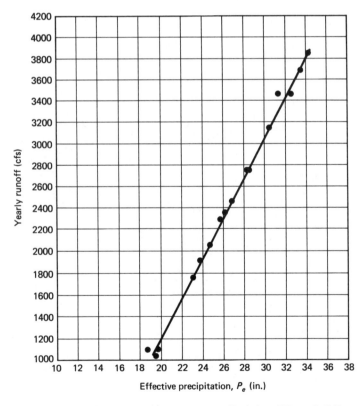

Figure 7.17 Effective precipitation and runoff relation of Example 7.12.

TABLE 7.17 REGRESSION ANALYSIS FOR PRECIPITATION–RUNOFF CORRELATION

No.	$Q = y$	$P_e = x$	x^2	xy ($\times 10^3$)	y^2 ($\times 10^6$)
1	3,450	31.29	979.1	107.95	11.90
2	3,450	32.29	1042.6	111.40	11.90
3	1,100	18.66	348.2	20.53	1.21
4	3,840	34.13	1,164.9	131.06	14.75
5	1,930	23.80	566.4	45.93	3.72
6	1,782	23.00	529.0	40.99	3.18
7	2,350	26.09	680.7	61.31	5.52
8	2,480	26.80	718.2	66.46	6.15
9	2,750	28.41	807.1	78.13	7.56
10	3,150	30.39	923.6	95.73	9.92
11	2,290	25.77	664.1	59.01	5.24
12	3,700	33.38	1,114.2	123.51	13.69
13	1,100	19.62	384.9	21.58	1.21
14	1,050	19.44	377.9	20.41	1.10
15	2,740	28.18	794.1	77.21	7.51
16	2,050	24.45	597.8	50.12	4.20
Σ	39,212	425.70	11,692.8	1,111.33	108.76

4. Regression analysis is arranged in Table 7.17.

From eq. (6.44a),

$$C = \frac{\left(\sum y\right)\left(\sum x^2\right) - \left(\sum x\right)\left(\sum xy\right)}{N\left(\sum x^2\right) - \left(\sum x\right)^2}$$

$$= \frac{39,212(11,692.8) - 425.7(1111.33 \times 10^3)}{16(11,692.8) - (425.7)^2}$$

$$= -2490$$

From eq. (6.44b),

$$m = \frac{N\left(\sum xy\right) - \left(\sum x\right)\left(\sum y\right)}{N\left(\sum x^2\right) - \left(\sum x\right)^2}$$

$$= \frac{16(1111.33 \times 10^3) - 39,212(425.7)}{16(11,692.8) - (425.7)^2}$$

$$= 185.6$$

From eq. (6.45a),

$$S_y^2 = \frac{\sum y^2 - \left(\sum y\right)^2 / N}{N - 1}$$

$$= \frac{(108.76 \times 10^6) - (39,212)^2/16}{15}$$

$$= 844,079$$

From eq. (6.45b),

$$S_{yx}^2 = \frac{\sum y^2 - C\sum y - m\sum xy}{N - 2}$$

$$= \frac{(108.76 \times 10^6) - (-2490)(39,212) - 185.6(1111.33 \times 10^3)}{14}$$

$$= 9645$$

From eq. (6.45c),

$$r = \left(1 - \frac{9645}{844,079}\right)^{1/2}$$

$$= 0.99 \quad \text{(good correlation)}$$

Thus the correlation equation is

$$Q = 185.6P_e - 2490 \quad \text{[unbalanced]} \tag{7.24}$$

(b) Extension of streamflow record

1. Using eq. (7.22), the average annual precipitation is converted to the effective precipitation as per column 2 of Table 7.18.
2. By eq. (7.24), the streamflow is ascertained from the effective precipitation as shown in column 3 of Table 7.18.

TABLE 7.18 COMPUTATION OF RUNOFF FROM PRECIPITATION

(1) Year	(2) Effective Precipitation (in.)	(3) Streamflow (cfs)
1976	19.58	1144
1977	27.70	2651
1978	18.46	936
1979	18.17	882
1980	28.93	2879
1981	24.61	2078
1982	19.65	1157
1983	17.94	840
1984	20.23	1265
1985	23.47	1866

7.16 CORRELATION OF GAGING-STATION RECORDS FOR ESTIMATION OF STREAMFLOW

The technique of correlation, which is a process of establishing a mutual relation between a variable and related variable(s), has been used in the rainfall–runoff relationship. Another convenient application is to correlate directly stream-gaging records between two or more stations, one with short-term data and others with long-term records. The relation thus established, based on the concurrent data corresponding to the short period of record, is used to obtain correlative estimates of monthly discharges of the short-duration station from the records of the long-duration station outside the common period.

The runoff depends on the climate and the basin characteristics. The complex interrelations of climate and drainage basin characteristics are integrated in the flow of the stream, and their aggregate effect is measured directly at the stream-gaging station (Searcy, 1960). Evaluating the effect of various drainage and climatic factors in order to predict the runoff is a difficult task. The direct comparison of their result (i.e., streamflows) is more convenient. However, there has to be a common basis for this comparison, which is provided by the similarity of the climate and certain common characteristics of the basins being compared. Thus a correlation between a mountain stream and a desert stream may not be satisfactory. The degree of reliability of the relation will decrease as the distance between two gaging stations increases. When two basins differ greatly in size (by more than 10 times), they can be poorly related.

7.16.1 Simple Correlation

Although the relations between the flow can be based on concurrent daily, weekly, monthly, or annual discharge streamflow, monthly flow is a commonly used duration. The relation between two stream-gaging records is usually expressed by two curves, one for the high flow and one for the low flow, and occasionally by a third curve at extremely low flows. Very seldom is only one curve adequate to represent the relation. It is thus desirable as a first step to plot the discharge of one station against the corresponding discharge at the other station to identify the position of changed relations (shift in the curve). The logarithmic plot (converting discharges to their log values or using log-log paper) is convenient, as it tends to transform common curvilinear relations to straight lines and to normalize the streamflow data.

The fitting of the curve (line) through the scatter of the data is performed either graphically or numerically. In the graphic procedure (Searcy, 1960), the plot is first divided into a number of slices of equal width using the vertical lines, and the median of the points in each slice is determined in both X and Y directions. Then, using the horizontal lines, the plot is divided into a number of equally spaced slices and the median of the points in each slice is determined again in both X and Y directions. The straight line(s) are drawn through the average of the median points, giving reduced weight to the endpoints.

In the numerical procedure, a linear regression for a short sequence (one station) and a long sequence (another station) of streamflows is used to lengthen the short sequence. If the observed events (streamflows) for the long and short sequences are represented as

$$x_1, x_2, \ldots, x_{N_1}, x_{N_1+1}, \ldots, x_{N_1+N_2}$$

$$y_1, y_2, \ldots, y_{N_1}$$

then the estimate of y_i beyond N_1 is obtained from the following regression of y on x, using the known values of x_i.

$$y_i = mx_i + C \qquad [\text{L}^3\text{T}^{-1}] \tag{7.25}$$

where the values of m and C are calculated from eqs. (6.44a) and (6.44b).

The estimated values of y_i from eq. (7.25) tend to yield a smaller variance than would the real observations. To "preserve" the variance inherent in the observed values, a random component is added to the regression estimates (Matalas and Jacobs, 1964). This component, referred to as noise, is normally distributed with zero mean and variance proportional to the variance of the short sequence. The noise is shown to have no effect on the reliability of the estimate of the mean, but improves the reliability of the variance for the lengthened sequence. Including the noise term, the equation for streamflow estimates is given by

$$y_i = mx_i + C + \sqrt{1 - r^2}\,S_y e_i \qquad [\text{L}^3\text{T}^{-1}] \tag{7.26}$$

where

r = product-moment correlation coefficient given by eq. (6.45c)

S_y = standard deviation of y from short sequence by eq. (6.45a)

e_i = random normal variable with zero mean and unit variance (Table 7.23)

The strength (goodness) of the regression analysis is measured by the product-moment correlation coefficient, r. Table 7.19 gives the minimum values of r for the means. When the computed r by eq. (6.45c) is equal to or larger than the value from the table for the length of the short sequence, by including estimated values, the lengthened sequence provides a better estimate of the mean than does the short sequence. Matalas and Jacobs (1964) suggested similar tables of minimum r for the variance with and without the noise component. If the correlation coefficient exceeds 0.8, the noise component need not be added to get a reliable estimate of the variance from the lengthened sequence. Thus eq. (7.25) can be used to estimate streamflows, since the use of pseudorandom numbers is not too appealing, with every investigator deriving different values of streamflows (although within the limits expected by chance).

TABLE 7.19 MINIMUM VALUES OF THE CORRELATION COEFFICIENT FOR THE MEAN

Period of short record	10	15	20	25	30	35
Minimum r	0.35	0.28	0.24	0.21	0.19	0.17

Source: Matalas and Jacobs (1964).

7.17 ESTIMATION OF STREAMFLOW BY COMPARISON OF FLOW–DURATION CURVES

The flow–duration curve and the procedure of constructing the curve may change depending on the length of the streamflow record. This characteristic is used to extend the data at a site with the short-term record, provided that the long-term data are available at any other site having similar hydrological conditions. The following procedure can be applied:

1. Construct the flow–duration curve for the site in question based on the short-term data.

2. For the adjacent site, construct two flow–duration curves, one based on the long-term data and one based on the data concurrent to the short-term record of the first site.

3. Compare the flow–duration curves based on the short-term record for the given site and the adjacent site and on the basis of this comparison produce a curve representing the long period of record at the site in question by proportionally adjusting the long-period flow–duration curve of the adjacent site.

4. From the long-term flow–duration curve of the adjacent site, for a recorded discharge on the adjacent site, read the percent time value exceeded from the abscissa. For this percent exceedence, the discharge at the given site can be ascertained from the adjusted curve. If the curves for the given site and the adjacent site are drawn on the same graph paper, this conversion can be obtained directly.

Example 7.13

Stream-gaging station A has a short-term streamflow record of 7 years, from 1947 to 1953, as compared with stream-gaging station B, which has an up-to-date record of 39 years. The mean monthly flow–duration curves prepared from the data of 1947 through 1953 for both the stations have been summarized in Table 7.20. The ordinates of the duration curve for station B based on the entire period of record are also indicated. Construct the flow–duration curve for site A representing the long period of record. If a flow of 2000 cfs was recorded in May 1970 for station B, what was the corresponding flow at site A?

TABLE 7.20 MEAN MONTHLY FLOW–DURATION CURVES

Percent Equaled or Exceeded	Streamflow Based on Short-Term Record (cfs)		Streamflow Based on Long-Term Record for Station B (cfs)
	Station A	Station B	
10	1500	2880	2520
20	1280	2500	2380
30	1160	2200	2250
40	1080	1960	2100
50	1000	1760	1960
60	960	1600	1800
70	900	1460	1660
80	850	1300	1520
90	800	1160	1380
95	760	1080	1300

Solution Refer to Table 7.21.

1. Column 4 determines the ratio of station A to station B for various exceedence.
2. This ratio is applied to the data of station B in column 5 to derive the adjusted flow of station A in column 6.

TABLE 7.21 CORRELATION OF LONG-TERM FLOW–DURATION CURVES

(1)	(2)	(3)	(4)	(5)	(6)
	Short-Term Record			Long-Term Record	
Percent Equaled or Exceeded	Station A	Station B	Ratio $\frac{A}{B}$	Station B	Station A (col. 5 × col. 4)
10	1500	2880	0.52	2520	1310
20	1280	2500	0.51	2380	1214
30	1160	2200	0.53	2250	1193
40	1080	1960	0.55	2100	1155
50	1000	1760	0.57	1960	1117
60	960	1600	0.60	1800	1080
70	900	1460	0.62	1660	1029
80	850	1300	0.65	1520	988
90	800	1160	0.69	1380	952
95	760	1080	0.70	1300	910

3. The flow–duration curves for Stations A and B, based on the data of columns 5 and 6, are plotted in Figure 7.18.
4. The discharge for station A, corresponding to a discharge of 2000 for station B for the same exceedence level of 47%, is read from the figure. Thus $Q = 1120$ cfs.

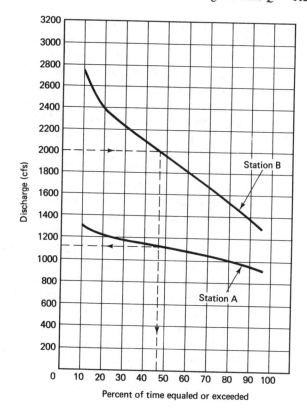

Figure 7.18 Long-term flow–duration curves.

Part C: Synthetic Technique

7.18 GENERATION OF SYNTHETIC STREAMFLOWS

A very common constraint encountered in the context of water resources planning is inadequacy of streamflow records. The available streamflows, known as historical records, are often quite short, generally less than 25 years in length. These do not cover even the economic life of a project of 50 to 100 years. The same flow sequence is not likely to be repeated within the span of 100 years. A system designed on the basis of the historical record only faces a chance of being inadequate for the unknown flow sequence that the system might experience. Further, the historical record comprising a single short series does not cover a sequence of low flows as well as high flows. The reliability of a system has to be evaluated under these conditions, which is not possible with the historical records alone. In a statistical sense, the historical record is a sample out of a population of natural streamflow process. If this process is considered stationary,* many series representing such samples can be formulated that will be statistically similar to the historical record. This is the basis of the synthetic technique. Thus the generated flows are neither historical flows nor a prediction of future flows but are representative of likely flows in a stream.

7.19 HYDROLOGIC TIME SERIES AND STOCHASTIC PROCESS

A sequence of values of any variable represented with respect to time is known as a time series. The time can be a discrete value, a time interval, or a continuous function. The hydrologic data of streamflows, precipitation, groundwater or lake levels, water temperatures, or oxygen concentration fall under the category of time series. These data can be deterministic, random, or a combination of the two. Streamflow, being a natural phenomenon, has a random component. However, it is not fully random, since it has been observed that a low flow tends to follow another low flow, and a high flow has a similar pattern. The word "stochastic" is used to denote the randomness in statistics, but in hydrology it refers to a partial random sequence as well. Thus the streamflow data represent a time series involving a stochastic process.

    The analysis of a time series, in the time domain, is performed by a parameter known as the serial correlation coefficient or the autocorrelation coefficient. This parameter indicates the dependence in successive values of a time series. This coefficient is determined not only for successive values (elements) of a time series but for elements that are various time intervals apart, known as the lag period. A graph of the autocorrelation coefficient against the lag period is known as the correlogram. If a correlogram shows zero or nearly zero values for all lag periods, the process is purely random. A value close to 1 will suggest a dominating deterministic process.

    The analysis of a time series in the frequency domain is done by the spectral density that identifies the cyclic nature or periodicity in the series. The density indicates the cycle in the deterministic data. In a purely random process it oscillates randomly. The purpose of streamflow synthesis, however, is not to analyze a time series

---

*The statistical properties of a series do not change with time.

but to generate the data based on the series. This does not require the decomposition of the time series by the analysis above but an understanding of its statistical properties to reproduce series of similar statistical characteristics. Various stochastic processes used for generating the hydrologic data are discussed below.

7.20 MARKOV PROCESS OR AUTOREGRESSIVE (AR) MODEL

The Markov process considers that the value of an event (i.e., streamflow) at one time is correlated with the value of the event at an earlier period (i.e., a serial or autocorrelation exists in the time series). In a first-order Markov process, this correlation exists in two successive values of the event. The first-order Markov model, which constitutes the classic approach in synthetic hydrology, states that the value of a variable x in one time period is dependent on the value of x in the preceding time period plus a random component. Thus the synthetic flows for a stream represent a sequence of numbers, each of which consists of two parts:

$$x_i = d_i + e_i \quad [L^3T^{-1}] \tag{7.25}$$

where

$$x_i = \text{flow at } i\text{th time } (i\text{th number of a time series})$$
$$d_i = \text{deterministic part at } i\text{th time}$$
$$e_i = \text{random part at } i\text{th time}$$

The values of x_i are tied up with the historical data by ensuring that they belong to the same frequency distribution and possess similar statistical properties (mean, deviation, skewness) as the historical series.

The general Markov procedure of data synthesis comprises: (1) determination of statistical parameters from the analysis of the historical record, (2) identifying the frequency distribution of the historical data, (3) generating random numbers of the same distribution and statistical characteristics, and (4) constituting the deterministic part considering the persistence (influence of previous flows) and combining with the random part. The various forms and combinations of deterministic and random components are recognized as different models. The simplest of these is a single season (annual) flow model of lag 1 which assumes that the magnitude of the current flow is significantly correlated with the previous flow value only. Multiple-season models divide the yearly flow into seasons or months.

Multilag models have a long memory, that is, consider the influence on the current flow to extend beyond the previous year alone. When the flows at the various locations are strongly correlated, multisite models are needed that involve comparatively complicated mathematics and often include simplifying assumptions.

7.20.1 Statistical Parameters of Historical Data

The four parameters that are important in a synthetic study are (1) mean flow, (2) standard deviation, (3) coefficient of skewness, and (4) correlation coefficient. The sample mean flow is*

*The units are the same as those of flow, X_i.

$$\overline{X} = \frac{1}{n} \sum_{i=1}^{n} x_i \qquad (7.26)$$

where

$\overline{X}$ = mean observed (historical) flow

n = total numbers (values) of flow

$\mu = E(\overline{X})$ (i.e., $\overline{X}$ is expected to be equal to the population mean μ as n tends to infinity)

x_i = ith number of observed flow

The sample estimate of the variance or standard deviation, S, which is a measure of the variability of the data, is given by[†]

$$S^2 = \frac{1}{n-1} \sum_{i=1}^{n} (x_i - \overline{X})^2 \qquad (7.27)$$

The sample coefficient of skewness, g, which is a measure of the lack of symmetry, is given by

$$g = \frac{n \sum_{i=1}^{n} (x_i - \overline{X})^3}{(n-1)(n-2)S^3} \qquad \text{[dimensionless]} \qquad (7.28)$$

The serial correlation coefficient is a measure of the extent to which a flow at any time is affected by the flow at another time. The K-lag coefficient, in which the effect extends by K time units is given by[‡]

$$r_K = \frac{\sum_{i=1}^{n-K} (x_i - \overline{X})(x_{i+K} - \overline{X})}{(n-K)S^2} \qquad \text{[dimensionless]} \qquad (7.29)$$

The one-lag serial coefficient, in which the current flow is affected only by the previous flow, can be obtained by substituting $K = 1$ in eq. (7.29). The additional lags should be included as long as they produce a model that explains more about the pattern of flows than one with fewer lags does (Fiering and Jackson, 1971).

7.20.2 Identification of Distribution

Theoretical frequency distributions are described in Chapter 8. In streamflow generation, the distributions generally used are normal, log-normal, and gamma families. The bell-shaped, or normal, distribution is most extensively used in statistical applications, because the sum of variables derived from any distribution tends to be distributed normally according to the central limit theorem. To test for normality, the historical values of flow are plotted against the percentage of values in the record that are equal to or greater than the plotted value on normal arithmetic probability paper. The flows are arranged in descending order. For each value x_i, the percent is computed

[†]The units are the flow units squared.
[‡]Equation (7.29) is slightly biased. It is, however, widely used for simplicity.

by $100(n - i + 1)/n$ where i is the rank of value x_i and n is the number of historic values. If the plot is a straight line, the distribution is normal. Also, the coefficient of skewness should be close to zero, since the normal distribution has no skewness.

The second distribution that has received wide usage in hydrology is the log-normal distribution, according to which the logarithms of variables (to any base) are normally distributed. The tables of standard normal distribution can be used to evaluate the log-normal distribution after logarithmic conversion of the variables. The log-normal distribution is positively skewed, which is a characteristic of many hydrologic variables. Since small changes in low values produce comparatively larger changes in their logarithmic values, this distribution is especially preferred in low-flow studies. To test for log-normality, the logarithms of the flows are plotted on normal arithmetic probability paper, as discussed above, or values are plotted directly on log probability paper. A straight-line plot indicates the log-normal distribution. Also, the skewness computed from the logarithms of the values should be close to zero.

When the historical records of flows, as also of logarithms of flows, show appreciable skewness, the gamma distribution is used, which has a distinctly positive skewness. This distribution should, however, be excluded when multiple lags exist (i.e., when a flow is affected by many previous flows). Often, the historical data do not clearly fit any of these distributions. The choice has to be made based on the purpose, economics, and any other considerations.

Example 7.14

The annual flows for the Lamprey River near Newmarket, New Hampshire, are given in column 2 of Table 7.22. Compute the statistical parameters for this flow sequence.

Solution From eq. (7.26): mean flow:

$$\overline{X} = \frac{1}{21} \sum_{i=1}^{21} x_i = \frac{1}{21} \sum \text{col. 2} = \frac{13{,}711}{21} = 653 \text{ cfs}$$

From eq. (7.27): variance:

$$S^2 = \frac{1}{21 - 1} \sum_{i=1}^{21} (x_i - \overline{X})^2 = \frac{1}{20} \sum \text{col. 4} = \frac{696{,}690}{20} = 34{,}834$$

$$S = 186.6$$

From eq. (7.28): coefficient of skewness:

$$g = \frac{21 \left(\sum \text{col. 5} \right)}{20(19)(186.6)^3} = \frac{21(11{,}661 \times 10^4)}{20(19)(186.6)^3} = 1$$

From eq. (7.29): lag 1 serial correlation coefficient: The value in column 6 is obtained by multiplying the number in column 3 by its successive value:

$$r_1 = \frac{\sum\limits_{1}^{21-1} (x_i - \overline{X})(x_{i+1} - \overline{X})}{(20)S^2}$$

$$= \frac{\sum \text{col. 6}}{(20)S^2} = \frac{2552.4 \times 10^2}{20(186.6)^2} = 0.37$$

TABLE 7.22 ANNUAL FLOWS AND COMPUTATION OF THE STATISTICAL PARAMETERS FOR THE LAMPREY RIVER, NEW HAMPSHIRE

(1) i	(2) Annual Flow, x_i (cfs)	(3) $x_i - \bar{X}$	(4) $(x_i - \bar{X})^2$	(5) $(x_i - \bar{X})^3$ ($\times 10^4$)	(6) $(x_i - \bar{X})(x_{i+1} - \bar{X})$ ($\times 10^2$)	(7) $(x_i - \bar{X})(x_{i+2} - \bar{X})$ ($\times 10^2$)
1	472	−181	32,761	−593	432.6	155.7
2	414	−239	57,121	−1,365.2	205.5	614.2
3	567	−86	7,396	−63.6	221	9.5
4	396	−257	66,049	−1,697.5	28.3	403.5
5	642	−11	121	−0.1	17.3	19
6	496	−157	24,649	−387	273.2	191.5
7	479	−174	30,276	−526.8	212.3	−276.7
8	531	−122	14,884	−181.6	−194	−297.7
9	812	159	25,281	402	388	−256.0
10	897	244	59,536	1,452.7	−392.8	−65.9
11	492	−161	25,921	−417.3	43.5	77.3
12	626	−27	729	−2.0	13	−32.7
13	605	−48	2,304	−11.1	−58.1	−249.1
14	774	121	14,641	177.2	628	119.8
15	1,172	519	269,361	13,979.8	513.8	980.9
16	752	99	9,801	97	187	16.8
17	842	189	35,721	675.1	32.1	139.9
18	670	17	289	0.5	12.6	−10.7
19	727	74	5,476	40.5	−46.6	75.5
20	590	−63	3,969	−25	−64.3	
21	755	102	10,404	106.1		
Σ	13,711		696,690	$11,661 \times 10^4$	$2,552.4 \times 10^2$	$1,614.8 \times 10^2$

Lag 2 serial correlation coefficient: The value in column 7 is obtained by multiplying the column 3 value by the value two time intervals following in the same column.

$$r_2 = \frac{\sum\limits_{i=1}^{21-2} (x_i - \overline{X})(x_{i+2} - \overline{X})}{(n-2)S^2}$$

$$= \frac{\sum \text{col. 7}}{19(186.6)^2} = \frac{1614.8 \times 10^2}{19(34,834)} = 0.24$$

Hence lag 1 is better related.

7.20.3 Generation of Random Numbers

The source of random numbers in a simulation study is either the computer-based pseudorandom-number generator or the random number tables. The tables provide a set of random numbers drawn from the uniform distribution or from the normal distribution. A set of normally distributed random numbers are given in Table 7.23. The random number should belong to the same distribution to which the historical record belongs for the generated flow to have similar characteristics. It is a simple matter to convert a sequence of uniformly distributed random numbers to normal random numbers if only a uniform distribution table exists. According to the central limit theorem, the numbers formed by summing up many (say, 12) consecutive uniformly distributed numbers will be approximately normally distributed. The normally distributed random numbers for normal or log-normal flow sequences are thus conveniently

TABLE 7.23 STANDARD NORMAL RANDOM SAMPLING DEVIATES

	0	1	2	3	4	5	6	7	8	9
0	−0.523	0.611	−0.359	−0.393	0.084	−0.931	−0.027	0.798	1.672	−1.077
1	−1.536	−0.454	0.071	−2.129	1.525	0.261	2.319	0.972	0.767	−2.849
2	−0.121	0.968	−1.943	0.581	−0.711	−0.060	−0.482	−0.746	−0.747	1.254
3	−0.542	−0.807	0.168	0.839	−0.756	−0.453	−1.912	0.766	−0.890	0.205
4	0.131	−0.859	−1.096	−0.785	0.310	1.314	−0.231	0.029	1.819	−1.602
5	−0.234	0.551	0.743	−0.900	0.435	−2.999	0.212	0.869	−0.716	−0.410
6	−1.010	1.347	0.230	0.009	−1.495	2.145	−1.033	0.729	0.309	0.920
7	0.273	−0.885	−0.016	0.775	−1.740	0.353	−1.519	0.958	−0.448	2.185
8	−0.102	−1.111	−0.585	1.461	−0.307	1.489	−0.196	0.506	−0.662	−1.175
9	0.368	−0.710	0.407	0.066	−0.617	−0.580	0.107	−2.247	1.616	−1.060
10	−1.762	1.382	1.142	−2.056	−0.400	−1.701	−0.914	−1.000	−0.172	0.903
11	0.306	−0.607	−0.324	1.171	1.016	−1.829	1.723	−0.513	−0.657	2.011
12	−0.465	−1.214	−0.174	0.894	0.245	−0.987	−1.155	0.592	−0.411	−0.109
13	−0.004	−0.029	−0.633	0.004	−0.603	1.104	−0.655	1.191	0.938	−0.805
14	0.593	0.252	−0.541	0.318	1.268	1.972	0.875	−1.030	−1.175	0.445
15	0.233	0.430	−0.331	−1.272	−0.289	−0.060	−0.754	0.789	0.546	0.687
16	0.571	−0.215	−1.090	0.610	−0.810	−0.364	−1.282	0.010	0.586	0.926
17	0.370	0.976	1.017	1.106	0.441	−2.376	0.793	0.016	−0.704	0.146
18	−0.009	−1.285	−0.346	−0.323	0.609	−0.373	0.078	−1.034	0.153	0.997
19	0.416	−0.131	0.668	0.662	−1.835	1.646	0.197	0.131	0.783	0.076

Source: Fiering and Jackson (1971).

available. For flows that are distributed as gamma variates, use the Wilson and Hilferty (Fiering and Jackson, 1971, p. 53) formula that incorporates the skewness coefficient and the serial correlation coefficient.

Further, the normal random numbers given in Table 7.23 have a zero mean and one standard deviation, known as the standard normal deviates. A transformed random variable with the zero mean and variance of random variables of S_e^2 can be given by

$$n_i = S_e t_i \quad \text{[dimensionless]} \tag{7.30}$$

where

t_i = standard normal variate (with mean zero and variance 1)

n_i = transformed random variable with zero mean and variance of random numbers of S_e^2

7.20.4 Form of Deterministic and Random Components

Streamflows show persistence as reflected in their flow pattern. The Markovian process or autoregressive (AR) model considers that this persistence is indicated through the serial correlation within the sequence. Thus in the p-order model the effect runs through p terms, and the autoregressive model, AR(p), takes the form

$$a_i = \phi_{p,1} a_{i-1} + \phi_{p,2} a_{i-2} + \phi_{p,3} a_{i-3} + \cdots + \phi_{p,p} a_{i-p} + n_i \quad \text{[dimensionless]} \tag{7.31}$$

where

a_i = ith variable of the sequence (stochastic component) of zero mean and unit variance

n_i = random number at ith time

$\phi_{p,1}, \phi_{p,2}, \ldots$ = autoregressive parameters or weights

The stochastic component can be related to the flow at any time by

$$x_i = \overline{X} + S a_i \quad [\text{L}^3\text{T}^{-1}] \tag{7.32}$$

where $\overline{X}$ and S are the mean and variance of the historical flow sequence.

The first-order autoregressive model AR(1), commonly known as a Markov model, reduces to the form

$$a_i = \phi_{i,1} a_{i-1} + n_i \quad \text{[dimensionless]} \tag{7.33}$$

When the conditions of mean $a_i = 0$, variance $a_i = 1$, and expectation $E(n_i a_{i-1}) = 0$ are included for the sequence a, the following relations are derived:

$$\phi_{i,1} = r_1 \quad \text{[dimensionless]} \tag{7.34}$$

where r_1 is the lag 1 autocorrelation coefficient and

$$S_e^2 = 1 - r_1^2 \quad \text{[dimensionless]} \tag{7.35}$$

7.20.5 Formulation of the Markov Model

When equations (7.32), (7.33), (7.30), (7.34), and (7.35) are combined, we arrive at the following Markov model for annual flow results comprising deterministic and random parts.

$$x_i = \overline{X} + \underbrace{r_1(x_{i-1} - \overline{X})}_{\text{deterministic}} + \underbrace{S\sqrt{(1 - r_1^2)}\,t_i}_{\text{random}} \quad [\text{L}^3\text{T}^{-1}] \tag{7.36}$$

where

$$x_i = \text{streamflow at } i\text{th time}$$

$$\overline{X} = \text{mean of recorded flow}$$

$$r_1 = \text{lag 1 serial or autocorrelation coefficient}$$

$$S = \text{standard deviation of recorded flow}$$

$$t_i = \text{random variate from an appropriate distribution}$$
$$\text{with a mean of zero and variance of unity}$$

$$i = i\text{th position in series from 1 to } N \text{ years}$$

The procedure to generate a series of flows is as follows:

1. Determine the mean flow, variance, coefficient of skewness, and lag 1 serial correlation coefficient from the historical record and identify the distribution.
2. Pick up the random numbers from a generator or a table. To use the table, close your eyes and place a pencil point at any number on any page. The random numbers are taken consecutively succeeding the number selected.
3. Starting with x_{i-1} equal to $\overline{X}$, determine x_i, which becomes x_{i-1} to compute the next value. This is a cyclic computation.
4. If the generated flow is negative, use this to generate the next flow and then discard it from the series.
5. To neglect the effect of the starting condition, discard the first 25 to 50 generated flows in the series.
6. If the distribution is log-normal, use the logarithm of the values and finally convert back the flows. Other procedures for generating log-normal flows are given by Beard (1965) and Matalas (1967). The gamma distribution needs conversion of random variates, as previously stated.

A model on the same lines for monthly flows, developed by Thomas and Fiering, has the following form (Maass et al., 1962, p. 467):

$$q_{i,j} = \overline{q}_j + b_j(q_{i-1,j-1} - \overline{q}_{j-1}) + t_{i,j}S_j(1 - r_j)^{1/2} \quad [\text{L}^3\text{T}^{-1}] \tag{7.37}$$

where

$$i = \text{month in series, measured from the beginning}$$

$$j = \text{month in year, } j = 1, 2, \ldots, 12 \text{ for January to December}$$

$$q_{i,j} = \text{flow in } i\text{th month from the beginning, for } j\text{th month of the year}$$

$q_{i-1,j-1}$ = immediate previous month

$\bar{q}_j$ = mean of flows of jth month (12 values)

b_j = regression coefficient of flows of jth month and flows of $(j - 1)$th month = $r_j S_j / S_{j-1}$ (12 values)

r_j = correlation coefficient between flows of jth month and $(j - 1)$th month (12 values)

S_j = standard deviation for jth month (12 values)

$t_{i,j}$ = random normal deviate of zero mean and unit standard deviation

Models of monthly flows are handled expediently by a computer. The Hydrologic Engineering Center of the U.S. Army Corps of Engineers has prepared a very versatile model, the HEC-4, that is capable of analyzing monthly streamflows at 10 stations simultaneously to produce a sequence of hypothetical flows with maximum, minimum, and average values for each month.

Example 7.15

Generate a synthetic annual series similar to the historical record of Example 7.14. Assume the normal distribution.

Solution

1. From Example 7.14, $\bar{X} = 653$ cfs, $S = 186.6$ cfs, and $r_1 = 0.37$.
2. Use Table 7.23 for standard normal deviates, t_i.
3. Using eq. (7.36), x_i are computed in Table 7.24, starting with $x_{i-1} = \bar{X}$, in column 2.

TABLE 7.24 COMPUTATION OF ANNUAL FLOW BY ONE-LAG MARKOV MODEL

(1)	(2)	(3)	(4)	(5) $\bar{X} + r_1(x_{i-1} - \bar{X})$ Deterministic Component	(6) t_i (Table 7.23)	(7) $t_i S \sqrt{1 - r_1^2}$ Random Component	(8)
i	x_{i-1}	$x_{i-1} - \bar{X}$	$r_1(x_{i-1} - \bar{X})$				x_i
1	653	0.0	0.0	653	0.131	22.7	675.7
2	675.7	22.7	8.4	661.4	−0.859	−148.9	512.5
3	512.5	−140.5	−52.0	601.0	−1.096	−190.0	411.0
4	411.0	−242.0	−89.5	563.5	−0.785	−136.1	427.4
5	427.4	−225.6	−83.5	569.5	0.310	53.7	623.2
6	623.2	−29.8	−11.0	642.0	1.314	227.8	869.8
7	869.8	216.8	80.2	733.2	−0.231	−40.0	693.2
8	693.2	40.2	14.9	667.9	0.029	5.0	672.9
9	672.9	19.9	7.3	660.3	1.819	315.3	975.6
10	975.6	322.6	119.4	772.4	−1.602	−277.7	494.7
11	494.7	−158.3	−58.6	594.4	−0.234	−40.5	553.9
12	⋮						

7.21 AUTOREGRESSIVE-MOVING AVERAGE (ARMA) MODEL

In the autoregressive models of Section 7.20, the flows generated depend on the pre-assigned number of past flow values and a random variate. There is another category, known as the moving average (MA) models, that considers a stochastic component (streamflow event) to be a constituent of a number of random variates, with the current variate assigned a weight of unity and random variates generated at antecedent times multiplied by the assigned factors. Such models are generally inappropriate for direct application to hydrology. A moving average model is, however, combined with an autoregressive model to produce a mixed model known as an ARMA model, which appropriately represents the effect of linear aquifer, linear reservoir, and independent rainfall pattern. One of the first applications of ARMA models in hydrography was made by Carlson et al. (1970).

ARMA (p, q) consists of two polynomials of order p and q, respectively, as follows:

$$a_i = \phi_{p,1}a_{i-1} + \phi_{p,2}a_{i-2} + \cdots + \phi_{p,p}a_{i-p} + n_i - \theta_{q,1}n_{i-1} - \theta_{q,2}n_{i-2} - \cdots$$
$$- \theta_{q,q}n_{i-q} \quad \text{[dimensionless]} \quad (7.38)$$

where $\theta_{q,1}, \theta_{q,2}, \ldots$ are the random variate weights.

Estimation of the parameters ϕ and θ is not a straightforward procedure. In the class of mixed models, the simplest is the ARMA (1,1) model given by

$$a_i = \phi_{1,1}a_{i-1} + n_i - \theta_{1,1}n_{i-1} \quad \text{[dimensionless]} \quad (7.39a)$$

or

$$a_i = \phi_{1,1}a_{i-1} + Se(t_i - \theta_{1,1}t_{i-1}) \quad (7.39b)$$

From the known statistical conditions of the series in eq. (7.39), the following relations are obtained, to be used in eq. (7.39b) and eq. (7.32) to generate flow:

$$\phi_{1,1} = \frac{r_2}{r_1} \quad \text{[dimensionless]} \quad (7.40)$$

$$S_e^2 = \frac{1 - \phi_{1,1}^2}{1 + \theta_{1,1}^2 - 2\phi_{1,1}\theta_{1,1}} \quad \text{[dimensionless]} \quad (7.41)$$

and

$$r_1 = \frac{(1 - \phi_{1,1}\theta_{1,1})(\phi_{1,1} - \theta_{1,1})}{1 + \theta_{1,1}^2 - 2\phi_{1,1}\theta_{1,1}} \quad \text{[dimensionless]} \quad (7.42)$$

where r_1 and r_2 are lag 1 and lag 2 serial correlation coefficients.

An initial estimate of $\phi_{1,1}$ is obtained from eq. (7.40) and of $\theta_{1,1}$ is obtained from eq. (7.42) by substituting $\phi_{1,1}$ and computed r_1. Final estimates of $\phi_{1,1}$ and $\theta_{1,1}$ are made by a least-squares fitting procedure (Kottegoda, 1980, p. 128).

If the value of parameter $\phi_{1,1}$ is too close to its limits of -1 and $+1$, nonstationary behavior of the historical hydrologic sequence is indicated. The nonstationariness is accounted for by means of a dth-order difference operator, which represents successive difference of d terms of stochastic variables (i.e., a_i values). This is known as an autoregressive-integrated moving average ARIMA (p, d, q) model. Box and Jenkins (1976) consider that the parameters p, d, and q need not be greater than 2 for practical purposes. By changing the form of variable a_i, an ARIMA (p, d, q) model can be transformed to an ARMA (p, q) type (Kottegoda, 1980, pp. 129).

The ARMA $(1, 1)$ model should be considered when an AR (1) type does not fit the data. Parameters $(p + q)$ should be minimum. For example, ARMA $(1, 1)$ is preferable to AR (3) if it is found that both fit an observed sequence.

Example 7.16

Solve Example 7.15 by the ARMA $(1, 1)$ model. Use the following values for standard normal deviates, t_i: 0.131, -0.859, -1.096 from Table 7.23.

Solution From Example 7.14, $\overline{X} = 653$, $S = 186.6$, $r_1 = 0.37$, and $r_2 = 0.24$. From eq. (7.40),

$$\phi_{1,1} = 0.24/0.37 = 0.65 .$$

From eq. (7.42),

$$\theta_{1,1} = 0.33 .$$

From eq. (7.41),

$$S_e^2 = 0.85 \qquad S_e = 0.92$$

$$a_1 = (0.65 \times 0) + 0.92(0.131 - 0.33 \times 0) = 0.121$$

$$a_2 = (0.65 \times 0.121) + 0.92(-0.859 - 0.33 \times 0.131) = -0.75$$

$$a_3 = (0.65 \times -0.75) + 0.92[-1.096 - (0.33 \times -0.859)] = -1.24$$

$$x_1 = \overline{X} + Sa_i = 653 + (186.6)(0.121) = 676 \text{ cfs}$$

$$x_2 = 653 + (186.6)(-0.75) = 513 \text{ cfs}$$

$$x_3 = 653 + (186.6)(-1.240) = 422 \text{ cfs}$$

7.22 DISAGGREGATION MODEL

In this type of model, the flows of a higher level (aggregate flows) are created by the sequential technique of Sections 7.20 and 7.21, which are distributed at a lower level by the disaggregation procedure. A disaggregation model divides annual flows into seasonal or monthly flows and the aggregate basin flows (monthly or annual) into flows at individual sites. Since introduction of the disaggregation approach by Valencia and Schaake (1973), many models of this type have been proposed. Some of these are by Mejia and Rousselle (1976), Lane (1979), Salas et al. (1980), and Stedinger and Vogel (1984). The advantage of a disaggregation model over a sequential model (AR or ARMA) results from (1) the sets of parameters required in the monthly sequential model are too large, and (2) relevant statistics of higher levels (annual) of flows are not necessarily preserved in lower-level (monthly) flows.

The disaggregation model takes the following mathematical form in matrix notation:

$$\mathbf{Y} = \overline{\mathbf{A}}\mathbf{X} + \overline{\mathbf{B}}\mathbf{V} \qquad [\text{L}^3\text{T}^{-1}] \tag{7.43}$$

where

$\mathbf{Y} = (n \times 1)$ vector of disaggregated variables (i.e., monthly flows)

$\overline{\mathbf{A}} = (n \times m)$ coefficient matrix

$\mathbf{X} = (m \times 1)$ vector of aggregated variables (i.e., annual flows)

$\overline{\mathbf{B}} = (n \times n)$ coefficient matrix

$\mathbf{V} = (n \times 1)$ vector of random standard normal deviates

It is assumed that the sample (historical) values of variables X and Y have been adjusted to have zero mean by subtracting their average values. Variables V, being the standard normal deviates, have zero mean, too.

Given generated values of X at m sites for a particular year (or for m years at one site), the seasonal monthly flows Y are generated at m sites for a particular year (or for m years at one site) by eq. (7.43), provided that the coefficient matrices $\overline{\mathbf{A}}$ and $\overline{\mathbf{B}}$ are estimated from historical data.

The criteria used in parameters (coefficients) estimation are that the expected means, variances, and covariances of generated data are equal to the historical means, variances, and covariances. By transposing the matrices and taking expected values of samples (historical data), the following relations for estimation of $\overline{\mathbf{A}}$ and $\overline{\mathbf{B}}$ are obtained (Valencia and Schaake, 1973):

$$\overline{\mathbf{A}} = E[\mathbf{YX^T}]E[\mathbf{XX^T}]^{-1} \qquad \text{[dimensionless]} \qquad (7.44)$$

and

$$\overline{\mathbf{BB}}^T = E[\mathbf{YY^T}] - E[\mathbf{YX^T}]E[\mathbf{XX^T}]^{-1}E[\mathbf{XY^T}] \qquad \text{[dimensionless]} \qquad (7.45)$$

where $E[\mathbf{XX^T}]$, $E[\mathbf{YX^T}]$ etc., are covariance matrices; equivalent to $\overline{\mathbf{S}}_{XX}$, $\overline{\mathbf{S}}_{YX}$ etc. and $\mathbf{X^T}$, etc., are transpose of $\mathbf{X}$, etc.

It is not necessary that $\overline{\mathbf{B}}$ have a solution. For a solution, $\overline{\mathbf{BB}}^T$ should be positive semidefinite, which is determined by the theory of Gramian matrix.

7.23 AUTORUN MODEL

Among the flow generation schemes, a comparatively recent methodology relates to run analysis, which has a distribution-free behavior (is independent of the probability distribution of the data). A run is defined as a succession of the same kinds of observations, preceded and succeeded by at least a single observation of a different kind. Thus a run is made up of a wet period (water surpluses) and a drought period (water deficits) with respect to a truncation level, x_0, often taken as the median value of the observations, as shown in Figure 7.19.

The models in previous sections are based on the sequential properties of the historical data. An autorun model reproduces the runs (i.e., preserves the lengths of wet and dry periods) as observed in the historical data. Two basic parameters of autorun analysis are (1) the autorun coefficient, and (2) the run length, such as positive run length, n_p, and negative run length, n_n, in Figure 7.19.

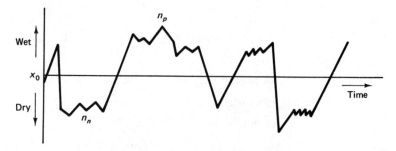

Figure 7.19 Runs in a hydrologic series.

On the lines of the autocorrelation coefficient of the Markovian process, Sen (1976) defined the "autorun coefficient" as a conditional probability of an observation being greater than the truncation level, given that the observation preceding the k lag is greater than the truncation level. The following relationship between the autocorrelation coefficient of Markovian nature and the autorun coefficient has been derived by Sen (1978):

$$r_k = \sin \pi[(r_0)_k - 0.5] \qquad \text{[dimensionless]} \qquad (7.46)$$

where

$$r_k = k\text{-lag autocorrelation coefficient}$$

$$(r_0)_k = k\text{-lag autorun coefficient}$$

A value of $(r_0)_k = 0.5$ means that the data are independent of each other, and $(r_0)_k = 1$ indicates perfectly correlated data.

The run length is a basic parameter to indicate the property of runs. It also serves as a test parameter. If the average positive or negative run length calculated from the available data is equal to 2, the historical sequence is independent; otherwise, it is dependent. In dependent hydrologic series, high values of streamflow tend to follow high values, and low values tend to follow low values.

The average positive (wet) and negative (dry) run lengths at any truncation level are calculated as follows:

$$\bar{n}_p = \frac{1}{m_p} \sum_{i=1}^{m_p} (n_p)_i \qquad \text{[L]} \qquad (7.47)$$

and

$$\bar{n}_n = \frac{1}{m_n} \sum_{i=1}^{m_n} (n_n)_i \qquad \text{[L]} \qquad (7.48)$$

where

$$m_p, m_n = \text{numbers of wet and dry periods}$$

$$(n_p)_i, (n_n)_i = i\text{th wet and dry period length in historical record}$$

For stationary process the run lengths are distributed according to the geometric distribution. The geometric PDF (probability distribution function) parameters $\bar{r}_p$ of wet periods and $\bar{r}_n$ of dry periods have been obtained by Sen (1985) as

$$\bar{r}_p = \frac{\bar{n}_p - 1}{\bar{n}_p} \qquad \text{[dimensionless]} \qquad (7.49a)$$

$$\bar{r}_n = \frac{\bar{n}_n - 1}{\bar{n}_n} \qquad \text{[dimensionless]} \qquad (7.49b)$$

The data generation mechanism first constructs alternately wet and dry periods and then within these periods determines the flow values of surplus and deficit magnitudes.

Geometrically distributed wet and dry periods are constructed with parameters $\bar{r}_p$ and $\bar{r}_n$ through

$$y = 1 + \frac{\log \varepsilon}{\log *} \qquad (7.50)$$

where

y = geometrically distributed random variable indicating a wet or dry period

ε = uniformly distributed random variable between 0 and 1 (from a table)

$*$ = assumes the value of $\bar{r}_p$ and $\bar{r}_n$ alternately

The initial wet or dry period can be selected according to the final period of the historical sequence.

Once the alternate sequence of wet and dry periods is determined, the flow values are selected randomly from the appropriate PDF corresponding to the historical sequence. The values of random variables below the threshold (lower than truncation level) are neglected in generating surpluses, and vice versa.

Part D: Estimation of Flow at Ungaged Site

7.24 INTERPOLATION OR EXTRAPOLATION OF DATA

Often a site for which streamflow data are needed is not gaged, but a gaging station exists on the same river upstream or downstream, or both upstream and downstream. The interpolation of the record provides a means of estimating the flow in such cases.

Consider that the gaging-station record is available at a site X having a drainage area A_x and an estimate has to be made for another site, Y, on the same river with a drainage area of A_y. The flow will be distributed in direct proportion of the drainage area; that is,

$$Q_y = \frac{Q_x}{A_x}A_y \qquad [\text{L}^3\text{T}^{-1}] \tag{7.51}$$

A better estimate is made when records at two gaging sites exist, preferably one upstream and one downstream of the ungaged site because between a gaged site and the ungaged site there might be some changes in the drainage pattern, such as the meeting of a tributary or extraction of water. Figure 7.20 shows that Bowie Creek meets the Leaf River between station A and station B. The variation in flow between the two gaged sites is adjusted either on the basis of the drainage area, as under

$$\frac{Q_y}{A_y} = \frac{Q_x}{A_x} + \left(\frac{\overline{Q}_z}{A_z} - \frac{\overline{Q}_x}{A_x}\right)\frac{A_y - A_x}{A_z - A_x} \qquad [\text{LT}^{-1}] \tag{7.52}$$

where

$$Q_x = \text{flow at gaged site X of drainage area A}_x$$
$$Q_y = \text{flow at ungaged site Y of drainage area A}_y$$
$$Q_z = \text{flow at gaged site Z of drainage area A}_z$$
$$\overline{Q}_x, \overline{Q}_z = \text{average of the entire record at X and Z}$$

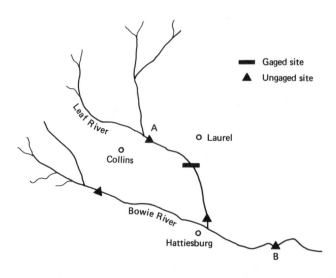

■ Gaged site
▲ Ungaged site

Figure 7.20 Section of Southern Mississippi Basin.

or on the basis of the distance between the sites, as follows:

$$\frac{Q_y}{A_y} = \frac{Q_x}{A_x} + \left(\frac{\overline{Q_z}}{A_z} - \frac{\overline{Q_x}}{A_x}\right)\frac{L_y}{L_z} \quad [LT^{-1}] \tag{7.53}$$

where

$$L_z = \text{distance between stations X and Z}$$
$$L_y = \text{distance between stations X and Y}$$

Example 7.17

The monthly mean discharge of the Leaf River near Collins, Mississippi, which has a drainage area of 752 mi², and the corresponding discharge of the same river at Hatties-burg, Mississippi, which has a drainage area of 1760 mi², are given in Table 7.25 for the year 1985. The distance between the two stations is 35 miles. Estimate the discharge for a site near Laurel that has a drainage area of 1035 mi² and is also located about 10 miles from the site near Collins.

TABLE 7.25 MONTHLY MEAN DISCHARGE OF LEAF RIVER, MISSISSIPPI

Month	Leaf River near Collins	Leaf River at Hattiesburg
October	312	1046
November	3651	6985
December	2524	6740
January	2803	5401
February	3592	8900
March	3329	7792
April	2944	7428
May	2029	4565
June	1003	2788
July	1237	3727
August	906	2796
September	858	2611

Solution

Average flow near Collins, $\overline{Q_x} = 2099$ cfs

Average flow at Hattiesburg, $\overline{Q_z} = 5065$ cfs

1. Adjustment based on the drainage area:

$$\left(\frac{\overline{Q_z}}{A_z} - \frac{\overline{Q_x}}{A_x}\right)\frac{A_y - A_x}{A_z - A_x} = \left(\frac{5065}{1760} - \frac{2099}{752}\right)\left(\frac{1035 - 752}{1760 - 752}\right)$$

$$= (2.88 - 2.79)\left(\frac{283}{1008}\right)$$

$$= 0.025$$

From eq. (7.52),

$$\frac{Q_y}{1035} = \frac{Q_x}{752} + 0.025 \quad [LT^{-1}] \tag{7.54}$$

Q_y values computed from eq. (7.54) are shown in column 2 of Table 7.26.

2. Adjustment according to the distance:

$$\left(\frac{\overline{Q}_z}{A_z} - \frac{\overline{Q}_x}{A_x}\right)\frac{L_y}{L_z} = (2.88 - 2.79)\left(\frac{10}{35}\right)$$
$$= 0.026$$

From eq. (7.53),

$$\frac{Q_y}{1035} = \frac{Q_x}{752} + 0.026 \quad [LT^{-1}] \tag{7.55}$$

Q_y values computed from eq. (7.55) are shown in column 3 of Table 7.26.

TABLE 7.26 ESTIMATED DISCHARGE OF THE LEAF RIVER NEAR LAUREL

(1) Month	(2) Discharge Based on the Drainage Area (cfs)	(3) Discharge Based on the Distance (cfs)
October	455	456
November	5051	5052
December	3500	3501
January	3884	3885
February	4970	4971
March	4608	4609
April	4078	4079
May	2818	2819
June	1406	1407
July	1728	1729
August	1273	1274
September	1207	1208

7.25 MODELS OF THE HYDROLOGIC CYCLE

Many models have been formulated incorporating elements of the hydrologic cycle. The depression storage and evaporation are reasonably fixed parameters during a specific period. The infiltration rate is thus a very important component of these models which relates the precipitation to the runoff. The infiltration depends on the moisture conditions of the drainage basin at the time of the precipitation. A considerable amount of research has been directed toward finding a parameter that can adequately represent the basin moisture conditions. An infiltration approach to the runoff estimate has been presented in Chapter 3.

7.26 STREAMFLOW FROM DRAINAGE-BASIN CHARACTERISTICS

The U.S. Geological Survey conducted the statistical multiple regression analysis to derive the generalized relations for the natural streamflows in four regions of the eastern, central, southern, and western United States (Thomas and Benson, 1970). The long-term streamflow records were used in the regression and a large number of topographic and climatic indices were included to define the drainage-basin characteristics. The regression considered various categories of flows, such as low flows, high flows, mean monthly flows, and annual flows. The multiple regression analyses defined the relation between each category of flow and the drainage basin characteristics. It was concluded that the flow can be defined more accurately in the humid eastern and southern regions than in the more arid western and central regions. Also, the mean flow can be more accurately defined than the high flows, and the low flows are very poorly defined.

The regression relation has the following form in which the constant, a, and coefficients, b_1, b_2, . . . , have different values for different regions and different categories of flows (i.e., high, low, mean, etc.).

$$Q = aA^{b_1} S^{b_2} L^{b_3} S_t^{b_4} E^{b_5} I_{24,2}^{b_6} P^{b_7} S_n^{b_8} F^{b_9} S_i^{b_{10}} t_1^{b_{11}} t_7^{b_{12}} E_v^{b_{13}} A_a^{b_{14}} \qquad \text{[unbalanced]}$$
$$(7.56)$$

where

Q = discharge, cfs

A = drainage area, mi^2

S = channel slope, ft/mi

L = channel length, miles

S_t = percent of total drainage area occupied by lakes, swamps, ponds

E = mean elevation of the basin, thousands of feet above sea level

$I_{24,2}$ = maximum 24-hour precipitation expected to be exceeded once every 2 years, in.

P = mean annual precipitation, in.

S_n = mean annual snowfall, in.

F = forest cover (i.e., percent of total area forested)

S_i = soil index for infiltration, in.

t_1 = mean of minimum January temperatures, °F

t_7 = mean of minimum July temperatures, °F

E_v = annual evaporation, in.

A_a = alluvial area in the basin, mi^2

For each category of flow characteristics, all of the indices above need not be included.

The indices most highly related to streamflow are drainage basin size and mean annual precipitation. The standard error of estimate for the mean annual flow ranged from 8.6% for the eastern region to 33% for the western region.

Example 7.18

The following relation was derived by the U.S. Geological Survey (Thomas and Benson, 1970) for the eastern region for the annual mean flow. Determine the annual mean flow for the Shetucket basin near Willimantic, Connecticut, which has a drainage area of

226 mi^2, a slope of 1.2 ft/mi, and an annual precipitation of 25 in. including 10 in. of snowfall.

$$Q_A = (2.89 \times 10^{-4})A^{1.06}S^{0.1}P^{1.87}S_n^{0.18} \qquad \text{[unbalanced]} \qquad (7.57)$$

Solution From eq. (7.57),

$$Q_A = (2.89 \times 10^{-4})(226)^{1.06}(1.2)^{0.1}(25)^{1.87}(10)^{0.18}$$

$$= 57.32 \text{ cfs}$$

7.27 STREAMFLOW RELATED TO CHANNEL GEOMETRY

For estimating the streamflow characteristics for an ungaged stream quickly and inexpensively, the channel-geometry relations have been derived by the USGS as an alternative approach which eliminates the need for the extensive input data required in other methods. The equations have been developed from the data collected at numerous stream-gaging sites in the arid to semiarid parts of the western United States to assess the mean annual flow and flood discharges with selected recurrence intervals from channel geometry (principally width) and channel material data (Hedman and Osterkamp, 1982).

In these relations the discharge (mean annual or flood flows) is directly related to the active channel width, which is indicative of relatively recent conditions of water and sediment discharge. It is subject to change by prevailing discharges. Its upper limit is defined by a break in the relatively steep bank slope of the active channel to a more gently sloping surface beyond the channel edge, as shown in Figure 7.21. The nature of the stream (i.e., perennial,* intermittent,† and ephemeral‡) affected the relations. The characteristics of the areas, such as mountains, plains, or deserts, also influenced the discharge. The studies indicated that width–discharge relations vary measurably with the channel material characteristics. The channel material has been grouped into three categories: (1) silt-clay channel, (2) sand channels, and (3) armored channels. The separate relations are accordingly derived for each of these characteristics, as shown in Table 7.27.

Example 7.19

Determine the mean annual flow for a stream that flows through a silty-clay bed in the desert of the southwest. The active-channel width has been measured to be 200 ft.

Solution

$$Q_A = 0.04W_{AC}^{1.75}$$

$$= 0.04(200)^{1.75}$$

$$= 425.5 \text{ acre-ft/year}$$

---

*Perennial: stream that has measurable discharge more than 80% of the time.

†Intermittent: stream that has discharge between 10 and 80% of the time.

‡Ephemeral: flows only in direct response to precipitation, with discharge occurring less than 10% of the time.

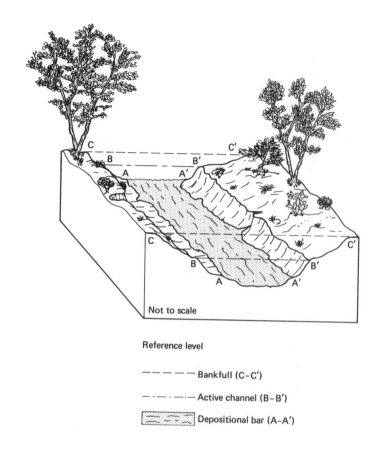

Reference level

‒ ‒ ‒ ‒ Bankfull (C–C′)

‒ · ‒ · ‒ Active channel (B–B′)

░░░░ Depositional bar (A–A′)

Figure 7.21 Reference points for active-channel geometry (from Hedman and Osterkamp, 1982).

Sec. 7.27   Streamflow Related to Channel Geometry

TABLE 7.27 EQUATIONS FOR DETERMINING MEAN ANNUAL RUNOFF FOR STREAMS IN THE WESTERN UNITED STATES

Flow Frequency	Areas of Similar Regional-Runoff Characteristics	Percentage of Time Having Discharge	Channel-Material Characteristics[a]	Equation[b]	Standard Error of Estimate (%)
Perennial	Alpine	More than 80	Silt-clay and armored	$Q_A = 64W_{AC}^{1.88}$	28
Intermittent	Plains north of latitude 39° N	10 to 80	Silt-clay and armored	$Q_A = 40W_{AC}^{1.80}$	50[c]
			Sand	$Q_A = 40W_{AC}^{1.65}$	50[c]
	Plains south of latitude 39° N	10 to 80	Silt-clay and armored	$Q_A = 20W_{AC}^{1.65}$	50[c]
			Sand	$Q_A = 20W_{AC}^{1.55}$	50[c]
Ephemeral	Northern and southern plains and intermontaine areas	6 to 9	Silt-clay and armored	$Q_A = 10W_{AC}^{1.55}$	d
			Sand	$Q_A = 10W_{AC}^{1.50}$	d
		2 to 5	Silt-clay and armored	$Q_A = 4.0W_{AC}^{1.50}$	40[c]
			Sand	$Q_A = 4.0W_{AC}^{1.40}$	40[c]
	Deserts of the southwest	1 or less	Silt-clay and armored	$Q_A = 0.04W_{AC}^{1.75}$	75[c]
			Sand	$Q_A = 0.04W_{AC}^{1.40}$	75[c]

[a] Silt-clay channels — bed material d_{50} less than 0.1 mm or bed material d_{50} equal to or less than 5.0 mm and bank silt-clay content equal to or greater than 70%. Sand channels — bed material $d_{50} = 0.1$–5.0 mm and bank silt-clay content less than 70%; armored channels — bed material d_{50} greater than 5.0 mm.

[b] Active-channel width, W_{AC}, in feet; discharge, Q_A, in acre-feet per year.

[c] Approximate — standard error of estimate of the basic regression equation.

[d] Standard error of estimate not determined; graphical analyses.

Source: Hedman and Osterkamp (1982).

or

$$Q_A = \left[425.5 \frac{(\text{acre})\,(\text{ft})}{\text{year}} \right] \left(\frac{\text{ft}^2}{\text{acre}} \right) \left(\frac{\text{year}}{\text{sec}} \right)$$

$$= 425.5 \times \frac{43{,}560}{1} \times \frac{1}{365 \times 24 \times 60 \times 60}$$

$$= 0.59 \text{ cfs}$$

7.28 VARIABILITY OF STREAMFLOW

The previous sections provide an indication that the streamflow is not constant in space or with respect to time, which makes it necessary to measure or estimate the data at a specific location for a long duration of time. From the available record at a site for a certain period, the trends in streamflows can be detected. Three devices used to study the variability of the flow that have direct applications in resource planning are frequency curve, mass curve, and duration curve. The frequency curve has been discussed in the context of peak flows in Chapter 8. The mass curve and duration curve are described below.

7.28.1 Flow–Mass Curve

A graph of the cumulated values of a hydrologic quantity as the ordinate, against time or date as the abscissa, is known as a mass curve. When the streamflow is taken as the hydrologic quantity, it is known as a flow–mass curve. The mass curve is a summation of the hydrograph that represents the area under the hydrograph from one time to another. Mathematically, a flow–mass curve can be expressed as

$$V = \sum_{t=t_1}^{t_2} Q_t \Delta t \qquad [\text{L}^3] \tag{7.58}$$

where

V = volume of streamflow

Q_t = discharge as a function of time or the hydrograph ordinate at time $t = (t_1 + t_2)/2$

Δt = period between time t_1 and t_2

This has been described in Chapter 9 in the context of reservoir storage requirements.

7.28.2 Flow–Duration Curve

To ascertain how often flow of a given magnitude occurred during the period of record, a flow–duration curve is prepared. From the available data, the discharge is plotted as the ordinate, against the percent of time that discharge is exceeded on the abscissa, as shown in Figure 7.22. This is referred to as a complete series analysis. Two other series, the annual series and the partial duration series, are described in the context of frequency analysis in Chapter 8. In a statistical sense, a duration curve represents

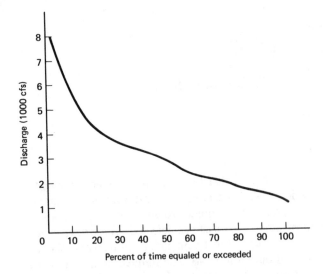

Figure 7.22 Flow–duration curve.

the cumulation of the frequency distribution curve. The duration curve can be of daily flows, mean monthly flows, or mean annual flows. The curve of mean annual flows is likely to be different from the monthly or daily flow curves. If the river flows fluctuate from month to month, but the total flow every year is nearly the same, the monthly duration curve will be similar to Fig. 7.22, but the mean annual duration curve will be nearly a horizontal line.

The procedure used to prepare a distribution curve is as follows, which has been demonstrated in Example 7.20.

1. The total range of discharge is divided into a number of classes. For instance, the discharge ranging from 0 to 10,000 cfs is divided into 20 classes of 500 cfs each.
2. The entire record is scanned day by day for daily flows, month by month for monthly flows, and year by year for yearly flows curves.
3. A mark is made in the appropriate class for each item in the record. The plot of the number of items in each class against the discharge value of the class will represent the frequency distribution, as shown in Figure 7.23a.
4. The number of items in each class are cumulated starting with the highest flow. The percent is determined of the accumulated number of items of each class with respect to total items of all classes.
5. The average discharge value of each class is plotted on the vertical scale against the percent (of time) determined in item 4, as the abscissa, as shown in Figure 7.23b.

Sometimes the ordinates of the flow–duration curve are plotted in dimensionless form, in terms of ratios, to the average discharge rather than actual discharges. When no flow data are available, a synthetic flow duration can be constructed by a procedure described by Chow (1964, pp. 14–43).

The flow–duration curve is a very important tool to appraise the flow values of various dependabilities, and thus is indispensable for water resources study. The slope

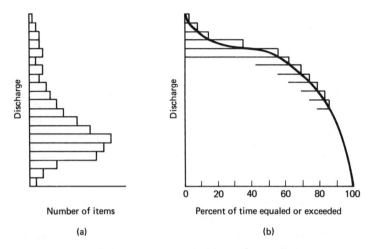

Figure 7.23 Frequency curve and flow–duration curve.

of the curve indicates streamflow characteristics. A flat-sloped curve suggests large natural storage in a stream and a steep slope indicates the flashy nature of a stream. The flow corresponding to 50% exceedence is called the median flow. This is usually different from the mean flow, which can be obtained by (1) computing the average of the data in the record, (2) dividing the duration curve in two equal parts by a horizontal line, or (3) using the following area summation equation:

$$Q_{\text{mean}} = 0.025(\tilde{Q}_0 + \tilde{Q}_{100}) + 0.05(Q_5 + Q_{95}) + 0.075(Q_{10} + Q_{90})$$
$$+ 0.10(Q_{20} + Q_{30} + Q_{40} + Q_{50} + Q_{60} + Q_{70} + Q_{80})$$
$$[\text{L}^3\text{T}^{-1}] \qquad (7.59)$$

where

$$Q_{\text{mean}} = \text{mean discharge}$$
$$Q_5, Q_{10}, \ldots = \text{discharge corresponding to 5\%, 10\%, etc., exceedence levels}$$
$$\tilde{Q}_0, \tilde{Q}_{100} = \text{discharge nearly 0 and 100\% of time (any discharge of less}$$
$$\text{than 5\%, and more than 95\%, respectively)}$$

The firm power of a run-of-river (without storage) power plant is estimated on the basis of the flow available at 90% of the time from an unregulated flow–duration curve. The minimum flow of a stream can be improved by creating reservoir storage, which will modify the duration curve to a flatter slope. This will enhance the firm power potential of a hydroelectric plant. The ordinate and abscissa of the curve can be replaced to represent kilowatts and hours, respectively, so that the area under the curve will provide the annual energy output of a plant in kilowatthours.

Example 7.20

Prepare the flow–duration curve from the mean monthly flow data given in Table 7.28. Determine the percent of time that a monthly flow of 650 cfs is equaled or exceeded. Also determine the median and mean monthly flows.

TABLE 7.28 MEAN MONTHLY FLOW IN CFS FOR EXAMPLE 7.20

Water Year	Oct.	Nov.	Dec.	Jan.	Feb.	Mar.	Apr.	May	June	July	Aug.	Sept.
1979	468	710	1462	841	899	735	971	1538	1247	839	576	481
1980	1114	1140	800	739	581	499	690	1592	2652	1088	691	467
1981	466	643	649	432	581	762	1121	2280	1762	1128	666	473
1982	414	960	769	655	694	924	892	1570	2660	1874	862	552
1983	818	1448	1536	727	1313	624	1169	1504	1484	1129	638	422
1984	556	544	501	330	642	428	970	1407	1249	1004	625	433
1985	341	227	243	1043	978	416	665	1149	1302	1404	715	500

Solution

1. The discharge, in the range 0 to 3000 cfs, has been divided into 15 classes of 200 cfs each, as shown in column 1 of Table 7.29.
2. The data are scanned through and each item is noted in the class group to which it belongs. The total in each class is shown in column 2.
3. Column 3 accumulates the number of items of column 2, starting from the bottom.
4. The items accumulated are shown as percents in column 4.
5. The plot of average value in each class in column 1 against column 4 is given in Figure 7.24.
6. From the plot, a flow of 650 cfs is equaled or exceeded 88% of the time.
7. The flow corresponding to 50% (i.e., $Q_{median} = 900$) cfs.
8. Mean flow, from eq. (7.59),

$$Q_{mean} = 0.025(2,700 + 300) + 0.05(1860 + 500) + 0.075(1580 + 520)$$
$$+ 0.10(1320 + 1160 + 1020 + 900 + 810 + 700 + 620)$$
$$= 75 + 118 + 158 + 653$$
$$= 1004 \text{ cfs}$$

TABLE 7.29 COMPUTATION OF FLOW–DURATION CURVE

(1) Class (Flow Range) (cfs)	(2) Numbers of Items	(3) Cumulated Number of Items	(4) Percent of Time
0–200	0	84	100
201–400	4	84	100
401–600	20	80	95.2
601–800	20	60	71.4
801–1000	11	40	47.6
1001–1200	10	29	34.5
1201–1400	4	19	22.6
1401–1600	10	15	17.8
1601–1800	1	5	6.0
1801–2000	1	4	4.8
2001–2200	0	3	3.6
2201–2400	1	3	3.6
2401–2600	0	2	2.4
2601–2800	2	2	2.4
2801–3000	0	0	0

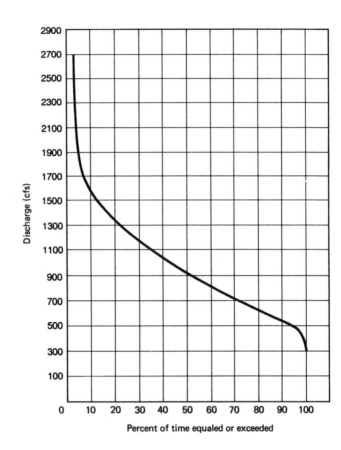

Figure 7.24 Flow–duration curve for Example 7.20.

Percent of time equaled or exceeded

7.29 COMPUTER APPLICATIONS

Hromadka et al. (1985) have formulated a Synthetic Runoff Hydrograph Program, the software of which is supported by the Advanced Engineering Software of Irvine, California. The program contains the following elements:

1. Development of a synthetic unit hydrograph using the watershed S-curve of the U.S. Army Corps of Engineers or the dimensionless curvilinear unit hydrograph of the U.S. Soil Conservation Service.
2. Estimation of area-averaged rainfall values for different durations and preparation of an intensity duration curve for the watershed.
3. Development of a synthetic critical storm pattern.
4. Assessment of effective rainfalls.
5. Computation of watershed runoff or a flood hydrograph.

A comprehensive modeling of precipitation–runoff process is offered through the Hydrologic Engineering Center (HEC)-1 Flood Hydrograph Package. The result of this modeling process is the computation of streamflow hydrographs at desired locations. Many previously developed individual programs have been combined and upgraded in the revised 1985 version of the program. The HEC-1 package includes the following component models or options.

1. Stream network model: This is the base model around which other capabilities of HEC-1 are built up. A basin is segmented into a number of subbasins. For each subbasin, the input is a precipitation hyetograph. Precipitation excesses are computed by subtracting the infiltration and detention losses. The resulting precipitation excesses are routed by the unit hydrograph or kinematic techniques to the outlet of the subbasin providing a runoff hydrograph. Runoff hydrographs from subbasins are routed through river reaches and combined together at control points. Reservoir routing functions similarly, wherein upstream inflows are routed using the storage routing method of Chapter 8.

2. Multiplan-multiflood analysis allows the simulation of up to nine design floods for up to five plans.

3. Dam-break simulation analyzes the consequences of dam overtopping and structural failures.

4. Depth-area option computes flood hydrographs from user-supplied precipitation depth-area relation.

5. Flood damage analysis provides the economic assessment of flood damage for damaged reaches.

6. Optimization procedure for optimal size of a flood control system.

Almost similar features are available in the U.S. Soil Conservation Service Technical Release (TR)-20 Program. Runoff hydrographs for each subbasin are generated by the SCS method that uses the SCS curve number, time of concentration, and standard rainfall pattern or user defined pattern for a storm. Runoff hydrographs are combined and routed through reaches or through detention ponds using the attenuation-kinematic method.

The HEC has also formulated the HEC-4 Monthly Streamflow Simulation Program related to statistical hydrology. It analyzes streamflows at a number of interrelated stations to determine their statistical characteristics and generates a sequence of hypothetical streamflows of any desired length.

The documents and manuals for the above programs are available for mainframe computers as well as microcomputers except the HEC-4 program, for which only a mainframe version is available at present.

PROBLEMS

7.1. From the following hourly streamflow record due to a storm, separate the baseflow by the recession curve technique. The drainage area is 30 acres. Also determine the runoff depth.

Time (hr)	Flow (cfs)	Time (hr)	Flow (cfs)
1	30	9	45
2	29.4	10	31.5
3	66	11	22.5
4	155	12	18
5	190	13	16
6	140	14	14.5
7	100	15	13
8	63		

7.2. Tabulated below are the ordinates of a hydrograph at a section of a stream having a drainage area of 250 km^2. Prepare the direct runoff hydrograph (DRH) after separating the baseflow using the recession curve technique. Determine the equivalent depth of runoff.

Time (hr)	Flow (m^3/s)	Time (hr)	Flow (m^3/s)
0	1.37	50	8.8
5	1.25	55	6.8
10	1.12	60	5.50
15	5.00	65	4.10
20	12.00	70	2.75
25	15.60	75	2.00
30	17.15	80	1.20
35	14.40	85	0.65
40	12.50	90	0.60
45	10.70		

7.3. Solve Problem 7.1 by all of the methods of the arbitrary approach.

7.4. Solve Problem 7.2 by the methods of the arbitrary approach.

7.5. If the hydrograph of Problem 7.1 resulted from an isolated storm of 2-hour duration, determine the unit hydrograph by the inverse procedure.

7.6. The hydrograph of Problem 7.2 was produced by an isolated storm of 5-hour duration. Derive the unit hydrograph by the inverse procedure.

7.7. Given below are the measured streamflows in cfs from a storm of 6-hour duration on a stream having a drainage area of 185 mi^2. Derive the unit hydrograph by the inverse procedure. Assume a constant baseflow of 550 cfs.

Hour	Day 1	Day 2	Day 3	Day 4
Midnight	550	5000	1900	550
6 a.m.	600	4000	1400	
Noon	9000	3000	1000	
6 p.m.	6600	2500	750	

7.8. In a basin of 80 mi^2, the following data were observed from a hydrograph and an isochronal map of the watershed. Determine **(a)** the instantaneous hydrograph and **(b)** the 1-hour unit hydrograph, using the Clark method.

$$\text{Time of concentration} = 8 \text{ hr}$$

$$\text{Attenuation constant} = 7.7 \text{ hr}$$

Area	Basin Area (mi^2)	Travel Time (hr)
A_1	3.3	1
A_2	9.3	2
A_3	19.4	3
A_4	14.3	4
A_5	9.4	5
A_6	10.8	6
A_7	10.0	7
A_8	1.5	8

7.9. The isochronal map of a basin is shown in Fig. P7.9. Assume that the hydrograph given in Problem 7.2 results from a storm of 5-hour duration in the basin on which the ordinate corresponding to a time scale of 40 hours represents the inflection point. Prepare **(a)** the time–area curve for the basin, **(b)** the instantaneous unit hydrograph, and **(c)** the 5-hour-duration unit hydrograph.

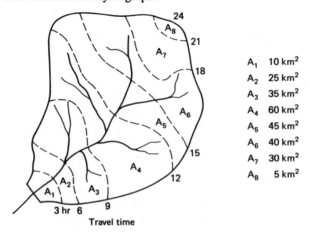

A_1	10 km²
A_2	25 km²
A_3	35 km²
A_4	60 km²
A_5	45 km²
A_6	40 km²
A_7	30 km²
A_8	5 km²

Figure P7.9

7.10. Given below is a 2-hour unit hydrograph. Derive an 8-hour unit hydrograph by the lagging method.

Time (hr)	0	1	2	3	4	5	6	7	8
Q (cfs)	0	100	200	400	300	200	100	50	0

7.11. The ordinates of a 4-hour unit hydrograph for a 42-mi² basin are given below. Find the 8-hour unit hydrograph by the lagging method.

Time (hr)	Q (cfs)	Time (hr)	Q (cfs)
0	0	11	1350
1	200	12	1100
2	1250	13	900
3	2200	14	700
4	3000	15	550
5	3500	16	400
6	3000	17	300
7	2600	18	200
8	2300	19	100
9	1900	20	50
10	1600	21	0

7.12. From the 2-hour unit hydrograph of Problem 7.10, determine the 8-hour unit hydrograph by the S-curve method.

7.13. From the 4-hour unit hydrograph of Problem 7.11, determine the 2-hour and 6-hour unit hydrographs by the S-curve method.

7.14. A basin of 140 mi² has a total stream length of 20 miles up to the upstream boundary of the basin and has a distance of 8 miles along the stream opposite the point of the basin centroid. Determine the 3-hour unit hydrograph using Snyder's method. Assume that $C_t = 2.0$ and $C_p = 400$.

7.15. For a basin of 400 km² having $L = 28$ km and $L_c = 12$ km, determine the 2-hour unit hydrograph using Snyder's method. $C_t = 1.5$ and $C_p = 0.20$.

7.16. Solve Problem 7.14 by the SCS method.

7.17. Solve Problem 7.15 by the SCS method.

7.18. Given a 500-km² basin with a lag of 10 hours, derive the 4-hour unit hydrograph using the SCS method.

7.19. For the following unit hydrograph and storm pattern, determine the composite direct runoff hydrograph. The unit hydrograph is a triangle as follows:

$$\text{Base length} = 4 \text{ time units}$$

$$\text{Time to peak} = 1 \text{ time unit}$$

$$\text{Peak flow} = \tfrac{1}{2} \text{ rainfall unit}$$

Storm pattern (time units)	1	2	3	4
Effective rainfall (rainfall units)	1.5	0	2.5	0.8

7.20. The baseflow in a stream and the 3-hour unit hydrograph for the basin are given below. Determine the total flow hydrograph for a storm of the pattern indicated.

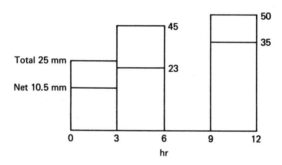

Figure P7.20

Time (hr)	Unit Hydrograph (m³/s)	Baseflow (m³/s)
1200	0	10
1500	4.7	10
1800	7.5	11
2100	5.7	11
2400	4.3	11
0300	3.1	12
0600	2.4	12
0900	1.4	12
1200	0.8	13
1500	0.2	13
1800	0	13

7.21. The following hydrograph resulted from an isolated 5-hour storm of 7.3 mm effective rainfall. Determine the streamflow hydrograph resulting from the storm sequence indicated.

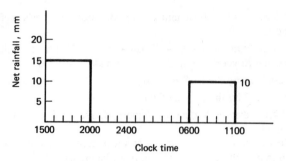

Figure P7.21

Time (hr)	Flow (m^3/s)
0	1.37
10	1.12
20	12.0
30	17.15
40	12.5
50	8.8
60	5.5
70	2.75
80	1.20
90	0.60

7.22. The rainfall excesses (effective rainfalls) derived from the weighted-average daily rainfall in a basin are indicated below. The 24-hour unit hydrograph from the observed data in the basin is also given. Estimate the daily streamflow. A baseflow of 20 m^3/s has been estimated.

Date, July 85	6	14	18
Rainfall excess (mm)	28	5.2	48.5

Unit Hydrograph:

Time (days)	0	1	2	3	4	5	6
Flow (m^3/s)	0	1.5	11.6	7.7	5.0	2.3	0

7.23. The yearly streamflow record of a stream and the corresponding average annual precipitation data from the basin are listed below. Assuming that these data have already been tested for consistency, determine the effective precipitation that contributes to the runoff.

Year:	1960	1961	1962	1963	1964	1965	1966	1967
Streamflow (m^3/s)	1.00	1.25	0.50	0.25	3.0	1.26	0.90	2.45
Precipitation (mm)	600	740	500	450	870	640	770	730

Estimation of Surface Water Flow   Chap. 7

Year:	1968	1969	1970	1971	1972	1973	1974	1975	1976
Streamflow (m^3/s)	1.55	1.65	1.00	1.10	2.00	0.60	2.70	1.20	1.90
Precipitation (mm)	680	670	590	570	900	420	830	620	820

Year:	1977	1978	1979	1980	1981	1982	1983	1984	1985
Streamflow (m^3/s)	0.4	1.15	0.50	—	—	—	—	—	—
Precipitation (mm)	480	610	760	560	450	730	490	500	620

7.24. (Adapted from Searcy and Hardison, 1960.) The following annual runoff and precipitation data relate to the Colorado River near the Grand Canyon, Arizona. Determine the effective precipitation that furnishes the annual runoff to the river.

Year:	1920	1921	1922	1923	1924	1925	1926	1927
Yearly runoff (million acre-ft)	—	21.27	17.84	17.05	13.01	11.74	14.42	17.26
Precipitation (in.)	16.68	16.52	14.16	16.30	11.56	15.15	13.78	20.36

Year:	1928	1929	1930	1931	1932	1933	1934	1935	1936
Runoff (million acre-ft)	15.63	19.43	13.42	6.74	15.97	10.01	4.66	10.22	12.32
Precipitation (in.)	12.49	18.43	14.00	9.94	13.85	12.12	10.09	13.12	12.72

Year:	1937	1938	1939	1940	1941	1942	1943	1944	1945	1946
Runoff (million acre-ft)	12.41	15.63	9.62	7.44	16.94	17.26	11.43	13.53	11.87	9.09
Precipitation (in.)	12.47	13.99	11.14	12.31	16.94	13.65	14.63	12.51	14.00	12.39

7.25. Determine the precipitation–runoff relation for Problem 7.23. Use this relation to extend the streamflow data up to 1985.

7.26. Determine the precipitation–runoff relation for Problem 7.24. Compute the runoff corresponding to all precipitation data using this relation.

7.27. A stream-gaging station X with the short-term record of 10 years has to be extended with the help of an adjacent station Y having a long period of record of 50 years. The flow–duration data based on the short period of record for station X and the corresponding duration for station Y are given below along with the flow–duration values based on the entire period of record for station Y. Estimate the flow at station X corresponding to a discharge of 680 cfs at station Y.

Percent of Time Equaled or Exceeded	Short-Term Flow (cfs)		Long-Term Flow at Station Y (cfs)
	Station X	Station Y	
10	2300	920	820
20	1920	700	720
30	1620	580	620
40	1380	500	550
50	1180	420	480
60	1020	380	400
70	880	320	350
80	720	270	300
90	580	220	250
95	500	180	230

7.28. The annual flows for the Warren River, Warren, Rhode Island, are given below. Determine the statistical parameters (mean, standard deviation, skew coefficient, and serial coefficient) and identify the frequency distribution of flow sequence.

Year	1959	1960	1961	1962	1963	1964	1965	1966
Streamflow (cfs)	2908	3450	3450	1100	3840	1930	1782	2350

Year	1967	1968	1969	1970	1971	1972	1973	1974	1975
Streamflow (cfs)	2480	2750	3150	2290	3700	1100	1050	2740	2050

7.29. The annual flow record of the Shetucket River near South Windham, Connecticut, is given below. Determine the statistical parameters of this series.

Year	1966	1967	1968	1969	1970	1971	1972	1973	1974	1975
Flow (m^3/s)	21.5	19.0	25.8	18	29.2	22.5	21.8	24.2	37	40.8

Year	1976	1977	1978	1979	1980	1981	1982	1983	1984	1985
Flow (m^3/s)	22.4	28.5	27.5	35.2	53.2	34.2	38.3	30.5	33	26.8

7.30. Generate a synthetic annual series using the AR(1) model for the data of Problem 7.28. For random numbers, use Table 7.23; start at column 0, row 0; continue along row.

7.31. Using the historical record of Problem 7.29, generate a sequence of annual flows for the Shetucket River by the AR(1) model. Assume a normal distribution. For random numbers, use Table 7.23 starting at column 1 of row 3.

7.32. Annual flow data of the Batten Kill River near Middle Falls, New York, provide a mean and standard deviation of 40.7 m^3/s and 9.58 m^3/s, respectively. The first and second serial correlation coefficients are computed to be 0.56 and 0.45, respectively. Estimate an initial set of parameters of an ARMA(1, 1) model. Generate three additional values of streamflows using the following standard normal deviates: 1.123, −0.821, −0.342.

7.33. Solve Problem 7.31 by the ARMA(1, 1) model.

7.34. The mean monthly discharges of the White River at Greenwater, Washington, of drainage area 220 mi^2, are given below for 1986. Estimate the discharge of the same river at Putnam, which has a drainage area of 75 mi^2.

Month	Oct.	Nov.	Dec.	Jan.	Feb.	Mar.	Apr.	May	June	July	Aug.	Sept.
Discharge (cfs)	1150	1200	900	760	591	511	702	1602	2680	1095	705	510

7.35. The mean monthly discharges of the Naugatuck River at Seymour, Connecticut, of drainage area 880 mi^2, are given below along with the discharges on the same river at Rosenfield, of drainage area 289 mi^2. The distance between the two sites is 45 mi. Determine the monthly discharge at site X, of drainage area 475 mi^2, which is located 16 miles from the Seymour site. Estimate both according to distance and drainage area.

Month:	Oct.	Nov.	Dec.	Jan.	Feb.	Mar.
Discharge at Seymour (cfs)	302	353	1140	730	1468	1510
Discharge at Rosenfield (cfs)	107	127	438	270	505	542

Month:	Apr.	May	June	July	Aug.	Sept.
Discharge at Seymour (cfs)	1040	900	340	261	268	223
Discharge at Rosenfield (cfs)	375	295	125	102	113	79

7.36. The regression relation developed by the U.S. Geological Survey for the mean January month flow in cfs in the eastern region is given below. Determine the mean January month flow for the Shetucket basin near Willimantic, Connecticut, which has a drainage area of 226 mi^2, a slope of 1.2 ft/mi, and an annual precipitation of 25 in., including 10 in. of snowfall. Assume that 25% of the drainage area is occupied by the water body (lakes, swamps, etc.). [Note: slope is not significant in this case.]

$$Q_1 = (4.61 \times 10^{-5})A^{1.03}S_t^{0.27}P^{2.30}S_n^{0.42}$$

7.37. Determine the mean annual flow for a perennial stream in the western United States that has an active channel width of 200 ft in silt-clay.

7.38. If the stream in Problem 7.37 is of an intermittent type in the plains south of latitude 39° N, what is its mean annual flow?

7.39. The mean monthly discharge data (cfs) for a river are given below. Prepare the flow–duration curve (perform the complete series analysis) for the stream. Determine the percent of time that a flow of 100 cfs is equaled or exceeded. Also determine the mean monthly and median monthly flows.

Year	Oct.	Nov.	Dec.	Jan.	Feb.	Mar.	Apr.	May	June	July	Aug.	Sept.
1979	77.2	163	533	263	245	270	341	408	255	111	58.1	47.1
1980	181	504	320	225	177	135	241	605	747	184	92.3	72.1
1981	86.3	156	159	102	171	282	433	833	425	160	67.3	45.3
1982	64.7	180	226	203	240	286	291	575	900	371	102	57.2
1983	94.0	328	400	235	327	175	411	528	319	114	57.5	43.1
1984	69.5	97.2	118	64.4	158	118	323	391	218	119	55.5	38.3
1985	32.1	32.9	35.0	276	297	124	206	364	357	191	71.9	47.7

7.40. The mean monthly flow data (m³/s) for the Menominee River below Koss, Michigan, are given below. Prepare the flow–duration curve for the stream. Determine whether the stream carries a 50-m³/s flow 90% of the time. Also determine the median flow.

Year	Oct.	Nov.	Dec.	Jan.	Feb.	Mar.	Apr.	May	June	July	Aug.	Sept.
1947	56.6	76.7	56.8	54.0	53.0	60.1	157.0	164.0	103.0	72.1	55.5	52.7
1948	51.0	59.3	47.1	49.2	35.4	72.5	103.0	82.3	48.5	42.2	45.8	39.6
1949	34.8	56.1	49.2	45.9	47.9	57.4	87.7	84.7	62.0	102.0	51.7	57.2
1950	57.0	57.8	58.6	59.1	57.6	61.1	199.0	243.0	101.0	69.2	65.8	49.0
1951	43.4	48.9	48.4	50.5	44.5	68.0	267.0	171.0	143.0	159.0	93.3	122.0
1952	147.0	117.0	87.0	76.9	75.6	66.5	219.0	94.0	86.7	153.0	97.0	58.6
1953	45.6	50.2	51.9	59.4	62.5	103.0	167.0	133.0	166.0	174.0	87.5	67.0
1954	56.1	53.4	63.6	57.0	66.7	68.5	182.0	184.0	138.0	73.2	61.0	91.1
1955	123.0	84.6	68.5	66.2	61.0	70.5	254.0	108.0	106.0	48.7	56.1	38.0
1956	59.7	63.0	57.4	59.0	54.8	49.7	270.0	108.0	88.4	116.0	83.4	61.5
1957	48.8	53.1	53.6	49.4	46.1	69.7	130.0	93.0	65.0	41.4	36.3	52.3
1958	52.7	73.3	59.8	54.0	51.3	61.8	123.0	65.9	62.0	132.0	47.0	58.5
1959	47.7	68.9	48.4	46.7	43.1	55.0	110.0	105.0	56.7	48.3	78.0	142.0
1960	155.0	122.0	78.2	82.3	71.0	62.4	242.0	373.0	135.0	83.4	72.1	80.8
1961	80.5	102.0	68.2	52.6	49.2	77.0	158.0	186.0	82.5	60.8	53.8	48.9
1962	57.9	67.6	63.1	53.9	52.4	69.2	168.0	168.0	107.0	59.3	52.1	73.8
1963	65.3	54.7	51.4	46.8	43.8	56.1	87.7	120.0	99.1	43.6	40.6	38.0
1964	34.2	35.6	35.7	37.2	33.8	41.8	70.2	131.0	63.2	42.1	56.9	65.1
1965	55.6	69.8	54.2	49.3	44.1	50.9	173.0	361.0	83.6	51.7	45.6	56.2
1966	67.5	76.6	87.2	78.5	66.6	131.0	157.0	117.0	113.0	44.6	63.0	43.5
1967	62.6	63.7	59.2	59.8	64.0	65.2	295.0	133.0	135.0	100.0	66.3	51.6
1968	86.0	108.0	63.3	50.5	58.0	72.6	134.0	108.0	168.0	141.0	78.1	155.0
1969	93.8	90.2	82.7	89.9	90.0	84.4	229.0	157.0	118.0	87.1	51.1	42.6
1970	65.3	68.9	58.2	62.7	51.0	58.8	114.0	109.0	157.0	56.9	46.6	49.1
1971	64.1	122.0	97.5	72.4	64.7	89.4	284.0	155.0	93.3	67.5	51.0	47.0
1972	92.4	88.2	80.2	65.9	56.7	68.5	194.0	254.0	88.6	68.2	108.0	91.0
1973	134.0	138.0	77.2	85.3	73.8	226.0	240.0	290.0	111.0	72.6	80.3	69.0
1974	66.9	82.5	66.3	62.8	65.4	73.4	142.0	107.0	111.0	61.6	86.4	77.8
1975	58.8	106.0	75.5	66.3	66.0	68.4	193.0	209.0	116.0	54.8	41.5	63.2
1976	43.4	69.2	86.3	67.5	69.2	96.9	298.0	147.0	79.1	41.2	36.9	30.3

Source: Davis and Cornwell (1985).

=====

COMPUTATION OF EXTREME FLOWS

8.1 METHODS OF COMPUTATION

The extreme hydrologic events are floods and droughts. Both of these are defined differently by various agencies throughout the world. In general qualitative terms, these refer to periods of unusually high and low water supplies. Any hydraulic structure in a river system, such as a dam, spillway, channel, road, or railway drainage, has to experience floods and droughts related to that system. The floods become relevant from the consideration of the capacities of these hydraulic structures. On the other hand, the depth of a navigable channel, the water supply during a dry period, and the quantity of flow below a regulatory structure are concerned with drought conditions. When the streamflow and/or precipitation records are available, these form the basis of estimation of flood and drought flows. These records are, however, not too long to provide the extreme values directly. The extrapolation is carried out by a statistical process or by the physical analysis of critical hydrometeorological events. Where streamflow data are not available, empirical and other indirect methods are applied.

Floods may arise from extreme rainstorms, or from the rapid melting of extensive snow deposits, or from a combination of the two. Where the records of streamflows are not available, the flood flow estimate is made from the data on extreme rainfall. The method of flood flow computation are summarized below.

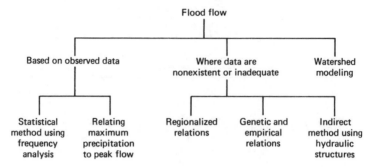

The following procedures are used in the study of droughts.

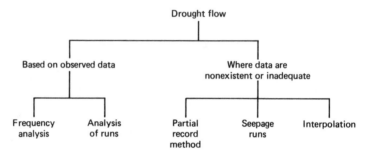

8.2 THE CONCEPT OF PROBABILITY IN HYDROLOGY

Since the magnitude of the flows recorded in the past will be repeated, a specific flood value will be equaled or exceeded (equaled or less in low flow analysis) in a period of time. The actual time between exceedences is called the recurrence interval. Statistical analysis of hydrological events considers the average elapsed time between occurrences of an event (i.e., flow of a certain magnitude or greater). This average recurrence interval for a certain event is also known as the return period of that event. The chance of occurence of a flood of a return period T, in a unit time, is $1/T$, called the probability of occurrence. Because the period is usually measured in years and the probability is expressed in percent, it is referred to as the percent probability of annual exceedence.

A flood discharge is a continuous variable that can acquire any value between two numbers. An individual observation or value of a variable, in this case flood flow, is known as a variate. An array of variates, constituting a time series, represents a sample from the population of peak discharges recorded in the past and to be observed in the future at the study site. The continuous series can be reduced to a discrete form by grouping the data into a number of classes of equal discharge interval, each class representing a discrete variate. The number of items in a class (number of occurrences of a variate) within the entire data base is called its frequency. A complete description of the frequency of all classes (variates), such as a plot of the num-

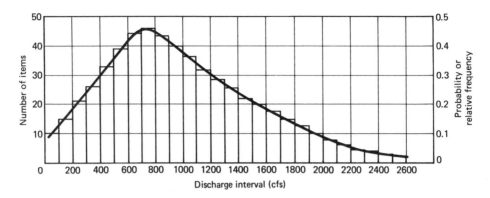

Figure 8.1 Frequency distribution curve.

ber of items in each class against the respective class interval as shown in Figure 8.1, is called a frequency distribution. When the number of items in a class is divided by the total number of items in all classes, the result is the probability of that class or variate, as defined earlier, that is,

$$p = \frac{n_i}{N} \quad \text{[dimensionless]} \tag{8.1}$$

where

p = probability of occurrence of flood flow of class i (variate i)

n_i = number of items in the ith class

N = total number of items in a series

The distribution of the probabilities of all classes (instead of their frequencies) is known as a probability distribution. The ordinates of the frequency distribution and the probability distribution are proportional to each other.

In the case of a continuous random variable, when a variate, x, takes a continuous value, the probability becomes a continuous function, p_x, called the probability density function (PDF). Statisticians have demonstrated that the distribution of a large number of natural phenomena, including hydrologic data, can be expressed by certain general mathematical equations. These are recognized to be theoretical probability distribution functions. There are many different types of probability distributions. Some of these, such as the bionomial, geometric, and Poisson distributions, consider the discrete process, while others, such as the uniform, normal, gamma, beta, Pearson, and extreme value distributions, are for the continuous random process, as described subsequently. These mathematical functions (equations) are very convenient in analysis because of their known solutions.

If the frequencies or probabilities, as shown in Figure 8.1, are successively summed up (accumulated) starting from the highest value, a curve of type (a) in Figure 8.2 results. This is known as the cumulative frequency or probability, P, which indicates the probability that a variable has a value equal or greater than a certain assigned value. This probability is designated as $P(X \geq x)$. When the probabilities of a variate are summed up starting with the lowest value, a curve of type (b) of

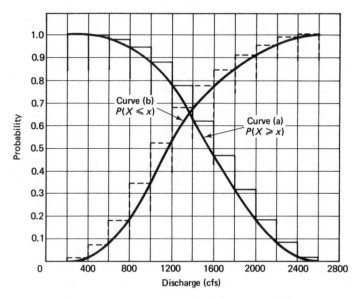

Curve (a): Cumulative frequency or probability of equal, or greater than, $P(X \geqslant x)$:
Curve (b): Probability of equal, or less than, $P(X \leqslant x)$:

Figure 8.2 Cumulative probability curve.

Figure 8.2 is obtained which indicates the cumulative probability, $P(X \leq x)$, that a variable has a value equal to or less than certain assigned value. In the case of a continuous variable, the summation for cumulative probability can be expressed by an integration of the probability distribution function (PDF), as follows:

$$P(X \leq x) = \int_{-\infty}^{x} p_x \, dx \qquad \text{[dimensionless]} \qquad (8.2)$$

This integration (area under the PDF) is called the cumulative distribution function (CDF). The total area must be equal to unity.

In eq. (8.1), the number of possible values for a continuous variable, and hence N, approaches infinity. Thus the probability that a variable will have an exact value, x, has no meaning [zero probability from eq. (8.1)] for a continuous function. Therefore, the probability of occurrence is expressed for a variate having a value greater than x or less than x. The probability estimates for continuous variables are, accordingly, related to areas under the PDF (i.e., CDF rather than the ordinates of the PDF). For the theoretical distribution functions, the tables are available for areas under the curves in standardized units. On graph paper specially constructed for a specific distribution, CDF plots as a straight line.

8.3 DESIGN FLOOD FOR HYDRAULIC STRUCTURES

The hydrologic design of a project is based on an optimum peak flood discharge. There are two approaches to estimating the optimum design discharge. The major hydraulic structures for flood control or other structures, where a high degree of protection is required due to the danger to human lives and extensive property damage, such as the spillway of an earthen dam, are designed on the basis of probable maximum flood and standard project flood associated with a critical combination of meteorological and hydrological conditions. No attempt is made to associate the design discharge with any specific probability of exceedence. The precipitation maximization method described subsequently is used for this purpose.

In the case of storage capacity of reservoirs, spillway design for concrete structures in remote areas, carrying capacity of channels and culverts, channel improvement schemes, and storm sewer systems, if the design capacity is exceeded, some damage will result, but not of a catastrophic nature. The optimum design flood of these structures is based on a certain probability of exceedence or return period. A frequency analysis is required in such cases. The design probability of flood discharge is determined from a consideration of (1) acceptable level of risk, (2) economic factors, and (3) standard practice.

8.3.1 Risk Basis for Design Flood

A structure designed for any level of peak discharge bears a certain risk of being overflowed in its lifetime. Consider that the return period of 10,000 cfs discharge is 100 years. This is referred to as a "100-year exceedence" flood. This means that the probability of exceeding 10,000 cfs in any one year is $1/100 = 0.01$, or the exceedence probability, P, is of 0.01 or 1%. The exceedence probability is defined with respect to a single trial (year). If it is desired to know the probability of exceeding 10,000 cfs in a total period of 100 years, the answer is not straightforward. A 100-year flood does not mean that it will definitely be exceeded exactly once in every 100 consecutive years, but it implies that in a very large number of occurrences it is the average return period.

Flood flows follow the Bernoulli's process according to which the probability of an event occurring is independent of time and independent of the past history of occurrences or nonoccurrences. In such a case, at any time, an event may either occur with probability P or not occur with probability, Q $(= 1 - P)$. The probability of one event in 3 years is $PQQ + QPQ + QQP$, which is equal to $3PQ^2$. Thus the probability of k events in n years is equal to the number of ways of arranging k values of P among n items. This is indicated in terms of the exceedence probability and is referred to as the bionomial probability distribution. It considers a discrete time scale.

$$f_x\{\text{exactly } k \text{ events in } n \text{ years}\} = C_k^n P^k (1 - P)^{n-k} \qquad \text{[dimensionless]} \qquad (8.3)$$

where

$$P = \text{exceedence probability of an event in any one year}$$
$$f_x = \text{probability of } k \text{ events (exceedences) in } n \text{ years}$$
$$C_k^n = \frac{n!}{k!\,(n - k)!}$$

In hydrologic study it is usually not important to know the probability that an event (e.g., flood) will exceed exactly k times, but to ascertain the probability that an event will occur once or more in n years. Thus

$$f_x\{1 \text{ or more event in } n \text{ years}\} = 1 - f_x\{\text{zero event in } n \text{ years}\}$$

From eq. (8.3),

$$f_x\{1 \text{ or more flood in } n \text{ years}\} = 1 - C_0^n P^0 (1 - P)^{n-0}$$

or

$$f_x\{\text{at least one flood in } n \text{ years}\} = 1 - (1 - P)^n \qquad \text{[dimensionless]} \qquad (8.4)$$

Equation (8.4) gives the probability, f_x, of a stucture overtopping at least once or the risk level, in n years, associated with a flood of any exceedence probability, P(return period, $T = 1/P$). Alternatively, for a project life, n, and an acceptable risk level, R (where $R = f_x \times 100$), the exceedence probability (P) and hence the return period $(1/P)$ of the design flood can be computed from eq. (8.4). The values are shown in Table 8.1 for various acceptable risk levels and a project life of 25, 50, and 100 years. When only a 1% chance (risk level) of a structure being overtopped in 50 years can be taken, it should be designed for a 5260-year-return-period flood (Table 8.1).

TABLE 8.1 RETURN PERIOD, $1/P$, FOR VARIOUS RISK LEVELS [EQ. (8.4)]

Acceptable Level of Risk, R (%)	Project Life, n (years)		
	25	50	100
1	2440	5260	9950
25	87	175	345
50	37	72	145
75	18	37	72
99	6	11	27

Example 8.1

A culvert has been designed for a "50-year exceedence interval." What is the probability that exactly one flood of the design capacity will occur in the 100-year lifetime of the structure?

Solution $n = 100$, $k = 1$. Exceedence probability, $P = \frac{1}{50} = 0.02$. From eq. (8.3),

$$f_x\{1 \text{ event in } 100 \text{ years}\} = C_1^{100} P^1 (1 - P)^{100-1}$$

$$= \frac{100!}{1!(100-1)!} (0.02)^1 (1 - 0.02)^{99}$$

$$= 0.27$$

Example 8.2

In Example 8.1, what is the probability that the culvert will experience the design flood one or more times (at least once) in its lifetime?

Solution From eq. (8.4),

$$f_x\{\text{at least once in } 100 \text{ years}\} = 1 - (1 - 0.02)^{100}$$

$$= 0.87$$

Computation of Extreme Flows   Chap. 8

Example 8.3

The spillway of a dam has a service life of 75 years. A risk of 5% for the failing of the structure (exceeding of the flood capacity) has been considered acceptable. For what return period should the spillway capacity be designed?

Solution

$$f_x = \frac{5}{100} = 0.05$$

$$n = 75$$

From eq. (8.4),

$$f_x = 1 - (1 - P)^n$$

$$0.05 = 1 - (1 - P)^{75}$$

$$P = 0.000684$$

$$T = \frac{1}{P} = 1460 \text{ years}$$

8.3.2 Economic Basis for Design Flood

From economic considerations the optimum design discharge is the peak flow rate corresponding to a return period whose use in the project design will minimize the average annual cost of the project. The average annual cost involves the following:

1. Annual cost allocated from the total cost of construction of a structure, apportioned over the economic life of the structure
2. Annual operation and maintenance cost of the structure
3. Annual flood damages in money terms with the proposed structure in position

A flood–frequency curve (a plot of flood magnitude versus exceedence probability) is necessary for this analysis, which has been described in Section 8.4. Various development levels for the project are considered. For each alternative, the costs of the first two items listed above are computed by standard procedures of engineering economics. The last item is computed by the following steps:

1. Prepare a flood–frequency curve of peak discharge (Q) versus exceedence probability, (P).
2. For a selected design alternative, collect data on flood damage from the field study, that is, monetary flood damages (J) for various flood stages (H) or flood discharges (Q).
    If it is in terms of flood stages, convert these stages into corresponding discharges, using the stage–discharge relation for the site of study.
3. Combining steps 1 and 2, prepare a damage–frequency relation or curve of damage (J) versus exceedence probability (P).
4. Determine the area under the damage–frequency curve graphically or arithmetically to obtain the annual flood damage under selected project conditions.

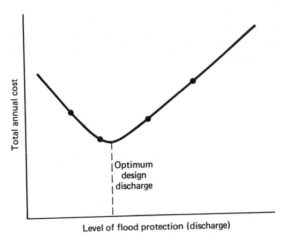

Figure 8.3 Determination of optimum design discharge on economic basis.

When the annual construction cost and the annual cost of operation and main-tenance of the selected design alternative are added to the annual damage cost computed above, the total annual cost is obtained. Perform a similar analysis with other alternative designs. Plot the computed data as shown in Figure 8.3. The optimum design discharge is the point at which the total cost is minimum. Determine the exceedence probability for this discharge from the flood–frequency curve.

8.3.3 Standard Practice for Design Exceedence Probabilities

The extensive analysis described above is justified for major projects. Moreover, the parameters used in the analysis are often not well defined. Consequently, it has become a practice to adopt a standard design exceedence probability based on (1) the type of structure, (2) the importance of the structure, and (3) the development of the area subject to flooding.

Large and flood-vulnerable hydraulic structures are designed for a recurrence interval of 1000 years or more. A common frequency level is 100 years, for which small to medium-sized hydraulic structures, navigable waterways, and river ports are designed. Minor structures, the culverts on highways and railway bridges are designed for a frequency of 10 to 50 years. The storm drainage in residential areas is designed using flood exceedence of 2.5 to 10 years.

8.4 FLOOD–FREQUENCY ANALYSIS

There are three applications of the statistical methods in hydrologic study. First, it is used in the regression analysis to determine the rating equation (Section 6.14.3) and to extend the short-duration record (Section 7.13). Second, it is applied for computing the statistic parameters in the synthetic streamflow generation (7.20.1). Third, it is used in the flood–frequency study to prepare a curve that indicates the magnitude of floods of various probabilities of occurrence. Once the selection of the design exceedence probability (or return period) has been made by the procedures of the preceding section, the peak discharge corresponding to that probability or return interval on the flood–frequency curve becomes the design discharge. The procedure of analy-

Computation of Extreme Flows   Chap. 8

sis and the reliability of a flood–frequency curve depend on the type and quality of observed flood flow series on which the curve is based. No amount of statistical sophistication can improve the quality of the data.

8.4.1 Type of Data

When a set contains all available data observed over a certain period of time, it constitutes a complete-duration series. An application of such a series is considered in the flow–duration analysis (Section 7.28.2). All of these data, however, have no significance in flood or drought estimation, which is governed by extreme flows. Accordingly, from a complete-duration series, two types of data are selected; annual series and partial or partial-duration series. The annual series includes the largest or smallest values recorded each year (or equal time intervals apart). The partial series contains all the data that have a magnitude greater than a certain base value irrespective of their year or duration of occurrence. If the base value selected is such that the number of values in the series is equal to the number of years of the record, the series is called an annual exceedence partial-duration series. The relationship between the probabilities of the annual exceedence partial-duration series and the annual series have been investigated by Langbein and Chow.

Where two types of flood peaks occur each year, such as spring snowmelt floods and winter rainstorm floods, or hurricane and nonhurricane floods, they are known as mixed population data that require special treatment, as discussed subsequently.

8.4.2 Quality of Data

For a reliable computation of a flood–frequency curve, the peak-discharge data should meet the following requirements:

1. *Stationariness of data*. The meaning of stationary data is that the properties or characteristics of the data do not change with time. As described in Section 7.14, the stationariness can be checked by dividing a long flood flow series into a number of segments (subsets). The statistical parameters of mean, standard deviation, and coefficient of skew should be comparable for each subset. For the data of short length, it is not feasible to perform this test, but their adequacy is checked.

2. *Homogeneity of data*. Homogeneity is an indicator that all data of a series belong to the same population. The homogeneity check may be performed qualitatively by studying the factors that have a disturbing effect on it. The quantitative analysis is made by the statistics theory of "test of hypothesis." According to this, it is hypothesized (assumed) that the data follow a certain distribution. An acceptable probability, or risk level, of making the wrong decision from the hypothesis is specified; this is known as the level of significance.

In flood flow computations the significance level is usually assumed to be 1, 2, or 5%. As an example, a 5% level of significance means that there is about a 5% chance that a hypothesis will be rejected when it should have been accepted. This represents an area at the extreme end of the probability distribution curve and is referred to as the critical region, as shown in Figure 8.4. Using the equation of the hy-

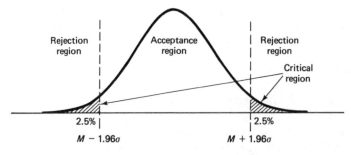

Figure 8.4 Critical region in hypothesis testing.

pothesized distribution and the statistical parameters derived from the sample, the value of the standard variate, called the test statistic, is computed. This is compared to the theoretical value of the variate obtained from a standard distribution table for the specified level of significance. If the computed value is more than the tabular or theoretical value, it is in the critical region or in the region of rejection of the hypothesis.

The distributions suitable for small samples of less than 30 values, as is usually the case with the flood flow series, are Student's distribution and the chi-square distribution. Both of these assume that the population, from which a sample has been derived, is normally distributed. Often the logarithmic values of peak discharges are used in analysis since the flood flows are normally distributed in that form.

To test the hypothesis that two samples of sizes N_X and N_Y come from the same population, the standard variate or test statistic of Student's distribution has the following form:

$$t = \frac{\overline{Y} - \overline{X}}{\sqrt{N_X S_X^2 + N_Y S_Y^2}} \sqrt{\frac{N_X N_Y \nu}{N_X + N_Y}} \qquad \text{[dimensionless]} \qquad (8.5)$$

where

$$\overline{X}, \overline{Y} = \text{mean values of two samples}$$

$$S_X, S_Y = \text{standard deviations of two samples}$$

$$\nu = \text{degrees of freedom}$$

$$= N_X + N_Y - 2$$

The theoretical standard variates for Student's distribution are given in Appendix E.

Equation (8.5) tests the homogeneity of the mean of the series. In its formulation it was considered that the two series are homogeneous with respect to the standard deviations. Hence, before performing the test above by Student's criterion, the homogeneity for the standard deviation is tested by the Fisher distribution, which is an extension of the chi-square distribution. The statistic, or variate, of χ^2 [χ is the Greek lowercase letter chi (pronounced "Kai") and χ^2 is chi-square] is given by $\chi^2 = NS^2/\sigma^2$, where σ (Greek lowercase letter sigma) is the standard deviation of the population, a sample of size N of which has the standard deviation S. The distribution has a degree of freedom, $\nu = N - 1$. For two samples, one of chi-square variate with $\nu = m$ degrees of freedom and the other of chi-square variate with $\nu = n$ degrees of freedom, the Fisher distribution is

$$F = \frac{\chi_m^2}{\chi_n^2}$$

or

$$F = \frac{S_X^2}{S_Y^2} \quad \text{[dimensionless]} \tag{8.6}$$

The cumulative F distribution with $\nu_1 = m$ and $\nu_2 = n$ degrees of freedom (m and n are considered as the numerator and denominator degrees of freedom, respectively) are given in Appendix F. As in the case of Student's criterion, when the computed F value is less than the theoretical value from the standardized statistical table at a specified level of significance, the hypothesis cannot be rejected (is accepted).

Example 8.4

The annual peak discharges of Lamprey River at Newmarket, New Hampshire, are given in Table 8.2. The discharges in parentheses have been contributed by snowmelts and the others by rainstorms. Determine whether the two type of peak discharges are a part of a single population of flood peaks. Adopt a 5% level of significance.

TABLE 8.2 ANNUAL PEAK DISCHARGE DATA FOR LAMPREY RIVER AT NEWMARKET, NEW HAMPSHIRE

Date	Discharge (cfs)	Date	Discharge (cfs)	Date	Discharge (cfs)
8.3.45	12,000	7.10.59	8,440	7.20.71	13,950
3.15.46	(5,050)	3.29.60	(6,995)	4.8.72	(12,680)
8.19.47	8,440	4.3.61	(5,298)	2.20.73	(2,730)
4.5.48	(4,026)	7.23.63	7,735	3.25.74	(14,000)
2.1.50	9,925	8.20.64	8,441	4.5.75	(4,840)
4.3.51	(6,145)	3.10.65	(7,840)	9.3.76	5,475
6.15.52	3,130	6.13.66	6,990	3.25.77	(7,275)
5.1.53	(15,750)	4.20.67	(4,026)	4.2.79	(5,050)
7.22.54	9,470	3.25.68	(7,880)	9.13.81	6,710
9.2.56	13,600	2.28.69	(3,285)	6.23.83	7,523
8.31.57	18,190	4.6.70	(10,560)	7.28.85	4,275
4.16.58	(11,655)				

Solution

1. The sequence above is divided into two series, one for rainstorm discharges and the other for snowmelt discharges.
2. The statistical parameters of mean and standard deviation for the two series, as computed by eqs. (7.26) and (7.27), are listed in Table 8.3.

TABLE 8.3 STATISTICAL PARAMETERS FOR SNOWMELT AND RAINSTORM SERIES

	Snowmelt	Rainstorm
1. Number of floods	18	16
2. Mean, $\overline{X}$ (cfs)	7505	9018
3. Standard deviation, S (cfs)	3871	3857

3. Since the homogeneity of standard deviations is a prerequisite for Student's criterion for the mean, the Fisher distribution is tested first. From eq. (8.6),

$$F(\text{computed}) = \frac{S_X^2}{S_Y^2} = \frac{(3857)^2}{(3871)^2} = 0.99$$

4. Using $(N_X - 1) = 15$ degrees freedom for S_X (numerator) and $(N_Y - 1) = 17$ degrees of freedom for S_Y (denominator), F(theoretical), from the table in Appendix F at 5% significance level, is*

$$F = 2.30$$

5. Since F(computed) $<$ F(theoretical), the hypothesis cannot be rejected. The data are homogeneous with respect to the standard deviations.

6. After the Fisher criterion is satisfied, Student's criterion is, from eq. (8.5),

$$t(\text{computed}) = \frac{\overline{Y} - \overline{X}}{\sqrt{N_X S_X^2 + N_Y S_Y^2}} \sqrt{\frac{N_X N_Y \nu}{N_X + N_Y}}$$

$$\nu = N_X + N_Y - 2 = 16 + 18 - 2 = 32$$

$$t(\text{computed}) = \frac{|7505 - 9018|}{\sqrt{16(3857)^2 + 18(3871)^2}} \sqrt{\frac{16(18)(32)}{16 + 18}}$$

$$= 1.11$$

7. From the table in Appendix E for the 5% level and 32 degrees of freedom,

$$t(\text{theoretical}) = 1.70$$

8. Since t(computed) $<$ t(theoretical), there is a homogeneity of means. Hence the data belong to a single homogeneous population.

8.4.3 Consistency of Data

The record of peak stages and discharges should be as complete as possible. The peak stage and discharge, when an instrument had failed, should be determined from high-water marks in the vicinity of the gaging station. If there is a reason for doubt that the data for the entire period are not related to the same datum, the streamflow data of a particular site should be checked for consistency and accuracy by comparison with several surrounding gaging sites by the mass curve analysis (Section 3.5.1). The inconsistencies should be reconciled and erroneous data should be recomputed or excluded. Natural flood flows are the basic data required for hydrologic design. Where regulated flows are available, the effect of regulation should be corrected by the flood-routing techniques (Section 8.18) to obtain natural discharge values.

8.4.4 Adequacy of Data

The length of record is an important factor since a short-duration record may not be representative of the true nature of peak flows at a site. The computed statistical parameters will thus not be reliable. A minimum period of record of 25 years has been considered desirable for the statistical analysis of peak flows. For a shorter observation period, the adequacy of the record may be evaluated by the analysis with respect to a long-term base gaging station in the region (Sokolov et al., 1976).

*For 5% or 0.05 significance, α in the table $= 1 - 0.05 = 0.95$.

Consider that the gaging station at the study site has a short record of S years. Also, that a base gaging station in the homogeneous region has a long-term record of L years that includes the period of S years. For the base gaging station, the statistical parameters (mean, $\overline{X}$; standard deviation, S_X; and coefficient of skewness, g) are computed for two sets of record of period S and L year, respectively. If the ratios of parameters for short and long length, $\overline{X}_S/\overline{X}_L$, $(S_X)_S/(S_X)_L$, and g_S/g_L, do not depart from unity by more than 15%, the short-term record of the study site is considered adequate. If these ratios show more than a 15% difference, the short-period record is extended by correlation analysis (Section 7.16) with the long-term base gage station.

8.5 METHODS OF FLOOD–FREQUENCY ANALYSIS

There are three methods to prepare a flood–frequency curve from the array of flood flow data; (1) graphical method, (2) empirical method, and (3) analytical method. The last of these has a wider application.

8.6 GRAPHICAL METHOD

In this method, the array of flood flows is divided into a number of class intervals of equal range in discharge. The number of occurrences of flood flows in each class interval are noted. The number of occurrences in each class interval are cumulated, starting with the highest value. The percentage of the accumulated number of items or occurrences of each class with respect to the total occurrences of all classes is determined. The computed percent is then plotted against the lower discharge limit of each class on a probability paper (Section 8.7.1), where it preferably plots a straight line.

This procedure can be applied only when the array consists of a very large number of flood events. The method has been described in detail in connection with the flow–duration curve or complete series analysis (Section 7.28.2). The peak-flow data are, however, not too long usually. As such, the other two methods are often used.

8.7 EMPIRICAL METHOD

This is also a graphic procedure. In this method, the plotting position of a magnitude of flood, is, however, determined by an empirical formula. If an array of n flood flow values is arranged in descending order of magnitude starting with the highest discharge, when n approaches infinity, a discharge ranking m in order of magnitude will have an exceedence probability as follows:

$$P_m = \left(\frac{m}{n}\right)_{n \to \infty} \quad \text{[dimensionless]} \quad (8.7)$$

Equation (8.7), when applied to a smaller sequence, however, will assign a probability of 100% to the lowest-valued flood of the sequence having a rank of n, which means that there is no possibility of a flow of less than that value. This is obviously erroneous. To remove the bias in plotting positions at two extreme ends, many empirical formulas have been suggested which are special cases of the following general formula:

$$P_m = \frac{m - a}{n + b}(100) \qquad \text{[dimensionless]} \qquad (8.8)$$

where a and b are constants. Several of these formulas are summarized by Adamowski (1981). All formulas give practically the same results in the middle range of discharge but produce different positions near the upper and lower tails of the distribution.

The following formula proposed by Weibull in 1939 is widely used:

$$P_m = \frac{m}{n + 1}(100) \qquad \text{[dimensionless]} \qquad (8.9)$$

For an annual series, the return period or recurrence interval T is the inverse of P_m.

8.7.1 Probability Graph Paper

The plot of peak-discharge magnitude against probability of exceedence (cumulative distribution function) is a curve on arithmetic paper. The purpose of a probability paper·is to linearize this plot, so that the extrapolation of the data, as often needed, is simplified.

The equation of the cumulative distribution function (CDF) has to be transformed to the form $Y = mX + C$ to plot as a straight line, where Y is a function of exceedence probability and statistical parameters and X indicates the peak flow. The frequency factors have this linear relation as described in Section 8.8.5. On graph paper, the linear transformation can be achieved by the distortion of the probability scale (abscissa). Since there are different equations of the CDF, separate graph paper has to be constructed for each theoretical distribution. Further, since the transformed function includes the statistical parameters, it is feasible to construct the paper for a distribution that is defined by two parameters only. Hence the probability papers have been designed for normal distribution, log-normal distribution (the ordinate is in log scale), type I extreme value or Gumbel distribution (two parameters and a fixed skew of 1.14), and type III extreme value or Weibull distribution, which is essentially a logarithmically transformed type I distribution.

The theoretical distributions will plot a straight line on respective papers. The natural floodflow data do not necessarily follow any exact theoretical distribution. Moreover, in the empirical method no distributional assumptions are made. Thus the observed data can be plotted on any kind of paper, and the best judgment is made in interpretation of past and future peak discharges. The plotting paper commonly used in the graphic method, however, is the logarithmic normal probability graph.

The steps in the empirical procedure are summarized below.

1. Rank the data from the largest to the smallest values. If two or more observations have the same value, assume that they have slightly different values and assign each a different rank.
2. Calculate the plotting position from eq. (8.9).
3. Do not omit any years during the period of record since it will have a biasing effect. If any data are missing, make their estimates (Section 7.13). The data

could be excluded when the cause of the interruption in data is known to be independent of the flow condition.

4. Often, data on one or more historical flood events that occurred prior to the period of record may be known. The plotting positions for the historical events and other peak flows higher than the historical events are determined based on the total period from the time of the first historical event to the end of the flood flow record. The plotting position of other peak flows is based on the period of record of the data only. For example, the flood flows are available for a period of 40 years, from 1946 to 1985. The magnitude of a historical peak flow that occurred in 1900 has been observed which is the highest of all recorded flows. The plotting position for historical flow will be based on 86 years (1900 to 1985) and that of other floods on 40 years of record. The modification of the plotting positions due to historical floods has been discussed by Dalrymple (1960, pp. 16–18).

5. Select the type of the probability paper to be used; the log-normal graph paper is common.

6. Plot the magnitude of flood on the ordinate and the corresponding plotting position on the abscissa, representing the probability of exceedence as one side of the scale and the return period on the other side.

Frequently, one or two extreme events may plot far off from other points as "outliers." This occurs because an extreme event in the recorded data may actually represent a much higher return period than the period of record. There is a probability of at least one occurrence of a T-year (higher) event in a n-year (smaller) period of record from eq. (8.4). If possible, the return period of these extreme events might be investigated based on available historical or regional information. A test for detection of high and low outliers and their treatment are described by the U.S. Water Resources Council (1981, pp. 17–18).

The extrapolation of the data for longer return periods should be done very cautiously because the probability distribution is very sensitive in the tail part of the curve.

Example 8.5

The maximum annual instantaneous flows of the Cedar River near Waterloo, Minnesota, are given in Table 8.4. From the historic record, a peak flow of 30,000 cfs was noted in 1905. (a) Prepare a flood-frequency curve. (b) Determine the probability of flow of 20,000 cfs. (c) Determine the magnitude of flow corresponding to an exceedence probability of 0.5. (d) Determine the magnitude of flow of a return period of 100 years.

Solution

(a) Flood–frequency curve
1. The rank of each flow value, starting with the highest flood, is indicated in Table 8.4. Many designers prefer to make a separate table by arranging values in descending order.
2. For historic flow, the base period is 68 years; hence the plotting position.

$$P = \tfrac{1}{69}(100) = 1.4\%$$

3. For each rank in Table 8.4, the plotting position is computed from eq. (8.9).
4. Peak discharge versus plotting position (exceedence probability) is plotted on log-normal paper in Figure 8.5.

TABLE 8.4 ANNUAL PEAK FLOWS OF THE CEDAR RIVER NEAR WATERLOO, MINNESOTA

Year	Flow (cfs)	Rank	Plotting Position (%)	Year	Flow (cfs)	Rank	Plotting Position (%)
1950	14,400	5	20	1962	6,240	17	68
1951	6,720	16	64	1963	22,700	1	4
1952	13,390˙	7	28	1964	11,140	10	40
1953	15,360	4	16	1965	4,560	21	84
1954	8,856	13	52	1966	5,376	19	76
1955	5,136	20	80	1967	12,480	9	36
1956	6,770	15	60	1968	19,200	3	12
1957	9,600	12	48	1969	12,984	8	32
1958	980	24	96	1970	5,450	18	72
1959	4,030	22	88	1971	13,440	6	24
1960	10,440	11	44	1972	22,680	2	8
1961	3,100	23	92	1973	8,400	14	56

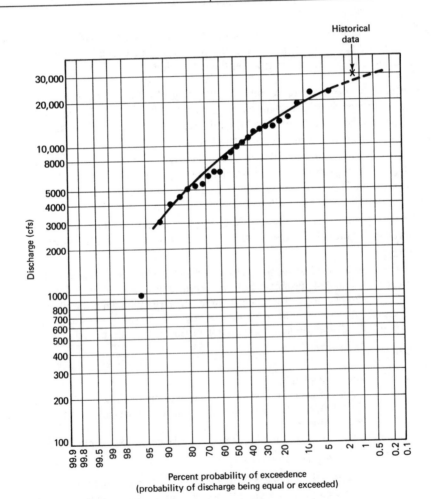

Figure 8.5 Flood–frequency curve of Cedar River (Example 8.5).

Computation of Extreme Flows   Chap. 8

5. The following values are read directly from the graph.

(b) Probability of flow of 20,000 cfs = 8% or 0.08.

(c) Flow of 0.5 (or 50%) probability = 9500 cfs.

(d) For $T = 100$ years, $P = \frac{1}{100}(100) = 1.0\%$. Flow for 1.0% probability = 28,000 cfs.

8.8 ANALYTICAL METHOD

This method makes use of the theoretical probability distribution functions. The plot of the cumulative density function (CDF), of any distribution, is a frequency curve by definition. The CDF equations contain statistical parameters which are computed from recorded data series. The requirement, however, is that the theoretical distribution should represent characteristics similar to those demonstrated by the natural recorded data. There are many distributions; four commonly applied for fitting the hydrologic sequences are described below. The features of these distributions, comprising PDF, range, mean, and variance, are summarized in Table 8.5. Of these, the log-Pearson type III has been widely adopted as a distribution of choice in flood flow analysis. The extreme value type III distribution is preferred in the study of low flows (droughts).

8.8.1 Normal (Gaussian) Distribution

This is a bell-shaped frequency function symmetrical about the mean value. It has very wide applications, although, due to its range from $-\infty$ to $+\infty$, it does not fit well to hydrologic sequences which do not have negative values. The distribution has two parameters, the mean, μ, and the standard deviation, σ ($\overline{X}$ and S for sample data), as shown in Table 8.5. It can be transformed to a single parameter function using a standard variate, Z, in terms of μ and σ.

8.8.2 Log-Normal Distribution

This is an extension of the normal distribution wherein the logarithmic values of a sequence are considered to be normally distributed. The PDF and all other properties of the normal distribution are applicable to this distribution when the data are converted to logarithmic form, $y = \ln x$, as indicated in Table 8.5. It is a two-parameter, bell-shaped, symmetrical distribution in this form. In terms of an untransformed variate, x, it is a three-parameter (skewed) distribution having a range from 0 to ∞.

This distribution suits to the hydrologic data and has the advantage of a link with the normal distribution.

8.8.3 Extreme Value Distribution

Consider n data series with m observations in each series. A largest or smallest (extreme) value is obtained out of m observations in each series. There will be n such extreme values. The probability distribution of these extreme values depends on the sample size, m, and the parent distribution of the series. Frechet, in 1927, and Fisher and Tippett, in 1928, found that the distribution of extreme values approaches an asymptotic form as m is increased indefinitely. The type of the asymptotic form is

TABLE 8.5 PROPERTIES OF COMMON DISTRIBUTIONS

Distribution	Probability Density Function (PDF), $p(X)$	Cumulative Density Function (CDF), $P(X \le x)$	Range	Mean μ or $\bar{X}$	Standard Deviation σ or S
1. Normal	$\dfrac{1}{\sqrt{2\pi}\sigma} e^{-(X-\mu)^2/2\sigma^2}$ or $\dfrac{1}{\sqrt{2\pi}} e^{-z^2/2}$ where $z = \dfrac{X-\mu}{\sigma}$	$\displaystyle\int_{-\infty}^{x} \dfrac{1}{\sqrt{2\pi}\sigma} e^{-(x-\mu)^2/2\sigma^2}\, dx$	$-\infty \le x \le \infty$	μ	σ
2. Log-normal $y = \ln x$	$\dfrac{1}{\sqrt{2\pi}\sigma_y} e^{-(y-\mu_y)^2/2\sigma_y^2}$	$\displaystyle\int_{-\infty}^{y} \dfrac{1}{\sqrt{2\pi}\sigma_y} e^{-(y-\mu_y)^2/2\sigma_y^2}\, dy$	$-\infty \le y \le \infty$ $0 \le x \le \infty$	μ_y	σ_y
3. Extreme value Type I, $y = (x-\beta)/\alpha$	$\dfrac{1}{\alpha} e^{-y-e^{-y}}$	$e^{-e^{-y}}$	$-\infty \le x \le \infty$	$\beta + 0.577\alpha$	1.283α
Type III	$\alpha x^{\alpha-1}\beta^{-\alpha} e^{-(x/\beta)^\alpha}$	$1 - e^{-(x/\beta)^\alpha}$	$x \ge 0$	$\beta\Gamma(1 + 1/\alpha)$	$\beta[\Gamma(1 + 2/\alpha) - \Gamma^2(1 + 1/\alpha)]^{1/2}$
4. Log Pearson Type III, $y = \ln x$	$p_0(1 + y/\alpha)^c e^{-cy/\alpha}$ where p_0 = prob. at the mode $= \dfrac{N}{\alpha}\dfrac{c^{c+1}}{e^c\Gamma(c+1)}$	$\displaystyle\int_{-\infty}^{y} p_0(1 + y/\alpha)^c e^{-cy/\alpha}\, dy$ (known as incomplete gamma function)	$-\infty \le y \le \infty$ $0 \le x \le \infty$	$(c + 1)\dfrac{\alpha}{c}$	$\sqrt{c+1}\,\dfrac{\alpha}{c}$

Γ is the gamma function; $\Gamma(n) = (n-1)!$.

dependent on the parent distribution from which the extreme values were obtained. Three types of asymptotic distributions have been developed based on different parent distributions.

The type I extreme value distribution, also known as the Gumbel distribution, results from the exponential-type parent distribution. The parent distribution is unbound (has no limit) in the direction of the extreme value. The density functions of type I distribution, which are in terms of parameters α and β, are given in Table 8.5. Using the mean and the standard deviation of the flood flow series, α and β can be evaluated from the relations given under mean and standard deviation in Table 8.5. The distribution has a constant skew coefficient of ± 1.14. Gumbel used the type I distribution first in an analysis of floods in 1941. He argued that the daily discharge of each year constituted a sufficiently large sample, with $m = 365$, from which an extreme value of flood flow was picked up.

The type II distribution orginates from the Cauchy-type distribution of the parent distribution, but it has little application in hydrologic events.

The type III or Weibull distribution also arises from an exponential-type parent distribution, but the parent distribution is limited in the direction of the extreme value (e.g., the low flows are bounded by zero on the left). The density functions are given in Table 8.5. This distribution is essentially a logarithmically transformed type I distribution. Gumbel applied this for low-flow analyses.

8.8.4 Log-Pearson Type III (Gamma-Type) Distribution

Karl Pearson proposed a general equation for a distribution that fits many distributions — including normal, beta, and gamma distributions — by choosing appropriate values for its parameters. A form of the Pearson function, similar to the gamma distribution, is known as the Pearson type III distribution. It is a distribution in three parameters with a limited range in the left direction, unbounded to the right, and has a large skew. Since the flood flow series commonly indicate considerable skew, this is used as the distribution of flood peaks. The distribution is usually fitted to the logarithms of flood values because this results in lesser skewness. The log-Pearson type III distribution has been adopted as a standard by U.S. federal agencies for flood analyses.

8.8.5 Approach to Analytical Analysis: Use of Frequency Factors

The CDF indicated in Table 8.5 for various probability distributions is of the type for which a direct integration is generally not possible. However, integration tables have been developed for various distributions from approximate or numerical analyses which are used to obtain the cumulative (exceedence) probability for a desired magnitude of flow corresponding to the statistical parameters from the sample data.

A simplified approach was suggested by Chow [1951]. He suggested that most frequency functions applicable to hydrologic sequences, including the four distributions above, can be resolved to the linearized form

$$X = \overline{X} + KS \qquad [\mathrm{L^3 T^{-1}}] \tag{8.10}$$

where

$$X = \text{flood of a specified probability}$$
$$\overline{X} = \text{mean of the sample (observed data)}$$
$$S = \text{standard deviation of the sample}$$
$$K = \text{frequency factor}$$

The frequency factor is a property of a specific probability distribution at a specified probability level. For a given distribution, the relationship has been developed between the frequency factor and the corresponding return interval, known as the $K–T$ relationship. For various distributions, these are expressed in mathematical terms, by tables or by curves called $K–T$ curves.

The procedure used for analysis is as follows:

1. Compute the statistical parameters (mean, standard deviation, and skewness coefficient, if necessary) from the flood flow series.
2. Use the $K–T$ relationship (commonly in the form of a table) for the proposed distribution.
3. For a given return interval, determine the corresponding frequency factor from the $K–T$ relation of step 2.
4. Compute the magnitude of flood by eq. (8.10).
5. Repeat steps 3 and 4 for various return intervals and make a frequency plot. The normal distribution is plotted on normal probability paper; the log-normal and log-Pearson III distributions are plotted on log-normal paper and the extreme value distribution is graphed on Gumbel extreme probability paper.

The $K–T$ relations for normal, log-Pearson type III and extreme value type I are given in Tables 8.6 through 8.8. The log-Pearson is a commonly used distribution in flood studies and the logarithmic form of extreme value type I (equivalent of

TABLE 8.6 FREQUENCY FACTOR FOR NORMAL DISTRIBUTION

Exceedence Probability	Return Period	K	Exceedence Probability	Return Period	K
0.0001	10,000	3.719	0.500	2.00	0.000
0.0005	2,000	3.291	0.550	1.82	−0.126
0.001	1,000	3.090	0.600	1.67	−0.253
0.005	200	2.576	0.650	1.54	−0.385
0.010	100	2.326	0.700	1.43	−0.524
0.025	40	1.960	0.750	1.33	−0.674
0.050	20	1.645	0.800	1.25	−0.842
0.100	10	1.282	0.850	1.18	−1.036
0.150	6.67	1.036	0.900	1.11	−1.282
0.200	5.00	0.842	0.950	1.053	−1.645
0.250	4.00	0.674	0.975	1.026	−1.960
0.300	3.33	0.524	0.990	1.010	−2.326
0.350	2.86	0.385	0.995	1.005	−2.576
0.400	2.50	0.253	0.999	1.001	−3.090
0.450	2.22	0.126	0.9995	1.0005	−3.291
0.500	2.00	0.000	0.9999	1.0001	−3.719

TABLE 8.7 FREQUENCY FACTORS FOR LOG-PEARSON TYPE III DISTRIBUTION

Skew Coefficient, g	Probability							
	0.99	0.80	0.50	0.20	0.10	0.04	0.02	0.01
	Return Period							
	1.0101	1.2500	2	5	10	25	50	100
3.0	−0.667	−0.636	−0.396	0.420	1.180	2.278	3.152	4.051
2.8	−0.714	−0.666	−0.384	0.460	1.210	2.275	3.114	3.973
2.6	−0.769	−0.696	−0.368	0.499	1.238	2.267	3.071	3.889
2.4	−0.832	−0.725	−0.351	0.537	1.262	2.256	3.023	3.800
2.2	−0.905	−0.752	−0.330	0.574	1.284	2.240	2.970	3.705
2.0	−0.990	−0.777	−0.307	0.609	1.302	2.219	2.912	3.605
1.8	−1.087	−0.799	−0.282	0.643	1.318	2.193	2.848	3.499
1.6	−1.197	−0.817	−0.254	0.675	1.329	2.163	2.780	3.388
1.4	−1.318	−0.832	−0.225	0.705	1.337	2.128	2.706	3.271
1.2	−1.449	−0.844	−0.195	0.732	1.340	2.087	2.626	3.149
1.0	−1.588	−0.852	−0.164	0.758	1.340	2.043	2.542	3.022
0.8	−1.733	−0.856	−0.132	0.780	1.336	1.993	2.453	2.891
0.6	−1.880	−0.857	0.099	0.800	1.328	1.939	2.359	2.755
0.4	−2.029	−0.855	−0.066	0.816	1.317	1.880	2.261	2.615
0.2	−2.178	−0.850	−0.033	0.830	1.301	1.818	2.159	2.472
0	−2.326	−0.842	0	0.842	1.282	1.751	2.054	2.326
−0.2	−2.472	−0.830	0.033	0.850	1.258	1.680	1.945	2.178
−0.4	−2.615	−0.816	0.066	0.855	1.231	1.606	1.834	2.029
−0.6	−2.755	−0.800	0.099	0.857	1.200	1.528	1.720	1.880
−0.8	−2.891	−0.780	0.132	0.856	1.166	1.448	1.606	1.733
−1.0	−3.022	−0.758	0.164	0.852	1.128	1.366	1.492	1.588
−1.2	−3.149	−0.732	0.195	0.844	1.086	1.282	1.379	1.449
−1.4	−3.271	−0.705	0.225	0.832	1.041	1.198	1.270	1.318
−1.6	−3.388	−0.675	0.254	0.817	0.994	1.116	1.166	1.197
−1.8	−3.499	−0.643	0.282	0.799	0.945	1.035	1.069	1.087
−2.0	−3.605	−0.609	0.307	0.777	0.895	0.959	0.980	0.990
−2.2	−3.705	−0.574	0.330	0.752	0.844	0.888	0.900	0.905
−2.4	−3.800	−0.537	0.351	0.725	0.795	0.823	0.830	0.832
−2.6	−3.889	−0.499	0.368	0.696	0.747	0.764	0.768	0.769
−2.8	−3.973	−0.460	0.384	0.666	0.702	0.712	0.714	0.714
−3.0	−4.051	−0.420	0.396	0.636	0.660	0.666	0.666	0.667

type III) is common in studies of droughts. However, the recorded data should be plotted along with the computed data for an assessment of the assumed distribution. The distribution should be changed if two sets of data do not compare reasonably.

8.8.6 Generalized Skew Coefficient

In the log-Pearson type III distribution, the frequency factor K is dependent on the skew coefficient, g. It is difficult to obtain accurate skew estimates from the sample data which are usually less than 100 events. The guide by the U.S. Water Resources Council (1981, pp. 10–13) recommends use of a weighted generalized skew coefficient. From the data of all nearby stations within a 100-mile radius, a regionalized skew coefficient for the specific site is estimated by the methods suggested in the

TABLE 8.8 FREQUENCY FACTORS FOR EXTREME VALUE TYPE I DISTRIBUTION

Sample Size, n	Probability								
	0.2	0.1	0.067	0.05	0.04	0.02	0.0133	0.01	0.001
	Return Period								
	5	10	15	20	25	50	75	100	1000
15	0.967	1.703	2.117	2.410	2.632	3.321	3.721	4.005	6.265
20	0.919	1.625	2.023	2.302	2.517	3.179	3.563	3.836	6.006
25	0.888	1.575	1.963	2.235	2.444	3.088	3.463	3.729	5.842
30	0.866	1.541	1.922	2.188	2.393	3.026	3.393	3.653	5.727
35	0.851	1.516	1.891	2.152	2.354	2.979	3.341	3.598	
40	0.838	1.495	1.866	2.126	2.326	2.943	3.301	3.554	5.576
45	0.829	1.478	1.847	2.104	2.303	2.913	3.268	3.520	
50	0.820	1.466	1.831	2.086	2.283	2.889	3.241	3.491	5.478
55	0.813	1.455	1.818	2.071	2.267	2.869	3.219	3.467	
60	0.807	1.446	1.806	2.059	2.253	2.852	3.200	3.446	
65	0.801	1.437	1.796	2.048	2.241	2.837	3.183	3.429	
70	0.797	1.430	1.788	2.038	2.230	2.824	3.169	3.413	5.359
75	0.792	1.423	1.780	2.029	2.220	2.812	3.155	3.400	
80	0.788	1.417	1.773	2.020	2.212	2.802	3.145	3.387	
85	0.785	1.413	1.767	2.013	2.205	2.793	3.135	3.376	
90	0.782	1.409	1.762	2.007	2.198	2.785	3.125	3.367	
95	0.780	1.405	1.757	2.002	2.193	2.777	3.116	3.357	
100	0.779	1.401	1.752	1.998	2.187	2.770	3.109	3.349	5.261
∞^{a}	0.719	1.305	1.635	1.866	2.044	2.592	2.911	3.137	4.936

[a]Additional data for $n = \infty$:

Probability	K
0.3	0.354
0.4	0.0737
0.5	−0.164
0.6	−0.383
0.8	−0.821
0.9	−1.100

guide. In the absence of detailed records, the regionalized skew coefficient is read from a map in the guide. The weighted generalized skew is obtained by the weighted average of the sample (station) skewness and the regional (map) skewness, as follows:

$$g = Wg_s + (1 - W)g_m \qquad \text{[dimensionless]} \qquad (8.11)$$

where

g = generalized skew coefficient

W = weighted factor

g_s = sample skew coefficient

g_m = map (regional) skew coefficient

Computation of Extreme Flows   Chap. 8

Tung and Mays (1981) and the U.S. Water Resources Council (1981) have suggested assigning of the weights in accordance with the variance of the sample skew and the variance of the regional skew. Accordingly,

$$W = \frac{V(g_m)}{V(g_s) + V(g_m)} \qquad \text{[dimensionless]} \qquad (8.12)$$

where V stands for the variance.

The value of $V(g_m)$ has to be estimated from the regional data. For the skew map in the guide of the Water Resources Council, $V(g_m)$ is estimated to be 0.3025.

Tung and Mays (1981) have discussed various methods to compute $V(g_s)$. The Water Resources Council suggests the following approximate equation:

$$V(g_s) = 10^{A - B \log N/10} \qquad (8.13)$$

where

$$N = \text{record length in years}$$
$$A = -0.33 + 0.08|g_s| \qquad \text{for } |g_s| \leq 0.9$$
$$A = -0.52 + 0.30|g_s| \qquad \text{for } |g_s| \geq 0.9$$
$$B = 0.94 - 0.26|g_s| \qquad \text{for } |g_s| \leq 1.5$$
$$B = 0.55 \qquad \text{for } |g_s| > 1.5$$
$$|g_s| = \text{absolute value of the station skew, } g_s$$

Equation (8.11) gives improper weights to the regional skew if the regional and sample skews differ by more than 0.5. In such a situation, the weights should be determined by studying the flood-producing characteristics of the watershed.

In Example 8.6, an application of the log-Pearson type III distribution has been made. The procedure is the same for other distributions except that different K–T tables have to be used.

Example 8.6

On the flood flow sequence of Example 8.5, perform the frequency analysis by the theoretical method, adopting the log-Pearson type III distribution. The generalized skew from the regional map is -0.7.

Solution

1. The data are converted to log form by $y = \log x$.
2. The statistical parameters computed by eqs. (7.26) through (7.28) are as follows:
   a. Mean, $\overline{X} = 3.920$.
   b. Standard deviation, $S = 0.308$.
   c. Coefficient of skewness, $g = -1.1$.

3. $V(g_m) = 0.3025$ and for $N = 24$,

$$A = -0.52 + 0.30(1.1) = -0.19$$

$$B = 0.94 - 0.26(1.1) = 0.654$$

From eq. (8.13),

$$V(g_s) = 10^{-0.19 - 0.654 \log 24/10} = 0.364$$

TABLE 8.9 COMPUTATION OF PEAK FLOWS OF DIFFERENT EXCEEDENCE PROBABILITY

(1) Percent Probability	(2) K	(3) $X = \overline{X} + KS$	(4) $Q = \log^{-1} X$ (cfs)
1	1.66	4.431	27,000
4	1.407	4.353	22,560
10	1.147	4.273	18,760
20	0.854	4.183	15,240
50	0.148	3.966	9,240
80	−0.769	3.683	4,820
90	−1.339	3.507	3,210
99	−2.957	3.009	1,020

From eq. (8.12),

$$W = \frac{0.3025}{0.3025 + 0.364} = 0.45$$

From eq. (8.11),

$$g = 0.45(-1.1) + 0.55(-0.7) = -0.88 \quad \text{or} \quad -0.9$$

4. For various percent exceedence probabilities, peak flows are computed in Table 8.9. The values of K (frequency factor) in column 2 are obtained from Table 8.7.

5. The data are plotted on log-normal paper in Figure 8.6.

6. For $T = 100$ or $P = 1\%$, flow = 27,000 cfs.

8.9 CONFIDENCE LIMITS AND PROBABILITY ADJUSTMENT

Frequency curve by the analytical method is only an estimate of the population curve. The uncertainty (error) of the estimated peak flow for a given probability is a function of the errors in estimating the mean and standard deviation for a known skew coefficient. The confidence limits provide a measure of uncertainty of the estimated peak flow. Beard (1962) proposed a method for constructing the error limit curves above and below a theoretically fitted frequency curve to form a reliability band. Table 8.10 provides the factors by which the standard deviation of flood series is multiplied to get the error limits. For a 5% error curve, the flood values from the fitted frequency curve are added to the computed error limits for the corresponding exceedence probabilities. For a 95% curve, the error limits are subtracted from the flood values at the same exceedence probabilities. There is a 90% probability that the true value lies between the 5% and 95% curves.

The probabilities computed from the theoretical distribution are of an infinite population. As given in Table 8.11, Beard (1962) proposed an adjustment to these probabilities to reflect the limited size of the sample. The table is based on the normal distribution but can be applied approximately to the Pearson type III distribution with small skew coefficients.

Example 8.7

Plot the 5% and 95% error limit curves (upper and lower confidence limits) on the flood–frequency curve of Example 8.6. Also determine the probabilities adjusted for the limited sample size.

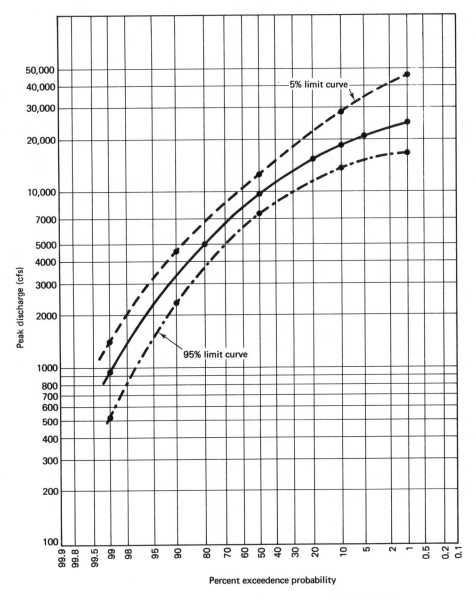

Figure 8.6 Log-Pearson type III frequency curve and a reliability band.

Solution

1. The log-Pearson type III curve of Example 8.6 is shown in Figure 8.6.
2. Error limit and probability adjustment computations are given in Table 8.12.
3. For a 5% limit curve, the values of row (e) are plotted against row (a) from Table 8.12 as shown in Figure 8.6 and for a 95% curve, the values of row (i) versus row (a) plotted on the same graph.

TABLE 8.10 ERROR LIMITS FOR FREQUENCY CURVE

Years of Record, N	Percent Exceedence Frequency (at 5% Level of Significance)[a]						
	0.1	1	10	50	90	99	99.9
5	4.41	3.41	2.12	0.95	0.76	1.00	1.22
10	2.11	1.65	1.07	0.58	0.57	0.76	0.94
15	1.52	1.19	0.79	0.46	0.48	0.65	0.80
20	1.23	0.97	0.64	0.39	0.42	0.58	0.71
30	0.93	0.74	0.50	0.31	0.35	0.49	0.60
40	0.77	0.61	0.42	0.27	0.31	0.43	0.53
50	0.67	0.54	0.36	0.24	0.28	0.39	0.49
70	0.55	0.44	0.30	0.20	0.24	0.34	0.42
100	0.45	0.36	0.25	0.17	0.21	0.29	0.37
	99.9	99	90	50	10	1	0.1

Percent Exceedence Frequency (at 95% Level of Significance)[a]

[a]Chance of true value being greater than the value represented by the error curve.

TABLE 8.11 P_n VERSUS P_∞ FOR NORMAL DISTRIBUTION (PERCENT)[a]

N − 1	P_∞						
	50	30	10	5	1	0.1	0.01
	Adjusted Probability, P_n						
5	50.0	32.5	14.6	9.4	4.2	1.79	0.92
10	50.0	31.5	12.5	7.3	2.5	0.72	0.25
15	50.0	31.1	11.7	6.6	1.96	0.45	0.13
20	50.0	30.8	11.3	6.2	1.7	0.34	0.084
25	50.0	30.7	11.0	5.9	1.55	0.28	0.06
30	50.0	30.6	10.8	5.8	1.45	0.24	0.046
40	50.0	30.4	10.6	5.6	1.33	0.20	0.034
60	50.0	30.3	10.4	5.4	1.22	0.16	0.025
120	50.0	30.2	10.2	5.2	1.11	0.13	0.017
∞	50.0	30.0	10.0	5.0	1.0	0.10	0.01

[a]Values for probability > 50 by subtraction from 100 [i.e., $P_{90} = (100 - P_{10})$].
Source: Beard (1962).

8.10 SPECIAL CASES OF FLOOD–FREQUENCY ANALYSIS

8.10.1 Combined-Population (Composite) Frequency Analysis

When the homogeneity test of Section 8.4.2 indicates nonhomogeneousness of the data, the series consists of events caused by different types of hydrologic phenomena, such as rainstorm floods and snowmelt floods or hurricane and nonhurricane events.

TABLE 8.12 ERROR LIMITS AND PROBABILITY ADJUSTMENT

	0.1	1	10	50	90	99	99.9
(a) P_α (%) (select)	0.1	1	10	50	90	99	99.9
(b) 5% level (from Table 8.10)	1.11	0.88	0.58	0.36	0.39	0.54	0.67
($N = 24$)							
(c) Error limit, $[(b) \times S]$	0.342	0.271	0.179	0.111	0.120	0.166	0.206
(d) Log value, $[X^a + (c)]$		4.702	4.452	4.077	3.627	3.175	
(e) Curve value, $[\log^{-1} (d)]$ (cfs)		50,350	28,310	11,940	4240	1496	
(f) 95% level (from Table 8.10)	0.67	0.54	0.39	0.36	0.58	0.88	1.11
(g) Error limit, $[(f) \times S]$	0.206	0.166	0.120	0.111	0.179	0.271	0.342
(h) Log value, $[X^a - (g)]$		4.265	4.153	3.855	3.328	2.738	
(i) Curve value, $[\log^{-1} (h)]$ (cfs)		18,410	14,220	7160	2130	550	
(j) P_N (for $N - 1 = 23$)	0.29	1.58	11.0	50.0	89.0	98.42	99.71
(from Table 8.11)							

S = Standard deviation of flood sample
a = Column 3 of Table 8.9

A composite flood–frequency curve is derived for such cases. The composite probability of events having individual probabilities of $P_1, P_2, P_3, \ldots, P_n$ is given by

$$P_c = 1 - (1 - P_1)(1 - P_2)\cdots(1 - P_n) \quad \text{[dimensionless]} \quad (8.14)$$

When only two types of events are involved in the observed peak discharges, eq. (8.14) reduces to

$$P_c = P_1 + P_2 - (P_1 P_2) \quad \text{[dimensionless]} \quad (8.15)$$

where

P_1 = exceedence probability of one type of phenomenon (i.e., rainstorm peak discharge)

P_2 = exceedence probability of other type of phenomenon (i.e., snowmelt peak discharge)

Equation (8.15) means that a given discharge may occur as an annual maximum in the form of either a rainstorm flood or a snowmelt flood $(P_1 + P_2)$ but not as both a rainstorm and a snowmelt flood [i.e., $(P_1 P_2)$].

The annual peak discharge series is separated into subseries each of which is homogeneous (i.e., rainstorm floods series and snowmelt floods series). Each of these subseries is analyzed separately, either by empirical (Section 8.7) or analytical (Section 8.8) procedures, and separate flood–frequency curves are derived. About 10 discharge values are selected and the exceedence probability of each of these discharges is obtained from each of the flood–frequency curves. The composite probability, P_c, for each of the discharges is obtained by eq. (8.14) or (8.15). The composite flood–frequency curve is plotted based on P_c values.

Example 8.8

From the annual peak flow record of a gaging site, the flood–frequency curves of rainstorm events and snowmelt events are derived separately, as shown in Figure 8.7. Prepare a composite flood–frequency curve for the flood flow sequence that includes both annual rainstorm and annual snowmelt discharges.

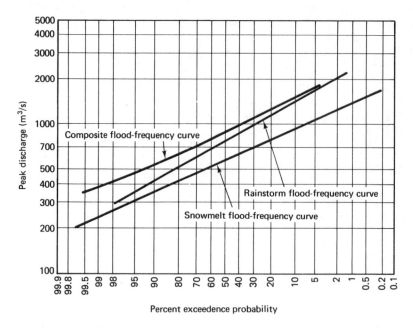

Figure 8.7 Flood–frequency curves of snow and rain events and the composite curve.

Solution

1. For selected discharges, the exceedence probabilities for rainstorm and snowmelt events are read from Figure 8.7, as shown in columns 2 and 3 of Table 8.13.
2. The composite exceedence probabilities of column 4 are computed by eq. (8.15).
3. The composite curve is plotted on Figure 8.7.

TABLE 8.13 COMPUTATION OF COMPOSITE PROBABILITIES

Q (m^3/s)	$P_{rainstorm}$ (%)	$P_{snowmelt}$ (%)	P_c [from eq. (8.15)] (%)
350	95	90	99.5
500	81	63	93.0
600	70	44	83.2
700	57	29	69.5
800	46	18	55.7
900	36	11	43.0
1000	28	7	33.0
1200	17	2.5	19.1
1400	10	0.9	10.8
1600	6	0.3	6.3

8.10.2 Frequency Analysis of Partial-Duration Series

A partial-duration series comprises all peak discharges greater than some arbitrary base discharge. The introduction of more than one discharge per year causes the problem of computation of probabilities of annual exceedences. As such, the partial series are not widely used in flood studies. When used, the empirical method is applied in their analysis, and the frequency plot is made on semilogarithm paper because the probability paper is generally designed for plotting hydrologic data of annual and complete series (Chow, 1964). The logarithmic scale is used for exceedence probability or return period.

When there are several peak discharges per year, the average number of flood peaks per year ($\overline{m}$) that exceed any given discharge, Q, is given by

$$\overline{m} = \frac{N_Q}{N} \quad \text{[dimensionless]} \quad (8.16)$$

where

N_Q = total number of peak discharges greater than the given discharge Q

N = total number of years of record

For each value of discharge considered, N_Q, and hence $\overline{m}$, will be different. The annual exceedence probability (P_m) of the given discharge Q has been related to $\overline{m}$ by the following (Sokolov et al., 1976):

$$P_m = 1 - e^{-\overline{m}} \quad \text{[dimensionless]} \quad (8.17)$$

For various values of Q, the values of $\overline{m}$ are computed by eq. (8.16) and then, by eq. (8.17), annual exceedence probabilities, P_m, are computed to make a frequency plot.

Equation (8.17) indicates that for a value of $\overline{m} < 0.1$ (recurrence interval of greater than 10 years), P_m is practically equal to $\overline{m}$ (i.e., both annual and partial series have almost the same probabilities).

8.10.3 Frequency Analysis of Flood Volume

Analysis until now considered instantaneous maximum flows (flood peaks). In a study like reservoir design, it is important to know the frequency of flood volumes comprising maximum one-day discharge (runoff), maximum 2-day (consecutive) discharge, and maximum 3-day, 7-day, . . . 90-day or any other period discharge in cfs-days. Such information is obtained from the daily discharge record at a gaging site. After flood-volume data are tabulated (the U.S. Geological Survey uses a special form for this purpose), the frequency analysis is performed exactly in a manner similar to that used for the flood peaks. The same procedures are applied for stage curves, rainfall curves, and other hydrologic factors as well.

8.10.4 Regional Frequency Analysis

The foregoing discussion on frequency analysis relates to the analysis at an individual site. The flood flow data at a single site involve a larger sampling error than a group of stations. Moreover, it is only rarely that the flood–frequency information is required exactly at the gage site. It is more often desired anywhere in a region, including ungaged locations. The U.S. Geological Survey accordingly recommends a study of frequency analysis on a regional basis. However, the region should be homogeneous with respect to flood-producing characteristics. A test has been developed by Langbein to define a homogeneous region (Dalrymple, 1960, p. 38).

There are many ways by which regional studies are performed. The procedure of the U.S. Geological Survey, described below, is very widely used (Dalrymple, 1960).

1. List maximum annual floods for all gaging stations in the region having a record of 5 years or more.
2. Select the longest period of record at any station as the base period.
3. Adjust all records to the base period by a correlation study.
4. Perform a test for homogeneity and exclude the stations that are nonhomogeneous.
5. For each station, arrange the data in descending order of magnitude.
6. Compute the plotting position and recurrence intervals.
7. Plot the frequency curves, one for each station.
8. For each station, determine the mean annual flood represented by the discharge corresponding to 2.33-year recurrence interval on the frequency curve.
9. Compute the ratios of floods of different recurrence intervals to the mean annual flood of each station.
10. Tabulate ratios of all stations in one table and select median ratios for various recurrence intervals.
11. Draw a regional frequency curve by plotting the median flood ratios against the corresponding recurrence intervals.
12. Plot the mean annual flood of each station against its drainage area.
13. For the known drainage area of any place in the region, the curve of item 12 will provide the mean annual flow. From the curve of item 11, the ratios of peak flow to mean can be converted to floodflows for various recurrence intervals.

The Hydrologic Engineering Center of the U.S. Army, Corps of Engineers (HEC, 1975), relates the mean of annual maximum flows with the drainage basin characteristics — such as drainage area, slope, surface storage, stream length, and infiltration characteristics — through a multiple linear regression analysis of all stations in the region. A similar regression equation is derived for the standard deviation of annual maximum flows also. From the known basin characteristics or a generalized map of regression constants, the values of the mean, X, and the standard deviation, S, are obtained using the established regression equations. For computed X and S, a frequency curve is prepared by the theoretical frequency factor method (Section 8.8.5).

In another method, the flood peaks of various return periods in the region are directly related to basin characteristics by multiple regression analysis. A different relationship exists for every return period.

8.11 COMPUTATION OF PEAK FLOW FROM PRECIPITATION

As mentioned earlier, in situations where the risk of human life is involved, the structures are not designed with respect to any specific frequency but for the worst expected conditions. Snyder (1964) suggested that the spillway of a major dam, having a storage capacity larger than 50,000 acre-ft, which involves considerable risk to life and excessive damage potential should be designed for the probable maximum flood (PMF). For intermediate dams of 1000 to 50,000 acre-ft capacity, where there is a possibility of loss of life, the spillway should be designed for the Standard Project Flood (SPF). The probable maximum flood is defined as the most severe flood considered reasonably possible in a region. The standard project flood excludes extremely rare storm conditions and thus is the most severe flood considered reasonably characteristic of the specific region. The peak discharge of a SPF is about 40 to 60% of that of a PMF for the same basin.

The floods result, directly or indirectly, from precipitation. The PMF is produced by the probable maximum precipitation (PMP), which is defined by the American Meteorological Society in 1959 as the theoretically greatest depth of precipitation for a given duration that is physically possible over a particular drainage area at a certain time of the year. The spatial and temporal distribution of the probable maximum precipitation, determined on the basis of maximization of the factors that operate to produce a maximum storm, leads to the development of the probable maximum storm (PMS).

A PMS thus developed can be used with a precipitation–runoff relation (Section 7.15) or a unit hydrograph method (Section 7.12) or a simulation model to compute a PMF hydrograph.

8.11.1 Estimation of PMP

There are two common approaches to determining the PMP: rational estimation and statistical estimation. The rational approach has the following basis:

$$\left(\frac{\text{precipitation}}{\text{moisture}}\right)_{\text{max}} \times \left(\frac{\text{moisture}}{\text{supply}}\right)_{\text{max}} = \text{PMP} \qquad [\text{L}] \qquad (8.18)$$

The first term, relating to the ratio of the observed rainfall value to the atmospheric moisture at the time of the actual storm, is available from the record of actual storms. The optimum moisture supply for the second term is obtained from the meteorological tables of the "effective precipitable water" based on the maximum persisting dew point for the basin. The maximization is achieved by considering all major storms in an area and transposed storms from the homogeneous region. For each of these storms the precipitation depth, computed by applying eq. (8.18), is plotted against the duration of the storm in a drainage area. The enveloping curve is the depth–duration curve for the given area. Such curves are prepared for different sizes of the drainage area in the study basin, thus providing depth–duration–area curves.

The statistical approach considers the linearized frequency distribution of the form of eq (8.10), as follows:

$$X_m = \overline{X}_m + K_m S_m \qquad [\text{L}] \qquad (8.19)$$

where

X_m = maximum observed rainfall (i.e., PMP of a given duration)

$\overline{X}_m$ = mean of a series of maximum annual rainfalls of a
specified duration from a drainage area

K_m = standard variate for probable maximum
precipitation; maximization of K value

S_m = standard deviation of annual maxima series

The value of K depends on the type of distribution and the recurrence interval. For an extreme value type I distribution which often adequately describes the rainfall extremes, K has a value of 3.5 for a 100-year rainfall (for the data size of 50). The value of K for the maximum probable precipitation is, obviously, higher than that of 100-year rain. It can be ascertained from the enveloping curve of extreme historic storms. Hershfield (1961) analyzed 24-hour rainfalls from 2600 stations, with a total of 95,000 station-years of data, and determined an enveloping K_m value of 15. From eq. (8.19), for the selected K_m value, the depth–area–duration curves can be prepared by considering storms of different durations from various drainage areas.

The generalized diagrams of PMP based on the rational approach and historic record have been prepared by the U.S. National Weather Service (formerly, U.S. Weather Bureau) for the entire country divided in two parts: east of the 105th meridian and west of the 105th meridian. All-season values of PMP covering west of the 105th meridian for areas to 400 mi^2 and duration to 24 hours were presented in 1960 (U.S. Weather Bureau, 1960). For east of the 105th meridian, the monthwise variations of PMP for areas from 10 to 1000 mi^2 and durations of 6, 12, 24, and 48 hours were prepared in 1956 (U.S. Weather Bureau, 1956). The all-season values for east of the 105th meridian of PMP have since been revised and extended in 1978 to 20,000 mi^2 and for durations of 6 to 72 hours (U.S. National Weather Service, 1978). Again, the seasonal variations for only 10 mi^2 area have been revised in 1980 (U.S. National Weather Service, 1980). Figure 8.8 shows the all-season envelope probable maximum 24-hour precipitation for a 200-mi^2 area for east of the 105th meridian. The PMP of a different duration and for a different-sized area can be obtained by applying the conversion factor of Table 8.14, corresponding to the relevant region indicated in Fig. 8.8. Thus Table 8.14 provides the generalized depth–area–duration relations of the PMP. The depth–area–duration analysis is described in Section 3.5.4.

To determine PMP for a size of area other than one listed in Table 8.14, the conversion factor of the table can be interpolated from a semilog plot of the factor (ordinary scale) versus areal size (log scale). Similarly, the value for an intermediate duration can be ascertained from a Cartesian or semilog plot of the factor versus duration. For both interpolations together, the areal interpolation is done first for different durations and then the durational interpolation is carried out.

8.11.2 Development of PMS

The temporal (time-varying) and spatial (isohyetal) distribution of the PMP estimated in the preceding section is necessary for determination of the probable maximum flood.

Temporal distribution. The U.S. National Weather Service (1982) considered 6-hour increments for sequencing a PMP. Based on the examination of 28 storm samples, the NWS recommended the following:

1. Arrange the individual 6-hour increments such that they decrease progressively to either side of the greatest 6-hour increment. This implies that the lowest

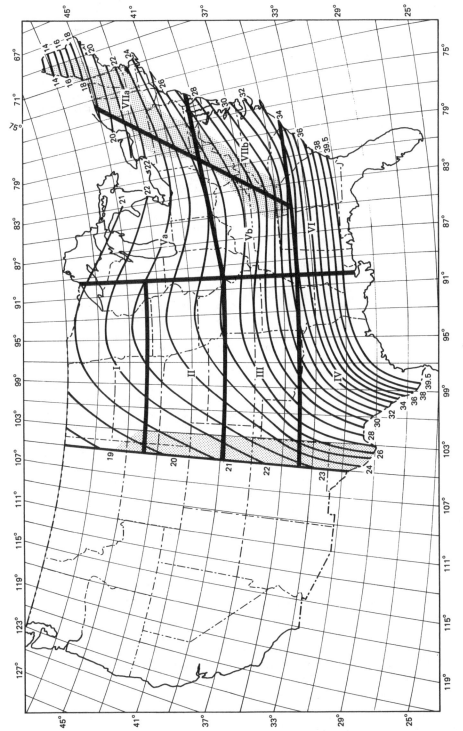

Figure 8.8 All-season PMP (in.) for 200 mi^2, 24 hr (based on US NWS, 1978).

Area mi²	km²	Duration (hr)	I	II	III	IV	Va	Vb	VI	VIIa	VIIb
10	26	6	1.00	1.09	1.03	0.93	1.04	1.01	0.90	1.04	1.00
		12	1.20	1.29	1.22	1.10	1.26	1.18	1.07	1.21	1.16
		24	1.28	1.38	1.31	1.25	1.34	1.31	1.25	1.34	1.33
		48	1.38	1.50	1.45	1.40	1.50	1.45	1.40	1.50	1.45
		72	1.47	1.60	1.55	1.50	1.52	1.53	1.50	1.52	1.53
200	518	6	0.75	0.78	0.74	0.66	0.76	0.72	0.67	0.73	0.68
		12	0.90	0.93	0.87	0.82	0.93	0.86	0.81	0.87	0.85
		24	1.00	1.00	1.00	1.00	1.00	1.00	1.00	1.00	1.00
		48	1.10	1.12	1.14	1.16	1.13	1.14	1.16	1.17	1.15
		72	1.15	1.20	1.20	1.22	1.22	1.23	1.23	1.23	1.25
1,000	2,590	6	0.57	0.56	0.54	0.50	0.56	0.52	0.50	0.52	0.52
		12	0.67	0.71	0.66	0.63	0.70	0.69	0.63	0.63	0.66
		24	0.77	0.80	0.79	0.79	0.80	0.79	0.83	0.80	0.80
		48	0.85	0.90	0.92	0.93	0.90	0.92	0.94	0.93	0.93
		72	0.96	0.97	0.98	1.00	0.97	0.98	1.04	0.98	0.98
5,000	12,950	6	0.36	0.36	0.31	0.28	0.36	0.31	0.28	0.33	0.31
		12	0.45	0.47	0.43	0.39	0.48	0.43	0.40	0.45	0.43
		24	0.52	0.54	0.54	0.55	0.54	0.54	0.55	0.56	0.56
		48	0.63	0.67	0.68	0.65	0.67	0.65	0.68	0.70	0.69
		72	0.70	0.74	0.76	0.76	0.74	0.76	0.78	0.74	0.76
10,000	25,900	6	0.26	0.27	0.23	0.21	0.28	0.23	0.22	0.28	0.23
		12	0.36	0.37	0.33	0.30	0.38	0.35	0.32	0.37	0.35
		24	0.42	0.45	0.43	0.43	0.47	0.44	0.45	0.47	0.45
		48	0.50	0.58	0.54	0.55	0.58	0.57	0.58	0.60	0.60
		72	0.60	0.62	0.64	0.65	0.66	0.64	0.67	0.67	0.65
20,000	51,800	6	0.18	0.20	0.17	0.16	0.20	0.17	0.16	0.20	0.16
		12	0.27	0.28	0.25	0.23	0.30	0.28	0.25	0.33	0.28
		24	0.35	0.36	0.35	0.32	0.38	0.36	0.36	0.40	0.37
		48	0.45	0.47	0.45	0.45	0.48	0.47	0.48	0.50	0.49
		72	0.50	0.55	0.55	0.55	0.56	0.55	0.56	0.57	0.55

[a]Factors derived by the author from the figures in U.S. National Weather Service (1978) to be applied to 24-hour values on 200-mi² area of Figure 8.8.

6-hour increment will be either at the beginning or the end of the sequence.

2. Place the four greatest 6-hour increments at any position in the sequence, except within the first 24-hour period of the storm sequence.

When it is necessary to consider values for a duration of less than 6 hours, the Hydrologic Engineering Center recommended the breakup of 6 hours according to the percentages given in Table 8.15.

In another procedure, the increments of precipitation are first aligned to match the ordinates of the unit hydrograph. The sequence of precipitation increments is then reversed to form the design hyetograph.

TABLE 8.15 HYETOGRAPH OF
6-HOUR PMP ACCORDING TO
THE HEC (1979)

Sequence (hr)	Percent of 6-hr depth
First hour	10
Second hour	12
Third hour	15
Fourth hour	38
Fifth hour	14
Sixth hour	11

Spatial distribution. For distributing the area-averaged PMP over a drainage area, the important considerations are (1) the shape of the isohyets, (2) the number of isohyets, (3) the magnitude of isohyets, and (4) the orientation of isohyetal pattern.

For drainage areas of less than approximately 1000 mi^2, a uniform depth of precipitation over the entire drainage basin is assumed (HEC, 1979). For larger areas, the U.S. National Weather Service (1982) has recommended an elliptical pattern with a major axis-to-minor axis ratio of 2.5:1 for the distribution of 6-hour increments of precipitation for the entire region east of the 105th meridian. This pattern contains 14 isohyets for areas up to 3000 mi^2 and 19 isohyets for the coverage of an area of 60000 mi^2. It is necessary to use only as many of the isohyets as needed to cover the drainage area. The U.S. National Weather Service has provided the nomograms (and tables) indicating the magnitude of each isohyet as a percent of the area-averaged PMP for various sizes of drainage area. For each of the first three greatest 6-hour PMP increment in the sequence, a separate table and nomogram are provided. For the fourth to twelfth 6-hour increments a common table and nomogram are given to obtain the magnitude of each isohyet as a percent of the area-averaged PMP for various sizes of drainage area. Since drainage basins have irregular shapes, there is mostly a disagreement between the shape of the isohyets and the shape of the drainage basin for which the adjustment factors are applied for each durational increment of the PMP. The U.S. National Weather Service (1982) has suggested that the isohyetal pattern be applied only to the three greatest 6-hour increments of PMP. For the nine remaining 6-hour increments of PMP in the 3-day storm, a uniform distribution of PMP throughout the area is adopted.

As regards the orientation, the outline of the given drainage is superimposed on the isohyetal map in a manner that gives the greatest volume of rainfall within the drainage area. This is done by trials and may result in a placement that does not coincide with the geographic center of the drainage. Each of the 6-hour PMP increments will be centered at the same point. This orientation is compared with the central orientation obtained from a generalized map prepared by the U.S. National Weather Service (1982, Fig. 8, p. 31). If the difference is more than 40°, the isohyet values for each increment of PMP are reduced by an adjustment factor obtained from the graph of the U.S. National Weather Service (1982, Fig. 10, p. 35).

Example 8.9

For the 890-mi^2 watershed of the Ponaganset River above Foster, Rhode Island, determine 6-hour PMP from the generalized chart. Prepare the hyetograph for this storm.

Solution

1. 24-hour PMP for 200 mi^2 (Fig. 8.8) = 24 in.
2. Conversion factor for 6-hour duration and different area:

Area (mi^2)	200	1000	5000
Factor (Table 8.14)	0.73	0.52	0.33

These values are plotted and for 890 mi^2 the factor is 0.54, from the plot.

3. 6-hour PMP = 24(0.54) = 12.96 in.
4. Distribution of PMP (based on Table 8.15):

	Percent	Rainfall (in.)
First hour	10	1.30
Second hour	12	1.56
Third hour	15	1.94
Fourth hour	38	4.92
Fifth hour	14	1.81
Sixth hour	11	1.43

8.11.3 Transformation of Design Storm to Flood Flow Hydrograph

The unit hydrograph method for developing design flood hydrograph is very widely used. The first step is to estimate the rainfall losses. The losses per unit duration of the rainfall are subtracted from each increment of total precipitation to give increments of rainfall excess (net rainfall). The computation of losses by infiltration capacity curve and index methods are discussed in Section 3.9.

The unit hydrograph (UHG) is derived from the rainfall–runoff data of a basin where such data are available; otherwise, it is formulated by the synthetic procedure from regional relations, as described in Section 7.11. However, an uncertainty arises from the fact that the design storm and resulting flood are invariably of greater magnitude than the storms and corresponding floods used to derive the UHG. The studies by the U.S. Army Corps of Engineers (1959) showed that the peak ordinate of a UHG derived from data for a major flood is 25 to 50% greater than the peak from data for a minor flood. Therefore, modification of UHG is made when the derived UHG does not represent conditions similar to those during the design storm. The modification procedure is as follows.

1. Read the widths of the UHG at ordinates equal to 50% and 75% of the peak discharge, referred to as W_{50} and W_{75}.
2. Plot the widths against the peak discharge on Figure 8.9, which is based on the study of a large number of drainage basins. Draw lines A–A' and B–B' through the plotted points parallel to curves W_{50} and W_{75}.

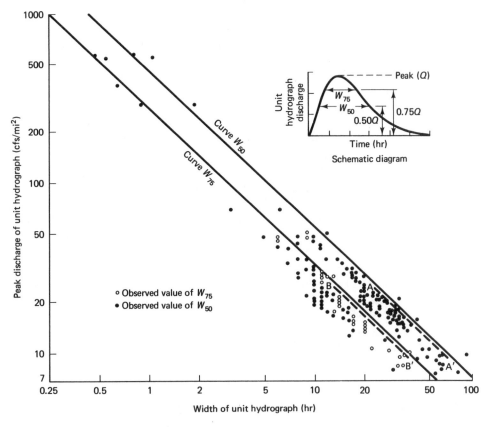

Figure 8.9 Width and peak discharge relation of the unit hydrograph (from Sokolov et al., 1976).

3. Increase the peak discharge of the UHG by 25 to 50%, depending on judgment.
4. Plot the revised peak discharge, Q_u, on Figure 8.9.
5. For the revised peak discharge, read W_{50} and W_{75} from lines A–A' and B–B'. Plot these values on the graph. The widths are plotted on each side of Q_u in the same ratio as existed on the original UHG.
6. The modified UHG is sketched on the same time base as the original UHG through the plotted points Q_u, W_{50}, and W_{75}.
7. The minor adjustment on the recession curve of the revised UHG can be made by (a) preparing an S-curve from the plotted modified UHG, (b) smoothening the S-curve, (c) deriving UHG from it, (d) adjusting the UHG ordinates, (e) drawing the S-curve again, and repeating the process until the most logical forms of both the UHG and S-curve are obtained. This is usually not done.

Once the UHG is derived, it is applied to the rainfall sequence of the design storm and the baseflow is added to produce the discharge hydrograph (Section 7.12). The application is shown in example 7.9.

8.12 PEAK SNOWMELT DISCHARGE

In some regions the contribution from snowmelt is vital. The maximum floods in many areas are either due to a combination of snowmelt and rainfall runoff or from snowmelt events alone. In the USSR, empirical equations similar to rainstorm floods (Section 8.13) are applied to compute the direct snowmelt peak discharge. In the United States the intensity (rate) of snowmelts in terms of the depth per unit time is computed from the equations that have been derived by combining the heat balance equation with empirical factors. These equations have been developed separately for plains and mountainous regions during rainfall and rain-free periods, as described in Section 3.10. The duration of a given intensity is the period for which the set of conditions, as applied in the equations of snowmelt, existed.

After the depth of melt has been estimated for the portion covered by the snowpack, it can be treated like the rainfall to ascertain the flow, that is, losses are deducted and melt excess is converted into a streamflow hydrograph by the application of the unit hydrograph or routing technique. The losses are treated in two ways. For a rain-on-snow event, the amount of water that is delayed very long in reaching a stream is considered lost. For a primarily snowmelt event, the losses consist of evapotranspiration, deep percolation, and permanent retention of water in the snowpack. The rates of melt excess are small but continue for a long period (are approximately continuous). Special long-tailed unit hydrographs or S-hydrographs are used for snowmelt streamflow computation.

8.13 FLOOD FLOW COMPUTATION BY GENETIC AND EMPIRICAL EQUATIONS

When the hydrologic data are insufficient to use the preceding methods, genetic and empirical equations are used. The genetic equations are based on physical considerations that embody the theoretical concepts of the runoff generation (Sokolov et al., 1976). On the other hand, empirical relations are not concerned with mechanisms of runoff generation but bring out the resultant effect of the relevant factors. For small basins of less than 20 mi^2, a genetic equation known as the rational method is commonly used. It is applied to the drainage analysis on a small watershed. The method has been described in Section 12.7. For rivers having a drainage area larger than 20 mi^2, empirical relations are used. The parameters of the empirical equations are determined from the regional analysis. An enormously large number of equations exist to determine the peak discharge from basin and climate parameters. Gray (1973) has presented about 50 empirical relations from the world experience. The two most commonly used relations are described below.

8.13.1 Creager Enveloping Curves

The general form of the Creager equation is

$$q = 46CA^{0.894A^{-0.048}-1} \quad \text{[unbalanced]} \quad (8.20)$$

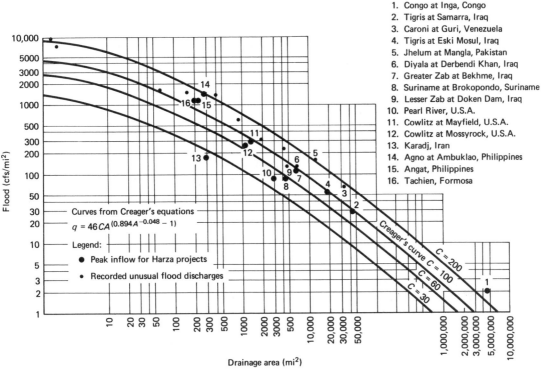

Figure 8.10 Creager envelope curves (from Creager, Justin, and Hinds, 1945).

The legend for the numbered points:

1. Congo at Inga, Congo
2. Tigris at Samarra, Iraq
3. Caroni at Guri, Venezuela
4. Tigris at Eski Mosul, Iraq
5. Jhelum at Mangla, Pakistan
6. Diyala at Derbendi Khan, Iraq
7. Greater Zab at Bekhme, Iraq
8. Suriname at Brokopondo, Suriname
9. Lesser Zab at Doken Dam, Iraq
10. Pearl River, U.S.A.
11. Cowlitz at Mayfield, U.S.A.
12. Cowlitz at Mossyrock, U.S.A.
13. Karadj, Iran
14. Agno at Ambuklao, Philippines
15. Angat, Philippines
16. Tachien, Formosa

where

$$q = \text{peak discharge, cfs/mi}^2$$
$$C = \text{coefficient depending on the type of drainage area}$$
$$A = \text{drainage area, mi}^2$$

Creager found that the curve of $C = 100$ enveloped all but a few of the known great floods in the United States. The curves for C equal to 30, 60, 100, and 200 are shown in Figure 8.10. The value of C to be used for a study basin can be selected by substituting in eq. (8.20) known information on floods in the region or else by comparing drainage characteristics with some of the known areas listed in Figure 8.10.

8.13.2 Myers–Jarvis Enveloping Curves

This relation has the simple form

$$Q = pA^{1/2} \qquad \text{[unbalanced]} \qquad (8.21)$$

where

$$Q = \text{peak discharge, cfs}$$
$$p = \text{Myers's rating, as given in Table 8.16}$$
$$A = \text{drainage area, mi}^2$$

TABLE 8.16 MYERS'S RATING FOR MAXIMUM FLOOD DISCHARGE

Region	Approximate Range of drainage Area (mi^2)	Myers's Rating
North Atlantic and Middle Atlantic	10 — 100	7,000
slope basins	500 — 80,000	5,500
South Atlantic slope and eastern		
Gulf of Mexico	10 — 10,000	6,000
Ohio River basin	10 — 900	7,000
	1,000 — 5,000	6,000
	5,000 — 200,000	4,300
Eastern Great Lakes and eastern	10 — 60	6,000
St. Lawrence River basins	70 — 1,000	3,500
Upper Mississippi River and	10 — 150	4,000
western Great Lakes basins	200 — 800	2,700
	900 — 15,000	1,800
Lower Missouri and western	10 — 5,000	7,500
Mississippi River tributaries	6,000 — 15,000	4,000
Western Gulf of Mexico basins	50 — 3,000	11,000
	1,000 — 9,000	10,000
	10,000	5,000
Pacific slope basins in California	10 — 3,000	6,000

On a log-log graph, eq. (8.21) plots a straight line with a slope of 1/2 (1 vertical to 2 horizontal) and an intercept of p. From the regional data, the peak discharge and the drainage area are plotted. A line as an envelope through the upper points is drawn with a slope of 1/2. Known as the Myers curve, this provides a value of p and an estimate of flood peaks that could occur anywhere in the region. When the regional data are not available to plot the curve, the value of p is ascertained guided by Table 8.16.

Example 8.10

The Miami River near Hamilton, Ohio, has a drainage area of 3630 mi^2. Determine the peak flood discharge. Assume that $C = 60$.

Solution

1. From Table 8.16, for a drainage area of 3000 mi^2 in the Ohio River Basin, $p = 6000$.

2. Substituting in eq. (8.21) yields

   $$Q = 6000(3630)^{1/2} = 361,500 \text{ cfs}$$

3. An actual peak flow of 352,000 cfs was recorded on the Miami River at Hamilton on March 16, 1913.

4. For $C = 60$ in the Creager equation,

$$0.894A^{-0.048} - 1 = 0.894(3630)^{-0.048} - 1 = -0.397$$

From eq. (8.20), $q = 46(60)(3630)^{-0.397} = 106.6$.

$$Q = qA = 106.6(3630) = 386,800 \text{ cfs}$$

8.14 MEASUREMENT OF PEAK DISCHARGE BY INDIRECT METHODS

In indirect methods, the hydraulic equations of flow applicable to different hydraulic systems are used. These relations, derived from the energy equation and continuity equation, are in terms of the head or its variation. The high-water marks left by the peak floods provide this information. For application of the method, a location of a particular structure is selected for which the specified hydraulic equation is applicable. The peak-flow measurements are commonly performed at the site of the following structures:

1. A reach of stream channel (slope–area method)
2. At width contractions of a bridge section
3. At dams
4. At culverts

8.14.1 Slope–Area Method

A suitable reach of a stream channel is selected from the primary consideration of good high-water marks. Manning's equation is applied, in which the slope (energy gradient) term is modified for the nonuniform condition. The following set of formulas relevant to channel flow (Chapter 10) are applicable:

$$K = \frac{1.486}{n} A R^{2/3} \qquad \text{[unbalanced]*} \tag{8.22}$$

$$\alpha = \frac{\sum Ki^3/Ai^2}{K_T^3/A^2} \qquad \text{[dimensionless]} \tag{8.23}$$

$$h_v = \alpha V^2/2g \qquad \text{[L]} \tag{8.24}$$

$$\Delta h_v = h_v \text{ (upstream)} - h_v \text{ (downstream)} \qquad \text{[L]} \tag{8.25}$$

When Δh_v is positive (i.e., expanding reach),

$$S = \frac{\Delta h + \Delta h_v/2}{L} \qquad \text{[dimensionless]} \tag{8.26a}$$

When Δh_v is negative (i.e., contracting reach), using Δh_v with negative sign

$$S = \frac{\Delta h + \Delta h_v}{L} \qquad \text{[dimensionless]} \tag{8.26b}$$

$$Q = \sqrt{K_1 K_2 S} \qquad \text{[unbalanced]} \tag{8.27}$$

*For English units.

where

Q = discharge between the reach

K = conveyance of the channel; K_1 at section 1 and K_2 at section 2

A = area of cross section at a selected section

R = hydraulic radius = area/wetted perimeter

n = Manning's roughness coefficient (Table 10.4)

α = velocity head coefficient; assumed to be 1.0 if the section is not subdivided; eq. (8.23) used for subdivided section

h_v = velocity head at a section

Δh_v = difference in velocity head at two sections

Δh = difference in water surface elevations (high-water marks)

L = length of the reach

S = friction slope

The following trial-and-error procedure is used since eq. (8.24) cannot be solved directly.

1. Determine K for upstream and downstream sections by eq. (8.22).

2. Determine α for subdivided sections by eq. (8.23); otherwise, $\alpha = 1$.

3. Determine S from eq. (8.26a) or eq. (8.26b), assuming that $\Delta h_v = 0$.

4. Compute Q from eq. (8.27). Consider it as an "assumed" value.

5. For assumed Q (and $v = Q/A$), determine h_v from eq. (8.24).

6. Determine revised S and Q from eqs. (8.26) and (8.27), respectively. Repeat steps 5 and 6 until Q stabilizes.

When more than two cross sections are selected in a reach, compute the discharge between each of them.

Example 8.11

From the following data obtained from the field measurements at two sections 129.5′ apart on Snake Creen near Connell, Washington, determine the peak discharge ($n = 0.045$).

	Section 1	Section 2
Area (ft^2)	208	209
Hydraulic radius (ft)	3.09	4.10
High-water mark (ft)	16.30	15.23

Solution

Trial	Reach	$K = \dfrac{1.486}{n}AR^{2/3}$	h	$h_v = \alpha \dfrac{v^2}{2g}$	Δh	Δh_v	S [eq. (8.26)]	Q [eq. (8.27)]	$v = \dfrac{Q}{A}$
1	Section 1	14,577	16.3	0					7.01
					1.07	0	0.00826	1459	
	Section 2	17,688	15.23	0					6.98
2	Section 1	14,577	16.3	0.763					
					1.07	0.007	0.00829	1462	
	Section 2	17,688	15.23	0.756					

8.14.2 Measurement at Width Contractions

The contraction of a stream channel at a roadway or railway crossing causes a sudden drop in the water level between an approach section and the contracted section under the bridge. From hydraulic considerations, the discharge is proportional to the square root of the head and the cross-sectional area of the contraction. The head is equal to the difference in water levels between the approach and contracted sections to which the approach velocity head is added and the friction head loss is subtracted. Expressing the velocity head and head loss in terms of discharge and conveyance, the U.S. Geological Survey (Matthai, 1967) indicated discharge as

$$Q = 8.02CA_3 \left[\frac{\Delta h}{1 - \alpha_1 C^2 (A_3/A_1)^2 + 2gC^2(A_3/K_3)^2(L + LwK_3/K_1)} \right]^{1/2} \qquad (8.28)$$

$$[L^3T^{-1}]$$

where

$\Delta h =$ difference in water levels between 1 and 3 [subscript 1 refers to the approach section, one b-width upstream (to the bridge opening of width b). Subscript 3 refers to the downstream side of the contraction]

$C =$ coefficient of discharge

$A =$ area of cross section

$\alpha =$ velocity head coefficient, given by eq. (8.23)

$K =$ conveyance, given by eq. (8.22)

$L =$ length of the bridge opening, Figure 8.11

$L_w =$ length of the approach reach from section 1 to water contact point on upstream of structure

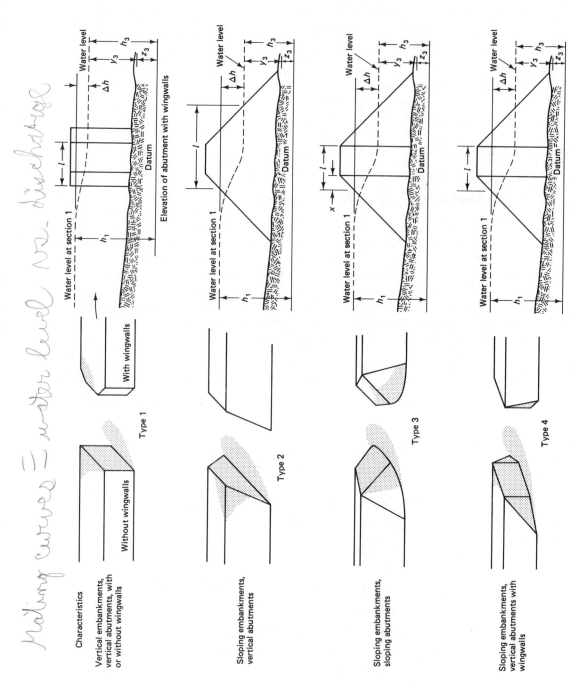

Figure 8.11 Classification of width contractions.

The discharge coefficient, C, in eq. (8.28) is a function of many parameters related to the geometry of the bridge opening and the flow pattern. The U.S. Geological Survey (Matthai, 1967) divided the bridge openings into four types, as shown in Figure 8.11, and provided separate discharge coefficients for each type. The basic discharge coefficients under each type have been indicated by the U.S. Geological Survey as a function of the L/b ratio and the channel-contraction ratio by a set of curves. The basic coefficients are further modified for the effect of the Froude number and angularity of flow. Generally, C varies between 0.6 and 0.9.

8.14.3 Measurement at Dams

The equation of flow over a weir, dam, or embankment is

$$Q = CbH^{3/2} \qquad [L^3T^{-1}] \qquad (8.29)$$

where

C = discharge coefficient

b = width of weir crest or flow section

H = total energy head ($h + v_1^2/2g$), h being the head over the structure and v_1 the approach velocity

The head, h, is determined on the basis of the field survey of high-water marks soon after the occurrence of the flood. The head is measured upstream from the crest at a distance of three to four times the head. The tailwater elevation is also noted. If the tailwater elevation is higher than the crest (submerged crest), the tailwater profile is also determined.

The two most important parameters in the computation of discharge over a weir, dam, or embankment are the head, h, and discharge. coefficient, C. The selection of a value of C affects the discharge computation. For some types of dam structures, the discharge coefficient can be determined more reliably than the others. The parameters that influence C, the computational aids available to determine C, and the effect of submergence on discharge for various types of dam structures are indicated in Table 8.17. For the determination of C, the U.S. Geological Survey (Hulsing, 1967) has prepared a series of curves for each type of structure, incorporating ratios of various factors that affect C in each case. For sharp-crested, broad-crested, and certain other shapes, the values of C can also be obtained from handbook tables (e.g., Brater and King, 1976, pp. 5-40 to 5-44).

8.14.4 Measurement at Culverts

The culverts are usually used to measure flood discharges from small drainage areas. The placement of a roadway embankment and culvert inside a stream channel causes a change in flow pattern resulting from acceleration of flow due to contraction of the cross-sectional area. By measuring high-water marks that define the headwater and tailwater elevations and from the flow conditions inside the culvert, which can be tranquil, critical, or rapid, the peak flow can be determined. The continuity equation and the energy equation between the approach section and a section at the culvert are used. This is described in section 12.13.3.

TABLE 8.17 DISCHARGE COEFFICIENT AND SUBMERGENCE EFFECT FOR WEIRS, DAMS, AND EMBANKMENTS

Type	C as a Function of:	Definition Sketch	Determination of C	Effect of Submergence
Rectangular sharp-crested weir	$f\left(\dfrac{h}{P}, \dfrac{b}{B}, E, \dfrac{r}{b}\right)$ where b = width of weir B = width of channel r = radius of abutment corners		USGS (Hulsing, 1967) provides curves of 1. h/P and E vs. C 2. h/P and b/B vs. factor, K_C 3. Adjustment of K_C for r/b values	1. USGS provides curves of h_t/H and H/P vs. factor, K_t. 2. C should be multiplied by the factor, K_t, for submergence effect
Broad-crested weir	$f\left(\dfrac{h}{L}, \dfrac{R}{h}, E_1, E_2, \dfrac{b}{B}, \dfrac{r}{b}\right)$		USGS provides 1. h/L and E_1 vs. C 2. h/P and b/B vs. factor, K_C 3. h/L and E_2 vs. factor K_S 4. R/h vs. factor K_R 5. Adjustment of K_C for r/b values	1. No effect on discharge if $h_t/h < 0.85$ 2. For $h_t/h > 0.85$, appreciable effect but no data available

			USGS provides

Round-crested weir

f(weir geometry, head)
For nappe fitting weir:

$$f\left(\frac{h_0}{P+e}, C_0, E, \frac{b}{B}, \frac{r}{b}\right)$$

$C_0 = Q^{3/2}/bh_0$

h_0 = Design head at which the nappe profile fits to the crest

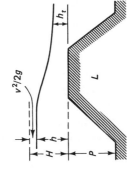

USGS provides

1. $\dfrac{h_0}{P+e}$ vs. C_0
2. h/h_0 vs. C/C_0
3. h/P and b/B vs. factor K_c
4. K_C adjustment for r/b

1. h_t/H and H_0 vs. factor, K_t
   where
   $H = h + V^2/2g$
   $H_0 = h_0 + V^2/2g$

Highway embankment

$f\left(\dfrac{h}{L}, n\right)$; for $\dfrac{h}{L} > 0.15$

$f(h, n)$, for $\dfrac{h}{L} < 0.15$

n = roughness of roadway

USGS provides

1. h/L and n vs. C
or
2. H and n vs. C

1. h_t/H and n vs. factor, K_t

Weir of unusual shapes

f(weir geometry, head)

Different shapes

USGS provides
tables of
1. E_1, E_2, and H vs. C
2. L and H vs. C

Not considered

8.15 COMPUTATION OF LOW FLOW

Low streamflows are significant because the conditions of drought upset the entire ecological balance; more specifically, low flows are important from a consideration of the adequacy of a stream to receive wastes, to supply municipal and industrial water requirements, to meet supplemental irrigation, or to maintain aquatic life. Of the following procedures of low-flow computation, the first two relate to the gaged sites and the others to ungaged locations.

1. *Frequency analysis.* This analysis is similar to peak-flow analysis (Sections 8.4 through 8.9).
2. *Analysis of runs.* The deficit in flow with respect to the baseflow is counted along with the duration (of the drought).
3. *Partial-record method.* A few baseflow measurements at a site are related to the concurrent discharges at a neighboring station for which a low-flow frequency curve is available.
4. *Seepage Runs.* During a period of baseflow, discharges at intervals along a channel reach are measured to identify the loss or gain in the flow along a river channel.
5. *Interpolation.* By plotting the low-flow characteristics at gage sites against the channel distance, the flows at intermediate points are interpolated.

The frequency analysis is a most common procedure in which an array of annual low flows is formulated from the records of daily discharge of a gaging station. The sequence can be prepared for the lowest daily discharge in each year or the lowest mean discharge for 2, 3, 7, 10, or more consecutive days. The frequency curves can be prepared from the extracted values of low flows for different periods of days. An empirical or analytical procedure, similar to the flood flow analysis, can be used.

8.15.1 Low-Flow Frequency Analysis by Empirical Method

1. Arrange the annual values in ascending order beginning with the smallest as number 1 (in peak flows these are arranged in descending order).
2. Compute the plotting position using eq. (8.9), that is, $P_m = [m/(n + 1)](100)$; the return period, T, is the inverse of P_m.
3. Plot each value on graph paper [log-normal or log-extreme value type I (Gumbel)]. Draw a smooth curve.

Example 8.12

The data for the annual 7-day minimum average flows at the Wallkill River near Walden, New York, are given in Table 8.18.

(a) Prepare the low-flow frequency curve.
(b) Determine (1) the probability of the 7-day low flow to be less or equal to 70 ft³/sec, (2) the return period of a flow of 100 ft³/sec, and (3) 10-year 7-day low flow.

TABLE 8.18 ANNUAL 7-DAY MINIMUM AVERAGE FLOW OF WALLKILL RIVER NEAR WALDEN, NEW YORK

(1) Year	(2) 7-Day Low-Flow (cfs)	(3) Rank	(4) Plotting Position (%)	(5) Year	(6) 7-Day Low-Flow (cfs)	(7) Rank	(8) Plotting Position (%)
1956	86.7	21	67.74	1971	83.3	18	58.06
1957	57.8	9	29.03	1972	99.0	27	87.10
1958	73.7	13	41.93	1973	66.2	12	38.71
1959	93.8	26	83.87	1974	88.6	22	70.97
1960	55.8	6	19.35	1975	111.6	29	93.55
1961	61.0	10	32.25	1976	46.3	1	3.23
1962	84.2	19	61.29	1977	85.5	20	64.52
1963	123.3	30	96.77	1978	80.1	17	54.84
1964	92.7	25	80.65	1979	55.3	5	16.13
1965	78.4	16	51.61	1980	57.2	8	25.81
1966	89.4	23	74.19	1981	63.0	11	35.48
1967	49.4	2	6.45	1982	105.3	28	90.32
1968	54.2	4	12.90	1983	91.8	24	77.42
1969	49.6	3	9.68	1984	76.3	15	48.39
1970	75.5	14	45.16	1985	56.5	7	22.58

Solution

1. The rank is shown in columns 3 and 7 of Table 8.18. It is advisable to arrange values in ascending order on a separate table.
2. The plotting position is computed by eq. (8.9) in columns 4 and 8 of the table.
3. The log of the flow values and the recurrence interval (100/plotting position) are plotted on extreme value (Gumbel) paper in Figure 8.12 [logs of the values are plotted since the logarithmic transform of type I (Gumbel) is essentially type III (Weibull) distribution, which is more suitable for low-flow sequence]
4. From Figure 8.12:
   (1) For $Q = 70$, $\log Q = 1.845$, $T = 2.25$, $P = 1/T = 0.444$.
   (2) For $Q = 100$, $\log Q = 2$, $T = 1.16$ years.
   (3) For $T = 10$ years, $\log Q = 1.71$, $Q = \log^{-1} 1.71 = 51$ cfs.

8.15.2 Low-Flow Frequency Analysis by Analytical Method

The analytical approach discussed in the context of flood flows (Section 8.8) is applicable to low flows. The linear equation (8.10), reproduced below, is used in theoretical analysis.

$$X = \overline{X} + KS \quad [L^3T^{-1}] \tag{8.10}$$

The distributions suitable for low flows are log-normal, log-Pearson type III, and extreme value type III distributions, with a preference for the latter. The frequency factors, K, are given in Tables 8.6 through 8.8. The K factors of the extreme value type I distribution in Table 8.8 are used with the logarithmic-flow values to fit to the extreme value type III distribution.

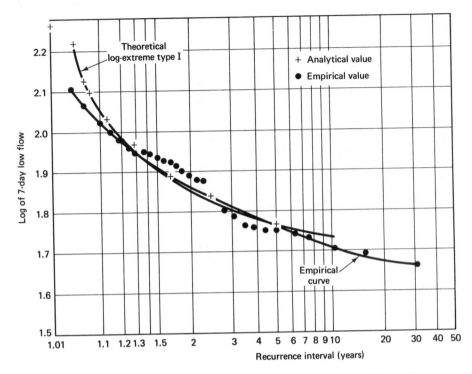

Figure 8.12 Frequency curve of 7-day annual minimum flow of Walkill River, New York, on extreme value probability-paper after Gumbel.

The frequency factors have been tabulated for use with the peak flows that indicate the probability of exceedence. However, the low-flow frequency curve indicates the probability of flow being equal to or less than the value indicated. The probability value of the table should thus be converted by subtracting from 1 (from 100 in the case of percent).

Example 8.13

Solve Example 8.12 by the analytical method using log-extreme value type I distribution.

Solution

1. The flow data are converted to log form.
2. The statistical parameters computed by eqs. (7.26) and (7.27) are as follows:
   a. Mean, $\overline{X}$ = 1.868.
   b. Standard deviation, S = 0.115.
3. For various probability levels, the flows are computed in Table 8.19. The values of K are obtained from Table 8.8.
4. The frequency curve has been plotted on Figure 8.12.
5. From the curve:
   (1) For Q = 70, log Q = 1.845, T = 2.05, P = 0.488.
   (2) For Q = 100, log Q = 2.0, T = 1.13 years.
   (3) For T = 10 years, log Q = 1.72, Q = $\log^{-1}$ 1.72 = 52 cfs.

TABLE 8.19 COMPUTATION OF LOW FLOW OF VARIOUS PROBABILITIES

$P(X>)$ Table 8.8 Value	$P(X<)$ $1 - P(X>)$	$T = \dfrac{1}{P(X<)}$	K	$X = \bar{X} + KS$	$Q = \log^{-1} X$
0.8[a]	0.2	5	−0.821	1.774	59.37
0.6[a]	0.4	2.5	−0.383	1.824	66.67
0.4[a]	0.6	1.67	0.0737	1.876	75.24
0.2	0.8	1.25	0.866	1.968	92.8
0.10	0.90	1.11	1.541	2.045	110.97
0.067	0.933	1.07	1.922	2.089	122.75
0.05	0.95	1.053	2.188	2.120	131.71
0.02	0.98	1.02	3.026	2.216	164.43
0.0133	0.987	1.013	3.393	2.258	181.22
0.01	0.99	1.010	3.653	2.288	194.13
0.001	0.999	1.001	5.727	2.527	336.2

[a]Based on sample size $n = \infty$.

8.15.3 Regionalization and Application to Ungaged Sites

The low-flow characteristics of streams in adjacent basins vary significantly. The geology-lithology and rock structure, influence low flows. It has not been possible to describe quantitatively the effect of geology on low flows by certain indexes. Hence the regionalization of low-flow characteristics has met with only limited success. Attempts have been made to regionalize by multiple regression on several basin characteristics, but the standard errors of regressions have been too large. The regional relations are developed only for limited conditions where all significant basin characteristics except drainage area in a region have extremely limited range. The lack of general regionalization does not permit estimates at ungaged sites by the extension of frequency relations.

8.16 FLOOD ROUTING

Flood routing is a process whereby the shape of a flood hydrograph is determined at a point in a reservoir, lake, or channel resulting from a known hydrograph at an upstream point. This is shown schematically in Figure 8.13.

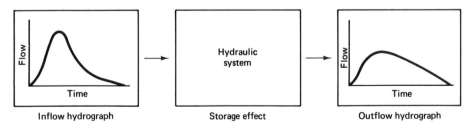

Figure 8.13 Flood routing process.

Such calculations serve the purpose of (1) determining the size of a spillway, (2) studying the effect of a reservoir on the modification of a flood peak, (3) developing design elevations of flood walls and levees, (4) derivation of unit hydrographs and synthetic hydrographs, and (5) any other flood-flow-related objectives, including forecasting of floods.

The three elements of the routing process are connected by the basic continuity equation

$$
\begin{bmatrix} \text{inflow volume in} \\ \text{time, } \Delta t \end{bmatrix} + \begin{bmatrix} \text{change in volume of} \\ \text{water stored by the} \\ \text{hydraulic system} \\ \text{during time, } \Delta t \end{bmatrix} = \begin{bmatrix} \text{outflow} \\ \text{volume} \\ \text{in time,} \\ \Delta t \end{bmatrix}
$$

or

$$ I\Delta t \qquad + \qquad \Delta S \qquad = \qquad O\Delta t \qquad [L^3] \qquad (8.30) $$

where I and O are the rates of inflow and outflow, respectively.

The hydraulic system in Figure 8.13 can be represented by a reservoir or by a stream-channel section, and the routing process is, accordingly, classified into two broad types: reservoir routing and channel, or streamflow, routing.

In eq. (8.30), I is a known input, O has to be determined, but S is also an unknown parameter. To solve the equation, either both O and S have to be related to a common unknown parameter or S has to be defined in terms of O. The former approach is applied to reservoir routing and the latter is adopted in streamflow routing.

Conceptually, eq. (8.30), expressed in differential form, can be integrated to provide the outflow as a function of time. Generally, however, the terms in the equation have a form that are not amenable to direct solution. The numerical solution is preferred. In terms of numerical approximation, eq. (8.30) can be written as

$$ \frac{I_1 + I_2}{2} + \frac{S_1 - S_2}{\Delta t} = \frac{O_1 + O_2}{2} \qquad [L^3 T^{-1}] \qquad (8.31) $$

where subscript 1 is at the beginning and subscript 2 at the end of the routing period, Δt. The procedure used to solve eq. (8.31) is described separately for reservoir and streamflow routings.

8.17 RESERVOIR ROUTING (THE PULS METHOD)

In the case of a reservoir, the volume of storage can be expressed as a function of water surface elevation by planimetering the reservoir surface area from the topographic map for successive elevations and multiplying the average area by the water depth. A typical relation is shown by curve (a) in Figure 8.14.

Also, the outflow of water through the reservoir (in addition to the controlled releases through sluices, turbines, etc.) depends on the depth of flow over the spillway and thus on the depth of water in the reservoir. A spillway rating curve of the relation between discharge and water surface elevation can be prepared as shown by curve (b) of Figure 8.14.

Since the outflow and the storage are both functions of water surface elevation or stage, the continuity equation becomes a relation between the known inflow and the unknown water stage, from which the stage can be computed as a function of time. These stages can readily be converted to outflows from the spillway rating curve. For this purpose, eq. (8.31), in numerical form, is rearranged as follows:

$$(I_1 + I_2) + \left(\frac{2S_1}{\Delta t} - O_1\right) = \left(\frac{2S_2}{\Delta t} + O_2\right) \qquad [\text{L}^3\text{T}^{-1}] \qquad (8.32)$$

At the initial time, $t = 0$ (start of the routing just before flood arrives), $I_1 = I_2 = O_1$ and S_1 corresponds to the storage at the spillway crest elevation. The left-hand side of the equation has known quantities that yield a value of $(2S_2/\Delta t + O_2)$, but still does not yield O_2 and S_2 separately. For computational expediency, by combining curves (a) and (b) of Figure 8.14, another curve of relation between $(2S/\Delta t + O)$ and surface elevation, or alternatively, $(2S/\Delta t + O)$ versus O, is constructed on the same paper for a selected value of Δt, as shown by curve (c) in Figure 8.14.

Using curve (c), for known $(2S_2/\Delta t + O_2)$, the elevation will be obtained which will provide S_2 and O_2 directly from curves (a) and (b), respectively. These values will be used as initial values on the left-hand side of eq. (8.32) for the next time step of the routing period.* The computation is repeated for the succeeding routing periods.

As a slight modification of the procedure above, only two curves, S versus O and $(2S/\Delta t + O)$ versus O, are constructed. From these curves it is possible to split $(2S/\Delta t + O)$ in O and S.

Example 8.14

Route the inflow hydrograph indicated below through a reservoir. The storage data (water surface elevation versus storage volume) for the reservoir are given below. The spillway discharge is $Q = 3LH^{3/2}$. The crest height of the spillway is 50 ft and the length of the spillway is 35 ft.

Inflow hydrograph:

Time (days)	0	0.5	1.0	1.5	2.0	2.5	3.0	3.5	4.0
Flow (cfs)	0	70	185	360	480	300	165	80	0

Storage data:

Elevation (ft)	50	50.5	51.0	51.5	52.0	52.5	53.0	53.5
Storage (acre-ft)	231	247	277	313	353	400	452	509

Solution

1. The discharge data computed from $Q = 3(35)H^{3/2}$ and the storage data are listed in Table 8.20.

---

*A variation of this procedure uses a $(2S/\Delta t + O)$ versus O curve. From known $(2S/\Delta t + O)$, the value of O becomes available from this curve for the next step. The subtraction of twice O provides $(2S/\Delta t - O)$ directly for application in eq. (8.32).

TABLE 8.20 STORAGE, DISCHARGE, AND $(2S/\Delta t + O)$ DATA

(1) Water Surface Elevation (ft)	(2) Head, H = (col. 1 − crest level)	(3) Storage, S (from given data) Acre-ft	(4) cfs-day	(5) Outflow, O (from formula) (cfs)	(6) $(2S/\Delta t + O)$ (cfs)
50	0	231	116.4	0	465.6
50.5	0.5	247	124.7	37.1	535.9
51.0	1.0	277	139.9	105.0	664.6
51.5	1.5	313	158.1	192.9	825.3
52.0	2.0	353	178.3	297.0	1010.2
52.5	2.5	400	202.0	415.0	1223.0
53.0	3.0	452	228.3	545.6	1458.8
53.5	3.5	509	257.1	687.5	1715.9

2. $(2S/\Delta t + O)$ has been calculated for $\Delta t = 0.5$ day, from the values of S and O in columns 4 and 5, respectively. These have been plotted in Figure 8.14.
3. The routing computations are performed in Table 8.21 and explained below.
>   Column 3: Addition of two successive values of column 2.
>   Columns 4 and 5: Obtained from the storage and the discharge curves (Figure 18.14), entering from the $(2S/\Delta t + O)$ curve corresponding to the value of column 7 in the previous line.
>   Column 6: Obtained from the values in columns 4 and 5.
>   Column 7: Right-hand side of eq. (8.32), column 3 + column 6.
>   Column 8: From Figure 8.14, for the value in column 7.

TABLE 8.21 RESERVOIR FLOOD ROUTING COMPUTATION

(1) Time (days)	(2) Inflow, I (cfs)	(3) $I_1 + I_2$	(4) LHS Outflow, O (cfs)	(5) LHS Storage, S (cfs-day)	(6) LHS $\dfrac{2S_1}{\Delta t} - O_1$ (cfs)	(7) RHS $\dfrac{2S_2}{\Delta t} + O_2$ (cfs)	(8) Water Elevation (ft)
Before flood arrives:	0	0	0	116.4	465.6	465.6	50.00
Inflow hydrograph:							
0	0	70	0	116.4	465.6	535.6	50.45
0.5	70	255	35	124.5	463	718	51.20
1.0	185	545	140	146.0	444	989	51.95
1.5	360	840	285	176.0	419	1259	52.6
2.0	480	780	450	210	390	1170	52.4
2.5	300	465	380	197	408	873	51.60
3.0	165	245	215	163	437	682	51.05
3.5	80	80	115	142	453	533	50.45
4.0	0	0	30	124	466	466	50.0
4.5	0	0	0	116.4	466	466	50.0

Computation of Extreme Flows   Chap. 8

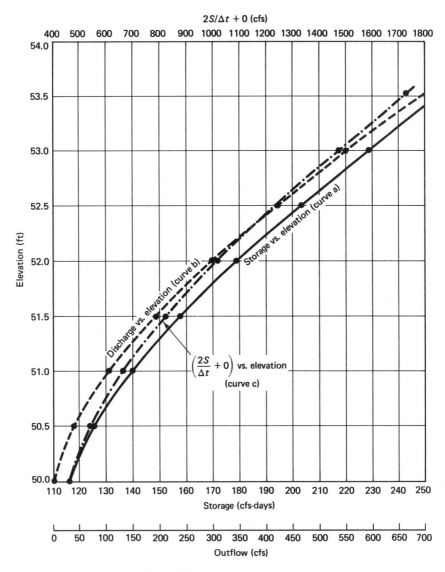

Figure 8.14 Reservoir routing curves.

4. The inflow hydrograph (column 2) and the outflow hydrograph (column 4) are plotted in Figure 8.15.

8.18 STREAMFLOW ROUTING (MUSKINGUM METHOD)

The approach of defining the storage and the outflow in terms of the stage is not applicable to streamflow routing because of the varied flow conditions in a river channel. The problem of storage and outflow, being two unknowns in eq. (8.30), is resolved by considering storage as being related to outflow (as well as to inflow). The storage

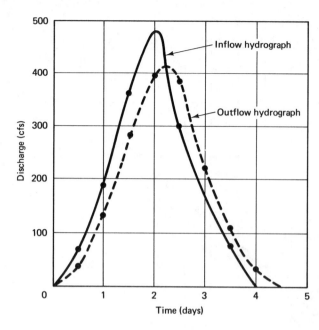

Figure 8.15 Inflow hydrograph and routed outflow hydrograph.

in a channel reach under varied flow conditions consists of two parts, as shown in Figure 8.16, the prism storage comprises the water below the line parallel to the channel bottom, and the wedge storage is the water between this line and the actual water surface. As seen from Figure 8.16, the wedge storage increases the storage volume during rising stages and reduces the volume in the falling stages for the same outflow. Thus, in a plot of storage versus outflow, a loop is observed due to the effect of the wedge storage. In simpler methods, the wedge storage is neglected and the channel storage is indicated in terms of the outflow only. (This is the case with reservoir routing, also.) To incorporate the effect of wedge storage, the inflow is also included as a parameter in the relation of storage. The inflow storage is related to the inflow rate and the outflow storage to the outflow rate, as follows:

$$S_I = KI^n \qquad \text{(a)}$$

$$S_O = KO^n \qquad \text{(b)}$$

where the subscripts I and O refer to inflow and outflow, n is an exponent, and K is a storage constant. If x is a weight factor to account for the relative effect of inflow and outflow on storage, then

$$S = xS_I + (1 - x)S_O \qquad \text{(c)}$$

Substituting eqs. (a) and (b) in (c) yields

$$S = K[xI^n + (1 - x)O^n] \qquad [L^3] \qquad (8.33)$$

Many applications of eq. (8.33) have been made. In a common case, the exponent n is taken to be unity, which results in

$$S = K[xI + (1 - x)O] \qquad [L^3] \qquad (8.34)$$

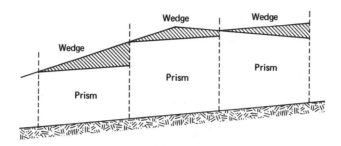

Figure 8.16 Components of channel storage.

Substituting, eq. (8.34) in eq. (8.31) gives us

$$\frac{I_1 + I_2}{2} + \frac{K[xI_1 + (1 - x)O_1] - K[xI_2 + (1 - x)O_2]}{\Delta t} = \frac{O_1 + O_2}{2} \tag{8.35}$$

$$[\text{L}^3\text{T}^{-1}]$$

Arranging the terms yields

$$O_2 = C_0 I_2 + C_1 I_1 + C_2 O_1 \qquad [\text{L}^3\text{T}^{-1}] \tag{8.36}$$

$$C_0 = \frac{0.5\,\Delta t - Kx}{K(1 - x) + 0.5\,\Delta t} \qquad [\text{dimensionless}] \tag{8.37a}$$

$$C_1 = \frac{0.5\,\Delta t + Kx}{K(1 - x) + 0.5\,\Delta t} \qquad [\text{dimensionless}] \tag{8.37b}$$

$$C_2 = \frac{K(1 - x) - 0.5\,\Delta t}{K(1 - x) + 0.5\,\Delta t} \qquad [\text{dimensionless}] \tag{8.37c}$$

Equation (8.36), known as the Muskingum equation, was used by McCarthy in a study of the Muskingum Conservancy District Flood Control Project in 1934–1935.

To use eq. (8.37), the values of K and x have to be established. This is done on the basis of actual observed inflow and outflow hydrographs, as described below.

8.18.1 Determination of Routing Constants

To determine the routing constants K and x, the storage represented by the area between the inflow and outflow hydrographs is computed for various time periods (Figure 8.17). The values of storage are graphed against the weight discharge $[xI + (1 - x)O]$ using different selected values of x, as shown in Figure 8.18. Since eq. (8.34) is a linear relation, the correct value of x is that which gives a straight-line plot, or the narrowest loop. The value of K is obtained automatically as the slope of the line.

The value of x will range from 0 to 0.5 with a value of 0.25 as an average for river reaches. Analysis of many flood waves indicates that the time required for the center of mass of the floodwave to pass from the upstream end to the downstream end of a reach is equal to the factor K. The time between the upstream and downstream peaks is approximately equal to K.

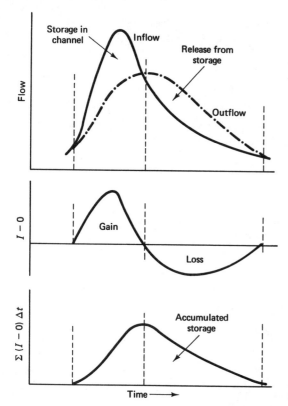

Figure 8.17 Computation of storage from hydrographs.

Example 8.15

The inflow and outflow hydrograph for a river reach are as shown in columns 2 and 3 of Table 8.22. Determine K and x for the reach.

Solution

1. The storage and the weighted discharge for values of x of 0.1, 0.2 and 0.3 have been computed in Table 8.22.
2. The storage versus weighted discharge for various x is plotted in Figure 8.18.
3. When x is increased from 0.2 to 0.3, the loop is reversed. A value of x between 0.2 and 0.3 (say, 0.25) will provide the best linear relation.
4. From Figure 8.18, for $x = 0.2$, $K = 24$ hr (slope of line); for $x = 0.3$, $K = 24.2$ hr. The factor K is practically the same for $x = 0.2$ and 0.3.

8.18.2 Application of the Muskingum Method

Once the values of constants K and x have been determined, the routing parameters C_0, C_1, and C_2 are computed from eqs. (8.37a) through (8.37c). The sum of these three parameters is equal to 1. The time interval, Δt, is selected so that all three parameters have positive values. The routing operation is simply a solution of eq. (8.36). The O_2 value of one routing period is used as O_1 for the succeeding period. Initially, outflow is equal to inflow.

TABLE 8.22 DETERMINATION OF STORAGE AND WEIGHTED DISCHARGE

(1) Time (hr)	(2) Inflow, I (cfs)	(3) Outflow, O (cfs)	(4) $I - O$ (cfs)	(5) S $\Delta t = 12$ hr (cfs-hr)[a]	(6) Cumulated S (cfs-hr)	(7) $x = 0.1$	(8) $x = 0.2$	(9) $x = 0.3$
						\multicolumn{3}{c}{$xI + (1 - x)O$}		
0	0	0	0					
				1.8	1.8	2.03	2.06	2.09
12	2.3	2.0	0.3					
				46.8	48.6	7.75	8.50	9.25
24	14.5	7.0	7.5					
				145.2	193.8	13.37	15.04	16.71
36	28.4	11.7	16.7					
				192.0	385.8	18.03	19.56	21.09
48	31.8	16.5	15.3					
				126.0	511.8	24.57	25.14	25.71
60	29.7	24.0	5.7					
				11.4	523.2	28.72	28.34	27.96
72	25.3	29.1	−3.8					
				−70.8	452.4	27.60	26.80	26.00
84	20.4	28.4	−8.0					
				−93.0	359.4	23.05	22.30	21.55
96	16.3	23.8	−7.5					
				−85.8	273.6	18.72	18.04	17.36
108	12.6	19.4	−6.8					
				−76.8	196.8	14.70	14.10	13.50
120	9.3	15.3	−6.0					
				−63.0	133.8	10.75	10.30	9.85
132	6.7	11.2	−4.5					
				−46.2	87.6	7.88	7.56	7.24
144	5.0	8.2	−3.2					
				−33.0	54.6	6.17	5.94	5.71
156	4.1	6.4	−2.3					
				−23.4	31.2	5.04	4.88	4.72
168	3.6	5.2	−1.6					

[a] $\frac{1}{2}$(sum of the value in column 4 with its previous value) $\times \Delta t$.

Example 8.16

For the inflow hydrograph indicated in columns 1 and 2 of Table 8.23, perform the routing through a river reach when $K = 20$ hours and $x = 0.25$.

Solution Consider $\Delta t = 12$ hours.

$$C_0 = \frac{0.5\,\Delta t - Kx}{K(1 - x) + 0.5\,\Delta t} = \frac{0.5(12) - 20(0.25)}{20(1 - 0.25) + 0.5(12)} = 0.05$$

$$C_1 = \frac{0.5\,\Delta t + Kx}{K(1 - x) + 0.5\,\Delta t} = \frac{0.5(12) + 20(0.25)}{20(1 - 0.25) + 0.5(12)} = 0.52$$

$$C_2 = \frac{K(1 - x) - 0.5\,\Delta t}{K(1 - x) + 0.5\,\Delta t} = \frac{20(1 - 0.25) - 0.5(12)}{20(1 - 0.25) + 0.5(12)} = 0.43$$

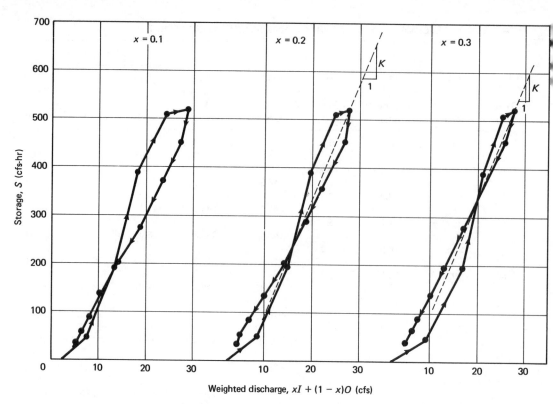

Figure 8.18 Determination of K and x in Muskingum Method.

TABLE 8.23 COMPUTATION BY THE MUSKINGUM METHOD

(1) Time (hr)	(2) Inflow (cfs)	(3) $C_0 I_2$	(4) $C_1 I_1{}^{\text{a}}$	(5) $C_2 O_1{}^{\text{b}}$	(6) O_2 (eq. 8.36)
12	100	—	—	—	100 (O_1 for next step)
24	300	15	52	43	110
36	680	34	156	47.3	237.3
48	500	25	353.6	102	480.6
60	400	20	260	206.7	486.7
72	310	15.5	208	209.3	432.8
84	230	11.5	161.2	186.1	358.8
96	180	9	119.6	154.3	282.9
108	100	5	93.6	121.6	220.2
120	50	2.5	52	94.7	149.2

[a] C_1 multiplied by the value of column 2 from the preceding step.
[b] C_2 multiplied by the value of column 6 from the preceding step.

PROBLEMS

8.1. A spillway is designed for a "100-year exceedence" flood. What is the probability of the design flood occurring exactly twice in a 50-year period?

8.2. What is the probability that five 50-year floods will occur in a 100-year period?

8.3. In Problem 8.1, what is the probability of the design flood occurring one or more times in 50 years?

8.4. Determine the probability of one or more floods of 50-year severity occurring in 100 years.

8.5. If a designer accepts a 5% chance of a flood control levee being overtopped in 25 years, what return period of flood should be used in the design?

8.6. For a flood control project having a service life of 100 years, only a 1% risk of overtopping can be taken. Determine the design exceedence level (required return period) of the flood.

8.7. A reinforced-concrete culvert has to be designed for a roadway. The flood–frequency data at the proposed site are given in the table below. The flood damages associated with various magnitudes of floods are also indicated in the table. A culvert of 30 in. diameter will pass a discharge of 40 cfs. Determine the total annual cost of this culvert including the cost of damage due to higher than the design level floods. Assume that the following relationships hold for annual direct cost and cost of operation and maintenance:

$$\text{Annual direct cost} = 750 + 70D \qquad (D \text{ is in inches})$$

$$\text{Annual O\&M cost} = 30\% \text{ of direct cost}$$

Discharge (cfs)	Annual Exceedence Probability	Estimated Damage (dollars)
20	0.46	0
24	0.36	600
32	0.22	4,000
40	0.14	10,000
60	0.06	22,500
74	0.03	75,000
108	0.01	110,000
200	0.001	150,000

8.8. Four alternative designs are proposed for the culvert of Problem 8.7. The cost of each alternative is analyzed per procedure adopted in Problem 8.7 and the total cost of each alternative is indicated below. The level of flood protection provided by each alternative is also given below. Determine the optimum design flood and its exceedence probability.

Culvert Size (in.)	Level of Protection (cfs)	Total Annual Cost (dollars)
27	30	10,100
30	40	9,450
40	70	10,250
50	100	11,800

8.9. The annual peak discharges of the Little Madawaska River near Fortfairfield, Maine, are shown below. The data in parentheses relate to snowmelt floods. Determine whether the peak discharges due to rainstorms are homogeneous to the peak discharges due to snowmelts at the 5% significance level.

Date	Discharge (m^3/s)	Date	Discharge (m^3/s)	Date	Discharge (m^3/s)
4/5/45	(223)	4/12/59	(84)	2/23/73	(269)
3/25/46	(194)	3/30/60	(200)	7/18/74	356
2/24/47	(266)	2/28/61	(103)	6/18/75	(323)
12/20/48	302	7/13/62	178	3/25/76	(123)
3/11/49	(129)	3/15/63	(205)	2/28/77	(70)
8/15/50	215	9/17/64	215	4/3/78	(350)
4/4/51	(103)	12/23/65	197	1/5/79	140
7/1/52	253	4/20/66	(135)	3/25/80	(97)
3/3/53	(157)	3/25/67	(178)	4/1/81	(185)
5/7/54	80	8/8/68	215	4/15/82	(130)
4/3/55	(401)	4/1/69	(297)	7/25/83	171
7/27/56	241	11/26/70	190	10/20/84	192
9/20/57	347	7/20/71	195	8/5/85	109
8/20/58	464	3/31/72	(215)	7/7/86	97

8.10. The annual maximum flow data for two different periods of record for Mill Creek near Los Molinos, California, are given below. Determine whether there is a homogeneity in the sequence of data of two periods, at the 2.5% level of significance.

Year	Discharge (cfs)	Year	Discharge (cfs)
1929	1,520	1950	4,430
1930	6,000	1951	3,870
1931	1,500	1952	4,930
1932	5,440	1953	7,710
1933	1,080	1954	4,910
1934	2,630	1955	2,480
1935	4,010	1956	9,180
1936	3,930	1957	6,140
1937	4,700	1958	3,060
1938	13,000	1959	3,800
1939	4,600	1960	7,520
1940	9,360	1961	4,010
1941	6,240	1962	6,100

Computation of Extreme Flows   Chap. 8

8.11. Prepare the flood–frequency curve by graphic procedure for the annual peak-discharge data of Problem 8.9 on log-normal probability paper. Determine the exceedence probability of a flood of 100 m³/s. What is the magnitude of flood of a 200-year return period?

8.12. The maximum annual instantaneous flows from Batten Kill at Greenwich, New York, are listed below. Plot the flood–frequency curve on log-normal probability paper by the graphic method. Determine the flood having a probability of exceedence of 0.5. Also determine the flood having a return period of 10 years.

Water Year	Flow (m³/s)	Water Year	Flow (m³/s)	Water Year	Flow (m³/s)
1960	224	1969	460	1978	223
1961	305	1970	1330	1979	218
1962	210	1971	340	1980	185
1963	333	1972	447	1981	256
1964	500	1973	298	1982	288
1965	98	1974	397	1983	483
1966	288	1975	195	1984	790
1967	370	1976	690	1985	500
1968	212	1977	410		

8.13. The following peak-discharge data pertain to the Juniata River near Millerstown, Pennsylvania. From the city records, a historic flood of 10,000 cfs was noted for the year 1900. Plot the flood–frequency curve on log-normal paper by the graphic method. Compute the flow of the 200-year return period.

Year	Peak Discharge (cfs)	Year	Peak Discharge (cfs)	Year	Peak Discharge (cfs)
1950	7420	1960	3560	1971	7420
1951	5290	1961	4680	1972	6944
1952	7880	1962	3560	1973	7250
1953	8800	1963	5550	1974	4290
1954	7590	1964	5770	1975	5385
1955	5550	1965	8260	1976	5984
1956	6175	1966	3430	1977	3040
1957	4170	1967	8060	1978	2064
1958	6215	1968	6855	1979	5150
1959	5010	1969	4370		
		1970	7345		

8.14. Perform a flood–frequency analysis of the annual maximum flow data of Problem 8.9 by the theoretical method, adopting the log-Pearson type III distribution. The regional skew coefficient from the map is 0.0.

8.15. On the flood flow sequence of Problem 8.12, perform a frequency analysis by the theoretical method, adopting the log-Pearson type III distribution. The regional skew coefficient from the map is 0.50.

8.16. Using the peak-discharge data of Problem 8.13, compute the flood–frequency curve based on log-Pearson type III distribution. The regional skew coefficient is 0.10.

8.17. Solve Problem 8.14 by the extreme value type I distribution.

8.18. Solve Problem 8.15 by the log-normal distribution.

8.19. Solve Problem 8.16 by the normal distribution.

8.20. Construct a reliability band of 0.05 and 0.95 error limits on the log-Pearson type III frequency curve of Problem 8.14.

8.21. On the frequency curve of Problem 8.16, plot the 5% and 95% error limit curves. Also compute the probability adjusted for the size of the flood flow series.

8.22. The observed peak discharges at a gaging site consist of hurricane and nonhurricane events for which separate flood–frequency curves have been computed. From these curves the following exceedence probabilities have been noted for two types of floods, corresponding to the discharge levels indicated. Prepare a composite flood–frequency curve for the combined series. What is the return period of a peak flow of 40,000 cfs?

Q (cfs)	P_h (for hurricane)	P_n (for nonhurricane)
26,500	0.048	0.002
25,000	0.051	0.009
20,000	0.064	0.153
17,500	0.072	0.308
15,000	0.084	0.52
12,500	0.098	0.727
10,000	0.119	0.883

8.23. Using the generalized map, determine the 18-hour PMP for a watershed of 320 mi² located in Boston, Massachusetts. Construct a hyetograph, assuming that the rainfall in each 3-hour duration is according to HEC distribution (Table 8.15).

8.24. The ordinates of a 4-hour unit hydrograph for a basin of 100 mi², which has resulted from an ordinary storm, are given below. Prepare the modified unit hydrograph that can be used for computing the design flood hydrograph. Assume that the peak of the major storm UHG is 130% of the ordinary storm.

Time (hr)	1	2	3	4	5	6	7	8
Flow (cfs)	475	2000	5240	7140	8300	7260	6190	5360

Time (hr)	9	10	11	12	13	14	15	16
Flow (cfs)	4520	3800	3200	2600	2140	1670	1300	950

Time (hr)	17	18	19	20
Flow (cfs)	710	475	240	120

8.25. A peak flow of 329,000 cfs was record on August 20, 1955, in the Delaware River near Trenton, New Jersey, having a drainage area of 6780 mi². Determine the error of estimation if the Creager's relation is used with $C = 50$.

8.26. If the peak flow of Problem 8.25 was estimated by the Myers method with a rating of 4000, what is the error of estimation?

8.27. A peak discharge of 1,110,000 cfs was recorded on the Ohio River near Louisville on January 26, 1937. The river has a drainage area of 91,170 mi². Estimate the peak-flow rates by the Creager method and the Myers method. For the Myers rating, refer to Table 8.16 and use a Creager coefficient of 60.

8.28. To determine the peak discharge by the slope–area method, the following field measurements were made at two sites that are 115 ft apart.

	Site 1	Site 2
1. Area (ft²)	225	209
2. Hydraulic radius (ft)	5.20	4.05
3. High-water marks (ft)	15.51	15.00
4. Roughness coefficient	0.045	0.030

8.29. Field investigations have been made of two adjacent sites of a stream. The fall of the water surface was 0.25 m in a reach of 35 m. The cross sections at both sides were subdivided. The area and hydraulic radius of subsections are indicated below. Determine the peak flow by the slope–area method.

Site	n	A (m²)	R (m)
1	0.075	0.60	0.25
	0.040	15.5	1.01
2	0.075	0.85	0.20
	0.045	20.5	1.20
	0.045	0.50	0.22

8.30. The following field observations were made just after a flood at the bridge opening of a stream shown in Fig. P8.30. Determine the peak discharge. Assume that $C = 0.91$.
1. Water elevation at approach section = 9.885 ft.
2. Water elevation at contracted section = 8.995 ft.
3. Approach section 1:

Subsection	n	Area (ft²)	Wetted Perimeter (ft)
1	0.05	15.9	30
2	0.035	116.0	38
3	0.045	12.5	25.2

4. Contracted section 3:

$$n = 0.035 \qquad A = 82 \text{ ft}^2 \qquad \text{wetted perimeter} = 33.3 \text{ ft}$$

5. Characteristics of constriction:

> Slopping embankment
>
> Vertical abutment
>
> Length of the opening, $L = 18.5$ ft
>
> Length of the approach reach, $L_w = 35$ ft

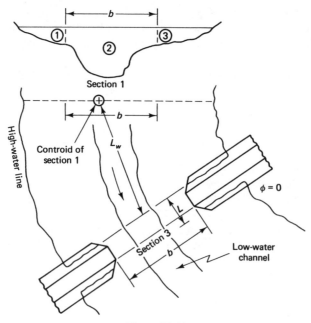

Figure P8.30

8.31. The data for 7-day low flows of the English River near Altona, New York, from the annual series are given below. Prepare the low-flow frequency curve on extreme value type III (log Gumbel) paper. Determine **(a)** the probability of a 7-day low flow of 7.5 m³/s, and **(b)** the return period for a flow of 4.5 m³/s.

Year	7-Day Low Flow (m³/s)	Year	7-Day Low Flow (m³/s)
1952	4.95	1969	4.15
1953	4.90	1970	6.00
1954	5.80	1971	5.80
1955	4.95	1972	4.90
1956	6.20	1973	7.25
1957	6.25	1974	10.10
1958	4.85	1975	6.65
1959	4.00	1976	5.55
1960	4.15	1977	5.05
1961	5.55	1978	3.60
1962	5.60	1979	4.00
1963	6.35	1980	3.55
1964	4.85	1981	4.20
1965	3.15	1982	2.80
1966	6.40	1983	5.40
1967	6.30	1984	6.15
1968	4.65	1985	6.10

8.32. The annual 7-day minimum average flows for the Buffalo River in Tennessee are given below. Plot the low-flow frequency curve on log-Gumbel paper. Determine 10-year 7-day low flows.

Year	Flow (cfs)	Year	Flow (cfs)
1926	99	1943	97
1927	145	1944	131
1928	130	1945	170
1929	159	1946	182
1930	110	1947	146
1931	96	1948	120
1932	168	1949	200
1933	149	1950	226
1934	125	1951	208
1935	122	1952	164
1936	115	1953	146
1937	146	1954	112
1938	155	1955	112
1939	149	1956	120
1940	124	1957	156
1941	93	1958	166
1942	102	1959	165

8.33. Solve Problem 8.31 by the Analytical Method using log-extreme value I distribution.

8.34. Solve Problem 8.32 by the Analytical Method using log-extreme value I distribution.

8.35. Route the inflow flood hydrograph indicated below through a reservoir. The storage (elevation versus volume) data obtained from the reservoir survey are also given. The spillway has the following characteristics:

$$\text{Flow} = 3LH^{3/2}$$
$$\text{Length} = 70 \text{ ft}$$
$$\text{Crest height} = 60 \text{ ft}$$

Inflow hydrograph:

Time (hr)	0	0.4	0.8	1.2	1.6	2.0	2.4	2.8	3.2	3.6
Flow (cfs)	0	600	2100	2500	1600	950	550	300	80	0

Storage data:

Elevation (ft)	60	61	62	63	64	65
Storage (acre-ft)	300	330	360	395	430	470

8.36. The reservoir storage data and the spillway rating data are given below. Route the following flood hydrograph through the reservoir.

Inflow hydrograph:

Time (hr)	0	0.5	1.0	1.5	2.0	2.50	3.0	3.50	4.0
Flow (cfs)	0	20	70	160	280	330	140	100	40

Storage and discharge data:

Elevation (ft)	15	16	17	18	19	20
Storage (1000 ft³)	180	252	414	655	990	1350
Outflow (cfs)	0	15	55	105	175	240

(1 cfs-hr = 3600 ft³.)

8.37. Given the following hydrographs at the upstream and downstream ends of a river reach, determine the Muskingum routing constants.

Date	Time (hr)	Inflow (cfs)	Outflow (cfs)	Date	Time (hr)	Inflow (cfs)	Outflow (cfs)
1	6	0	0	3	6	25	48.5
	12	75	32.3		12	25	37.7
	18	170	66.3		18	25	31.8
	24	125	107.8		24	25	28.7
2	6	100	112.3				
	12	77.5	103.3				
	18	57.5	88.2				
	24	25.0	69.3				

8.38. From the following hydrographs, compute the Muskingum routing constants.

Time	Inflow (m³/s)	Outflow (m³/s)	Time	Inflow (m³/s)	Outflow (m³/s)
Midnight	17.3	3.1	Noon	12.6	21.9
Noon	28.8	7.5	Midnight	9.8	17.0
Midnight	35.9	16.3	Noon	7.7	13.6
Noon	37.0	26.8	Midnight	6.2	10.9
Midnight	30.6	33.7	Noon	5.0	9.0
Noon	22.0	34.0	Midnight	4.2	7.5
Midnight	16.4	28.0	Noon	3.6	6.2

8.39. The routing constants for a reach of a river have been found to be 24 hours and $x = 0.2$. Route the following inflows through the reach of the river by the Muskingum method.

Time (hr)	12	24	36	48	60	72	84	96
Flow (cfs)	300	450	750	825	740	600	400	270

8.40. Determine the hydrograph at the downstream section if a storm produced the following hydrograph at the upstream section. The Muskingum constants are $K = 8$ hours and $x = 0.15$ for the reach.

Time	Inflow (cfs)	Time	Inflow (cfs)
6 A.M.	150	Midnight	300
Noon	180	6 A.M.	200
6 P.M.	300	Noon	150
Midnight	750	6 P.M.	100
6 A.M.	850	Midnight	100
Noon	550	6 A.M.	80
6 P.M.	400	Noon	50

CHAPTER
NINE

STORAGE AND CONTROL STRUCTURES

9.1 HYDRAULIC STRUCTURES

There are a large variety of hydraulic structures to serve the many purposes for which water resources are put to use. A classification based on the function performed by the structure is given in Table 9.1. This lists only the common structures. There are other specialized hydraulic structures also, such as hydrofoils, offshore structures, and hydrodynamic transmissions.

Having estimated the water requirements for an intended project (Chapter 2) and having assessed the available water resources at a prospective site (Chapter 3 through 7), a planning engineer is faced with one of the three situations:

1. The rate at which the resources are available is always in excess of the requirements.
2. The total quantity of available resources over a period of time is equal to or in excess of the overall requirements, but the rate of requirement at times exceeds the available rate.
3. The total available resources are less than the overall requirements.

In the first case, a run-of-river project can be formulated in which water can be used directly from the stream as need arises. A storage reservoir is the solution to the second case. Under the third condition, a supplemental source or an alternative site has to be explored. Run-of-river projects primarily incorporate conveyance structures. Reservoir projects include storage, control, and conveyance structures. This chapter deals with major storage structures and control works, such as dams and spillways. Conveyance structures are covered in Chapters 10 and 11.

446

TABLE 9.1 CLASSIFICATION OF HYDRAULIC STRUCTURES

Type	Purpose	Structures
Storage structures	To store water	Dams, tanks
Flow control structures	To regulate the quantity and pass excess flow	Spillways, outlets, gates, valves
Flow measurement structures	To determine discharge	Weirs, orifices, flumes
Diversion structures	To divert the main course of water	Coffer dams, weirs, canal headworks, intake works
Conveyance structures	To guide flow from one location to another	Open channels, pressure conduits, pipes, canals, sewers
Collection structures	To collect water for disposal	Drain inlets, infiltration galleries, wells
Energy dissipation structures	To prevent erosion and structural damage	Stilling basins, surge tanks, check dams
Shore protection structures	To protect the banks	Dikes, groins, jetties, rivetments, breakwaters, seawalls
River training and waterway stabilization structures	To maintain a river channel and water transportation	Levees, cutoffs, locks, piers, culverts
Sediments and quality control structures	To control or remove sediments and other pollutants	Racks, screens, traps, sedimentation tanks, filters, sluiceways
Hydraulic machines	To convert energy from one form to other	Pumps, turbines, rams

9.2 REQUIRED STORAGE CAPACITY

Storage is of two types. When the demands for water can be satisfied by holding some of the high flow each year for release during a later period of low flow, it is a seasonal or "within-year" storage. However, if there is not enough high flow every year to raise the flow to a desired level, extra water must be stored during wetter years to release during dry years. This is termed "over-year" or "carryover" storage.

There are two approaches for determining the size of reservoir storage required. The simplified methods, which are commonly used in planning stage studies, comprise mass curve analyses. The detailed methods, commonly used at the time of developing reservoir operating plans, perform a sequential reservoir routing of the historical flow record. With the availability of computer program packages, the sequential routing study is also being used increasingly in the planning stage. The theory of reservoir routing is discussed in Section 8.17. The simplified approach is still very valuable for planning a single project when demands for water are relatively simple.

9.2.1 Simplified Procedure for Reservoir Storage Capacity

There are two methods of analyses: (1) the sequential mass-curve method, and (2) the nonsequential mass-curve method.

A sequential mass-curve method known as the Rippl method considers the most critical period of recorded flow, which might be a severe drought period. The cumulative differences between inflow to the reservoir and outflow (draft) during successive periods are evaluated, the maximum value of which is the required storage:

$$S = \text{maximum } \sum (I_t - O_t) \qquad [\text{L}^3] \qquad\qquad (9.1)$$

where

S = required storage capacity

O_t = reservoir output or draft (yield) during period Δt

I_t = inflow during period Δt

Equation (9.1) can be solved either graphically or analytically. There are two graphic procedures. In the first method, I_t is accumulated separately as a mass inflow curve (curve a, Figure 9.1) and O_t as a mass yield curve (curve b, Figure 9.1). Yield represents the total demand for water and evaporation. For a constant draft rate, the yield curve is a straight line having a slope equal to the draft rate. At each high point on the mass inflow curve (curve a), a line is drawn parallel to the yield curve (curve b) and extended until it meets the inflow curve. The maximum vertical distance between the parallel yield line and the mass inflow curve (i.e., FD) represents the required storage. Assuming that the reservoir is full at A in Figure 9.1, going from A to E along the mass curve (ABCDE) represents the same volume of water as going from A to E along the straight line (AFE). From A to B, the draft is more than the inflow, resulting in a lowering of the reservoir; from B to C, the inflow is higher than the draft, but not enough to refill the reservoir; from C to D, the draft is more, once again causing a further drop in the level; from D to E, however, the inflow is very high, thus filling the reservoir at E.

The second graphical procedure plots the difference of successive accumulated values of the inflow and yield $\Sigma (I_t - O_t)$ against time as shown in Figure 9.2. The maximum vertical difference is the storage. In the analytical procedure, the maximum accumulated difference is determined arithmetically to calculate the reservoir size.

The storage computed from the sequential analysis of historic flow data does not indicate the frequency (return period) associated with the selected size though it identifies within-year or over-year storage. Low flows of different durations can be analyzed nonsequentially to prepare flow–duration–frequency curves. From these curves, a mass inflow curve for any selected frequency can be prepared and used for determining the storage corresponding to that frequency. However, it does not make a distinction between within-year or over-year storage. The methods have been described by the HEC (1967) and Riggs and Hardison of the U.S. Geological Survey (1983).

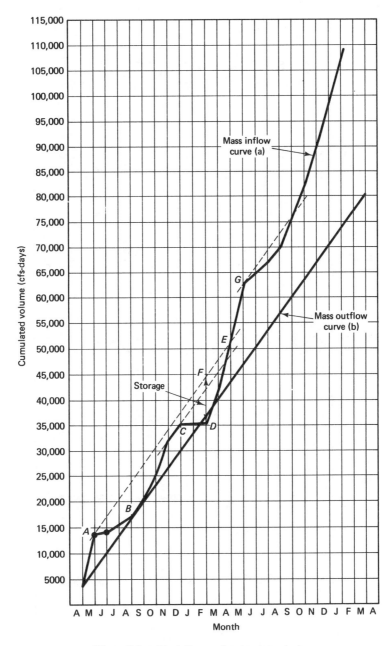

Figure 9.1 Rippl diagram for storage analysis.

Example 9.1

During a critical flow period, the monthly inflows at the site of a proposed dam are given in Table 9.2. The water supply requirements to be met from the storage are also shown in the table. A uniform release of 8 cfs has to be maintained to meet local requirements. Average monthly evaporation rate is 12 cfs. Determine the required capacity of the reservoir.

TABLE 9.2 DATA AND COMPUTATIONS FOR THE RESERVOIR CAPACITY

(1) Month	(2) Inflow (cfs)	(3) Outflow (cfs)	(4) Total Outflow (cfs)	(5) Inflow Volume, I_t (cfs-day)	(6) Outflow Volume, O_t (cfs-day)	(7) Cumulative Inflow ΣI_t (cfs-day)	(8) Cumulative Outflow ΣO_t (cfs-day)	(9) Difference $\Sigma I_t - \Sigma O_t$ (cfs-day)
Apr.	141	90	110	4,230	3,300	4,230	3,300	+930
May	310	92	112	9,610	3,472	13,840	6,772	7,068
June	18	92	112	540	3,360	14,380	10,132	4,248
July	56	93	113	1,736	3,505	16,116	13,637	2,479
Aug.	40	90	110	1,240	3,410	17,356	17,047	309
Sept.	135	90	110	4,050	3,300	21,406	20,347	1,059
Oct.	160	90	110	4,960	3,410	26,366	23,757	2,609
Nov.	221	89	109	6,630	3,270	32,996	27,027	5,969
Dec.	85	89	109	2,635	3,379	35,631	30,406	5,225
Jan.	0	89	109	0	3,379	35,631	33,785	1,846
Feb.	0	91	111	0	3,108	35,631	36,893	-1,262
Mar.	241	90	110	7,471	3,410	43,102	40,303	2,799
Apr.	359	90	110	10,770	3,300	53,872	43,603	10,269
May	312	92	112	9,672	3,472	63,544	47,075	16,469
June	75	92	112	2,250	3,360	65,794	50,435	15,359
July	50	93	113	1,550	3,505	67,344	53,940	13,404
Aug.	82	93	113	2,542	3,505	69,886	57,445	12,441
Sept.	247	90	110	7,410	3,300	77,296	60,745	16,551
Oct.	198	90	110	6,138	3,410	83,434	64,155	19,279
Nov.	268	90	110	8,040	3,300	91,474	67,455	24,019
Dec.	266	89	109	8,246	3,379	99,720	70,834	28,886
Jan.	305	89	109	9,455	3,379	109,175	74,213	35,962

Solution

1. *Rippl or mass diagram.* Inflows are cumulated in column 7 of Table 9.2 and plotted (curve a) in Figure 9.1. Total outflows are determined by adding evaporation and releases (and seepage, if any). Outflows are added in column 8 and plotted (curve b) in Figure 9.1.

Storage = 8400 cfs-days or 16,630 acre-ft

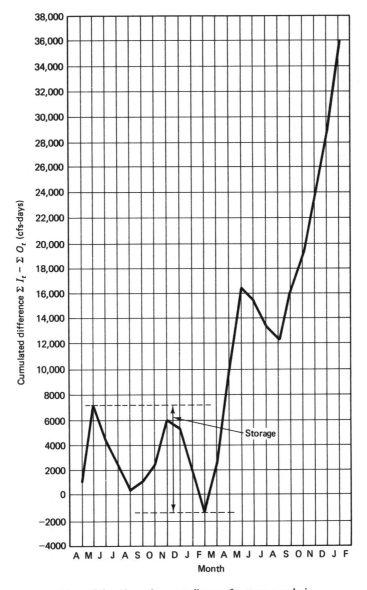

Figure 9.2 Alternative mass diagram for storage analysis.

2. *Modified mass diagram.* The difference of accumulated inflows (column 7) and outflows (column 8) are computed in column 9 and plotted on Figure 9.2.

$$\text{Storage} = 8350 \text{ cfs-days or } 16,530 \text{ acre-ft}$$

3. *Analytical method.* The maximum difference between the highest value and the subsequent lowest value of the accumulated difference (col. 9) is the required storage.

$$\text{Storage} = 7068 - (-1262) = 8330 \text{ cfs-days or } 16,490 \text{ acre-ft}$$

9.2.2 Capacity of Water Supply Tank

When widely fluctuating demands are imposed on a water supply distribution system, a distribution reservoir or a service tank is provided to accommodate an operating storage equivalent to a 24-hour demand on a maximum day. The requirements for fire-fighting purposes — sufficient to provide flow for 10 to 12 hours in large communities and for at least 2 hours in small communities — are added to this value. Emergency requirements for 2 or 3 days are also included.

The operating storage is filled in by a uniform 24-hour pumping at a rate of average hourly demand or at double this rate in 12 hours. In this case, the cumulated inflow is represented by a straight line corresponding to the uniform pumping into the tank and the cumulated outflow or demand is an undulating curve. To determine the storage, the construction of the parallel lines on the mass curve is reversed [i.e., the lines are drawn at the lowest and highest points of the demand curve parallel to the cumulated inflow (pumping) plot]. The vertical distance between these lines is the storage. For 12-hour pumping, a parallel line on the demand is drawn at a point corresponding to the time at start of pumping and continued until pumping ceases. The vertical distance between the demand curve and the parallel pumping line at the end of pumping time is the required storage. For variable demands it is convenient to compute storage by the method of difference of cumulated inflows and outflows.

Example 9.2

The hourly demand rates for the maximum day are shown in Table 9.3. Determine the required storage for (a) a uniform 24-hour pumping, and (b) a pumping period of 6 a.m. to 6 p.m.

Solution

(a) 24-hour pumping

1. Average hourly demand $= \dfrac{6182.5(1000)}{24} = 257.6 \times 10^3$ gal
   or pumping rate

2. Hourly demands are given in column 3 of Table 9.3, which is column 2 $\times$ 60.

3. Column 4 accumulates the demand of column 3, and column 5 accumulates pumping at 257.6×10^3 gal/hr.

4. The difference of the cumulated pumping and demands is given in column 6 and plotted in Figure 9.3(a).

5. Storage = maximum difference = 1319×10^3 gal or 1.32 MG.

(b) 12-hour pumping

1. Average hourly demand $= \dfrac{6182.5(1000)}{12} = 515.2 \times 10^3$ gal/hr
   or pumping rate.

TABLE 9.3 DATA AND COMPUTATIONS FOR THE TANK CAPACITY

(1)	(2)	(3)	(4)	(5) 24-hr Pumping	(6) 24-hr Pumping	(7) 12-hr Pumping	(8) 12-hr Pumping
Time	Hourly Demand Rate (gpm)	Hourly Demand (gal × 1000)	Cumulative Hourly Demand (gal × 1000)	Cumulative 24-hr Pumping (gal × 1000)	Cumulative Difference (col. 5 − col. 4) (gal × 1000)	Cumulative 12-hr Pumping (gal × 1000)	Cumulative Difference (col. 7 − col. 4) (gal × 1000)
12 night	2061	123.7	123.7	257.6	133.9	0	−123.7
1 am	1953	117.2	240.9	515.2	274.3	0	−240.9
2	1890	113.4	354.3	772.8	418.5	0	−354.7
3	1818	109.1	463.4	1030.4	567.0	0	−463.4
4	1773	106.4	569.8	1288.0	718.2	0	−569.8
5	1782	106.9	676.7	1545.6	868.9	0	−676.7
6	1872	112.3	789.0	1803.2	1014.2	515.2	−273.8
7	3267	196.0	985.0	2060.8	1075.8	1030.4	45.4
8	4671	280.3	1265.3	2318.4	1053.1	1545.6	280.3
9	5058	303.5	1568.8	2576.0	1007.2	2060.8	492.0
10	5310	318.6	1887.4	2833.6	946.2	2576.0	688.6
11	5436	326.2	2213.6	3091.2	877.6	3091.2	877.6
12	5688	341.3	2554.9	3348.8	793.9	3606.4	1051.5
1 pm	5796	347.8	2902.7	3606.4	703.7	4121.7	1219.0
2	5733	344.0	3246.7	3864.0	617.3	4636.9	1390.2
3	5688	341.3	3588.0	4121.6	533.6	5152.1	1564.1
4	5706	342.4	3930.4	4379.2	448.8	5667.3	1736.9
5	5976	358.6	4289.0	4636.8	347.8	6182.5	1893.5
6	6588	395.3	4684.3	4894.4	210.1	6182.5	1498.2
7	8400	504.0	5188.3	5152.0	−36.3	6182.5	994.2
8	7488	449.3	5637.6	5409.6	−228.0	6182.5	544.9
9	4545	272.7	5910.3	5667.2	−243.1	6182.5	272.2
10	2313	138.8	6049.1	5924.8	−124.3	6182.5	133.4
11	2223	133.4	6182.5	6182.5	0.0	6182.5	0
Total		6182.5					

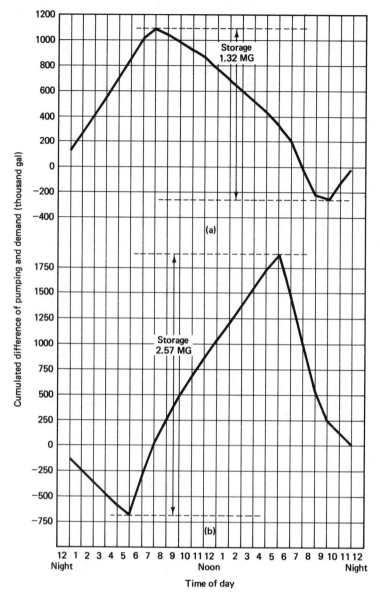

Figure 9.3 (a) Storage for 24-hour pumping; (b) storage for 12-hour pumping for Example 9.2.

2. Accumulated values of pumping are given in column 7 of Table 9.3.
3. Accumulated difference of pumping and demand ascertained in column 8 and plotted in Figure 9.3(b).
4. Storage = 2.57 MG.

9.3 RESERVOIR STORAGE–YIELD RELATIONS

The yield of a reservoir is the amount of water delivered by the reservoir in a pre-scribed interval of time. In a broad sense it is the rate of draft from the storage. In the preceding section, the storage required to meet a given yield is determined. In some cases it is necessary to determine the yield for a fixed storage. For different yields (rates of draft), required storages are determined from the methods of the preceding section. A storage–yield relation for a given site is plotted as illustrated by curve *a* in Figure 9.4. A flat curve indicates that only a small improvement in yield takes place with a large increase in storage. A steep slope has an opposite effect. For reservoir planning, a storage–elevation curve (curve *b*, Figure 9.4) is also plotted by planime-tering on the contour map the area impounded at different elevations and determining the incremental volume between the elevations.

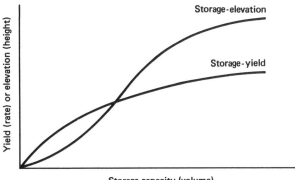

Figure 9.4 Reservoir planning curves.

9.4 RESERVOIR FEATURES

The storage capacity determined in Section 9.2 and related to yield in Section 9.3 refers to the active storage that is required to meet the demand of water for intended uses. An additional storage capacity known as the dead storage is provided to collect sediment and to maintain a minimum pool level. Also, if flood control is one of the purposes of the reservoir, an extra space above the active storage is provided to ac-commodate flood flows.

A reservoir section with related pool levels is shown in Figure 9.5. Many consid-erations are involved in the selection of a reservoir site. The suitability of the dam site and the adequacy of storage capacity are primary factors.

9.5 SEDIMENTATION OF RESERVOIRS

Sediment loading in a reservoir is measured by the trap efficiency, which is defined as the percent of incoming sediment that is retained in a reservoir. The trap efficiency of a reservoir depends on many factors. These include (1) the storage capacity, (2) the

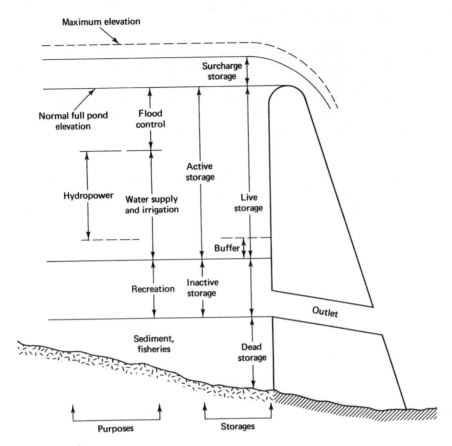

Figure 9.5 Reservoir features.

runoff from the watershed (inflow), (3) the age of the reservoir, (4) the shape of the basin, (5) the type of outlets and methods of operation, and (6) the grade-size characteristics of the sediment. These factors have the same effect, regardless of the size of the reservoir. Brune (1953), after analyzing records of 44 reservoirs, found that a parameter defined by the ratio of the reservoir storage capacity (C) to the annual runoff from the watershed [i.e., inflow (I)] offers a good correlation with the trap efficiency. The relationship between the capacity–inflow ratio (C/I) and the trap efficiency is indicated in Figure 9.6 for normally ponded reservoirs, which are conventional reservoirs, as distinguished from desilting basins and dry reservoirs. The desilting basins, because of shape and method of operation, have a very high trap efficiency, over 90%. On the other hand, the semidry reservoirs have a low efficiency of about 60%, and the dry reservoirs in the range 10 to 40%, depending on the C/I ratio.

The volume occupied by the sediment in a reservoir is obtained by multiplying the trap efficiency with the incoming sediment in a selected period. The incoming sediment or the sediment production in a watershed is estimated by the methods of sediment determination. As sediment deposits, the reservoir capacity (C) is reduced

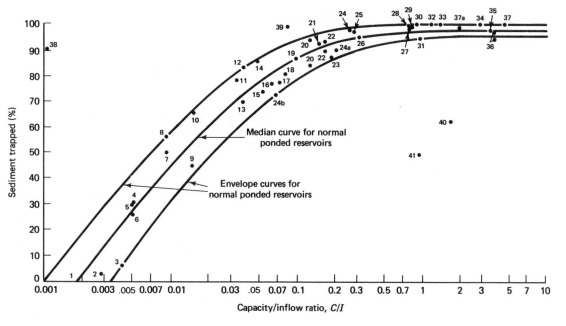

Figure 9.6 Reservoir trap efficiency (after Brune, 1953).

with time and hence trap efficiency. The computations are made in steps as illustrated in Example 9.3. Sediment deposition may be reduced by venting gravity underflow through the sluiceways during floods when the sediment concentration is high.

Example 9.3

A reservoir has a drainage area of 49,000 acres. The reservoir has a capacity of 10,000 acre-ft. Streamflow runoff averages 15.5 in. per year. The annual sediment production is 9.5×10^6 ft^3. Determine the life of the reservoir, assuming that the life is over when 80% of the original capacity is lost.

Solution

1. Annual inflow, $I = (49,000) \left(\dfrac{15.5}{12} \right) = 63,290$ acre-ft.

2. Annual sediment production = 9.5×10^6 ft^3 or 218.5 acre-ft.

3. Unfilled capacity (20%) = $\left(\dfrac{20}{100} \right) (10,000) = 2000$ acre-ft.

4. Computations are arranged in Table 9.4.

9.6 DAMS

There are three common classification schemes for dams. According to the function performed, dams are classified into (1) storage dams for impounding water for developmental uses, (2) diversion dams for diverting streamflow into canals or other con-

TABLE 9.4 RESERVOIR SEDIMENTATION COMPUTATION

(1) Reservoir Capacity (acre-ft)	(2) Capacity Lost (acre-ft)	(3) Capacity Inflow (C/I) Ratio[a]	(4) Trap Efficiency from Fig. 9.6 (%)	(5) Average Trap Efficiency (%)	(6) Annually Trapped Sediment (acre-ft) (col. 5 × Production Rate)	(7) Time to Fill (years) (col. 2/col. 6)
10,000		0.16	92			
	2000			90	197	10.1
8,000		0.13	88			
	2000			86.5	189	10.6
6,000		0.09	85			
	2000			82.5	180	11.1
4,000		0.06	80			
	2000			74	162	12.3
2,000 (80% filled)		0.03	68			
Total						44.1

[a]col. 1/I; I = 63,290 in this example.

veyance system, and (3) detention dams to hold the water temporarily to retard flood flows. From hydraulic design considerations, dams are classified as (1) overflow dams to carry discharge over their crests, and (2) nonoverflow dams, which are not designed to be overtopped. However, the most common classification is based on the materials of which dams are made. This classification makes further subclassification by recognizing the basic type of design, such as concrete gravity or concrete arch dams. Types of dams include:

1. Earthfill dams
2. Rockfill dams
3. Concrete dams
   a. Concrete gravity dams
   b. Concrete arch dams
   c. Concrete buttress dams
4. Stone masonry
   a. Stone-masonry gravity dams
   b. Stone-masonry arch dams
5. Timber dams
6. Steel coffer dams

9.6.1 Selection of Type of Dam

Physical factors important in the choice of type of dam to be constructed are discussed briefly below. Topographically, a narrow stream section with high rocky walls suggests a suitable site for a concrete dam. Where the walls are strong to resist arch thrust, a concrete arch dam is adaptable. Low-rolling-plains country suggests an earthfill or rockfill dam.

When the geologic characteristics of foundation are comprised of solid rock, any type of dam can be constructed, although concrete gravity and arch dams are favorable. Gravel foundations are suitable for earthfill, rockfill, and low concrete gravity dams. Silt and fine sand foundations are used to support earthfill and low concrete gravity dams but are not suitable for rockfill dams. Clay foundations in general are not suitable for the construction of dams. However, earthfill dams can be constructed with special treatment.

Availability of certain materials close to the site will effect a considerable reduction in cost if the type of dam selected utilizes these materials in sufficient quantity.

Size, type, and natural restrictions in location of a spillway influence the choice of dam. A large spillway requirement indicates the adoption of a concrete gravity dam. A small spillway requirement favors the selection of an earth or rockfill dam. When the excavated material from a side channel spillway can be used in dam embankment, an earthfill dam is advantageous. Side channel spillways, irrespective of its materials use, can be used with any type of dam.

The factors listed above, and others, such as the cost of diverting the stream, availability of labor, and traffic requirements on the top of the dam, favor one type over the other, but there is no unique choice for a given dam site. Several types have to be considered with their preliminary designs and estimates before making a final choice, mostly from cost considerations.

9.7 EARTHFILL DAMS

Most common among type of dams, earth dams (built to a height of several hundred feet) are considered as safe as any other type. These are designed as a nonoverflow section with a separate spillway. The foundation requirements are not as rigorous as for concrete dams. Since all earth material is pervious to some extent, problems of movement of water through the body and under the dam are intimately associated with the design of earth dams, in addition to the structural stability of embankment and foundation. The criteria for the design of earth dams are stated below.

1. Sufficient spillway capacity and freeboard are provided so that there is no danger of overtopping of the dam.
2. Seepage flow through the embankment is controlled so that the amount lost does not interfere with the objective of the dam and there is no erosion or sloughing of soil.
3. Uplift pressure due to the seepage underneath is not enough to cause piping.
4. The slopes of the embankment are stable under all conditions of reservoir operation, including rapid drawdown.
5. The stresses imposed by the embankment upon the foundation are less than the strength of material in the foundation with a suitable factor of safety.
6. The upstream face is properly protected against wave action, and the downstream face is protected against the action of rain.

Design criteria 2 and 3, which relate to hydraulic aspects, are discussed here. The structural analysis involved in criteria 4 and 5 has been summarized. One frequent cause for the failure of earth dams is the flow of water over the top of the dam. Once the downstream embankment is washed, the core, whether of impervious soil or concrete, will break or tip over. Safety against overtopping is provided by (1) a spillway of a design capacity from 100-year flood to maximum possible flood, and (2) a freeboard from the consideration of wind setup, height of tides, wave uprush, and a margin of safety. Item 6 is self-explanatory.

9.7.1 Seepage Through the Dam

Consider a dam of homogeneous material situated on an impervious foundation. The seepage pattern through the dam is shown in Figure 9.7. This pattern is the same irrespective of the material (sand, clay, loam) of the dam. Of course, the rate of seepage depends on the type of soil. The emergence of seepage lines on the downstream slope tends to make the downstream face unstable. Either the downstream slope has to be made very flat or the seepage must be diverted away from the downstream slope. The second alternative, being more economical, is favored and widely applied in dam design.

Seepage with horizontal drainage blanket (filter). One way to keep the seepage lines off the downstream slope is to provide a gravel filter on the base of the dam at the downstream side. The seepage net for this case is shown in Figure 9.8.

The flow lines enter the blanket vertically. Cassagrande (1937) has shown that the phreatic line, which is the topmost seepage line, quite closely approximates a parabola. The parabola intersects the water surface at A such that $AB = 0.3CB$. Near the upstream face, the phreatic line diverges from the parabola to join B perpendicularly.

The focus of the parabola is at F and the directrix passes through D. Every point on the parabola is equidistant from the focus and directrix. The equation of the parabola with origin at focus F (Figure 9.8) can be given by

$$x = \frac{y^2 - y_0^2}{2y_0} \quad \text{[L]} \tag{9.2}$$

At point A, $x = d$ and $y = H$. Substituting in eq. (9.2) yields

$$y_0 = \sqrt{H^2 + d^2} - d \quad \text{[L]} \tag{9.3}$$

y_0, which is also the distance of FD, is determined from eq. (9.3). FE is $\frac{1}{2}y_0$. The phreatic line can be drawn by eq. (9.2). The phreatic line acts as an upper boundary of

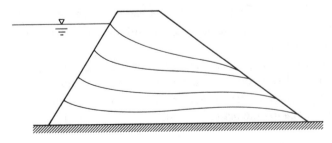

Figure 9.7 Seepage pattern through a homogeneous section.

Storage and Control Structures   Chap. 9

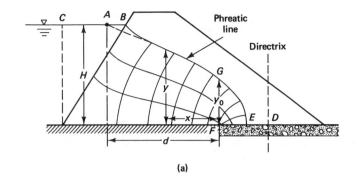

(a)

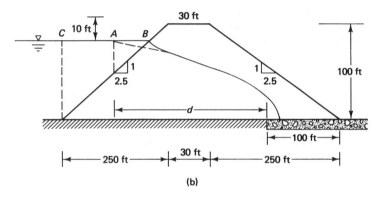

(b)

Figure 9.8 (a) Seepage pattern through a homogeneous section with a filter blanket; (b) seepage for Example 9.4.

the flow net. The flow net can be drawn following the procedure of Section 5.1 and the quantity of seepage computed by eq. (5.1).

For an approximate estimate of seepage, without reference to the flow net, apply Darcy's law at point G, per unit length of dam.

$$q = KiA = K\left(\frac{dy}{dx}\right)y_0 \tag{a}$$

From eq. (9.2),

$$\frac{dy}{dx} = \frac{y_0}{\sqrt{2xy_0 + y_0^2}} \tag{b}$$

since at G, $x = 0$ and $y = y_0$, or

$$\frac{dy}{dx} = 1 \tag{c}$$

Hence

$$q = K(1)(y_0) \tag{d}$$

Sec. 9.7   Earthfill Dams

461

The flow through the section at G is the same as at any other section. Consequently, the total seepage through length L of the dam

$$Q = K(y_0)(L) \tag{e}$$

Substituting y_0 from eq. (9.3) gives us

$$Q = K(\sqrt{H^2 + d^2} - d)L \qquad [L^3T^{-1}] \tag{9.4}$$

Where vertical and horizontal permeabilities differ, a transformed section of the embankment is used. All horizontal distances are distorted by a factor of $\sqrt{K_V/K_H}$, and a mean permeability value of $\sqrt{K_V K_H}$ is used for K in eq. (9.4).

Example 9.4

A homogeneous earthen dam has a top width of 30 ft and a height of 100 ft with a freeboard of 10 ft. The side slopes are 1 (vertical) : 2.5 (horizontal). It has a horizontal drainage blanket at the base that extends from the downstream toe to a distance of 100 ft. The embankment has a permeability of 1.5×10^{-5} ft/sec. Determine the seepage through the dam.

Solution

$H = 90$ ft

$CB = 90(2.5) = 225$ ft

$AB = 0.3(225) = 67.5$ ft and $CA = 157.5$ ft

$d =$ base length $-$ blanket length $- CA$

$\quad = 530 - 100 - 157.5 = 272.5$ ft

From eq. (9.3),

$$y_0 = \sqrt{90^2 + 272.5^2} - 272.5 = 14.5 \text{ ft}$$

From eq. (9.4),

$$q = (1.5 \times 10^{-5})(14.5)(1) = 2.18 \times 10^{-4} \text{ cfs}$$

$$\text{or } 18.8 \text{ ft}^3/\text{day per foot of dam.}$$

Seepage in composite section.   Another method used to prevent the seepage lines from emerging on the downstream face is to use an impervious core. This reduces the amount of seepage substantially. With relatively pervious shoulders, most of the vertical drop in the seepage line occurs in the impervious core, as shown in Figure 9.9. Since the outer shell material is several hundred times as pervious as the center core, the shell on the upstream side will have little effect on the position of the seepage lines. The small amount of water that seeps through the central core will flow in the lower layer of the downstream shell (rising only slightly above the tailwater level) to emerge at the toe, where a gravel and rock filter is provided.

Thus for seepage, only the central core section may be considered, through which seepage lines may be drawn. The central portion of Figure 9.9 has been replotted along with the phreatic line in Figure 9.10. As previously, the seepage line is a parabola. However, when the seepage line meets the discharge face at an angle α other than 180° (emerges in other than a vertical direction), it diverges from the parabola (point D_0) near the lower end to emerge tangent (at point D) to the discharge

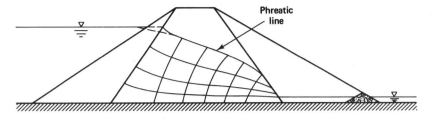

Figure 9.9 Seepage net in a dam of composite section.

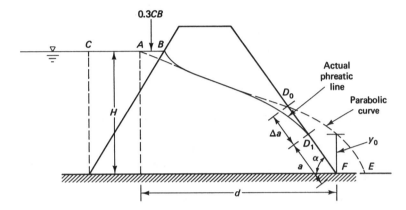

Figure 9.10 Central core portion of a composite section.

face. As the angle increases, the divergence from a true parabola becomes smaller. The horizontal drainage blanket in the preceding case corresponds to $\alpha = 180°$. From the property of a parabola, the distance to any point on the parabola from focus F, and thus FD_0, is given by

$$a + \Delta a = \frac{y_0}{1 - \cos \alpha} \qquad [\text{L}] \qquad (9.5)$$

The distance Δa varies with the slope angle α, becoming zero when $\alpha = 180°$ (Cassagrande, 1937). According to an empirical formula,

$$\Delta a = (a + \Delta a)\left(0.5 - \frac{\alpha}{360°}\right) \qquad [\text{L}] \qquad (9.6)$$

Combining eqs. (9.5) and (9.6) yields

$$a = \frac{y_0}{1 - \cos \alpha}\left(0.5 + \frac{\alpha}{360°}\right) \qquad [\text{L}] \qquad (9.7)$$

Another alternative approximate formula for a is

$$a = \sqrt{H^2 + d^2} - \sqrt{d^2 - H^2 \cot^2 \alpha} \qquad [\text{L}] \qquad (9.8)$$

The location of the breakout point, D, is determined by eq. (9.7) or (9.8), and a short transition curve from D to the base parabola is drawn by a free hand. This is the upper boundary of the flow net. The net is prepared, as discussed earlier, to determine the seepage.

For an approximate solution without reference to the flow net, apply Darcy's law at a section through D. The depth of the section is $a \sin \alpha$. The slope of the phreatic line is slightly flatter than the slope of the downstream face or approximately $\sin \alpha$ for α of less than 60°. Thus, for <60°, by Darcy's law,

$$Q = KaL \sin^2\alpha \qquad [L^3T^{-1}] \qquad (9.9)$$

where

$$a = \text{breakout point from eq. (9.7) or (9.8)}$$
$$L = \text{length of dam}$$
$$\alpha = \text{angle of downstream face}$$

For $\alpha > 60°$, eq. (9.4) can be used with sufficient accuracy.

Example 9.5

In the dam of Example 9.4, a center core of permeability 3×10^{-6} ft/sec in the horizontal direction and 1×10^{-6} ft/sec in the vertical direction is inserted. (There is no drainage blanket.) The core is symmetrical about the center of the dam. It has a top width of 30 ft and the two faces are at an angle of 40° from horizontal. Determine the seepage through the dam.

Solution

1. The section of the dam is shown in Figure 9.11(a).
2. The horizontal distances for the core to be modified by $\sqrt{K_V/K_H} = \sqrt{1/3} = 0.58$. The distorted core section is shown in Figure 9.11(b).
3. Revised slope angle $\tan \alpha = \dfrac{100}{69.1} = 1.45$ or $\alpha = 55.35°$.
4. $CB = \dfrac{90}{\tan 55.35} = 62.2$ ft, $CA = 43.5$ ft

$$d = (69.1 + 17.4 + 69.1) - 43.5 = 112.2 \text{ ft}$$
$$y_0 = \sqrt{(112.1)^2 + (90)^2} - 112.1 = 31.66 \text{ ft}$$
$$a = \dfrac{31.66}{1 - \cos 55.35}\left(0.5 + \dfrac{55.35}{360}\right) = 48 \text{ ft}$$
$$K = \sqrt{K_V K_H} = \sqrt{(3 \times 10^{-6})(1 \times 10^{-6})} = 1.73 \times 10^{-6} \text{ ft/sec}$$
$$q = (1.73 \times 10^{-6})(48) \sin^2 55.35 = 5.6 \times 10^{-5} \text{ cfs}$$

$$\text{or } 4.86 \text{ ft}^3/\text{day per foot.}$$

9.7.2 Seepage under the Dam

Instead of the impervious foundation assumed in the preceding section, when the dam is located on a relatively pervious material such as alluvial sand or gravel, seepage occurs underneath the dam, as shown in Figure 9.12(a). If the upward seepage pressure of water near the toe is greater than the effective weight of the soil, the surface of the soil will rise at a point of least resistance, and water and soil will start flow-

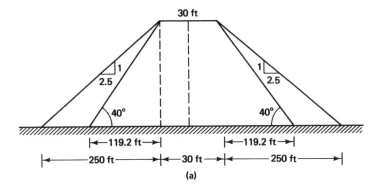

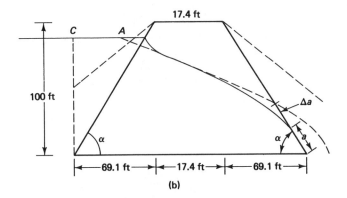

Figure 9.11 Composite anisotropic dam section.

ing away from the dam. This is known as piping and can result in the sliding of the toe or the settling of the whole dam.

The submerged unit weight of soil is given by

$$\gamma_{sub} = \frac{G_s - 1}{1 + e}\gamma_w \qquad [FL^{-3}] \qquad (9.10)$$

where

$$G_s = \text{specific gravity of soil}$$

$$e = \text{void ratio}$$

For a seepage line at a hydraulic gradient, i, the upward seepage force per unit volume is $i\gamma_w$. When the two forces are in balance,

$$i = \frac{G_s - 1}{i + e} \qquad [\text{dimensionless}] \qquad (9.11)$$

This is known as the critical gradient. Typically, the right side of eq. (9.11) has a value of unity. A gradient of slightly higher than unit value will cause piping or sand particles to be in an unstable condition known as the quicksand. The actual gradient at the downstream end of the dam is evaluated from the flow net by dividing the head

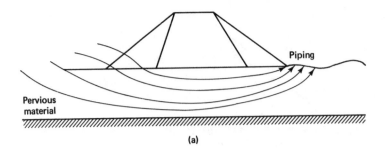

(a)

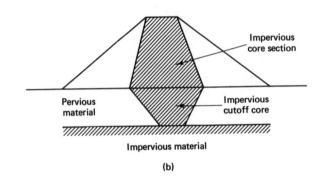

(b)

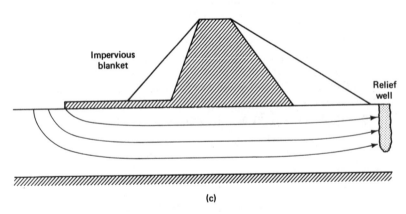

(c)

Figure 9.12 (a) Piping phenomenon in dam; (b) cutoff trench on pervious foundation; (c) impervious blanket to relieve pressure.

difference between the last two potential lines by the distance between these potential lines. This should be less than unity.

In an empirical approach, the creep ratio, L/H, is computed; here L is the length along the surface of contact between the soil and the base of the structure. This ratio is kept sufficiently large (4 for gravel and 18 for fine sand and silt) to prevent piping (Terzaghi and Peck, 1967).

When the pervious layer in foundation is thin, one way to prevent the danger of piping is to excavate a trench to the depth of the impervious layer and to extend the center core of the dam into this trench, as shown in Figure 9.12b. Alternatively, a row of sheet piles can be driven under the dam; or a combination of both is adopted.

When the pervious layer under the dam is too thick, so that a trench or pile cannot reach the impervious stratum, the upward pressure or exit gradient of the underneath seepage can be reduced by lengthening the seepage path. For this purpose, an impervious blanket is provided in front of the embankment, as shown in Figure 9.12c. If the seepage pressure is still high despite the blanket, relief wells are also provided as shown. The approximate quantity of seepage under the dam (through the foundation) can be obtained by the application of Darcy's law.

Example 9.6

The dam in Example 9.5 is located on a pervious stratum of 40 ft depth, having a permeability of 1.0×10^{-5} ft/sec. Determine the seepage through the dam. Assume that the average length of the seepage path is equal to the base of the dam.

Solution

$$L = 530 \text{ ft}$$

$$i = \frac{H}{L} = \frac{90}{530} = 0.17$$

$$q = (1.0 \times 10^{-5})(0.17)(40) = 6.8 \times 10^{-5} \text{ cfs}$$

$$\text{or } 5.88 \text{ ft}^3/\text{day per foot.}$$

9.7.3 Structural Stability of Earth Dam

Possible failure of an earth embankment can occur by a slide on the slope of the dam. A Swedish National Commission concluded in 1922 that in most of the slides that actually occurred, the line of failure approached the circumference of a circle. The approach of analysis consists of locating a segment of circle through the slope, or through the slope and foundation, and taking moments about the center of all the forces on the segment. The forces comprise the weight of the segment of the earth and the shear force along the circumference. The ratio of resisting moment to overturning moment is the factor of safety. The analysis is repeated until the most dangerous circle is found. The subject of slope stability is covered in textbooks on soil mechanics (e.g., McCarthy, 1982, pp. 446–493).

9.8 CONCRETE AND MASONRY GRAVITY DAMS

The following forces that act on a dam section have to be considered in the analysis and design of gravity dams: (1) weight of the dam, (2) hydrostatic force, (3) ice force, (4) uplift force, (5) earthquake force, and (6) the reaction of the first five forces. The line of action of these forces is shown in Figure 9.13, with the circled number referring to the force number.

A dam segment of unit length (perpendicular to the paper) is considered in analysis and design. The force due to the weight of the dam acting through the centroid of the section is the cross-sectional area of the dam section multiplied by the specific weight of the concrete or masonry.

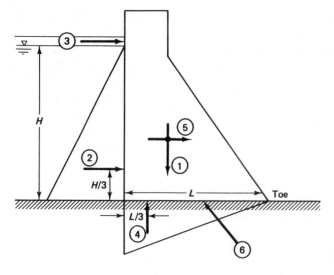

Figure 9.13 Forces acting on a gravity dam.

It is also relatively simple to determine the hydrostatic force upon the face of the dam. The pressure diagram is a triangle of base width $\gamma_w H$, where H is the water depth. The total force is the area of the triangle (i.e., $\frac{1}{2}\gamma_w H^2$) that acts horizontally at a height of $H/3$ from the base. When the lower part of the upstream face is slanted, the weight of the column of water resting on the surface is included as a force, with its action line from the centroid of the mass of the water column on the face.

The quantitative determination of the ice force is uncertain. The ice sheet expands or contracts with change of temperature. A slow rise of temperature, and hence pressure, causes the ice to deform, thus distributing the pressure. This is less effective in terms of the buildup of force as compared to a sudden rise of temperature. When there is no opportunity for the ice sheet to expand, a dam section experiences a greater force. Also, the force is greater when the ice sheet becomes thicker. However, a thickness beyond 2 to 3 feet may not be expected. The ice force is assumed arbitrarily based on practice and experience. The assumed force varies from 10,000 lb per linear foot in Canada to 60,000 lb per linear foot of dam in the northern United States. Its line of action is from the center of thickness of the ice sheet. The ice sheet force can be eliminated by attaching a horizontal perforated steel pipe to the face of the dam near the bottom from which warm air is blown. It is a matter of economics as to whether a bubbler system is more feasible than the added cost to withstand the ice force.

The uplift force is due to the pressure of water acting upward from the pores in the foundation. Given sufficient time, even a dam on solid rock will be subject to an uplift, which equals the water head at the upstream and downstream faces and with a linear variation between in the form of a trapezoid. The total uplift force is given by the area of the trapezoid, and the resultant acts through its centroid. The uplift force can be reduced by (1) constructing a grout curtain wall underneath the dam near the upstream face and (2) drilling a system of drainage holes from a gallery behind the cutoff wall. For preliminary design, full uplift pressures equivalent to headwater and tailwater levels are assumed.

The earthquake causes acceleration of the earth crust along faults. The dam should move with the sudden movement of the earth crust if a rupture is to be avoided. The acceleration of the mass of dead load (weight of dam) introduces an inertia force given by

$$P_e = Ma = \frac{W}{g}\alpha g = W\alpha \qquad [F] \qquad (9.12)$$

where

W = weight of dam

α = ratio of earthquake horizontal acceleration to the acceleration due to gravity (g).

The earthquake force, P_e, acts horizontally through the center of gravity of the dam section.

The most severe earthquake ever recorded has been on the order of $\alpha = 0.5$. A severity of 0.1 is considered disastrous. Most dams in seismically active regions are designed with $\alpha = 0.1$.

During an earthquake the inertia force is also caused by the acceleration of the mass of water in the reservoir. For increased hydro pressure, elliptical and parabolic distributions have been suggested (Hinds et al., 1945). In a simplistic approach, the earthquake acceleration, α, is increased by 50%, to 0.15, and the force is still applied at the dam section.

Earthquake movement may take place in any direction. A horizontal acceleration in the upstream direction is most critical when the inertia force acts downstream. An upward acceleration opposes the acceleration of gravity. This momentarily reduces the weight of the concrete and the water in the same ratio, thus not affecting the position of the resultant. However, the stresses are changed slightly. The deviation of the earthquake force from the horizontal is ignored because of uncertainties in the value of α.

Three basic rules of statics are applied to determine the reaction of the foundation on the dam: Σ vertical force = 0 and Σ horizontal force = 0 provide the magnitude of the vertical and horizontal components of the reaction, and Σ moment = 0 with respect to toe or heel provides the location of the reaction from that point.

Example 9.7

For the concrete dam section shown in Figure 9.14, determine the magnitude of all forces with their line of actions. The dam is located in seismically active regions of earthquake acceleration of $0.1g$.

Solution   Computations are given in Table 9.5.

1. Quantity in column 2 = volume per foot = area of indicated dam section or force diagram.

2. Force in column 3 = specific weight × quantity in column 2.

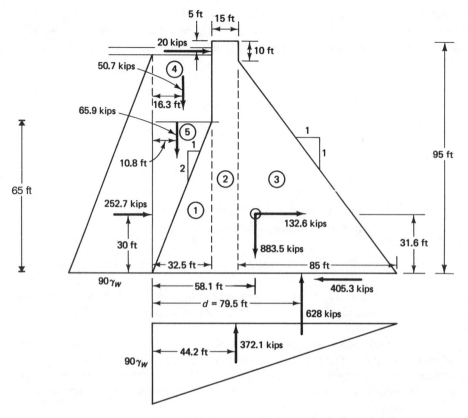

Figure 9.14 Forces on the dam section in Example 9.7.

3. For forces broken down into elements, the centroid of

$$\text{total force} = \frac{\sum(\text{element force, col. 3})(\text{CG of element, col. 4})}{\sum(\text{element force, col. 3})}$$

4. Seismic force $= \alpha W = 0.15(\text{weight of section})$.
5. Computation of reactions:

$$\sum V = 0: \quad \uparrow V + 372.1 - 116.6 - 883.5 = 0, \quad \uparrow V = 628 \text{ kips}$$

$$\sum H = 0: \quad \overleftarrow{H} - 20 - 252.7 - 132.6 = 0, \quad \overleftarrow{H} = 405.3 \text{ kips}$$

$$\sum M \text{ at heel} = 0:$$
$$628(\overset{\frown}{d}) + 372.1(\overset{\frown}{44.2}) - 883.5(\overset{\frown}{58.1}) - 132.6(\overset{\frown}{31.6})$$
$$- 20(\overset{\frown}{89}) - 252.7(\overset{\frown}{30}) - 116.6(\overset{\frown}{13.2}) = 0$$

TABLE 9.5 ANALYSIS OF FORCES ON A GRAVITY DAM[a]

(1) Description	(2) Quantity (ft^3)	(3) Force (kips)	(4) CG from Heel (ft)
Weight of dam			
Section 1	1056.3	153.1	21.7
Section 2	1425.0	206.6	40.0
Section 3	3612.5	523.8	75.8
Total weight of dam		883.5	58.1
Hydrostatic force			
Horizontal force	4050	252.7	30
Vertical force			
Section 4	812.5	50.7	16.3
Section 5	1056.3	65.9	10.8
Total vertical hydrostatic		116.6	13.2
Ice force (assumed)		20.0	89
Uplift force	5962.5	372.1	44.2
Seismic force (horizontal)			
Section 1		23.0[b]	21.7
Section 2		31.0	47.5
Section 3		78.6	28.3
Total Seismic force		132.6	31.6
Reaction			
Vertical		628	79.5
Horizontal		405.3	0

[a]For a section of dam, 1 ft long.
[b]Force $= \alpha \cdot$ weight of section $= 0.15W$.

Therefore,

$$d = 79.5 \text{ ft}$$

Resultant reaction,

$$R = \sqrt{(628)^2 + (405.3)^2} = 747.4 \text{ kips}$$

9.8.1 Requirements for Stability

There are three ways in which a gravity dam may fail:

1. By sliding on a horizontal, or nearly horizontal, plane (a) above the foundation or (b) along the foundation
2. By overturning on a horizontal plane (a) within the dam or (b) at the base
3. By the compressive stresses in the dam or in the foundation exceeding the permissible values

    The methods of computation and the safety requirements of these are discussed below.

9.8.2 Resistance to Sliding

Resistance to sliding at any joint or at the base is offered by a combination of static friction induced by the resultant of the vertical forces and the shearing strength of the joint or the base. For all conditions of loading, this resistance must exceed the total horizontal force above the joint by a factor of safety of 4 for usual loading conditions and 1.3 for an extreme loading combination. This relation can be expressed as

$$(SF)_S = \frac{\mu V + f_v A}{H} \qquad \text{[dimensionless]} \qquad (9.13)$$

where

$(SF)_S$ = factor of safety against sliding

μ = coefficient of friction between the materials on each side of the joint or the base (usually between 0.5 and 0.75)

V = resultant of vertical forces

H = resultant of horizontal forces

f_v = average shear stress of the material

A = area of the joint or base

The distribution of shear along a horizontal plane is from about zero at the heel to a maximum value near the toe. The average shear is roughly one-half of the maximum shear strength. The strength of concrete/masonry is considered for the joint and of the foundation rock for the analysis at the base.

In the past, the practice has been to make another check for the ratio of the horizontal reaction to the vertical reaction without uplift to be less than the coefficient of internal friction.

Example 9.8

For Example 9.7, determine the factor of safety against sliding at the base. Assume a coefficient of internal friction of 0.65 and the shear strength of the foundation of 400 psi.

Solution

1. From Example 9.7: V = 628 kips, H = 405.3 kips, A = (base width) (1 ft length of dam) = 132.5 ft^2.
2. $f_v = \frac{1}{2}$ ultimate strength = 200 psi or 28.8 kips/ft^2.
3. From eq. (9.13),

$$(SF)_S = \frac{0.65(628) + 28.8(132.5)}{405.3} = 10.4$$

9.8.3 Overturning Stability

The dam will overturn if the resultant of all horizontal and vertical forces acting on the dam above any horizontal plane, including uplift, passes outside the limits of the dam section. When the resultant intersects the downstream face, overturning may be prevented, but tensile stress is created at the upstream face. For concrete and ma-

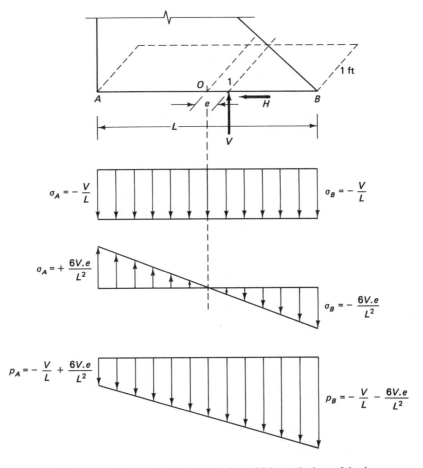

Figure 9.15 Stress distribution along a horizontal joint or the base of the dam.

sonry, the tensile strength is always considered unreliable and is disregarded. The location of the resultant should be such that no tension exists in any joint or base.

In Figure 9.15, point 0 coincides with the center of gravity of the dam section of 1 ft length (perpendicular to the paper) at any joint or base. Point 1 is the location of the resultant of forces. Per foot length of the dam of a rectangular base: $A = L$, $I = L^3/12$, $M = Ve$, distance of centroid to extreme fiber, $c = 0.5L$. Using a negative sign for compression gives us

$$\text{axial stress} = \frac{-V}{A} = \frac{-V}{L} \tag{a}$$

$$\text{flexural stress} = \pm \frac{Mc}{I} = \pm \frac{6Ve}{L^2} \tag{b}$$

combining yields

$$p_A = \frac{-V}{L}\left(1 - \frac{6e}{L}\right) \qquad [\text{FL}^{-2}] \tag{9.14a}$$

$$p_B = \frac{-V}{L}\left(1 + \frac{6e}{L}\right) \qquad [\text{FL}^{-2}] \qquad (9.14b)$$

where V is the vertical component of reaction.

Equation (9.14) provides the vertical stresses (bearing pressures) at the heel and toe of a dam section. If the entire section has to be free of tensile stress, neither p_A nor p_B should be positive, which means that e should not exceed $\frac{1}{6}L$ [eq. (9.14a)] and also that e should not exceed $-\frac{1}{6}L$ [eq. (9.14b)] (i.e., the resultant for all conditions of loading shall fall within the middle third of the base). This is known as the law of the middle third.

Example 9.9

For the dam of Example 9.7, show that it is safe from overturning and that the base will be free of tension. Determine the vertical stresses at the base.

Solution

1. When the resultant of all forces intersects within the base, it is safe against overturning. There will be no tension if the middle-third law is satisfied.

2. $\dfrac{L}{3} = \dfrac{132.5}{3} = 44.2$ ft, $\dfrac{2L}{3} = 88.3$ ft. Location of reaction, $d = 79.5$ (i.e., between the middle third), $e = 79.5 - \dfrac{132.5}{2} = 13.25$ ft.

3. Bearing pressures, from eq. (9.14), in kips per square foot:

$$P_{\text{heel}} = -\frac{628}{132.5}\left[1 - \frac{6(13.25)}{132.5}\right] = -1.90 \text{ ksf (compression)}$$

$$P_{\text{toe}} = -\frac{628}{132.5}\left[1 + \frac{6(13.25)}{132.5}\right] = -7.58 \text{ ksf (compression)}$$

9.8.4 Governing Compressive Stresses

The vertical compressive stresses on a horizontal joint or on the base as given by eq. (9.14) are not the maximum stresses that occur in the structure. Since there is no stress on the face of a dam (when the reservoir is empty) or the hydrostatic stress acts normal (or nearly normal) to the face, that face and a plane perpendicular to it represent the principal planes. Maximum stress occurs on the principal plane normal to the face.

Determination of stresses in the body of a dam is complicated. Among the number of methods used, the simplest is the gravity method, which considers a dam to be composed of a series of vertical cantilever beams of 1 ft width, each acting independent of each other. The theory of principal stresses is used to determine the maximum stress. In Figure 9.16, if σ_x and σ_y denote the principal stresses, the normal and shear stress on any inclined plane ab can be given by (refer to a strength-of-materials text)

$$\sigma_n = \sigma_x \cos^2\phi + \sigma_y \sin^2\phi \qquad [\text{FL}^{-2}] \qquad (9.15a)$$

or

$$\sigma_x = \frac{\sigma_n}{\cos^2\phi} - \sigma_y \tan^2\phi \qquad [\text{FL}^{-2}] \qquad (9.15b)$$

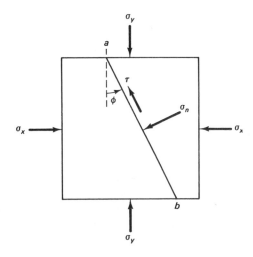

Figure 9.16 Stresses on an inclined plane due to principal stresses.

and

$$\tau = (\sigma_x - \sigma_y) \sin \phi \cos \phi \quad [\mathrm{FL}^{-2}] \tag{9.16}$$

Now refer to Figure 9.17; if the stress σ_n is represented by the vertical pressure, p, obtained from eq. (9.14), the stress σ_y corresponds to the external pressure on the face of the dam, p_E, and the stress σ_x is represented by the maximum normal stress, p_{max}, then by substitution in eq. (9.15b) and (9.16),

$$p_{max} = \frac{p}{\cos^2\phi} - p_E \tan^2\phi \quad [\mathrm{FL}^{-2}] \tag{9.17a}$$

$$\tau = (p - p_E) \tan \phi \quad [\mathrm{FL}^{-2}] \tag{9.17b}$$

For maximization, p_E may be taken as zero (empty reservoir condition); then the equations above simplify to

$$p_{max} = \frac{p}{\cos^2\phi} \quad [\mathrm{FL}^{-2}] \tag{9.18}$$

and

$$\tau = p \tan \phi \quad [\mathrm{FL}^{-2}] \tag{9.19}$$

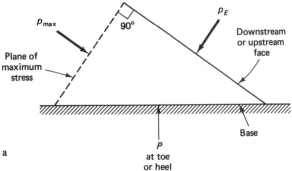

Figure 9.17 Stress element at the end of a joint or the base.

Sec. 9.8   **Concrete and Masonry Gravity Dams**   475

where

$$p = \text{vertical stress at toe or heel from eq. (9.14)}$$

$$\phi = \text{angle of the face from the vertical}$$

9.9 ARCH DAMS

An arch dam is a relatively thin concrete section curved upstream in plan. Unlike the gravity dam, in which the force of water is mainly held back by the weight of the dam, an arch dam obtains a large measure of stability by transmitting the water and other forces by arch action into the canyon walls. It is, as such, suitable to sites where the abutments are solid rocks capable of resisting arch thrust and the canyon is narrow.

There are two types of arch dams: (1) massive arch, where a single curved wall spans the full width between abutments, and (2) multiple arch, in which a number of (usually inclined) smaller arches are supported on piers and buttresses. The second category is classified into the buttress dam.

For massive arch dam design, the following theories are used in the computation of stresses:

1. *Cylinder theory.* This is the simplest theory, in which a dam is assumed to act similar to a thin cylinder. The stresses computed are only approximate. This theory is used for small dams in simple conditions and for preliminary studies.

2. *Elastic arch theory.* Cylinder theory cannot be applied to irregular arch forms and for special loading such as silt and earthquake forces, and also when temperature variations and abutment yielding are involved. Additional stresses due to shear and moment are created on the arch. The elastic theory gives a better idea of actual stress. By the theory of structures, each slice of dam is analyzed as an elastic arch section for normal thrust, shear force, and moment at the base of the arch (at the foundation with canyon wall) and at different cross sections of the arch. The combined stresses are computed therefrom.

3. *Composite system theory.* Elastic arch theory considers that each arch acts independent of its neighbors. Actually, adjacent arches restrain each other. As such, the distribution of stress in an arch dam is complex. In the frequently used trial-load method, the dam is assumed to be made up of two systems of elements: horizontal arches and vertical cantilevers. By successive trials, the total load is distributed between these two sets of elements in such a manner that the computed deflection for any point in the dam, considered as a point in the arch system, is identical with its computed deflection, considered as a point in the cantilever system. The U.S. Bureau of Reclamation (1977a) describes in detail the trial-load method and the finite element method of arch dam analysis.

Due to the complexity of the trial-and-error method, the arch dams have been analyzed by measuring the stresses on scale models. This at least serves as a check on design.

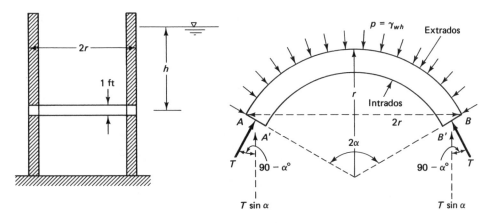

Figure 9.18 Hydrostatic pressure and the resulting thrust on a thin circular arc.

9.9.1 Design of Arch Dams

The design of a dam according to the cylinder theory is discussed below. Consider a slice of dam of 1 ft height at a depth of h feet below the reservoir water level, as shown in Figure 9.18. The hydrostatic pressure acting radially on the arch,

$$p = \gamma_w h \tag{a}$$

$$\text{total water force perpendicular to the transverse section } AB = (\gamma_w h)\,(2r \sin \alpha)\,(1) \tag{b}$$

Balancing the normal forces gives us

$$2T \sin \alpha = 2\gamma_w hr \sin \alpha \quad \text{or} \quad T = \gamma_w hr \tag{c}$$

For a small thickness, t, compared to the radius, r, the stress distribution on section AA and BB can be considered uniform, given by

$$f_c = \frac{T}{t} = \frac{\gamma_w hr}{t} \quad [\text{FL}^{-2}] \tag{9.20}$$

or

$$t = \frac{\gamma_w hr}{f_c} \quad [\text{L}] \tag{9.21}$$

where

t = thickness of the arch dam

γ_w = unit weight of water

h = depth below the reservoir level

r = radius of extrados (outside) of the arch

f_c = allowable compression stress on concrete (up to 1000 psi)

For a given span, the concrete volume of an arch section is minimum when

$$2\alpha = 133°34' \qquad (9.22)$$

or when

$$r = 0.544L \qquad [\text{L}] \qquad (9.23)$$

where

$$r = \text{outside, inside, or central radius}$$

$$L = \text{span for any of the radius selected above}$$

Topography seldom permits 2α to be greater than 150°. The design is controlled in the following three ways:

1. *Constant radius.* The radius is fixed according to an increased angle (say $2\alpha = 150°$) at the top span, so that the average angle for the entire dam height is nearly at the best angle of $133°34'$. The thickness at various depths is calculated by eq. (9.21), which varies as a straight line for the fixed radius.
2. *Constant angle.* The angle is fixed at the fixed value given by eq. (9.22). At various depths, the span length is determined and the radius r is computed by eq. (9.23). For this radius, the thickness, t, is computed by eq. (9.21). However, this produces overhang (lower arches project outside the upper ones on the upstream side).
3. *Variable radius.* As a compromise between the two methods above, the radius is kept fixed for a few sections and varied in others so as to reduce the overhang and also affect the volume savings.

9.10 SPILLWAYS

A spillway is a passageway to convey past the dam flood flows that cannot be contained in the allotted storage space or which are in excess of those turned into the diversion system. Ordinarily, a reservoir is operated to release the required quantity for usage through headworks or through outlets in the dam. Thus a spillway functions only infrequently, at times of floods or sustained high runoffs, when other facilities are inadequate. But its ample capacity is of prime importance for the safety of dams and other hydraulic structures. Determination of design flood for the spillway capacity which comprises the peak, volume, and variation or a hydrograph of flow is a problem of hydrology for which a reference is made to the study of flood flows (Chapter 8). After the hydrograph of spillway design flood is established, flood routing (Section 8.17) through a selected size and type of spillway provides the maximum reservoir water level. Estimates of various combinations of spillway discharge capacity and reservoir height for alternative sizes and types of spillways provide a basis for selection of the economical spillway. Since innumerable spillway arrangements could be considered, a judgment on the part of the designer is required to select only those alternatives that have adaptability to the site and show definite advantages.

Hydraulic aspects of spillway design extend beyond determination of inflow design flood and routing of flood. These relate to design of the three spillway components: control structure, discharge channel, and terminal structure. The control structure regulates outflows from the reservoir and may consist of a sill, weir section, orifice, tube, or pipe. Design problems relate to determining the shape of the section and computing discharge through the section. The flow released through the control structure is conveyed to the streambed below the dam in a discharge channel. This can be the downstream face of the overflow section, a tunnel excavated through an abutment, or an open channel along the ground surface. The channel dimensions are fixed by the hydraulics of channel flow. An estimate of the loss of energy through the channel section is also important to design the terminal structure. Terminal structures are energy-dissipating devices that are provided to return the flow to the river without serious scour or erosion at the toe of the dam. These comprise a hydraulic jump basin, a roller bucket, a sill block apron, or a basin with impact baffles and walls. The hydraulic aspects relating only to control structures are discussed in the chapter.

9.10.1 Types of Spillways

Spillways are classified according to their most prominent feature as it pertains to the control structure or to the discharge channel. The two most common types of spillways are the concrete overflow spillway associated with gravity dams and the chute spillway often used with earth dams. For dams in narrow canyons, the spillway inlets are placed upstream of the dam in the form of either a side channel or a shaft (morning glory) spillway. Spillways are usually referred to as controlled or uncontrolled, depending on whether or not they are equipped with gates.

9.11 OVERFLOW SPILLWAYS

This is a special form of a weir whose shape is made to conform closely to the profile of the lower nappe of a ventilated sheet of water falling from a sharp-crested weir. In high-overflow spillways, the velocity of approach is negligible,* whereas low spillways have appreciable velocity of approach, which affects both the shape of the crest and the discharge coefficients. In low spillways the spillway crest curve is continuous with the toe curve, forming an S-shape or ogee profile. However, the name "ogee spillways" is also applied to high spillways which have a straight tangent section between the crest curve and the toe curve.

9.11.1 Crest Shape of Overflow Spillways

The lower surface of a nappe from a sharp-crested weir is a function of (1) the head on the weir, (2) the slope or inclination of the weir face, and (3) the height of the crest, which influences the velocity of approach.

On the crest shape based on a design head H_d, when the actual head is less than H_d, the trajectory of the nappe falls below the crest profile, creating positive pressures on the crest, thereby reducing the discharge (coefficient). On the other hand, with a

---

*The effect of the velocity of approach is negligible for a ratio of crest height to head on a weir greater than 1.33.

higher-than-design head, the nappe trajectory is higher than the crest, which creates negative pressure pockets and results in increased discharge. Accordingly, it is considered desirable to underdesign the crest shape of a high overflow spillway for a design head, H_d, less than the head on the crest corresponding to the maximum reservoir level, H_e. However, with too much negative pressure, the cavitation may occur. The previous recommendation (U.S. Bureau of Reclamation, 1977c) has been that H_e/H_d should not exceed 1.33. Vacuum tank observations have, subsequently, indicated that cavitation on the crest would be incipient at an average pressure of about -25 ft of water. The Corps of Engineers (COE) has, accordingly, recommended (U.S. Department of the Army, 1986) that a spillway crest be designed so that the maximum expected head will result in an average pressure on the crest no lower than -15 ft of water. Based on model studies by Murphy in 1970 and Maynord in 1985, design curves have been prepared that show a relationship between H_e and H_d for a pressure of -15 ft and -20 ft of water on the crest. The curves, corresponding to -15 ft pressure, can be approximated by the following equations (Reese and Maynord, 1987):

$$H_d = 0.33H_e^{1.22} \qquad \text{(without piers)} \qquad (9.24a)$$

$$H_d = 0.30H_e^{1.26} \qquad \text{(with piers)} \qquad (9.24b)$$

where

H_e = maximum total head on the crest, ft

H_d = design head to be adopted, ft

Equations (9.24a) and (9.24b) apply over a limit. When the total head on the spillway, H_e, is less than about 30 ft, H_e/H_d should be set to a value of 1.42 for the crest without piers and to 1.35 with piers in place of the above equations.

Numerous crest profiles have been proposed by various investigators.* The U.S. Bureau of Reclamation played a leading role in investigations of the shape of the nappe. The Bureau described the complete shape of the lower nappe by separating it into two quadrants, one upstream and one downstream from the crest axis (apex), as shown in Figure 9.19. The equation for the downstream quadrant is expressed as

$$\frac{y}{H_d} = \frac{1}{K}\left(\frac{x}{H_d}\right)^n \qquad \text{[dimensionless]} \qquad (9.25a)$$

where

H_d = design head excluding the velocity of approach head

x, y = coordinates of the crest profile, with the origin at the highest point (O), as shown in Figure 9.19

K, n = constants that depend on upstream inclination and velocity of approach

Murphy (1973) suggested that n can be set equal to 1.85 for all cases and K can be varied from 2.0 for a deep approach to 2.2 for a very shallow approach, as shown in Figure 9.20(a).

*For an account of various profiles, refer to Grzywienski (1951).

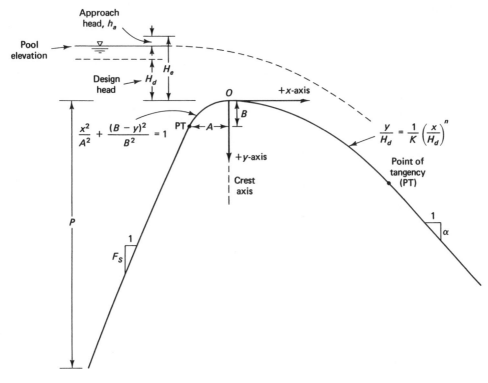

Figure 9.19 Definition sketch of overflow spillway section.

In a high-overflow section, the crest profile merges with the straight downstream section of slope α, as shown in Figure 9.19 (i.e., $dy/dx = \alpha$). Differentiation of eq. (9.25a) and expressing that in terms of x yield the distance to the position of downstream tangent as follows:

$$\frac{X_{DT}}{H_d} = 0.485(K\alpha)^{1.176} \qquad \text{[dimensionless]} \qquad (9.25b)$$

where

X_{DT} = horizontal distance from the apex to the downstream tangent point

α = slope of the downstream face

The discharge efficiency of a spillway is highly dependent on the curvature of the crest immediately upstream of the apex. To fit a single equation to the upstream quadrant had proven more difficult. Many compound curves have been proposed, including the tricompound circle for which the U.S. Bureau of Reclamation provided a tabular solution (1977c). Investigations by Murphy (1973), as confirmed on model studies by Maynord (Reese and Maynord, 1987), suggested that an ellipse, of which both the major and minor axes vary systematically with the depth of approach, can closely approximate the lower nappe surfaces generated by the U.S. Bureau of Reclamation method. Furthermore, any sloping upstream face could be used with little loss of accuracy if the slope became tangent to the ellipse calculated for a vertical upstream face.

Sec. 9.11   Overflow Spillways   **481**

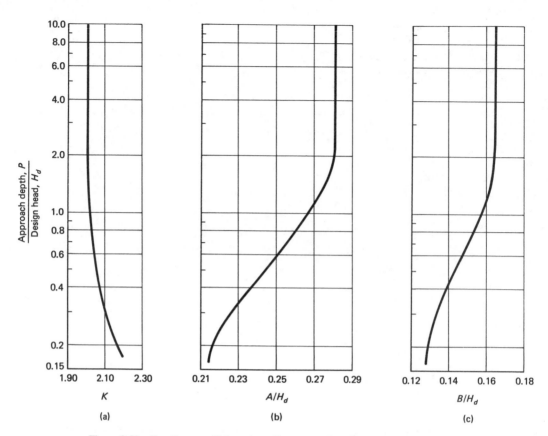

Figure 9.20 Coordinate coefficients for spillway crest (from U.S. Department of the Army, 1986).

With respect to origin at the apex, the equation of the elliptical shape for upstream quadrant is expressed as

$$\frac{x^2}{A^2} + \frac{(B-y)^2}{B^2} = 1 \qquad \text{[dimensionless]} \qquad (9.26a)$$

where

x = horizontal coordinate, positive to the right

y = vertical coordinate, positive downward

A, B = one-half of the ellipse axes, as given in Figure 9.20(b) and (c) for various values of approach depth and design head

For an inclined upstream face of slope F_s, the point of tangency with the elliptical shape can be determined by the following equation, obtained by differentiation of eq. (9.26a):

$$X_{\text{UT}} = \frac{A^2 F_s}{(A^2 F_s^2 + B^2)^{1/2}} \qquad \text{[L]} \qquad (9.26b)$$

where

$$X_{UT} = \text{horizontal distance of upstream tangent point}$$
$$F_s = \text{slope of upstream face}$$

9.11.2 Discharge for Overflow Spillways

The following equation of flow through a weir with a consolidated coefficient, as derived in Section 6.12.2, applies to an overflow spillway (in FPS units).

$$Q = CL_e H_e^{3/2} \qquad [L^3T^{-1}] \qquad (9.27a)$$

where

$Q = $ discharge

$C = $ variable coefficient of discharge

$L_e = $ effective length of crest

$H_e = $ total head, including velocity of approach, h_a (Figure 9.19)

Where crest piers and abutments cause side contractions of the overflow, the effective length is less than the crest length, as follows:

$$L_e = L - 2(NK_p + K_a)H_e \qquad [L] \qquad (9.27b)$$

where

$L_e = $ effective length

$L = $ length of crest

$N = $ number of piers

$K_p = $ pier contraction coefficient; $K_p = 0.02$ for square-nose piers and 0 for pointed-nose piers (for details, see U.S. Department of the Army, 1986)

$K_a = $ abutment contraction coefficient; $K_a = 0.2$ for square abutment and 0 for rounded abutment of radius greater than one-half of the total head (for details, see U.S. Department of the Army, 1986)

The effect of the velocity of approach is negligible when the height of the crest is greater than 1.33 times the design head, H_d.

The discharge coefficient of overflow spillways is influenced by a number of factors. These include (1) crest height-to-head ratio or the velocity of approach, (2) the actual head being different from the design head, (3) the upstream face slope, and (4) the downstream submergence. When a spillway is being underdesigned, three graphs (Figures 9.21 through 9.23) developed by the U.S. Bureau of Reclamation (1977c) may be used to assess the coefficient of discharge, as follows:

1. Figure 9.21 provides the basic coefficient for the case when the crest pressures are essentially atmospheric in a vertical-faced spillway. For the known ratio of the crest height to the design head, P/H_d, the basic coefficient, C_o, is determined from this figure.

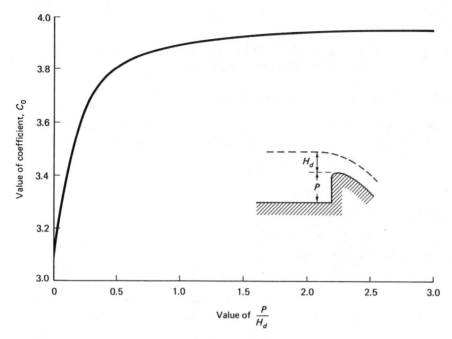

Figure 9.21 Basic discharge coefficient for vertical-faced section with the atmospheric pressure on the crest (U.S. Bureau of Reclamation, 1977c).

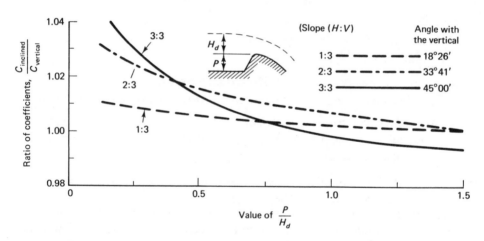

Figure 9.22 Correction factor for slopping upstream face (from U.S. Bureau of Reclamation, 1977c).

2. From Figure 9.22, the correction factor is determined for spillways having a sloping upstream face.
3. Figure 9.23 indicates the correction factor for the actual head being different than the design head. For the ratio of maximum pool level to design head (H_e/H_d), a factor is determined from this figure.

Storage and Control Structures   Chap. 9

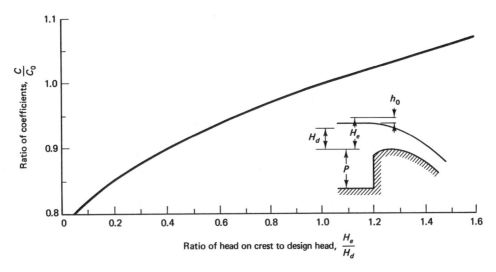

Figure 9.23 Correction factor for other than the design head (from U.S. Bureau of Reclamation, 1977c).

The basic coefficient multiplied by the two factors provide the corrected value of the coefficient in eq. (9.27a). The coefficient of discharge decreases under the condition of submergence.

Example 9.10

Design an overflow spillway section for a design discharge of 50,000 cfs. The upstream water surface is at El. 800 and the channel floor is 680. The spillway, having a vertical face, is 180 ft long.

Solution

1. Assume a high overflow spillway section ($C = 3.95$).
2. From the discharge equation,

$$H_e^{3/2} = \frac{Q}{CL} = \frac{50,000}{3.95(180)} = 70.32 \quad \text{or} \quad H_e = 17.1$$

3. Depth of water upstream $= 800 - 680 = 120$ ft.

$$\text{Velocity of approach, } v_0 = \frac{50,000}{120(180)} = 2.31 \text{ ft/sec}$$

$$\text{Velocity head} = \frac{(2.31)^2}{2(32.2)} = 0.08 \text{ ft}$$

4. Maximum water head $= 17.1 - 0.08 = 17.0$ ft.
5. Height of crest, $P = 120 - 17.0 = 103$ ft.
6. Since $H_e < 30$ ft, design head, $H_d = \dfrac{17.1}{1.42} = 12.0$ ft.

7. $\dfrac{P}{H_d} = \dfrac{103}{12} = 8.58 > 1.33$, high overflow section.

8. *Downstream quadrant:* From eq. (9.25a) of the crest shape,

$$\frac{y}{12} = \frac{1}{2}\left(\frac{x}{12}\right)^{1.85}$$

or

$$y = 0.06X^{1.85}$$

Coordinates of the shape computed by this equation are as follows:

x (select) (ft)	y (computed) (ft)
5	1.18
10	4.25
15	9.00
20	15.31
30	32.42

9. *Point of tangency:* Assume a downstream slope of 2:1, from eq. (9.25b).

$$X_{DT} = 0.485[2(2)]^{1.176}(12) = 30 \text{ ft}$$

10. *Upstream quadrant:* From Figure 9.20(b) and (c), $A/H_d = 0.28$, $B/H_d = 0.165$.

$$A = 0.28(12) = 3.36 \qquad B = 0.165(12) = 2.00 \text{ ft}$$

From eq. (9.26a),

$$\frac{x^2}{(3.36)^2} + \frac{(2.0 - y)^2}{(2.0)^2} = 1$$

The coordinates are computed as follows:

x (select) (ft)	y (computed) (ft)
1.0	0.09
2.0	0.39
3.0	1.10
3.36	2.00

11. The crest shape has been plotted in Figure 9.24.

9.12 CHUTE OR TROUGH SPILLWAYS

Chute spillway derives its name from the shape of the discharge channel component of the spillway. In this type of spillway, the discharge is conveyed from a reservoir to the downstream river level through a steep open channel placed either along the dam abutment or through a saddle. The designation of the channel as a chute implies that the velocity of flow is greater than critical. This name of the spillway applies regardless of the control device used at the head of the chute, which can be an overflow crest, a gated orifice (acting as a large orifice), or a side-channel crest. This type of structure consists of four parts as shown in Figure 9.25: an entrance channel, a control

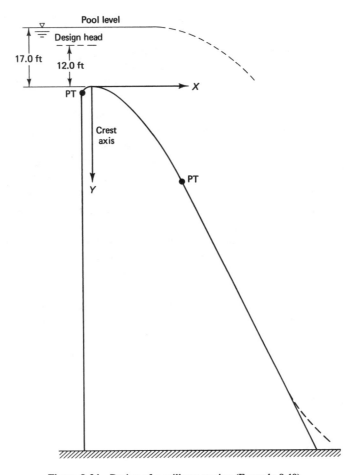

Figure 9.24 Design of a spillway section (Example 9.10).

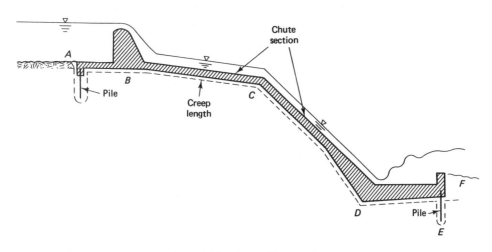

Figure 9.25 Chute spillway section.

structure or crest, the sloping chute, and a terminal structure. The entrance channel at *A* is a relatively wide channel of subcritical flow. More often, the axis of the entrance channel is curved to fit the alignment to the topography, because of the low-energy losses in the approach channel. The control section at *B* is placed in line with or upstream from the centerline of the dam. The critical velocity occurs when the water passes over the control.

Flows in the chute are ordinarily maintained at supercritical stage until the terminal structure *DE* is reached. Economy of excavation generally makes it desirable that from *B* to *C*, where a heavy cut is involved, the chute may be placed on a light slope. From *C* to *D* it follows the steep slope on the side of the river valley. An energy-dissipating device is placed at the bottom of the valley at *D*. The axis of the chute is kept straight as far as practicable; otherwise, the floor has to be superelevated to avoid the piling up of the high-velocity flow around the curvature. The velocity of flow increases rapidly in the chute with drop in elevation. The curvature should be confined to the upper reaches, where the velocity is comparatively low. In the lower reaches, the alignment can be curved only if the chute floor and walls are shaped adequately to force water into a turn without overtopping the walls.

It is preferable that the width of the control section, the chute, and the stilling basins are the same. Quite often, these widths are not the same, because of the design requirements of the spillway and stilling basin. Extreme care must be taken that the transitions take place very gradually, or undesirable standing waves may develop.

To prevent hydrostatic uplift under the chute, a cutoff wall is provided under the control structure and a drainage system of filters and pipes is provided, as shown in Figure 9.25. When the stilling basin is operating, there is a substantial uplift under the lower part of the chute and upstream part of the stilling basin floor. The floor must be made sufficiently heavy or anchored to the foundation.

9.12.1 Slope of Chute Channel

It is important that the slope of the chute in the upstream section *BC* should be sufficiently steep to maintain a supercritical flow to avoid formation of a hydraulic jump in the chute. Flow through a channel is given by the following Manning's formula, derived in Chapter 10:

$$Q = \frac{1.486}{n} AR^{2/3} S^{1/2} \tag{9.28}$$

where

Q = discharge, cfs

n = roughness coefficient given in Table 10.4

A = cross-section area of channel, ft^2

R = hydraulic radius, A/P, ft

S = slope of bottom or water surface of channel

For a rectangular channel, $A = by$ and $R \approx y$ for a wide section. Under the condition of critical flow, $y_c = (q^2/g)^{1/3}$, where $q = Q/b$ (i.e., discharge per foot width). Thus eq. (9.28) reduces to*

$$S_c = \frac{21.3n^2}{q_c^{0.222}} \quad \text{[dimensionless]} \quad (9.29)$$

The subscript c refers to the critical stage. Since the reliable information on the value of n is difficult, a conservatism is indicated in the selection of n. The slope of the chute should be more than S_c for a supercritical flow. As the spillway must function properly from small to very large discharges, the critical slope has to be investigated for the entire range of flow. The required S_c from eq. (9.29) is normally a fraction of 1%.

A review of existing spillways indicate that the actual slopes of the upstream section of the chute are 1 to 2% or more. Research indicated that an unstable rapid flow occurs when the Froude number exceeds 1.56 to 1.64. It is therefore likely that chutes designed with a conservative slope have a Froude number well over this limit and an unstable flow will occur with "bumpy" surface.

Example 9.11

Determine the minimum slope in the upper reach of chute section of 100 ft width. The range of discharge is 5000 to 70,000 cfs. $n = 0.015$.

Solution

1. Under minimum flow conditions, $q = \dfrac{5000}{100} = 50$.

2. From the slope equation,

$$S_c = \frac{21.3(0.015)^2}{(50)^{0.222}} = 0.002$$

3. Under maximum flow conditions,

$$70,000 = \frac{1.486}{0.015}[(100)y](y)^{2/3}(0.002)^{1/2}$$

or

$$y = 20.86 \text{ ft}$$

4. Velocity of flow for maximum discharge:

$$v = \frac{Q}{A} = \frac{70,000}{100(20.86)} = 33.56 \text{ ft/sec}$$

5. Froude number,

$$\text{Fr} = \frac{v}{\sqrt{gy}} = \frac{33.56}{\sqrt{32.2(20.86)}} = 1.29$$

Since $1.29 < 1.56$, we have stable rapid flow.

*In SI units, this has the form $S_c = \cdot 12.6n^2/q_c^{0.222}$.

9.12.2 Chute Sidewalls

Except for converging and diverging sections, chute channels are designed with parallel vertical sidewalls, commonly of reinforced concrete 12 to 18 in. thick. The sidewalls are designed as the retaining walls. The height of the walls is designed to contain the depth of flow for the spillway design flood. The water surface profile from the control section downward is determined for this purpose. The allowance is made for pier end waves, roll waves, and air entrainment.

The water surface profile is computed by the methods of varied flow discussed in Section 10.11.4. The initial values of discharge, velocity, and depth at the entering section are known. Since the flow in the chute is supercritical, computation proceeds in a downstream direction. It may be noted that in the steep channel, either the S_2 curve (when $y_c > y > y_n$) or the S_3 curve (when $y_c > y_n > y$) is involved (see Section 10.11.3).

In view of uncertainties involved in the evaluation of surface roughness, pier end waves, roll waves, and air entrainment buckling, a freeboard given by the following empirical equation is added to the computed depth of the water surface profile.

$$\text{Freeboard (ft)} = 2.0 + 0.002Vd^{1/3} \qquad \text{[unbalanced]} \qquad (9.30)$$

where

V = mean velocity in chute section under consideration, ft/sec

d = mean depth, ft

The converging and diverging transitions in the sidewalls must be made gradual. The U.S. Bureau of Reclamation experiments have shown that the angular variation (flare angle) not exceeding the following value will provide an acceptable transition for either a contracting or an expanding channel.

$$\tan \alpha = \frac{1}{3\text{Fr}} \qquad \text{[dimensionless]} \qquad (9.31)$$

where

α = angular variation of the sidewall with respect to the channel centerline

Fr = Froude number = $V/\sqrt{gy}$

V, y = averages of the velocities and depths at the beginning and the end of the transition

9.13 SIDE-CHANNEL SPILLWAYS

In side-channel spillway, the overflow weir is placed along the side of the discharging channel, so that the flow over the crest falls into a narrow channel section (trough) opposite the weir, turns a right angle, and then continues in the direction approximately parallel to the weir crest. A plan and a cross section is shown in Figure 9.26.

This type of spillway is adaptable to certain special conditions, such as when a long overflow crest is desired but the valley is narrow, or where the overflows are most economically passed through a deep narrow channel or a tunnel. The crest of the spillway is similar to an overflow or other ordinary weir section. Downstream

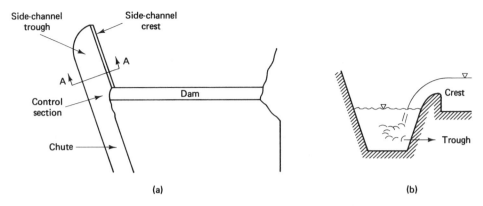

Figure 9.26 Side-channel spillway: (a) side channel plan; (b) cross section at AA.

from the side channel trough, a control section is achieved by constricting the channel sides or elevating the channel bottom to induce the critical flow. Flows upstream from this section are at subcritical stage. Downstream from this section functions similar to a chute-type spillway. Thus the side-channel design is concerned only with the hydraulic action in the trough upstream of the control section, where spatially varied flow takes place. Flow in the trough should be at a sufficient depth so as to carry away the accumulated flow and not to submerge the flow over the crest. Hydraulic aspect is concerned with the water surface profile in the trough, which is determined from the momentum principle (not by the energy principle as applied in the gradual flow in Section 10.11 because of excessive energy losses due to high turbulence). A trapezoidal section is a most common section for the side-channel trough. The bottom width is kept to a minimum. The trough is placed on the rock foundation and a concrete lining is provided.

In Figure 9.27, consider a short reach of distance Δx. The rate of change of momentum in the reach is equal to the external force acting in the reach.

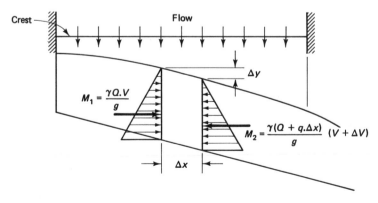

Figure 9.27 Analysis of side-channel flow.

Rate of momentum at upper section $= \dfrac{\gamma}{g} Q_1 v_1$

Rate of momentum at lower section $= \dfrac{\gamma}{g} Q_2 v_2$

Change of rate of momentum $= \dfrac{\gamma}{g}(Q_2 v_2 - Q_1 v_1)$ (c)

Forces comprise the hydrostatic pressures, component of weight in the direction of flow and the friction in the channel section. Neglecting the weight component and the channel friction that tend to compensate each other, the resultant hydrostatic force

$$P_2 - P_1 = \gamma \cdot A \cdot dy$$

Representing the average area of cross section $A = Q_1 + Q_2/(v_1 + v_2)$

$$\text{Hydrostatic force} = \gamma \left(\frac{Q_1 + Q_2}{v_1 + v_2} \right) \Delta y \qquad (d)$$

Equating the two forces (c) and (d) and rearranging, the change in water elevation can be expressed:

$$\Delta y = \frac{-Q_2}{g} \frac{v_1 + v_2}{Q_1 + Q_2} \left[(v_2 - v_1) + \frac{v_1 (Q_2 - Q_1)}{Q_2} \right] \quad [L] \qquad (9.32)$$

Equation (9.32) is solved by a trial-and-error procedure. For a reach of length Δx, Q_1 and Q_2 are known. Starting from the control point where the critical depth exists, a trial depth at the other end is found which will satisfy the equation.

Example 9.12

Design a side-channel trough for a spillway of 100 ft length for a maximum discharge of 2500 cfs. The side-channel trough has a length of 100 ft and a bottom slope of 1 ft in 100 ft. A control section of 10 ft width is placed downstream from the trough with the bottom of the control at the same elevation as the bottom of the trough floor at the downstream end.

Solution

1. Critical depth at the control section, $y_c = \left(\dfrac{q_1^2}{g} \right)^{1/3}$.

2. $q_1 = \dfrac{2500}{10} = 250 \text{ cfs/ft.}$

3. $y_c = \left(\dfrac{250^2}{32.2} \right)^{1/3} = 12.44 \text{ ft.}$

4. $v_c = \dfrac{q_1}{y_c} = \dfrac{250}{12.44} = 20.1 \text{ ft/sec.}$

5. Velocity head, $h_c = \dfrac{v_c^2}{2g} = \dfrac{(20.1)^2}{2(32.2)} = 6.27 \text{ ft.}$

6. For a side-channel trough, assume a trapezoidal section with 2(vertical): 1(horizontal) slope and a 10-ft bottom width. Also, assume that the transition loss from the end of side channel trough to the control section is equal to 0.2 of the difference in velocity heads between the ends of the transition.

7. The following energy equation may be written between the trough end and the control section:

$$y_{100} + h_{100} = y_c + h_c + 0.2(h_c - h_{100})$$

or

$$y_{100} + 1.2h_{100} = 12.44 + 1.2(6.27) = 19.96 \qquad (9.33)$$

where the subscript c refers to "critical" and "100" refers to the distance of the trough from the upstream end of the spillway (i.e., at the end of the trough). Equation (9.33) is solved by trial and error. Assume that $y_{100} = 19.2$; then $A = 376.3$ ft^2, $v = Q/A = 2500/376.3 = 6.64$ ft/sec, and $h_{100} = (6.64)^2/2(32.2) = 0.68$. Thus Eq. (9.33) is satisfied.

8. The known values above relate to section 1 at the downstream end of the trough. Section 2 is taken at the upper end of a selected increment Δx, 25 ft in this case. A value of the change in water level, Δy, is assumed for the reach, all terms of eq. (9.32) are evaluated, and Δy is computed, as illustrated in Table 9.6. The assumed and computed values of Δy should match; otherwise, a new value is assumed for Δy.

9. The process is repeated until the upstream end of the channel is reached.

9.14 MORNING GLORY OR SHAFT SPILLWAYS

This type of spillway consists of four parts: (1) a circular weir at the entry, (2) a flared transition conforming to the shape of the lower nappe of a sharp-crested weir, (3) a vertical drop shaft, and (4) a horizontal or near-horizontal outlet conduit or tunnel. This spillway is used at dam sites in narrow canyons or where a diversion conduit/tunnel of the dam is available to be utilized as an outlet conduit. As the head increases, the control shifts from weir crest, to drop shaft, and to outlet conduit. Three control conditions are indicated in Figure 9.28 and listed in Table 9.7.

Condition 1 of a free-discharging weir prevails as long as the nappe forms to converge into the shape of a solid jet. Under condition 2, the weir crest is drowned out. The U.S. Bureau of Reclamation (1977c) indicated that this condition is approached when $H_d/R_s > 1$, where H_d is the design head and R_s is the radius of the crest. Further increase in head leads to condition 3 when the spillway is flooded out, showing only a slight depression and eddy at the surface. Under condition 3, the head rises rapidly for a small increase in discharge. Thus the design is not recommended under this condition (i.e., under the design head the outlet conduit should not flow more than 75% full). Thus the discharge through a shaft spillway is limited. The U.S. Bureau of Reclamation suggested that the following weir formula may be used for the flow through the shaft spillway entrance regardless of the submergence, by adjusting the coefficient to reflect the flow conditions (U.S. Bureau of Reclamation, 1977c, p. 416).

$$Q = C(2\pi R_s)H^{3/2} \qquad [\text{L}^3\text{T}^{-1}] \qquad (9.37)$$

TABLE 9.6 SIDE-CHANNEL SPILLWAY COMPUTATIONS[a]

(1)	(2)	(3)	(4)	(5)	(6)	(7)	(8)	(9)	(10)	(11)	(12)	(13)	(14)
Δx (select)	Bottom Level[b]	Δy (Assume)	Water Level[c]	y (col. 4 − col. 2)	A[d]	Q = $q(L - \Delta x)$[e]	v = Q/A	$Q_1 + Q_2$[f]	$Q_2 - Q_1$	$v_1 + v_2$	$v_2 - v_1$	Δy Computed Eq. 9.32	Remarks on Assumed Δy
D/S end	100.0		119.2	19.20	376.3	2500	6.64						
25	100.25	1.0	120.2	19.95	398.5	1875	4.71	4375	−625	11.35	−1.93	0.63	High
		0.62	119.82	19.57	387.2		4.84	4375	−625	11.48	−1.80	0.61	ok
25	100.50	0.50	120.32	19.82	394.6	1250	3.17	3125	−625	8.01	−1.67	0.41	High
		0.40	120.22	19.72	391.6		3.19	3125	−625	8.03	−1.65	0.41	ok
25	100.75	0.25	120.47	19.72	391.6	625	1.60	1875	−625	4.79	−1.59	0.24	ok
15	100.90	0.10	120.57	19.67	390.2	250	0.64	875	−375	2.24	−0.96	0.07	High
		0.07	120.54	19.64	389.3		0.64	875	−375	2.24	−0.96	0.07	ok

[a] $q = Q/L = 2500/100 = 25$ cfs/ft, bottom slope = 1 in 100 ft given.
[b] (Slope × channel length) + datum.
[c] Final water level at the preceding station (section) + assumed Δy.
[d] Area of cross section of trough computed for depth y in Column 5, $A = (10 + 0.5y)y$.
[e] $q(L - \Sigma \Delta x)$, L = crest length.
[f] Column 7 + value in column 7 at the preceding station (section).

494

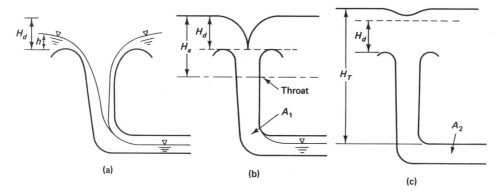

Figure 9.28 Flow conditions of a shaft spillway: (a) condition 1: crest control; (b) condition 2: tube control; (c) condition 3: pipe control.

TABLE 9.7 DISCHARGE CHARACTERISTICS OF SHAFT SPILLWAY

Control Point	Condition	Characteristics	Relation	
Crest	Unsubmerged flow	Weir flow	$Q = CLh^{3/2}$	(9.34)
Throat of drop shaft	Partially submerged	Orifice flow	$Q = C_d A_1 \sqrt{2gH_a}$ $C_d = 0.95$	(9.35)
Downstream of outlet conduit	Submerged flow	Pipe flow	$Q = A_2 \sqrt{\dfrac{2gH_T}{\Sigma K}}$ $\Sigma K =$ loss coefficients through pipe flow	(9.36)

where

C = discharge coefficient related to H_d/R_s and P/R_s from model tests, where H_d = design head and P = crest height from the outlet pipe; values of C are given in Figure 9.29

R_s = radius to the circular crest

H = head over the weir

Alternatively, eq. (9.37) can be used to determine the crest size (radius), R_s, for a given design discharge under the maximum head. This is a minimum radius required when small subatmospheric pressures along the overflow crest can be tolerated. The U.S. Bureau of Reclamation (1977c) have provided the tables to compute the crest profile. Similarly, eq. (9.35) can be used to determine the shape of the transition (drop shaft) that is required to pass the design discharge with the maximum head over the crest.

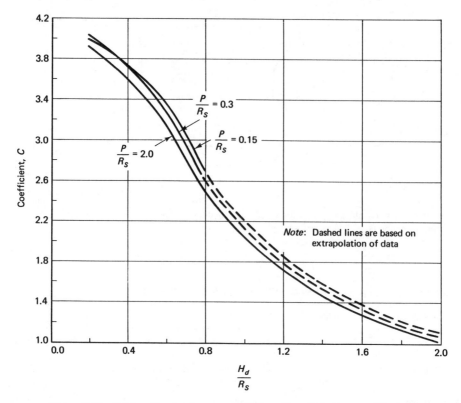

Figure 9.29 Relation for circular crest coefficient (from U.S. Bureau of Reclamation, 1977c).

Example 9.13

A shaft spillway is to discharge 2000 cfs under a design head of 10 ft. Determine the minimum size of the overflow crest. Also determine the shape of the transition if the control section is 4 ft below the crest level.

Solution

1. Since the coefficient C is related to P and R_s, assume that $P/R_s > 2$ and determine R_s by trial and error.
2. Try $R_s = 7$ ft.

$$\frac{H_d}{R_s} = \frac{10}{7} = 1.43$$

From Figure 9.29, $C = 1.44$.
3. From eq. (9.37),

$$Q = C(2\pi R_s)H_d^{3/2}$$
$$= 1.44(2\pi)(7)(10)^{3/2} = 2002 \text{ cfs}$$

This is practically the same as the required discharge. Hence the crest radius = 7 ft.

4. Depth of the control section from the water surface:

$$H_a = 10 + 4 = 14 \text{ ft}$$

5. From eq. (9.35),

$$Q = 0.95(\pi R^2)\sqrt{2gH_a}$$

$$R^2 = \frac{2000}{0.95\pi\sqrt{2gH_a}}$$

or

$$\text{Drop shaft radius } R = \frac{9.14}{H_a^{1/4}}$$

H_a (select) (ft)	R (ft)
14	4.73
16	4.57
18	4.44
20	4.32

PROBLEMS

9.1. At a proposed dam site for a water supply project on the Battenkill River near Greenwich, New York, the mean monthly flows are given below. If the demand of water is at a uniform rate of 60 cfs, determine the size of storage required. An allowance of 15 cfs is made to account for the seepage, evaporation losses, and downstream releases. Solve by **(a)** the graphic method, and **(b)** the analytic method.

Month	Discharge (cfs)	Month	Discharge (cfs)
Oct.	130	May	195
Nov.	70	June	100
Dec.	50	July	65
Jan.	41	Aug.	85
Feb.	30	Sept.	157
Mar.	95	Oct.	300
Apr.	230	Nov.	346

9.2. It is proposed to locate an impounding reservoir on a streambed of a 150-mi^2 drainage basin. Monthly runoffs in inches for a critical year are given in the table below. The estimated demands in mgd for each month are also given. Monthly evaporation losses from the reservoir pool and the net precipitation on the pool are also indicated. If the reservoir occupies 5% of the drainage area, estimate within-year storage requirement for the project.

Month	Runoff (in.)	Demand (mgd)	Evaporation (in.)	Net Precipitation (in.)
Oct.	1.54	125	2.25	4.10
Nov.	1.62	127	1.50	3.75
Dec.	1.99	143	1.0	4.20
Jan.	2.99	137	1.05	4.0
Feb.	2.05	150	1.75	4.6
Mar.	2.10	145	3.0	3.5
Apr.	3.20	153	4.3	3.2
May	1.50	155	5.5	3.0
June	0.36	161	5.95	4.0
July	2.78	158	5.0	4.5
Aug.	3.05	152	4.10	3.2
Sept.	3.01	153	3.20	3.9

(*Hint:* Convert all data to millions of gallons. Inflow is the sum of runoff plus net precipitation, and outflow is the sum of demand and evaporation.)

9.3. For the following hourly demand rates, determine the operating storage capacity for (a) uniform 24-hour pumping, and (b) pumping from 6 a.m. to 6 p.m. only.

Time	Demand Rate (gpm)	Time	Demand Rate (gpm)
1 a.m.	1710	1 p.m.	5850
2 a.m.	1620	2 p.m.	5814
3 a.m.	1616	3 p.m.	5787
4 a.m.	1530	4 p.m.	5850
5 a.m.	1620	5 p.m.	6030
6 a.m.	1719	6 p.m.	6407
7 a.m.	2880	7 p.m.	5000
8 a.m.	4500	8 p.m.	4836
9 a.m.	5085	9 p.m.	4092
10 a.m.	5400	10 p.m.	4803
11 a.m.	5589	11 p.m.	2500
12 noon	5670	12 night	3187

9.4. A reservoir with an initial capacity of 30,000 acre-ft has an average annual water inflow of 150,000 acre-ft. A sediment inflow of 220 acre-ft per year is expected from the watershed. How long will it take for the reservoir to become 75% filled with sediment? Use the mean curve.

9.5. A reservoir has a capacity of 5×10^6 m³ and a drainage area of 190 km². The average annual runoff from the watershed is 390 mm, which brings in a sediment quantity of 600 m³/km². Determine the time required to reduce the reservoir capacity to 1×10^6 m³. Use the mean curve.

9.6. An earth dam has a crown (top) width of 20 ft. The height is 70 ft with a freeboard of 10 ft. The slopes are 1 (vertical):3 (horizontal). It rests on an impervious foundation having a drainage blanket extending back 50 ft from the toe. The permeability of embankment material is 3.2×10^{-5} ft/sec. Determine the seepage per foot length of the dam.

9.7. An earth dam on an impervious foundation has a crown width of 25 ft, a total height of 80 ft, and a freeboard of 10 ft. It has an upstream slope of 20° and a downstream slope of 30°. A horizontal blanket extends back 80 ft from the back toe. The dam has a permeability of 4×10^{-4} cm/s. Determine the seepage through the dam.

9.8. In the dam of Problem 9.7, if the material has a permeability of 4×10^{-4} cm/s horizontally and 1×10^{-4} cm/s vertically, determine the seepage through the dam.

9.9. An earth dam of crown width 30 ft and height 85 ft rests on an impervious foundation. It has side slopes of 1 (vertical):4 (horizontal) and a permeability of 2×10^{-4} cm/s. A center core is provided symmetrically about the vertical axis of the dam. The core has a top width of 20 ft, side slopes of 1 (vertical) to 1 (horizontal), and a permeability of 2×10^{-5} cm/s. The water depth in the reservoir is 75 ft. Determine the quantity of seepage through the dam.

9.10. If the center core in Problem 9.9 has a permeability of 2×10^{-5} cm/s in the horizontal direction and 1×10^{-5} cm/s in the vertical direction, determine the seepage through the dam.

9.11. If the center core in Problem 9.10 has vertical sides (rectangular shape), determine the seepage.

9.12. An earth dam has a crown width of 25 ft and a height of 80 ft with a freeboard of 8 ft. The side slopes are at angle of 20° with the horizontal. It has a permeability of 3×10^{-5} ft/sec horizontally and 1×10^{-5} ft/sec vertically. It rests on an impervious foundation and has no drainage blanket (the seepage line emerges on the downstream slope). Determine the seepage per foot of length of the dam.

9.13. If the dam of Problem 9.9 is located on a pervious foundation of 30 ft depth with a permeability of 1×10^{-4} cm/s, determine the quantity of seepage through the foundation assuming that the average length of seepage is equal to the base length.

9.14. For the dam of Problem 9.12, determine the seepage through the foundation if it is a pervious stratum of 40 ft depth and permeability 1.5×10^{-5} ft/sec. Assume that the seepage length is equal to the base length.

9.15. A concrete gravity dam has a maximum height of 110 ft and a crest width of 20 ft. The upstream face has a slope of 5 (vertical):1 (horizontal) and the downstream face slope is 1.5 (vertical):1 (horizontal). The dam rests directly on the foundation without a cutoff wall. The tailwater depth is negligible. The earthquake acceleration is 0.1g. The ice force is 10,000 lb/ft with a 2-ft depth of ice sheet. The specific weight of concrete is 150 lb/ft^3. Compute **(a)** all the forces acting on the dam, **(b)** vertical and horizontal components of the foundation reaction, and **(c)** line of action water depth 100 ft.

9.16. For a masonry gravity dam section shown in Fig. P9.16, compute all the forces and the reaction from the foundation with the line of its action. Consider the seismic acceleration of 0.05g for the dead weight with 50% increase for water acceleration. Full uplift water pressure acts from the foundation. Assume a specific gravity of masonry of 2.5. Neglect the dynamic force of moving water.

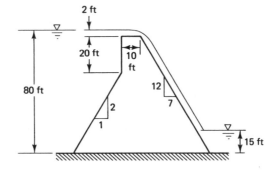

Figure P9.16

9.17. For the gravity dam in Problem 9.15, determine the factor of safety against sliding at the base. The coefficient of internal friction is 0.6 and the foundation rock has a shear strength of 500 psi.

9.18. For the gravity dam in Problem 9.16, determine the factor of safety against sliding at the base. The coefficient of friction is 0.7. The shear resistance by the entire section is 1,000,000 lb.

9.19. A triangular masonry dam section of a vertical face has a total height of 200 ft, and the downstream slope is 1.5 (vertical):1(horizontal). Water is filled to the top of the dam. The masonry weighs 135 lb/ft^3 and has a shear strength of 600 psi. For a friction coefficient of 0.7, determine the stability against sliding 50 ft above the base of the dam. Neglect the ice and earthquake forces.

9.20. Show that the dam in Problem 9.15 is safe against overturning, and indicate whether any tensile stress develops at the base. Determine stresses at the base.

9.21. Demonstrate that the dam section of Problem 9.16 will not overturn. Show it is free from tension at the base. Determine the bearing stresses in the foundation.

9.22. On a gravity dam the loading consists of the dam weight, water pressure, and uplift only, with no ice, tailwater, or earthquake loads. Show that the dam section, from a consideration of no tension at the base, can be represented by a simple triangle of apex angle given by $\tan \phi = 1/\sqrt{s - 1}$, where s is specific gravity of the dam material.

9.23. For the dam of Problem 9.15, determine the maximum compressive stress and the maximum horizontal shear stress in the dam.

9.24. If the dam in Problem 9.16 is built of rubble material having an allowable compression strength of 20,000 lb/ft^2, is the dam safe from failure by crushing?

9.25. The profile of a dam site is shown in Fig. P9.25. Design a constant-radius arch dam with a minimum crest thickness of 2.5 ft. The allowable stress of concrete is 500 psi.

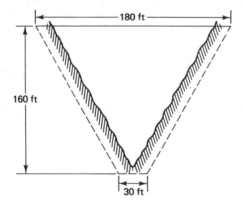

Figure P9.25

9.26. Using the data of Problem 9.25, design a constant-angle arch dam.

9.27. A high-overflow spillway has a maximum head of 12.0 ft. Determine the profile of the spillway crest having a vertical upstream face and a 1 (horizontal):2 (vertical) downstream slope. Take the design head to be 0.75 times the maximum head.

9.28. An overflow spillway has a design head of 2.5 m. Determine the crest profile for a spillway of 2 m height having both the upstream and downstream slopes of 1:1.

9.29. A vertical-faced overflow spillway of 10 ft height is designed to discharge 8000 cfs at the pool level of 7 ft. Determine the length of the spillway. Underdesign the spillway.

9.30. An underdesigned overflow spillway of 1 m height and 40 m width has an upstream slope of 2 (horizontal):3 (vertical). Determine the discharge for a design head of 1.8 m.

9.31. A high 100-ft-wide overflow section of 1 (horizontal):3 (vertical) upstream slope is designed for a head of 10 ft. Determine the discharge when the actual total head is 12 ft.

9.32. For Problem 9.31, determine the discharge if the crest supports four square-nosed piers and has square abutments.

9.33. An overflow is to be designed to carry a peak flow of 70,000 cfs. The upstream reservoir level is at El. 1000 ft and the average channel floor is at El. 850. The design head over the spillway is 20 ft. Determine the length of the spillway and define the crest profile for a vertical upstream section and 1 (horizontal):2 (vertical) downstream section.

9.34. An overflow spillway is to be designed having a upstream slope of 1 (horizontal):1.5 (vertical) and a downstream slope of 1 (horizontal):2 (vertical). It has to carry a peak flow of 2000 m^3/s. The depth of the reservoir upstream is 40 m. The crest length is 75 m. Determine the crest height and the shape of the overflow section.

9.35. Determine the minimum slope and the type of flow (stable or unstable) in the upper part of a chute section of 150 ft width. The discharge range is 1000 to 100,000 cfs. $n = 0.015$.

9.36. For a 50-m-wide chute, having a discharge range of 30 to 500 m^3/s, determine the minimum slope of the chute section and the condition of flow in the chute. $n = 0.016$.

9.37. The control section of a chute spillway of design capacity 50,000 cfs is 150 ft wide, whereas the rectangular chute has a width of 100 ft. The depths of flow at the control section and at the beginning of the 100-wide section are 17.5 ft and 12.0 ft, respectively. Design the converging section.

9.38. At the bottom of the valley, where the depth is 10 ft, the chute section is to be expanded again to a width of 200 ft to join the stilling basin. Design the diverging section of 9.37.

9.39. An overflow section of 120 ft length discharges 4000 cfs into a side channel. Determine the water surface profile in a side-channel trough having a bottom slope of 0.004. The channel trough and chute have a uniform rectangular section of 20 ft width. The control section is formed by raising the channel bottom opposite and perpendicular to the lower end of the crest.

9.40. In Problem 9.39, if the trough has a trapezoidal shape of 20 ft width and a side slope of 1 (vertical):2 (horizontal), compute the water surface profile. Assume the transition losses between the trough and the chute section at the end of the crest to be 0.2 of the difference in velocity heads in the two sections.

9.41. A 40-m-long spillway of design discharge 40 m^3/s spills into a side channel of 1% bottom slope. The channel trough has a bottom width of 5 m and side slopes of 3 (vertical):2 (horizontal) and joins a chute section of 4 m width at the control. Compute the water surface profile in the channel trough. Assume the transition losses between the trough and the chute to be 0.2 of the difference in the velocity heads at two ends of the transition.

9.42. A morning glory spillway is to discharge 3500 cfs under a head of 14 ft. Determine the minimum size of the overflow crest. For a control section in the drop shaft at a depth of 20 ft below the water surface, draw the shaft profile.

9.43. A morning glory spillway has a crest diameter of 10.5 m. The design discharge is 775 m^3/s. Determine the design head over the spillway. If the control section is 2 m below the crest, determine the size of the drop shaft at the control section.

9.44. A morning glory spillway is so designed that the crest is just drowned under the maximum head of 11 ft. Compute the crest size (radius at the crest) and the shaft profile for a design discharge of 200 cfs. [*Hint:* Since the orifice control condition prevails, apply eq. (9.35) for the head versus radius computations starting at the crest level.]

CONVEYANCE SYSTEM:
Open Channel Flow

10.1 INTRODUCTION

The two modes of transporting flowing water from one point to another are by pipes and open channels. In pipes, the flow of water is under pressure, whereas the open channel flow has a free water surface. The basic theory is the same for both kinds of flow, but there is an important difference in the boundary condition. The open channel flow is more difficult to deal with, due to the presence of the free water surface, various possible configurations of the channel section, and the changing position of the water surface with respect to time and space. The treatment of open channel flow is somewhat empirical in nature because of these factors.

An open channel can be a natural stream or a river. It can also be an artificial channel in the form of a canal, flume, chute, culvert, tunnel, ditch, partly filled pipe (conduit), or an aqueduct of any shape. An artificial channel is commonly used to convey the water from the source of supply to a distribution point. Further distribution beyond this point is made by a network of pipes. An artificial channel in the form of a canal is excavated either in a firm foundation such as a rock bed or in the erodible materials. Irrigation canals, formed in alluvial and other granular material, are erodible unless lined with nonerodible materials. In designing erodible channels, consideration has to be given to the stability of the channel geometry such that substantial scouring does not take place. This criterion is not applicable to nonerodible channels. Two kinds of channels are described separately.

10.2 ELEMENTS OF CHANNEL SECTION

The flow in an open channel is due to the gravitational force; hence the channel bottom should have a slope in the direction of flow. A channel having unvarying cross section and constant bed slope throughout its length is known as a prismatic channel. The cross section of a channel taken normal to the direction of flow at any point is referred to as the channel section. A section and a longitudinal profile of a channel are shown in Figure 10.1.

The definition of various geometric elements is given below:

1. *Depth of flow, y*: vertical distance from the channel bottom to the free surface
2. *Depth of flow section, d*: depth of flow normal to the direction of flow; $d = y \cos \theta$, but the terms d and y are used interchangeably
3. *Top width, T*: width at the free surface
4. *Flow area, A*: cross-sectional area of the flow normal to the direction of flow; $A = bd$ for a rectangle
5. *Wetted perimeter, P*: across a channel section, the length of the channel surface in contact with water; $P = b + 2d$ for a rectangle
6. *Hydraulic radius, R* = ratio of the flow area to the wetted perimeter; $R = A/P$
7. *Hydraulic depth, D* = ratio of the flow area to the top width; $D = A/T$
8. *Section factor for critical flow, $Z_c = A\sqrt{D}$*
9. *Section factor for uniform flow, $Z_n = AR^{2/3}$*

For a circular section, the geometric elements in the dimensionless form, as a ratio with an appropriate power of the diameter, d_0 of the section, are given in Table 10.1, which provides a convenient means of determining the geometric elements for various depths of flow. For a depth of flow of $0.94d_0$, the section factor $AR^{2/3}$ has a maximum value of $0.335d_0^{8/3}$ in a circular section (i.e., discharge is maximum at this depth). Between the range of the water depth of $0.82d_0$ and d_0, there are two depths corresponding to the same level of discharge — one above $0.94d_0$ and one below it.

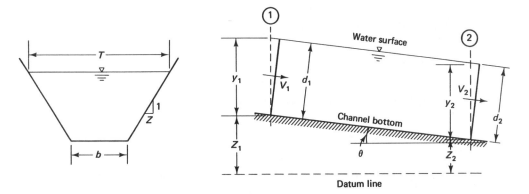

Figure 10.1 Channel section and longitudinal profile.

TABLE 10.1 GEOMETRIC ELEMENTS FOR CIRCULAR SECTION

$\dfrac{y}{d_0}$	$\dfrac{A}{d_0^2}$	$\dfrac{P}{d_0}$	$\dfrac{R}{d_0}$	$\dfrac{T}{d_0}$	$\dfrac{D}{d_0}$	$\dfrac{A\sqrt{D}}{d_0^{5/2}}$	$\dfrac{AR^{2/3}}{d_0^{8/3}}$
0.01	0.0013	0.2003	0.0066	0.1990	0.0066	0.0001	0.0000
0.05	0.0147	0.4510	0.0326	0.4359	0.0336	0.0027	0.0015
0.10	0.0409	0.6435	0.0635	0.6000	0.0682	0.0107	0.0065
0.15	0.0739	0.7954	0.0929	0.7141	0.1034	0.0238	0.0152
0.20	0.1118	0.9273	0.1206	0.8000	0.1398	0.0418	0.0273
0.25	0.1535	1.0472	0.1466	0.8660	0.1774	0.0646	0.0427
0.30	0.1982	1.1593	0.1709	0.9165	0.2162	0.0921	0.0610
0.35	0.2450	1.2661	0.1935	0.9539	0.2568	0.1241	0.0820
0.40	0.2934	1.3694	0.2142	0.9798	0.2994	0.1603	0.1050
0.45	0.3428	1.4706	0.2331	0.9950	0.3446	0.2011	0.1298
0.50	0.3927	1.5708	0.2500	1.0000	0.3928	0.2459	0.1558
0.55	0.4426	1.6710	0.2649	0.9950	0.4448	0.2949	0.1825
0.60	0.4920	1.7722	0.2776	0.9798	0.5022	0.3438	0.2092
0.65	0.5404	1.8755	0.2881	0.9539	0.5666	0.4066	0.2358
0.70	0.5872	1.9823	0.2962	0.9165	0.6408	0.4694	0.2608
0.75	0.6318	2.0944	0.3017	0.8660	0.7296	0.5392	0.2840
0.80	0.6736	2.2143	0.3042	0.8000	0.8420	0.6177	0.3045
0.85	0.7115	2.3462	0.3033	0.7141	0.9964	0.7098	0.3212
0.90	0.7445	2.4981	0.2980	0.6000	1.2408	0.8285	0.3324
0.94[a]	0.7662	2.6467	0.2896	0.4750	1.6130	0.9725	0.3353
0.95	0.7707	2.6906	0.2864	0.4359	1.7682	1.0242	0.3349
1.00	0.7854	3.1416	0.2500	0.0000	∞	∞	0.3117

[a]Maximum flow occurs at 0.94 full depth.

Example 10.1

For the channel section shown in Figure 10.2, determine the geometric elements.

Solution

$$y = 36 \text{ in. or } 3 \text{ ft} \qquad d_0 = 60 \text{ in. or } 5 \text{ ft}$$

$$\frac{y}{d_0} = \frac{3}{5} = 0.6$$

From Table 10.1:

$$\frac{A}{d_0^2} = 0.492, \qquad A = 0.492(5)^2 = 12.3 \text{ ft}^2$$

$$\frac{P}{d_0} = 1.7722, \qquad P = 1.7722(5) = 8.861 \text{ ft}$$

$$\frac{R}{d_0} = 0.2776, \qquad R = 0.2776(5) = 1.39 \text{ ft}$$

$$\frac{Z_c}{d_0^{5/2}} = \frac{A\sqrt{D}}{d_0^{5/2}} = 0.3438, \qquad A\sqrt{D} = 0.03438(5)^{5/2} = 1.92$$

$$\frac{Z}{d_0^{8/3}} = \frac{AR^{2/3}}{d_0^{8/3}} = 0.2092, \qquad AR^{2/3} = 0.2092(5)^{8/3} = 15.29$$

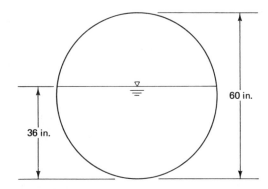

Figure 10.2 Partial full circular channel section of Example 10.1.

10.3 TYPES OF FLOW

The flow in an open channel is classified according to the change in the depth of flow with respect to space and time. If the depth of flow remains the same at every section of the channel, the flow is known as the uniform or normal flow. In the varied or nonuniform flow, the depth changes along the length of the channel. When the change in the depth occurs abruptly over a short distance, it is a rapidly varied flow; otherwise, it is a gradually varied flow.

If the depth of flow does not change during the time interval under consideration, it is referred to as the steady flow. It is unsteady if the depth changes with time. Combining the space and time criteria, the flow in an open channel can be classified as follows:

Type of Flow	Example
Steady uniform flow	Laboratory channel
Steady gradually varied flow	Irrigation, navigation channel
Steady rapidly varied flow	Flow over a weir, hydraulic jump
Unsteady gradually varied flow	Streamflow, flood wave
Unsteady rapidly varied flow	Surges, pulsating flow

For an unsteady uniform flow, the depth should vary from time to time while always remaining parallel to the channel bottom. This is not a practically feasible condition. Even the steady uniform flow is difficult to obtain in natural channels due to irregular section and in artificial channels because of the existence of the controls. The steady uniform flow, however, is a fundamental type of flow that is considered in all channel design problems. The effect of the varied flow is superimposed over the uniform flow condition to determine the channel section requirements. In the computation of flow in natural streams, the steady flow condition is assumed during the time interval under consideration. The unsteady flow relates to the propagation of a wave in the channel which is outside the scope of the book in the context of a conveyance channel.

10.4 STATE OF FLOW

The viscosity and gravity affect the state of flow in an open channel. The Reynolds number and the Froude number are both relevant in the channel flow. The Reynolds number, a ratio of the inertia force to the viscous force, is expressed as follows for an open channel:

$$R_e = \frac{VR}{\nu} \quad \text{[dimensionless]} \tag{10.1}$$

where

$$R_e = \text{Reynolds number}$$
$$V = \text{mean velocity of flow, ft/sec}$$
$$R = \text{hydraulic radius, ft}$$
$$\nu = \text{kinematic viscosity of water, ft}^2/\text{sec}$$
$$= \frac{\text{dynamic viscosity, lb-sec/ft}^2}{\text{mass density, slugs/ft}^3}$$

The flow is laminar when the viscous forces are dominating, resulting in a Reynolds number of less than 500. It is turbulent if the viscous forces are weak and the Reynolds number is higher than 2000. Between 500 to 2000 is the transitional range of R_e. The experiments on smooth channels and rough channels indicated the following characteristics (Chow, 1959, pp. 9–12):

1. The Darcy–Weisbach formula of flow in pipes is applicable to uniform flow in open channels.
2. In the laminar region, the friction factor relation, $f = K/R_e$ of pipe flow, is applicable to both smooth and rough channels. The value of K varies with channel shape and is higher for rough channels.
3. In the turbulent region, the friction factor relation of smooth pipes (Blasius and Prandt–von Kármán) is approximately representative of smooth channels.
4. In the turbulent region of rough channels, the channel shape, roughness, and the Reynolds number have a pronounced effect on the friction factor. The friction factor relation deviates from the pipe flow relation.

The common type of flow pertains to item 4, thus necessitating a separate relation for the channel flow.

The gravity effect is incorporated in the Froude number, which is represented by a ratio of inertia force to gravity force, as follows:

$$Fr = \frac{V}{\sqrt{gD}} \quad \text{[dimensionless]} \tag{10.2}$$

where

$$Fr = \text{Froude number}$$
$$V = \text{mean velocity of flow, ft/sec}$$
$$D = \text{hydraulic depth, ft}$$

Conveyance System: Open Channel Flow   Chap. 10

When Fr = 1, the flow is in a critical state; when F < 1, the flow is subcritical or tranquil having a low velocity; and when F > 1, the flow is supercritical or shooting, having a high velocity.

The friction factor characteristics of laminar and turbulent flow as discussed above relate to the subcritical flow. In the supercritical turbulent regime of flow, the friction factor becomes larger with an increase in the Froude number. Up to a value of 3, the Froude number has a negligible effect on the friction factor.

10.5 VELOCITY DISTRIBUTION COEFFICIENTS

The velocity in a channel section is not uniformly distributed due to the presence of a free water surface and the friction along the wall of the channel. The mean velocity, V, is used in computing the velocity head or fluid momentum at a channel section. The values thus computed are generally lower than with the actual distribution.

When the energy principle is used, the corrected velocity head is given by $\alpha(V^2/2g)$. The value of α, known as the energy coefficient, or Coriolis coefficient, is expressed as follows, based on the total kinetic energy at a section:

$$\alpha = \frac{\sum v^3 \Delta A}{V^3 A} \qquad \text{[dimensionless]} \qquad (10.3)$$

where

$$v = \text{velocity in elementary area, } \Delta A$$
$$V = \text{mean velocity in the section}$$
$$A = \text{area of the section}$$

When the momentum principle is used, the true momentum is expressed by $\beta \gamma QV/g$. The value β, known as the momentum, or Boussinesq coefficient, is computed based on the momentum passing through the section.

$$\beta = \frac{\sum v^2 \Delta A}{V^2 A} \qquad \text{[dimensionless]} \qquad (10.4)$$

Table 10.2 can be used to select the values of the coefficients where eqs. (10.3) and (10.4) are not used. In practical applications, the effect of the energy coeffi-

TABLE 10.2 VELOCITY DISTRIBUTION COEFFICIENTS FOR ENERGY AND MOMENTUM

Channels	Value of α			Value of β		
	Min.	Av.	Max.	Min.	Av.	Max.
Regular channels, flumes, spillways	1.10	1.15	1.20	1.03	1.05	1.07
Natural streams and torrents	1.15	1.30	1.50	1.05	1.10	1.17
Rivers under ice cover	1.20	1.50	2.00	1.07	1.17	1.33
River valleys, overflooded	1.50	1.75	2.00	1.17	1.25	1.33

cient, α, on computations, and hence on designs, is quite significant and therefore should not be overlooked, even though the value of the coefficient may not always be determined accurately (Chow, 1959).

Example 10.2

From the discharge measurements in a channel, the values as indicated in Figure 10.3 have been ascertained for the velocity and area of the different sections. Determine the energy and momentum coefficients.

Solution

$$Q = AV = \Delta A_1 v_1 + \Delta A_2 v_2 + \Delta A_3 v_3 + \cdots$$

$$(3820)V = (120)1.2 + (540)1.43 + (880)2.30 + (920)2.42$$
$$+ (800)2.52 + (480)1.92 + (80)0.95$$
$$= 8180.2$$

$$V = \frac{8180.2}{3820} = 2.14$$

Section	Area, ΔA (ft^2)	Velocity, V (ft/sec)	ΔAV^3	ΔAV^2
1	120	1.2	207.36	172.80
2	540	1.43	1,579.07	1,104.25
3	880	2.30	10,706.96	4,655.20
4	920	2.42	13,038.69	5,387.89
5	800	2.52	12,802.41	5,080.32
6	480	1.92	3,397.39	1,769.47
7	80	0.95	68.59	72.20
Total	3820		41,800.5	18,242.13

$$\alpha = \frac{\sum V^3 \Delta A}{V^3 A} = \frac{41,800.5}{3820(2.14)^3} = 1.11$$

$$\beta = \frac{\sum V^2 \Delta A}{V^2 A} = \frac{18,242.13}{3820(2.14)^2} = 1.04$$

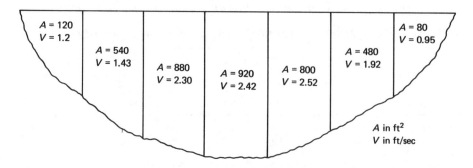

Figure 10.3 Discharge measurement at a channel section.

Conveyance System: Open Channel Flow   Chap. 10

10.6 PRINCIPLES OF CONTINUITY, ENERGY, AND MOMENTUM

These principles based on the three conservation laws of physics form the basis to describe all types of flow problems in hydraulic engineering. The principle of continuity is the law of conservation of mass. Since water is considered to be incompressible, its volume can be used interchangeably with mass. The continuity equation, where water does not run in or out along the course of flow, is

$$Q = V_1 A_1 = V_2 A_2 = \cdots \qquad [\text{L}^3\text{T}^{-1}] \qquad (10.5)$$

where

$$Q = \text{rate of flow or discharge}$$
$$V_1, V_2, \ldots = \text{velocity of flow at section } 1, 2, \ldots$$
$$A_1, A_2, \ldots = \text{cross-sectional area of section } 1, 2, \ldots$$

In hydraulics, the law of conservation of energy is expressed as the total head of water, which is equal to the sum of the elevation above a datum, the pressure head, and the velocity head at any point of flow. According to the principle of energy, the total energy head at a point in the upstream section 1 in Figure 10.4 is equal to the total energy head at a point in the downstream section 2, including the internal loss of energy in the mass of the water between the two sections, or

$$Z_1 + y_1 + \alpha_1 \frac{V_1^2}{2g} = Z_2 + y_2 + \alpha_2 \frac{V_2^2}{2g} + h_f \qquad [\text{L}] \qquad (10.6)$$

Equation (10.6) is known as the energy equation, or Bernoulli's equation.*

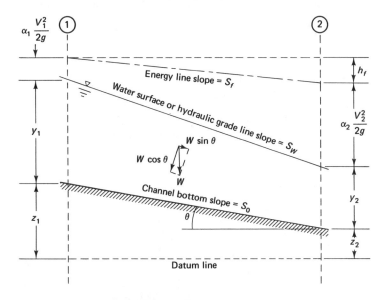

Figure 10.4 Various forms of energy in open channel flow.

*In Bernoulli's equation, $\alpha_1 = \alpha_2 = 1$ and $h_f = 0$.

Example 10.3

Determine the discharge per unit width of broad-crested weir in a rectangular channel as shown in Figure 10.5. Neglect the losses and assume that $\alpha_1 = \alpha_2 = 1$.

Solution Applying the energy equation at point 1 and at point 2 in the center of the weir. Neglecting the velocity of approach at point 1, we have

$$10 = 4 + 3 + \frac{V^2}{2g}$$

$$V = 13.90 \text{ ft/sec}$$

From the continuity equation,

$$q = AV$$

$$= (3 \times 1)(13.90) = 42 \text{ cfs/ft width}$$

(The coefficient of discharge, C_d, has not been included in the computation.)

The third conservation law relates to the momentum. Momentum should not be lost in a hydraulic system. Whenever a change occurs in momentum, it is converted into an impulse force (force multiplied by the time). According to Newton's second law of motion, the resultant external force acting on any body in any direction is equal to the rate of change of momentum of the body in that direction. The rate of momentum is defined as the product of mass flow rate (ρQ) and its velocity (V). In Figure 10.4,

$$\frac{\text{change of rate of momentum}}{\text{between sections 1 and 2}} = \frac{\text{resultant force}}{\text{between the sections}}$$

$$\rho Q(\beta_2 V_2 - \beta_1 V_1) = P_1 - P_2 + W \sin \theta - F_f \qquad [\text{MLT}^{-2}] \qquad (10.7)$$

where

$$P_1, P_2 = \text{resultant pressure forces at sections 1 and 2}$$

$$W = \text{weight of water between sections 1 and 2}$$

$$F_f = \text{force of friction acting along the surface of contact between the water and the channel}$$

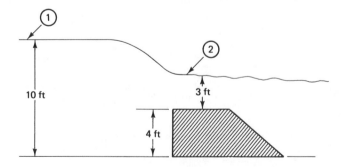

Figure 10.5 Application of energy principle on broad-crested weir.

Equation (10.7) is the momentum equation. The term, F_f, measures the force exerted by the water on the channel walls. It is not related to the internal losses. In many applications, F_f can be omitted.

The momentum equation accounts for all forces in a hydraulic system. It is distinct from the energy equation since energy is a scalar quantity, whereas momentum is a vector quantity. In certain problems, both lead to the similar results. In other problems, only one of them can be used conveniently. The energy equation is simpler to use and has a wider application. However, in problems involving unknown energy losses such as a hydraulic jump, the momentum equation has a direct application.

Example 10.4

Solve Example 10.3 by the momentum principle. Assume that $F_f = 0, \beta_1 = \beta_2 = 1$.

Solution   Refer to Figure 10.6.

1. Considering the horizontal bed, the weight component in the direction of flow is zero.
2. Pressure forces acting are:
   a. Upstream pressure at section 1,
   $$P_1 = \tfrac{1}{2}\gamma b y_1^2 = \tfrac{1}{2}\gamma(1)(10)^2$$
   b. Pressure at section 2,
   $$P_2 = \tfrac{1}{2}\gamma b y_2^2 = \tfrac{1}{2}\gamma(1)(3)^2$$
   c. Water pressure on the weir face at section 2,
   $$\begin{aligned} P_3 &= \tfrac{1}{2}\gamma b h[(y_1 - h) + y_1] \\ &= \tfrac{1}{2}\gamma(1)(4)[(10 - 4) + 10] \\ &= 32\gamma \end{aligned}$$

3. Applying the momentum equation between sections 1 and 2:
   $$\frac{\gamma}{g} q(V_2 - V_1) = \frac{1}{2}(100)\gamma - \frac{1}{2}(9)\gamma - 32\gamma$$

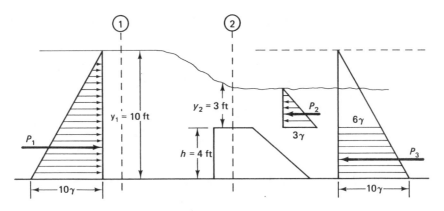

Figure 10.6   Application of the moment principle.

4. From the continuity equation,

$$q = A_1V_1 \quad \text{or} \quad V_1 = \frac{q}{(1)(10)} = \frac{q}{10}$$

and

$$q = A_2V_2 \quad \text{or} \quad V_2 = \frac{q}{(1)(3)} = \frac{q}{3}$$

5. Substituting in step 3 yields

$$\frac{1}{g} q^2 \left(\frac{1}{3} - \frac{1}{10} \right) = 13.5$$

$$q = 43 \text{ cfs/ft width}$$

A slight difference in the answer compared to Example 10.3 is due to the assumptions specifically neglecting the velocity of approach in Example 10.3.

10.7 CRITICAL FLOW CONDITION

10.7.1 Concept of Specific Energy

The energy in a channel section measured with respect to the channel bottom as the datum is known as the specific energy. In eq. (10.6), with $Z = 0$ and assuming that $\alpha = 1$, the specific energy is given by

$$E = y + \frac{V^2}{2g} \quad [L] \tag{10.8a}$$

Since $V = Q/A$, the equation may be written as

$$E = y + \frac{Q^2}{2gA^2} \quad [L] \tag{10.8b}$$

The first term on the right side relates to the static energy and the second to the kinetic energy. These have been plotted separately and then combined in Figure 10.7 for a graph between the depth against the specific energy for a constant discharge.

The combined curve (3) indicates that at point O, the specific energy is minimum. It will be demonstrated that this corresponds to the critical state of flow. The flow below this point is supercritical (low depth, high velocity). For a given specific energy, there are two alternate depths, one in the supercritical range and one in the subcritical range. At the critical state they merge into one depth, y_c.

For the condition of a minimum specific energy, $dE/dy = 0$. Differentiating eq. (10.8b), for a constant Q,

$$\frac{dE}{dy} = 1 - \frac{2Q^2}{2gA^3} \frac{dA}{dy} = 0 \tag{a}$$

Since $dA/dy = T$ and $A/T = D$, substituting in (a) gives us

$$\frac{Q^2}{gA^2} \frac{1}{D} = 1 \tag{b}$$

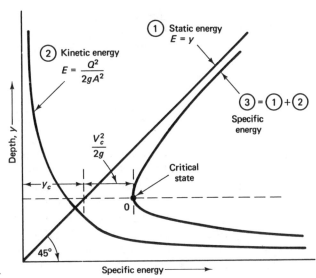

Figure 10.7 Specific energy plot.

Also, since $Q/A = V$,

$$\frac{V^2}{gD} = 1$$

or

$$\frac{V}{\sqrt{gD}} = 1 \tag{c}$$

The term on the left side of eq. (c) is the Froude number, Fr. As stated in section 10.5, Fr = 1 is the condition of the critical flow. Hence the specific energy is a minimum at the critical flow.

10.7.2 Computation of Critical Flow

From eq. (b) in Section 10.7.1,

$$A\sqrt{D} = \frac{Q}{\sqrt{g}} \tag{d}$$

Since $Z_c = A\sqrt{D}$ and including the energy coefficient,

$$Z_c = \frac{Q}{\sqrt{g/\alpha}} \qquad [\text{L}^{5/2}] \tag{10.9}$$

Equation (10.9) is used in the following two ways:

1. *Critical depth, y_c, given, to compute Q.* Compute the section factor, Z_c, for known y_c. Determine Q using eq. (10.9).
2. *To compute the critical depth for a given Q.* Calculate $Q/\sqrt{g/\alpha}$, which is equal to Z_c. For simple geometric sections, Z_c is expressed in terms of y_c in the form of an algebraic equation. The value of y_c is solved from the equation. For

complicated sections, the graphic procedure is used in which a curve of depth (y) versus $Z \, (= A\sqrt{D})$ is constructed. Corresponding to the Z_c value equal to $Q/\sqrt{g/\alpha}$, the critical depth is obtained directly from the curve.

Dimensionless curves and tables are available in handbooks (e.g., Chow, 1959, pp. 65, 625–627).

In Figure 10.7, curve (3) is almost vertical at the critical depth. A slight change in energy can cause substantial variation in depth. The flow at the critical state is thus unstable at which the water surface appears wavy. In the channel design, if the depth is found at or near the critical state for a great length of the channel, the shape or slope of the channel should be altered (Chow, 1959). The critical flow, however, serves the purpose of a control section. It is useful in defining the flow conditions and developing the surface water profiles, as explained subsequently.

The slope of a channel at which the computed uniform or normal depth of flow is equal to the critical depth is known as the critical slope, S_c. A slope less than the critical slope is the mild or subcritical slope, and a slope greater than critical is the steep or supercritical slope. These are important in flow profiles, as discussed subsequently.

Example 10.5

(a) Determine the specific energy of water in a rectangular channel 25 ft wide having a flow of 500 cfs at a velocity of 5 ft/sec.
(b) What is the critical depth of water in the channel?
(c) What is the critical velocity?

Solution

(a) 1. From the continuity equation,

$$A = \frac{Q}{V} = \frac{500}{5} = 100 \text{ ft}^2$$

$$by = 100 \quad \text{or} \quad y = \frac{100}{25} = 4 \text{ ft}$$

2. Specific energy

$$E = y + \frac{V^2}{2g} = 4 + \frac{(5)^2}{2(32.2)} = 4.39 \text{ ft}$$

(b) $Z_c = \dfrac{Q}{\sqrt{g}} = \dfrac{500}{\sqrt{32.2}} = 88.11$

If y_c is the critical depth, then for a rectangular channel,

$$A = by_c = 25y_c$$

$$D = \frac{A}{T} = \frac{25y_c}{25} = y_c$$

$$Z_c = A\sqrt{D} = 25y_c\sqrt{y_c} = 25y_c^{3/2}$$

or

$$25y_c^{3/2} = 88.11, \qquad y_c = 2.32 \text{ ft}$$

(c) At the critical flow,

$$V_c = \sqrt{gD} = \sqrt{gy_c} = \sqrt{32.2(2.32)} = 8.64 \text{ ft/sec}$$

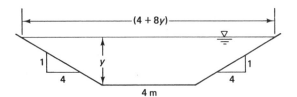

Figure 10.8 Channel section for Example 10.6.

Example 10.6

A trapezoidal channel with a bottom width of 4 m and side slopes of 4:1 carries a discharge of 30 m³/s (Figure 10.8). Determine the (a) critical depth, (b) critical velocity, and (c) minimum specific energy.

Solution

(a) 1. From eq. (10.9), assuming that $\alpha = 1$, $Z_c = Q/\sqrt{g}$.

2. $\dfrac{Q}{\sqrt{g}} = \dfrac{30}{\sqrt{9.81}} = 9.58$.

3. For any depth y,

$$A = \tfrac{1}{2}[4 + (4 + 8y)]y = (4 + 4y)y$$

$$T = 4 + 8y$$

$$Z_c = A\sqrt{D} = A\left(\frac{A}{T}\right)^{1/2} = \frac{A^{3/2}}{T^{1/2}} = \frac{[(4 + 4y)y]^{3/2}}{(4 + 8y)^{1/2}}$$

4. Using this formula, Z_c is computed for the various selected values of y in Table 10.3.

TABLE 10.3 SECTION FACTOR FOR A TRAPEZOIDAL CHANNEL

y (m) Select:	$A = (4 + 4y)y$	$A^{3/2}$	$T = (4 + 8y)$	$T^{1/2}$	$Z_c = \dfrac{A^{3/2}}{T^{1/2}}$
1.0	8	22.63	12	3.46	6.54
1.5	15	58.09	16	4.0	14.52
2.0	24	117.58	20	4.47	26.30
2.5	35	207.06	24	4.90	42.26

5. y versus Z_c has been plotted in Figure 10.9.
6. For Z_c corresponding to $Q/\sqrt{g}$ of 9.58, the value of y_c has been read from Figure 10.9 as

$$y_c = 1.22 \text{ m}$$

(b) At critical depth, $V_c = \sqrt{gD}$:

$$D = \frac{A}{T} = \frac{(4 + 4y_c)y_c}{4 + 8y_c} = \frac{[4 + 4(1.22)]1.22}{4 + 8(1.22)} = 0.787$$

$$V_c = \sqrt{9.81(0.787)} = 2.78 \text{ m/s}$$

(c) Specific energy at critical state

$$E = y_c + \frac{V_c^2}{2g} = 1.22 + \frac{(2.78)^2}{2(9.81)} = 1.61 \text{ m}$$

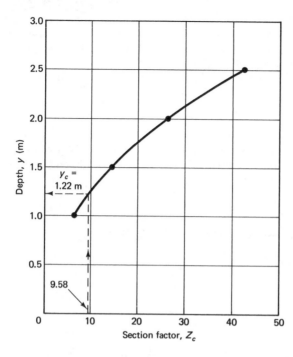

Figure 10.9 Depth versus section factor curve for critical depth for Example 10.6.

10.8 UNIFORM CHANNEL FLOW

Uniform flow occurs in a steady state only. In uniform flow, the depth, area of cross section, velocity of flow, and discharge are the same at every section of a channel. Such a condition develops when the force in the direction of the flow is fully balanced by the resistance encountered by the water as it moves downstream. This does not occur frequently. The long channels may have only small reaches of uniform flow. However, the uniform flow is a basic flow in channel hydraulics. The designs are based on the consideration of the uniform flow.

10.8.1 Hydraulics of Uniform Flow

As stated earlier, the Darcy–Weisbach (1845)* equation of pipe flow is applicable to the uniform channel flow. The commonly encountered flow in channels is rough turbulent flow, for which the friction factor relations of pipe, by Blasius (1913)* and Prandtl and von Kármán (1935)*, are not directly applicable. In 1939, Colebrook and White suggested an equation (Hydraulic Research Station, 1983) for open channel flow which is, however, not directly solvable because of its implicit form, in which the term of channel slope appears on both sides of the equation. Hydraulic Research Station Ltd. in the United Kingdom have prepared charts for the application of this equation.

The common approach in the United States is to use the Chezy or Manning formula developed in 1769 and 1889, respectively. There is a conformity in the concept of Chezy and Darcy–Weisbach that the head loss varies as the square of the velo-

*Referenced in Chapter 11.

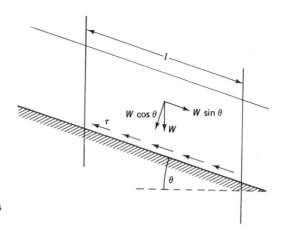

Figure 10.10 Derivation of Chezy's formula.

city. Chezy's formula can be derived by equating the propulsive force due to the weight of water in the direction of flow with the retarding shear force at the channel boundary. From Figure 10.10,

$$\text{propulsive force,} \quad F = W \sin \theta = \gamma AL \sin \theta$$

$$\text{resisting force,} \quad R = \tau PL$$

Since $F = R$,

$$\gamma AL \sin \theta = \tau PL$$

For turbulent flow,

$$\tau \propto V^2 \quad \text{or} \quad \tau = KV^2$$

and for small slope,

$$\sin \theta = \tan \theta = S$$

Hence

$$V = \left(\frac{\gamma}{K}\frac{A}{P}S\right)^{1/2}$$

or

$$V = C\sqrt{RS} \quad [\text{LT}^{-1}] \tag{10.10}$$

where

V = mean velocity, ft/sec

R = hydraulic radius, ft

S = slope of energy line, which is equal to channel bottom for uniform flow

C = Chezy's constant

Three formulas by Ganguillet and Kutter, Bazin, and Powell are commonly used to determine the Chezy constant, C. Of these the first formula is most satisfactory. It uses the roughness coefficient, known as Kutter's n, which is almost equal to Manning's coefficient n.

If $C = R^{1/6}/n$ is substituted in Chezy's formula, Manning's formula results. Manning's formula has proven most reliable in practice. This empirical formula, suitable for a fully rough turbulent flow, is given by

$$V = \frac{1.486}{n} R^{2/3} S^{1/2} \qquad \text{(English units)} \qquad [LT^{-1}] \qquad (10.11)$$

$$V = \frac{1}{n} R^{2/3} S^{1/2} \qquad \text{(Metric units)} \qquad [LT^{-1}] \qquad (10.12)$$

where n is the Manning's roughness coefficient. It depends on channel material, surface irregularities, variation in shape and size of the cross section, vegetation and flow conditions, channel obstruction, and degree of meandering. Chow (1959) has provided a detailed table and photographs of channels for values of n in different conditions. Typical values are summarized in Table 10.4.

TABLE 10.4 VALUES OF MANNING'S ROUGHNESS COEFFICIENT (For sheet flow see eq. 12.8)

Conduit material	Manning n
Closed conduits	
Asbestos-cement pipe	0.011–0.015
Brick	0.013–0.017
Cast iron pipe	
Cement-lined and seal coated	0.011–0.015
Concrete (monolithic)	
Smooth forms	0.012–0.014
Rough forms	0.015–0.017
Concrete pipe	0.011–0.015
Corrugated-metal pipe	
(½-in. × 2⅔-in. corrugations)	
Plain	0.022–0.026
Paved invert	0.018–0.022
Spun asphalt lined	0.011–0.015
Plastic pipe (smooth)	0.011–0.015
Vitrified clay	
Pipes	0.011–0.015
Liner plates	0.013–0.017
Open channels	
Lined channels	
Asphalt	0.013–0.017
Brick	0.012–0.018
Concrete	0.011–0.020
Rubble or riprap	0.020–0.035
Vegetal	0.030–0.40
Excavated or dredged	
Earth, straight and uniform	0.020–0.030
Earth, winding, fairly uniform	0.025–0.040
Rock	0.030–0.045
Unmaintained	0.050–0.14
Natural channels (minor streams,	
top width at flood stage < 100 ft)	
Fairly regular section	0.03–0.07
Irregular section with pools	0.04–0.10

10.8.2 Computation of Uniform Flow

Combining the continuity equation $Q = AV$ with eq. (10.11) or (10.12), the Manning formula is obtained in terms of discharge as follows:

$$Q = \frac{1.486}{n} AR^{2/3}S^{1/2} \qquad \text{(English units)} \qquad [L^3T^{-1}] \qquad (10.13a)$$

$$Q = \frac{1}{n} AR^{2/3}S^{1/2} \qquad \text{(Metric units)} \qquad [L^3T^{-1}] \qquad (10.13b)$$

The terms on the right side, excluding the slope, are grouped into a single term known as the conveyance, K. Thus the formula is also stated as $Q = K\sqrt{S}$. As defined earlier, $AR^{2/3}$ is called the section factor, Z_n, for uniform or normal flow, which is a function of depth for a given channel section. The depth of flow corresponding to uniform channel flow is known as the normal depth, y_n.

Three variables are involved in eq. (10.13): (1) discharge or velocity, (2) slope, and (3) section factor (a function of depth). For any two of these known, the third one can be computed. Three cases are described below:

1. *Normal depth and slope are known; compute the discharge.* Determine the section factor, $AR^{2/3}$, for a given normal depth. The application in eq. (10.13) is direct to compute Q.
2. *Discharge and normal depth are known; compute the slope.* Again the application is direct in eq. (10.13).
3. *Discharge and slope are known; compute the normal depth.* Equation (10.13) is rearranged as

$$AR^{2/3} = \frac{Qn}{1.486S^{1/2}} \qquad [L^{8/3}] \qquad (10.14)$$

The right side of eq. (10.14) is evaluated from the known variables. For simple channel geometry, $AR^{2/3}$ is expressed in terms of y_n and solved for directly. For other cases, a plot of y versus $AR^{2/3}$ is prepared. From this the value of y_n is obtained for the computed value of $AR^{2/3}$ equal to $Qn/1.486S^{1/2}$.

For a circular channel section running full, the application of the Manning formula is direct in all three cases above. In a circular section that is flowing only partially full, the geometric relations in terms of the diameter given in Table 10.1 are very useful. For an example to determine the depth of flow, the section factor $AR^{2/3}$ is computed from the discharge. Entering Table 10.1 from the last column of $AR^{2/3}/d_0^{8/3}$, the value of y/d_0, and hence y, is obtained. Figure 10.11 can also be used for quick computation. To use this figure it is necessary first to find the values when the section is flowing full. As stated earlier, the maximum flow in a circular section occurs at the depth of 0.94 of the diameter.

Example 10.7

Calculate the discharge through a 3-ft-diameter circular clean earth channel running half full. The bed slope is 1 in 4500.

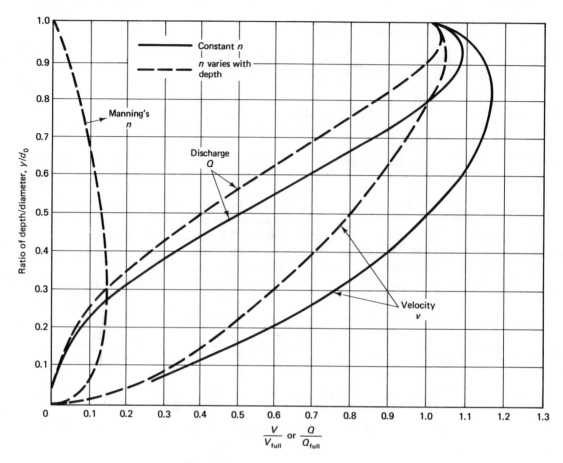

Figure 10.11 Hydraulic elements of a circular section.

Solution

1. From Table 10.4, $n = 0.018$.

2. $S = \dfrac{1}{4500} = 2.22 \times 10^{-4}$.

3. $\dfrac{y_n}{d_0} = \dfrac{1.5}{3} = 0.5$.

4. From Table 10.1, $\dfrac{AR^{2/3}}{d_0^{8/3}} = 0.1558$.

5. $AR^{2/3} = 0.1558(3)^{8/3} = 2.92$.

6. $Q = \dfrac{1.49}{0.018}(2.92)\,(2.22 \times 10^{-4})^{1/2} = 3.60$ cfs.

Alternative Solution   From Figure 10.11 for constant n, y/d_0 of 0.5, $Q/Q_{\text{full}} = 0.5$,

$$Q_{\text{full}} = \frac{1.49}{0.018}\left[\frac{\pi}{4}(3)^2\right]\left(\frac{3}{4}\right)^{2/3}(2.22 \times 10^{-4})^{1/2} = 7.19 \text{ cfs}$$

$$Q = 0.5Q_{\text{full}} = 0.5(7.19) = 3.60 \text{ cfs}$$

Example 10.8

A trapezoidal channel of bottom width 25 ft and side slope 1:2.5 carries a discharge of 450 cfs with a normal depth of 3.5 ft. The elevations at the beginning and end of the channel are 685 and 650 ft, respectively. Determine the length of the channel if $n = 0.02$.

Solution   Refer to Figure 10.12.

1. This is a problem of the determination of S.

2. $A = \frac{1}{2}(25 + 42.5)3.5 = 118.13$

$$P = 25 + 9.42 + 9.42 = 43.84$$

$$R = \frac{A}{P} = \frac{118.13}{43.84} = 2.69$$

3. $S = \left[\dfrac{Q}{(1.49/n)AR^{2/3}}\right]^2 = \left[\dfrac{450}{(1.49/0.02)(118.13)(2.69)^{2/3}}\right]^2 = 0.0007$

4. $\dfrac{H_1 - H_2}{L} = S$ or

$$L = \frac{H_1 - H_2}{S} = \frac{685 - 650}{0.0007} = 50,000 \text{ ft}$$

Example 10.9

The channel of Example 10.6 has the bottom of 0.1% and $n = 0.025$. Determine the (a) normal depth, (b) critical slope, and (c) state of flow in the channel.

Solution

(a) 1. $S = \dfrac{0.1}{100} = 0.001$

2. $\dfrac{Qn}{S^{1/2}} = \dfrac{30(0.025)}{(0.001)^{1/2}} = 23.72$

3. $A = (4 + 4y)y$

$$P = 4 + 8.24y$$

$$R = \frac{(4 + 4y)y}{4 + 8.24y}$$

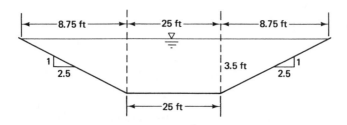

Figure 10.12   Channel section of Example 10.8.

4. For selected values of y, Z_n is computed below.

Selected y	$A = (4 + 4y)y$	$P = 4 + 8.24y$	$R = \dfrac{A}{P}$	$Z_n = AR^{2/3}$
1.0	8	12.24	0.65	6.00
1.5	15	16.36	0.92	14.19
2.0	24	20.48	1.17	26.65
2.5	35	24.60	1.42	44.22
3.0	48	28.72	1.67	67.57

5. y versus Z_n has been plotted in Figure 10.13. From this graph, for Z_n of 23.72, $y_n = 1.9$ m.

(b) 1. Critical depth, $y_c = 1.22$ m (from Example 10.6). For critical depth,

$$A = 10.83$$

$$P = 14.05$$

$$R = \frac{A}{P} = \frac{10.83}{14.05} = 0.77$$

2. $S_c = \left[\dfrac{30}{(1/0.025)(10.83)(0.77)^{2/3}} \right]^2 = 0.0068$

(c) Since the channel bottom slope < the critical slope, it is a mild slope and subcritical flow.

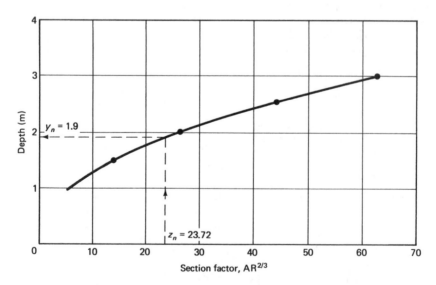

Figure 10.13 Depth versus section factor for normal depth for Example 10.8.

10.9 DESIGN OF RIGID-BOUNDARY CHANNELS

This is based on uniform flow. In the preceding section on hydraulic computations, while calculating the depth of flow, the channel dimensions and the longitudinal bed slope are taken as known values. The channel design comprises sizing up these variables in addition to the total channel depth. Obviously, additional relations and criteria are needed for this purpose. These are based on the considerations discussed below.

10.9.1 Permissible Velocity of Flow

There is an upper and a lower limit for the channel velocity. In a rigid-boundary channel, the maximum permissible velocity is not a problem. As a minimum limit it is necessary to choose a velocity that will not start sediment deposit, induce aquatic growth, or start sulfide formation in the case of sanitary sewers. The velocity required to transport material in sewers is only slightly dependent on conduit shape and depth of flow, and primarily dependent on the particle size and specific weight. Generally, a minimum velocity of 2 to 3 ft/sec (0.61 to 0.91 m/s) is used for open channels and sanitary and storm sewers. A velocity of 2 ft/sec will be sufficient to move a 15.0-mm-diameter organic or 2.0-mm sand particle (ASCE and Water Pollution Control Federation, 1982).

10.9.2 Bottom Longitudinal Slope

This is governed by the topography and head requirements. When the two ends of a channel are fixed and the channel has to be laid on a predetermined alignment, the slope gets fixed accordingly. The conveyance channels for water supply, irrigation, and hydropower require a higher level at the point of delivery and therefore have a relatively small slope.

In circular pipes, the slopes required for a minimum velocity of 2 ft/sec (0.6 m/s) at a value of n of about 0.015* for various flows are indicated in Table 10.5

TABLE 10.5 SLOPES REQUIRED FOR VARIOUS FLOWS FOR CIRCULAR CONDUIT

Flow (cfs)	m³/s	Slope (ft/1000 ft)
0.1	2.8×10^{-3}	9.2
0.2	5.7×10^{-3}	6.1
0.3	8.5×10^{-3}	4.8
0.4	1.1×10^{-2}	4.1
0.6	1.7×10^{-2}	3.22
0.8	2.3×10^{-2}	2.73
1.0	2.8×10^{-2}	2.39
1.5	4.2×10^{-2}	1.89
2.0	5.7×10^{-2}	1.59
3.0	8.5×10^{-2}	1.26
4.0	1.1×10^{-1}	1.06

*For clay, concrete, cast iron, and plastic pipes, n is considered to be 0.015. For smooth joints and good construction, it is 0.013.

according to a study by Pomeroy (1967). These are valid for any conduit size if the indicated flow is not less than 0.1% or more than 95% of the full conduit capacity.

10.9.3 Channel Side Slopes

The channel side slopes depend on the type of material of the channel. A nearly vertical slope in rocks, to 1 (vertical):3 (horizontal) in sandy soil, are recommended slopes. The Bureau of Reclamation prefers a 1:1.5 slope for usual sizes of lined canals.

10.9.4 Freeboard

This is the vertical distance from the water surface to the top of the channel. It should be sufficient to prevent the overtopping of the channel by waves or fluctuating water surface. The Bureau of Reclamation has recommended the following formula:

$$u = \sqrt{cy} \qquad \text{[unbalanced]} \qquad (10.15)$$

where

u = freeboard, ft

c = coefficient varying from 1.5 for a capacity of
       20 cfs to 2.5 for a capacity of 3000 cfs or more

y = water depth in the canal, ft

10.9.5 Hydraulic Efficient Sections

For a given slope and roughness coefficient, the discharge increases with an increase in the section factor. For a given area, the section factor is highest for the least wetted perimeter. The expressions for the wetted perimeter can be written in terms of the depth for various channel shapes. Its minimization by differentiating and equating to zero provides the depth relation of the best hydraulic section. From hydraulic efficiency considerations a semicircle is the best of all sections for an open channel. The best closed section flowing full is a circle. Any reasonably shaped open section is more efficient than the closed (conduit) sections flowing full. The dimensions of a channel are not governed entirely by the hydraulic efficiency but by practical and cost considerations as well. A trapezoidal section is very common. The following properties are related to the best hydraulic section except where these have to be changed from practical considerations.

1. *Trapezoidal section*. The best hydraulic section never has a base width larger than the depth of water.
2. *Rectangular section*. The width is twice the depth in the best hydraulic section.
3. *Triangular section*. Side slopes are selected by practical consideration.
4. *Circular section*. A semicircle is a best section for channels open at top, and a circle is best as a closed section.

10.9.6 Design Procedure

1. Select S and estimate n from available data.
2. Substituting in the right side of eq. (10.14), determine the section factor $AR^{2/3}$.
3. Select the side slope z and assume b/y as necessary. Express $AR^{2/3}$ in terms of the depth. Solve for the depth as in the preceding section.
4. Assuming several values of the unknowns, a number of section dimensions can be obtained to make a cost comparison.
5. Check for the minimum velocity.
6. Add a freeboard to the water depth for an open section.

Storm sewers and wastewater sewers are designed by the procedure above except for the computation of the quantity of flow. The storm discharge is computed based on the drainage area and wastewater flow from the quantity of water supply, as discussed in Chapter 2.

Example 10.10

Design a rigid-boundary earth channel to carry a discharge of 1.08 m³ per second.

Solution

1. Based on the topography and channel alignment, $S = 0.001$.

2. For a clean earth channel, $n = 0.018$.

3. Designing a trapezoidal channel: Based on the material, the side slope $z = 1$.

4. $AR^{2/3} = \dfrac{Qn}{S^{1/2}} = \dfrac{1.08(0.018)}{(0.001)^{1/2}} = 0.615$.

5. Assume that $b/y = 1$.

$$A = 2y^2$$
$$P = 3.83y$$
$$R = \frac{2y^2}{3.83y} = 0.522y$$

6. $AR^{2/3} = (2y^2)(0.522y)^{2/3} = 1.30y^{8/3}$ or

$$1.30y_n^{8/3} = 0.615, \qquad y_n = 0.75 \text{ m}$$

7. Check for velocity:

$$V = \frac{Q}{A} = \frac{1.08}{2(0.75)^2} = 0.96$$

Since V of 0.96 > V_{min} of 0.61, it is OK.

8. For freeboard, values to be used:

$$y = 0.75 \text{ m} \quad \text{or} \quad 2.46 \text{ ft}$$
$$u = \sqrt{1.5(2.46)} = 1.92 \text{ ft} \quad \text{or} \quad 0.59 \text{ m}$$

9. Total channel depth = $0.75 + 0.59 = 1.34$ m.

Example 10.11

A district has a drainage area of 2500 acres with a population of 20 persons per acres. The daily water supply to the district is 40 gallons per head. It has been observed that 10% of this flow passes along the sewer between the hours of 7 to 8 a.m. If the sewer consists of vitrified clay laid to 0.1% grade, design the sewer.

Solution $n = 0.013$ for vitrified clay, $S = 0.1/100 = 0.001$.

$$\text{Total water supply} = (\text{area})(\text{person/acre})(\text{supply/head})$$

$$= 2500(20)(40) = 2 \times 10^6 \text{ gpd}$$

$$\text{Flow passing to the sewer/hr} = 0.1(2 \times 10^6) = 0.2 \times 10^6 \text{ gph} \quad \text{or} \quad 7.4 \text{ cfs}$$

$$AR^{2/3} = \frac{Qn}{1.486S^{1/2}} = \frac{(7.4)(0.013)}{(1.486)(0.001)^{1/2}} = 2.05$$

Maximum flow at $0.94d_0$, for which $AR^{2/3} = 0.3353d_0^{8/3}$; from Table 10.1, hence $0.3353d_0^{8/3} = 2.05$. Therefore, $d_0 = 1.97$ ft.

The minimum velocity should be checked for some minimum rate of flow which is not specified. Small sized sewers are usually designed to flow partially full, as discussed in Sections 12.4 and 12.5.

10.10 DESIGN OF LOOSE-BOUNDARY CHANNELS

Channels formed in erodible or alluvial material such as irrigation canals are subject to erosion by the following water. The uniform flow formula used in the design of rigid-boundary channels does not provide a sufficient condition for the design of erodible channels because the stability of such channels depends on the properties of the channel material. The erodible channels have to be designed for stability, which has been defined by Lane as a condition in an unlined channel, the bed and banks of which are not scoured by the moving water and in which objectionable deposits of sediment do not occur. A channel may scour or form deposits to some extent, but the long-term effect over a yearly period should be zero. Further, the channels are operated over a wide range of discharge conditions. It is not possible to design a channel to be stable for more than one discharge. Since the sediment deposited at low flows will be carried away during high flows, it should be designed for maximum discharge. The physical factors and their behavior on flow in erodible channels have not been exactly identified. The precise design on an analytical basis is not possible. Field observations and experience play an important part. There are two approaches in design: the regime theory and the tractive force theory. The regime theory, based on empirical relations, is suitable to channels carrying the sediment load. The earlier regime theory by Kennedy (1895), Lindley (1919), Lacey (1930), Inglish (1941), and Blench (1952) was developed for channels formed in the silt-sand range. Recent studies by Kellerhals (1967) and Charlton et al. (1978) have been made for gravel-bed channels (Richards, 1982). The tractive force theory is convenient for (1) silt and sand-bed channels carrying clean water without sediment, (2) gravel-bed channels carrying fine sediment, and (3) rigid-boundary channels carrying sediment of any size.

10.10.1 Regime Theory

The regime theory was started by R. G. Kennedy, who introduced a formula in 1895 of nonscouring nonsilting velocity. E. S. Lindley, however, advanced the concept of regime in 1919, according to which all channel dimensions and slope are fixed by nature for the given conditions. Based on the extensive study of the data and following the regime concept, Lacey produced a set of formulas in 1929, which were modified by him in 1953. Refinements and extensions of Lacey's work were made by Inglish during 1941–1947 and Blench during 1955–1966, but Lacey's formulas are still popular in design. Lacey's three basic relations to achieve true regime are:

$$\text{Velocity–depth relation:} \quad V = 1.15\sqrt{fR} \qquad \text{[unbalanced]} \qquad (10.16)$$

$$\text{Velocity–slope relation:} \quad V = 16R^{2/3}S^{1/3} \qquad \text{[unbalanced]} \qquad (10.17)$$

$$\text{Width–discharge relation:} \quad P = 2.67Q^{1/2} \qquad \text{[unbalanced]} \qquad (10.18)$$

where

$$V = \text{mean velocity, ft/sec}$$
$$R = \text{hydraulic mean radius, ft}$$
$$P = \text{wetted perimeter, ft}$$
$$S = \text{longitudinal slope}$$
$$f = \text{sediment factor given as follows:}$$
$$f = 1.76D_o^{1/2} \qquad \text{[unbalanced]} \qquad (10.19)$$

where D_o is the mean grain diameter, mm. The values of f for various materials are indicated in Table 10.6.

TABLE 10.6 SILT FACTOR AND PERMISSIBLE UNIT TRACTIVE FORCE

Material	Size (mm)	Silt factor, f	Average permissible unit tractive force (lb/ft^2)
Small/medium boulders, cobbles, and shingles	64–256	6.12–9.75	0.92
Coarse gravel	8–64	4.68	0.48
Fine gravel	4–8	2.00	80×10^{-3}
Coarse sand	0.5–2	1.44–1.56	50×10^{-3}
Medium sand	0.25–0.5	1.31	35×10^{-3}
Fine sand	0.06–0.25	1.1–1.3	25×10^{-3}
Silt (colloidal)		1.00	0.2–0.3
Fine silt (colloidal)		0.4–0.9	
Compact clay (colloidal)			0.3–0.4
Loose clay (colloidal)			0.05

Combining eqs. (10.16) through (10.18), the following equations emerge that are used in the design:

$$R = 0.47\left(\frac{Q}{f}\right)^{1/3} \quad \text{[unbalanced]} \tag{10.20}$$

$$S = \frac{f^{5/3}}{1859Q^{1/6}} \quad \text{[unbalanced]} \tag{10.21}$$

The design steps are as follows:

1. Design discharge, Q, and usually the slope, S, are known.
2. Select a value of silt factor, f. It can be taken from Table 10.6 or computed by eq. (10.19) using the mean diameter of bed material. If the sediment load exceeds 2000 ppm with significant coarse material, the value of f should be more than from eq. (10.19).
3. Determine P from eq. (10.18).
4. Determine R from eq. (10.20).
5. Using the expressions for P and R in terms of channel width and depth, solve simultaneously for b and y.
6. Compute S by eq. (10.21). Compare with the available slope. Where the computed slope is more than the available slope, the channel could be widened or excess head can be absorbed in drop structures. Where the computed slope is less than the available slope, the realignment may be necessary. Thus the slope by eq. (10.21) is the minimum slope.
7. Add a proper freeboard.

Example 10.12

A canal for a maximum discharge of 2000 cfs is to be formed in alluvial material of the mean diameter of 0.2 mm. (8×10^{-3} in.). The channel alignment has a slope of 1 in 10,000. The sediment load is 1000 ppm. Design the channel.

Solution

1. From eq. (10.19), $f = 1.76\sqrt{0.2} = 0.79$. Since the sediment load < 2000 ppm, the calculated f can be used.
2. From eq. (10.18), $P = 2.67(2000)^{1/2} = 119$ ft.
3. From eq. (10.20), $R = 0.47\left(\frac{2000}{0.79}\right)^{1/3} = 6.4$ ft.
4. $A = PR = 119\,(6.4) = 762$ ft².
5. Assuming the side slope of $1:1$:

$$A = (b + y)y \quad \text{and} \quad P = b + 2.83y$$

Hence

$$(b + y)y = 762 \tag{a}$$
$$b + 2.83y = 119 \tag{b}$$

Solving eqs. (a) and (b), $b = 98$ ft, $y = 7.2$ ft.

6. From eq. (10.21),

$$S_{min} = \frac{(0.79)^{5/3}}{1859(2000)^{1/6}} = 0.0001$$

Available $S = 1/10{,}000 = 0.0001 =$ computed slope; hence OK.

7. Adding a freeboard of 3.0 ft, total depth $= 10.2$ ft.

10.10.2 Tractive Force Theory

The force exerted by the moving water on the wetted surface is the tractive or drag force, which is equal to the component of the weight of water in the direction of flow. From Figure 10.14,

$$\text{tractive force,} \quad F = W \sin \theta = \gamma ALS$$

$$\text{boundary shear stress} = \frac{\text{tractive force}}{\text{contact area}}$$

$$= \frac{\gamma ALS}{PL}$$

$$= \gamma RS \quad [ML^{-1}T^{-2}] \quad (10.22)$$

The boundary shear stress is, however, not uniformly distributed on the bed and the slopes. For trapezoidal sections, the maximum stress, referred to as the theoretical stress or theoretical unit tractive force, may be taken as

$$\text{For bed:} \quad \tau_0 = \gamma y S \quad [ML^{-1}T^{-2}] \quad (10.23)$$

$$\text{For sides:} \quad \tau_s = 0.76 \gamma y S \quad [ML^{-1}T^{-2}] \quad (10.24)$$

If the theoretical shear stresses above are less than the stresses that will cause the material of the channel boundary to move, the channel will be stable. The stress at which the channel material moves is known as the critical or permissible stress (permissible unit tractive force) and is a function of the size of material and the sediment content of water. The U.S. Bureau of Reclamation has prepared the graphs of the permissible stresses for noncohesive and cohesive materials (Chow, 1959, pp. 173–174). Typical values for straight channels with clear water are given in Table 10.6.

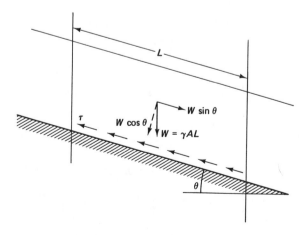

Figure 10.14 Theory of tractive force.

The permissible stresses for the sides of the channel are only a fraction of the permissible values for the bed, since a particle on the slope is also subject to the gravity force down the slope in addition to the shear force of the flowing water. The reduction factor is given by

$$K = \sqrt{1 - \frac{\sin^2\theta}{\sin^2\phi}} \qquad \text{[dimensionless]} \qquad (10.25)$$

where

$$K = \text{factor for permissible stress on sides}$$
$$\theta = \text{slope of the side to the horizontal}$$
$$\phi = \text{angle of repose of the material}$$

The design procedure comprises proportioning of the section based on the maximum stress on the sides and checking for the maximum stress on the bottom. The steps are as follows:

1. Select a value of the permissible stress (critical) for the channel material.
2. Determine K for a predecided side slope. Multiply the value of step 1 by K to obtain the permissible stress for sides of the channel.
3. Equate the permissible stress on sides to the theoretical value (i.e., $0.76\gamma yS$) and solve for y.
4. Substitute in Manning's equation to obtain the width, b.
5. Determine the theoretical shear stress on the bed = γyS. This should be less than the permissible value of step 1.

Example 10.13

Design a trapezoidal channel to carry a discharge of 300 cfs. The channel laid on a slope of 0.0015 is excavated in coarse gravels of effective diameter of 30 mm (1.2 in.). Consider a side slope of $1:2$, $n = 0.030$, and angle of repose = $35°$.

Solution

1. For coarse noncohesive material of 30 mm, critical stress = 0.48 lb/ft^2.
2. For a $1:2$ slope, $\theta = 26.57°$.
3. $K = \sqrt{1 - \dfrac{(\sin 26.57)^2}{(\sin 35)^2}} = 0.63$
4. Permissible stress on sides = 0.63(0.48) = 0.30.
5. $0.76\gamma yS = 0.30$, or

$$y = \frac{0.3}{0.76(62.4)(0.0015)} = 4.22 \text{ ft}$$

6. From Manning's equation,

$$300 = \frac{1.49}{0.03}[b + 8.44](4.22)\left[\frac{(b + 8.44)4.22}{b + 4.47(4.22)}\right]^{2/3}(0.0015)^{1/2}$$

or

$$14.15 = \frac{(b + 8.44)^{5/3}}{(b + 18.86)^{2/3}}$$

By trial and error, $b = 11$ ft.

7. Theoretical shear stress on the bed:

$$\tau_0 = \gamma y S = 62.4(4.22)(0.0015) = 0.39 < 0.48 \qquad \text{OK}$$

8. Add a freeboard of 2.5 ft.

10.11 GRADUALLY VARIED FLOW

10.11.1 Dynamic Equation of Gradually Varied Flow

When the gravity force causing the flow is not balanced with the resisting drag force, the depth varies gradually along the length of the channel. The dynamic equation of gradually varied flow is derived from the energy principle and indicates the slope of the water surface in the channel.

In Figure 10.15, the total energy at point 1 is

$$H = Z_1 + y + \alpha \frac{V^2}{2g} \tag{a}$$

Differentiating with respect to the channel bottom as the x-axis:

$$\frac{dH}{dx} = \frac{dZ_1}{dx} + \frac{dy}{dx} + \alpha \frac{d}{dx}\left(\frac{V^2}{2g}\right) \tag{b}$$

If the level increasing in the direction of flow is assumed positive then $dH/dx = -S_f$, $dZ_1/dx = -S_0$, and

$$\frac{d}{dx}\left(\frac{V^2}{2g}\right) = \frac{d}{dy}\left(\frac{V^2}{2g}\right)\frac{dy}{dx}$$

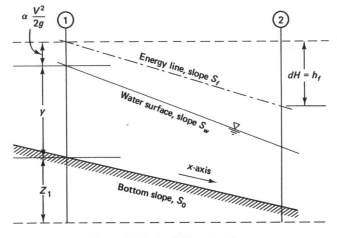

Figure 10.15 Gradually varied flow.

Equation (b) becomes

$$\frac{dy}{dx} = \frac{S_0 - S_f}{1 + \alpha[d(V^2/2g)/dy]} \qquad \text{[dimensionless]} \qquad (10.26)$$

This is the equation of gradually varied flow. To reduce the equation further, it is considered that with the energy grade, S_f, used for the slope term in Manning's equation, that formula can be used for the gradually varied flow through a section, that is,

$$Q = \frac{1.49}{n} AR^{2/3} S_f^{1/2} \qquad (c)*$$

or

$$Q = K\sqrt{S_f} \qquad (d)$$

where K is the general expression for the conveyance. Also, in the case of uniform flow,

$$Q = K_n \sqrt{S_0} \qquad (e)$$

where K_n is the normal flow conveyance. From eqs. (d) and (e),

$$\frac{S_f}{S_0} = \frac{K_n^2}{K^2} \qquad (f)$$

The denominator term of eq. (10.26) may be developed as follows:

$$\alpha \frac{d}{dy}(V^2/2g) = \alpha \frac{d}{dy}\left(\frac{Q^2}{2gA^2}\right) = -\alpha \frac{Q^2}{g}\left(\frac{1}{A^3}\right)\frac{dA}{dy} \qquad (g)$$

Since $dA/dy = T$ and in general terms, $Z = \sqrt{A^3/T}$,

$$\alpha \frac{d}{dy}\left(\frac{V^2}{2g}\right) = -\alpha \frac{Q^2}{gZ^2} \qquad (h)$$

For the critical flow, $Z_c = Q/\sqrt{g/\alpha}$; hence

$$\alpha \frac{d}{dy}\left(\frac{V^2}{2g}\right) = -Z_c^2/Z^2 \qquad (i)$$

Substituting eqs. (f) and (i) in eq. (10.26), we have

$$\frac{dy}{dx} = S_0 \frac{1 - (K_n/K)^2}{1 - (Z_c/Z)^2} \qquad \text{[dimensionless]} \qquad (10.27)$$

Equation (10.27) is another form of the gradually varied flow equation, which is convenient for the evaluation.

10.11.2 Types of Flow Profile Curves

The integration of eq. (10.27) will represent the surface curve of the flow. The shape or profile of the surface curve depends on (1) the slope of the channel, and (2) the

*In metric units $Q = \frac{1}{n} AR^{2/3} S_f^{1/2}$

depth of flow compared to the critical and normal depths. The classification is as follows:

Sign convention

1. If the water surface is rising in the direction of flow, the curve, known as the backwater curve, is positive.
2. If the water surface is dropping, the curve, known as the drawdown curve, is negative.

Channel slopes

1. When $y_n > y_c$, the slope is mild.
2. When $y_n < y_c$, the slope is steep.
3. When $y_n = y_c$, the slope is critical.
4. When $S_0 = 0$, the slope is horizontal. For horizontal slope, $y_n = \infty$.
5. When $S_0 < 0$, the slope is adverse. For adverse slope, y_n is negative or nonexistent.

Flow profiles. If the lines are drawn at the critical depth and the normal depth parallel to the channel bottom, three zones are formed. Zone 1 is the space above the upper line, zone 2 is the space between the two lines, and zone 3 is the space between the lower line and the channel bottom. The zone in which the water surface lies determines the flow profile, as shown in Figure 10.16.

10.11.3 Flow Profile Analysis

The analysis predicts the general shape of the flow profile in a longitudinal section of a channel without performing the quantitative analysis. For a channel of constant slope, the conditions described in the preceding section determine the type of flow profile. A break in the slope of a channel results in a change in the flow condition as well. A surface curve is often formed to negotiate the change of pattern of flow. Chow (1959, p. 232) has indicated 20 typical flow profiles for a combination of different types of channel slopes. Certain points in the channel reach serve as a control section where the depth of flow is fixed (i.e., either it is y_c or y_n or has some other known value).

10.11.4 Computation of Flow Profile

The analytical determination of the shape of flow profile essentially is a solution of eq. (10.27). Since the variables on the right side of the equation cannot be expressed explicitly in terms of y, the exact integration of the equation is not practically possible. There are four approaches to computing the surface profile:

1. Graphical or numerical integration method
2. Analytical or direct integration method
3. Direct step method
4. Standard step method

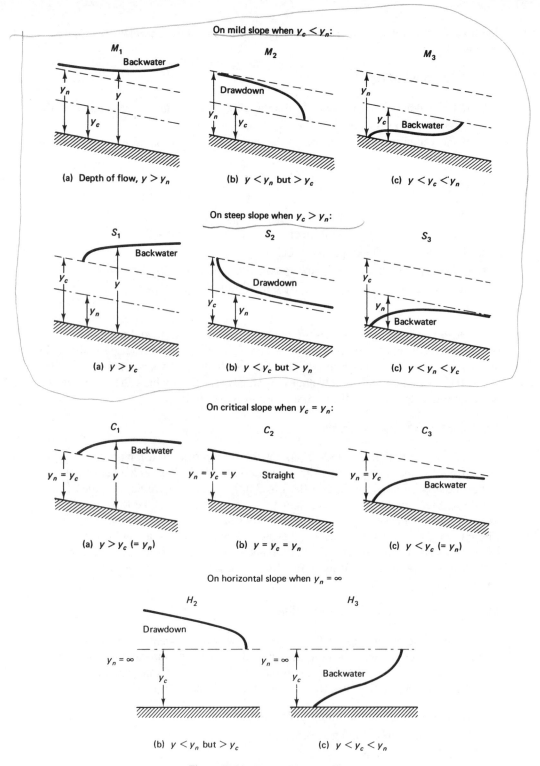

Figure 10.16 Types of flow profiles.

On adverse slope when y_n nonexistent:

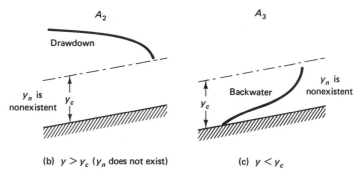

(b) $y > y_c$ (y_n does not exist)   (c) $y < y_c$

Figure 10.16   (continued)

Two of these are described in detail. The water surface elevation at the start of the curve from which the computation starts is a control section. The computation should proceed upstream from the control section in subcritical flow and in the downstream direction for supercritical flow.

Numerical integration method.   Consider a channel section having a depth y_1 at x_1 and y_2 at x_2 as shown in Figure 10.17(a).

$$x_2 - x_1 = \int_{x_1}^{x_2} dx$$

or

$$x_2 - x_1 = \int_{y_1}^{y_2} \frac{dx}{dy} dy$$

The right-hand side indicates the area under the dx/dy versus y curve, as shown in Figure 10.17(b). If a to b is considered a straight line for a small difference in y_1 and y_2, then

$$x_2 - x_1 = \frac{[(dx/dy)_1 + (dx/dy)_2]}{2}(y_2 - y_1) \qquad [\text{L}] \qquad (10.28)$$

The procedure comprises solving eq. (10.28) by the following steps:

1. Select several values of y starting from the control point.
2. For each value of y, calculate dx/dy by the inverse of eq. (10.27).
3. Using eq. (10.28), calculate x for two successive values of y.

For a backwater curve, the y values should be selected at close intervals near the tail part of the curve.

Example 10.14

A trapezoidal channel with a bottom width of 4 m and side slopes of 4:1 carries a discharge of 30 m³/s. The channel has a constant bed slope of 0.001. A dam backs up the water to a depth of 3.0 m just behind the dam. Compute the backwater profile to a depth 5% greater than the normal channel depth. $n = 0.025$, $\alpha = 1.0$.

Sec. 10.11   **Gradually Varied Flow**   **535**

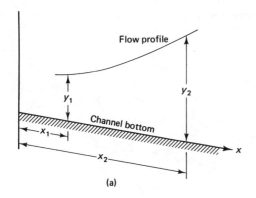

(a)

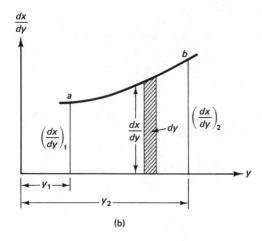

(b)

Figure 10.17 Derivation of numerical integration method.

Solution

1. The channel is the same as given in Examples 10.6 and 10.9.
2. From Example 10.6, critical depth $y_c = 1.22$ m.
3. From Example 10.9, normal depth $y_n = 1.90$ m.
4. Since $y_n > y_c$, the channel has a mild slope.
5. At the control point y of 3 ft is greater than y_n, thus, the profile is of M_1 type curve.
6. Section factor for critical flow $Z_c = \dfrac{Q}{\sqrt{g/\alpha}} = \dfrac{30}{\sqrt{9.81}} = 9.58$.
7. Conveyance for uniform flow, $K_n = \dfrac{Q}{\sqrt{S_0}} = \dfrac{30}{\sqrt{0.001}} = 948.68$
8. At the starting point of the curve, the control section depth = 3 m. The last computed point which is 5% greater than $1.90 = 1.05(1.9) = 2.0$ ft.
9. Since the flow is subcritical, computation proceeds upstream from the dam as the origin. The computations are arranged in Table 10.7.
10. The profile has been shown in Figure 10.18 by a plot between y (column 1) and x (column 10).

TABLE 10.7 COMPUTATION OF THE FLOW PROFILE BY THE NUMERICAL INTEGRATION METHOD

(1) y Select:	(2) T^a	(3) A^b	(4) R^c	(5) $R^{2/3}$	(6) $K = \frac{1}{n}AR^{2/3\,d}$	(7) $Z = \sqrt{\frac{A^3}{T}}^{\,d}$	(8) $\frac{dx}{dy}$ [Inverse of eq. (10.27)]^e	(9) Δx [eq. (10.28)] (m)	(10) x Cumulated (m)
3.0	28.0	48.0	1.67	1.41	2707.2	62.85	1113.50		
2.8	26.4	42.56	1.57	1.35	2298.2	54.04	1167.50	228	228
2.6	24.8	37.44	1.47	1.29	1931.9	46.00	1260.60	243	471
2.4	23.2	32.64	1.37	1.23	1605.89	38.72	1442.03	270	741
2.2	21.6	28.16	1.27	1.17	1317.89	32.15	1891.18	333	1074
2.1	20.8	26.04	1.22	1.14	1187.42	29.14	2466.0	218	1292
2.0	20.0	24.0	1.17	1.11	1065.60	26.29	4181.25	332	1624

$^a T = 4 + 8y.$

$^b A = (4 + 4y)y.$

$^c R = \dfrac{A}{P} = \dfrac{(4 + 4y)y}{4 + 8.24y}.$

^{d} K and Z calculated for each y selected.

$^e \dfrac{dx}{dy} = \dfrac{1}{S_0} \cdot \dfrac{1 - \left(\dfrac{Z_c}{Z}\right)^2}{1 - \left(\dfrac{K_n}{K}\right)^2}.$

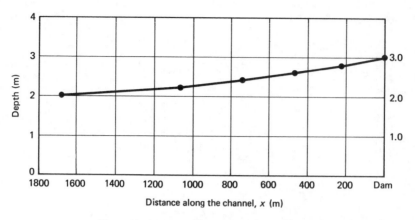

Figure 10.18 Backwater profile for Example 10.14.

Direct step method. This method directly uses the energy principle from section to section in the entire reach of the channel. It is applicable to prismatic channels. Applying the energy principle at points 1 and 2 of Figure 10.15 gives

$$Z_1 + y_1 + \alpha_1 \frac{V_1^2}{2g} = Z_2 + y_2 + \alpha_2 \frac{V_2^2}{2g} + h_f \tag{a}$$

or

$$(Z_1 - Z_2) - h_f = \left(y_2 + \alpha_2 \frac{V_2^2}{2g} \right) - \left(y_1 + \alpha_1 \frac{V_1^2}{2g} \right) \tag{b}$$

or

$$S_0 \Delta x - S_f \Delta x = E_2 - E_1 \tag{c}$$

where Δx is the distance between the two sections. E_1 and E_2 are specific energies and $h_f = S_f \Delta x$, or

$$\Delta x = \frac{E_1 - E_2}{S_f - S_0} \tag{d}$$

If the energy grade between the two sections is considered to be the average of the grade at sections 1 and 2, then

$$\Delta x = \frac{E_1 - E_2}{\overline{S_f} - S_0} \quad [\text{L}] \tag{10.29a}$$

with

$$\overline{S_f} = \frac{S_{f1} + S_{f2}}{2} \quad [\text{dimensionless}] \tag{10.29b}$$

$$S_f = \frac{V^2 n^2}{2.22 R^{4/3}} \quad [\text{dimensionless}] \tag{10.29c}$$

Equation (10.29c) in metric units is $S_f = V^2 n^2 / R^{4/3}$.

The steps of the procedure are as follows:

1. Select several values of y starting from the control point.
2. For a selected y, calculate A, R, $R^{4/3}$, and V $(= Q/A)$.
3. For a selected y, also calculate the velocity head $[\alpha (V^2/2g)]$, the specific energy, E $(= y + \alpha V^2/2g)$, and the energy slope, S_f $(= n^2V^2/2.22R^{4/3})$.
4. For two successive values of y, determine the difference between the specific energy, ΔE, and the average of the energy slope, $\overline{S}_f$.
5. Compute Δx from eq. (10.29a).

The analytical integrating method requires use of the varied flow function tables (see Appendix D-2 of Chow, 1959). Many alternative computation procedures have been proposed under this method. The standard step method, also based on the energy principle, is a trial-and-error procedure wherein the depth of flow is determined for a given channel distance and not the inverse as in the other methods.

Example 10.15

Determine the flow profile using the data of example 10.14 by the direct step method.

Solution The computations are arranged in Table 10.8.

10.12 RAPIDLY VARIED FLOW

This involves a sharp change in the curvature of the water surface, sometimes to the extent of discontinuity in the flow profile. The streamlines are so disturbed that the pressure distribution is not hydrostatic. The rapid variation in the flow conditions occurs within a short reach. As a result, the energy loss due to boundary friction is negligible in the rapidly varied flow, which is dominant in the gradually varied flow conditions. The overall energy losses are substantial due to turbulent conditions. The problems related to rapidly varied flow are usually studied on an individual basis, with each phenomenon given a specific treatment. Flow over spillways and hydraulic jumps are two common cases of the rapidly varied flow. The latter is described below briefly.

10.12.1 Hydraulic Jump

When a shallow stream of high velocity impinges on water of sufficient depth, the result is usually an abrupt rise in the surface in the region of impact. This phenomenon is known as the hydraulic jump. A similar phenomenon takes place when the flow passes over from a steep slope to a mild slope or when certain obstruction is met in the passage of a supercritical flow. For formation of a jump, the flow should be supercritical, which converts into subcritical flow after the jump.

In the process, a substantial loss of energy takes place. Since unknown energy losses are involved in the jump, the use of the principle of energy is not practical. The principle of momentum is used instead, as described in Section 10.6. By this principle, the change in the forces between two sections are equated with the change in the rate of momentum, which is equal to the mass of water times change in the velocity. Commonly, the relation is developed for the horizontal or slightly inclined

TABLE 10.8 COMPUTATION OF THE FLOW PROFILE BY THE DIRECT STEP METHOD

(1) y Select:	(2) A	(3) $R = \dfrac{A}{P}$	(4) $R^{4/3}$	(5) $V = \dfrac{Q}{A}$	(6) $\alpha\dfrac{V^2}{2g}$	(7) $E = y + \alpha\dfrac{V^2}{2g}$	(8) $\Delta E_S = E_1 - E_2$	(9) $S_f = \dfrac{n^2 V^2}{R^{4/3}}$ $(\times 10^3)$	(10) $\overline{S}_f$ $(\times 10^3)$	(11) $\overline{S}_f - S_0$ $(\times 10^3)$	(12) Δx [eq (10.29a)]	(13) x Cumulated
3.0	48.0	1.67	2.00	0.625	0.020	3.020		0.122				
2.8	42.56	1.57	1.82	0.705	0.025	2.825	0.195	0.171	0.147	−0.853	228	228
2.6	37.44	1.47	1.66	0.801	0.033	2.633	0.192	0.242	0.207	−0.793	242	470
2.4	32.64	1.37	1.51	0.919	0.043	2.443	0.190	0.350	0.296	−0.704	270	740
2.2	28.16	1.27	1.37	1.065	0.058	2.258	0.185	0.517	0.433	−0.567	326	1066
2.1	26.04	1.22	1.30	1.152	0.068	2.168	0.090	0.638	0.578	−0.422	213	1279
2.0	24.0	1.17	1.23	1.250	0.080	2.080	0.088	0.794	0.716	−0.284	310	1589

channels in which the weight of water between the section and the boundary friction are neglected. It is applicable to most field channels.

The following formulas are derived for the rectangular channels from the momentum principle on the basis described above:

$$\frac{D_2}{D_1} = \frac{1}{2}(\sqrt{1 + 8Fr^2} - 1) \qquad \text{[dimensionless]} \qquad (10.30)$$

where

$$D_1 \text{ and } D_2 = \text{depth before and after the jump}$$
$$Fr_1 = \text{Froude number before the jump} = V_1/\sqrt{gD_1}$$

The depths D_1 and D_2 are referred to as conjugate depths. Equation (10.30) can be used to ascertain D_2 when D_1 is known. In the equation, subscripts 1 and 2 can be replaced by each other. Thus the equation can also be used to determine the pre-jump depth, D_1, for a known post-jump depth, D_2.

The discharge through the jump where b is the width of a rectangular channel can be given by

$$Q = b\left[(gD_1D_2)\frac{D_1 + D_2}{2}\right]^{1/2} \qquad [L^3T^{-1}] \qquad (10.31)$$

The energy dissipated in a jump is computed from

$$E_{loss} = \frac{(D_2 - D_1)^3}{4D_1D_2} \qquad [L] \qquad (10.32)$$

There are many applications of the hydraulic jump. The main use is to dissipate energy in water flowing over spillways or weirs to prevent scouring downstream of the structure.

Example 10.16

Water flows at a rate of 360 cfs in a rectangular channel of 18 ft width with a depth of 1 ft. (a) Is a hydraulic jump possible in the channel? (b) If so, what is the depth of flow after the jump? (c) How much energy is dissipated through the jump?

Solution

(a) $A = 18 \times 1 = 18$

$$V = \frac{Q}{A} = \frac{360}{18} = 20$$

$$Fr_1 = \frac{V_1}{\sqrt{gD_1}} = \frac{20}{\sqrt{3.22(1)}} = 3.52$$

Since $Fr_1 > 1$, supercritical flow, jump can form.

(b) $\frac{D_2}{D_1} = \frac{1}{2}(\sqrt{1 + 8(3.52)^2} - 1) = \frac{9.0}{2}$

$$D_2 = \frac{1}{2}(9.0)(1) = 4.50 \text{ ft}$$

(c) Loss of energy,

$$E_{\text{loss}} = \frac{(D_2 - D_1)^3}{4D_1 D_2}$$

$$= \frac{(4.5 - 1.0)^3}{4(1.0)(4.5)}$$

$$= 2.38 \text{ ft-lb/lb}$$

PROBLEMS

10.1. Compute the hydraulic radius, hydraulic depth, and section factors Z_c and Z_n for the trapezoidal channel section shown in Fig. P10.1.

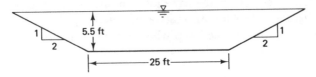

Figure P10.1

10.2. In a conduit of diameter of 4.5 ft, the depth of flow is 4.0 ft. **(a)** Determine the hydraulic radius, hydraulic depth, and section factors for critical and normal flows. **(b)** Also determine the alternate depth of flow that will carry the same discharge.

10.3. Discharge measurements provided the following velocity and area values for various sections of a channel. Determine the energy and momentum coefficients.

Section	1	2	3	4	5	6
Area (m²)	75	401	521	492	387	125
Velocity (m/s)	0.18	0.31	0.55	0.58	0.40	0.22

10.4. For the 50-ft-wide rectangular channel section shown in Fig. P10.4, determine the depth of flow and the velocity at section 2 using the energy principle. Neglect the energy losses. Assume that $\alpha_1 = \alpha_2 = 1$.

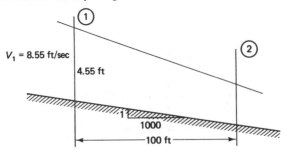

Figure P10.4

10.5. Solve Problem 10.4 by the momentum principle.

10.6. A rectangular channel section increases in width from 40 ft to 50 ft in a length of 100 ft. The channel slope is 0.1%. If the discharge and the velocity of flow at section 1 are 2950 cfs and 9.8 ft/sec, respectively, determine the depths of flow at sections 1 and 2 and the velocity at section 2. Use the energy principle. Neglect the losses. Assume that $\alpha_1 = \alpha_2 = 1$.

10.7. Solve Problem 10.6 by the momentum principle.

10.8. A trapezoidal channel with a side slope of 1(vertical):2(horizontal) and a bottom width of 10 ft carries a discharge of 300 cfs. **(a)** Plot the specific energy curve for the channel. **(b)** At what depth will the critical flow occur? **(c)** Determine the alternate depth to a 4.0-ft depth of flow. **(d)** What is the state of flow at the alternate depth?

10.9. Prove that for a rectangular channel at the critical state of flow, **(a)** the depth of flow is equal to two-thirds of the minimum specific energy, and **(b)** the velocity head is equal to one-third of the minimum specific energy.

10.10. A right-angled triangular channel carries a flow of 20 m³/s. Determine the critical depth and the critical velocity of flow.

10.11. A 40-in. conduit carries a discharge of 25 cfs. Determine the critical depth using the geometric elements in Table 10.1.

10.12. A trapezoidal channel has a bed width of 3.5 m and the side slope of 30° from the horizontal. Determine the critical depth and the critical velocity for a flow of 22 m³/s.

10.13. Determine the discharge through the following sections for a normal depth of 5 ft, $n = 0.013$, and $S = 0.2\%$.
  (a) A rectangular section 20 ft wide.
  (b) A circular section 20 ft in diameter.
  (c) A right-angled triangular section.
  (d) A trapezoidal section with a bottom width of 20 ft and side slope of 1 (vertical):2 (horizontal).
  (e) A parabolic section having a top width of 20 ft for a 5-ft depth. [*Hint:* $A = \frac{2}{3}Ty$, $P = T + \frac{8}{3}(y^2/T)$, where T is the top width and y is the depth.]

10.14. In a 3.0-m-wide rectangular channel of bed slope 0.0015, a discharge of 4 m³/s is observed at a depth of 0.8 m. Estimate the discharge when the depth is doubled.

10.15. Determine the conveyance of the channel in Problem 10.14.

10.16. A long, rectangular channel of 15 ft width, lined with concrete, is supplied by a reservoir as shown in Fig. P10.16. Neglecting the entry losses into the channel, determine the depth of flow and discharge through the channel. $n = 0.015$.

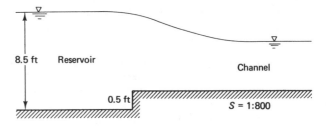

Figure P10.16

10.17. A discharge of 5.2 m³/s occurs in a rectangular channel of 2 m width having a bed slope of 1:625. Determine the **(a)** normal depth of flow, **(b)** critical depth of flow, and **(c)** state of flow. $n = 0.013$.

10.18. The channel in Problem 10.12 is excavated in smooth rock to a bed slope of 0.2%. Determine the **(a)** normal depth of flow, **(b)** critical slope, and **(c)** state of flow in the channel.

10.19. What diameter of a circular conduit flowing full would be required to carry the same quantity of flow as in a concrete trapezoidal channel of 20 ft width and 45° side slopes, running at a depth of 3.0 ft?

10.20. A concrete-lined trapezoidal channel has to be constructed to carry a discharge of 500 cfs. Design the channel. Assume that the following values were fixed based on the site conditions:

$$\text{Bed slope} = 0.002$$
$$n = 0.015$$
$$\text{Side slope, } z = 1.5(\text{horizontal}) : 1(\text{vertical})$$
$$b/y \text{ ratio} = 2.0$$

10.21. Design the channel of Problem 10.20 as a most hydraulic efficient channel (of semi-circular section).

10.22. Design the channel of Problem 10.20 as a best hydraulic rectangular section.

10.23. Design a storm sewer laid on a minimum grade to carry a peak flow of 10 cfs. Consider that the depth at the peak flow is 0.7 times the sewer diameter. The minimum flow to be maintained is 3.0 cfs. $n = 0.013$.

10.24. A circular sewer laid on a 1% grade is to carry 3.0 cfs when full. **(a)** Design the sewer. **(b)** At a dry weather flow of 0.6 cfs, what is the depth and velocity of flow? **(c)** What is the depth of flow at minimum velocity of 2 ft/s? ($n = 0.015$.)

10.25. A water supply conduit of 36 in. diameter was laid 20 years ago on a slope of 0.1% with $n = 0.015$. At the present time the conduit carries a flow of 15 cfs. Determine the **(a)** original capacity and velocity of flow, **(b)** present velocity when full, and **(c)** present value of n.

10.26. A sewer of vitrified clay has been laid on a gradient of 1:100. It receives the flow from 300 houses. The per capita daily water supply is 200 liters/day. The population density is 3.5 persons per house. Design the sewer. Assume that (1) the sewage quantity is equal to the water consumption; (2) the sewer is to be designed for the maximum hourly flow, which is 400% of the average daily flow; and (3) the minimum hourly flow is 50% of the average flow.

10.27. A trapezoidal irrigation canal is excavated in silty-sand to convey a discharge of 10 m³/s on a bed slope of 1:10,000. The side slope is 1 (vertical):2 (horizontal). The mean grain diameter is 0.3 mm. The sediment load is 3000 ppm, for which the silt factor is increased by 10%. Design the channel by the regime theory.

10.28. Design the channel with the data of Problem 10.27 by the tractive force method. Angle of friction = 30° and $n = 0.02$. Permissible tractive force = 2.4 N/m².

10.29. A trapezoidal concrete channel has a constant bed slope of 0.0015, a bed width of 3.0 m, and side slopes of 1:1. It carries a discharge of 20 m³/s. The channel is a tributary to a river in which the existing flood level is 3.5 m above the channel bottom. Compute the water surface profile by the numerical integration method to a depth 5% greater than the uniform flow depth. $\alpha = 1.1$ and $n = 0.025$.

10.30. Water flows under a gate opening (sluice) into a trapezoidal channel having a bed slope of 0.35%, a width of 20 ft, and side slopes of 1 (vertical):2 (horizontal). The sluice gate is regulated for a discharge of 400 cfs with a depth of opening of 0.8 ft. Compute the flow profile by the numerical integration method. Take $n = 0.025$ and $\alpha = 1.10$. Consider the control point at the vena contracta. (The distance from the gate opening to the vena contracta is approximately equal to the height of opening of 0.8 ft, and the depth at vena contracta may be taken as 0.6 ft. The computations may be performed in the downstream direction from the gate.)

10.31. Determine the flow profile using the data of Problem 10.29 by the direct step method.

10.32. Determine the flow profile using the data of Problem 10.30 by the direct step method.

10.33. A rectangular channel of 20 ft width has a depth of 4 ft and velocity of 60 ft/sec. Determine **(a)** whether a jump can form in the channel, **(b)** the downstream depth needed to form the jump, and **(c)** the loss of energy through the jump.

10.34. In a rectangular channel of 12 m width, water flows at a rate of 150 m³/s. At the end of the channel there is a horizontal concrete apron of 12 m width, on which the water depth is 3 m. Will a hydraulic jump be formed in the channel? What is the pre-jump depth? What is the loss of energy through the jump?

10.35. A rectangular section of a stream has a width of 50 ft and a depth of 5 ft. It has a slope of $1:1000$ and $n = 0.015$. The flow into the stream merges from a steep channel through a sluice of 1.50 ft depth. Determine whether a hydraulic jump is going to be formed. If so, what is the pre-jump depth?

PRESSURE FLOW SYSTEM:
Pipes and Pumps

The transmission and feeder mains of a waterworks from the treatment plant to the distribution system and the distribution mains consisting of an interconnecting pipe network up to the source point are the principal components of a water system that carries water under pressure. The sewage force mains that receive discharge from a pumping station also carry flows under pressure. There is a range of minimum to maximum pressure in which these components operate. The pressure is reduced as water flows through the system, due to frictional resistance by the pipe walls and fittings. This is measured in terms of the energy loss. The energy equation is thus appropriate in all pipe flow problems.

11.1 ENERGY EQUATION OF PIPE FLOW

Figure 11.1 shows a pipeline segment. The total energy at any point consists of potential or elevation head, pressure head, and velocity head. The hydraulic grade line shows the elevation of pressure head along the pipe (i.e., it is a line connecting the points to which the water will rise in piezometric tubes inserted at different sections of a pipeline). This concept is similar to the water surface in open channel flow. The energy grade line represents the total head at different points of a pipe section. In a uniform pipe, the velocity head is constant. Thus the energy grade line is parallel to the hydraulic grade line.

Applying the energy equation between points 1 and 2 gives us

$$Z_1 + \frac{p_1}{\gamma} + \alpha\frac{V_1^2}{2g} = Z_2 + \frac{p_2}{\gamma} + \alpha\frac{V_2^2}{2g} + h_f \qquad [\text{L}] \qquad (11.1)$$

In eq. (11.1), h_f is the loss of head along the pipeline due to friction. The energy gradient $S_f = h_f/L$. Additional losses resulting from valves, fittings, bends, and so

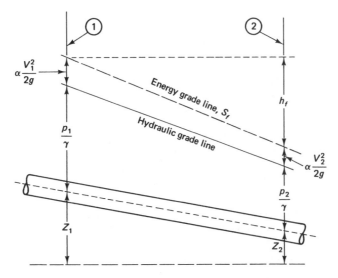

Figure 11.1 Hydraulic grade line and energy grade line in a pipe flow.

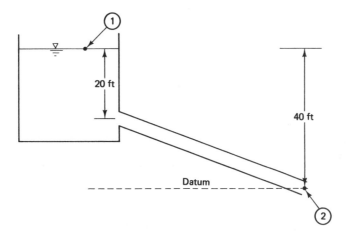

Figure 11.2 Headloss through a pipeline in Example 11.1.

on, are known as the minor losses, h_m, and have to be included when present. Then, in eq. (11.1) the term h_f will be replaced by the total head loss, h_{loss}. Since the minor losses are localized, the energy grade line, represented by h_f/L, will have breaks wherever the minor losses occur. If a mechanical energy is added to the water by a pump or taken off by a turbine between the two points of interest, it should be added or subtracted from the left side of eq. (11.1). In a uniform pipe, $V_1 = V_2$, and elevations Z_1 and Z_2 are generally known. To ascertain the pressure reduction, it is necessary to evaluate the head loss (and minor losses if present).

Example 11.1

From a reservoir, water flows at a rate of 10 cfs through a pipe of 12 in. diameter, as shown in Figure 11.2. Determine the loss of head in the system.

Solution

$$\text{Area of cross section of pipe, } A = \frac{\pi}{4}\left(\frac{12}{12}\right)^2 = 0.785 \text{ ft}^2$$

From a continuity equation, $Q = AV$ or

$$V = \frac{Q}{A} = \frac{10}{0.785} = 12.74 \text{ ft/sec}$$

Consider that the datum passes through point 2. Applying the energy equation between points 1 and 2 gives us

$$Z_1 + \frac{p_1}{\gamma} + \frac{V_1^2}{2g} = Z_2 + \frac{p_2}{\gamma} + \frac{V_2^2}{2g} + h_f$$

Since the pressure is atmospheric at points 1 and 2 and water is practically stationary (i.e., $V = 0$ at point 1),

$$40 + 0 + 0 = 0 + 0 + \frac{(12.74)^2}{2(32.2)} + h_f$$

and

$$h_f = 37.50 \text{ ft}$$

11.2 EVALUATION OF HEAD LOSS DUE TO FRICTION

11.2.1 Darcy–Weisbach Equation

The Darcy–Weisbach equation (1845) is the most general formula in the pipe flow application. It was obtained experimentally. However, Chezy's equation (1769), derived in Section 10.8.1 from balancing the motivating and drag forces on the moving water, can be reduced to the Darcy–Weisbach equation.

According to Chezy's formula,

$$V = C\sqrt{RS} \qquad [LT^{-1}] \tag{10.10}$$

Since $S = h_f/L$, $R = d/4$ for pipe and treating $C = \sqrt{8g/f}$, eq. (10.10) reduces to

$$h_f = \frac{fL}{d}\frac{V^2}{2g} \qquad [L] \tag{11.2}$$

where

h_f = loss of head due to friction in pipe, ft or m
f = friction factor, dimensionless
L = length of pipe, ft or m
d = internal diameter of pipe, ft or m
V = mean velocity of flow in pipe, ft/sec or m/s

The solution of eq. (11.2) requires the interim step of ascertaining an appropriate value of the friction factor, f, to be used in the equation.

11.2.2 Friction Factor for Darcy–Weisbach Equation

The friction factor relation depends on the state of flow, which is classified according to the Reynolds number. For pipes, the diameter is used as a characteristic dimension and the Reynolds number is given by

$$\text{Re} = \frac{Vd}{\nu} \quad \text{[dimensionless]} \tag{11.3}$$

where

$$V = \text{average velocity of flow, ft/sec or m/s}$$
$$d = \text{internal diameter of pipe, ft or m}$$
$$\nu = \text{kinematic viscosity of fluid, ft}^2/\text{s or m}^2/\text{s}$$

The flow is classified as follows:

Type of Flow	Value of Re
Laminar	<2000
Transition to turbulent (critical region)	2000–4000
Turbulent	>4000

For laminar flow, the friction factor is a function of the Reynolds number only. It is given by the following relation, based on the Hagen–Poiseuille equation:

$$\text{(For laminar flow)} \quad f = \frac{64}{\text{Re}} \quad \text{[dimensionless]} \tag{11.4}$$

In the critical region of Re between 2000 and 4000 the flow alternates between the laminar and turbulent regimes. Any friction factor relation cannot be applied with certainty in this region.

In turbulent regime, the friction factor is a function of the Reynolds number as well as the relative roughness of the pipe surface. During 1932 and 1933, Nikuradse published the results of now famous experiments on smooth (uncoated) and rough pipes coated with sand grains of uniform size. The experiments' results, plotted as the friction factor versus the Reynolds number, are shown in Figure 11.3. The roughness is characterized by a parameter, K/d, where K is the average diameter of the sand grains and d is the internal diameter of the pipe. In contrast to the Nikuradse sand roughness, the roughness of commercial pipe is not uniform. As a means of differentiating, the nonuniform roughness of commercial pipe is designated ϵ and is given in equivalent sand roughness. Table 11.1 indicates equivalent roughness for pipe of different material. Based on the results of these experiments, the turbulent flow is further classified in three zones as follows:

1. Flow in smooth pipe, where the relative roughness ϵ/d is very small
2. Flow in fully rough pipe
3. Flow in partially rough pipe where both the relative roughness and viscosity are significant

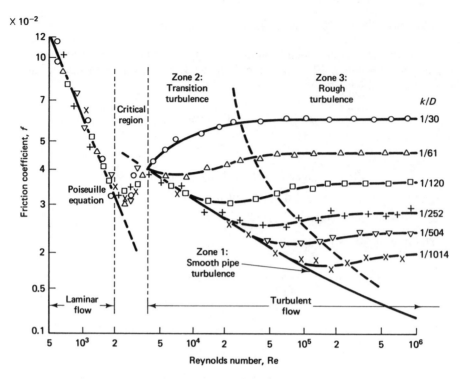

Figure 11.3 Nikuradse's experiment on smooth and sand-coated pipes.

TABLE 11.1 ROUGHNESS VALUES FOR PIPES

Pipe Material	Equivalent Roughness, ϵ (ft)	Hazen–Williams Coefficient, C
Brass, copper, aluminum	Smooth	140
PVC, plastic	Smooth	150
Cast iron		
new	8.0×10^{-4}	130
old	—	100
Galvanized iron	5.0×10^{-4}	120
Asphalted iron	4.0×10^{-4}	—
Wrought iron	1.5×10^{-4}	—
Commercial and welded steel	1.5×10^{-4}	120
Riveted steel	60.0×10^{-4}	110
Concrete	40.0×10^{-4}	130
Wood stave	20.0×10^{-4}	120

Nikuradse's experiments permitted Prandtl and von Kármán to establish the following formulas for smooth and fully rough pipes of categories 1 and 2 above. For flow in smooth pipe in a turbulent regime:

$$\frac{1}{\sqrt{f}} = -2 \log\left(\frac{2.51}{Re\sqrt{f}}\right) \quad \text{[dimensionless]} \quad (11.5)$$

For flow in fully rough pipe in a turbulent regime:

$$\frac{1}{\sqrt{f}} = -2 \log\left(\frac{\epsilon}{3.7d}\right) \quad \text{[dimensionless]} \quad (11.6)$$

However, most hydraulic problems and flow in commercial pipes relate to the flow of the third category of partial rough pipe. With a specific reference to this transition zone between smooth and rough pipes, Colebrook, in collaboration with White, in 1939, combined eqs. (11.5) and (11.6). The formula thus developed covers the entire turbulent regime.

For all types of flow in a turbulent regime,

$$\frac{1}{\sqrt{f}} = -2 \log\left(\frac{\epsilon}{3.7d} + \frac{2.51}{Re\sqrt{f}}\right) \quad \text{[dimensionless]} \quad (11.7)$$

For smooth pipes, when ϵ/d is very small, eq. (11.7) reduces to eq. (11.5). For rough pipes at very high Reynolds number, it takes the form of eq. (11.6). However, eq. (11.7) is implicit since the friction factor appears on both sides of the equation. As such, it involves trial-and-error solution.

By fitting the curve to the Colebrook relation of smooth pipe and combining the equation of rough pipe, Jain (1976) has suggested the following explicit equation for the entire turbulent regime, which gives results within 1% of the Colebrook equation:

$$\frac{1}{\sqrt{f}} = -2 \log\left(\frac{\epsilon}{3.7d} + \frac{5.72}{Re^{0.9}}\right) \quad \text{[dimensionless]} \quad (11.8)$$

From the implicit relations of Prandtl, von Kármán and Colebrook–White, Moody (1944) has prepared a diagram between the friction factor versus the Reynolds number and the relative roughness as shown in Figure 11.4. The diagram can conveniently be used instead of eq. (11.8) to determine the friction factor. The equivalent roughness is obtained from Table 11.1.

For the application of eq. (11.8), or the Moody diagram, the velocity of flow and the diameter of the pipe should be known so that the Reynolds number can be determined. When the velocity or diameter is unknown, the procedure has been discussed in the next section.

Example 11.2

Determine the friction factor for water flowing at a rate of 1 cfs in a cast iron pipe 2 in. in diameter at 80°F.

Solution

1. Area of cross section of pipe $= \frac{\pi}{4}\left(\frac{2}{12}\right)^2 = 0.022 \text{ ft}^2$.

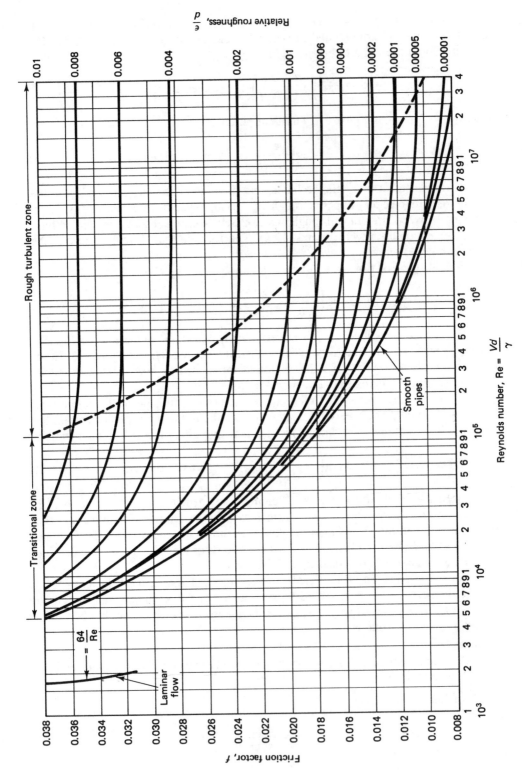

Figure 11.4 Moody diagram for friction factor for pipes.

552

2. Velocity of flow, $V = \dfrac{Q}{A} = \dfrac{1}{0.022} = 45.45$ ft/sec.

3. At 80°F, kinematic viscosity, $\nu = 0.93 \times 10^{-5}$ ft²/sec.

4. The Reynolds number,

$$\text{Re} = \frac{Vd}{\nu} = \frac{45.45(2/12)}{0.93 \times 10^{-5}} = 8.1 \times 10^5$$

5. Since Re > 4000, turbulent flow.

6. Equivalent roughness, $\epsilon = 8.0 \times 10^{-4}$ ft (Table 11.1).

7. Relative roughness, $\epsilon/d = \dfrac{8 \times 10^{-4}}{2/12} = 0.005$.

8. From eq. (11.8),

$$\frac{1}{\sqrt{f}} = -2 \log \left[\frac{0.005}{3.7} + \frac{5.72}{(8.1 \times 10^5)^{0.9}} \right] = 5.721$$

$$f = 0.031$$

Alternative Solution   On the Moody diagram, the point of intersection of Re = 8.1 × 10^5 and $\epsilon/d = 0.005$ is projected horizontally to the left to read

$$f = 0.03$$

11.2.3 Application of Darcy–Weisbach Equation

Darcy–Weisbach equation (11.2) has three applications:

1. It is used to compute the head loss, h_f, in a given size pipe, d, that carries a known flow, V or Q.
2. It is used to ascertain the flow, V (or Q), through a given size pipe, d, in which the head loss, h_f, is known.
3. It is used to determine the pipe size, d, to pass a given rate of flow, Q, within a known limit of head loss, h_f.

In the first case, the application of eq. (11.2) is direct. From known V and d values, Re can be computed and then f can be determined by eq. (11.4) for laminar flow and eq. (11.8) for turbulent flow, or by the Moody diagram. Thus eq. (11.2) can be solved for h_f.

However, in the second and third cases, Re, and hence f, cannot be determined. Since Re is unknown, many researchers have prepared special diagrams between certain groups of variables in nondimensional form, instead of Re versus f, that enable direct determination of pipe size or flow. The common procedure is of trial and error, comprising the following steps:

1. A value of f is assumed near the rough turbulence if ϵ/d is known.
2. Using the Darcy–Weisbach equation (11.2), either V or d, as desired, is computed.
3. Re and revised f are determined.
4. Steps 2 and 3 are repeated until the value of f stabilizes.

Example 11.3

Water is delivered at a rate of 0.80 cfs by a 6-in. cast iron pipe at 80°F between two points A and B that are 1000 ft apart. If point A is 100 ft higher than point B, what is the pressure difference between two points?

Solution

1. Computing the head loss in the pipe, at 80°F, $\nu = 0.93 \times 10^{-5}$ ft²/sec.

$$\text{Velocity of flow} = \frac{Q}{A} = \frac{0.80}{(\pi/4)(0.5)^2} = 4.08 \text{ ft/sec}$$

$$\text{Re} = \frac{Vd}{\nu} = \frac{4.08(0.5)}{0.93 \times 10^{-5}} = 2.2 \times 10^5$$

$$\text{Relative roughness,} \frac{\epsilon}{d} = \frac{0.0008}{0.5} = 0.0016$$

From the Moody diagram (Figure 11.4), $f = 0.023$. Hence

$$h_f = \frac{fL}{d}\frac{V^2}{2g} = (0.022)\left(\frac{1000}{0.5}\right)\frac{(4.08)^2}{2(32.2)} = 11.9 \text{ ft.}$$

2. Applying the energy equation between points A and B gives us

$$Z_1 + \frac{p_1}{\gamma} + \frac{V_1^2}{2g} = Z_2 + \frac{p_2}{\gamma} + \frac{V_2^2}{2g} + h_f$$

$$100 + \frac{p_1}{\gamma} = 0 + \frac{p_2}{\gamma} + 11.9$$

$$\frac{p_2 - p_1}{\gamma} = 100 - 11.9 = 88.1 \text{ ft}$$

$$p_2 - p_1 = 88.1(62.4) = 5500 \text{ psf or } 38.2 \text{ psi}$$

Example 11.4

A 2000-m-long commercial steel pipeline of 200 mm diameter conveys water at 20°C between two reservoirs, as shown in Figure 11.5. The difference in water level between the reservoirs is maintained at 50 m. Determine the discharge through the pipeline. Neglect the minor losses.

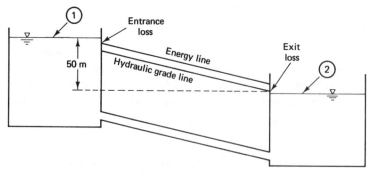

Figure 11.5 Pipe connecting two reservoirs in Example 11.4.

Solution

1. Consider the datum at the water level of the second reservoir. Apply the energy equation at points 1 and 2:

$$Z_1 + \frac{p_1}{\gamma} + \frac{V_1^2}{2g} = Z_2 + \frac{p_2}{\gamma} + \frac{V_2^2}{2g} + h_f$$

$$50 + 0 + 0 = 0 + 0 + 0 + h_f$$

$h_f = 50$ m, the difference in water level.

2. For commercial steel, $\epsilon = 1.5 \times 10^{-4}$ ft or 4.6×10^{-2} mm. Hence

$$\frac{\epsilon}{d} = \frac{4.6 \times 10^{-2}}{200} = 2.3 \times 10^{-4}$$

In the rough turbulence, when f is independent of Re, $f = 0.014$ for $\epsilon/d = 2.3 \times 10^{-4}$.

3. For the first trial, assume that $f = 0.014$. From the Darcy–Weisbach equation,

$$h_f = \frac{fL}{d}\frac{V^2}{2g}$$

or

$$50 = (0.014)\left(\frac{2000}{0.2}\right)\frac{V^2}{2(9.81)}$$

or $V = 2.65$ m/s.

4. At 20°C, $\nu = 1.02 \times 10^{-6}$ m²/s.

$$\text{Re} = \frac{Vd}{\nu} = \frac{2.65(0.2)}{1.02 \times 10^{-6}} = 5.2 \times 10^5$$

For Re $= 5.2 \times 10^5$ and $\epsilon/d = 2.3 \times 10^{-4}$, from the Moody diagram the value of f is 0.016.

5. Recompute V for revised $f = 0.016$.

$$50 = (0.016)\left(\frac{2000}{0.2}\right)\frac{V^2}{2(9.81)}$$

$$V = 2.48 \text{ m/s}$$

6. Revised Re $= \dfrac{(2.48)(0.2)}{(1.02)(10^{-6})} = 4.9 \times 10^5$.

For Re $= 4.9 \times 10^5$ and $\epsilon/d = 2.3 \times 10^{-4}$, from the Moody diagram, $f = 0.016$. Since the value of f stabilizes, no further trials are necessary.

7. Hence,

$$V = 2.48 \text{ m/s}$$

$$Q = AV = \frac{\pi}{4}(0.2)^2(2.48) = 0.078 \text{ m}^3/\text{s}$$

11.2.4 Hazen–Williams Equation for Friction Head Loss

Another common formula for head loss in pipes that has found almost exclusive usage in water supply engineering is the Hazen–Williams equation:

$$V = 1.318CR^{0.63}S^{0.54} \quad \text{(English units)} \quad \text{[unbalanced]} \quad (11.9)$$

where

V = mean velocity of flow, ft/sec

C = Hazen–Williams coefficient of roughness given in Table 11.1

R = hydraulic radius, ft

S = slope of energy gradient = h_f/L.

There is a direct comparison between Chezy's formula and the Hazen–Williams formula for $R = 1$ ft and $S = 1/1000$, the Hazen–Williams coefficient C being the same as Chezy's coefficient, C. Thus the Hazen–Williams formula is accurate within a certain range of diameters and friction slopes, although it is used indiscriminately in pipe designs. Jain et al. (1978) indicated that an error of up to 39% can be involved in the evaluation of the velocity by the Hazen–Williams formula over a wide range of diameters and slopes. Two sources of error in the Hazen–Williams formula are: (1) the multiplying factor 1.318 should change for different R and S to be comparable with Chezy's formula for the same value of C, and (2) the Hazen–Williams coefficient C is considered to be related to the pipe material only as shown in Table 11.1, whereas it must also depend on pipe diameter, velocity, and viscosity, similar to the friction factor of Darcy–Weisbach. Jain et al. suggested a modified formula that incorporates the kinematic viscosity and contains a coefficient the values of which varies with material, pipe diameter, and velocity of flow. The Hazen–Williams formula, however, has a wide application because of its simplicity.

Equation (11.9) can be written as follows in terms of discharge for a circular pipe by substituting $V = Q/A$, $A = (\pi/4)d^2$, and $R = d/4$:

$$Q = 0.432Cd^{2.63}S^{0.54} \quad \text{(English units)} \quad \text{[unbalanced]} \quad (11.10a)$$

$$Q = 0.278Cd^{2.63}S^{0.54} \quad \text{(metric units)} \quad \text{[unbalanced]} \quad (11.10b)$$

A nomogram based on eq. (11.10) is given in Figure 11.6 to facilitate the solution. Equation (11.10) and the nomogram provide a direct solution to all types of pipe problems, as mentioned in Section 11.2.3: (1) computation of head loss, (2) assessment of flow, and (3) determination of pipe size.

The nomogram in Figure 11.6 is based on the coefficient $C = 100$. For pipes of a different coefficient, the adjustments are made as follows:

To adjust discharge: $\quad Q = Q_{100}\left(\dfrac{C}{100}\right) \quad \quad [L^3T^{-1}] \quad (11.11a)$

To adjust diameter: $\quad d = d_{100}\left(\dfrac{100}{C}\right)^{0.38} \quad \quad [L] \quad (11.11b)$

To adjust friction slope: $\quad S = S_{100}\left(\dfrac{100}{C}\right)^{1.85} \quad \quad \text{[dimensionless]} \quad (11.11c)$

Where the subscript 100 refers to the value obtained from the nomogram.

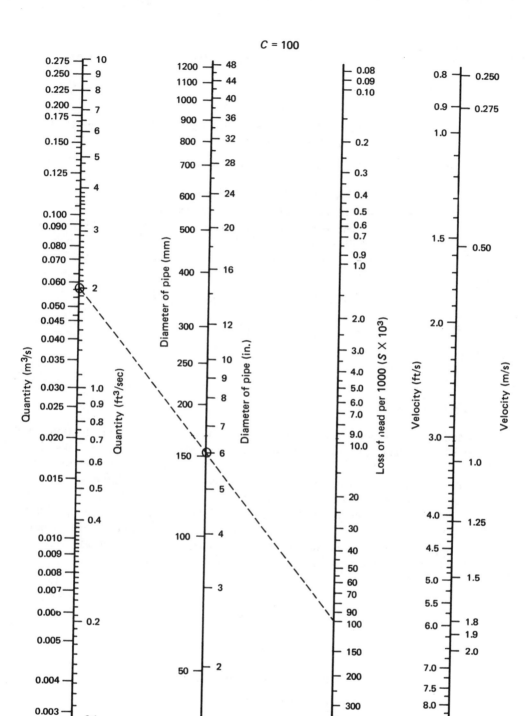

Figure 11.6 Nomogram based on the Hazen–Williams equation, for C = 100.

Example 11.5

Recompute the head loss in Example 11.3 by the Hazen–Williams formula.

Solution

1. For a new cast iron, $C = 130$.
2. From eq. (11.10), $0.80 = 0.432(130)(0.5)^{2.63}S^{0.54}$ or $S = 0.011$. Hence $h_f = SL = 0.011(1000) = 11.0$ ft.

Alternative Solution From the nomogram (Figure 11.6):

1. A point is marked at 0.80 cfs on the discharge scale.
2. Another point is marked at 6 in. on the diameter scale.
3. The straight line joining these points meets the head loss for 1000 scale at 18. Thus $S_{100} \times 10^3 = 18$ or $S_{100} = 0.018$.
4. Adjustment of S for $C = 130$.

$$S = S_{100}\left(\frac{100}{C}\right)^{1.85} = (0.018)\left(\frac{100}{130}\right)^{1.85} = 0.011$$

$$h_f = SL = 0.011(1000) = 11.0 \text{ ft}$$

11.3 MINOR HEAD LOSSES

In addition to the continuous head loss along the pipe length due to friction, local head losses occur at changes in pipe section, at bends, valves, and fittings. These losses may be neglected for long pipes but are significant for less than 100-ft-long pipes. Since pipe lengths in water supply and wastewater plants are generally short, minor losses are important. There are two ways to compute these losses. In the equivalent-length technique, a fictitious length of pipe is estimated that will cause the same pressure drop as any fitting or change in a pipe cross section. This length is added to the actual pipe length. In the second method, the loss is considered proportional to the kinetic energy head given by the following formula:

$$h_m = K\frac{V^2}{2g} \qquad [\text{L}] \qquad (11.12)$$

where

$$h_m = \text{minor loss of head, ft or m}$$
$$K = \text{loss coefficient}$$
$$V = \text{mean velocity of flow, ft/sec or m/s}$$

Some typical values of the loss coefficient are given in Table 11.2.

TABLE 11.2   MINOR HEAD LOSS COEFFICIENTS

Item	Loss Coefficient, K
Entrance loss from tank to pipe	
Flush connection	0.5
Projecting connection	1.0
Exit loss from pipe to tank	1.0
Sudden contraction	
$d_1/d_2 = 0.5$	0.37
$d_1/d_2 = 0.25$	0.45
$d_1/d_2 = 0.10$	0.48
Sudden enlargement	
$d_1/d_2 = 2$	0.54
$d_1/d_2 = 4$	0.82
$d_1/d_2 = 10$	0.90
Fittings	
90° bend — screwed	0.5–0.9
90° bend — flanged	0.2–0.3
Tee	1.5–1.8
Gate valve (open)	0.19
Check valve (open)	3.00
Glove valve (open)	10.00
Butterfly valve (open)	0.30

11.4 PIPELINE ANALYSIS AND DESIGN

The analysis involves determination of the head loss or the rate of flow through a pipeline of a given size. The design situation relates to selection of a pipe size that will carry a design discharge between two points with specified reservoir elevations or a known pressure difference. The problems can be solved by the Darcy–Weisbach equation as explained in Section 11.2.3. If minor losses are neglected, the Hazen–Williams equation leads to the direct solution of both the analysis and design problems. The problems in different pipeline systems are discussed in sections 11.5 through 11.9.

11.5 SINGLE PIPELINES

The application is made of the (1) energy equation, (2) Darcy–Weisbach or Hazen–Williams equation, and (3) minor losses relation, as demonstrated in Example 11.6.

Example 11.6

Two reservoirs are connected by a 200-ft-long cast iron pipeline, as shown in Figure 11.7. If the pipeline is to convey a discharge of 2 cfs at 60°F, what is the size of the pipeline required?

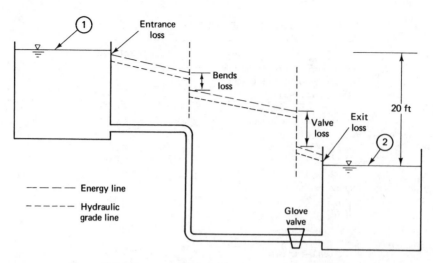

Figure 11.7 Pipe system connecting two reservoirs in Example 11.6.

Solution

1. Apply the energy equation between points 1 and 2 with respect to point 2 as the datum:

$$Z_1 + \frac{p_1}{\gamma} + \frac{V_1^2}{2g} = Z_2 + \frac{p_2}{\gamma} + \frac{V_2^2}{2g} + h_{\text{loss}}$$

$$20 + 0 + 0 = 0 + 0 + 0 + h_{\text{loss}}$$

$$h_{\text{loss}} = 20 \tag{a}$$

2. Friction loss,

$$h_f = \frac{fL}{d}\frac{V^2}{2g} = \frac{fL}{d}\frac{Q^2}{[(\pi/4)d^2]^2 2g}$$

$$= \frac{fLQ^2}{39.68d^5} \tag{b}$$

3. Minor losses,

$$h_m = \sum \frac{KV^2}{2g} = \sum K \frac{Q^2}{[(\pi/4)d^2]^2 2g} = \frac{\sum KQ^2}{39.68d^4} \tag{c}$$

Item	K
Entrance loss	0.5
Exit loss	1.0
Two 90° bends at 0.9	1.8
Glove valve	10.0
Total	13.3

4. $h_{\text{loss}} = h_f + h_m$ \hfill (d)

5. Substituting eqs. (a), (b) and (c) in (d) above:

$$\frac{fLQ^2}{39.68d^5} + \frac{\sum KQ^2}{39.68d^4} = 20$$

$$\frac{f(200)(2)^2}{(39.68)d^5} + \frac{13.3(2)^2}{(39.68)d^4} = 20$$

$$20d^5 - 1.34d - 20.16f = 0 \qquad\qquad (e)$$

6. In the first trial, assume that $f = 0.03$. Substitute f in eq. (e): $20d^5 - 1.34d - 0.605 = 0$. Solve by trial and error: $d = 0.59$ ft. Thus

$$V = \frac{4Q}{\pi d^2} = \frac{4(2)}{\pi(0.59)^2} = 7.32 \text{ ft/sec}$$

$$\text{Re} = \frac{Vd}{\nu} = \frac{7.32(0.59)}{1.217 \times 10^{-5}} = 3.5 \times 10^5$$

$$\frac{\epsilon}{d} = \frac{8 \times 10^{-4}}{0.59} = 0.0014$$

$$f = 0.0215 \quad \text{(from the Moody diagram)}$$

7. First revision: Substitute f in eq. (e): $20d^5 - 1.34d - 0.433 = 0$. Solve by trial and error: $d = 0.57$ ft. Thus

$$V = \frac{4(2)}{\pi(0.57)^2} = 7.84 \text{ ft/sec}$$

$$\text{Re} = \frac{7.84(0.57)}{1.217 \times 10^{-5}} = 3.7 \times 10^5$$

$$\frac{\epsilon}{d} = \frac{8 \times 10^{-4}}{0.57} = 0.0014$$

$$f = 0.0215 \quad \text{(from the Moody diagram)}$$

Since f stabilizes, $d = 0.57$ ft or 6.8 in.

11.6 SINGLE PIPELINES WITH PUMPS

The pumps are common in waterworks and wastewater systems at the source to lift the water level and at intermediate points to boost the pressure. Figure 11.8 indicates a situation of water supplied from a lower reservoir to an upper-level reservoir.

To analyze the system, the energy equation is applied between upstream and downstream ends of the pipe:

$$Z_1 + \frac{p_1}{\gamma} + \frac{V_1^2}{2g} + H_p = Z_2 + \frac{p_2}{\gamma} + \frac{V_2^2}{2g} + h_f + h_m \qquad\qquad (a)$$

Treating $V_1 = V_2$ yields

$$H_p = \left(Z_1 + \frac{p_1}{\gamma}\right) - \left(Z_2 + \frac{p_2}{\gamma}\right) + h_f + h_m \qquad\qquad (b)$$

$$H_p = \Delta Z + h_{\text{loss}} \qquad [\text{L}] \qquad\qquad (11.13)$$

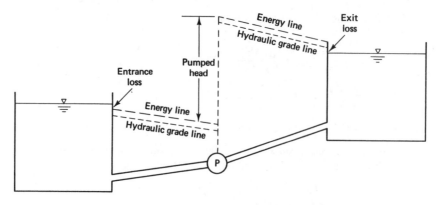

Figure 11.8 Pumped pipeline system.

where

$$H_p = \text{energy added by the pump, ft or m}$$

ΔZ = difference in downstream and upstream piezometric
heads or water levels, or total static head, ft or m

h_f = friction head loss = $(fL/d)(V^2/2g)$, ft or m

h_m = minor head losses = $\Sigma KV^2/2g$, ft or m

h_{loss} = total of friction and minor head losses

The energy head, H_p, and the brake horse power of the pump are related as

$$\text{BHP} = \frac{\gamma Q H_p}{550\eta} \qquad [\text{FLT}^{-1}] \qquad (11.14)$$

where

$$\text{BHP} = \text{brake horsepower}$$

Q = discharge through pipe, cfs

H_p = pumphead, ft

η = overall pump efficiency

Example 11.7

Water has to be transported at a rate of 1 cfs from a reservoir of water elevation 1000 ft to a reservoir at water elevation of 1100 ft through a 4000-ft-long 6-in.-diameter steel pipeline at 50°F. Determine the horsepower of the pump required having an efficiency of 70%. Neglect the minor losses.

Solution From eq. (11.13), $H_p = Z + h_{\text{loss}}$.

$$Z = 1100 - 1000 = 100$$

$$h_m = 0$$

Friction loss h_f:

$$V = \frac{Q}{A} = \frac{1}{\pi/4(0.5)^2} = 5.1 \text{ ft/sec}$$

$$\text{Re} = \frac{Vd}{\nu} = \frac{5.1(0.5)}{1.4 \times 10^{-5}} = 1.82 \times 10^5$$

$$\frac{\epsilon}{d} = \frac{0.00015}{0.5} = 0.0003$$

$$f = 0.018 \text{ (from the Moody diagram)}$$

$$h_f = (0.018)\left(\frac{4000}{0.5}\right)\frac{(5.1)^2}{2(32.2)} = 58.2 \text{ ft}$$

$$H_p = 100 + 58.2 + 0 = 158.2 \text{ ft}$$

$$\text{BHP} = \frac{\gamma Q H_p}{550\eta} = \frac{62.4(1)(158.2)}{550(0.70)} = 25.6 \text{ hp}$$

11.7 PIPES IN SERIES

Pipes in series or a compound pipeline consists of several pipes of different sizes connected together as shown in Figure 11.9. According to the continuity and the energy equations, the following relations apply to the pipes in series:

$$Q = Q_1 = Q_2 = Q_3 = \cdots \qquad [L^3 T^{-1}] \qquad (11.15a)$$

$$h_f = h_{f1} + h_{f2} + h_{f3} + \cdots \qquad [L] \qquad (11.15b)$$

For analysis purpose, the different-size pipes are replaced by a pipe of a uniform diameter of a length that will pass a discharge, Q, with the total head loss, h_f, given by eq. (11.5). This is known as the equivalent pipe. The procedure will be illustrated by an example.

Example 11.8

In Figure 11.10, cast iron pipes 1, 2, and 3 are 1000 ft of 6-in. diameter, 500 ft of 3-in. diameter, and 1800 ft of 4-in. diameter, respectively. If the difference in head is 50 ft, determine the discharge at 50°F.

Solution The nomogram of Figure 11.6 for $C = 100$ can be used for converting a series of pipes of any material (any value of C) into an equivalent length. The steps are as follows:

1. Assume a discharge through the series of pipes: say, $Q = 0.5$ cfs.
2. For each pipe, for the assumed discharge and known diameter, compute the fric-

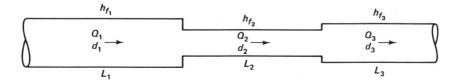

Figure 11.9 Compound pipeline.

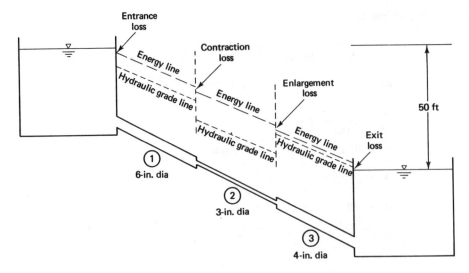

Figure 11.10 Compound pipe system connecting two reservoirs.

tion slope by the Hazen–Williams equation (11.10)* or the nomogram, as shown in column 4 of Table 11.3.

3. Multiply the friction slope (column 4) by the pipe length (column 5) to obtain the head loss (column 6).

4. The addition of col. 6 provides the total head loss per eq. (11.15b). In this case, $h_f = 213$ ft.

5. Select any desired size of uniform pipe. Selected $d = 4$ in.

6. For the selected diameter and the assumed Q of step 1, compute S, again by eq. (10.10) or the nomogram. In this case $S = 0.055$.

7. The required length of uniform pipe, L = step 4/step 6:

$$L = \frac{h_f}{S} = \frac{218}{0.055} = 3964 \text{ ft}$$

TABLE 11.3 COMPUTATION FOR PIPES IN SERIES

(1) Pipe	(2) Pipe Size (ft)	(3) Discharge, Q, Assumed	(4) Friction Slope, S, Computed	(5) Pipe Length, L (ft)	(6) Head Loss, h_f, (ft) (SL)
1	0.5	0.5	0.008	1000	8.00
2	0.25	0.5	0.222	500	111.00
3	0.333	0.5	0.055	1800	99.00
	0.333 (select)	0.5 (assumed)	0.055	?	218

*Alternatively, the Darcy–Weisbach formula can be used to determine h_f for steps 2 and 3 above.

Thus a uniform pipe of 4 in. diameter and 3964 ft length is equivalent to the three pipes in series.

The discharge can be determined by the method of a single pipe using the Darcy–Weisbach or Hazen–Williams equation.

$$\text{Given, } h_f = 50 \text{ ft, thus, } S = \frac{50}{3964} = 0.0126$$

For $d = 0.333$, and $S = 0.0126$, by the Hazen–Williams equation (11.10a), $Q = 0.23$ cfs.

11.8 PIPES IN PARALLEL

For the parallel or looping pipes of Figure 11.11, the continuity and energy equations provide the following relations:

$$Q = Q_1 + Q_2 + Q_3 + \cdots \qquad [L^3 T^{-1}] \qquad (11.16a)$$

$$h_f = h_{f_1} = h_{f_2} = h_{f_3} = \cdots \qquad [L] \qquad (11.16b)$$

A procedure similar to that used for pipes in series is also used in this case, as illustrated in the following example.

Example 11.9

A welded steel pipeline of 2 ft diameter is 1 mile long. To augment the supply, a pipe of the same diameter is attached in parallel to the first in the middle half of the length as shown in Figure 11.12. The head above the outlet is 100 ft. Find the discharge through the pipe. Neglect the minor losses.

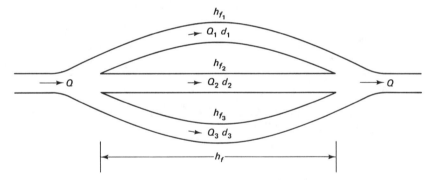

Figure 11.11 Parallel pipe system.

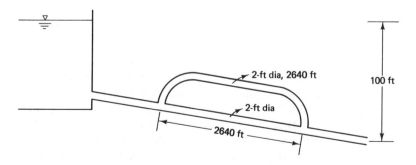

Figure 11.12 Pipes in parallel for Example 11.9.

Solution The following steps are followed to convert the parallel pipes into a single pipe of a uniform size.

1. Assume a head loss through the system: say, $h_f = 100$ ft.
2. For each pipe, compute $S = h_f/L$ (column 5, Table 11.4).
3. From the known values of diameter (column 2) and computed slope (column 5) for each pipe, compute the discharge (column 6) by the Hazen–Williams equation (11.10) or the nomogram.
4. The addition of column 6 is the total discharge, Q, per eq. (11.16a). In this case, $Q = 91.34$ cfs.
5. Select any desired size of a uniform pipe. The selected $d = 2$ ft.
6. For the selected diameter and total Q of step 4, compute the friction slope, S, again by eq. (11.10) or the nomogram.
7. The required length of the uniform pipe

$$L = \frac{\text{step 1}}{\text{step 6}} = \frac{h_f}{S} = \frac{100}{0.137} = 730 \text{ ft}$$

A single pipe of 2 ft diameter and 293 ft length is equal to the parallel portion of the pipe.

TABLE 11.4 COMPUTATION OF PARALLEL PIPE SYSTEM

(1) Pipe	(2) Pipe Diameter (ft)	(3) h_f (ft), Assumed	(4) Length (ft)	(5) $S = h_f/L$ (col. 3/col. 4)	(6) Discharge Q (cfs) Computed
1	2	100	2640	0.0379	45.67
2	2	100	2640	0.0379	45.67
1	2 (selected)	100 (assumed)	?	0.137 (computed)	91.34

Total length of uniform pipe of 2 ft diameter = $2640 + 730 = 3370$ ft. For steel, $C = 120$. Given $h_f = 100$ ft,

$$S = \frac{100}{3370} = 0.03$$

From eq. (11.10a)

$$Q = 0.432(120)(2)^{2.63}(0.03)^{.054} = 48.3 \text{ cfs}$$

11.9 PIPE NETWORKS

An extension of pipes in parallel is a system in which the pipes are interconnected to form a complex loop configuration. The flow to an outlet comes from several paths. The analytical solution of such systems, referred to as pipe networks, is quite complicated. Three simple methods are the Hardy Cross method, the linear theory method, and the Newton–Raphson method. Of these, the Hardy Cross method, which involves a series of successive approximations and corrections to flows in individual pipes, is a most popular procedure of analysis.

According to the Darcy–Weisbach equation,

$$h_f = \frac{fL}{d} \frac{V^2}{2g} = \frac{16}{\pi^2} \frac{fL}{d^5} \frac{Q^2}{2g} \tag{a}$$

As per the Hazen–Williams equation,

$$Q = 0.432 C d^{2.63} \left(\frac{h_f}{L}\right)^{0.54} \tag{b}$$

or

$$h_f = \frac{4.727L}{C^{1.85} d^{4.87}} Q^{1.85} \tag{c}$$

Both (a) and (c) can be expressed in the general form

$$h_f = KQ^n \tag{11.17}$$

where

K = equivalent resistance as given in Table 11.5

n = 2.0 for Darcy–Weisbach equation and 1.85 for Hazen–Williams equation.

TABLE 11.5   EQUIVALENT RESISTANCE, K, FOR PIPE

Formula	Units of Measurement	K
Hazen–Williams	Q, cfs; L, ft; d, ft; h_f, ft	$\dfrac{4.73L}{C^{1.85} d^{4.87}}$
	Q, gpm; L, ft; d, in.; h_f, ft	$\dfrac{10.44L}{C^{1.85} d^{4.87}}$
	Q, m^3/s; L, m; d, m; h_f, m	$\dfrac{10.70L}{C^{1.85} d^{4.87}}$
Darcy–Weisbach	Q, cfs; L, ft; d, ft; h_f, ft	$\dfrac{fL}{39.70 d^5}$
	Q, gpm; L, ft; d, in.; h_f, ft	$\dfrac{fL}{32.15 d^5}$
	Q, m^3/s; L, m; d, m; h_f, m	$\dfrac{fL}{12.10 d^5}$

The sum of head losses around any closed loop is zero: that is,

$$\sum h_f = 0 \tag{d}$$

Consider that Q_a is an assumed pipe discharge that varies from pipe to pipe of a loop to satisfy the continuity of flow. If δ is the correction made in the assumed flow of all pipes of a loop to satisfy eq. (d), then by substituting eq. (11.17) in eq. (d),

$$\sum K(Q_a + \delta)^n = 0 \tag{e}$$

Expanding eq. (e) by the binomial theorem and retaining only the first two terms yields

$$\delta = -\frac{\sum KQ_a^n}{n \sum KQ_a^{n-1}} \tag{f}$$

or

$$\delta = -\frac{\sum h_f}{n \sum |h_f/Q_a|} \qquad [\text{L}^3\text{T}^{-1}] \tag{11.18}$$

Equations (11.17) and (11.18) are used in the Hardy Cross procedure. The values of n and K are obtained based on the Darcy–Weisbach or Hazen–Williams equations from Table 11.5. The procedure is summarized as follows:

1. Divide the network into a number of closed loops. The computations are made for one loop at a time.
2. Compute K for each pipe using the appropriate expression from Table 11.5 (column 3 of Table 11.6).
3. Assume a discharge, Q_a, and its direction in each pipe of the loop (column 4). At each joint, the total flow in should equal the flow out. Consider the clockwise flow to be positive and the counterclockwise flow to be negative.
4. Compute h_f in column 5 for each pipe by eq. (11.17), retaining the sign of column 4. The algebraic sum of column 5 is h_f.
5. Compute h_f/Q for each pipe without regard to the sign. The sum of column 6 is $\sum |h_f/Q|$.
6. Determine the correction, δ, by eq. (11.18). Apply the correction algebraically to the discharge of each member of the loop.
7. For common members among two loops, both δ corrections should be made, one for each loop.
8. For the adjusted Q, steps 4 through 7 are repeated until δ becomes very small for all loops.

Lyle and Weinberg (1957) and Watters (1984) have given computer programs for the Hardy Cross analysis.

Example 11.10

Find the discharge in each pipe of the welded steel pipe network shown in Figure 11.13. All pipes are 4 in. in diameter. The pressure head at A is 50 ft. Determine the pressure at the different nodes.

Solution Refer to Tables 11.6 through 11.8.

1. In the first trial, the flow in each pipe is assumed as indicated in Figure 11.13.
2. $K = \dfrac{4.73L}{d^{4.87}C^{1.85}}$

$$K_{AB} = K_{AC} = K_{BD} = K_{CE} = \frac{4.73(100)}{(0.333)^{4.87}(120)^{1.85}} = 14.25$$

$$K_{BC} = K_{DE} = \frac{4.73(50)}{(0.333)^{4.87}(120)^{1.85}} = 7.12$$

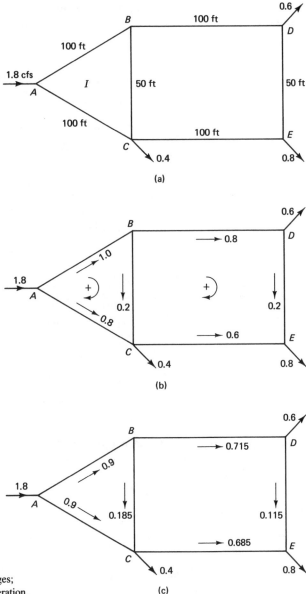

Figure 11.13 (a) Pipe network of
Example 11.10; (b) assumed discharges;
(c) discharge after second and final iteration.

11.10 HYDRAULIC TRANSIENTS

The interim stage when a flow changes from one steady-state condition to another
steady-state condition is known as the transient state of flow. In conduits and open
channels, such conditions occur when the flow is decelerated or accelerated due to
sudden closing or opening of the control valves, starting or stopping of the pumps,
rejecting or accepting of the load by a hydraulic turbine, or similar situations of sudden increased or decreased inflows. The variations in velocity result in change of

TABLE 11.6 ITERATION 1 OF HARDY CROSS PROCEDURE

(1) Loop	(2) Pipeline	(3) K	(4) Q_a (cfs)	(5) $h_f = KQ_a^{1.85}$	(6) $\left|\dfrac{h_f}{Q}\right|$	(7) Q corrected $= Q_a + \delta$
1	AB	14.25	+1.0	+14.25	14.25	+0.9
	BC	7.12	+0.2	+0.36	1.80	+0.185
	CA	14.25	−0.8	−9.43	11.79	−0.9
				+5.18	27.84	
2	BD	14.25	+0.8	+9.43	11.79	+0.715
	DE	7.12	+0.2	+0.36	1.80	+0.115
	EC	14.25	−0.6	−5.54	9.23	−0.685
	CB	7.12	−0.2	−0.36	1.80	−0.185
				+3.89	24.62	

For loop 1,

$$\delta = \frac{-5.18}{1.85(27.84)} = -0.10$$

Adjusted $Q_{AB} = +1.00 + (-0.1) = 0.9$.

For loop 2,

$$\delta = \frac{-3.89}{1.85(24.62)} = -0.085$$

Adjusted $Q_{BC} = 0.2 + (-0.1) - (-0.085) = +0.185$ (common pipe in two loops).

TABLE 11.7 ITERATION 2 OF HARDY CROSS PROCEDURE

(1) Loop	(2) Pipeline	(3) K	(4) Q_a	(5) $h_f = KQ_a^{1.85}$	(6) $\left|\dfrac{h_f}{Q_a}\right|$	(7) Q corrected $= Q_a + \delta$
1	AB	14.25	0.9	11.73	13.03	0.894
	BC	7.12	0.185	0.31	1.68	0.188
	CA	14.25	−0.9	−11.73	13.03	−0.906
				0.31	27.74	
2	BD	14.25	0.715	7.66	10.71	0.706
	DE	7.12	0.115	0.13	1.13	0.106
	EC	14.25	−0.685	−7.08	10.34	−0.694
	CB	7.12	−0.185	−0.31	1.68	−0.188
				0.40	23.86	

Loop 1: $\delta = \dfrac{-0.31}{1.85(27.74)} = -0.006$

Loop 2: $\delta = \dfrac{-0.40}{1.85(23.86)} = -0.009$

TABLE 11.8 FINAL FLOWS AND PRESSURE HEADS

Pipe	Flow (cfs)	Head Loss (ft)	Node	Pressure Head (ft)
AB	0.9	$14.25(.894)^{1.85} = 11.6$	A	50 ft (given)
BC	0.185	0.3	B	$50 - h_{AB} = 38.4$
AC	0.9	11.9	C	$50 - h_{AC} = 38.1$
BD	0.715	7.5	D	$h_B - h_{BD}$
				$38.4 - 7.5 = 30.9$
DE	0.115	0.1	E	$h_C - h_{CE}$
				$38.1 - 7.3 = 30.8$
EC	0.685	7.3		

momentum. The fluid is subject to an impulse force equivalent to the rate of change of the momentum according to Newton's second law. An appreciable increase of pressure occurs with respect to time due to this impulse force. This pressure fluctuation is called water hammer (or oil hammer) because a hammering noise is usually associated with this phenomenon. More commonly, this is now referred to as hydraulic transients. The system design should be adequate to withstand both the normal static pressure and the maximum rise in pressure due to hydraulic transients.

11.10.1 Equations of Transient Flow

The conservation of momentum equation describes the transient state of flow. This equation is also referred to as the dynamic equation or equation of motion. Along with the momentum equation, the continuity (conservation of mass) equation is used to fully describe the transient flow phenomenon. Since the velocity and pressure are functions of time as well as distance in a transient state, these equations are expressed as partial differential equations.

In simplified form, after dropping the convective acceleration terms, which are very small compared to the other terms, the two equations applied to the element Δx in Figure 11.14 yield*

$$\text{Continuity equation:} \quad \frac{\partial h}{\partial t} + \frac{c^2}{g}\frac{\partial v}{\partial x} = 0 \qquad [LT^{-1}] \qquad (11.19)$$

$$\text{Momentum equation:} \quad \frac{\partial h}{\partial x} + \frac{1}{g}\frac{\partial v}{\partial t} + \frac{fv|v|}{2Dg} = 0 \qquad [\text{dimensionless}] \qquad (11.20)$$

where

h = pressure head above a specified datum

c = speed of propagation of pressure wave in a specified fluid and specific conduit material

v = flow velocity; $|v|$ is absolute value of v

f = friction factor

D = conduit diameter

*For derivation of these equations, refer to a fluid mechanics or fluid transients textbook (e.g., Chaudhry, 1987, pp. 30–39).

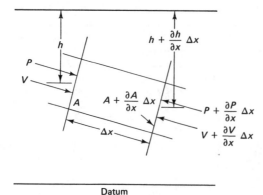

Figure 11.14 Element of fluid for continuity and momentum equations.

Datum

The speed of the pressure wave is related to the medium of flow (fluid), conduit properties, and type of anchoring. A general expression is as follows (Halliwell, 1963):

$$c = \sqrt{\frac{K}{\rho[1 + (K/E)\psi]}} \qquad [LT^{-1}] \qquad (11.21)$$

where

K = bulk modulus of elasticity of fluid

ρ = density of fluid

E = modulus of elasticity of conduit material

ψ = dimensionless parameter that depends on the elastic properties of conduit and the type of anchoring, as follows:

1. For rigid conduits:

$$\psi = 0 \qquad (11.22)$$

2. For a thin-walled conduit without expansion joints, anchored throughout its length:

$$\psi = \frac{D}{e}(1 - \nu^2) \qquad (11.23)$$

where

D = conduit diameter

e = conduit wall thickness

ν = Poisson's ratio

3. For a thin-walled conduit without expansion joints, anchored at the upper end:

$$\psi = \frac{D}{e}(1 - 0.5\nu) \qquad (11.24)$$

4. For a thin-walled conduit with frequent expansion joints:

$$\psi = \frac{D}{e} \qquad (11.25)$$

An important factor that affects the pressure rise is the time of closure (stoppage) of flow, t_c. For a very slow closure, when $t_c > 20L/c$, where L is the conduit length, the conduit acts as a rigid body and the fluid behaves as an incompressible mass. The entire column of fluid is subjected to a uniform deceleration. The rigid column theory applies and the transient or surge pressure can be computed from the momentum principle [eq. (11.20)] applied to the rigid column of fluid.

As closing time, t_c, decreases, the inertia force increases. A point is reached when the force (pressure) is sufficient to cause the fluid to compress and the pipe material to expand. Under this condition, the transient process is radically changed. The kinetic energy of flow is converted into the elastic energy. The elastic wave theory applies, in which the shock or pressure wave travels through the fluid. The continuity and momentum equations [eqs. (11.19) and (11.20)] are solved simultaneously to compute the pressure fluctuations. There are two cases of this category: (1) when the time of closure, t_c, is not more than $2L/c$ — this is known as a "rapid" closure, and (2) when t_c is greater than $2L/c$, but less than $20L/c$ — this is known as a "slow" closure. These are discussed subsequently.

11.10.2 Mechanism of Pressure-Wave Propagation

Consider the piping system shown in Figure 11.15(a), in which a steady-state flow is taking place from a reservoir at a velocity, V_0, and head, H_0, with no friction losses. When the valve at the downstream end of the pipe is instantly closed, the following sequence of events takes place.

1. Immediately following the valve closure, the fluid in proximity of the valve is brought to rest. This causes a local pressure increase. Due to the increased pressure, the liquid is compressed and the pipe walls expand. This provides a little extra volume in which the liquid enters to come to a stop. An instant later, the next section immediately upstream follows the same process. Thus a wave of increased pressure propagate upstream toward the reservoir. The process which is started in Figure 11.15(b) completes in Figure 11.15(c). When the wave reaches the reservoir, the entire pipe is expanded; the liquid column is compressed and comes to a complete stop. The time taken by the wave to reach the reservoir is L/c.

2. At the end of step 1, the pressure in the pipe is much higher than the pressure in the reservoir. Hence the halted water begins to flow from the pipe into the reservoir. This relieves the pressure in the pipe. The liquid column decompresses and the pipe material contracts. The process starts at the reservoir end of the pipe [Figure 11.15(d)] and continues toward the valve [Figure 11.15(e)]. The time in this step is L/c, and the total time taken by the wave to return to the valve is $2L/c$, which is known as the water-hammer period.

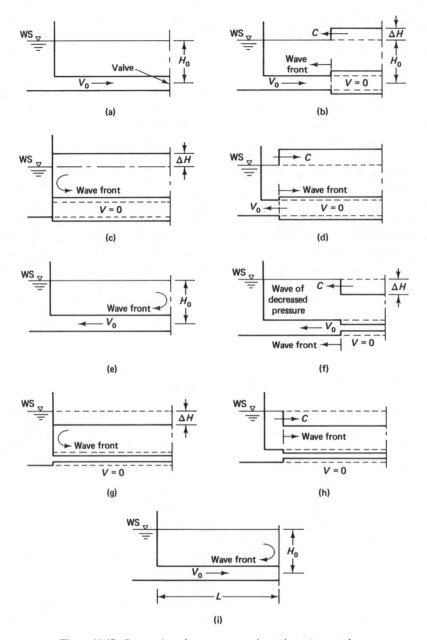

Figure 11.15 Propagation of pressure wave due to instantaneous closure.

3. The wave of backwater motion cannot go past the closed valve. The sudden stoppage (inertia) of this moving fluid at the valve causes the pressure to drop below the normal level. This sends a wave of negative pressure again upstream toward the reservoir. The process starts at Figure 11.15(f) and ends at (g).

4. At the end of step 3, since the pressure at the reservoir end of the pipe is negative (less than that in the reservoir), liquid starts flowing from the reservoir into

the pipe. The process starts at Figure 11.15(h) and ends at (i). At the end of this step, the conditions are identical to the beginning of step 1 (i.e., one cycle is completed in a period of $4L/c$).

This process could have continued indefinitely as shown in Figure 11.16(a), but the friction losses in the system diminishes the pressure successively, as shown in Figure 11.16(b), until the waves finally die out.

11.10.3 Very Slow Closure of Valve

For $t_c > 20L/c$, the fluid and pipe material act as solid media. There is little or no pressure-wave propagation, as illustrated in Section 11.10.2. The pressure surge acts due to uniform deceleration of the column of water in the pipe, which can be ascertained from the momentum (or energy) principle. Neglecting the friction losses and expressing the terms in the lumped finite forms in eq. (11.20), the maximum pressure difference is given by

$$\Delta h_m = \frac{L}{g} \frac{V_0}{t_c} \qquad [\text{L}] \tag{11.26}$$

where

Δh_m = maximum pressure rise

L = length of pipeline

V_0 = initial velocity of flow

t_c = time of closure (velocity changes from V_0 to 0 in time t_c)

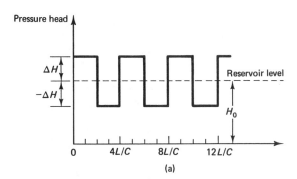

(a)

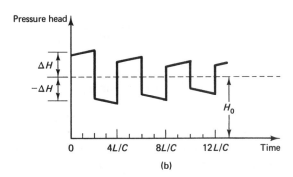

Figure 11.16 Pressure variation at the valve: (a) friction losses neglected; (b) friction losses considered.

(b)

Example 11.11

From a reservoir, water at a rate of .744 m³/s flows through a 1000-m-long horizontal cast iron pipeline discharging into the atmosphere through a valve at the downstream end. The pipe is 500 mm in diameter and 15 mm thick. The valve is fully closed in 20 seconds. Calculate the transient pressure at the valve and at 400 m upstream of the valve. The pipe is without expansion joints and anchored throughout. $K = 2.2$ GPa, E (cast iron) $= 150$ GPa, $\nu = 0.25$.

Solution

1. For steady-state flow when the value is open:

$$Q = A_0 V_0 = 0.744$$

or

$$\frac{\pi}{4}(0.5)^2 V_0 = 0.744$$

or

$$V_0 = 3.79 \text{ m/s}$$

2. Speed of wave, c:

$$\psi = \frac{D}{e}(1 - \nu^2) = \frac{500}{15}[1 - (0.25)^2] = 31.3$$

$$c = \sqrt{\frac{2.2 \times 10^9}{1000[1 + (2.2 \times 10^9/150 \times 10^9) \times 31.3]}} = 1230 \text{ m/s}$$

3. Wave travel time:

$$\frac{L}{c} = \frac{1000}{1230} = 0.8 \text{ sec}$$

Since a t_c of 20 seconds $> 20L/c$, this is a case of very slow valve closing.

4. At the valve when the valve is fully closed:

$$\Delta h_m = \frac{LV_0}{gt_c} = \frac{1000(3.79)}{9.81(20)} = 19.3 \text{ m}$$

5. At 400 m upstream from the valve when the valve is closed,

$$\Delta h_m = \frac{600(3.79)}{9.81(20)} = 11.6 \text{ m}$$

11.10.4 Rapid Closure of Valve

When the time of closure, t_c, is not more than $2L/c$, the return pressure wave finds the valve closed. The maximum transient pressure in such a case is the same as for the instantaneous closing of the valve. Analytical solution of this elastic wave phenomenon is obtained by simplifying the momentum equation (11.20) by dropping the friction loss term and solving the simultaneous differential equations (11.19) and (11.20). Riemann expressed the solution as

$$h_{tx} = h_0 + F\left(t - \frac{x}{c}\right) + f\left(t + \frac{x}{c}\right) \qquad [L] \qquad (11.27)$$

$$V_{tx} = V_0 - \frac{g}{c}\left[F\left(t - \frac{x}{c}\right) + f\left(t + \frac{x}{c}\right)\right] \qquad [LT^{-1}] \qquad (11.28)$$

where

h_0, V_0 = initial head and velocity in steady state

h_{tx}, V_{tx} = head and velocity at a section x from the valve at time t

$F(t - x/c)$ = pressure wave propagated upstream at a speed c

$f(t + x/c)$ = reflected pressure wave propagated downstream

Denoting i as the time interval equal to the water-hammer period $(2L/c)$, the reflected wave, f, at any instant will have the same magnitude (and opposite sign) as the direct wave, F, one time interval earlier (i.e., $f_i = -F_{i-1}$).

At the valve ($x = 0$), writing the head relations [from eq. (11.27)] and the velocity relations [from eq. (11.28)] for the two successive time periods $i - 1$ and i and by combining them, the following general relation results:

$$h_i + h_{i-1} - 2h_0 = \frac{c}{g}(V_{i-1} - V_i) \qquad [L] \qquad (11.29)$$

For rapid and instantaneous valve closure, $i - 1$ denotes the initial condition at time 0 and i denotes the next period when the valve is closed; thus $V_i = 0$. Substituting in eq. (11.29) yields

$$h_1 + h_0 - 2h_0 = \frac{c}{g}(V_0 - 0) \qquad (a)$$

or

$$\Delta h_m = \frac{cV_0}{g} \qquad [L] \qquad (11.30)$$

where $\Delta h_m = h_1 - h_0$ = maximum transient head.

If a pipeline is operating at partial opening and the complete closure is attained starting from this opening, the maximum pressure rise can still be given by eq. (11.30), when V_0 will correspond to the velocity at partial opening.

Example 11.12

In Example 11.11, the valve is closed in 1.5 seconds. Determine the maximum transient pressure at the valve.

Solution

1. Since $t_c < 2L/c$, this is a case of rapid closure.

2. From eq. (11.30),

$$\Delta h_m = \frac{1230(3.79)}{9.81} = 475.2 \text{ m}$$

11.10.5 Slow Closure of Valve

When the time of closure exceeds $2L/c$ (but is less than $20L/c$, so that the elastic theory is applicable), the reflection or negative wave returning at the valve partially compensates the pressure rise and the transient pressure is not as high as for the rapid closure. There are many analytical methods (besides the numerical procedures) to ascertain the pressure variation with time. The classic procedure of Allievi (1925) is described here.

Using the equation of discharge through a valve (sluice), we have

$$V_0 = C_{d,0} \frac{A_0}{A_p} \sqrt{h_0} \tag{a}$$

$$\frac{V_i}{V_0} = \frac{C_{d,i}}{C_{d,0}} \frac{A_i}{A_0} \sqrt{\frac{h_i}{h_0}} \tag{b}$$

where

$$A_0, A_i = \text{areas of value initially and at time } i = 0$$
$$A_p = \text{area of pipe}$$

If C_d is a constant, $n_i = A_i/A_0$ and $\xi_i = h_i/h_0$, then $V_i = n_i \xi_i V_0$ and $h_i = h_0 \xi_i^2$. Substituting in eq. (11.29) gives

$$h_0 \xi_i^2 + h_0 \xi_{i-1}^2 - 2h_0 = \frac{cV_0}{g}(n_{i-1}\xi_{i-1} - n_i \xi_i) \tag{c}$$

Denoting $N = cV_0/2gh_0$, we obtain

$$\xi_i^2 + \xi_{i-1}^2 - 2 = 2N(n_{i-1}\xi_{i-1} - n_i \xi_i) \quad \text{[dimensionless]} \tag{11.31}$$

Equation (11.31) is solved at different time intervals, $i = 1, 2, 3, ...$, to calculate the head variation.

In practice, the slower rate of closing is selected wherever possible to prevent the excessive rise in pressure associated with the instant closing.

Example 11.13

In Example 11.11, the valve is fully closed in 8 sec. The area of the valve opening varies as follows. Calculate the variation of pressure head at the valve during closure. Neglect the friction losses. $C_d = 0.6$.

Time (s)	0	1.6	3.2	4.8	6.4	8.0
Valve area (m)	0.06	0.048	0.036	0.024	0.012	0

Solution

1. From Example 11.11, $L/c = 0.8$ second, time interval $= 2L/c = 1.6$ seconds, $V_0 = 3.79$ m/s.
2. For continuity, $A_p V_0 = C_d A_0 \sqrt{2gh_0}$. Hence

$$h_0 = \left(\frac{A_p V_0}{C_d A_0 \sqrt{2g}}\right)^2 = \left[\frac{0.196(3.79)}{0.6(0.06)\sqrt{2(9.81)}}\right]^2 = 21.7 \text{ m}$$

3. $N = \dfrac{cV_0}{2gh_0} = \dfrac{1230(3.79)}{2(9.81)(21.7)} = 10.95.$

4. At $i = 1$ ($t = 1.6$ seconds), from eq. (11.31)

$$\xi_1^2 + \xi_0^2 - 2 = 2N(n_0\xi_0 - n_1\xi_1)$$

$$\xi_1^2 + 1 - 2 = 2(10.95)\left[1(1) - \left(\frac{0.048}{0.06}\right)\xi_1\right]$$

or

$$\xi_1^2 + 17.52\xi_1 - 22.90 = 0$$

or

$$\xi_1 = \sqrt{\frac{h_1}{h_0}} = 1.22, \quad \text{hence } h_1 = \xi_1^2 h_0 = (1.22)^2 (21.7) = 32.3 \text{ m}$$

and

$$V_1 = n_1\xi_1 V_0 = \left(\frac{0.048}{0.060}\right)(1.22)(3.79) = 3.70 \text{ m/s}$$

5. At $i = 2$ ($t = 3.2$ sec),

$$\xi_2^2 + (1.22)^2 - 2 = 2(10.95)\left[\left(\frac{0.048}{0.060}\right)(1.22) - \left(\frac{0.036}{0.06}\right)\xi_2\right]$$

$$\xi_2^2 + 13.14\xi_2 - 21.89 = 0$$

or

$$\xi_2 = 1.50$$

Hence

$$h_2 = (1.50)^2 (21.7) = 48.83 \text{ m}$$

and

$$V_2 = n_2\xi_2 V_0 = 0.6(1.50)(3.79) = 3.41 \text{ m/s}$$

6. The other steps are completed in Table 11.9.

TABLE 11.9 COMPUTATION OF TRANSIENT PRESSURE BY ALLIEVI METHOD

i	$t = 2L/c$	$n_i = \dfrac{A_i}{A_0}$	ξ_i [eq. (11.31)]	$h_i = \xi_i^2 h_0$	$V_i = n_i\xi_i V_0$
0	0	1	1	21.7	3.79
1	1.6	0.8	1.22	32.3	3.70
2	3.2	0.6	1.50	48.83	3.41
3	4.8	0.4	1.84	73.47	2.79
4	6.4	0.2	2.23	107.91	1.69
5	8.0	0	2.61	147.82	0

11.10.6 Numerical Methods for Transient Pressure

A major simplification in the analytical procedures above has been made when the frictional losses are ignored. On the other hand, the numerical methods consider the entire partial differential equations and hence involve substantive computations. The differential equations are expressed in numerical (finite difference or finite element) form in these methods, to be conveniently solved on a computer. With the common use of computers these days, the simplified analytical procedures, as discussed above, are rarely used now. However, the classic methods are important for an understanding of the basic principles. Among the numerical methods, perhaps the most popular is the method of "characteristics." In this method, the pipe length is divided into a number of sections and the time of closure into a number of time increments. The grid formed by these divisions on the X–t plane is used to indicate the progress of events (head and velocity distribution) at various points in different time intervals. Equations similar to eq. (11.29) but including the friction term is written for two successive time increments for each section (various X increments) along the pipe. Starting from the known steady-state conditions at time t_0, successive solution of these equations leads to the values for the time increment t_1. Using these as initial conditions for the next step, the values for the subsequent time period are obtained. Thus the solution marches in the direction of t. For detailed treatment of the method of characteristics, refer to a fluid transients textbook (e.g., Wylie and Streeter, 1983).

11.11 PUMPS

In the design of a pumping station, a water resources engineer is concerned with the selection of a pump based on its performance information. Therefore, the components of a pump and its design details have not been considered. Karassik et al. (1976) is recommended reading. In water and wastewater works, the centrifugal pumps are most common in application. In the following sections the characteristics are specifically described with respect to this type of pump.

11.12 PUMP CLASSIFICATION: SPECIFIC SPEED

The pump performance parameters comprise (1) the rotational speed, (2) the discharge capacity, (3) the pumping head, (4) the power applied, and (5) the efficiency. For each pump, these parameters have certain relationships to each other that vary from pump to pump. By combining the three main parameters of speed, discharge, and head, a single term known as the specific speed has been evolved that is fixed for all pumps operating under dynamic conditions that are geometrically similar (homologous) to one another. The term is thus suitable to group the pumps with respect to the similarity of their design and to compare the performance of the pumps of different designs.

The specific speed, expressed as follows, is measured in inconsistent but standard units in the United States and thus is used as an index.

$$N_S = \frac{N\sqrt{Q}}{H^{3/4}} \quad [\ ]* \tag{11.32}$$

*In principle, the term is dimensionless in the form $NQ^{1/2}/(gH)^{3/4}$.

where

$$N_S = \text{specific speed}$$
$$N = \text{rotational speed, rpm}$$
$$Q = \text{discharge-capacity, gpm}$$
$$H = \text{total head, ft}$$

Under any similar operating conditions of head and capacity (similar conditions established by the laws of similarity explained in the next section), the specific speed is the same for all pumps of geometrically similar designs. For any pump the value of the specific speed, however, changes under different operating conditions. In the classification of a pump, the specific speed corresponding to the operating condition at the maximum efficiency, called the type specific speed, is used.

Certain features of the specific speed are as follows:

1. The efficiency starts dropping drastically when lowering the specific speed below 1000.
2. High specific speeds above 5000 also have a lower efficiency than the medium specific speed range.
3. At all specific speeds, smaller pump capacities have lower efficiencies than those of pumps of higher capacities.

Radial-flow centrifugal pumps have a specific speed between 500 and 3500 and are suitable for low discharge under relatively high pressure. Mixed-flow pumps of specific speed in the range 3500 to 7500 are used for flows of more than 1000 gpm. Axial-flow pumps with specific speed between 7500 and 15,000 deliver a high discharge of over 5000 gpm.

11.13 RELATIONS FOR GEOMETRICALLY SIMILAR PUMPS

The relations that relate the parameters of geometrically similar pumps are known as the affinity laws. These are useful in predicting the performance of a pump from pumping tests on a model pump or homologous* pump.

Since $Q = AV$, where the area is proportional to the square of the impeller diameter, D^2, and the velocity is proportional to the impeller diameter, D, and angular speed, N, hence $Q \propto ND^3$. Since $V = \sqrt{2gH}, H \propto V^2$ or $H \propto D^2N^2$. Also, the power is the multiplication of Q and H and hence $P \propto N^3D^5$.

Thus, from dimensional analysis considerations, the affinity laws are

$$\frac{Q_2}{Q_1} = \frac{N_2}{N_1}\left(\frac{D_2}{D_1}\right)^3 \qquad \text{[dimensionless]} \qquad (11.33a)$$

$$\frac{H_2}{H_1} = \left(\frac{N_2}{N_1}\right)^2\left(\frac{D_2}{D_1}\right)^2 \qquad \text{[dimensionless]} \qquad (11.33b)$$

$$\frac{P_2}{P_1} = \left(\frac{N_2}{N_1}\right)^3\left(\frac{D_2}{D_1}\right)^5 \qquad \text{[dimensionless]} \qquad (11.33c)$$

*Two units that are geometrically similar and have similar vector diagrams are said to be homologous.

where

$$Q = \text{capacity}$$
$$H = \text{head}$$
$$P = \text{power}$$

where the subscript 1 refers to the parameters at which characteristics are known and the subscript 2 refers to the unit for which values to be predicted.

Example 11.14

A model pump of 5 in. diameter develops 0.25 hp at a speed of 800 rpm under a head of 2.5 ft. A geometrically similar pump 15 in. in diameter is to operate at the same efficiency at a head of 49.0 ft. What speed and power should be expected?

Solution From eq. (11.33b),

$$\frac{H_2}{H_1} = \left(\frac{N_2}{N_1}\right)^2 \left(\frac{D_2}{D_1}\right)^2$$

$$\frac{49}{2.5} = \left(\frac{N_2}{800}\right)^2 \left(\frac{15}{5}\right)^2$$

$$N_2 = 1181 \text{ rpm}$$

From eq. (11.33c),

$$\frac{P_2}{P_1} = \left(\frac{N_2}{N_1}\right)^3 \left(\frac{D_2}{D_1}\right)^5$$

$$P_2 = \left(\frac{1181}{800}\right)^3 \left(\frac{15}{5}\right)^5 (0.25) = 195 \text{ hp}$$

11.14 RELATIONS FOR ALTERATIONS IN THE SAME PUMP

For a given pump operating at a given speed, there are definite relationships among parameters, known as the performance characteristics. If the pump size is altered or the speed is changed, the same relations do not hold. It has been found that for the velocity triangle at the exit from the impeller to remain the same before and after the alteration in the pump diameter, the following relations should apply:

$$\frac{Q_2}{Q_1} = \frac{N_2}{N_1} \frac{D_2}{D_1} \qquad \text{[dimensionless]} \qquad (11.34a)$$

$$\frac{H_2}{H_1} = \left(\frac{N_2}{N_1}\right)^2 \left(\frac{D_2}{D_1}\right)^2 \qquad \text{[dimensionless]} \qquad (11.34b)$$

$$\frac{P_2}{P_1} = \left(\frac{N_2}{N_1}\right)^3 \left(\frac{D_2}{D_1}\right)^3 \qquad \text{[dimensionless]} \qquad (11.34c)$$

The efficiency is considered constant, with change in speed and diameter in the relations above.

When only the speed is changed, the relations are as follows:

$$\frac{N_2}{N_1} = \frac{Q_2}{Q_1} = \left(\frac{H_2}{H_1}\right)^{1/2} = \left(\frac{P_2}{P_1}\right)^{1/3} \qquad \text{[dimensionless]} \qquad (11.35)$$

For a changed diameter at the same speed, the relations are as follows:

$$\frac{D_2}{D_1} = \frac{Q_2}{Q_1} = \left(\frac{H_2}{H_1}\right)^{1/2} = \left(\frac{P_2}{P_1}\right)^{1/3} \qquad \text{[dimensionless]} \qquad (11.36)$$

Equations (11.34), (11.35), and (11.36) are used to determine the revised characteristics of a pump for a desired change in speed or diameter or both. Alternatively, the equations are used to determine the speed or diameter to produce a desired change in the discharge capacity or the head of the pump without changing the efficiency.

Example 11.15

A pump tested at 1800 rpm gives the following results: capacity = 4000 gpm, head = 157 ft, power = 190 hp. (a) Obtain the performance of this pump at 1600 rpm. (b) If along with the speed the diameter of the impeller is reduced from 15 in. to 14 in., obtain the revised pump characteristics.

Solution

(a) $N_2/N_1 = 1600/1800 = 0.89$. From eq. (11.35),

$$\frac{Q_2}{Q_1} = 0.89 \qquad \text{or} \qquad Q_2 = 0.89(4000) = 3560 \text{ gpm}$$

$$\frac{H_2}{H_1} = (0.89)^2 \qquad \text{or} \qquad H_2 = (0.89)^2(157) = 124 \text{ ft}$$

$$\frac{P_2}{P_1} = (0.89)^3 \qquad \text{or} \qquad P_2 = (0.89)^3(190) = 134 \text{ hp}$$

(b) $D_2/D_1 = 14/15 = 0.933$. From eq. (11.34),

$$\frac{Q_2}{Q_1} = (0.933)(0.89) \qquad \text{or} \qquad Q_2 = 0.83(4000) = 3320 \text{ gpm}$$

$$\frac{H_2}{H_1} = (0.933)^2(0.89)^2 \qquad \text{or} \qquad H_2 = 0.69(157) = 108 \text{ ft}$$

$$\frac{P_2}{P_1} = (0.933)^3(0.89)^3 \qquad \text{or} \qquad P_2 = 0.57(190) = 108 \text{ hp}$$

11.15 HEAD TERMS IN PUMPING

For the following definitions, refer to Figure 11.17.

Static suction lift The vertical distance from the water level in the source tank to the centerline of the pump. If the pump is located at a lower level than the source tank [Figure 11.17(b)], the static suction lift is negative.

Static discharge head The vertical distance from the centerline of the pump to the water level in the discharge tank.

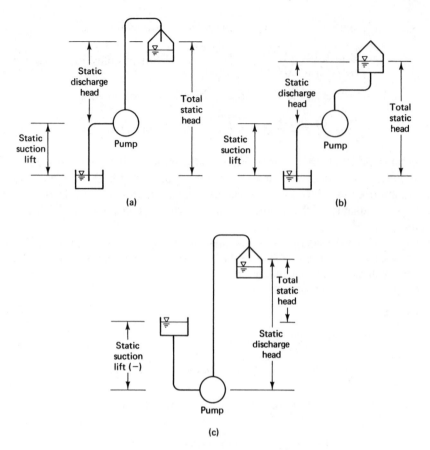

Figure 11.17   Head terms in pumping.

Total static head   The sum of the static suction lift and the static discharge head, which is equal to the difference between the water levels of discharge and source tanks.

Total dynamic head (TDH)   The sum of the total static head and the friction and minor losses. This term is now known as the total head. The relation for the total head was developed in Section 11.6, given by eq. (11.13):

$$H_p = \Delta Z + h_{loss} \quad [\text{L}]. \tag{11.13}$$

where

$$H_p = \text{total dynamic head (TDH)}$$
$$\Delta Z = \text{total static head}$$
$$h_{loss} = h_f + h_m \text{ i.e. friction and minor losses}$$

The total dynamic head, H_p, is used to calculate the horsepower requirement for the pump as given by eq. (11.14).

11.16 SYSTEM HEAD CURVE

For any piping system, the friction loss, $h_f [= (fL/d)(V^2/2g)]$, and the minor loss, h_m $(= \Sigma KV^2/2g)$, can be expressed in terms of the flow through the system. Thus eq. (11.13) can be expressed as

$$H_p = \Delta Z + \frac{0.81}{g}\left(\frac{fLQ^2}{d^5} + \frac{\Sigma KQ^2}{d^4}\right) \qquad [L] \qquad (11.37)$$

The plot of eq. (11.37) between H_p versus Q, as shown in Figure 11.18, is known as the system head curve. This curve, representing the behavior of the piping system, is important in the selection of a pump.

11.17 PUMP CHARACTERISTIC CURVES

As stated earlier, for a given pump at a given speed, there are definite relationships among the pump discharge capacity, head, power, and efficiency. These relations are derived from the actual tests on a given pump or a similar unit and are usually depicted graphically by the pump characteristics or performance curves, comprising the following:

1. Pumping head versus discharge
2. Brake horsepower versus discharge
3. Efficiency versus discharge

Figure 11.19 illustrates the typical characteristic curves. The general shape of the curves varies with the size, speed, and design of a particular pump. The important feature of the curves is that an increase in the head reduces the capacity. These curves are supplied by the manufacturer of the pump. In fact, since a pump casing can accommodate impellers of several sizes, the manufacturer supplies a series of sets of curves drawn on the same graph, corresponding to various sizes of the impellers, which can be derived by use of the laws explained in Section 11.14. A set of characteristic curves represents the behavior of a given-size pump operating at a given speed, in the same manner as a system head curve represents the behavior of a piping system. At a given speed, a pump is rated at the head and discharge, which gives the

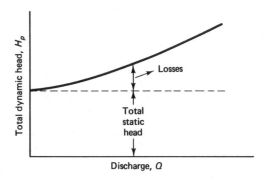

Figure 11.18 Typical system head curve.

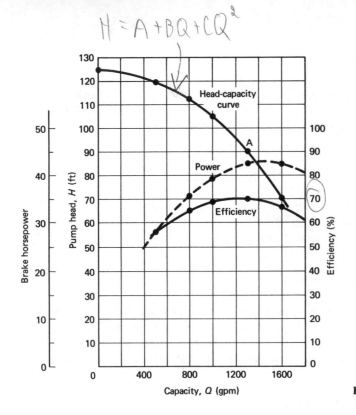

$$H = A + BQ + CQ^2$$

Figure 11.19 Pump characteristic curves.

maximum efficiency, referred to as the best efficiency point, shown by point A in Figure 11.19. The characteristic curves, particularly the head–discharge curve, are important in pump selection.

Example 11.16

The characteristic data as supplied by the manufacturer for a 8-in. pump rotating at 1750 rpm are given in Table 11.10. (a) Plot the pump characteristic curves. (b) Determine the type specific speed. (c) If the pump speed is reduced to 1450 rpm, determine the pump characteristics. (d) If the diameter of the pump is reduced to 6 in., determine the pump characteristics. (e) If a similar unit of 6 in. diameter is used, determine its characteristics.

TABLE 11.10 DATA FOR 8-IN. PUMP AT 1750 RPM

Q (gpm)	H (ft)	P (hp)	Efficiency (%)
0	124		0
500	119	27.75	54
800	112	35.34	64
1000	104	38.69	68
1300	90	42.30	70
1600	70	42.20	67

Solution

(a) The characteristic curves are plotted in Figure 11.19.

(b) At best (maximum) efficiency point, $Q = 1300$ gpm, $H = 90$ ft.

$$N_S = \frac{N\sqrt{Q}}{H^{3/4}} = \frac{1750\sqrt{1300}}{90^{3/4}} = 2159$$

(c) For a change in the pump speed, from eq. (11.35),

$$Q_2 = \frac{N_2}{N_1}Q_1 = \frac{1450}{1750}Q_1 = 0.83Q_1$$

$$H_2 = \left(\frac{N_2}{N_1}\right)^2 H_1 = (0.83)^2 H_1 = 0.69 H_1$$

$$P_2 = \left(\frac{N_2}{N_1}\right)^3 P_1 = (0.83)^3 P_1 = 0.57 P_1$$

The values of Table 11.10 are adjusted by these factors in columns 1, 2, and 3 of Table 11.11.

(d) For the reduced diameter of the same pump, from eq. (11.36)

$$Q_2 = \frac{D_2}{D_1}Q_1 = \frac{6}{8}Q_1 = 0.75Q_1$$

$$H_2 = \left(\frac{D_2}{D_1}\right)^2 H_1 = (0.75)^2 H_1 = 0.56 H_1$$

$$P_2 = \left(\frac{D_2}{D_1}\right)^3 P_1 = (0.75)^3 P_1 = 0.42 P_1$$

TABLE 11.11 ADJUSTED PUMP CHARACTERISTICS

8-in. at 1450 rpm Same Pump			6-in. at 1750 rpm, Same Pump			6-in. at 1750 rpm, Similar Unit		
(1)	(2)	(3)	(4)	(5)	(6)	(7)	(8)	(9)
Q	H	P	Q	H	P	Q	H	P
$0.83Q_1$	$0.69H_1$	$0.57P_1$	$0.75Q_1$	$0.56H_1$	$0.42P_1$	$0.42Q_1$	$0.56H_1$	$0.237P_1$
0	85.6		0	69.4		0	69.4	
415	82.1	15.82	375	66.6	11.66	210	66.6	6.58
664	77.3	20.14	600	62.7	14.84	336	62.7	8.38
830	71.8	22.05	750	58.2	16.25	420	58.2	9.17
1079	62.1	24.11	975	50.4	17.77	546	50.4	10.03
1328	48.3	24.05	1200	39.2	17.72	672	39.2	10.00

(e) For a homologous unit, from eq. (11.33),

$$Q_2 = \left(\frac{D_2}{D_1}\right)^3 Q_1 = (0.75)^3 Q_1 = 0.42 Q_1$$

$$H_2 = \left(\frac{D_2}{D_1}\right)^2 H_1 = (0.75)^2 H_1 = 0.56 H_1$$

$$P_2 = \left(\frac{D_2}{D_1}\right)^5 P_1 = (0.75)^5 P_1 = 0.237 P_1$$

The final values are computed in Table 11.11 by the factors above.

11.18 SINGLE PUMP AND PIPELINE SYSTEM

The suitability of a given pump for a certain known piping system is determined by superimposing the system head curve of the piping system on the head-capacity characteristic curve of the pump. The intersection point of two curves indicates the operating point (i.e., the head and discharge of the given pump). If the efficiency of the pump is too low at this point, another pump should be considered. This is illustrated in Example 11.17.

Example 11.17

The pump, having the characteristic curves given in Table 11.10, is to be used in the pipeline system shown in Figure 11.20. Determine the (a) operating head and discharge, (b) efficiency of the pump and hence its suitability, and (c) input power (brake horsepower) for the pump. $f = 0.02$.

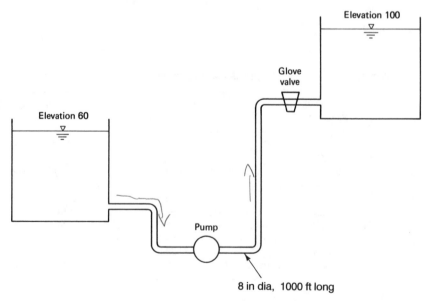

Figure 11.20 Piping system for Example 11.17.

Solution

1. For the system head curve, from eq. (11.37),

$$H_p = \Delta Z + \frac{0.81}{g}\left(\frac{fLQ^2}{d^5} + \frac{\sum KQ^2}{d^4}\right)$$

	K
One entrance	0.5
One exit	1.0
Four 90° bends at 0.9	3.6
One glove valve	10.0
	15.1

$$H_p = (100 - 60) + \frac{0.81}{32.2}\left[\frac{0.02(1000)Q^2}{(8/12)^5} + \frac{(15.1)Q^2}{(8/12)^4}\right]$$

or

$$H_p = 40 + 5.74Q^2 \quad [\text{L}] \tag{11.38}$$

2. The system curve between the capacity and the head is plotted in Figure 11.21, based on Table 11.12.

3. On the same graph, the head-capacity curve of the pump characteristics has been plotted from the data in Table 11.10.

4. The operating head and capacity at the intersection point. $Q = 1300$ gpm, $H = 89$ ft.

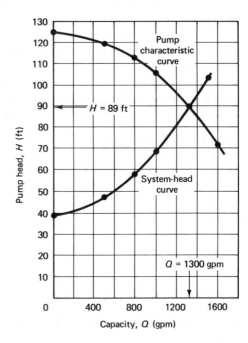

Figure 11.21 Determination of pump operating condition.

TABLE 11.12 COMPUTATION FOR SYSTEM HEAD CURVE

Capacity (gpm) (Select)	$Q = \dfrac{\text{Capacity}}{449}$ (cfs)	Hp [by eq. (11.38)] (ft)
0	0	40
500	1.11	47.07
800	1.78	58.20
1000	2.23	68.54
1500	3.34	104.03

5. At this point, the efficiency = 70% which is the maximum efficiency; hence the pump is OK.

6. Input power $= \dfrac{\gamma QH}{550\eta} = \dfrac{(62.4)(1300/449)(89)}{550(0.70)} = 41.8$ hp.

11.19 MULTIPLE PUMP SYSTEM

A single pump is suitable within a narrow range of head and discharge in proximity of the optimum pump efficiency. However, in a piping system the discharge and head requirements may vary considerably at different times. Within a certain range, these fluctuations in head and discharge can be accommodated by adopting variable-speed motors. Pump characteristic curves can be altered by suitable adjustment of the speed, as discussed in Section 11.14. When the fluctuations are considerable, or either the head or capacity requirement is too high for a single pump, two or more pumps are used in series or in parallel. It is advantageous both from hydraulic and economic considerations to use pumps of identical size to match their performance characteristics. The pumps are used in series in a system where substantial head changes take place without appreciable difference in the discharge (i.e., the system head curve is steep). In series, each pump has the same discharge. The parallel pumps are useful for systems with considerable discharge variations with no appreciable head change. In parallel, each pump has the same head.

11.19.1 Pumps in Series

The following relations apply:

$$H = H_A + H_B + \cdots \qquad [\text{L}] \qquad (11.39)$$

$$Q = Q_A = Q_B = \cdots \qquad [\text{L}^3\text{T}^{-1}] \qquad (11.40)$$

$$\eta = \frac{H_A + H_B + \cdots}{H_A/\eta_A + H_B/\eta_B + \cdots} \qquad [\text{dimensionless}] \qquad (11.41)$$

$$P = \frac{\gamma Q(H_A + H_B + \cdots)}{550\eta} \qquad [\text{ML}^2\text{T}^{-3}] \qquad (11.42)$$

where $A, B, \cdots$ refer to different pumps.

The composite characteristic curves of pumps in series can be prepared by the equations above. The composite head characteristic curve is prepared by adding the ordinates (heads) of all the pumps for the same values of discharge [eq. (11.39)], as shown in Figure 11.22. The intersection point of the composite head characteristic curve and the system head curve provides the operating condition.

Example 11.18

Pumps A and B have the following characteristics:

Pump A: 8-in., 1450 rpm			Pump B: 10-in., 1750 rpm		
Q (gpm)	H (ft)	Efficiency	Q (gpm)	H (ft)	Efficiency
0	186	0	0	172	0
500	179	54	400	166	59
1000	158	70	800	140	77
1500	112	67	1200	90	74

These pumps are arranged in series in a system having a static lift of 80 ft. The pipeline comprises 6-in.-diameter pipe of 1200 ft length, with the minor losses 20 times the velocity head. Determine the operating condition and the power input. $f = 0.022$.

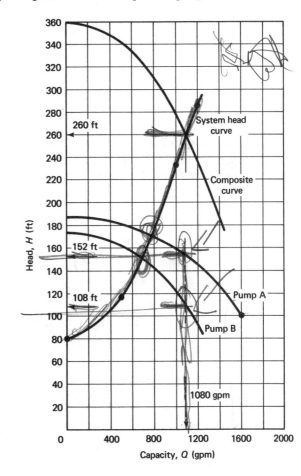

Figure 11.22 Head characteristics and operating condition for pumps in series.

Solution

1. The characteristic curves for pumps A and B are plotted in Figure 11.22. The composite head characteristics are computed below.

Capacity (gpm)	Head for Pump A (ft)	Head for Pump B (ft)	Total head $H_A + H_B$ (ft)
0	186	172	358
400	182	166	348
800	168	140	308
1200	144	90	234

The composite curve is illustrated in Figure 11.22.

2. The system head curve is computed below:

$$H_p = \Delta Z + \frac{0.81}{32.2}\left[\frac{0.022(1200)Q^2}{(0.5)^5} + \frac{20Q^2}{(0.5)^4}\right]$$
$$= 80 + 29.3Q^2$$

Capacity (gpm)	Q = Capacity/449 (cfs)	$H_p = 80 + 29.3Q^2$
0	0	80.0
500	1.11	116.1
1000	2.23	225.7
1200	2.67	288.9

The system head curve has been plotted on Figure 11.22.

3. The discharge and head at the operating condition from the intersection of the characteristic and the system curves are

$$Q = 1080 \text{ gpm}, \qquad H = 260 \text{ ft}$$

4. Corresponding to Q = 1080 gpm:

$$\text{For Pump A: } H_A = 152 \text{ ft}, \qquad \eta_A = 69\%$$
$$\text{For Pump B: } H_B = 108 \text{ ft}, \qquad \eta_B = 75\%$$

5. $\eta = \dfrac{H_A + H_B}{H_A/\eta_A + H_B/\eta_B} = \dfrac{260}{152/69 + 108/75} = 71.4\%.$

6. Input power $= \dfrac{\gamma Q(H_A + H_B)}{550\eta} = \dfrac{62.4(1080/449)\,(260)}{550(0.714)} = 99.4 \text{ hp.}$

11.19.2 Pumps in Parallel

For parallel pumps, the relations are as follows:

$$H = H_A = H_B = \cdots \qquad [\text{L}] \qquad (11.43)$$
$$Q = Q_A + Q_B + \cdots \qquad [\text{L}^3\text{T}^{-1}] \qquad (11.44)$$

$$\eta = \frac{Q_A + Q_B + \cdots}{Q_A/\eta_A + Q_B/\eta_B + \cdots} \qquad [\text{dimensionless}] \qquad (11.45)$$

$$P = \frac{\gamma H(Q_A + Q_B + \cdots)}{550\eta} \qquad [\text{ML}^2\text{T}^{-3}] \qquad (11.46)$$

The composite head characteristic curve is obtained by summing up the abscissas (discharges) of all the pumps for the same values of head [eq. (11.44)], as shown in Figure 11.23.

Example 11.19

The two pumps of Example 11.18 are arranged in parallel for the same pipeline system, in which the pipe diameter only has been changed to 9 in. Determine the operating condition (a) combined for two pumps and (b) for each pump. (c) Also determine the power input. The static head is 40 ft.

Solution

1. The characteristics curve for pumps A and B are plotted in Figure 11.23. The composite characteristics are computed below.

Head (ft)	Capacity of Pump A (gpm)	Capacity of Pump B (gpm)	Total Capacity $Q_A + Q_B$ (gpm)
172	760	0	760
160	980	520	1500
140	1230	800	2030
120	1430	990	2420

The composite curve is plotted in Figure 11.23.

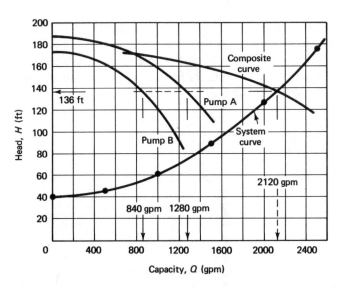

Figure 11.23 Head characteristics and operating condition for pumps in parallel.

2. The system head curve is computed as follows:

$$H_p = 40 + \frac{0.81}{32.2}\left[\frac{0.022(1200)Q^2}{(0.75)^5} + \frac{20Q^2}{(0.75)^4}\right]$$

$$= 40 + 4.39Q^2$$

Capacity, Q (gpm)	Q = capacity/449 (cfs)	$H_p = 40 + 4.39Q^2$
0	0	40
500	1.11	45.4
1000	2.23	61.8
1500	3.34	89.0
2000	4.45	126.9
2500	5.57	176.2

The system curve is plotted in Figure 11.23.

3. The operating condition at the intersection of the characteristic and system curves:

$$Q = 2120 \text{ gpm}, \qquad H = 136 \text{ ft}$$

4. Corresponding to the head of 136 ft:

For Pump A: $Q_A = 1280$ gpm, $\eta_A = 68\%$

For Pump B: $Q_B = 840$ gpm, $\eta_B = 77\%$

5. Overall $\eta = \dfrac{Q_A + Q_B}{Q_A/\eta_A + Q_B/\eta_B} = \dfrac{2120}{1280/68 + 840/77} = 71.3\%$

6. Input power $= \dfrac{(62.4)\,(2120/449)\,(136)}{550(0.713)} = 102$ hp.

11.20 LIMIT ON PUMP LOCATION

The pressure on the suction side of a pump which is located above the supply tank is below the atmospheric (vacuum) pressure. If the absolute pressure at the suction inlet (point s of Figure 11.24) falls below the vapor pressure of the water at the operating temperature, vapor pockets are formed which can damage the pump. This phenomenon is known as cavitation.

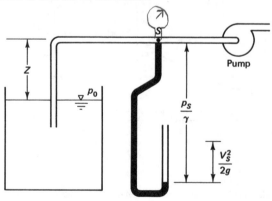

Figure 11.24   Pressure head at suction inlet.

The absolute pressure at the suction intake of a pump is referred to as the net positive suction head (NPSH). It should not fall below a certain minimum value which is influenced by the further reduction in pressure within the impeller. For each pump, the manufacturer indicates the *required* NPSH based on the pump performance test. The *available* NPSH for a given system should be more than the required NPSH. This places a limitation on the elevation for the location of a pump.

In terms of the pressure head at the suction inlet, the available NPSH is

$$\text{NPSH} = \frac{p_s}{\gamma} + \frac{V_s^2}{2g} - \frac{e_w}{\gamma} \qquad [\text{L}] \qquad (11.47)$$

where the subscript refers to values at suction inlet and e is the vapor pressure, given in Appendices C-1 and C-2.

It is not convenient in many instances to measure p_s and V_s at the suction inlet. In such cases it is preferable to express NPSH in terms of pressure on the reservoir surface.

$$\text{NPSH} = \frac{p_0}{\gamma} - Z - h_L - \frac{e_w}{\gamma} \qquad [\text{L}] \qquad (11.48)$$

where

p_0 = absolute pressure at the reservoir surface; atmospheric for open reservoir
Z = elevation of suction intake from the reservoir surface
h_L = friction and local head losses up to the suction inlet

The cavitation parameter, σ, is given by

$$\sigma = \frac{\text{NPSH}}{\text{total pump head}} \qquad [\text{dimensionless}] \qquad (11.49)$$

The computed σ should be higher than the critical σ furnished by the manufacturer. The cavitation parameters range from 0.05 for a specific speed of 1000 to 1.0 for a specific speed of 8000.

Example 11.20

The pumping system shown in Figure 11.24 is to deliver 1 cfs of water at 60°F. The suction line is 6 in. in diameter in a 300-ft-long cast iron pipe. The suction inlet is 20 ft above the reservoir level. The atmospheric pressure of 14.7 psi absolute exists over the reservoir. The required NPSH of the pump is 7. Determine whether the system will have a cavitation problem.

Solution

$$V = \frac{Q}{A} = \frac{1}{(\pi/4)(0.5)^2} = 5.1 \text{ ft/sec}$$

$$\text{Re} = \frac{Vd}{\nu} = \frac{5.1(0.5)}{1.21 \times 10^{-5}} = 2.1 \times 10^5$$

$$\frac{\epsilon}{d} = \frac{0.0008}{0.5} = 0.0016$$

From the Moody diagram, $f = 0.023$:

$$h_f = \frac{fL}{d}\frac{V^2}{2g} = (0.023)\left(\frac{300}{0.5}\right)\frac{(5.1)^2}{2(32.2)} = 5.57 \text{ ft}$$

Exit loss $\quad K = 0.5$

Bend $\qquad \dfrac{K = 0.9}{\Sigma K = 1.4}$

$$hm = \frac{1.4(5.1)^2}{2(32.2)} = 0.57 \text{ ft}$$

At 60°F, e_w (vapor pressure) = 0.26 psi (Appendix C-1). From eq. (11.48),

$$\text{NPSH} = \frac{p_0}{\gamma} - Z - h_L - \frac{e_w}{\gamma}$$

$$= \frac{14.7(144)}{62.4} - 20 - (5.57 + 0.57) - \frac{0.26(144)}{62.4}$$

$$= 7.18 \text{ ft}$$

Since the available NPSH of 7.18 is greater than the required NPSH 7.0, there is no cavitation problem.

PROBLEMS

11.1. Water flows from a reservoir at a rate of 2.5 cfs through a system of pipes as shown in Fig. P11.1. Determine the total energy lost in the system.

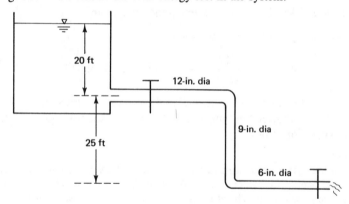

Figure P11.1

11.2. Crude oil of density 925 kg/m³ flows from a closed tank which has a pressure 70 kPa above the atmosphere to an open tank. If the oil level in the open tank is 2 m higher than the closed tank, determine the total loss of energy between the two tanks.

11.3. At the inlet point of a 1000-ft-long pipeline 6 in. in diameter, the energy head is 80 ft from a reference datum. Pipe carries water flow of 0.6 cfs. The pressure at the outlet is 18 psi and the elevation of the pipe at the end is 12 ft from the datum. Determine the loss of energy through the pipe.

11.4. Determine the slope of energy grade line for Problem 11.3. Determine the piezometric height (height of the hydraulic grade line) above the reference datum at a point 300 ft from the inlet.

11.5. Water flows at a velocity of 0.3 ft/sec in a cast iron pipe of 1 in. inside diameter at 40°F. Determine the friction factor.

11.6. Determine the friction factor for a 100-mm-inside-diameter commercial steel pipe in which water flows at 60°C at a rate of 0.1 m³/s.

11.7. The discharge through a concrete pipe of 1 ft diameter and 3000 ft length is 1.0 cfs. What is the loss of energy if the water temperature is 60°F? Apply the Darcy–Weisbach equation.

11.8. Water at 80°F flows from a storage tank through a 80-ft-long galvanized iron pipe of 4 in. in diameter laid horizontally. Calculate the depth of water required in the tank above the centerline of the pipe to carry a discharge of 2 cfs. Neglect the minor losses. Apply the Darcy–Weisbach equation.

11.9. A fire hydrant is supplied through a 6-in. cast iron pipe of 16,000 ft length. The total drop in pressure is limited to 35 psi. What is the discharge through the pipe for a water temperature of 80°F? Apply the Darcy–Weisbach equation.

11.10. Water is transferred at a rate of 2.1 cfs from an upper reservoir to a lower reservoir by a concrete pipe 3000 ft long. The difference in water level is 60 ft. Determine the size of pipe for a water temperature of 60°F. Apply the Darcy–Weisbach equation.

11.11. An upper reservoir A is connected to a lower reservoir B by a 10,000-ft-long, 2-ft-diameter cast iron pipe. The difference in water elevations of the two reservoirs is 100 ft. There is a hill between the two reservoirs whose summit is 15 ft above the upper reservoir. The pipeline has to cross the hill when its length is 3000 ft. Determine the **(a)** discharge of water through the pipeline, and **(b)** minimum depth below the summit to which the pipeline should be laid if the pressure in the pipe is not to fall below atmospheric pressure. Assume a water temperature of 50°F. Neglect the minor losses. Apply the Darcy–Weisbach equation.

11.12. For Problem 11.7, determine the loss of energy by the Hazen–Williams formula. $C = 120$.

11.13. Rework Problem 11.8 by the Hazen–Williams formula. $C = 120$.

11.14. For Problem 11.9, determine the discharge by the Hazen–Williams formula.

11.15. For Problem 11.10, determine the pipe size by the Hazen–Williams formula.

11.16. A pipeline 200 m long delivers water from an impounding reservoir to a service reservoir with the difference in water levels of 20 m. The pipeline of commercial steel is 400 mm in diameter. It has two 90° flanged elbows, a check valve, and an orifice ($K = 2.5$). Determine the flow through the pipe. Take a water temperature of 20°C.

11.17. In a cast iron piping system of 450 ft length shown in Fig. P11.17, the rate of flow is 1 cfs at 70°F. Determine the diameter of the pipe.

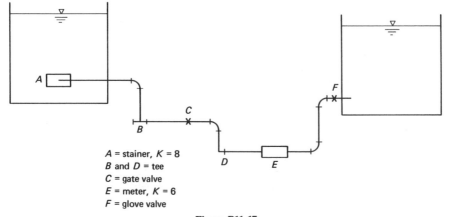

A = stainer, $K = 8$
B and D = tee
C = gate valve
E = meter, $K = 6$
F = glove valve

Figure P11.17

11.18. A 300-mm-diameter cast iron pipeline 1000 m long delivers water from an upstream reservoir to a downstream reservoir. The difference in water levels of the two reservoirs is 30 m. **(a)** Determine the discharge through the pipe. **(b)** If the discharge has to be maintained at 0.3 m³/s without a change in the pipe, determine the horsepower of the pump to be installed at an efficiency of 70%. Consider the entrance and exit losses and the water temperature of 10°C.

11.19. In the supply system shown in Fig. P11.19, the pressure required at the delivery end is 60 psi. Determine **(a)** the pumping head, and **(b)** the power delivered by the pump. Consider the entrance and exit losses and a water temperature of 70°F.

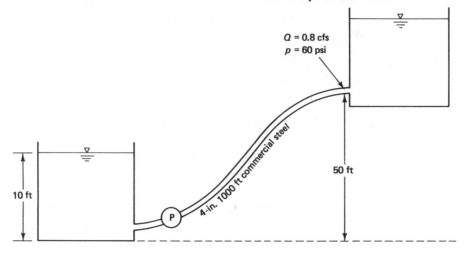

$Q = 0.8$ cfs
$p = 60$ psi

4-in. 1000 ft commercial steel

50 ft

10 ft

Figure P11.19

11.20. A fire hydrant is supplied through three welded steel pipelines arranged in series. The total drop in pressure due to friction in the pipeline is limited to 50 psi. What is the discharge through the hydrant? Neglect the minor losses.

Pipe	Diameter (in.)	Length (ft)
1	6	500
2	8	2,000
3	12	16,000

11.21. Find the elevation of the downstream reservoir for a flow of 5 cfs in the system shown in Fig. P11.21. Consider the minor losses. The water temperature = 50°F.

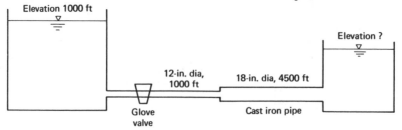

Elevation 1000 ft

Elevation ?

12-in. dia, 1000 ft

18-in. dia, 4500 ft

Glove valve

Cast iron pipe

Figure P11.21

11.22. A flow of 1 m³/s is divided into three parallel cast iron pipes of diameters 500, 250, and 400 mm and lengths 400 m, 100 m, and 250 m, respectively. Determine the head loss and flow through each pipe.

11.23. Two reservoirs having a difference of 40 ft in water elevations are connected by a 1.5-mile-long cast iron pipe of 1 ft in diameter. A second pipe 1.5 ft in diameter is laid alongside the first one for the last half-mile. How is the discharge affected?

11.24. For the pipe system shown in Fig. P11.24, determine the rate of flow. Neglect the minor losses.

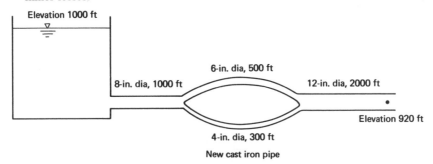

Figure P11.24

11.25. In the network shown in Fig. P11.25, determine the (a) flows in the pipe, and (b) pressure heads at the nodes. $C = 130$.

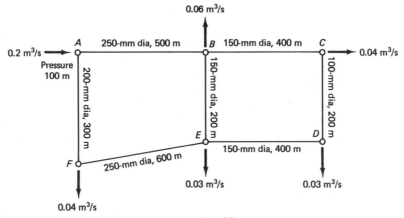

Figure P11.25

11.26. The length of each pipe in the network shown in Fig. P11.26 is 1000 ft. Determine

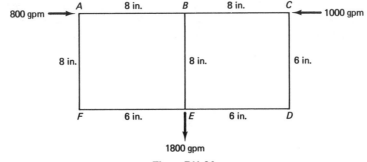

Figure P11.26

the **(a)** flows in the pipe, and **(b)** pressure heads at the nodes. The pressure at point C is 40 ft. The elevations of different points are given below. $C = 100$.

Node	A	B	C	D	E	F
Elevation (ft)	200	150	300	150	200	150

11.27. In the pipe network shown in Fig. P11.27, determine the discharge in each pipe. Assume that $f = 0.015$.

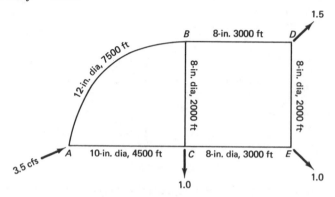

Figure P11.27

11.28. A steel pipeline 1600 m long discharges 150 liters per second of water from a reservoir to the atmosphere through a valve at the downstream end. The pipeline is 250 mm in diameter with a wall thickness of 10 mm. The valve is fully closed in 30 seconds. Determine the maximum transient pressure at the valve and 500 m upstream of the valve. The pipeline is without expansion joints and anchored at the upstream end only. $K = 2.1$ GPa, $E = 210$ GPa, $\nu = 0.3$.

11.29. A pipe system operates under a head of 50 ft at the valve to carry a discharge of 3.6 cfs. The cast iron pipe has a length of 4000 ft, a diameter of 9 in., and a wall thickness of 0.375 in. If the valve at the downstream end is fully closed in 20 seconds, determine the maximum transient pressure at the center of the pipe. The pipe has frequent expansion joints. $K = 320$ ksi, $E = 20 \times 10^3$ ksi, and $\nu = 0.25$.

11.30. In Problem 11.28, the valve is fully closed in 2 seconds. Determine the maximum pressure rise at the valve.

11.31. In Problem 11.29, the valve is fully closed in 1.9 seconds. Determine the maximum transient pressure at the valve.

11.32. A steel pipe 1625 m long discharges water through a valve under a head of 10 m. The pipe has a diameter of 300 mm. The speed of the pressure wave is 1300 m/s. The area of the valve opening varies as shown below. Determine the variation of head at the valve if it is fully closed in 20 seconds. Neglect the friction losses. $C_d = 0.6$.

Time (s)	0.0	2.5	5.0	7.5	10.0	12.5	15.0	17.5	20.0
Valve area (m²)	0.03	0.0234	0.0165	0.011	0.008	0.006	0.004	0.002	0

11.33. In Problem 11.29, if the valve is fully closed in 15.2 seconds, calculate the pressure variation at the valve. The area of the valve opening varies linearly with the time of closure. $C_d = 0.65$.

11.34. A 6-in. pump operating at 1760 rpm delivers 1480 gpm for a head of 132 ft at its maximum efficiency. Determine the specific speed. Make a recommendation on the type of pump to be used.

11.35. A centrifugal pump discharges 300 gpm against a head of 55 ft when the speed is 1500 rpm. The diameter of the impeller is 12.5 in. and the brake horsepower is 6.0. **(a)** Determine the efficiency of the pump. **(b)** What should be the speed of a geometrically similar pump of 15 in. diameter running at a capacity of 400 gpm?

11.36. If the rotational speed of a pump motor is reduced by 35%, what is the effect on the pump performance in terms of capacity, head, and power requirements?

11.37. The following performance characteristics are obtained for a pump tested at 1800 rpm. Determine the performance data of this pump operating at 1400 rpm.

Capacity (gpm)	Head (ft)	Power (hp)	Efficiency (%)
3000	200	175	87
2000	221	143	78.5
1000	229	107	54

11.38. The speed of the pump in Problem 11.37 is not changed, but the impeller diameter is reduced from 14.75 in. to 14.0 in., determine the performance data of the pump.

11.39. If both the speed and the impeller diameter are changed for the pump in Problems 11.37 and 11.38, determine the pump characteristic data.

11.40. From the manufacturer's data, a pump of 10 in. impeller diameter has a capacity of 1200 gpm at a head of 61 ft when operating at a speed of 900 rpm. It is desired that the capacity be about 1500 gpm at the same efficiency. Determine the adjusted speed of the pump and the corresponding head.

11.41. The characteristic data for a 14.75-in. pump rotating at 1800 rpm are given below. **(a)** Plot the pump characteristic curves. **(b)** Determine the type specific speed. **(c)** If the pump speed is changed to 1600 rpm, determine the performance characteristics of the pump. **(d)** If the diameter of the pump is reduced 14 in. (speed retained at 1800 rpm), determine the performance characteristics. **(e)** If another similar pump of 14 in. diameter is used, determine the pump characteristics.

Capacity (gpm)	Head (ft)	Power (hp)	Efficiency (%)
0	230	76.5	0
1000	228.5	107	54
2000	221.0	142.3	78.4
3000	200.5	174.5	87.0
3500	183.5	185.0	87.6
4000	157.0	189.5	83.7

11.42. The pump with the characteristics given in Problem 11.41 is to be used for the piping system shown in Fig. P11.42. Determine the **(a)** pump operation point, **(b)** operation efficiency, **(c)** suitability of the pump, and **(d)** input horsepower to the pump. $f = 0.022$.

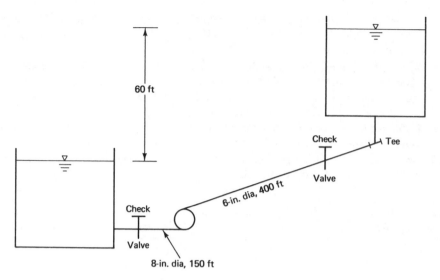

Figure P11.42

11.43. Water is pumped from a lower tank to a higher tank having a difference in elevation of 10 m. The piping system comprises a 200-mm-diameter pipe of 2000 m length with minor losses = 6.2 times of the velocity head. Determine the rate of flow. $f = 0.02$. Pump characteristics:

Discharge (liters/s)	0	10	20	30	40	50
Total head (m)	25	23.2	20.8	17.0	12.4	7.3
Efficiency (%)	—	45	65	71	65	45

11.44. Each of the two identical pumps has the following characteristics:

Q (gpm)	H (ft)	Efficiency (%)
0	124	0
500	119	54
800	112	64
1000	104	68
1300	90	70
1600	70	67

The pumps are used in series to supply water between two tanks with a static lift of 40 ft. The pipeline is 6 in. in diameter of 1200 ft in length, with the minor losses 20 times the velocity head. Determine the **(a)** operating condition, **(b)** head developed by each pump, and **(c)** input power. $f = 0.022$.

11.45. The two pumps of Problem 11.44 are arranged in parallel to supply water for the piping system of Problem 11.44. The diameter of the pipe has been changed to 10 in. Determine the operating condition and the input power. $f = 0.022$.

11.46. The suction side of a pipe system is as shown in Fig. P11.46. The pump discharges 495 gpm of water under a total head of 100 ft at 40°F. Atmospheric pressure = 32.7 ft of water. Determine the **(a)** available NPSH, and **(b)** cavitation parameter. If the required NPSH is 9, will there be a cavitation problem?

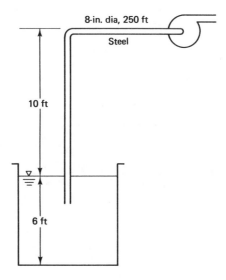

8-in. dia, 250 ft

Steel

10 ft

6 ft

Figure P11.46

11.47. A pump delivers water under a head of 130 ft at 100°F. The barometric (atmospheric) pressure is 14.3 psi absolute. At the suction intake the pressure is a vacuum (below atmosphere) of 17 in. Hg and the velocity is 12 ft/sec. Determine the NPSH and the cavitation parameter. (Head of water = head of Hg × sp. gr. of Hg.)

CHAPTER TWELVE

DRAINAGE SYSTEM

12.1 TYPES OF DRAINAGE SYSTEM

The term "drainage" applies to the process of removing excess water to prevent public inconvenience and to provide protection against loss to property and life. In an undeveloped area, drainage occurs naturally as a part of the hydrologic cycle. This natural drainage system is not static but is constantly changing with environmental and physical conditions. Development of an area interferes with nature's ability to accommodate severe storms without significant damage, and a man-made drainage system becomes necessary. A drainage system can be classified according to the following categories:

1. urban drainage system,
2. agriculture land drainage system,
3. roadway drainage system, and
4. airport drainage system.

12.2 URBAN DRAINAGE SYSTEM

In an urbanized area, the runoff is contributed by (1) excess surface water after a rainfall, from roofs, yards, streets, etc., and (2) wastewater* from households, commercial establishments, and industries. Past practice was to convey the entire runoff

---

* Wastewater is the spent or used water of a community, comprising water-carried wastes from residences, institutions, commercial buildings, and industries. Sewage is the liquid waste of a community conveyed by a sewer. Thus both terms have the same meaning. In recent usage the word "wastewater" has taken precedence. "Sewerage" implies collection, treatment, and disposal of sewage.

through a single system known as the combined sewer system. Combined sewers are no longer built; the present practice is to construct a system to discharge rainfall excess only, and a separate system to transport wastewater, also referred to as the dry-weather flow. The former is known as the stormwater sewer system and the latter as the sanitary sewer system.

In a combined sewer system the ratio of maximum flow rate (stormwater) to minimum flow rate (dry-weather flow) is over 20, even exceeding 100 during severe storms. The flow through the sewer system is carried to treatment facilities. During severe storms, however, it is not practical to carry the entire quantity to the treatment works. Excess diluted flow is passed into a stream at the nearest discharge point through "storm sewer overflows." The combined overflows induce large quantities of polluting materials into receiving waters. These should be kept to a minimum and the "first flushes," which are more polluted, should be passed entirely to the treatment facilities. Combined sewers tend to get silted up during dry-weather flow because the combined flow capacity is large and the dry-weather flow is comparatively small. These are flushed at the time of a storm.

In a separate system, sewers are designed to maintain a self-cleansing velocity at higher discharges. For storm sewers, comparatively shorter lengths are needed because the stormwater can be discharged directly to the nearest point in a water course. There is no need for overflow structures. Combined sewers are less costly to construct but have additional operational cost for treatment of larger quantities. Generally, a separate system is more favorable.

12.3 LAYOUT OF URBAN DRAINAGE SYSTEM

There are perpendicular, zonal, fan, and ring patterns of sewer systems, as shown in Figure 12.1. Among the sewer components, a lateral sewer is the unit in which no other common sewer discharges. A submain receives the discharge of a number of laterals. A main or trunk sewer receives the discharge from one or more submains. In a storm system, main sewers outfall to receiving waters. In sanitary or combined sewer systems, an intercepting sewer receives flow from a number of mains and conducts it to a point of treatment. Excess water is allowed to overflow in the water course through outfalls.

For the layout in a horizontal plane, the sewer lines are laid alongside the streets or utility easement. A layout is produced of the main sewer leaving the area at its lowest point and submains and laterals radiating to the outlying areas. The tributary area to each sewer is drawn from the ground contour map.

The vertical layout is influenced by the ground surface profile. A cover of about 10 ft is provided on sewer pipe in northern states to prevent freezing. A minimum cover of 3 ft is adopted in the south from imposed load considerations. For a minimum excavation, a sewer line is laid parallel to the ground surface. However, this is not practical in very steep or in flat terrain. For sanitary sewers, a slope has to be provided to maintain a self-cleansing velocity of minimum of 2 ft/sec (0.6 m/s). Table 10.5 indicates the required slopes for various flows. In flat areas all sewer lines drain to a collection point for pumping to a gravity main. A separate sewer system is constructed for a higher-elevation area.

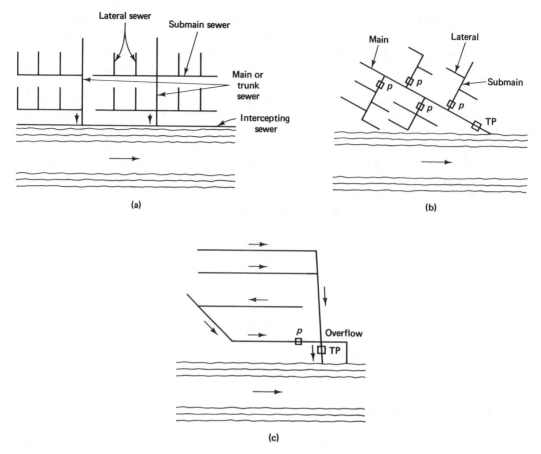

Figure 12.1 Layout of sanitary sewers: (a) perpendicular pattern; (b) fan pattern; (c) zone pattern. P, Pumping station; T, treatment plant.

Manholes, for service and maintenance of sewers, are located at (1) all sewer intersections, (2) major changes in slope of sewers, (3) changes in size of sewers, (4) changes in horizontal layout of sewers, (5) drop in vertical layout of sewers, and (6) along straight sewer runs at a spacing of 300 to 500 ft (90 to 150 m); this is 500 to 1000 ft (150 to 300 m) for large-diameter sewers.

A section of sewer invert entering a manhole and another section leaving the manhole are not always at the same elevation. A drop as shown at manhole 19 in Figure 12.2 is sometimes needed to adjust the slope of a sewer section without resorting to a steep gradient or deep cut. In a sewer on a short radius curve or where there is a bend or change in the sewer direction, a noticeable loss of head (energy) is involved. For the usual velocities, this justifies a drop of about 0.1 ft (30 mm) in the manhole invert. When the sewer increases in size to overcome the head loss, a drop in the invert elevation is made so that either the tops, or crowns, of the two sewer sections or the 0.8-depth points of the two sewers remain at the same elevation.

For each sewer line, a profile is drawn that shows the (1) ground level, (2) location of borings, (3) rock levels, (4) underground structures, (5) elevations of foundations and cellars, (6) cross streets, (7) location and number of manholes,

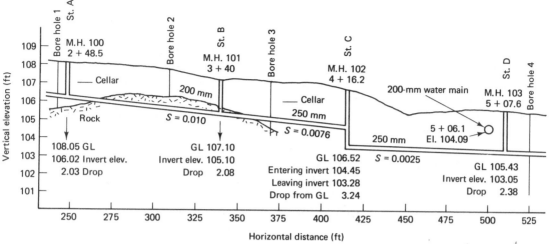

Figure 12.2 Profile of a sewer section (from Steel and McGhee, 1979). *gravity pipe system*

(8) the elevations of the sewer invert entering and leaving the manhole at each man-hole, and (9) slopes and sizes of sewer lines. A profile is illustrated in Figure 12.2. The profile assists in the design, cost estimation, and construction of a sewer.

12.4 DESIGN OF SANITARY SEWER SYSTEM

A design period throughout which a sewer capacity will be adequate is established prior to the design of a sewer system. This is between 25 and 50 years. A sanitary sewer performs two main functions: carrying the peak sanitary discharge, and transporting suspended solids without deposition. It is designed according to the following considerations:

1. To determine the hydraulic capacity of the sewer for the estimated peak flow at the end of the design period, known as the design discharge.
2. To check for a minimum cleansing velocity of 2 ft/sec for the peak rate of flow at the beginning of the design period (at the present time).

Sanitary sewers are designed with one-half to full depth of flow at design discharge. Usually, pipes up to 16 in. (400 mm) in diameter are designed to flow half full, pipes between 16 and 35 in. (900 mm) to flow two-thirds full, and pipes over 35 in. to flow at three-fourths to full depth of flow. An upper limit on the velocity is often taken to be about 10 ft/sec (3 m/s) in sanitary sewers.

The design involves fixing the diameter and slope of the sewer sections. Since the trench excavation is a major cost component, the aim is to lay the sewers to a minimum grade still meeting the criteria above.

Two steps in the design are (1) determination of the peak rate of sewage flow, and (2) selection of the sewer for the estimated flow. The second step is based on Manning's equation and has been described in Chapter 10. Peak-flow computations are discussed below.

12.4.1 Quantity of Wastewater

For sewer design, an estimate is made of present and future quantities of domestic, commercial, institutional, and industrial wastewaters, groundwater infiltration, and any stormwater entering the system. The assessment of wastewater quantity is made from the quantity of water consumed in a city. It is frequently assumed that the average rate of wastewater flow, including a moderate amount of infiltration, is equal to the average rate of water consumption in the city. The domestic component of water consumption is estimated by multiplying the population by the per capita rate of consumption. To this are added the contributions from commercial buildings, institutions, industries, and extraneous sources. The accuracy of population at present and projected at the end of the design period is important. Population forecasting techniques are discussed in Chapter 2. Since the system is designed in sections to serve segments of area, population density is used to determine the quantity of flow, as follows:

$$Q = pDA \qquad [L^3T^{-1}] \tag{12.1}$$

where

Q = domestic average daily flow
p = water usage per person per day
D = population density, persons per unit area
A = tributary area

The density is estimated by dividing the total population by the total area of the community with proper deductions for parks, playgrounds, swamps, lakes, ponds, and rivers. Main and intercepting sewers are designed for future density at the end of the design period. Laterals and submains are designed on the basis of saturation density for the area.

Domestic water usage per capita depends on the living conditions of consumers and tends to rise continuously. On an average it is 60 gallons (230 liters) per person per day. The average projected domestic consumption in the year 2000 is 79 gallons (300 liters) per person per day.

The quantity of commercial usage in smaller communities is assumed to be included within the domestic contribution. In large cities, a separate allowance from 4500 to 160,000 gpd per acre of floor area per building served is included.

Industrial wastewater quantities vary over a wide range depending on the type and size of industry. Data on wastes from various industries are available in certain guides. The average contribution from industry may vary from 8 to 25 gpd per person per shift.

Institutions such as hospitals, jails, and schools, and public services such as fire protection agencies, have a fixed pattern of demand. Their requirements amount to 13 to 26 gpd per person.

For infiltration, a moderate allowance has been included by assuming the rate of sewage equal to the rate of water consumption. When a separate estimate has to be made, an allowance of 30,000 gpd per mile (71000 liters/day per kilometer) is made for infiltration in small to medium-sized sewers up to 24 in.

In the average city, the total consumption is expected to be over 200% of the domestic consumption [i.e., 150 to 180 gpd per person (570 to 700 liters/day per person)].

The capacity is designed for the peak wastewater flow, which varies from about 150% of the average daily flow for high-population areas to about 450% for low-population areas, as shown in Figure 12.3. Thus, for average conditions, the peak flow is estimated at 360 to 720 gpd per capita. Many state regulatory agencies have set a lowest design rate of 400 gpd per capita (1500 liters/day per capita) for laterals and 250 gpd per capita (950 liters/day per capita) for mains where no actual measurements and other pertinent data are available (American Society of Civil Engineers and Water Pollution Control Federation, 1982).

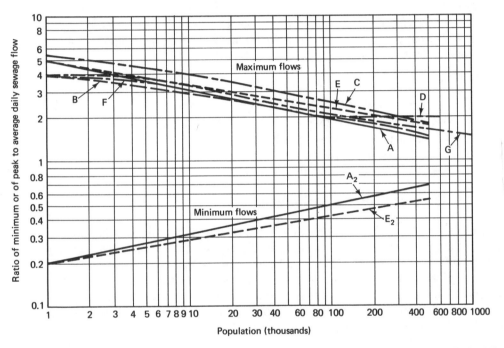

Curve A source: H. E. Babbitt, *Sewerage and Sewage Treatment,* 7th ed., John Wiley & Sons, Inc., New York, 1953.
Curve A_2 source: H. E. Babbitt and E. R. Baumann, *Sewerage and Sewage Treatment,* 8th ed., John Wiley & Sons, Inc., New York, 1958.
Curve B source: W. G. Harman, "Forecasting Sewage at Toledo under Dry-Weather Conditions," *Eng. News Rec.,* v. 80, p. 1233, 1918.
Curve C source: Youngstown, Ohio, report.
Curve D source: Maryland State Department of Health curve prepared in 1914, in *Handbook of Applied Hydraulics* (ed.) Davis, C. V., 2nd ed., McGraw-Hill Book Company, New York, 1952.
Curve E source: H. M. Gifft, "Estimating Variations in Domestic Sewage Flows," *Waterworks and Sewerage,* v. 92, p. 175, 1945.
Curve F source: Manual of Military Construction, U.S. Army Corps of Engineers, Washington, D.C., 19
Curve G source: G. M. Fair and J. C. Geyer, *Water Supply and Waste-Water Disposal,* John Wiley & Sons, Inc., New York, 1954. Curves A_2, B, and G were constructed as follows: curve A_2, $5/P^{0.107}$; curve B, $[14/(4 + \sqrt{P})] + 1$; curve G, $(18 + \sqrt{P})/(4 + \sqrt{P})$, where P is the population in thousands.

Figure 12.3 Ratio of peak flow to average daily flow (from American Society of Civil Engineers and Water Pollution Control Federation, 1982).

Example 12.1

Determine the maximum hourly (peak) rate of dry-weather flow from an area of 25 km² if the present population density is 8000 persons per square kilometer. Assume a domestic water consumption of 300 liters per day per capita and total consumption, including infiltration, to be 200% of domestic use.

Solution

1. Population = density × area = 8000(25) = 200×10^3 persons.
2. Domestic rate of flow = $300(200 \times 10^3) = 60 \times 10^6$ liters/day or 0.69 m³/s.
3. Total average daily flow = $2(60 \times 10^6) = 120 \times 10^6$ liters/day or 1.39 m³/s.
4. From Figure 12.3, ratio of peak to average flow = 2.0.
5. Peak wastewater flow = 2(1.39) = 2.78 m³/s.

Example 12.2

The future population density at the end of 15 years for Example 12.1 is projected to be 11,000 persons per square kilometer. Calculate the design flow (peak discharge at the end of the design period).

Solution

1. Future population = 11,000(25) = 275×10^3 persons.
2. Domestic flow = $300(275 \times 10^3) = 82.5 \times 10^6$ liters/day or 0.95 m³/s.
3. Total average daily flow = 2(0.95) = 1.9 m³/s.
4. Ratio of maximum to average, from Figure 12.3 = 1.9.
5. Peak flow in future = 1.9(1.9) = 3.61 m³/s.

12.4.2 Friction Coefficient for Sanitary Sewer

Experiments on small-diameter pipes of different materials indicated that after a short period of use, the capacity of a sewer depends on the characteristics of the slime grown on the pipe wall. Thus, to a large extent, roughness values become independent of the nature of the pipe material. The friction coefficient, n, of 0.012 in Manning's formula is considered satisfactory for pipes up to 35 in. (900 mm) in diameter, from this consideration. For larger pipes, the values of n are based on the type of material of the pipe, as given in Table 10.4.

12.4.3 Design Procedure for Sanitary Sewers

Based on the criteria stated in Section 12.4, the design procedure is as follows:

1. Compute the present peak sewage flow, Q_{peak}.
2. Compute the ultimate (at the end of the design period) peak sewage flow, Q_{design}.
3. Determine the minimum slope from Table 10.5 for Q_{design}; ascertain the ground-level gradient. The critical (steeper) of the two slopes is taken as the sewer slope.
4. For Q_{design}, calculate the pipe diameter by Manning's formula for the slope of step 3 and the partial flow condition at $\frac{1}{2}$ full for a lateral, $\frac{2}{3}$ full for a submain, and $\frac{3}{4}$ full for a main or an interceptor.

5. From Figure 10.11 on hydraulic elements, determine the full discharge capacity for the pipe size of step 4; From Table 10.1 obtain the depth of flow corresponding to the flow ratio Q_{peak}/Q_{full}.

6. For this depth of flow and a velocity of 2 ft/sec, recompute the required slope from Manning's formula. Select the steepest of the slopes from steps 3 and 6 as the design slope.

7. If the slope of step 3 is in considerable excess (over 10 times higher) than that of step 6,* the velocity of flow for design discharge should be checked to confirm that it is less than the scour velocity of 10 ft/sec. If necessary, a larger diameter should be selected.

8. Fix the invert levels accordingly.

Example 12.3

The ground-surface elevations at the location of two manholes 300 ft apart are 101.2 ft and 100 ft, respectively. The present rate of peak sewage flow is 1 mgd, which is expected to increase to 1.5 mgd at the end of 25 years. Design the sewer.

Solution

1. Q_{design} = 1.5 × 1.547 = 2.32 cfs.

2. Q_{peak} = 1 × 1.547 = 1.55 cfs.

3. Surface grade = $\dfrac{101.2 - 100.0}{300}$ = 0.004.

4. For 2.32 cfs, minimum slope (from Table 10.5) = 0.0016.

5. Critical slope = 0.004.

6. Designing for 1/2-full condition. From Table 10.1, for y/d_0 = 0.50, $AR^{2/3}$ = $0.156d_0^{8/3}$.

7. Using Manning's equation (10.13),

$$2.32 = \left(\frac{1.49}{0.012}\right)(0.156d_0^{8/3})(0.004)^{1/2}$$

$$d_0 = 1.27 \text{ ft or } 15.25 \text{ in.}$$

Use d_0 = 15 in. or 1.25 ft.

8. From Figure 10.11, for y/d_0 = 0.5 (half-full condition), Q/Q_{full} = 0.5.

$$Q_{full} = \frac{Q_{design}}{0.5} = \frac{2.32}{0.5} = 4.64 \text{ cfs}$$

9. Q_{peak}/Q_{full} = 1.55/4.64 = 0.33. From Figure 10.11, for Q/Q_{full} of 0.33, y/d_0 = 0.4. From Table 10.1, for y/d_0 of 0.4, R/d_0 = 0.214.

10. From Manning's equation (10.11),

$$V = \frac{1.49}{n}R^{2/3}S^{1/2}$$

---

* When the slope from step 6 is steeper than that from step 3, the velocity of flow at Q_{design} will go up above 2 ft/sec, but it is unlikely to exceed 10 ft/sec, since the velocity variation beyond 0.5 depth is not substantial.

or

$$2 = \frac{1.49}{0.012}(0.214 \times 1.25)^{2/3}S^{1/2}$$

$$S = 0.0015 < 0.004$$

11. Invert elevations

$$\begin{aligned} \text{Upstream end} &= \text{ground level} - \text{cover} - \text{diameter} \\ &= 101.2 - 10 - 1.25 = 89.95 \text{ ft} \\ \text{Downstream end} &= 100 - 10 - 1.25 = 88.75 \text{ ft} \end{aligned}$$

12.4.4 A Sanitary Sewer Project

An example illustrates the procedure. This may be varied to suit the requirements of local and state regulatory agencies.

Example 12.4

Design a sanitary sewer system for a part of the city shown in Figure 12.4. The present density of population is 8000 persons per square kilometer. It is estimated that the maximum density will be 12,000 persons per square kilometer. Assume that the maximum rate of sewage flow is 1500 liters/day per person for all the sewers. The ground-level gradient for various sewer segments is given in column 15 of Table 12.1.

Solution   Guided by contours, the directions of flow are indicated with arrows on the map. The manholes, provided at intersections, at changes in direction, and at intervals not exceeding 150 m, have been numbered from the upper end of the tributary lateral. The area tributary to each line segment from manhole to manhole is sketched by the dashed lines on the map and indicated in column 6 of Table 12.1. Design proceeds from the uppermost point of the system downward through the laterals. Where a branch joins a line already designed, the computation is made for the branch at the intersection (i.e., manhole 4) and restarted from the uppermost lateral meeting the other end of the branch (i.e., from manhole 5).

The procedure has been explained in Tables 12.1 and 12.2. The first table computes the sewage quantity and the second completes the hydraulic design. Computation is started from manhole 1 to 2 in Table 12.1. The tributary area of .007 km² is multiplied by the present and maximum density to obtain the incremental population. The total population is obtained by adding the incremental population from each manhole, until the submain 4 − 8 is met. The total population at the submain at manhole 4 equals the total of the laterals from manholes 1 through 4. After this, the laterals from manholes 5 to 8 and 10 to 8 are investigated as shown in lines 5 through 9 of Table 12.1. Their total is added to that of line 4 of the table to obtain the total population contributing to the submain at manhole 8 (line 10). This procedure is repeated until the point of connection to the existing system is reached. The quantity of wastewater flow is obtained by multiplying the total population by per capita consumption. Surface elevation data are known from a land survey. The street slope is the difference in elevation of the upper and lower manholes divided by the length of the line. Table 12.1 indicates the slopes directly.

Once the design flow (column 12 of Table 12.1) and present peak flow (column 11) have been determined, the hydraulic design is accomplished by the method of Example 12.3. This is arranged in Table 12.2. For the stipulated depth of flow of column 6, the value of parameter $AR^{2/3}/d_0^{8/3}$ is read from Table 10.1 and substituted in Manning's equation. In the equation, using Q_{design} (column 12 of Table 12.1) for the

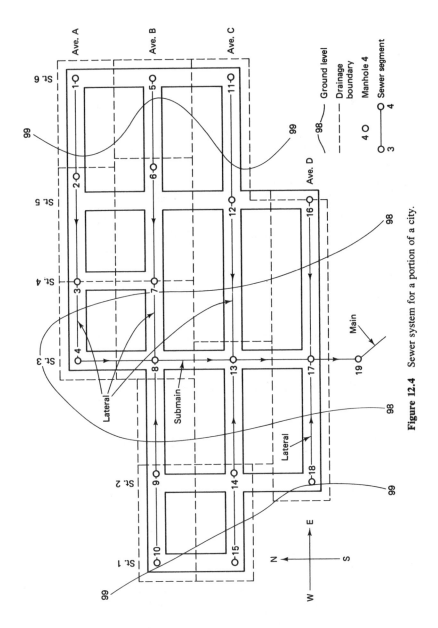

Figure 12.4 Sewer system for a portion of a city.

TABLE 12.1 COMPUTATION OF SEWAGE QUANTITY

(1) No.	(2) Location	(3)(4) Manhole From	To	(5) Length of Line (m)	(6) Area Drained (km²)	(7)(8) Increment of Population Present	Ultimate	(9)(10) Total Population Present	Ultimate	(11)(12) Wastewater Flow (m³/s) Present	Ultimate	(13) Surface Elevation Upper Manhole	(14) Lower Manhole	(15) Street Slope
1	Avenue A	1	2	100	0.007	56	84	56	84	.0010	.0015			0.007
2	Avenue A	2	3	100	0.007	56	84	112	168	.0019	.0029			0.008
3	Avenue A	3	4	90	0.005	40	60	152	228	.0026	.0040			0.009
4	Street 3	4	8	90	—	—	—	152	228	.0026	.0040			0.002
5	Avenue B	5	6	100	0.014	112	168	112	168	.0019	.0029			0.005
6	Avenue B	6	7	100	0.014	112	168	224	336	.0039	.0058			0.006
7	Avenue B	7	8	90	0.010	80	120	304	456	.0053	.0079			0.006
8	Avenue B	10	9	100	0.006	48	72	48	72	.0008	.0013			0.008
9	Avenue B	9	8	120	0.008	64	96	112	168	.0019	.0029			0.006
10	Street 3	8	13	97	—	—	—	568	852	.0099	.0148			0.002
11	Avenue C	11	12	140	0.008	64	96	64	96	.0011	.0017			0.009
12	Avenue C	12	13	150	0.016	128	192	192	288	.0033	.0050			0.010
13	Avenue C	15	14	100	0.006	48	72	48	72	.0008	.0013			0.012
14	Avenue C	14	13	120	0.014	112	168	160	240	.0028	.0042			0.009
15	Street 3	13	17	93	—	—	—	920	1,380	0.016	.0240			0.003
16	Avenue D	16	17	150	0.008	64	96	64	96	.0011	.0017			0.009
17	Avenue D	18	17	130	0.007	56	84	56	84	.0010	.0015			0.015
18	Street 3	17	19	95	—	—	—	1,040	1,560	.0181	.0270			0.004

TABLE 12.2 HYDRAULIC DESIGN OF SANITARY SEWERS

(1) No.	(2) Manhole From	(3) Manhole To	(4) Slope Street	(5) Slope Minimum (Table 10.5)	(6) Flow Depth, $\frac{y}{d_0}$	(7) Design for Q_{design} $AR^{2/3}/d_0^{8/3}$ (Table 10.1)	(8) Diameter, d_0 (mm)[a] [Eq. (10.13)]	(9) $\frac{Q_{design}}{Q_{full}}$ (Fig. 10.11)	(10) Check for $V=0.6$ m/s (2 ft/sec) at Q_{peak} Q_{full}	(11) $\frac{Q_{peak}}{Q_{full}}$	(12) $\frac{y}{d_0}$ (Fig. 10.11)	(13) Slope, S R/d_0 (Table 10.1)	(14) required[b] [Eq. (10.11)]	(15) Adopted Sewer Slope
1	1	2	0.007	0.009	0.5	0.156	85[c]	0.5	.003	0.33	0.4	0.214	0.011	0.011
2	2	3	0.008	0.009	0.5	0.156	105	0.5	.0058	0.33	0.4	0.214	0.008	0.009
3	3	4	0.009	0.007	0.5	0.156	120	0.5	.0080	0.33	0.4	0.214	0.007	0.009
4	4	8	0.002	0.007	0.67	0.24	105	0.8	.005	0.52	0.51	0.253	0.007	0.007
5	5	6	0.005	0.009	0.5	0.156	105	0.5	.0058	0.33	0.4	0.214	0.008	0.009
6	6	7	0.006	0.006	0.5	0.156	145	0.5	.0116	0.34	0.4	0.214	0.005	0.006
7	7	8	0.006	0.005	0.5	0.156	160	0.5	.0158	0.34	0.4	0.214	0.005	0.006
8	10	9	0.008	0.009	0.5	0.156	80	0.5	.0026	0.33	0.4	0.214	0.012	0.012
9	9	8	0.006	0.009	0.5	0.156	105	0.5	.0058	0.33	0.4	0.214	0.008	0.009
10	8	13	0.002	0.004	0.67	0.24	190	0.8	.0185	0.52	0.51	0.253	0.003	0.004
11	11	12	0.009	0.009	0.5	0.156	85	0.5	.0034	0.32	0.4	0.214	0.011	0.011
12	12	13	0.010	0.007	0.5	0.156	125	0.5	.0100	0.33	0.4	0.214	0.007	0.010
13	15	14	0.012	0.009	0.5	0.156	75	0.5	.0026	0.33	0.4	0.214	0.013	0.013
14	14	13	0.009	0.007	0.5	0.156	120	0.5	.0082	0.33	0.4	0.214	0.007	0.009
15	13	17	0.003	0.003	0.67	0.24	240	0.8	.0300	0.53	0.51	0.253	0.002	0.003
16	16	17	0.009	0.009	0.5	0.156	85	0.5	.0034	0.32	0.4	0.214	0.011	0.011
17	18	17	0.015	0.009	0.5	0.156	75	0.5	.003	0.34	0.4	0.214	0.013	0.015
18	17	19	0.004	0.002	0.67	0.24	240	0.8	.034	0.53	0.51	0.253	0.002	0.004

[a] $d_0 = \left[\dfrac{n \cdot (\text{col. 12 of Table 12.1})}{(\text{col. 13})(\text{col. 4 or 5})^{1/2}} \right]^{3/8}$

[b] $S = 0.36 n^2 / [(\text{col. 13})(\text{col. 8 in m})]^{4/3}$ For English units, 1.49 appears in denominator within parenthesis in both eq.

[c] There may be a minimum limit on the size of sewer, say 200 mm. In the worst case it will slightly reduce the velocity at Q peak which can be offset by increasing the sewer slope 33% as compared to col. 14.

discharge, and the critical of the values in columns 4 and 5 for the slope, and $n = 0.012$, the diameter, d_0, is computed. For the depth of flow in column 6, the discharge ratio of column 9 is obtained from Figure 10.11. Full-pipe flow (column 10) is thereby obtained by dividing the design flow by the ratio of column 9. The ratio in column 11 is determined by dividing peak flow (column 11 of Table 12.1) by column 10. For the discharge ratio in column 11, the depth ratio is ascertained in column 12 from Figure 10.11. Table 10.1 provides the hydraulic radius parameter of column 13 corresponding to the depth ratio of column 12. Using Manning's equation in terms of the velocity [eq. (10.11) or (10.12)], the required slope for a velocity of 0.6 m/s (2 ft/sec) is computed. The steeper of the slopes in columns 4, 5, and 14 is adopted to lay the sewer line. This is a general procedure, in which variations are made to suit the local conditions and preferences of a designer.

Another tabulation, as shown in Table 12.3, is added to determine the elevations of the sewers. Columns 1 through 8 of the table are completed from information in Tables 12.1 and 12.2. Column 9 is equal to column 5 times column 8. Invert elevations are computed from surface elevation, fall of sewer, cover, and sewer size. Recommendations regarding matching of the crown levels and the drops to accommodate losses, as discussed earlier, are kept in consideration.

Example 12.5

In the layout of Fig. 12.4, determine the sewer arrangement for manholes 1 through 4 if the ground level is 100 m at manhole 1. Provide a cover of 2 m.

Solution Invert elevations:
Manhole 1:

$$\text{surface elevation} - \text{cover} - \text{diameter} = 100 - 2 - 0.085 = 97.92$$

Manhole 2:

$$\text{Lower end of sewer } 1\text{–}2 = \text{upper end} - \text{sewer fall}$$
$$= 97.92 - 1.10 = 96.82$$

Upper end of sewer 2–3: The crown levels of sewers at the manhole should match. The crown level of sewer 1–2 $= 96.82 + 0.085 = 96.91$. The invert of sewer 2–3 $= 96.91 - 0.105 = 96.80$. Hence there is a drop of 0.02 m in manhole 2.
Manhole 3:

$$\text{Lower end of sewer } 2\text{–}3 = 96.80 - 0.9 = 95.90$$

Upper end of sewer 3–4: The crown level of sewer 2–3 $= 95.90 + 0.105 = 96.01$. The invert of sewer 3–4 $= 96.01 - 0.12 = 95.89$.
Manhole 4:

$$\text{Lower end of sewer } 3\text{–}4 = 95.89 - 0.81 = 95.08$$

Upper end of sewer 4–8: The crown level should match. Also, since a change in direction is involved, a drop of 30 mm should be provided. The crown level of sewer 3–4 $= 95.08 + 0.12 = 95.20$. The invert of sewer 4–8 $= 95.20 - 0.11 = 95.09$. However to provide a drop of 30 mm at manhole 4, the invert at upper end of sewer 4–8 $= 95.08 - 0.03 = 95.05$

12.5 DESIGN OF A STORM SEWER SYSTEM

Similar to the sanitary sewer system, the design process involves two steps: (1) determination of the quantity of stormwater, and (2) establishing a sewer capacity to

TABLE 12.3 SEWER ELEVATION ANALYSIS

(1) No.	(2) Manhole From	(3) Manhole To	(4) Diameter of Sewer (mm)	(5) Length of Sewer (m)	(6) Surface Elevation (m) Upper Manhole	(7) Surface Elevation (m) Lower Manhole	(8) Grade of Sewer	(9) Fall of Sewer	(10) Invert Elevation Upper Manhole	(11) Invert Elevation Lower Manhole
1	1	2	85	100	100	99.30	0.011	1.10	97.92	96.82
2	2	3	105	100	99.30	98.50	0.009	0.9	96.80	96.00
3	3	4	120	90	98.50	97.69	0.009	0.81	95.99	95.18
4	4	8	105	90	97.69	97.51	0.007	0.63	95.15	

pass this quantity. The quantity pertains to the peak rate of runoff within the urban or residential drainage area produced by a precipitation storm of a certain specified return period referred to as the design frequency. For this design rate of flow, the storm sewers are designed for just flowing full at the grade of ground surface slope by applying Manning's equation. A minimum velocity of 3 ft/sec (0.9 m/s) is maintained when flowing full in order to produce a minimum nondepositing slope for silt and grit particles that are heavier than the sewage solids. For this purpose, the slope at times increased in excess of the surface grade. The upper limit on velocity is about 15 ft/sec (5 m/s) from scour considerations.

12.6 QUANTITY OF STORMWATER

There are two common methods of computing peak stormwater flows from urban watersheds which are comparatively of smaller sizes (usually less than 20 mi^2). The rational method, used for the first time in the United States by Kuichling in 1889, is still very popular for estimating stormwater quantity. The U.S. Soil Conservation Service has evolved a procedure to determine the peak discharge by making use of the soil cover complex curves and extended it to apply to urban watersheds. With respect to the rational method, McPherson (1969) demonstrated that extensive variability of results is inherent in application of the method due to considerable variations in the interpretation of the variables and methodology of use. A wide latitude of subjective judgment is involved in the method. Despite its shortcomings, the rational method is the preferred technique in storm drainage design practice. Improved methods are evolving very slowly because of a dearth of rainfall–runoff field measurements that have a national transferability of results. McPherson pointed out that until improved methods are developed, the rational method is as satisfactory as any other oversimplified empirical approach.

12.7 RATIONAL METHOD

The basic equation in the rational method has the form

$$Q = C_f CiA \qquad [L^3T^{-1}] \qquad (12.2)$$

where

Q = peak rate of flow

C_f = frequency factor

C = runoff (rational) coefficient

i = intensity of precipitation for a duration equal to time of concentration, t_c, and a return period, T

A = drainage area

Equation (12.2) is dimensionally homogeneous (i.e., if A is in ft^2, i is in ft/sec, and Q is in cfs). However, this equation also yields correct values for Q in cfs, i in in./hr and A in acres. C and C_f are dimensionless coefficients. Their product should not exceed 1.

12.7.1 Frequency Correction factor, C_f

In a common form of eq. (12.2), C_f is taken as unity, which applies to design storms of a 2- to 10-year recurrence interval — a representative frequency for residential sewers. For storms of higher return periods, the coefficients are higher because of smaller infiltration and other losses, as shown in Table 12.4.

TABLE 12.4 FREQUENCY FACTOR

Return Period (years)	C_f
2–10	1.0
25	1.1
50	1.2
100	1.25

The U.S. Department of Transportation (1979) has proposed the curves of Figure 12.5 to determine the frequency correction factor. Application of the curves is explained in the following example.

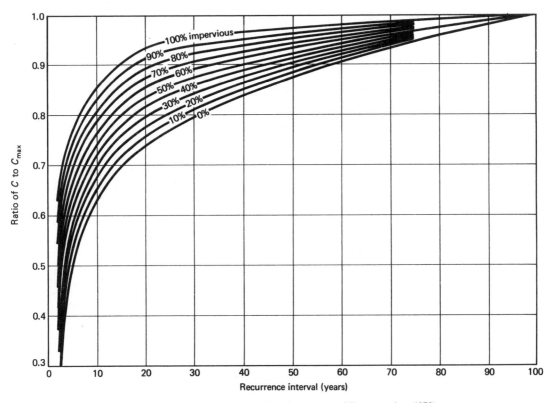

Figure 12.5 Correction for design storm frequency (from U.S. Department of Transportation, 1979).

Example 12.6

For a drainage area comprising the paved surface ($C = 0.9$), determine the runoff coefficient corresponding to a 50-year design frequency.

Solution From Table 12.4, $C_f = 1.2$.

$$CC_f = 0.9(1.2) = 1.08$$

Since CC_f cannot be greater than 1.0, $CC_f = 1.00$. From Figure 12.5, at imperviousness (C value) of 90% and for a 10-year frequency,

$$\frac{C}{C_{max}} = 0.83$$

For a 50-year frequency,

$$\frac{C}{C_{max}} = 0.96$$

$$C_f = \frac{\text{ratio for 50 yr}}{\text{ratio for 10 yr}} = \frac{0.96}{0.83} = 1.16$$

$$CC_f = 1.16(0.9) = 1.04 > 1.0$$

Hence $CC_f = 1.00$.

12.7.2 Runoff Coefficient, *C*

This is a highly critical element that serves the function of converting the average rainfall rate of a particular recurrence interval to the peak runoff intensity of the same frequency. Therefore, it accounts for many complex phenomena of the runoff process. Its magnitude will be affected by antecedent moisture condition, ground slope, ground cover, depression storage, soil moisture, shape of drainage area, overland flow velocity, intensity of rain, and so on. Yet its value is generally considered fixed for any drainage area, depending only on the surface type. This simplistic approach is a major cause of criticism for the rational method. The values of the coefficient are given in Table 12.5.

For an area having different types of surfaces, a composite coefficient is determined by estimating the fraction of each type of surface within the total area, multiplying each fraction by the appropriate coefficient for that type of surface, and then summing up the product for all types of surfaces. The coefficients are selected so as to reflect the conditions that are expected at the end of the design period.

The U.S. Department of Transportation (1979) has included a set of curves based on the following formula of Mitci, which relates the runoff coefficient to the degree of imperviousness (type of surface) and the antecedent rainfall (time from beginning of a rainfall to the occurrence of the design intensity rain of the duration of the time of concentration within the overall rainfall period).

$$C = \frac{0.98t}{4.54 + t} P + \frac{0.78t}{31.17 + t} (1 - P) \qquad \text{[unbalanced]} \qquad (12.3)$$

TABLE 12.5 RATIONAL RUNOFF COEFFICIENT

a. Urban Catchments

General Description	C	Surface	
City	0.7–0.9	Asphalt paving	0.7–0.9
Suburban business	0.5–0.7	Roofs	0.7–0.9
Industrial	0.5–0.9	Lawn heavy soil	
		>7° slope	0.25–0.35
Residential multiunits	0.6–0.7	2–7°	0.18–0.22
Housing estates	0.4–0.6	<2°	0.13–0.17
Bungalows	0.3–0.5	Lawn sandy soil	
		>7°	0.15–0.2
Parks, cemeteries	0.1–0.3	2–7°	0.10–0.15
		<2°	0.05–0.10

b. Rural Catchments (less than 10 km^2)

Ground Cover	Basic Factor	Corrections: Add or Subtract
Bare surface	0.40	Slope < 5%: −0.05
Grassland	0.35	Slope > 10%: +0.05
Cultivated land	0.30	Recurrence interval < 20 yr: −0.05
Timber	0.18	Recurrence interval > 50 yr: +0.05
		Mean annual precipitation < 600 mm: −0.03
		Mean annual precipitation > 900 mm: +0.03

(*Source*: Stephenson, 1981)

where

C = runoff coefficient corrected for antecedent rain condition

t = time, in minutes, from beginning of rainfall to the occurrence of the design intensity rain of short duration

P = percent of impervious surface (coefficient from Table 12.5)

Within the long-period rainfall, since the time sequence of the design rainfall intensity used in the rational method is not fixed, eq. (12.3) or the curves of the U.S. Department of Transportation cannot be used directly. Usually, the short-duration intense design storm (conforming to the time of concentration at the point under consideration) is placed at the midpoint of the longer-duration storm. This gives the time position to determine C for known imperviousness (type of surface). A common practice is to use the runoff coefficient from Table 12.5 and assume it not to vary through the duration of storm.

Example 12.7

A drainage area consists of 42% of turf ($C = 0.3$) and 58% of paved surface ($C = 0.9$). The point under design has a time of concentration of 20 minutes. The total duration of the rainstorm is 3 hours. Determine the value of C corrected for antecedent rainfall.

Solution

1. Composite $C = \dfrac{0.42(0.3) + 0.58(0.9)}{0.42 + 0.58} = 0.65.$

2. Midpoint of longer-duration storm = 1.5 hr or 90 min.

3. Time to start of 20-minute design rainfall from beginning of 3-hour storm, $t = 90 - \frac{1}{2}(20) = 80$ min.

4. Time to end of 20-minute design rainfall from beginning of 3-hour storm, $t = 90 + \frac{1}{2}(20) = 100$ min.

5. $C = \dfrac{0.98(80)}{4.54 + 80}(0.65) + \dfrac{0.78(80)}{31.17 + 80}(1 - 0.65) = 0.80.$

6. $C = \dfrac{0.98(100)}{4.54 + 100}(0.65) + \dfrac{0.78(100)}{31.17 + 100}(1 - 0.65) = 0.82.$

7. Average $C = 0.81.$

12.7.3 Drainage Area, A

Area represents the drainage area for a site under consideration. For a natural system it represents the watershed. For a sewer system it is the area tributary to a point of inlet. If a system consists of a number of sewers, the complete area is subdivided into component parts, separating a tributary area to each inlet point of every sewer segment. Many arrangements of sewer layout and inlet locations are tried before adopting a final one.

12.7.4 Rainfall Intensity, i

Intensity of rainfall is dependent on the duration of rainfall (short-duration storms are more intense) and the storm frequency or recurrence interval (less frequent storms are more intense). Rainfall intensity–duration–frequency (IDF) relation for a gaging site is developed from the data of a recording rain gage. The procedure has been explained in Section 3.5.3. Since the point rainfalls or observations at a gaging site are considered representative of a 10-mi^2 drainage area, IDF analysis for a station or combined for two stations is adequate for application in the small-area urban drainage design. Typical IDF curves are shown in Figure 12.6.

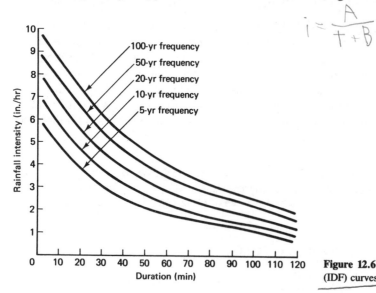

Figure 12.6 Intensity–duration–frequency (IDF) curves for Bridgewater, Conn.

Drainage System   Chap. 12

The collection and analysis of rainfall data and preparation of IDF curves for the site condition are done only for extensive projects. Generally, the rainfall data and the maps prepared by the National Weather Service and other government agencies are used in place of the local statistical analysis, to prepare the IDF curves for the locality selected.

In the absence of data and maps to prepare the curves, empirical relations are used for the duration of less than 2 hours. For any given frequency, the intensity is related to the duration by a relation of eq. (3.15):

$$i = \frac{A}{t + B} \quad \text{[unbalanced]} \tag{3.15}$$

where

$\qquad i$ = intensity, in./hr

$\qquad t$ = duration, min

$\qquad A, B$ = constants that depend on the frequency and climatic conditions

The values for constants A and B are obtained using observed rainfall or the National Weather Service data for the locality selected. General values of the constants for the different parts of the country (Figure 3.7) are given in Table 3.8 for frequency levels of 2, 5, 10, 25, 50, and 100 years.

For application in the rational formula, the extreme (probable maximum, etc.) value of the rainfall intensity is not used because nearly complete protection of the area is not justified. The following range of design frequency is commonly used:

1. 2 to 15 years for sewers in residential areas, most commonly 10 years
2. 10 to 50 years for sewers in commercial and high-value areas
3. 50 years or more for flood-protection works

12.7.5 Time of Concentration, t_c

With regard to storm duration to be considered for runoff assessment, a term "time of concentration, t_c," is relevant. It is defined as the time required for runoff from the hydraulically most remote part of the drainage area to reach the point of reference. There is another definition of this term as well, as stated in Section 7.8.1. For various routes of flow, t_c is taken as the longest time of travel to the point of reference. Since rainfall intensity reduces with increase in storm duration, the duration should be as short as possible. However, if the rainfall duration is less than t_c, then only a part of the drainage area will be contributing to the runoff. For an entire area to contribute, the shortest storm duration should equal the t_c. Thus the time of concentration is used as a unit duration for which the rainfall intensity is determined.

In storm sewer design, in addition to the time required for the rain falling on a most remote point of the tributary area to flow across the ground surface, along streets and gutters, to the point of entry to a sewer, the time of flow through the sewer line is also important. Either the surface and sewer flow times are added together (rational method) or they are considered separately (SCS-55 method). There are many ways to estimate t_c. Some of these are designed primarily for overland flow, some primarily for channel flow, and a few for both overland and channel

TABLE 12.6 EMPIRICAL RELATIONS FOR TIME OF OVERFLOW, t_i

Name	Formula for t_i	Remarks	Number
1. Kerby	$3.03\left(\dfrac{rL^{1.5}}{H^{.5}}\right)^{0.467}$	Applicable to $L < 0.4$ km $r = 0.02$ smooth pavement 0.1 bare packed soil 0.3 rough bare or poor grass 0.4 average grass 0.8 dense grass, timber	(12.4)
2. Izzard	$\dfrac{(0.024i^{0.33} + 878k/i^{0.67})L^{0.67}}{(CH^{0.5})^{0.67}}$	Applicable to $iL < 3.8$ $k = 0.007$ smooth asphalt 0.012 concrete pavement 0.017 tar and gravel pavement 0.046 closely clipped sod 0.060 dense bluegrass turf	(12.5)
3. Brasby–Williams	$\dfrac{0.96L^{1.2}}{H^{0.2}A^{0.1}}$		(12.6)
4. Aviation Agency	$\dfrac{3.64(1.1 - C)L^{0.83}}{H^{0.33}}$		(12.7)

where A = drainage area, km^2; i = rainfall intensity, mm/hr; H = difference of elevation, m; L = length of flow path, km; t_i = time of overflow, hr.

flows. Many formulas are summarized in Table 12.6, which can be used when overflow conditions dominate. The specific condition for which a formula applies is indicated in the table. To apply the Izzard formula, rainfall intensity must be known. A suitable procedure is to assume a time of concentration, determine the intensity from eq. (3.15), and calculate the time of concentration from the Izzard formula. If the initially assumed value was inconsistent, the process above is repeated. According to the U.S. Soil Conservation Service (1986), water moves through a watershed as sheet flow, shallow concentrated flow, open channel flow, or some combination of these before it enters the sewer line. The types that occur depend on the drainage area and can best be determined by field inspection. Time of concentration is the sum of travel time (T_t) values by sheet flow, shallow concentrated flow, and channel flow, whichever occur.

The sheet flow in the form of a thin layer can occur for a maximum length of 300 ft. The travel time is given by Manning's kinematic solution (Overton and Meadows, 1976) as follows:

$$T_{t1} = \frac{0.007(nL)^{0.8}}{(P_2)^{0.5}s^{0.4}} \quad \text{[unbalanced]} \tag{12.8}$$

where

Tt_1 = time, hr, P_2 = 2-yr, 24-hr rainfall, in., L = length, ft
n = Manning's coefficient, smooth surfaces 0.011, fallow 0.05,
    cultivated soil: 0.06 (residue cover < 20%), 0.17 (cover > 20%)
    grass: 0.15 (short), 0.24 (dense grass), 0.41 (Bermuda grass)
    woods: 0.4 (light underbrush) to 0.8 (dense underbrush)

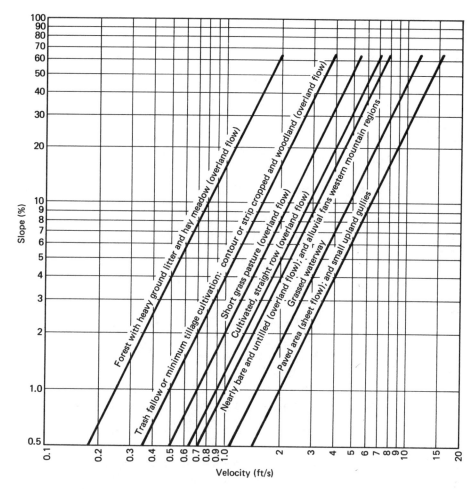

Figure 12.7 Average velocity of overland flow (from U.S. SCS, 1975b).

After a maximum of 300 ft, sheet flow usually becomes shallow concentrated flow. The average velocity for this flow can be determined from Figure 12.7 using the land slope and the type of soil cover. The travel time for shallow concentrated flow is the length divided by the average velocity.

Open channel is assumed to begin where a channel form is visible from field investigations or on aerial photographs. Manning's equation of open channel flow is used to determine the average velocity and the travel time therefrom.

Whenever a drainage area consists of several types of surfaces, the time of concentration is determined by adding the times for different surfaces.

Example 12.8

An urbanized watershed in Providence, Rhode Island, is shown in Figure 12.8. Determine the time of concentration to point C by the various methods. The average velocity of flow in storm drain = 1 m/s.

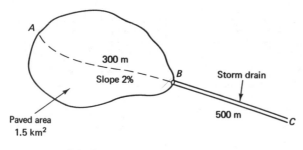

A

300 m

Slope 2%

B

Storm drain

500 m

C

Figure 12.8 Urbanized watershed for Example 12.8.

Paved area
1.5 km²

Solution

A. Inlet time (time of overland flow):
  1. *Kerby method*

$$r = 0.02, \qquad H = (\text{slope} \times \text{distance}) = \frac{2}{100} \times 300 = 6 \text{ m}$$

$$t_i = 3.03 \left[\frac{0.02(0.3)^{1.5}}{6^5} \right]^{0.467}$$

$$= 0.138 \text{ hr} \quad \text{or} \quad 8.3 \text{ min}$$

  2. *Bransby–Williams method*

$$t_i = \frac{0.96(0.3)^{1.2}}{(6)^{0.2}(1.5)^{0.1}} = 0.152 \text{ hr} \quad \text{or} \quad 9.1 \text{ min}$$

  3. *Federal Aviation Agency method*

$$C = 0.9 \text{ for asphalt paving}$$

$$t_i = 3.64 \left[\frac{(1.1 - 0.9)(0.3)^{0.83}}{(6)^{0.33}} \right] = 0.148 \text{ hr} \quad \text{or} \quad 8.9 \text{ min}$$

  4. *Izzard method.* Assume that the time of concentration = 15 min. For Providence, Rhode Island (area 3 in Fig. 3.7) and 5-year frequency,

$$i = \frac{131.1}{t + 19} = \frac{131.1}{15 + 19} = 3.85 \text{ in./hr} \quad \text{or} \quad 98 \text{ mm/hr}$$

$iL = 98(0.3) = 29.4 > 3.8$; thus the formula is not applicable.

  5. *SCS method.* Consider that there is no channel flow segment. Sheet flow for first 300 ft

$$P_2 = 3.5,^* \qquad n = 0.011 \quad \text{(smooth surface)}$$

$$T_{t1} = \frac{0.007[0.011(300)]^{0.8}}{(3.5)^{0.5}(0.02)^{0.4}} = 0.047 \text{ hr} \quad \text{or} \quad 2.8 \text{ min}$$

Shallow concentrated flow. From Figure 12.7, $v = 2.8$ ft/sec or 0.85 m/s.

$$L = 300 - \frac{300}{3.281} = 208.6 \text{ m}$$

$$T_{t2} = \frac{208.6}{0.85} = 245.4 \text{ s} \quad \text{or} \quad 4.1 \text{ min}$$

$$T_t = 2.8 + 4.1 = 6.9 \text{ min}$$

*From a map (Figure B-3) in the TR-55 (U.S. SCS, 1986).

B. Sewer flow time:

$$t_f = \frac{\text{sewer length}}{\text{velocity}} = \frac{500}{1} = 500 \text{ sec} \quad \text{or} \quad 8.3 \text{ min}$$

Adding inlet and flow time, t_c varies from 15.2 min to 17.4 min, depending on the method of computation.

12.7.6 Application of the Rational Method

The drainage area usually consists of more than one type of surface. Equation (12.2) is then applied in the following form:

$$Q = iC_f \sum_{j=1}^{n} C_j a_j \quad [\text{L}^3\text{T}^{-1}] \tag{12.9}$$

where

Q = peak discharge

C_j = runoff coefficient of subdrainage area a_j

$A = \sum_{j=1}^{n} a_j$

i = rainfall intensity for the time of concentration, which is equal to the longest total (inlet plus flow) time to the point where the value of Q is desired

The equation above provides the design flow at the mouth of the composite drainage area, as shown in Example 12.9. However, a storm drainage system consists of many segments of sewer drains. The tributary area and the amount of flow entering each drain is different. To determine the peak discharge, not only at the outlet but at interim points of entry to each drain, a step-by-step application of eq. (12.9), known as the Lloyd–Davies method, is made as illustrated in Example 12.10.

Example 12.9

An urban watershed is shown in Figure 12.9 along with the travel paths from the most remote points in each subarea. The details of the subareas are given in Table 12.7. Assuming that Figure 12.6 reflects the intensity–duration–frequency for the site, determine the 20-year peak flow at the drainage outlet G.

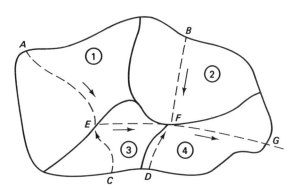

Figure 12.9 Watershed for Example 12.9.

TABLE 12.7 DETAILS OF SUBAREAS OF WATERSHED IN FIGURE 12.9

No.	Area Drained (acres)	Type of Surface	Path	Length (ft)	Slope (%)
1	14	Lawn	AE	1600	4.0
2	12.5	Bare surface	BF	1490	3.0
3	11.1	Asphalt paved	CE	1280	2.0
			EF	1300	1.5
4	8.5	Concrete paved	DF	1250	2.0
			FG	1510	1.5

Solution

1. Travel time by paths:

Path	Length (ft)	Slope (%)	Average Velocity (ft/sec) (Fig. 12.7)	$\text{Time} = \dfrac{\text{Length}}{\text{Velocity}} \times \dfrac{1}{60}$ (min)
AE	1600	4	3.0	8.9
BF	1490	3	1.8	13.8
CE	1280	2	2.9	7.4
EF	1300	1.5	2.5	8.7
DF	1250	2	2.9	7.2
FG	1510	1.5	2.5	10.1

2. Possible routes:
   (a) $AE + EF + FG = 8.9 + 8.7 + 10.1 = 27.7$ min $\leftarrow t_c$
   (b) $CE + EF + FG = 7.4 + 8.7 + 10.1 = 26.2$ min
   (c) $BF + FG \quad\quad = 13.8 + 10.1 \quad\quad = 23.9$ min
   (d) $DF + FG \quad\quad = 7.2 + 10.1 \quad\quad = 17.3$ min

3. From Figure 12.6, for 20-year frequency and a t_c of 27.7 min, $i = 4.6$ in./hr.

4.

Area	Area Drained	C	aC	$\sum aC$
1	14	0.2	2.8	2.8
2	12.5	0.3	3.75	6.55
3	11.1	0.8	8.88	15.43
4	8.5	0.9	7.65	23.08

5. $Q = i \sum aC = 4.6(23.08) = 106.2$ cfs.

Example 12.10

A storm drainage system comprises the four areas (Figure 12.10) with the data given in Table 12.8. Determine the 5-year design flow for each section of the sewer line.

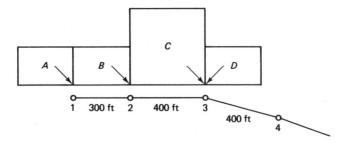

Figure 12.10 Storm drainage system.

TABLE 12.8 DETAILS OF DRAINAGE AREA

Unit	Area (acres)	Runoff Coefficient	Inlet time (min)
A	12	0.8	10
B	12	0.8	10
C	34	0.6	28
D	10	0.9	8

$$\text{Average velocity in sewer} = 4.5 \text{ ft/sec}$$

$$\text{5-year rainfall intensity, } i = \frac{105}{t_c + 15}$$

Solution

1. Flow time in sewers:

$$\text{Sewer 1–2} = \frac{300}{4.5} = 66.6 \text{ sec} \quad \text{or} \quad 1.1 \text{ min}$$

$$\text{Sewer 2–3} = \frac{400}{4.5} = 88.89 \text{ sec} \quad \text{or} \quad 1.5 \text{ min}$$

$$\text{Sewer 3–4} = \frac{400}{4.5} = 88.89 \text{ sec} \quad \text{or} \quad 1.5 \text{ min}$$

2. Computations are arranged in Table 12.9.

TABLE 12.9 STEP METHOD OF COMPUTATION

(1) Manhole	(2) Area, a (acres)	(3) Coefficient, C	(4) aC	(5) $\sum aC$	(6) Routes	(7) Inlet	(8) In Sewer	(9) Total Time	(10) Intensity (in./hr)	(11) Q_{peak} (cfs)
1	12	0.8	9.6	9.6	A–1	10	—	10	4.20	40.32
2	12	0.8	9.6	19.2	B–2	10	—	10		
					A–1–2	10	1.1	11.1	4.02	77.24
3	34	0.6	20.4	39.6						
	10	0.9	9.0	48.6	C–3	28	—	28	2.44	118.67
					D–3	8	—	8		
					2–3	11.1	1.5	12.6		

Note: Columns (7), (8), (9) are grouped under "Travel Time (min)".

Consider manhole 3: Areas C and D are contributing at the manhole as listed in column 2 of the table. The coefficients for these areas and the product aC are given in the next two columns. Column 5 accumulates the product; it sums up the value of column 4 to the value in the previous line of column 5 if the flow of the previous area passes through the manhole in question. That is, all areas whose flow passes through the manhole in consideration are added together. In column 6 all possible routes for the flow to reach the manhole in question (i.e., manhole 3) are listed along with their inlet and sewer flow time in columns 7 and 8, respectively. For the route of flow from the previous manhole (i.e., 2–3), the highest value of the total time (column 9) of the preceding manhole computation is taken as the inlet time. The flow time in a sewer section is determined from the average velocity. In a design problem, this is established from the design of the pipe size of the preceding section. The maximum of the total time in column 9 from all routes is underlined and denotes the time of concentration, t_c. Corresponding to this t_c, intensity is determined either from the intensity–duration–frequency curve or the empirical relation. Peak discharge is the multiplication of the last value in columns 5 and 10.

12.8 THE SCS TR-55 METHOD

In 1964 the SCS developed a computerized watershed model known as the TR-20. It is a very versatile model that has the capability of solving many hydrologic problems comprising the formulation of runoff hydrographs; routing hydrographs through channels and reservoirs, thus providing discharges at selected locations; combining or separating of hydrographs at confluences; and determining peak discharges and their time of occurrences for individual storm events. The model is widely used in small watershed projects and floodplain studies. When the sole purpose is to assess the peak discharge or peak flow hydrograph for drainage design, a method simplified from TR-20 is used. This is referred to as TR-55 (after the document Technical Release 55, in which the method is contained). There are two approaches in the TR-55 method, known as the graphical method and the tabular method.

The graphical method uses the following equation to determine the peak flow:

$$q_p = q_u A_m Q F_p \qquad [L^3T^{-1}] \qquad (12.10)$$

where

q_p = peak discharge, cfs

q_u = unit peak discharge, cfs/mi^2/in.

A_m = drainage area, mi^2

Q = runoff, in.

F_p = pond or swamp adjustment factor

Runoff (Q) is determined by the procedure of Section 3.9.6 from the data of 24-hour rainfall of the desired design frequency. Generalized 24-hour rainfall data for various frequency are available in the TR-55 of the U.S. Soil Conservation Service (1986). Adjustment factor (F_p) is obtained from Table 12.10. Unit peak discharge (q_u) is determined from the graphs contained in TR-55. A sample graph is shown in Figure 12.11. The information required to use the graph to ascertain q_u is as follows:

1. Time of concentration as calculated from Section 12.7.5
2. The ratio I_a/P, I_a being obtained from Table 12.11 and P is the 24-hr rainfall
3. Rainfall distribution type

TABLE 12.10 ADJUSTMENT FACTOR (F_p) FOR POND AND SWAMP AREAS THAT ARE SPREAD THROUGHOUT THE WATERSHED

Percentage of Pond and Swamp Areas	F_p
0	1.00
0.2	0.97
1.0	0.87
3.0	0.75
5.0	0.72

Source: U.S. SCS (1986).

The SCS has developed four synthetic 24-hour rainfall distribution from the National Weather Service duration–frequency data to represent various regions of the United States:

Type I: Hawaii, Alaska

Type IA: coastal side of Sierra Nevada, and Cascade Mountains in southern and northern California, Oregon, and Washington

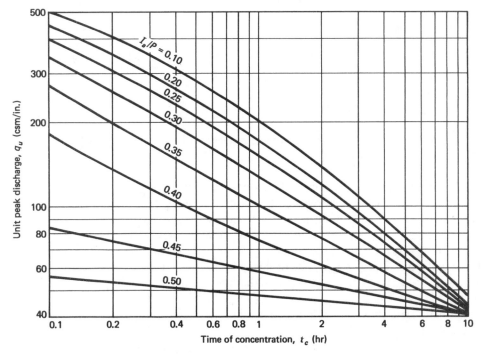

Figure 12.11 Unit peak discharge (q_u) for SCS type I rainfall distribution (from U.S. SCS, 1986).

Sec. 12.8   The SCS TR-55 Method

631

TABLE 12.11 I_a VALUES FOR RUNOFF CURVE NUMBERS

Curve Number	I_a (in.)	Curve Number	I_a (in.)
40	3.000	70	0.857
41	2.878	71	0.817
42	2.762	72	0.778
43	2.651	73	0.740
44	2.545	74	0.703
45	2.444	75	0.667
46	2.348	76	0.632
47	2.255	77	0.597
48	2.167	78	0.564
49	2.082	79	0.532
50	2.000	80	0.500
51	1.922	81	0.469
52	1.846	82	0.439
53	1.774	83	0.410
54	1.704	84	0.381
55	1.636	85	0.353
56	1.571	86	0.326
57	1.509	87	0.299
58	1.448	88	0.273
59	1.390	89	0.247
60	1.333	90	0.222
61	1.279	91	0.198
62	1.226	92	0.174
63	1.175	93	0.151
64	1.125	94	0.128
65	1.077	95	0.105
66	1.030	96	0.083
67	0.985	97	0.062
68	0.941	98	0.041
69	0.899		

Source: U.S. SCS (1986).

Type III: Gulf of Mexico and east coast area from Maryland to Maine

Type II: rest of the United States

The tabular method is suitable when a complete hydrograph is desired instead of peak flow only or when subdivision of a watershed into subareas is involved. For each subarea, the following information is ascertained: (1) weighted curve number; (2) runoff, Q; (3) ratio I_a/P; (4) time of concentration within the subarea, t_c; and (5) travel time downstream of the subarea to the outlet, T_t. For the selected rainfall distribution type and known I_a/P, one of the tables in TR-55 provides the hydrograph ordinates for the subarea that correspond to the time of concentration t_c and routed to the outlet for the travel time T_t. Once the routed ordinates for all subareas at different times are tabulated, their summation at each time yields the composite hydrograph.

Example 12.11

Solve Example 12.9 by the SCS TR-55 method. The rainfall distribution is type II and the 20-year 24-hour rainfall is 4 in.

Solution Computations are given in Tables 12.12 and 12.13.

TABLE 12.12 COMPUTATION OF RUNOFF AND INITIAL ABSTRACTION

Area	Drainage Area, A_m (mi²)	24-Hr Rainfall, P (in.)	Curve Number, CN (Table 3.19)	Runoff, Q (in.) (Table 3.21)	Area × Runoff, $A_m Q$ (mi² − in.)	I_a (in.) (Table 12.11)	I_a/P
1	0.0219	4	68	1.20	0.026	0.94	0.23
2	0.0195	4	75	1.67	0.033	0.67	0.17
3	0.0173	4	98	3.77	0.065	0.04	0.01
4	0.0133	4	98	3.77	0.050	0.04	0.01

12.9 A STORM SEWER DESIGN PROJECT

Design flows for various sewer sections are estimated by the procedure of the preceding section. The design sequence is as follows:

1. For Q_{design} and a velocity of 3 ft/sec (0.9 m/s), determine the diameter of the sewer by the continuity equation, $Q = VA$. This is the maximum size.
2. For Q_{design} and the surface grade, determine the diameter by Manning's equation. For a roughness coefficient, use Table 10.4.
3. If the diameter in step 2 is smaller than step 1, select it (rounded to a next-higher standard size) as the sewer size and the surface grade as the sewer slope.
4. If the diameter of step 2 is bigger than step 1, recompute the slope by Manning's formula adopting the sewer diameter of step 1. In this case a steeper slope than street grade is required. In certain circumstances it is necessary to use a grade lower than this. A provision should then be made for flushing of the pipe.

An example illustrates the procedure.

Example 12.12

Design a storm drainage system for the section of a city shown in Figure 12.12 (the city for which a sanitary sewer system is designed in Example 12.4). Design for the following conditions:

1. The coefficient of runoff, C, at the time of maximum development will be as shown in Figure 12.12.
2. The inlet time, which can be calculated by the method of Section 12.7.5, is assumed to be 15 minutes for each inlet.
3. The system is to be designed for 5-year peak flows. The rainfall intensity in mm/hr is given by $i = 3330/(t + 19)$, where t is in minutes.
4. Manning's n is 0.013.

Solution Unlike sanitary sewers, the storm sewer line need not run through individual lots because the connections from housing units are not required. Thus the drains can be laid by the shortest route. However, the arrangement will be governed by the topography (contour pattern) of the drainage area that dictates the direction of runoff and hence the positioning of the inlets and laying of sewer lines. Also, relatively larger areas can be

TABLE 12.13 HYDROGRAPH COMPUTATION

Area	Time of Conc., t_c (hr) (Example 12.9)	Downstream Travel Route	D/S Travel Time, ΣT_t (hr) (Example 12.9)	I_a/P (rounded)	$A_m Q$ (mi²-in.)	11.9	12.0	12.1	12.2	12.3	12.4	12.5	12.6	12.7	12.8	13.0
						Hydrograph Ordinates in cfs = (Value from TR – 55[a] × $A_m Q$)										
1	0.15	EF + FG	0.31	0.2	0.026	.6[b]	.9	1.6	3.6	8.1	12.7	13.9	12.0	9.2	6.9	3.9
2	0.23	FG	0.17	0.2	0.033	.9[b]	1.6	3.8	9.3	16.3	19.4	16.7	12.2	8.7	6.3	3.9
3	0.12	EF + FG	0.31	—	0.065	3.6[c]	6.0	11.3	21.9	37.8	43.0	35.4	25.3	17.5	12.4	7.1
4	0.12	FG	0.17	—	0.050	3.1[c]	5.5	10.8	20.9	35.2	35.1	24.3	15.6	10.5	7.6	4.7
						8.2	14.0	27.5	55.7	97.4	110.2	90.3	65.1	45.9	33.2	19.6

Hydrograph Times (hr)

[a] From Exhibit 5-II (U.S. Soil Conservation Service, 1986, pp. 5-29 and 5-30).

Rounding of t_c and T_t:

Area	t_c	T_t	Sum
1	0.2	0.3	0.5
2	0.2	0.2	0.4
3	0.1	0.3	0.4
4	0.1	0.2	0.3

[b] Table values at I_a/P of 0.1 and 0.3 are averaged.
[c] Table values at I_a/P of 0.1 are used.

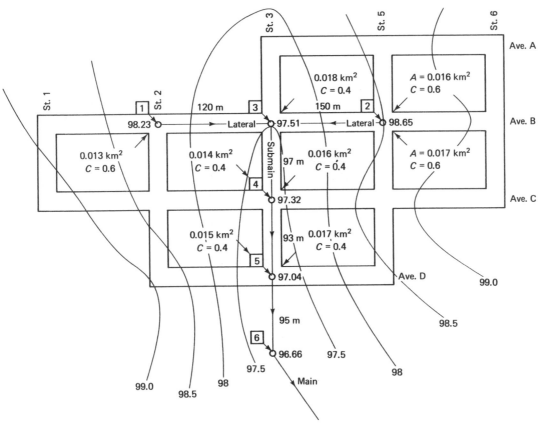

Figure 12.12 Storm drains layout for a section of a city.

covered by each section of the storm drain. For the section of the city in Figure 12.12, the arrow indicates the lowest point in each block. The general direction of flow of the block will be toward the arrow. A layout of drains has been arranged keeping this in mind. A minimum number of drains are included. The drainage area tributary to each intercept point determined on the basis of the contours is indicated in the figure along with the runoff coefficient, which has been taken as 0.6 for the commercial district and 0.4 for the residential area.

The peak flows at each intercept point are computed in Table 12.14 by the step method of Section 12.7.6. To calculate the flow at intercept 3, the time of concentration is to be determined which involves computation of the time of travel in sewer 1–3 and sewer 2–3 (column 11). This requires a determination of the size and the flow velocity (design) of the two sewer sections. Thus the design of section proceeds simultaneously with estimation of peak flow. The peak flows at the head of section 1–3 (intercept 1) and section 2–3 (intercept 2) are computed in Table 12.14. For these flows, pipe sections 1–3 and 2–3 are designed in lines 1 and 2 in Table 12.15. This provides the average velocity of flow (column 13) and the travel time through the sewer (column 14) in Table 12.15. This time of flow is included in column 11 of Table 12.14 to determine the time of concentration. The peak flow is then determined for the next pipe to be designed. Thus the computation alternates in Tables 12.14 and 12.15.

TABLE 12.14 COMPUTATION OF PEAK DISCHARGE

(1)	(2)	(3)	(4)	(5)	(6)	(7)	(8)	(9)	(10)	(11)	(12)	(13)	(14)
		Intercept		Tributary					Travel Time (min)				
Line	Location	From	To	Area, a (km²)	Coefficient, C	aC (m²)	ΣaC (m²)	Route	Inlet	In Sewer	Total	Intensity (mm/hr)	Q (m³/s)
1	Avenue B	1	3	0.013	0.6	7,800	7,800	TA–1[a]	15	0	15	97.9	0.212
2	Avenue B	2	3	0.016	0.6	9,600	9,600	TA–2	15	0	15	97.9	0.538
3	Street 3	3	4	0.017	0.6	10,200	19,800	TA–3	15	0	15		
				0.018	0.4	7,200	34,800[b]	1–3	15	1.47	16.47	93.9	0.908
								2–3	15	1.31	16.31		
4	Street 3	4	5	0.014	0.4	5,600	40,400	TA–4	15	0	15		
				0.016	0.4	6,400	46,800	3–4	16.47	1.23	17.70	90.7	1.180
5	Outside	5	6	0.015	0.4	6,000	52,800	TA–5	15	0	15		
				0.017	0.4	6,800	59,600	4–5	17.70	0.95	18.65	88.4	1.464

[a]TA – 1 = Tributary area to intercept 1.
[b]TA + area from 1–3 + area from 2–3 = 7200 + 7800 + 19,800 = 34,800.

TABLE 12.15 STORM SEWER DESIGN COMPUTATIONS

(1) Line	(2) (3) Manhole		(4) Design Flow, Q_{design} (m³/s)	(5) Length of Sewer (m)	(6) (7) Surface Elevation		(8) Street Slope	(9) Maximum Diameter for Velocity of 0.9 m/s[a] (mm)	(10) Diameter for Street Grade[b] (mm)	(11) (12) (13) Design Parameters			(14) Travel Time (min) $\left(\dfrac{\text{col. }5}{\text{col. }13}\times\dfrac{1}{60}\right)$
	From	To			Upstream	Downstream				Diameter (mm)	Sewer Grade	Velocity at Full[c] (m/s)	
1	1	3	0.212	120	98.23	97.51	0.006	550	445	445	0.006	1.36	1.47
2	2	3	0.538	150	98.65	97.51	0.0076	875	600	600	0.0076	1.90	1.31
3	3	4	0.908	97	97.51	97.32	0.002	1135	940	940	0.002	1.31	1.23
4	4	5	1.180	93	97.32	97.04	0.003	1290	960	960	0.003	1.63	0.95
5	5	6	1.464	95	97.04	96.66	0.004	1440	990	990	0.004	1.90	0.83

[a] $D = (1.274Q/v)^{1/2}$ (continuity equation).

[b] $D = \left[\dfrac{(3.211)nQ}{s^{1/2}}\right]^{0.375}$ (Manning's equation). For FPS units, use 2.16 in place of 3.211.

[c] $v = 1.274\left(\dfrac{Q}{D^2}\right)$ (continuity equation).

Sec. 12.9   A Storm Sewer Design Project

The design procedure of Table 12.15 is as follows. The value in column 4 is taken from column 14 of Table 12.14. Columns 5, 6, and 7 are based on the layout plan. Column 8 is the difference between columns 6 and 7, divided by column 5. In column 9, the maximum sewer size for a minimum velocity of 0.9 m/s is determined using the continuity equation, $Q = AV$. In column 10, the diameter corresponding to the street slope of column 8 is computed from Manning's equation. The design diameter in column 11 is the minimum of columns 9 and 10 (rounded to a standard size). If this pertains to column 10, the sewer grade in column 12 is equivalent to the street grade. If the design diameter is based on column 9, the sewer grade is computed from Manning's equation. The velocity of flow in column 13 is determined by the continuity equation for known flow (column 4) and diameter (column 11). When the velocity is excessive, it is reduced to a limiting value of 5.0 m/s and for the known design flow, the diameter is recomputed by the continuity equation and the slope from Manning's equation.

12.10 DETENTION BASIN STORAGE CAPACITY

Urbanization of rural areas increases peak discharges that adversely affect downstream floodplains. Many local governments are adopting ordinances which require that the postdevelopment discharge not exceed the predevelopment discharge, i.e. zero excess runoff, for a defined storm frequency at a development area. The detention basin is the most widely used measure to control the peak discharge. When a detention basin is installed, the reservoir routing procedure can be used to estimate the effect on hydrographs. The size of the detention basin can be adjusted to maintain a required level of outflow discharge. A quick method of estimation has been included in TR-55 that relates the ratio of peak outflow to peak inflow discharge (q_o/q_i) with the ratio of detention storage volume to runoff volume (V_s/V_r), as illustrated in Figure 12.13. This figure is used to estimate the detention storage volume (V_s) from the known information of runoff volume (V_r), peak outflow discharge (q_o), and peak inflow discharge (q_i) or to estimate q_o from the known values of V_r, V_s, and q_i. q_o is

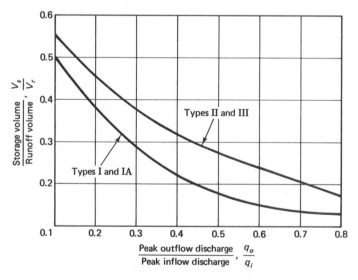

Figure 12.13 Detention basin storage volume (from U.S. SCS, 1986).

the predevelopment level of peak flow or a desired level of discharge from the drainage area. q_i is the peak discharge from the developed area computed by the TR-55 method of Section 12.8. While using the tabular method to estimate q_i for a subarea, the peak discharge associated with $T_{t=0}$ is used. V_r is the drainage area times the runoff, which was determined when computing q_i. The computed V_s is adequate for preliminary designs.

Another simplified approach based on the Rational Method follows.

Example 12.13

For watershed in Figure 12.9, if the peak rate of outflow of 5-year frequency is 50 cfs,* determine the size of the detention basin required. The intensity–duration is given by $i = 315 / (t_c + 25)$, where t_c is the rainfall duration in minutes.

Solution

1. From Example 12.9, $\Sigma ac = 23.08$.
2. Refer to Table 12.16.
3. Detention Basin Capacity (Table 12.16, col. 6) $= 149.5 \times 10^3$ cft.

TABLE 12.16 COMPUTATION OF DETENTION STORAGE

(1) Rainfall Duration min	(2) Intensity $i = 315 / (t_c + 25)$ in. per hr	(3) Peak Inflow $Q_i = i \Sigma ac$ cfs	(4) Peak Outflow Q cfs	(5) Rate of Flow Detained cfs	(6) Detention Capacity 1000 cft
26	6.18	142.63	50	92.63	144.5
35	5.25	121.17	50	71.17	149.5 ←
50	4.20	96.94	50	46.94	140.8
56	3.89	89.76	50	39.76	133.6
65	3.50	80.78	50	30.78	120.0
80	3.0	69.24	50	19.24	92.4

column 5 = column 3 − column 4
column 6 = column 5 × column 1 × 60

12.11 STRUCTURAL STRENGTH OF SEWERS

For both sanitary and storm sewers, the size, slope, and other characteristics of sewer pipes are determined from hydraulic considerations, as explained in previous sections. There is another phase of design that relates to the strength of sewers to withstand forces. The designer's task involves (1) computation of the loads on buried sewer lines due to overburden earth forces and superimposed traffic loads, and (2) design of proper bedding for the sewer with due regard to the crushing strength of the material, position of the conduit from the ground, type of soil, width of the excavated trench, and condition of the traffic. This aspect of structural design is not covered in the book. The reader is referred to a manual of the American Society of Civil Engineers and Water Pollution Control Federation (1969).

*This may represent runoff at the predevelopment level.

12.12 AGRICULTURE DRAINAGE SYSTEM

The removal of water from the surface of land is surface drainage. Similarly, the removal or control of water beneath the land surface is termed subsurface drainage. Urban storm drainage is concerned with surface drainage. Agriculture drainage, on the other hand, deals with both the removal of excess precipitation and irrigation surface waste as a part of surface drainage and removal and control of groundwater percolated from precipitation or irrigation or leaked from canals, through the subsurface drainage. The latter component, termed the land drainage, is achieved by the flow of water through the porous soil medium by gravity to the natural outlet. If the water is added through irrigation or heavy precipitation at a faster rate than it can travel to the outlets, the water table rises and can approach the surface to waterlog the land. In such cases, additional man-made outlets are provided in the form of drains. The installation of the drains or man-made subdrainage system has been found essential for the agriculture land because of the rapid buildup of the water table. For highway and airport pavement structures also, some provision of subsurface drains is required, in addition to major surface drainage facilities.

12.12.1 Surface Drainage for Agriculture Land

Surface flow that should be carried away from agriculture lands comprises precipitation excess and farm irrigation surface waste (excess). In a humid region, the former constitutes almost the entire surface flow. In arid regions, the irrigation waste is a major constituent. Surface runoff from agriculture land is much less than urban runoff because of the perviousness of land surface. The procedures to determine surface flows due to precipitation and irrigation are described below. Once the quantity of runoff (from storm and irrigation) at various points of interception is known, the surface drainage system is designed (1) as a separate system for the large land area on the line of an urban storm drain system, or (2) as a system of open drains comprising the laterals or field drains, submains, and mains. The submains and mains include contributions from the subsurface drains. The drains are sized for the combined surface and subsurface quantity of flow by Manning's equation.

The peak rate of surface runoff due to a rainstorm of specified frequency can be determined by the rational method using an appropriate value of the runoff coefficient for the agriculture area. Surface drains are designed to handle flows from 5- to 15-year storm frequencies. Where damages can be expensive, a more conservative design frequency of 25 years should be used. However, the Soil Conservation Service (SCS) procedure of soil and cover conditions, as described in Section 12.8, is more appropriate for agriculture land.

Surface runoff produced by waste from irrigation varies with many factors, including soil texture, land slope, length of irrigation run, and irrigation efficiency. This may amount to as much as 50% of the water applied to any farm unit. The total amount of farm (irrigation) waste at any point depends on the amount that is wasted from a unit area times the total irrigable area up to that point. From the data of an irrigation canal in a particular location, a canal capacity curve can be prepared as shown in Figure 12.14 that indicates the required capacity of the canal to irrigate various sizes of areas. Unless a better estimate of farm waste is available, a factor is applied to the canal capacity to determine the farm waste from irrigated area. The drain is

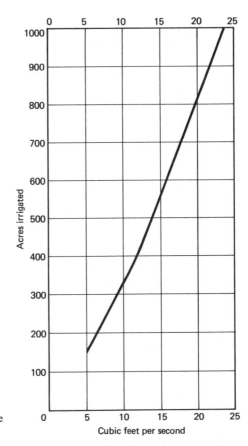

Figure 12.14 Typical canal capacity curve (from U.S. Bureau of Reclamation, 1984).

located on a topographic map. For any point on the drain, the total irrigated acreage is determined. The canal capacity for that acreage is read from Figure 12.14. By applying a factor between 15 and 25%, the irrigation waste for drainage design is computed. For example, assume that a topographic map shows an irrigable area of 500 acres at certain point on the drain in question. From Figure 12.14, the canal capacity is 14 cfs for 500 acres. The irrigation waste will be 15% of this, or 2.1 cfs.

12.12.2 Subsurface Drainage for Agriculture Land

In terms of agriculture requirements, the major objectives of subsurface drainage are (1) to maintain the water table below the plant root zone, which will otherwise rise close to the land surface due to excess infiltrated precipitation in humid regions or application of irrigation water in arid areas, and (2) to leach adequate quantity of water through the root zone of the plants to keep the salinity* from exceeding a specified limit.

The drainage system consists of either open drains or pipe drains or their combination. Open drains or ditches, with an exposed water surface, are used both for surface drainage, and subsurface drainage. They are used as field drains, branches,

---

* Irrigation water contains salt that gets deposited in the root zone.

mains, and intercepting drains. Their main advantages are the ability to carry a large quantity of water and the low initial cost, which is, however, partly offset by a high maintenance cost. Their principal disadvantages arise from the loss of land they occupy, which could otherwise be cultivated, and difficulty in farming operations. The size is determined from the theory of open channel flow by using the Manning's formula to carry the subsurface drained flow alone or along with the surface flow, depending on the intended use. The shape, depth of flow, and grade of ditch are the factors that enter in the design. A semicircle is a most efficient section. A trapezoidal section is also commonly used. Depth of ditches is usually 6 to 12 ft. Ditches should be deep enough to receive the discharge from the drains emptying into them. The slope is determined by the topography of the land, which is very small. A minimum slope should be 0.005%. The maximum grade should not induce velocity more than scour velocity, which ranges from 2 to 6 ft/sec, depending on the soil. Lateral ditches are rarely placed closer than 0.3 mile (0.5 km). A spacing of 0.6 mile (1 km) is satisfactory for the lateral for favorable slopes.

Pipe drains or tube drains are buried beneath the surface of soil. The modern tendency is in favor of pipe drains. Ditches are used for main and intercepting drains into which the pipe drains empty. Pipe drains are also designed by Manning's formula (open channel flow) for carrying the design flow with just flowing full condition. The smallest size in general use in the United States is 4 in. (100 mm) [plastic tubing of 3 in. (80 mm) has been installed in some places]. The grade for the pipe is provided to maintain a velocity of 1 to 1.5 ft/sec when running full to carry the small sediment that enter into the pipe. A minimum grade of 0.15% is recommended for 4-in. pipe and 0.005% for 12-in. or larger pipe. The spacing of 50 to 150 ft are usually adopted. The spacing requirement for a parallel drain system is discussed subsequently. Where very close spacing is required, mole drains are provided, which are the round channels formed by pulling a steel lug (cutting edge) through the subsoil at shallow depth.

Pipe drains comprise unglazed clay tile, concrete pipe, or corrugated plastic pipe. The pipes are placed with the ends of pipes butted together. Water enters through the space between abutted sections. If the space between pipe sections is $\frac{1}{8}$ in. or larger, the joints are covered with a filter material or special joints are used.

Based on the function performed, there are five types of drains designated as relief, interceptor, collector, suboutlet, and outlet drains. Relief drains are used to affect a lowering of groundwater over relatively large flat areas where the gradient of both the water table and subsurface strata do not permit sufficient lateral movement of the groundwater (U.S. Bureau of Reclamation, 1984). Interceptor drains cut off or intercept groundwater moving at a steeper slope down the hill. Collector drains receive water from subsurface relief and interceptor drains and from surface drains carrying irrigation surface waste and precipitation surface runoff. These can be open or pipe drains. Suboutlets receive inflows from a number of collector drains and convey to outlets. They are located in topographic lows such as draws and creeks. Outlet drains take away water from the drainage area. They are usually natural water courses but can be man-made structures.

12.12.3 Layout of Pipe (Tube) Drainage System

The arrangement of drains is mostly determined by the topography. The common types of layout are shown in Figure 12.15 and described briefly on the next page.

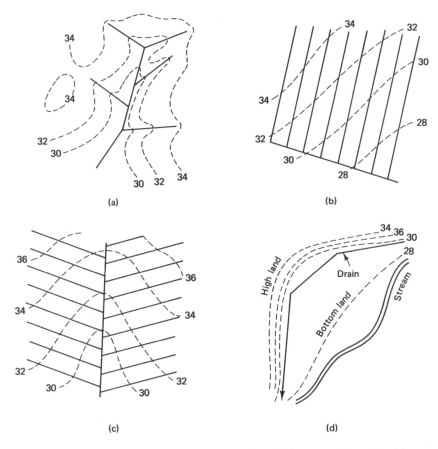

Figure 12.15 Arrangements of tile (subsurface) drains: (a) random; (b) gridiron; (c) herringbone; (d) intercepting (from Linsley and Franzini, 1979).

1. *Random system.* This system is used where the topography is undulating and drainage is required in isolated areas or in small swales and valleys.

2. *Gridiron system.* The parallel laterals enter the submain from one side. It is suitable for flat land or where the land slopes away on one side.

3. *Herringbone system.* The parallel laterals enter the submain at an angle, usually from both sides. It is suitable for a valley-shaped land where the submain is placed in the depression and better grades for laterals are obtained by angling them upslope.

4. *Interception drains.* Seepage moving down a slope (hill) is collected by drains placed along the toe of the slope.

The design of subsurface drainage system involves the layout of drains and the determination of depth, spacing, and size of drains, together with outlet and appurtenant works. Several layouts and tentative designs are worked out before adopting a final arrangement most suitable technically and economically.

12.12.4 Depth and Spacing of Drains

Methods for estimating the depth and spacing of drains have been developed based on the drainage theory, which is essentially the theory of movement of groundwater through the porous medium. As discussed in Chapter 4, the governing partial differential equation of groundwater flow is too complex to be solved for real field conditions. In the drainage theory, an idealized soil-water system is considered and a practical judgment is used in the application. The depth and spacing requirements as determined from mathematical analyses are verified from operating systems in similar conditions, if possible. Where wide variations exist between field observations and mathematical solution, field data should be checked to justify adoption of field-observed values. The simplest drainage theory, developed by the Dutch engineer Hooghoudt in 1940 and still very popular, considers a steady-state condition of a stabilized water table. It applies Darcy's law to derive an expression for the spacing of drains. Kirkham, in 1958, derived the spacing equation based on the steady-state groundwater equation. The U.S. Bureau of Reclamation extended the theory to unsteady-state conditions of falling and rising water table by applying the linearized differential equation of groundwater flow. Some other theories considered the nonlinear form of the groundwater equation. The unsaturated flow theories have also been developed. An idea of extensive research in the theory of drainage can be derived from the fact that over 60 drain spacing formulas have been reported in the category of steady- and unsteady-state flows.

The Bureau of Reclamation is a leader in the field of irrigation drainage. The validity of the Bureau's method has been demonstrated from the field tests. According to the Bureau, the height of the water table at the midpoint between the drains is given by

$$y = \frac{192}{\pi^3} \sum_{n=1,3,5,\ldots}^{\infty} (-1)^{(n-1)/2} \frac{n^2 - 8/\pi^2}{n^5} e^{-\pi^2 n^2 \alpha t / L^2} \qquad [L] \qquad (12.11)$$

where

$\alpha = KD/S$

K = hydraulic conductivity

D = average depth of flow region = $d + y_0/2$

S = specific yield (% by volume) from Figure 12.16

L = drain spacing

y = water-table height above drain at midpoint at the end of drain period, t

y_0 = water-table height above drain at midpoint at the beginning of drainout

d = depth from drain to barrier (impermeable surface)

The Bureau of Reclamation presented the solution in the form of the curves given in Figure 12.17 in terms of dimensionless parameters, y/y_0 versus KDt/SL^2 for the case where drains are located above a barrier and in terms of y/y_0 versus Ky_0t/SL^2 for drains located on the barrier.

The water level reaches its highest position after the last irrigation (at the end of the peak period of irrigation) or after recharge. The water table recedes during

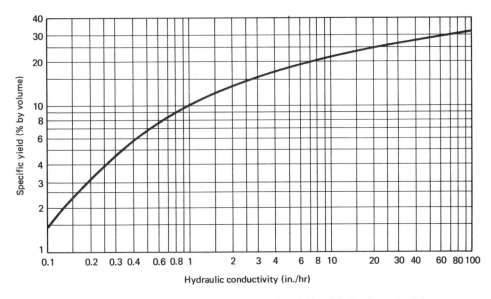

Figure 12.16 General relationship between specific yield and hydraulic conductivity (from U.S. Bureau of Reclamation, 1984).

a nonirrigation or slack period. It starts rising again with the beginning of irrigation or recharge. If annual discharge (drainage) from an area is less than annual recharge (from precipitation and irrigation), the water table will progressively rise upward from year to year. When the annual discharge and recharge are about equal, the range of the cyclic annual water-table fluctuation becomes reasonably constant. This condition is referred to as "dynamic equilibrium." The Bureau's method determines the drains spacing that will produce a dynamic equilibrium for a specified water-table depth.

For application of the method, it is necessary that the initial water-table condition be known. When the drains are being planned on an operating project, field measurements will provide information on buildup in the water table due to irrigation/precipitation. For a new project, the amount of deep percolation reaching the drain is determined as a percentage of the irrigation net input of water into the soil. These percentages are given in Table 12.17. The buildup in the water table is computed by dividing the amount of deep percolation by the specific yield of the soil within the zone of fluctuation of the water table, as given in Figure 12.16. In humid and semihumid areas, the infiltration due to rainfall should also be considered. Due to rainfall, the fraction of infiltration going into the deep percolation and contributing to the water-table buildup is computed by the same procedure.

12.12.5 Application of U.S. Bureau of Reclamation Method

The flow converges toward the drain resulting in the loss of head. To account for this convergency, the depth from drain to barrier, d, is converted to an equivalent depth by applying the Hooghoudt correction, as given in Figure 12.18. This equivalent depth is used to determine D. This correction is not required when the drains are located on the barrier.

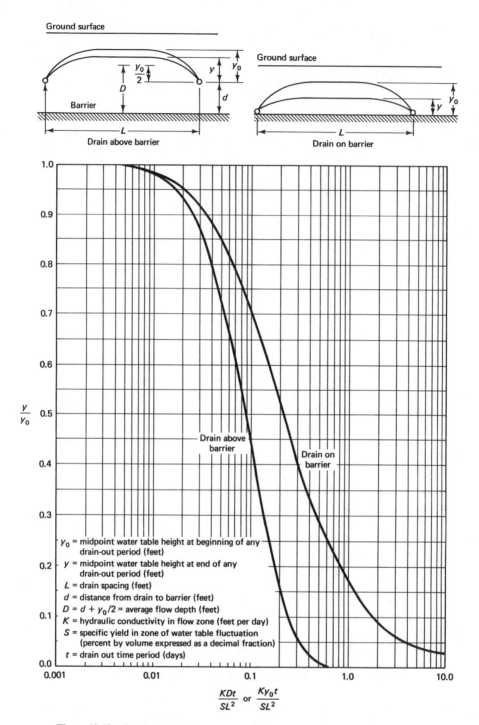

Figure 12.17 Calculation of drain spacing by the transient-flow theory (from U.S. Bureau of Reclamation, 1984).

Drainage System   Chap. 12

TABLE 12.17 DEEP PERCOLATION OF IRRIGATION INPUT OF WATER

a. By Texture

Texture	Percent	Texture	Percent
LS	30	CL	10
SL	26	SiCL	6
L	22	SC	6
SiL	18	C	6
SCL	14		

b. By Infiltration Rate

Infiltration Rate (in./hr)	Deep Percolation (%)	Infiltration Rate (in./hr)	Deep Percolation (%)
0.05	3	1.00	20
0.10	5	1.25	22
0.20	8	1.50	24
0.30	10	2.00	28
0.40	12	2.50	31
0.50	14	3.00	33
0.60	16	4.00	37
0.80	18		

Source: U.S. Bureau of Reclamation (1984).

The method involves a trial-and-error procedure as follows:

1. Estimate the initial (maximum) water-table height, y_0, after the last irrigation of the season or recharge.
2. Assume a drain spacing L.
3. Calculate the successive positions of the water table during the nonirrigation (drainout) period.
4. Calculate the buildup and drainout of the water table from each irrigation for the next season.
5. If the water-table height at the end of the series of calculations is not the same as the initial height, y_0, repeat the procedure with a different L until a dynamic equilibrium is achieved.

Example 12.14

Determine the drain spacing for the following conditions:

1. Depth from the surface to impervious layer = 30 ft.
2. Depth of the drain from the surface = 10 ft.
3. Root zone (water table below surface) requirement = 5 ft.
4. Average hydraulic conductivity = 4 in./hr or 8 ft/day.
5. The spring snowmelt and irrigation schedule are as follows. The runoff is about 20% and the infiltration rate in the root zone is 1.25 in./hr.

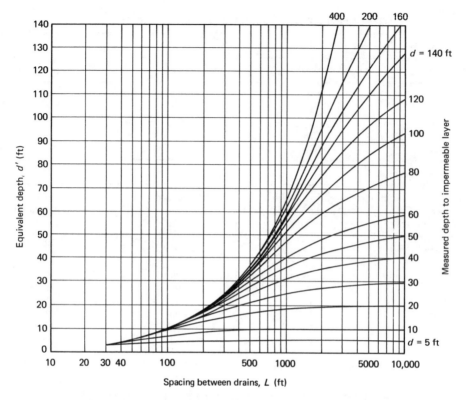

140
130
120
110
100
90
80
70
60
50
40
30
20
10
0

400 200 160

Equivalent depth, d' (ft)

Measured depth to impermeable layer

$d = 140$ ft

120

100

80

60

50

40

30

20

10

$d = 5$ ft

10   20 30 40   100   500   1000   5000 10,000

Spacing between drains, L (ft)

Figure 12.18 Hooghoudt's correction for convergency for drains of radius 0.6 ft (from U.S. Bureau of Reclamation, 1984).

Event	Water Application (in.)	Date	Time between Events (days)
Snowmelt	6	April 15	
			47
First irrigation	5	June 1	
			30
Second irrigation	7	July 1	
			29
Third irrigation	7.1	July 30	
			29
Fourth irrigation	7	August 28	
			135

Solution

1. Maximum allowable water-table height above the drain, $y_0 = 10 - 5 = 5$ ft.
2. For hydraulic conductivity of 4 in./hr, from Figure 12.16, $S = 17\%$.
3. Assume that the drainage spacing, $L = 1600$ ft.
4. Drain to barrier depth = 20. The corrected depth from Figure 12.18 = 18.5 ft.

5. For infiltration rate of 1.25 in./hr, the deep percolation from Table 12.17 is 22%.

6. Water-table buildup from the snowmelt and each irrigation is computed in Table 12.18.

TABLE 12.18 WATER-TABLE BUILDUP

(1)	(2)	(3)	(4)	(5)	(6)	(7)
					Water-Table Increment	
Event	Water Applied (in.)	Runoff (in.) (0.2 × col. 2)	Net Input (col. 2 − col. 3)	Deep Percolation (in.) (0.22 × col. 4)	(in.) (col. 5/S)	Feet
Snowmelt	6	1.20	4.80	1.06	6.24	0.52
First irrigation	5	1.00	4.00	0.88	5.17	0.43
Second irrigation	7	1.40	5.60	1.23	7.24	0.60
Third irrigation	7.1	1.42	5.68	1.25	7.35	0.61
Fourth irrigation	7	1.40	5.60	1.23	7.24	0.60

7. The starting point of the computation is the end of the last irrigation season when the water table is at its maximum allowable height of 5 ft above the drain. Then during the nonirrigation period of 230 days (365 − 135), the water table recedes. It builds up again with recharge/irrigation as per the schedule indicated. If the assumed spacing is correct, the water table should rise up to 5 ft again after application of the last irrigation.

8. The nonirrigation period is divided into two periods of 115 days each.

9. The computations are arranged in Table 12.19.

12.12.6 Design Discharge for Determining Subsurface Drain Pipe Size

Discharge into the drains takes place due to (1) deep percolation of the water from the surface, and (2) underground flow from upslope irrigated areas, and from canals, streams, and other water bodies. Thus

$$q = q_p + q_u \qquad [L^2T^{-1}] \tag{12.12}$$

where

$$q = \text{total flow per unit length of drain}$$

$$q_p = \text{flow due to deep percolation}$$

$$q_u = \text{flow from upslope sources}$$

Flow of parallel spaced drains can be computed using the following formulas:

$$q_p = \frac{2\pi K y_0 D}{L} \qquad \text{(for drains above a barrier)} \qquad [L^2T^{-1}] \tag{12.13}$$

$$q_p = \frac{4K y_0^2}{L} \qquad \text{(for drains on a barrier)} \qquad [L^2T^{-1}] \tag{12.14}$$

All terms are as defined previously.

TABLE 12.19 BUREAU OF RECLAMATION METHOD

(1) Event	(2) Period, t (days)	(3) Water-Table Buildup (ft) (Table 12.18)	(4)[a] Initial Height, y_0 (ft)	(5) $D = d' + \dfrac{y_0}{2}$ (ft)	(6)[b] $\dfrac{kDt}{SL^2}$	(7)[c] $\dfrac{y}{y_0}$ (Fig. 12.17)	(8)[d] Height after the Period, y (ft)
Last season							5.00
Nonirrigation 1	115		5.00	21.00	0.044	0.73	3.65
Nonirrigation 2	115		3.65	20.33	0.043	0.74	2.70
Snowmelt		0.52					
	47		3.22	20.11	0.017	0.95	3.06
First irrigation		0.43					
	30		3.49	20.25	0.011	0.97	3.39
Second irrigation		0.60					
	29		3.99	20.50	0.011	0.97	3.87
Third irrigation		0.61					
	29		4.48	20.74	0.011	0.97	4.35
Fourth irrigation		0.60					
			4.95				

[a]Col. 4 = col. 8 of preceding period plus col. 3 during the period.
[b]Col. 6 = K/SL^2 × col. 2 × col. 5.
[c]Col. 7 = From Figure 12.17, corresponding to the value in col. 6.
[d]Col. 8 = col. 7 × col. 4.

Subsurface flow from upslope is given by Darcy's law for the saturated portion above the drain. Hence

$$q_u = KiA \frac{y_0}{y_0 + d} \qquad [\text{L}^2\text{T}^{-1}] \qquad (12.15)$$

where

i = slope of water table obtained from a water-table contour map along a line normal to the contours

A = saturated area along the plane parallel to the contours or normal to the direction of flow, for a unit length of drain

The total design discharge is obtained from $Q = qX$, where a pipe is X units long. The formula is applied for a length X which serves an area that can be irrigated probably within about 2 days (U.S. Bureau of Reclamation, 1984).

Flow q in the equations above is the maximum rate of discharge. For a collector drain receiving water from a number of drains, each branch will not deliver at the maximum rate at the same time. The Bureau of Reclamation has suggested that the following equations will provide a reasonable design capacity for most collector drains:

$$Q = C \frac{2\pi K y_0 D}{L} \left(\frac{A}{L}\right) \qquad \text{(drains above barrier)} \qquad [\text{L}^3\text{T}^{-1}] \qquad (12.16)$$

$$Q = C \frac{4K y_0^2}{L} \left(\frac{A}{L}\right) \qquad \text{(drains on barrier)} \qquad [\text{L}^3\text{T}^{-1}] \qquad (12.17)$$

where

A = area drained

C = area discharge factor given in Table 12.20

TABLE 12.20 AREA DISCHARGE
FACTORS

Area Drained (acres)	Factor C
0–40	1.0
40–160	1.0–0.82
160–320	0.82–0.72
320–640	0.72–0.60
640–960	0.60–0.54
960–1280	0.54–0.50
1280–5000	0.50

12.13 ROADWAY DRAINAGE SYSTEM

Roadways occupy a narrow strip of land but stretch lengthwise through many watersheds of different characteristics. Two different types of drainage problems are associated with roadways:

1. It is necessary to take away precipitation falling on the road surface and to divert stormwater approaching the road. The facility alongside the road or the longitudinal system takes care of this.
2. A roadway crosses many natural water courses and channels in valley areas. Water carried by these channels has to be conveyed across the road smoothly. Cross-drainage works comprising culverts and bridges are provided for this purpose.

12.13.1 Longitudinal Drainage System

Roads are designed with a crown in the center and cross slopes in both directions away from the centerline. On rural roads, water falling on roads flows laterally off the road surface into the countryside or into shoulder drains. On city streets and urban highways, water falling on or near pavements and sidewalks is directed by the cross slopes to the gutters formed between the edge of the road surface and the vertical curb. It flows along gutters to curb or gutter inlets and from them into underground storm drains. Thus the longitudinal system consists of (1) collector structures such as gutters, gutter inlets, curb opening inlets, grate for inlets, and so on, and (2) underground drains (conduits) that conduct water to the outfall. A detailed treatment of the design of collector structures is presented by the U.S. Department of Transportation (1979). The design of drains is discussed here.

The drainage of city streets is mostly a part of the city storm sewer system, which is based on 5- or 10-year frequency. Highways use high drainage standards since they need urgent water removal from high-traffic pavements; the main storm drains of freeways are based on 50 to 100-year storms. For highway sections that traverse through city areas, a coordination of local drainage authority and highway authority is required since they utilize the local drainage facilities, especially the outfall facilities, to the extent possible. For a highway section at grade, the primary need is a provision of the adequate number of suitably located inlets for rapid removal of water into the existing urban drainage system facilities. When the section is in a cut, there will be sumps or low points at which excess runoff will collect and pumping will be needed. On an elevated highway section, connection to the existing facility is easier.

Highways traversing off city areas need separate drainage systems. The provision of detention storage to handle runoff from highways is considered part of a stormwater management plan. The acquisition of suitable sites and the cooperation of other agencies are required for such storage facilities.

12.13.2 Design Flows for Longitudinal Drainage

Drains are designed to carry the design flows by the procedure of storm drains of Section 12.9. The design flows for longitudinal drainage components — gutters, inlets, spillways, underground drains — are determined by the rational method (Section 12.7). For a direct application to the roadways, the rational formula is expressed as follows:

$$q = C_f C i L \qquad [\mathrm{L}^2\mathrm{T}^{-1}] \qquad (12.18)$$

where

q = peak flow per unit length of pavement

C_f = frequency correction factor (Section 12.7.1)

C = runoff coefficient (Section 12.7.2)

i = rainfall intensity of the design frequency for the time of transverse flow across the pavement

L = length of overland flow normal to contours

Like the original rational equation, eq. (12.18) is dimensionally homogeneous (i.e., if i is in ft/sec and L is in ft, then q is in cfs/ft). Usually, the unit for i of in./hr is used; then eq. (12.18) is divided by 43,200 to get q in cfs/ft.

The length of overland flow, L, is approximated by the formula

$$L = W\sqrt{r^2 + 1}/r \qquad [\mathrm{L}] \qquad (12.19)$$

where

W = roadway width from the center

r = ratio of cross slope to longitudinal slope

The time of overland flow for ascertaining the rainfall intensity is determined by the methods of Section 12.7.5, commonly by the Izzard method, eq. (12.5).

For short distances and steep slopes, the total discharge at an inlet is taken to be q multiplied by the length between inlets. For long distances and flat slopes, the discharge at an inlet is determined either by the routing procedure to provide allowance for the channel storage, or the original rational eq. (12.2) is used, incorporating the entire area tributary up to the point of the inlet. The original rational equation is also used when many interconnecting drains are involved and the travel time through them has to be considered. The U.S. Department of Transportation (1979) designed the underground drains considering them to be running under pressure (applying the pipe flow formula). However, the drains are commonly designed as a nonpressure system by the method of Section 12.9.

The Department of Transportation further recommends that any pipe wholly or partly under a roadbed should have a minimum diameter of 18 in. Elsewhere, it should be a minimum of 15 in. in diameter.

The erosion control of the slopes of highway embankments requires a serious consideration. An easy grade for the slope, sod and grass covers, and intercepting dikes or ditches are some of the measures of erosion control as discussed in Section 12.14 in context of airport drainage.

A subsurface (flow) drainage system, to remove the infiltrated water and to lower the high water table from all important highway pavement structures, is an essential part of a highway design. The requirements and the guidelines for the design of subsurface drainage for highways are given by the U.S. Department of Transportation (1973).

Example 12.15

A 30-ft-wide road section has a longitudinal slope of 0.013. It has a cross slope of $\frac{1}{4}$ in./ft. Determine the peak flow at the gutter inlet if the spacing of inlets is 150 ft. The 10-year rainfall intensity in in./hr is given by $i = 170/(t + 23)$, when t is in minutes.

Solution

1. Using eq. (12.19),

$$r = \frac{0.25}{12(0.013)} = 1.60$$

$$L = \frac{15[(1.6)^2 + 1]^{1/2}}{1.6} = 17.7 \text{ ft}$$

2. To determine the time of overflow by the Izzard method:
   a. Assume that $t_c = 5$ min.

   b. $i = \dfrac{170}{5 + 23} = 6.07$ in./hr or 154.18 mm/hr

   c. Using eq. (12.5),

$$L = 17.7 \text{ ft} \quad \text{or} \quad 5.4 \times 10^{-3} \text{ km}$$

$$H = \frac{0.25}{12}(17.7) = 0.37 \text{ ft} \quad \text{or} \quad 0.112 \text{ m}$$

$$iL = 154.18(5.4 \times 10^{-3}) = 0.84 < 3.8 \quad \text{OK}$$

$$t_i = \frac{[0.024(154.18)^{0.33} + 878(0.017)/(154.18)^{0.67}](5.4 \times 10^{-3})^{0.67}}{[0.8 \times 0.112^{0.5}]^{0.67}}$$

$$= 0.046 \text{ hr} \quad \text{or} \quad 2.8 \text{ min}$$

d. Repeat with $t = 3$ min.

$$i = \frac{170}{3.0 + 23} = 6.54 \text{ in./hr} \quad \text{or} \quad 166.1 \text{ mm/hr} \quad 6.54/43200 \text{ ft./s}$$

$$t_i = \frac{[0.024(166.1)^{0.33} + 878(0.017)/(166.1)^{0.67}](5.4 \times 10^{-3})^{0.67}}{[0.8 \times 0.112^{0.5}]^{0.67}}$$

$$= 0.045 \text{ hr} \quad \text{or} \quad 2.7 \text{ or } 3 \text{ min}$$

3. Using eq. (12.18),

$$q = \frac{(1)(0.8)(6.54)}{43,200}(17.7) = 0.002 \text{ cfs/ft}$$

4. $Q = 0.002(150) = 0.32$ cfs.

12.13.3 Cross Drainage System: Culverts

Bridges and culverts are two cross drainage works that pass stream channels under roadways. The hydraulics of bridge opening have been discussed in Section 8.14.2. The distinction between a bridge and a culvert on the basis of size is arbitrary, with a structure whose span is in excess of 20 ft being classed as a bridge. However, a distinctive feature is that the culverts can be designed to flow with a submerged inlet. A culvert acts as a control structure. In the hydraulic sense, a device is said to control flow if it limits the flow of water which would otherwise be exceeded under existing upstream and downstream conditions. In a control device, the head adjustment across the control section takes place until a balance is achieved between the inflow and the discharge through the section. In the case of culverts, a difficulty arises because the control section can be at the inlet or at the outlet, depending on the type of flow. In supercritical flow the flow velocity is faster than the velocity of a wave, so that the water waves cannot travel upstream, and hence control cannot be exercised from downstream (i.e., there is an inlet control). In the subcritical flow condition, control from downstream will backup water until an equilibrium profile is achieved upstream of the control (i.e., the outlet control exists).

Inlet control means that conditions at the entrance—depth of headwater and entrance geometry—control the capacity of the culvert. An orifice type of flow takes place at the entrance. A culvert runs part full (atmospheric pressure). Thus the barrel size beyond the inlet can be reduced without affecting the discharge, or the capacity can be increased by improving the inlet conditions. The detailed design of improved inlets have been discussed by the U.S. Department of Transportation (1972). These improvements comprise provision of wingwalls; beveling or rounding of culvert edges; tapering the sides of the inlet, including slope tapering; and providing a drop inlet. The geometry of the top and sides of the inlet is important, but not as important as that of the culvert floor. The inlet geometry and channel contraction affect the coefficient of discharge as discussed by Bodhaine (1982).

In outlet control, the culvert can flow full or part full, depending on headwater and tailwater levels. The friction head in the barrel of a culvert affects the headwater or the total energy to pass the discharge through the culvert.

Discharge through culverts depends not only on the type of control but on different types of flow in each control. A general classification of flow through culverts is shown in Table 12.21, separated into two groups: unsubmerged and submerged flow. For submerged flow, the headwater-to-barrel diameter ratio should exceed approximately 1.2. The features of each type of flow with respect to culvert slope, flow depth, and control section are indicated in the table. The first three types relate to unsubmerged flow, with the first one relating to inlet control condition. The other three types in the table relate to submerged culvert flow. The discharge equations given in the table for each type neglect entrance losses.

12.13.4 Design of Culverts

Some box culverts are designed such that their top forms the base of the roadway. These are unsubmerged culverts that belong to types 1, 2, and 3. For a trial selected size, the type of flow can be determined as follows:

1. For the design flow, determine the critical depth, d_c (Section 10.7.2), and the normal depth, d_n (Section 10.8.2).
2. Compare the depths above with the tailwater, h_4.
3. If $d_n < d_c$ and $h_4 < d_c$, it is type 1 flow. If $d_n > d_c$ and $h_4 < d_c$, it is type 2 flow. If $d_n > d_c$ and $h_4 > d_c$, it is type 3 flow.

The appropriate discharge equation of Table 12.21 is used to confirm the size and type. If not proper, then guided by the computed size, the trial may be repeated. In eqs. (12.20) through (12.25) in Table 12.21, the friction head loss between indicated sections is determined by Manning's equation, arranged as follows:

$$h_{ab} = \frac{n^2 V^2 L}{2.22 R^{4/3}} \quad \text{[unbalanced]} \quad (12.26a)$$

or

$$h_{ab} = \frac{29 n^2 L V^2}{R^{4/3} 2g} \quad \text{[unbalanced]} \quad (12.26b)$$

where

h_{ab} = friction head loss between a and b, ft

V = velocity of flow, ft/sec

L = length of section ab, ft

R = hydraulic radius, A/P, ft

The majority of culverts are designed for submerged conditions (types 4, 5, and 6), since the entrance is submerged at least with the peak rate of flow.

When a culvert is submerged by both headwaters and tailwaters, it is a type 4 condition in which eq. (12.23) is applicable. However, the distinction between types 5 and 6 when the tailwater is low is not as obvious.

To classify type 5 or 6 flow, the curves of Figures 12.19 and 12.20, which are adopted from Bodhaine (1982), are used. Figure 12.19 is applicable to a concrete barrel box or pipe culverts of square, rounded, or beveled entrances with or without wingwalls. Figure 12.20 is for rough (corrugated) pipes of circular or arch sections

TABLE 12.21 CLASSIFICATION OF CULVERT FLOW

(1) Category	(2) Type	(3) Culvert Slope	(4) Flow	(5) Control Section	(6) Discharge, Q	(7) Illustration
Unsubmerged, $H/D \leqslant 1.2$	1	Steep	Part full	Inlet	$C_d A_c \sqrt{2g(H + V_1^2/2g - d_c - h_{1,2})}$ (12.20)	
	2	Mild	Part full	Outlet	$C_d A_c \sqrt{2g(H + z + V_1^2/2g - d_c - h_{1,2} - h_{2,3})}$ (12.21)	
	3	Mild	Part full	Outlet	$C_d A_3 \sqrt{2g(H + z + V_1^2/2g - h_3 - h_{1,2} - h_{2,3})}$ (12.22)	

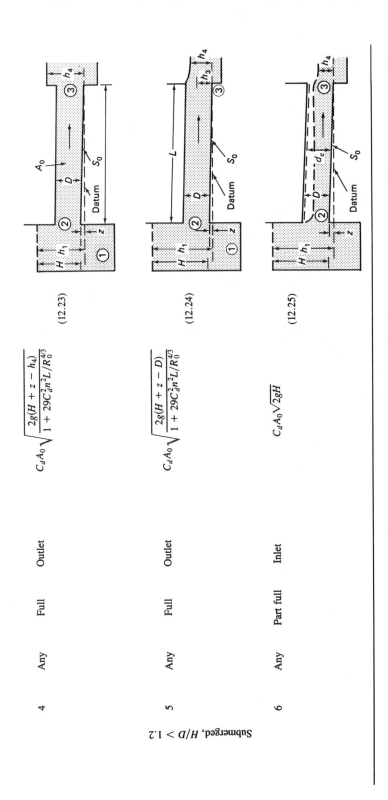

Submerged, $H/D > 1.2$

4	Any	Full	Outlet	$C_d A_0 \sqrt{\dfrac{2g(H + z - h_4)}{1 + 29C_d^2 n^2 L / R_0^{4/3}}}$	(12.23)
5	Any	Full	Outlet	$C_d A_0 \sqrt{\dfrac{2g(H + z - D)}{1 + 29C_d^2 n^2 L / R_0^{4/3}}}$	(12.24)
6	Any	Part full	Inlet	$C_d A_0 \sqrt{2gH}$	(12.25)

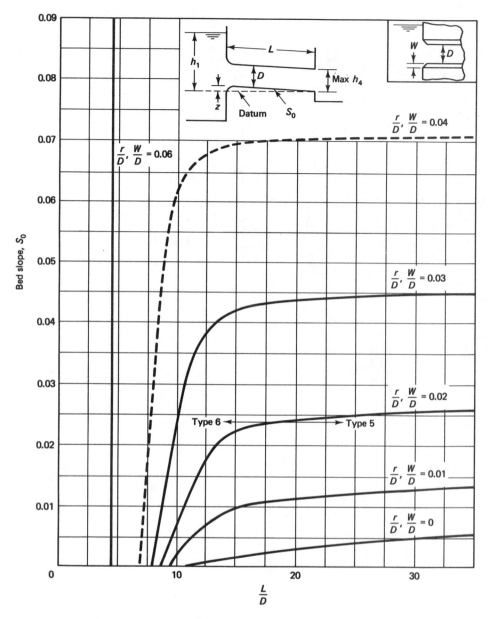

Figure 12.19 Criterion for classifying types 5 and 6 flow in box or pipe culverts with concrete barrels and square, rounded, or beveled entrances, either with or without wingwalls (from Bodhaine, 1982).

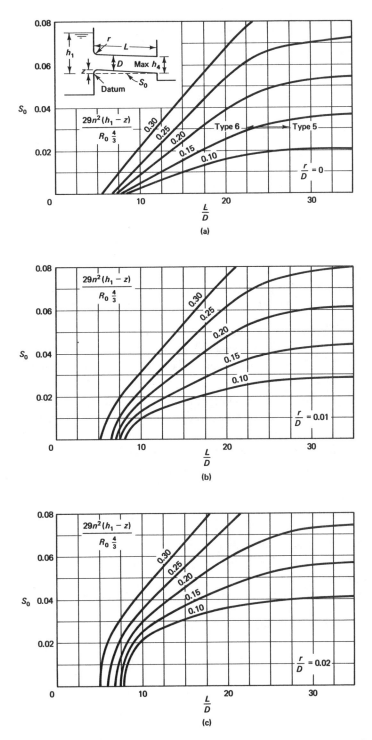

Figure 12.20 Criterion for classifying types 5 and 6 flow in pipe culverts with rough barrels (from Bodhaine, 1982).

Sec. 12.13   Roadway Drainage System

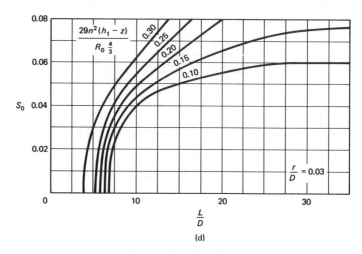

Figure 12.20 (contd.)

mounted flush in a vertical headwall with or without wingwalls. The procedure to classify type 5 or 6 is as follows:

1. Compute the ratios L/D, r/D or w/D, S_0, and $(29n^2H/R_0^{4/3}$ for rough pipes). r is the radius of rounding and w is the effective bevel.

2. For concrete pipes, select the curve of Figure 12.19 corresponding to r/D or w/D for the culvert.

3. For rough pipes, select from Figure 12.20 the graph corresponding to the value of r/D for the culvert and then select the curve corresponding to the $29n^2H/R_0^{4/3}$ computed for the culvert.

4. Plot the point defined by the computed values of S_0 and L/D for the culvert.

5. If the point plots to the right of the curve in step 2 or 3, the flow is type 5. If it plots to the left, the flow is type 6.

As in the case of bridge openings, the coefficient of discharge, C_d, is a function of many variables relating to type of flow, degree of channel contraction, and the geometry of the culvert entrance. The coefficient varies from 0.4 to 0.98. A systematic presentation has been made by Bodhaine (1982).

Example 12.16

A culvert section is shown in Figure 12.21 with upstream and downstream water levels. Design the culvert for a peak discharge of 120 cfs. For the culvert section, the corrugated metal pipe ($n = 0.024$) is to be used without rounding.

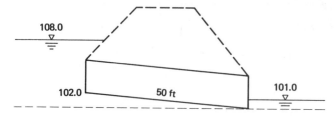

Figure 12.21 Submerged culvert section.

Solution

1. Consider a pipe section 4 ft in diameter.
2. $r/D = 0$, $S_0 = 2/50 = 0.04$, $L/D = 50/4 = 12.5$.

$$R_0 = D/4 = \frac{4}{4} = 1, \qquad \frac{29n^2H}{R_0^{4/3}} = \frac{29(0.024)^2(6)}{1} = 0.10$$

From Figure 12.20, flow is type 6.

3. Apply eq. (12.25):

$$A_0 = \frac{\pi}{4}(4)^2 = 12.56 \text{ ft}^2$$

$$H = 108 - 102 = 6$$

$$Q = 0.5(12.56)\sqrt{2(32.2)(6)}$$

$$= 123 \text{ cfs} \approx 120 \text{ cfs} \qquad \text{OK}$$

Example 12.17

Design a box culvert of concrete section ($n = 0.020$) to carry a discharge of 520 cfs for the condition shown in Figure 12.22. It has a square-edged entrance. The approach stream has a rectangular section of width 40 ft. $C_d = 0.95$.

Solution

1. Assume a 8-ft square section. The approach section will be one width (8 ft) upstream.
2. $H/D = 8/8 = 1.0 < 1.2$; unsubmerged case.
3. Determination of the critical depth:
   From eq. (10.9),

$$Z_c = \frac{Q}{\sqrt{g}} = 520/\sqrt{32.2} = 91.64$$

Since

$$Z_c = A^{3/2}/T^{1/2} = (8d_c)^{3/2}/(8)^{1/2} = 8.0d_c^{3/2},$$

$$8.0d_c^{3/2} = 91.64 \quad \text{or} \quad d_c = 5.09 \text{ ft}$$

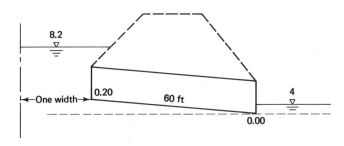

Figure 12.22 Unsubmerged box culvert section.

4. Determine the normal depth. From eq. (10.14),

$$AR^{2/3} = \frac{Qn}{1.49S^{1/2}} = \frac{520(0.015)}{1.49(0.0033)^{1/2}} = 91.13$$

But $AR^{2/3} = (8d_n)^{5/3}/(8 + 2d_n)^{2/3}$. Hence

$$\frac{(8d_n)^{5/3}}{(8 + 2d_n)^{2/3}} = 91.13$$

Solving either by trial and error or by plotting d versus $AR^{2/3}$, $d_n = 6.3$ ft.

5. Since $d_n > d_c$ but $h_4 < d_c$, it is type 2 flow.

6. $V_1 = \dfrac{Q}{A} = \dfrac{520}{8.2(40)} = 1.59$ ft/sec

$$\frac{V_1^2}{2g} = \frac{(1.59)^2}{2(32.2)} = 0.04 \text{ ft}$$

7. $h_{1,2} = \dfrac{n^2 V^2 L}{2.22 R_1^{4/3}}$

$$R_1 = \frac{40(8.2)}{40 + 2(8.2)} = 5.82 \text{ ft}$$

$$h_{1,2} = \frac{(0.013)^2(1.59)^2(8)}{2.22(5.82)^{4/3}} = \text{negligible}$$

8. $A_2 = 8(6.3) = 50.4$ ft^2

$$R_2 = \frac{50.4}{8 + 2(6.3)} = 2.45 \text{ ft}$$

$$V_2 = \frac{Q}{A_2} = \frac{520}{50.4} = 10.32 \text{ ft/sec}$$

$$h_{2-3} = \frac{(0.020)^2(10.32)^2(60)}{2.22(2.45)^{4/3}} = 0.36$$

9. From eq. (12.21), $A_c = 8(5.09) = 40.72$ ft^2.

$$Q = C_d A_c \sqrt{2g\left(H + z + \frac{V_1^2}{2g} - d_c - h_{1-2} - h_{2-3}\right)}$$

$$= 0.95(40.72)\sqrt{2(32.2)(8 + 0.2 + 0.04 - 5.09 - 0 - 0.36)}$$

$$= 519 \approx 520 \text{ cfs} \qquad \text{OK}$$

12.14 AIRPORT DRAINAGE SYSTEM

The objectives of the airport drainage system are (1) to collect and drain off surface water runoff, (2) to remove excess groundwater and lower the water table where it is too high, and (3) to protect all slopes from erosion.

The first objective is met by (1) properly grading the airport area so that all shoulders and slopes drain away from runways, taxiways, and all paved areas; (2) providing field storm drainage system serving all the depressed areas; and (3) constructing peripheral and other ditches to convey the outfall from the drainage system, to collect surface flows from the airport and adjoining sites, and to intercept ground-water flow from higher adjacent areas. Proper coordination of grading and draining is most desirable since a drainage system cannot function effectively unless the area is graded correctly to divert the surface into the drainage system. Similarly, ditches form an integral part of the drainage system.

Subsurface drainage is provided to take care of the second objective of diverting subterranean flows, lowering the water table, and controlling the moisture in the base and subbase of the pavements. Intercepting ditches or intercepting drainlines are provided to collect flows through the porous water-bearing stratum. For draining off the moisture pocketed in pervious soils over impervious stratum or in the low-lying areas of undulating impervious stratum, the subsurface drains are placed within wet masses of soil. It is desirable to place the best drainable soils adjacent to and beneath the paved areas to provide drainage away from the pavement. Less-drainable soils are placed in nontraffic areas. The draining of large areas through subsurface drainage system is usually not required on airports since it can be done more efficiently be grading properly and installing surface drainage (Federal Aviation Agency, 1965).

The cut and fill slopes face the problem of erosion. As a first step of protection, these slopes are made as flat as possible. Deep-cut slopes of over 10 ft, with higher ground above them, are provided with a cutoff ditch running back to the top-of-cut line and set back a few feet from the top of the bank to intercept the water flowing down from the higher ground. A ditch is constructed at the base of the bank to collect runoff. The cut slopes are protected by riprap, sod, grass, or vegetation. The fill slopes above 5 ft high are protected by constructing beams and gutters along the top of the slope to prevent water from running down the slope.

Only the surface storm drainage is discussed here. The design starts with a comprehensive study of the topography of the site and surrounding areas to identify surface and subsurface direction of flow, natural water courses, and outfalls. The topography affects the layout of the runways, taxiways, aprons, and buildings. The outline of the boundary of the airport is superimposed on the map. A plan is prepared from the topographic map, showing the contours of the finished grade and the location of such features as runways, taxiways, aprons, buildings, and roads. This is known as the drainage working drawing.

On the plan, the entire surface drainage system is sketched, showing all laterals, submains, and main storm sewers; direction of flow; gradients; and identifying each subarea, catch basin, inlet, gutter, shallow channel, manhole, and peripheral and outfall ditches.

The layout should cover all depressed areas in which overland flow will accumulate. Inlet structures are located at the lowest points within each field area. Each inlet is connected to the drainage line. The pipelines lead to the major outfalls.

Once a layout of inlets, manholes, and storm pipes has been made, determination of the area contributing to each inlet, tabulation of data, and computations of peak flows and drains capacity proceed in exactly the manner described in Section 12.9 for urban storm sewer design. This is illustrated in Example 12.18. Several different drainage layouts are necessary to select the most economic and effective system.

The rate of outflow from a drainage area is controlled by the capacity of the drain pipe. Whenever the rate of runoff to an inlet exceeds the drain capacity, ponding or temporary storage occurs. Where considerable low-lying flat field areas exist away from the pavements, the desirability of using ponding facility should be considered. This will reduce the size and/or number of drains. Also, this will act as a safety factor in the case of heavier-than-design storms. The volume of storage in ponding is determined by plotting against time the cumulated runoff corresponding to the design frequency and the cumulated discharge capacity of the drain pipe used (both in terms of volume). The optimum difference between the two curves is the storage volume of ponding. The time it takes to discharge the volume of ponding corresponds to the intersection point of the two curves.

Example 12.18

A surface (storm) drainage system for a part of an airport is shown in Figure 12.23. The finished contours, drainage layout, and length and slope of drains are marked on the figure. The computed tributary area, the composite runoff coefficient, and the time of inlet flow to each intercept are given in Table 12.22. The 5-year rainfall intensity in in./hr is given by $190/(t + 25)$, where t is in minutes. Manning's coefficient $n = 0.015$.

TABLE 12.22 DRAINAGE DATA FOR EXAMPLE 12.18

Intercept	Tributary Area (acres)	Weighted Runoff Coefficient	Time of Inlet Flow (min)
1	14.2	0.65	25.0
2	16.3	0.65	28.0
3	20.7	0.35	35.0
4	13.5	0.35	35.0
5	25.0	0.35	40.0

Solution

1. The design is performed in exactly the same manner as for the storm design in Section 12.9.
2. Computations for peak flows by the Rational Method and size of drains are arranged in Tables 12.23 and 12.24, respectively.

12.15 COMPUTER APPLICATIONS

The following list of software, along with their supporting documents, are related to drainage design and are available in microcomputer version for IBM compatibles.

The U.S. Soil Conservation Service has formulated a software based on their document TR-55. The software has a menu-driven interface to calculate the peak-flow by the Graphical Method or to generate a peak inflow hydrograph by the Tabular Method, as discussed in section 12.8, for small watersheds from 1 to 200 acres.

To design a network of storm sewer systems, the Texas State Department of Highways and Public Transportation has sponsored and Haestad Methods of Water-

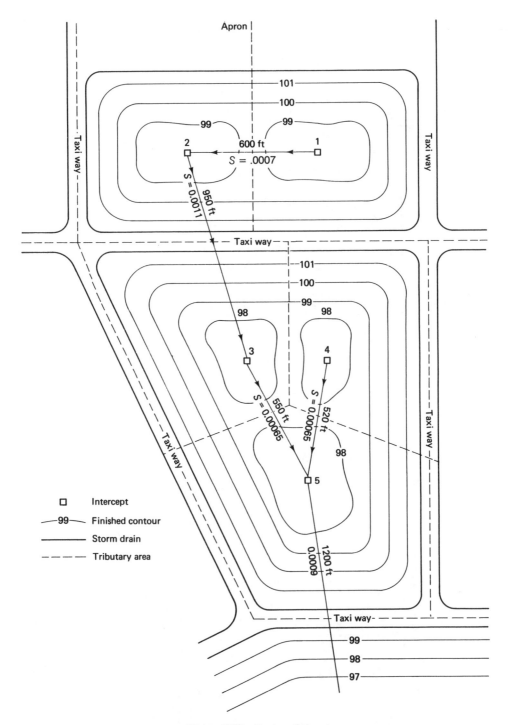

Figure 12.23 Section of airport.

TABLE 12.23 COMPUTATION OF PEAK DISCHARGE

(1) Line	(2) Location	(3) Intercept From	(4) Intercept To	(5) Tributary Area, a (acres)	(6) Coefficient, C	(7) aC (acres)	(8) ΣaC (acres)	(9) Route	(10) Travel Time (min) Inlet	(11) Travel Time (min) In Sewer[b]	(12) Total	(13) Intensity (in/hr)	(14) Q (cfs)
1		1	2	14.2	0.65	9.23	9.23	TA–1	25	0	25	3.8	35.1
2		2	3	16.3	0.65	10.60	19.83	TA–2	28	0	28	3.57	70.8
								1–2	25	3.18	28.18		
3		3	5	20.7	0.35	7.25	27.08	TA–3	35	0	35	3.17	85.8
								2–3	28.18	4.15	32.33		
4		4	5	13.5	0.35	4.73	4.73	TA–4	35	0	35	3.17	15.0
5		5	outlet	25.0	0.35	8.75	40.56[a]	TA–5	40	0	40	2.92	118.4
								4–5	35	2.89	37.89		
								3–5	35	2.85	37.85		

[a]Areas relating to (TA–5) + (drain 3–5) + (drain 4–5) = 8.75 + 27.08 + 4.73 = 40.56.
[b]From column 14 of Table 12.24.

TABLE 12.24 STORM SEWER DESIGN COMPUTATIONS

(1) Line	(2) (3) Intercept		(4) Design Flow (cfs)	(5) Length of Sewer (ft)	(6) (7) Surface Elevation		(8) Street Slope	(9) Maximum Diameter for Velocity of 3 ft/s[a] (in.)	(10) Diameter for Street Grade[b] (in.)	(11) Design Parameters Diameter (in.)	(12) Sewer Grade	(13) Velocity at Full[c] (ft/sec)	(14) Travel Time (min)[d]
	From	To			Upstream	Downstream							
1	1	2	35.1	600			0.0007	46	49	45	0.0011	3.14	3.18
2	2	3	70.8	950			0.0011	66	58	60	0.0011	3.82	4.15
3	3	5	85.8	550			0.00065	72	69	69	0.00065	3.22	2.85
4	4	5	15.0	520			0.00065	30	36	30	0.0017	3.0	2.89
5	5	outlet	118.4	1200			0.0009	85	74	75	0.0009	4.0	5.0

[a]$D = (1.274 Q/v)^{1/2}$ (continuity eq.).

[b]$D = \left(\dfrac{2.155 nQ}{S^{1/2}} \right)^{0.375}$ (Manning's eq.).

[c]$v = \dfrac{0.59}{n} D^{2/3} S^{1/2}$ (Manning's eq.).

[d]column 5/column 13.

bury, CT has marketed a program THYSYS that has a component to select inlet and pipe sizes of circular, arch or box type sections with up to 100 junctions in the network. Dodson and Associates, Inc., of Houston, TX has also prepared a software PIBS that computes runoff hydrographs, routes the hydrographs through a network of up to 500 storm sewer pipes and selects the pipe diameter for each sewer segment as a pressurized system. Both of these programs have separate components to design and analyze bridges and culverts. Programs for sewer designs are available from many other agencies as well. For detention pond design and analysis, POND-2 program of Haestad Methods has two options: 1) to route a specified inflow hydrograph through any shape and size of a pond and formulate an outflow hydrograph, and 2) to overlay specified inflow and outflow hydrographs and estimate the storage requirements by numerically integrating the area between the two hydrographs.

The Storm Water Management Model (SWMM) of the Environmental Protection Agency has a diverse array of routines. It can develop runoff hydrographs from rain or snowfall. These hydrographs are for analysis and design of a storm sewer network. SWMM can employ land use and population statistics to estimate the rate of sewage flow and to analyze a sanitary sewer network. It can model runoff pollutants and the treatment of sewage through the system.

PROBLEMS

12.1. From an area of 10 km^2 having a present population density of 8000 persons/km^2, determine the peak dry-weather flow if the domestic water consumption is 670 liters/person per day. The domestic consumption is expected to be 50% of the average total consumption. Determine the design flow if the density increases to 13,000 persons/km^2 at the end of the design period.

12.2. Wastewater to a sewer is contributed by two 500-acre areas. One area is sparsely populated, with 15 persons/acre, and the other is an apartment district with a heavy population of 150 persons/acre. For the total consumption rate of 160 gal/person per day in both areas, calculate the peak rate of sewage flow.

12.3. The composition of a district of 30 km^2 size is 60% residential area, 25% commercial zone, and 15% industrial zone. Twenty percent of the commercial zone is estimated to be covered by the buildings. The residential section has a density of 10,000 persons/km^2 and industrial zone of 50,000 persons/km^2. Determine the peak rate of flow. Assume the following parameters:

Domestic consumption = 300 liters/person per day

Commercial consumption = 12.2 liters/day per sq. meter of building area

Industrial consumption = 60 liters/person per shift

Number of shifts = 2

Length of sewer line = 10.2 km

12.4. A sewer has to be laid in a place where the ground has a slope of 1.75 in 1000 m. If the present and ultimate peak sewage discharge are 40 and 165 liters/sec, respectively, design the sewer section. Consider 2/3 full condition and $n = .012$.

12.5. Between two manholes 500 ft apart, the ground elevations are 100 ft and 99.25 ft, respectively. The present peak rate of sewage flow is 2.75 cfs, which is estimated to go up to 10.0 cfs at the end of 25 years. Design the sewer section.

12.6. The layout of a sanitary sewer system is as shown in Fig. P12.6. Data on area, length, and elevations are given below. The present population density, 40 persons/acre, is expected to rise to 100 per acre by conversion of the dwellings to apartments. The peak rate of sewage flow is 400 gpd per person. Design the sewer system, $n = .012$.

Block	Area (acres)	Length (ft)	Elevation (ft) Upstream	Downstream
A	2.0	390	101.50	97.17
B	2.5	350	100.67	97.17
C	1.5	330	97.17	93.29
D	1.3	230	98.69	97.54
E	1.2	295	100.5	97.54
F	5.7	650	97.54	94.29
G	2.1	300	94.29	93.29
H	3.5	550	93.29	86.42

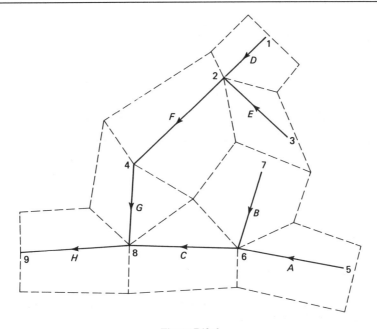

Figure P12.6

12.7. Design the sanitary sewer system for the city apartment district shown in Fig. P12.7. The length of the sewer segments, the tributary area of each, and the ground-level elevations are shown in the figure. The present density of population is 100,000 persons/km², which is expected to rise to 150,000/km² by the end of the design period. The maximum rate of sewage flow is 1500 liters/person per day.

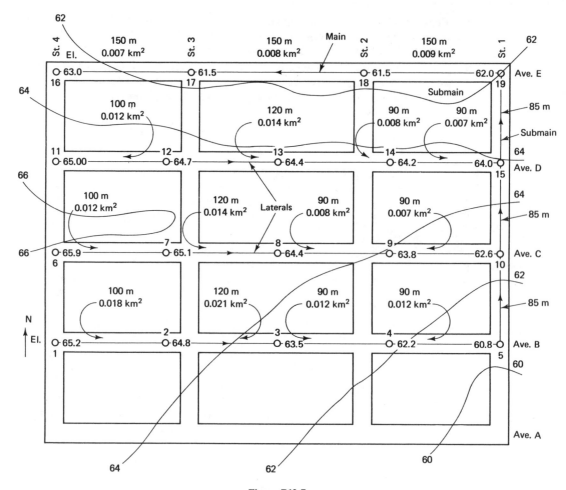

Figure P12.7

12.8. Determine the sewer arrangement of manhole 5 through 8 of Problem 12.6. Provide a cover of 10 ft.

12.9. Determine the arrangement of sewers at manholes 1 through 5, in Problem 12.7 allowing a cover of 2 m.

12.10. A drainage area consists of 30% turf ($C = 0.3$), 35% bare surface ($C = 0.4$), and 35% paved surface ($C = 0.9$). The time of concentration at the inlet point under consideration is 12 minutes. The total duration of the rainstorm is 3 hours. Determine the value of runoff coefficient corrected for antecedent rainfall condition.

12.11. A storm drain system is as shown in Fig. P12.11. For the flow conditions indicated, determine the time of concentration by the different methods. Assume that $A = 5230$ and $B = 30$ for intensity relation.

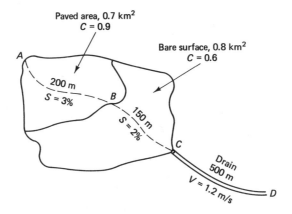

Figure P12.11

12.12. For the storm drain system shown in Fig. P12.12, determine the time of concentration at point C using Figure 12.7.

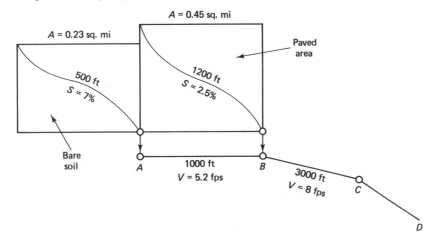

Figure P12.12

12.13. An urban watershed has a main ditch as shown in Fig. P12.13. Paths from remote points to the main ditch are also indicated. The details of each subarea are listed below. Determine the peak flow at the outlet ditch by the rational method, assuming that the 5-year rainfall intensity is given by $170/(t_c + 23.0)$.

Area	Drainage Area (acres)	Type of Surface	Path	Length (ft)	Slope (%)
1	12.0	Bare surface, $C = .4, CN = 75$	AB	1300	6.0
2	13.5	Asphalt paved, $C = 0.8$	BD	1250	1.5
			CD	1420	2.0
3	11.8	Lawn, $C = 0.3$	ED	1800	1.0
4	14.1	Concrete paved, $C = 0.9$	DG	1510	1.5
			FG	1660	2.5

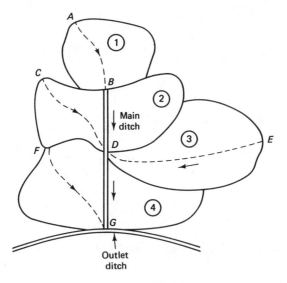

Figure P12.13

12.14. Assuming that Figure 12.6 reflects the intensity–duration–frequency relation for the watershed of Problem 12.13, determine the peak rate of runoff for 5-, 10-, and 20-year frequency.

12.15. Determine by the rational method the peak flow at the outfall of the watershed shown. The 5-year intensity relation is $190/(t_c + 25.0)$.

Given:

1. Area, $A_1 = 13.0$ acres, $C_1 = 0.6$.
    Area, $A_2 = 20.0$ acres, $C_2 = 0.4$.
    Area, $A_3 = 18.5$ acres, $C_3 = 0.5$.
2. Time of travel to inlet points I within each area $= 10$ min.
3. Inlet to manhole times (min): $I_1M_1 = 6.5$, $I_2M_2 = 12.1$, $I_3M_4 = 13.5$ min.
4. Average velocity of flow between manholes $= 4.1$ ft/sec.

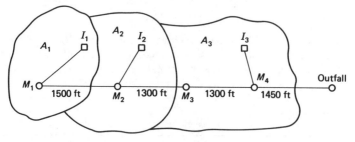

Figure P12.15

12.16. Solve Problem 12.13 by the SCS TR-55 method to determine the peak flow hydrograph. Assume type II rainfall distribution and the 5-year 24-hour rainfall of 3.5 in. Poor state soil group is A. Obtain tables of hydrograph unit discharges from the U.S. SCS publication *Urban Hydrology for Small Watersheds*, Technical Release 55 (1986), in your library.

Drainage System   Chap. 12

12.17. Solve Problem 12.15 by the SCS TR-55 method to determine the peak flow hydrograph. The entire area comprises the urban business district of hydrologic group A. Assume type III rainfall distribution and the 5-year 24-hour rainfall of 4 in. Obtain the tables from the source indicated in Problem 12.16.

12.18. A storm system consists of four areas with details as shown in Fig. P12.18. The direction of flow from each area and between the manholes is given by an arrow. Determine the 10-year peak rate of flow for each sewer section by the rational method, assuming that Figure 12.6 reflects the intensity–duration–frequency relation for the area. The travel time between each manhole = 5 min.

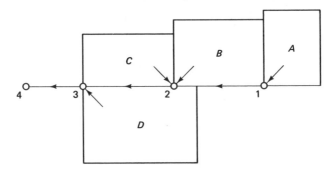

Figure P12.18

Drainage Unit	Area (acres)	Coefficient, C	Inlet Time (min)
A	5	0.6	8.0
B	5.4	0.7	9.2
C	2.5	0.7	7.0
D	6.5	0.5	14.5

12.19. For the drainage system shown in Fig. P12.19, determine the design flow for each sewer section by the rational method. The rainfall intensity (in./hr) is represented by $i = 100/(t_c + 15)$; t_c is in minutes. Flow from each area is shown by an arrow.

(a) Unit	Area (acres)	C	Inlet Time (min)
A	0.4	0.7	10
B	0.5	0.7	10
C	0.3	0.80	5
D	0.3	0.80	5
E	0.4	0.65	10
F	0.5	0.60	10
G	0.6	0.60	10

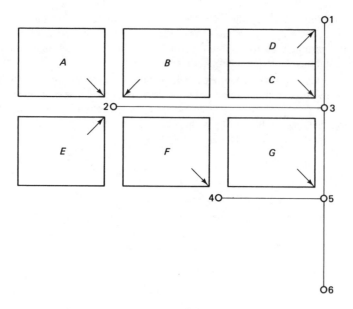

Figure P12.19

(b) Pipe Lengths	Feet	Slope
1–3	500	0.008
2–3	1500	0.01
3–5	600	0.0075
4–5	750	0.01
5–6	600	0.0075

(c) Average velocity through pipes = 4 ft/sec.

12.20. Determine the design discharge for the storm sewers between the manholes as shown

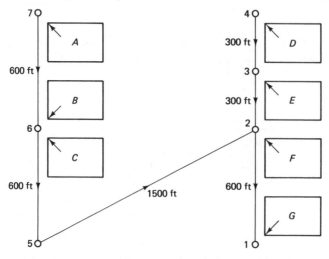

Figure P12.20

in Fig. P12.20. The design rainfall intensity is given by $i = 96/(t_c + 16)$ and the average velocity of flow is 3.5 ft/sec.

Block:	A	B	C	D	E	F	G
Area (acres)	10	9	8	10	9	8	12
C	0.65	0.7	0.55	0.6	0.5	0.45	0.4
Inlet Time (min)	10	9	7	11	10	8	10

12.21. Design the storm sewer system of Problem 12.19. Assume that $n = 0.013$.

12.22. For the part of a city in Problem 12.7, design the storm sewer system. The layout is shown in Fig. P12.22. Assume the following conditions.

1. The tributary areas and ground surface elevations are given in the figure.
2. The runoff coefficients are indicated in the figure.
3. The inlet time from each area is 20 minutes.
4. The design frequency is 2 years, for which the rainfall intensity (mm/hr) is given by $i = 2590/(t + 17)$, where t is in minutes.
5. Manning's $n = 0.013$.

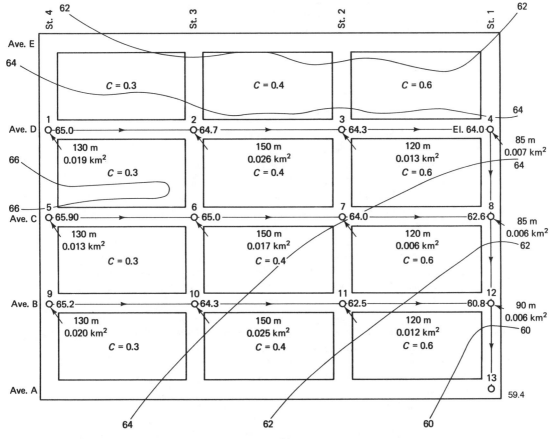

Figure P12.22

12.23. In an agriculture area, 6 in. of water is applied per irrigation application, 20% of which runs off. The infiltration rate of the soil in the upper root zone is 1.25 in./hr. If the hydraulic conductivity of the soil is 0.8 in./hr, determine the rise in the water table after the irrigation.

12.24. The irrigation application in an area is 5.0 in. The soil in the root zone has a sandy loam texture. About 30% of the irrigation water runs off. For a hydraulic conductivity of 0.4 in./hr, determine the water-table buildup.

12.25. For an agriculture drainage system, design the drains (determine the drain spacing) for the following conditions:

1. Depth from surface to impervious layer = 30 ft.
2. Depth of drains from surface = 8 ft.
3. Root zone or water-table requirement = 4 ft below the surface.
4. Uniform hydraulic conductivity = 10 ft/day.
5. Water application is as follows:

Event	Date	Time between Events (days)
Snowmelt	Apr. 22	
First irrigation	June 6	45
Second irrigation	July 1	25
Third irrigation	July 21	20
Fourth irrigation	Aug. 4	14
Fifth irrigation	Aug. 18	14
Sixth irrigation	Sept. 1	14

6. Each irrigation application, as well as spring snowmelt, contributed to a deep percolation of 1 in. (Adapted from the Bureau of Reclamation, 1984.)

12.26. In Problem 12.25, if the depth from the surface to the impervious layer is only 8 ft (i.e., the drains are located on the barrier), determine the drain spacing.

12.27. A 60-ft-wide asphalt-paved highway section has a longitudinal slope of 1% and a cross slope of $\frac{1}{4}$ in. to 1 ft. If the gutter inlet spacing is 180 ft, determine the peak flow at the inlet. The rainfall intensity (in./hr) is given by $i = 180/(t + 25)$, where t is in minutes. $C = 0.8$.

12.28. A highway section of 100 ft width traverses a suburb, where the intensity–duration–frequency curves of Figure 12.6 apply. The maximum 10-year intensity is 7 in./hr. A grated inlet located at station 285 + 95 has an elevation of 551.45 ft. Another inlet is located at station 284 + 05, with an elevation of 549.60 ft. The cross slope is 0.0208. Determine the inlet peak flow of 10-year frequency. $C = 0.7$.

12.29. A highway section is as shown in Fig. P12.29. The runoff is caught by the grated inlets at station 204 + 00 and at the gutter sumps at station 205 + 78.1. In addition, the runoff from the drainage area to the south of the highway is collected in the two intercepts. One intercept connects to the inlet at 204 + 00 and the other to a manhole at 205 + 95. The runoff from the south side of the highway is then conveyed under the highway, where the north side inlets are picked up. The accumulated runoff is discharged into a natural water course.

The tributary areas and their breakdown between pervious and impervious parts, length of drains, and surface slopes are indicated in the figure. The inlet time to

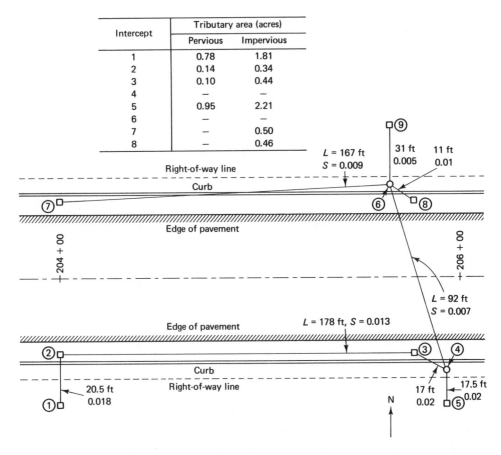

Intercept	Tributary area (acres)	
	Pervious	Impervious
1	0.78	1.81
2	0.14	0.34
3	0.10	0.44
4	–	–
5	0.95	2.21
6	–	–
7	–	0.50
8	–	0.46

Figure P12.29

the intercepts has been considered to be 5 min. The 10-year rainfall intensity (in./hr) is given by $i = 149/(t + 15.7)$, where t is in minutes. Design the longitudinal drainage system. $C = 0.3$ for pervious areas and 0.95 for impervious areas. Use concrete pipe ($n = 0.013$).

12.30. A culvert section is as shown in Fig. P12.30. Design the culvert of a beveled concrete pipe section ($n = 0.012$) to carry a peak discharge of 210 cfs. $C_d = 0.96$, $w = 0.15$ ft.

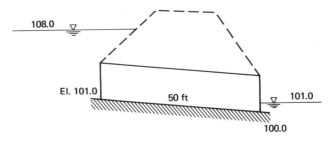

Figure P12.30

12.31. A culvert section is as shown in Fig. P12.31. Design a culvert of a rounded corrugated metal pipe ($n = 0.024$) to carry a peak flow of 125 cfs. $C_d = 0.5$, $r = 0.05$ ft.

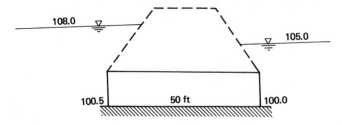

Figure P12.31

12.32. Design a circular corrugated-metal pipe ($n = 0.024$) culvert to carry a discharge of 250 cfs for the condition shown in Fig. P12.32. The approach stream has a width of 40 ft. $C_d = 0.90$.

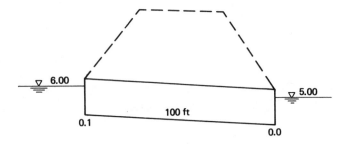

Figure P12.32

12.33. Design a surface drainage system for a part of the airport shown in Fig. P12.33. The finished contours, drains layout, and length and slope of the drains are as shown. The tributary area, weighted runoff coefficient, and time of inlet flow to each intercept point are listed below. The 5-year rainfall intensity (in./hr) is given by $i = 96/(t + 16)$, where t is in minutes. Use $n = 0.015$.

Intercept	Tributary Area (acres)	Weighted Coefficient, C	Time to Inlet (min)
1	15.0	0.40	25.0
2	16.5	0.40	26.0
3	25.0	0.40	30.0
4	12.0	0.40	20.0
5	30.0	0.35	40.0

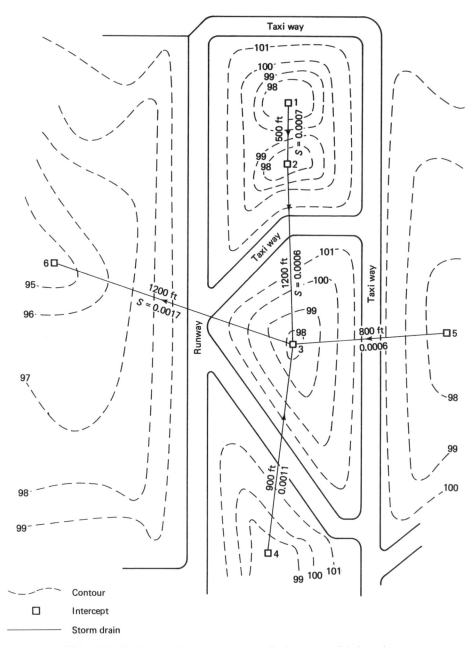

Figure P12.33 Portion of an airport showing final contour and drainage layout.

12.34. Determine the capacity of a detention basin if the peak rate of outflow for watershed in Problem 12.13 is limited to 50 cfs.

12.35. Determine the capacity of a detention basin in Problem 12.14 if the outflow from the basin is not to exceed 50 cfs at all levels of frequency.

12.36. Determine the size of detention basin for watershed of Problem 12.15 for restricting the outflow to 40 cfs.

Appendix A

TABLE A-1 LENGTH EQUIVALENTS

Unit	Equivalent					
	mm	m	in.	ft	yd	mi
Millimeter	1	10^{-3}	0.0394	0.00328	0.00109	6.214×10^{-7}
Meter	10^3	1	39.37	3.281	1.0936	6.214×10^{-4}
Inch	25.4	0.0254	1	0.0833	0.02778	1.578×10^{-5}
Foot	304.8	0.3048	12	1	0.333	1.894×10^{-4}
Yard	914.4	0.9144	36	3	1	5.682×10^{-4}
Mile	1.609×10^6	1.609×10^3	6.336×10^4	5280	1760	1

TABLE A-2 AREA EQUIVALENTS

Unit	Equivalent				
	in.^{2}	ft^2	m^2	acre	mi^2
Square inch	1	6.944×10^{-3}	6.452×10^{-4}	1.59×10^{-7}	2.491×10^{-10}
Square foot	144	1	0.0929	2.3×10^{-5}	3.587×10^{-8}
Square meter	1550	10.764	1	2.5×10^{-4}	3.861×10^{-7}
Acre	6.27×10^6	43,560	4047	1	1.56×10^{-3}
Square mile	4.014×10^9	2.788×10^7	2.59×10^6	640	1

TABLE A-3 VOLUME EQUIVALENTS

Unit			Equivalent			
	in^3	gal	ft^3	m^3	acre-ft	cfs-day
Cubic inch	1	0.00433	5.79×10^{-4}	1.64×10^{-5}	1.33×10^{-8}	6.70×10^{-9}
Gallon	231	1	0.134	0.00379	3.07×10^{-6}	1.55×10^{-6}
Cubic foot	1728	7.48	1	0.0283	2.30×10^{-5}	1.16×10^{-5}
Cubic meter	61,000	264	35.3	1	8.11×10^{-4}	4.09×10^{-4}
Acre-foot	7.53×10^7	3.26×10^5	43,560	1233	1	0.504
Cubic foot per second-day	1.49×10^8	6.46×10^5	86,400	2447	1.98	1

TABLE A-4 VELOCITY EQUIVALENTS

Unit			Equivalent		
	ft/sec	mi/hr	m/s	km/hr	kn
Feet per second	1	0.6818	0.3048	1.097	0.5925
Miles per hour	1.467	1	0.4470	1.609	0.8690
Meters per second	3.281	2.237	1	3.600	1.944
Kilometers per hour	0.9113	0.6214	0.2778	1	0.5400
Knots	1.688	1.151	0.5144	1.852	1

TABLE A-5 DISCHARGE EQUIVALENTS

Unit			Equivalent			
	gal/day	ft^3/day	gal/min	acre-ft/day	cfs	m^3/s
U.S. gallon per day	1	0.134	6.94×10^{-4}	3.07×10^{-6}	1.55×10^{-6}	4.38×10^{-8}
Cubic foot per day	7.48	1	5.19×10^{-3}	2.30×10^{-5}	1.16×10^{-5}	3.28×10^{-7}
U.S. gallon per minute	1440	193	1	4.42×10^{-3}	2.23×10^{-3}	6.31×10^{-5}
Acre · foot per day	3.26×10^5	43,560	226	1	0.504	0.0143
Cubic foot per second	6.46×10^5	86,400	449	1.98	1	0.0283
Cubic meter per second	2.28×10^7	3.05×10^6	15,800	70.0	35.3	1

TABLE A-6 PRESSURE EQUIVALENTS

Unit	Equivalent						
	ft H$_2$O	in. Hg	mm Hg	mbar	kPa	psi	kg/m^2
Feet of water (32°F)	1	0.883	22.42	29.89	2.989	0.4335	304.8
Inch of mercury (32°F)	1.133	1	25.40	33.86	3.386	0.4912	345.3
Millimeter of mercury (0°C)	0.0446	0.03937	1	1.333	0.1333	0.01934	13.60
Millibar	0.0335	0.02953	0.7501	1	0.1000	0.01450	10.20
Kilopascal (N/m$^2 \times 10^3$)	0.335	0.2953	7.501	10.00	1	0.1450	102.0
Pounds per square inch	2.307	2.036	51.71	68.95	6.895	1	703.1
Kilograms per square meter	0.00328	0.002896	0.07356	0.09807	0.009807	0.001422	1

TABLE A-7 ENERGY EQUIVALENTS

Unit	Equivalent					
	Btu	cal	J	kW-hr	ft-lb	hp-hr
British thermal unit (60°F)	1	252.0	1055	0.0002930	777.9	0.0003929
Calorie (15°C)	0.003969	1	4.186	1.163×10^{-6}	3.087	1.559×10^{-6}
Joule	0.0009482	0.2389	1	2.778×10^{-7}	0.7376	3.725×10^{-7}
Kilowatt-hour	3413	860,100	3.600×10^6	1	2.655×10^6	1.341
Foot-pound	0.001286	0.3239	1.356	3.766×10^{-7}	1	5.051×10^{-7}
Horsepower-hour	2545	641.300	2.685×10^6	0.7457	1.980×10^6	1

Appendix B

TABLE B-1 SOME OTHER USEFUL CONVERSION FACTORS

Multiply:	by:	to obtain:
Mass (kg)	9.81	Weight in newton
Pound	4.448	Newton (N)
Liter	1000	Cubic centimeter
Pound per ft^2	47.88	N/m^2 *or* pascal
Horsepower	745.7	Watt
	550	Foot-lb/sec
Std atmosphere	101.325	Kilopascal (kPa)
U.S. ton	2000	Pound
Nautical mile	1852	Meter
U.S. mile	1609	Meter
Square mile	2.59	Square kilometer
Square kilometer	100	Hectare (ha)
°F	$\frac{5}{9}(°F - 32)$	°C
Log to base e (i.e., $\log_e$, where $e = 2.718$)	0.434	Log to base 10 (i.e., $\log_{10}$)

Appendix C

TABLE C-1 PHYSICAL PROPERTIES OF WATER IN ENGLISH UNITS

Temp. (°F)	Specific gravity	Specific weight (lb/ft³)	Surface Tension (lb/ft)	Heat of vaporization (Btu/lb)	Viscosity Dynamic (lb-sec/ft²)	Viscosity Kinematic (ft²/sec)	Bulk modulus of elasticity (psi)	Vapor pressure in. Hg	Vapor pressure Millibar	Vapor pressure lb/in.²
32	0.99986	62.418	0.518×10^{-2}	1075.5	3.746×10^{-5}	1.931×10^{-5}	293×10^{3}	0.180	6.11	0.089
40	0.99998	62.426ᵃ	0.514	1071.0	3.229	1.664	294	0.248	8.39	0.122
50	0.99971	62.409	0.509	1065.3	2.735	1.410	305	0.362	12.27	0.178
60	0.99902	62.366	0.504	1059.7	2.359	1.217	311	0.522	17.66	0.256
70	0.99798	62.301	0.500	1054.0	2.050	1.058	320	0.739	25.03	0.363
80	0.99662	62.216	0.492	1048.4	1.799	0.930	322	1.032	34.96	0.507
90	0.99497	62.113	0.486	1042.7	1.595	0.826	323	1.422	48.15	0.698
100	0.99306	61.994	0.480	1037.1	1.424	0.739	327	1.933	65.47	0.950
120	0.98856	61.713	0.473	1025.6	1.168	0.609	333	3.448	116.75	1.693
140	0.98321	61.379	0.454	1014.0	0.981	0.514	330	5.884	199.26	2.890
160	0.97714	61.000	0.441	1002.2	0.838	0.442	326	9.656	326.98	4.742
180	0.97041	60.580	0.426	990.2	0.726	0.386	318	15.295	517.95	7.512
200	0.96306	60.121	0.412	977.9	0.637	0.341	308	23.468	794.72	11.526
212	0.95837	59.828	0.404	970.3	0.593	0.319	300	29.921	1013.25	14.696

ᵃMaximum specific weight is 62.427 lb/ft³ at 39.2°F.

TABLE C-2 PHYSICAL PROPERTIES OF WATER IN METRIC UNITS[a]

Temp. (°C)	Specific gravity	Density (g/cm³)	Surface tension (N/m)	Heat of vaporization (cal/g)	Viscosity Dynamic (poise)[b]	Viscosity Kinematic (stokes)[c]	Bulk modulus of elasticity (N/m²)	Vapor pressure mm Hg	Vapor pressure Millibar	Vapor pressure g/cm²
0	0.99987	75.6×10^{-3}	0.99984	597.3	1.79×10^{-2}	1.79×10^{-2}	2.02×10^{9}	4.58	6.11	6.23
5	0.99999	0.99996	74.9	594.5	1.52	1.52	2.06	6.54	8.72	8.89
10	0.99973	0.99970	74.2	591.7	1.31	1.31	2.10	9.20	12.27	12.51
15	0.99913	0.99910	73.5	588.9	1.14	1.14	2.14	12.78	17.04	17.38
20	0.99824	0.99821	72.8	586.0	1.00	1.00	2.18	17.53	23.37	23.83
25	0.99708	0.99705	72.0	583.2	0.890	0.893	2.22	23.76	31.67	32.30
30	0.99568	0.99565	71.2	580.4	0.798	0.801	2.25	31.83	42.43	43.27
35	0.99407	0.99404	70.4	577.6	0.719	0.723	2.27	42.18	56.24	57.34
40	0.99225	0.99222	69.6	574.7	0.653	0.658	2.28	55.34	73.78	75.23
50	0.98807	0.98804	67.9	569.0	0.547	0.554	2.29	92.56	123.40	125.83
60	0.98323	0.98320	66.2	563.2	0.466	0.474	2.28	149.46	199.26	203.19
70	0.97780	0.97777	64.4	557.4	0.404	0.413	2.25	233.79	311.69	317.84
80	0.97182	0.97179	62.6	551.4	0.355	0.365	2.20	355.28	473.67	483.01
90	0.96534	0.96531	60.8	545.3	0.315	0.326	2.14	525.89	701.13	714.95
100	0.95839	0.95836	58.9	539.1	0.282	0.294	2.07	760.00	1013.25	1033.23

[a]SI units:

Density: $kg/m^3 = g/cm^3 \times 10^3$.

Specific weight: $N/m^3 = $ density in $kg/m^3 \times 9.81$.

Dynamic viscosity: $N \cdot s/m^2 = $ poise $\times 10^{-1}$.

Kinematic viscosity: $m^2/s = $ stokes $\times 10^{-4}$.

Vapor pressure: $N/m^2 = $ millibar $\times 10^2$ or $g/cm^2 \times 98.1$.

[b]poise $= (g/cm \cdot s)$;

[c]stokes $= (cm^2/s)$.

Appendix D

TABLE D-1 PHYSICAL PROPERTIES OF AIR AT STANDARD ATMOSPHERIC PRESSURE

Temperature °F	Density, slugs/ft^3	Specific weight, lb/ft^3	Dynamic viscosity, lb-sec/ft^2	Kinematic viscosity, ft^2/sec
0	0.00268	0.0862	3.28×10^{-7}	1.26×10^{-4}
20	0.00257	0.0827	3.50	1.36
40	0.00247	0.0794	3.62	1.46
60	0.00237	0.0763	3.74	1.58
80	0.00228	0.0735	3.85	1.69
100	0.00220	0.0709	3.96	1.80
120	0.00215	0.0684	4.07	1.89
150	0.00204	0.0651	4.23	2.07
200	0.00187	0.0601	4.49	2.40
°C	kg/m^3	N/m^3	N · s/m^2	m^2/s
−20	1.39	13.6	1.56×10^{-5}	1.13×10^{-5}
−10	1.34	13.1	1.62	1.21
0	1.29	12.6	1.68	1.30
10	1.25	12.2	1.73	1.39
20	1.20	11.8	1.80	1.49
40	1.12	11.0	1.91	1.70
60	1.06	10.4	2.03	1.92
80	0.99	9.71	2.15	2.17
100	0.94	9.24	2.28	2.45

Appendix E

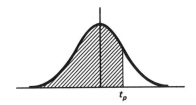

t_p

TABLE E-1 CUMULATIVE STUDENT t DISTRIBUTION

v	$t_{.999}$	$t_{.975}$	$t_{.95}$	$t_{.90}$	$t_{.75}$	$t_{.70}$	$t_{.60}$	$t_{.55}$
10	3.17	2.23	1.81	1.37	.700	.542	.260	.129
11	3.11	2.20	1.80	1.36	.697	.540	.260	.129
12	3.06	2.18	1.78	1.36	.695	.539	.259	.128
13	3.01	2.16	1.77	1.35	.694	.538	.259	.128
14	2.98	2.14	1.76	1.34	.692	.537	.258	.128
15	2.95	2.13	1.75	1.34	.691	.536	.258	.128
16	2.92	2.12	1.75	1.34	.690	.535	.258	.128
17	2.90	2.11	1.74	1.33	.689	.534	.257	.128
18	2.88	2.10	1.73	1.33	.688	.534	.257	.127
19	2.86	2.09	1.73	1.33	.688	.533	.257	.127
20	2.84	2.09	1.72	1.32	.687	.533	.257	.127
21	2.83	2.08	1.72	1.32	.686	.532	.257	.127
22	2.82	2.07	1.72	1.32	.686	.532	.256	.127
23	2.81	2.07	1.71	1.32	.685	.532	.256	.127
24	2.80	2.06	1.71	1.32	.685	.531	.256	.127
25	2.79	2.06	1.71	1.32	.684	.531	.256	.127
26	2.78	2.06	1.71	1.32	.684	.531	.256	.127
27	2.77	2.05	1.70	1.31	.684	.531	.256	.127
28	2.76	2.05	1.70	1.31	.683	.530	.256	.127
29	2.76	2.04	1.70	1.31	.683	.530	.256	.127
30	2.75	2.04	1.70	1.31	.683	.530	.256	.127
40	2.70	2.02	1.68	1.30	.681	.529	.255	.126
60	2.66	2.00	1.67	1.30	.679	.527	.254	.126
120	2.62	1.98	1.66	1.29	.677	.526	.254	.126
∞	2.58	1.96	1.645	1.28	.674	.524	.253	.126

Appendix F

TABLE F-1 CUMULATIVE F DISTRIBUTION (m NUMERATOR AND n DENOMINATOR DEGREES OF FREEDOM)

α	n	$m \longrightarrow$ 10	12	15	20	30	60	120	∞
.90		2.32	2.28	2.24	2.20	2.15	2.11	2.08	2.06
.95		2.98	2.91	2.84	2.77	2.70	2.62	2.58	2.54
.975	10	3.72	3.62	3.52	3.42	3.31	3.20	3.14	3.08
.99		4.85	4.71	4.56	4.41	4.25	4.08	4.00	3.91
.995		5.85	5.66	5.47	5.27	5.07	4.86	4.75	4.64
.90		2.19	2.15	2.10	2.06	2.01	1.96	1.93	1.90
.95		2.75	2.69	2.62	2.54	2.47	2.38	2.34	2.30
.975	12	3.37	3.28	3.18	3.07	2.96	2.85	2.79	2.72
.99		4.30	4.16	4.01	3.86	3.70	3.54	3.45	3.36
.995		5.09	4.91	4.72	4.53	4.33	4.12	4.01	3.90
.90		2.06	2.02	1.97	1.92	1.87	1.82	1.79	1.76
.95		2.54	2.48	2.40	2.33	2.25	2.16	2.11	2.07
.975	15	3.06	2.96	2.86	2.76	2.64	2.52	2.46	2.40
.99		3.80	3.67	3.52	3.37	3.21	3.05	2.96	2.87
.995		4.42	4.25	4.07	3.88	3.69	3.48	3.37	3.26
.90		1.94	1.89	1.84	1.79	1.74	1.68	1.64	1.61
.95		2.35	2.28	2.20	2.12	2.04	1.95	1.90	1.84
.975	20	2.77	2.68	2.57	2.46	2.35	2.22	2.16	2.09
.99		3.37	3.23	3.09	2.94	2.78	2.61	2.52	2.42
.995		3.85	3.68	3.50	3.32	3.12	2.92	2.81	2.69
.90		1.82	1.77	1.72	1.67	1.61	1.54	1.50	1.46
.95		2.16	2.09	2.01	1.93	1.84	1.74	1.68	1.62
.975	30	2.51	2.41	2.31	2.20	2.07	1.94	1.87	1.79
.99		2.98	2.84	2.70	2.55	2.39	2.21	2.11	2.01
.995		3.34	3.18	3.01	2.82	2.63	2.42	2.30	2.18
.90		1.71	1.66	1.60	1.54	1.48	1.40	1.35	1.29
.95		1.99	1.92	1.84	1.75	1.65	1.53	1.47	1.39
.975	60	2.27	2.17	2.06	1.94	1.82	1.67	1.58	1.48
.99		2.63	2.50	2.35	2.20	2.03	1.84	1.73	1.60
.995		2.90	2.74	2.57	2.39	2.19	1.96	1.83	1.69
.90		1.65	1.60	1.54	1.48	1.41	1.32	1.26	1.19
.95		1.91	1.83	1.75	1.66	1.55	1.43	1.35	1.25
.975	120	2.16	2.05	1.94	1.82	1.69	1.53	1.43	1.31
.99		2.47	2.34	2.19	2.03	1.86	1.66	1.53	1.38
.995		2.71	2.54	2.37	2.19	1.98	1.75	1.61	1.43
.90		1.60	1.55	1.49	1.42	1.34	1.24	1.17	1.00
.95		1.83	1.75	1.67	1.57	1.46	1.32	1.22	1.00
.975	∞	2.05	1.94	1.83	1.71	1.57	1.39	1.27	1.00
.99		2.32	2.18	2.04	1.88	1.70	1.47	1.32	1.00
.995		2.52	2.36	2.19	2.00	1.79	1.53	1.36	1.00

Glossary

Affinity laws   Equations that relate the performance of geometrically similar pumps.

Antecedent moisture   The degree of wetness of soil at the beginning of a runoff, determined by summation of weighted daily rainfall amounts for a period preceding the runoff.

Antecedent precipitation index   An index of moisture stored in a basin before a storm.

Aquiclude   Formation which, although porous and capable of absorbing water, does not transmit water at a sufficient rate to furnish an appreciable supply for a well or spring.

Aquifer   Porous water-bearing formation of permeable rock, sand, or gravel capable of yielding significant quantities of water.

Aquifer, leaky (semiconfined)   Aquifer overlain and/or underlain by a thin semipervious layer through which flow into or out of the aquifer can take place.

Aquifer, perched   Groundwater unit, generally of moderate dimensions, that occurs whenever a groundwater body is separated from the main groundwater supply by a relatively impermeable stratum and by the zone of aeration above the main water body.

Aquifuse   Formation that has no interconnected openings and hence cannot absorb or transmit water.

Aquitard   Formation of a rather impervious and semiconfining nature which transmits water at a very slow rate compared to an aquifer.

Area of influence   Area around a pumping well in which the water table (in an unconfined aquifer) and the piezometric surface (in a confined aquifer) are lowered by pumping.

Baseflow   Streamflow during dry periods which is contributed to the stream channel by the groundwater.

Basin   Drainage area of a stream or a lake.

Capacity (field or effective or water-holding or maximum field carrying capacity)   Amount of water held in a soil sample after the excess gravitational water has drained away.

Catchment area (basin)   *See* Drainage basin.

Cavitation   Formation of cavities filled with air and water vapor due to internal pressure reduced below atmosphere.

Cone of depression   Depression in the shape of an inverted cone, of the piezometric surface which defines the area of influence of a well.

Consumptive use   Use of water that results in a loss in the original quantity of water, such as absorption by crops.

Correlation coefficient   A measure of the goodness of fit of an equation assumed between the variables, on a scale of 0 to 1.

Critical depth; critical flow (discharge)   Flow conditions at which the specific energy for a given discharge is minimum (the Froude number is unity).

Depression storage   Volume of water that is required to fill small natural depressions to their full levels.

Detention (or **retarding**) **reservoir**   Reservoir with uncontrolled outlets to store floodwater temporarily.

Detention storage (or **surface detention**)   That part of precipitation which is stored temporarily en route to the stream; it includes surface and channel detention but does not include depression storage.

Direct flow   Flow of water promptly entering the stream channel, which includes the surface flow and interflow.

Discharge   Volume of fluid flowing through a cross section of a stream per unit time.

Double-mass curve   Plot of successive accumulated values of one variable against the contemporaneous accumulated values of another variable.

Drainage basin (area)   Drainage basin with reference to a point (section) represents an area from which runoff collects at that point; entire area having a common outlet for its runoff.

Duration curve   Graph representing the percentage of time during which the value of a given parameter (e.g., water level, discharge, etc.) is equaled or exceeded.

Equipotential line (or surface)   Line along which a potential (head) is constant.

Evaporation   Amount of water evaporated from a free water surface at a temperature below the boiling point.

Evapotranspiration   Amount of water transferred from the soil to the atmosphere by evaporation and plant transpiration.

Evapotranspiration, potential   Maximum quantity of water capable of being evaporated from the soil and transpired from the vegetation of a specified region in a given time interval under existing climatic conditions, expressed as depth of water.

Extreme value series   Hydrological series which includes the largest or smallest values, with each value selected from an equal time interval in the record.

Flood, maximum probable   Greatest flood that may be expected at a place, taking into account all pertinent factors of location, meteorology, hydrology, and terrain.

Flood routing   Process of determining progressively the timing and shape of a flood wave at successive points along a river or through a reservoir.

Flood, standard project   Discharge that may be expected from the most severe combination of meteorological and hydrological conditions that are considered reasonably characteristic of the geographical region involved, excluding extremely rare combinations.

Flow, laminar   Flow in well-defined flow lines in which the viscous force is predominant; in channels it occurs at Reynolds number smaller than 500–2000 and through porous media at Reynolds Number smaller than 1–10.

Flow net   A system of streamlines and orthogonal equipotential lines used to compute seepage.

Flow, nonuniform   Flow in which the velocity vector (magnitude and direction) is not constant from point to point.

Flow, steady   Flow in which the velocity vector does not change with time.

Flow, subcritical   Flow at velocity less than critical; at a Froude number less than unity.

Flow, supercritical   Flow with velocity greater than the critical velocity; at a Froude number greater than unity.

Flow, turbulent   Flow with turbulence; it occurs in channels at a Reynolds number larger than 5000.

Flow, uniform   Flow in which the velocities are the same in both magnitude and direction from point to point along the conduit or channel.

Flow, varied   Flow occurring in streams having a variable cross section or slope (space is the criterion).

Frequency analysis   Statistical procedure involved in interpreting the past record of a hydrological event to occurrences of that event in the future (e.g., estimates of frequencies of floods, droughts, storages, rainfall, water quality).

Frequency distribution   An arrangement of quantities pertaining to a single event, in order of magnitude and frequency of occurrence.

Friction losses   Total energy losses in the flow of water due to friction between the water and the walls of a conduit, channel, or porous medium, usually expressed in units of height.

Head, static discharge   Vertical distance between the centerline of a pump and the outlet (discharge flange) level.

Head total or energy head   The sum of the pressure, velocity and position heads above a datum.

Head, total dynamic   Total static head plus the friction head plus the velocity head.

Head, total static   The sum of the static suction lift and static discharge head.

Hydraulic conductivity   The capability of a rock or soil to transmit water under pressure; it defines the relationship, called Darcy's law, between the discharge and the hydraulic gradient causing it.

Hydraulic gradient   In a closed conduit, the slope of the line joining the elevations to which water would rise in piezometric pipes; in an open channel, the slope of the water surface; in a porous medium, the decrease in piezometric head per unit distance in the direction of flow.

Hydraulic radius   The wetted area of the cross section of a stream or conduit divided by its wetted perimeter.

Hydraulic transients   Interim stage when a flow changes from one steady-state condition to another steady-state condition because of a sudden acceleration or deceleration of flow.

Hydrograph (discharge hydrograph, flood hydrograph, sediment hydrograph, stage hydrograph, well hydrograph)   A graph showing hydrologic observed data against time at a given point (e.g., stage, discharge, sediment load).

Hydrograph, composite unit   Superposition of unit hydrographs for various subdivisions of a large area, with the times of beginning of rise lagged by the times of travel from the outlets of the subdivisions to a major gaging site.

Hydrograph, compound   Hydrograph from a storm which is made up of more than one substorm.

Hydrograph, instantaneous unit   Unit hydrograph resulting from the net precipitation (of 1 in.) applied to the basin instantly (in an infinitesimally short time).

Hydrograph, S-curve   Hydrograph that would result from an infinite series of runoff increments of unit rate (e.g. 1 inch in *t* hours), obtained by adding a series of unit hydrographs each lagged by *t* hours with respect to the preceding one.

Hydrograph, unit   Discharge hydrograph resulting from a unit runoff (net rain of 1 in.) of a specified duration over the watershed.

Hydrologic cycle   Succession of stages through which water passes from the atmosphere to the earth and returns back to the atmosphere.

Hydrometric network   Network of stations at which measurement of hydrological parameters is performed.

Hyetograph   Chart displaying temporal distribution of precipitation during a storm; also, graph displaying the intensity of precipitation versus time.

Infiltration capacity   Maximum rate at which water can be absorbed for a given soil per unit surface under given conditions.

Initial detention   That part of precipitation that does not appear either as infiltration or runoff at the time active runoff begins. It includes interception by vegetal cover, depression storage, and evaporation during precipitation, but does not include surface detention.

Interception   Precipitation that is caught and held by vegetation or structures and then lost by evaporation without reaching the ground.

Interflow   That portion of the precipitation which has not passed down to the water table but is discharged from the area as subsurface flow into streams.

Lag time   Time between center of mass of rainfall to center of mass of runoff or to the peak of runoff.

Limb (rising or **falling)**   The part of the hydrograph in which the discharge is steadily increasing or decreasing.

Net positive suction head   Force available to drive the flow into the pump, which is equal to the total head at the suction side of a pump minus the vapor pressure head of the liquid being pumped.

Net (storm) rain   Portion of rainfall during a storm which reaches a stream as direct surface flow.

Partial duration series   Events, such as floods, occurring above a selected base value without regard to the number within a given period.

Permeability, intrinsic   Property of a porous medium allowing for the movement of liquids and gases through the medium under the combined action of gravity and pressure; it is related to the coefficient of permeability by properties of the fluid.

Phreatic surface (or **groundwater surface**)   *See* Water table.

Piezometric head (surface)   An imaginary surface coinciding with the static level to which water from a confined aquifer will rise if a vertical opening is made through the confining layer.

Plant capacity factor   Ratio of the average energy output of an electric generating plant to the installed capacity.

Precipitation, effective   In agriculture, that portion of the rainfall that remains in the .soil and contributes to crop growth.

Precipitation, initial   Precipitation at the beginning of a storm before the depression storage is fully filled up.

Precipitation, intensity   Amount of precipitation collected in a unit time interval.

Precipitation, probable maximum   Amount of precipitation that is the physical upper limit for a given duration over a particular basin.

Probabilistic process   Process in which the probability of occurrence of the variables is taken into consideration (their sequence of occurrence is ignored).

Probability distribution   Frequency distribution divided by the total number of occurrences, provided that the latter tends to infinity.

Pump characteristics curves   A group of curves showing head-capacity, efficiency-capacity, and power input-capacity relationships for a pump. These are supplied by the manufacturer.

Pump (head) curve   Relationship between the head developed by a pump and the capacity (flow) for a constant rotative speed.

Rainfall (Precipitation), Excess   Rainfall available for direct runoff.

Rainfall (Precipitation) intensity–area curve   Curve showing the relationship between average rainfall depth (or rate) and the area over which it occurs for a given storm duration.

Rainfall (Precipitation) intensity–duration curve   Curve showing the relationship between average rainfall depth (or rate) and storm duration in a given area.

Rainfall (Precipitation) intensity–duration–frequency curves   Curves showing relationship between rainfall intensity and duration for different levels of frequency; each curve represents the rainfall intensity–duration which will be equaled or exceeded once in a certain number of years, indicated as the frequency of that curve.

Rainfall (Precipitation) intensity frequency   Average time interval between the occurrence of rainfall of given or greater intensity.

Rainfall, maximization   (1) **Moisture Maximization:** The process of adjusting precipitation upward to a theoretical value that would have pertained if the moisture content of the air had been at the maximum with other storm conditions remaining unchanged. (2) **Sequential Maximization:** Reducing the observed elapsed time between storms to develop a hypothetical severe precipitation sequence. (3) **Spatial Maximization:** Reducing the distance between precipitation storms for a hypothetical severe sequence.

Rain (precipitation), residual   Rain that falls at the end of a storm at a rate less than the infiltration capacity.

Rating curve (stage–discharge relation)   Curve showing the relation between stage and discharge of a stream at a given gaging station.

Recession hydrograph (curve)   Falling limb of a hydrograph; two stages are distinguished: (a) recession curve during surface runoff; (b) recession curve when the surface runoff is stopped (groundwater recession curve), which reflects the decrease of subsurface flow, and in later stages, the exhaustion of groundwater storage.

Regression analysis   A procedure to establish relationships between variables.

Retention   That part of the precipitation falling on a drainage area that does not escape as a surface streamflow, during a given period.

Return period (or **recurrence interval**)   Average interval of the time or number of years within which an event will be equaled or exceeded (e.g., flood peak discharge).

Runoff   The part of precipitation that flows toward a stream on the ground surface (surface runoff) or within the soil (subsurface runoff).

Saturation vapor pressure   Maximum possible partial pressure of water vapor in the atmosphere at a given temperature.

Similarity, dynamic   Exists between two geometrically similar units when the ratios of inertial force to the individual forces in the first unit are the same as the corresponding ratios in the second unit at the corresponding points in space.

Similarity, geometric   Exists between two units if the ratios of all corresponding dimensions of the two units are equal.

Soil-moisture deficiency (or **deficit**)   Difference between the water-holding capacity of the soil and the instantaneous soil moisture.

Soil-moisture retention   Part of soil moisture retained by surface tension and molecular forces against the influence of gravity.

Specific capacity   Ratio of discharge of a well to drawdown.

Specific discharge   Rate of flow of groundwater divided by the normal cross-section area of flow section.

Specific drawdown   Drawdown in a well per unit discharge.

Specific energy   Sum of the piezometric head and the velocity head; total energy, with respect to the bottom of a conduit or channel as a datum.

Specific speed   Parameter used to define the pump type; it is the speed at which a geometrically scaled-down model will discharge a unit flow under a unit head at maximum efficiency.

Specific storage Volume of water removed or added within the unit volume of an aquifer per unit change of head.

Specific yield Amount of water removed from the soil per unit horizontal area and unit drawdown.

Stage–discharge relation *See* Rating curve.

Static suction lift Vertical distance between the source (suction level) and the centerline of a pump.

Stochastic process Process in which both the probability and sequence of occurrence of the variables are taken into account.

Storage coefficient (or **storativity**) Volume of water removed from or added to an artesian aquifer per unit horizontal area and per unit change of head.

Storm, design Rainfall amount and distribution adopted over a given drainage area, used in determining the design flood.

Streamline (flowline) The path followed by a particle of water as it moves through a saturated soil mass.

Submergence Condition of a weir when the elevation of the water surface on the downstream side is equal to or higher than that of the weir crest.

Surface detention That part of the rain that remains on the ground surface during rain and either runs off or infiltrates after the rain ends, not including depression storage.

System head curve A curve of system head comprising total static head and head loss in the system versus flow through the system.

Terminal velocity Final velocity of falling soloid particles in water or in air or of raindrops in air.

Time of concentration Period of time required for storm runoff to flow from the most remote point of a drainage basin to the outlet.

Transmissivity (also, **coefficient of transmissivity**) Rate at which water is transmitted through a unit width of the aquifer under a unit hydraulic gradient; it is expressed as the product of the hydraulic conductivity and thickness of the saturated portion of the aquifer.

Transpiration Process by which water from vegetation is transferred into the atmosphere in the form of vapor.

Velocity potential Mathematical scalar function at a point such that its positive or negative gradient at a point yields the velocity vector at that point.

Vena contracta Minimum cross section of a jet of fluid discharging from an orifice or over a weir.

Wastewater Water containing liquid or solid matter discharged as waste.

Water hammer Very rapid pressure wave in a conduit due to a sudden change in flow.

Water loss (1) Sum of water lost from a given land area during a specified time by transpiration, evaporation, and interception. (2) In irrigation, seepage and evaporation from land and ditches; excess water drained from the land surfaces and the deep percolation.

Water table The surface separating the upper unsaturated soil from the lower saturated soil.

Water yield Total runoff from a drainage basin through surface channels and aquifers.

Well capacity (or **potential yield**) Maximum rate at which a well will yield water under a stipulated set of conditions, such as given drawdown.

Well function Mathematical function by means of which the unsteady drawdown can be computed at a given point in an aquifer at a given time due to a given constant rate of pumping from a well.

Wetted perimeter For a channel cross-section perpendicular to flow, the length of channel surface in contact with water.

Yield (1) Streamflow in a given time interval derived from a unit area of drainage basin. (2) Quantity of water that can be collected for a given use from surface or groundwater sources in a basin in a given time interval.

References

CHAPTER ONE

AMERICAN SOCIETY OF CIVIL ENGINEERS, *Principles of Project Formulation for Irrigation and Drainage Projects,* Technical Committee of the Irrigation and Drainage Division, ASCE, New York, 1982.

CLARK, J. W., VIESSMAN, W., and HAMMER, M. J., *Water Supply and Pollution Control,* 3rd ed., IEP, A Dun-Donnelley Publisher, New York, 1977.

GOODMAN, A. S., *Principles of Water Resources Planning,* Prentice-Hall, Inc., Englewood Cliffs, N.J., 1984.

GRIGG, N. S., *Water Resources Planning,* McGraw-Hill Book Company, New York, 1985.

KUIPER, E., *Water Resources Development: Planning, Engineering and Economics,* Butterworth & Company (Publishers) Ltd., London, 1965.

LINSLEY, R. K., and FRANZINI, J. B., *Water Resources Engineering,* 3rd ed., McGraw-Hill Book Company, New York, 1979.

UNITED NATIONS, *Water Resources Project Planning,* Economic Committee for Asia and Far East (Bangkok), Water Resources Series No. 41, U.N., New York, 1972.

U.S. INTER-AGENCY COMMITTEE ON WATER RESOURCES, *Proposed Practices for Economic Analysis of River Basin Projects,* May 1950.

U.S. NATIONAL WATER COMMISSION, *Water Resources Planning,* Consulting Panel on Water Resources Planning, 1972.

U.S. PRESIDENT'S WATER RESOURCES COUNCIL, *Policies, Standards and Procedures in the Formulation, Evaluation and Review of Plans for Use and Development of Water and Related Land Resources,* Sen. Doc. 97, 1962.

U.S. WATER RESOURCES COUNCIL, "Principles and Standards for Planning Water and Related Land Resources," *Fed. Reg.,* v. 38, n. 174, Sept. 10, 1973.

U.S. WATER RESOURCES COUNCIL, "Procedures for Evaluation of National Economic Development Benefits and Cost in Water Resources Planning," *Fed. Reg.,* v. 44, n. 242, Dec. 14, 1979; revised, v. 45, n. 190, Sept. 29, 1980.

U.S. WATER RESOURCES COUNCIL, "Economic and Environmental Principles and Guidelines for Water and Related Land Resources Implementation Studies," *Fed. Reg.,* v. 47, n. 55, Mar. 22, 1982.

U.S. WATER RESOURCES COUNCIL, "Economic and Environmental Principles and Guidelines for Water and Related Land Resources Implementation Studies," *Fed. Reg.,* v. 48, n. 48, Mar. 10, 1983.

VIESSMAN, W., and WELTY, C., *Water Management Technology and Institutions,* Harper & Row, Publishers, Inc., New York, 1985.

CHAPTER TWO

AMERICAN WATER WORKS ASSOCIATION, *Energy and Water Use Forecasting,* AWWA, Denver, Colo., 1980.

AMERICAN SOCIETY OF CIVIL ENGINEERS, *Consumptive Use of Water and Irrigation Water Requirements,* Technical Committee of the Irrigation and Drainage Division, ASCE, New York, 1973.

AMERICAN SOCIETY OF CIVIL ENGINEERS, "Describing Irrigation Efficiency and Uniformity," On-Farm Irrigation Committee of the Irrigation and Drainage Division, *ASCE J. Irrig. Drain. Div.,* 104, n. IR1, pp. 35–42, 1978.

AMERICAN SOCIETY OF CIVIL ENGINEERS, *Principles of Project Formulation for Irrigation and Drainage Projects,* Technical Committee of the Irrigation and Drainage Division, ASCE, New York, 1982.

BLANEY, H. F., "Monthly Consumptive Use Requirements for Irrigated Crops," *ASCE J. Irrig. Drain. Div.,* v. 85, n. IR1, pp. 1–12, 1959.

BLANEY, H. F., and CRIDDLE, W. D., *Determining Water Requirements in Irrigated Areas from Climatological Data,* U.S. Dept. of Agriculture, Soil Conservation Service, Washington, D.C., 1945.

BLANEY, H. F., and CRIDDLE, W. D., *Determining Consumptive Use and Irrigation Water Requirements,* Tech. Bull. 1275, U.S. Dept. of Agriculture, Washington, D.C., 1962.

CARL, K. J., YOUNG, R. A., and ANDERSON, G. C., "Guidelines for Determining Fire-Flow Requirements," *J. Am. Water Works Assoc.,* v. 65, May 1973.

CRIDDLE, W. D., "Methods of Computing Consumptive Use of Water," *ASCE J. Irrig. Drain. Div.,* v. 84, n. IR1, pp. 1–27, 1958.

DAVIS, D. W., *Technical Factors in Small Hydropower Planning,* Tech. Paper 61, Hydrologic Engineering Center, U.S. Army Corps of Engineers, Davis, Calif., 1979.

DAVIS, D. W., and SMITH, B. W., *Feasibility Analysis in Small Hydropower Planning,* Technical Committee of the Irrigation and Drainage Division, ASCE, New York, 1982.

DAVIS, C. V., and SORENSON, K. E. (eds.), *Handbook of Applied Hydraulics,* 3rd ed., Secs. 30, 31, 33, and 36, McGraw-Hill Book Company, New York, 1969.

FAIR, G. M., GEYER, J. C., and OKUN, D. A., *Water and Wastewater Engineering,* vol. 1, *Water Supply and Wastewater Removal,* John Wiley & Sons, Inc., New York, 1966.

FAO/UNENSCO, *Irrigation, Drainage, and Salinity: An International Source Book,* Hutchinson Publishing Group Ltd., London, 1973.

FREDRICH, A. J., *Hydroelectric Power Analysis in Reservoir Systems,* Tech. Paper 24, Hydrologic Engineering Center, U.S. Army Corps of Engineers, Davis, Calif., 1970.

FRITZ, J. J., *Small and Mini Hydropower Systems,* McGraw-Hill Book Company, New York, 1984.

GOODMAN, A. S., *Principles of Water Resources Planning,* Prentice-Hall, Inc., Englewood Cliffs, N.J., 1984.

HANSEN, V. E., ISRAELSON, O. W., and STRINGHAM, G. E., *Irrigation Principles and Practices,* 4th ed., John Wiley & Sons, Inc., New York, 1980.

HARGREAVES, G. H., "Water Requirements and Man-Induced Climate Change," *ASCE J. Irrig. Drain. Div.,* v. 107, n. IR3, pp. 247–255, 1981.

HART, F. C., "Experience with Seepage Control in the Pacific Northwest," *Proc. of the Seepage Symposium,* ARS 41–90, pp. 93–97, Phoenix, Ariz., 1963.

HELWEG, O. J., and ALVAREZ, D., "Estimating Irrigation Water Quantity and Quality," *ASCE J. Irrig. Drain. Div.,* v. 106, n. IR3, pp. 175–188, 1980.

HOLTZ, D., and SEBASTIAN, S., *Municipal Water Systems,* Indiana University Press, Bloomington, Ind., 1978.

JENSEN, M. E., and HAISE, H. R., "Estimating Evapotranspiration from Solar Radiation," *ASCE J. Irrig. Drain. Div.*, v. 89, n. IR4, pp. 15–41, 1963.

KUIPER, E., *Water Resources Development: Planning, Engineering and Economics,* Butterworth & Company (Publishers) Ltd., London, 1965.

LINSLEY, R. K., "Two Centuries of Water Planning Methodology," *ASCE J. Water Resour. Plan. Manage. Div.*, v. 105, n. WR1, pp. 39–45, 1979a.

LINSLEY, R. K., "Hydrology and Water Resources Planning 1776–1976," *ASCE J. Water Resour. Plan. Manage. Div.*, v. 105, n. WR1, pp. 113–121, 1979b.

LINSLEY, R. K., and FRANZINI, J. B., *Water Resources Engineering,* 3rd., McGraw-Hill Book Company, New York, 1979.

MCCLEAN, J. E., "More Accurate Population Estimates by Means of Logistic Curves," *Civ. Eng.*, pp. 35–37, Feb. 1952.

MCJUNKIN, F. E., "Population Forecasting by Sanitary Engineers," *ASCE J. Sanit. Eng. Div.*, v. 90, n. SA4, pp. 31–55, 1964.

MUNSON, W. C., "Method for Estimating Consumptive Use of Water for Agriculture," *ASCE J. Irrig. Drain. Div.*, v. 88, n. IR4, pp. 45–57, 1960.

NATIONAL RESEARCH COUNCIL, *Water Resources,* Publication 1000-B, National Academy of Sciences, National Research Council, Washington, D.C., 1962.

STEEL, E. W., and MCGHEE, T. J., *Water Supply and Sewerage,* 5th ed., McGraw-Hill Book Company, New York, 1979.

UNITED NATIONS, *The Demand for Water: Procedures and Methodologies for Projecting Water Demands in the Context of Regional and National Planning,* Water Resources Series No. 3, U.N., New York, 1976.

UNITED NATIONS, *Water as a Factor in Energy Resources Development,* Economic and Social Comm. for Asia and the Pacific (Bangkok, Thailand), Water Resources Series No. 60, U.N., New York, 1985.

U.S. NATIONAL WATER COMMISSION, *Forecasting Water Demands,* compiled by R. G. Thompson and others, Report PB 206 491, U.S. NWC, Springfield, Va., 1971.

U.S. SOIL CONSERVATION SERVICE, *Irrigation Water Requirements,* Tech. Release 21, U.S. Dept. of Agriculture, Washington, D.C., 1964.

U.S. SOIL CONSERVATION SERVICE, *Crop Consumptive Irrigation Requirements and Irrigation Efficiency Coefficients for the U.S.,* U.S. Dept. of Agriculture, Washington, D.C., 1976.

U.S. WATER RESOURCES COUNCIL, *The Nation's Water Resources,* U.S. Government Printing Office, Washington, D.C., 1968.

VIESSMAN, W., and HAMMER, M. J., *Water Supply and Pollution Control,* 4th ed., Harper & Row, Publishers, Inc., New York, 1985.

VIESMAN, W., and WEILTY, C., *Water Management, Technology and Institutions,* Harper & Row, Publishers, Inc., New York, 1985.

WARNICK, C. C., *Hydropower Engineering,* Prentice-Hall, Inc., Englewood Cliffs, N. J., 1984.

WORSTELL, R. V., "Estimating Seepage Losses from Canal Systems," *ASCE J. Irrig. Drain. Div.*, v. 102, n. IR1, pp. 137–147, 1975.

CHAPTER THREE

ALLEY, W. M., "Treatment of Evapotranspiration, Soil Moisture Accounting and Aquifer Recharge in Monthly Water Balance Models," *Water Resour. Res.*, v. 20, n. 8, pp. 1137–1149, Aug. 1984.

BOUWER, H., "Rapid Field Measurement of Air-Entry Value and Hydraulic Conductivity of Soil as Significant Parameters in Flow System Analysis," *Water Resour. Res.*, v. 2, n. 4, pp. 729–738, 1966.

BUTLER, S. S., *Engineering Hydrology*, Prentice-Hall, Inc., Englewood Cliffs, N.J., 1957.

CHOW, V. T. (ed.), *Handbook of Applied Hydrology*, Secs. 9–12, 14, McGraw-Hill Book Company, New York, 1964.

CHU, S. T., "Infiltration during Unsteady Rain," *Water Resour. Res.*, v. 14, n. 3, pp. 461–466, 1978.

COOK, H. L., "The Infiltration Approach to the Calculation of Surface Runoff," *Trans. Am. Geophys. Union*, v. 27, n. V, pp. 726–743, Oct. 1946.

CRIDDLE, W. D., "Methods of Computing Consumptive Use of Water," *ASCE J. Irrig. Drain. Div.*, v. 84, n. IR1, 1958.

EAGLESON, P. S., *Dynamic Hydrology*, McGraw-Hill Book Company, New York, 1970.

GRAY, D. M. (ed.), *Handbook on the Principles of Hydrology*, National Research Council of Canada, Water Information Center, Inc., Port Washington, N.Y., 1973.

GREEN, W. H., and AMPT, G., "Studies of Soil Physics, Part I: The Flow of Air and Water through Soils," *J. Agric. Sci.*, v. 4, n. 1, pp. 1–24, 1911.

HAAN, C. T., "A Water Yield Model for Small Watersheds," *Water Resour. Bull.*, v. 8, n. 1, pp. 58–69, 1972.

HANKS, R. J., "Model for Predicting Plant Yield as Influenced by Water Use," *Agron. J.*, v. 66, pp. 600–665, 1974.

HARBECK, G. E., *A Practical Field Technique for Measuring Reservoir Operation Utilizing Mass Transfer Theory*, Professional Paper 272-E, U.S. Geological Survey, Washington, D.C., 1962.

HARBECK, G. E., and CRUSE, R. R., *Evaporation Control Research, 1955–58*, Water Supply Paper 1480, U.S. Geological Survey, Washington, D.C., 1960.

HJELMFELT, A. T., and CASSIDY, J. J., *Hydrology for Engineers and Planners*, Iowa State University Press, Ames, Iowa, 1975.

HOLTON, H. N., et al., *USDAHL-74 Revised Model of Watershed Hydrology*, Tech. Bull. 1518, Agriculture Research Service, U.S. Dept. of Agriculture, Washington, D.C., 1975.

HORTON, R. E., "Analyses of Runoff-Plat Experiments with Varying Infiltration Capacity," *Trans. Am. Geophys. Union*, pt. 4, pp. 693–711, 1939.

HORTON, R. E., "A Simplified Method of Determining the Constants in the Infiltration-Capacity Equation," *Trans. Am. Geophys. Union*, pp. 575–577, 1942.

HUGGINS, L. F., and MONKE, E. J., *The Mathematical Simulation of the Hydrology of Small Watersheds*, Tech. Rep. 1, Water Resources Research Center, Purdue Univ., West Lafayette, Ind., 1966.

HYDROLOGIC ENGINEERING CENTER, *Hydrograph Analysis*, Hydrologic Engineering Methods for Water Resources Development, Vol. 4, U.S. Army, Corps of Engineers, Davis, Calif., 1973.

KOHLER, M. A., and PARMELE, L. H., "Generalized Estimates of Free Water Evaporation," *Water Resour. Res.*, v. 3, n. 4, pp. 997–1005, 1967.

KOHLER, M. A., NORDENSON, T. J., and FOX, W. E., *Evaporation From Pans and Lakes*, Research Paper 38, U.S. Weather Bureau, Washington, D.C., 1955.

LANGBEIN, W. B., et al., *Annual Runoff in the United States*, Circular 52, U.S. Geological Survey, Washington, D.C., 1949.

LARSON, L. W., and PECK, E. L., "Accuracy of Precipitation Measurements for Hydrologic Modeling," *Water Resour. Res.*, v. 10, n. 4, Aug. 1974.

LINSLEY, R. K., "A Simple Procedure for the Day-to-Day Forecasting of Runoff from Snow-Melt," *Trans. Am. Geophys. Union*, pp. 62–66, 1943.

LINSLEY, R. K., KOHLER, M. A., and PAULHUS, J. L. H., *Hydrology for Engineers*, 3rd ed., McGraw-Hill Book Company, New York, 1982.

MATHER, J. R., "Using Computed Streamflow in Watershed Analysis," *Water Resour. Bull.*, v. 17, n. 3, pp. 474–482, 1981.

References

MEIN, R. G., and LARSON, C. L., *Modeling Infiltration Component of the Rainfall-Runoff Process*, Bull. 43, Water Resources Research Center, Univ. of Minnesota, Minneapolis, Minn., 1971.

MEIN, R. G., and LARSON, C. L., "Modeling Infiltration during a Steady Rain," *Water Resour. Res.*, v. 9, n. 2, pp. 384–394, 1973.

MILLER, D. H., *Water at the Surface of the Earth*, Academic Press, Inc., New York, 1977.

MOREL-SEYTOUX, H. J., "Two Phase Flows in Porous Media," *Adv. Hydrol.*, v. 9, 1973.

MOREL-SEYTOUX, H. J., and KHANJI, J., "Derivation of an Equation of Infiltration," *Water Resour. Res.*, v. 10, n. 4, pp. 795–800, 1974.

MUSTONEN, S. E., and MCGUINNESS, J. L., "Lysimeter and Evapotranspiration," *Water Resour. Res.*, v. 3, n. 4, pp. 989–996, Feb. 1967.

OSBORN, H. B., "Estimating Precipitation in Mountainous Regions," *ASCE J. Hydraul. Div.*, v. 110, n. 12, pp. 1859–1863, 1984.

PALMER, W. C., *Meteorological Drought*, Research Paper 45, U.S. Weather Bureau, Washington, D.C., 1965.

PHILIP, J. R., "Theory of Infiltration," *Adv. Hydrosci.*, v. 5, 1969.

RAWLS, W. J., BRAKENSIEK, D. L., and MILLER, N., "Green-Ampt Infiltration Parameters from Soil Data," *ASCE J. Hydraul. Div.*, v. 109, n. 1, pp. 62–70, 1983.

REINHART, K. G., and TAYLOR, R. E., "Infiltration and Available Water Storage Capacity in the Soil," *Trans. Am. Geophys. Union*, v. 35, n. 5, pp. 791–794, 1954.

RIPPLE, C. D., RUBIN, J., and VAN HYLCKAMA, T. E. A., *Estimating Steady-State Evaporation Rates from Bare Soils under Conditions of High Water Table*, Water Supply Paper 2019-A, U.S. Geological Survey, Washington, D.C., 1972.

ROBINSON, T. W., and JOHNSON, A. I., *Selected Bibliography on Evaporation and Transpiration*, Water Supply Paper 1539-R, U.S. Geological Survey, Washington, D.C., 1961.

SEARCY, J. K., and HARDISON, C. H., *Double Mass Curves*, Water Supply Paper 1541-B, U.S. Geological Survey, Washington, D.C., 1960.

SHERMAN, L. K., "Derivation of Infiltration Capacity from Average Loss Rates," *Trans. Am. Geophys. Union*, v. 21, pp. 541–550, 1940.

SHERMAN, L. K., and MUSGRAVE, G. W., "Infiltration" in *Physics of the Earth, IX: Hydrology*, ed. O. E. Meinzer, McGraw-Hill Book Company, New York, 1942.

SKAGGS, R. W., and KHALEEL, R., "Infiltration," in *Hydrologic Modeling of Small Watersheds*, ed. C. T. Haan, H. P. Johnson, and D. L. Brakensiek, American Society of Agricultural Engineers, St. Joseph, Mich., 1982.

SMITH, R. E., and PARLANGE, J. Y., "A Parameter-Efficient Hydrologic Infiltration Model," *Water Resour. Res.*, v. 14, n. 3, pp. 533–538, 1978.

SOKOLOV, A. A., and CHAPMAN, T. G., *Methods for Water Balance Computations*, Studies and Reports in Hydrology, UNESCO, Paris, 1974.

STEEL, E. W., and MCGHEE, T. J., *Water Supply and Sewerage*, 5th ed., McGraw-Hill Book Company, New York, 1979.

STURROCK, A. M., *Evaporation and Radiation Measurements at Salton Sea, California*, Water Supply Paper 2053, U.S. Geological Survey, Washington, D.C., 1978.

THOMAS, H. A., *Improved Methods for National Water Assessment*, Report, Contract WR 15249270, Harvard Water Resources Group, Harvard Univ., Cambridge, Mass., 1981.

THORNTHWAITE, C. W., and MATHER, J. R., *The Water Balance*, publication of the Climatology Lab., *Climatol. Drexel Inst. Technol*, v. 8, n. 1, 1955.

U.S. ARMY CORPS OF ENGINEERS, *Runoff from Snowmelt*, Engineering Manual 1110-2-1406, Washington, D.C., 1960.

U.S. ARMY CORPS OF ENGINEERS, *Snow Hydrology*, North Pacific Div., Portland, Oreg., 1965.

U.S. BUREAU OF RECLAMATION and U.S. FOREST SERVICE, *Factors Affecting Snowmelt and Streamflow*, U.S. Dept. of Interior, Washington, D.C., 1958.

U.S. GOVERNMENT, *National Handbook of Recommended Methods for Water Data Acquisition,* Chap. 8, "Evaporation and Transpiration," 1982; prepared by agencies of the U.S. government under the sponsorship of the Water Data Coordinator, U.S. Geological Survey, Virginia, 1977.

U.S. SOIL CONSERVATION SERVICE, *National Engineering Handbook,* Sec. 4, *Hydrology,* U.S. Dept. of Agriculture, Washington, D.C., 1972.

U.S. SOIL CONSERVATION SERVICE, *Engineering Field Manual for Soil Conservation Service Practice,* U.S. Dept. of Agriculture, Washington, D.C., 1975.

U.S. SOIL CONSERVATION SERVICE, *Urban Hydrology for Small Watersheds,* Tech. Release 55, U.S. Dept. of Agriculture, Washington, D.C., 1986.

U.S. WEATHER BUREAU, *Rainfall Frequency Atlas of the United States,* Tech. Paper 40, U.S. Weather Bureau, Washington, D.C., 1961.

VIESSMAN, W., KNAPP, J. W., and LEWIS, G. L., *Introduction to Hydrology,* 2nd ed., IEP, A Dun-Donnelley Publisher, New York, 1977.

WEBB, E. K., "A Pan-Lake Evaporation Relationship," *J. Hydrol.,* v. 4, pp. 1–11, 1966.

WEISS, L. L., and WILSON, W. T., "Evaluation of Significance of Slope Changes in Double-Mass Curves," *Trans. Am. Geophys. Union,* v. 34, n. 6, pp. 893–896, Dec. 1953.

WINTER, T. C., "Uncertainties in Estimating the Water Balance of Lakes," *Water Resour. Bull.,* v. 17, n. 1, pp. 82–115, Feb. 1981.

WORLD METEOROLOGICAL ORGANIZATION, *Manual for Depth–Area–Duration Analysis of Storm Precipitation,* WMO No. 237, Tech. Paper 129, WMO, Geneva, 1969.

WORLD METEOROLOGICAL ORGANIZATION, *Preparation of Maps of Precipitation and Evaporation with Special Regard to Water Balance,* WMO, Geneva, 1970.

CHAPTER FOUR

BENNETT, G. D., *Introduction to Groundwater Hydraulics,* Techniques of Water Resources Investigations, Chap. B2, Book 3, U.S. Geological Survey, Washington, D.C., 1976.

BOWEN, R., *Groundwater,* Applied Science Publishers Ltd., Barking, Essex, England, 1980.

COOPER, H. H., "The Equation of Groundwater Flow in Fixed and Deforming Coordinates," *J. Geophys. Res.,* v. 71, n. 20, pp. 4785–4790, 1966.

DAVIS, S. N., and DEWIEST, R. J. M., *Hydrogeology,* John Wiley & Sons, Inc., New York, 1966.

DEWIEST, R. J. M., *Geohydrology,* John Wiley & Sons, Inc., New York, 1965.

DEWIEST, R. J. M., "On the Storage Coefficient and Equation of Groundwater Flow," *J. Geophys. Res.,* v. 71, n. 4, pp. 1117–1122, 1966.

DEWIEST, R. J. M. (ed.), *Flow through Porous Media,* Academic Press, Inc., New York, 1969.

FREEZE, A. R., and CHERRY, J. A., *Groundwater,* Prentice-Hall, Inc., Englewood Cliffs, N.J., 1979.

GUPTA, R. S., *Groundwater Reservoir Operation for Drought Management,* Ph.D. dissertation, Polytechnic Institute of New York, 1983.

HANTUSH, M. S., "Depletion of Storage, Leakage, and River Flow in Gravity Wells in Sloping Sand," *J. of Geophys. Res,* v. 69, n. 12, p. 2551–2560, 1964.

HEATH, R. C., *Basic Ground-Water Hydrology,* Water Supply Paper 2220, U.S. Geological Survey, Washington, D.C., 1983.

JACOB, C. E., "Flow of Groundwater" in *Engineering Hydraulics,* ed. H. Rouse, John Wiley & Sons, Inc., New York, 1950.

JORGENSEN, D. G., *Relationship between Basic Soil Engineering Equations and Basic Ground-Water Flow Equations,* Water Supply Paper 2064, U.S. Geological Survey, Washington, D.C., 1980.

LINSLEY, R. K., KOHLER, M. A., and PAULHUS, J. L. H., *Hydrology for Engineers,* 3rd ed., McGraw-Hill Book Company, New York, 1982.

LOHMAN, S. W., *Ground-Water Hydraulics,* Professional Paper 708, U.S. Geological Survey, Washington, D.C., 1972.

MARIÑO, M. A., and LUTHIN, J. N., *Seepage and Groundwater,* Developments in Water Science Series No. 13, Elsevier Science Publishing Co., Inc., New York, 1982.

McWHORTER, D. B., and SUNADA, D. K., *Ground-Water Hydrology and Hydraulics,* Water Resources Publications, Fort Collins, Colo., 1977.

MEINZER, O. E., *Outline of Groundwater Hydrology,* Water Supply Paper 494, U.S. Geological Survey, Washington, D.C., 1923 (reprint 1960).

PINNEKER, E. V. (ed.), *General Hydrogeology,* Cambridge University Press, Cambridge, 1983.

TODD, D. K., *Groundwater Hydrology,* 2nd ed., John Wiley & Sons, Inc., New York, 1980.

UNESCO/WMO, *International Glossary of Hydrology,* Secretariat of World Meteorological Organization, Switzerland, 1974.

U.S. DEPARTMENT OF INTERIOR, Water and Water Resources Service, *Groundwater Manual,* U.S. Government Printing Office, Washington, D.C., 1977 (reprint 1981).

VERRUIJT, A., *Theory of Groundwater Flow,* Gordon and Breach, Science Publishers, Inc., New York, 1970.

VIESSMAN, W., KNAPP, J. W., and LEWIS, G. L., *Introduction to Hydrology,* 2nd ed., IEP, A Dun-Donelley Publisher, New York, 1977.

WALTON, W. C., *Groundwater Resources Evaluation,* McGraw-Hill Book Company, New York, 1970.

CHAPTER FIVE

BENNETT, G. D., *Introduction to Ground-Water Hydraulics,* Techniques of Water Resources Investigations, Chap. B2, Book 3, U.S. Geological Survey, Washington, D.C., 1976.

BENTALL, RAY (compiler), *Methods of Determining Permeability, Transmissibility and Drawdown,* Water-Supply Paper 1536-I, U.S. Geological Survey, Washington, D.C., 1963.

BOWEN, R., *Ground Water,* Applied Science Publishers Ltd., Barking, Essex, England, 1980.

BROWN, R. H., et al. (eds.), *Ground-Water Studies*: *An International Guide for Research and Practice,* UNESCO, Paris, 1977.

CHOW, V. T., "On the Determination of Transmissibility and Storage Coefficients from Pumping Test Data," *Trans. Am. Geophys. Union,* v. 33, pp. 397–404, 1952.

DAVIS, S. N., and DEWIEST, R. J. M., *Hydrogeology,* John Wiley & Sons, Inc., New York, 1966.

DEWIEST, R. J. M., *Geohydrology,* John Wiley & Sons, Inc., New York, 1965.

DEWIEST, R. J. M. (ed.), *Flow through Porous Media,* Academic Press, Inc., New York, 1969.

FREEZE, A. R., and CHERRY, J. A., *Groundwater,* Prentice-Hall, Inc., Englewood Cliffs, N.J., 1979.

GUPTA, R. S., *Groundwater Reservoir Operation for Drought Management,* Ph.D., dissertation, Polytechnic Institute of New York, 1983.

HAMMER, M. J., and MACKICHAN, K. A., *Hydrology and Quality of Water Resources,* John Wiley & Sons, Inc., New York, 1981.

HANTUSH, M. S., "Hydraulics of Wells," in *Advances in Hydroscience,* Vol. I, ed. V. T. Chow, Academic Press, Inc., New York, 1964a.

HANTUSH, M. S., "Depletion of Storage, Leakage and River Flow by Gravity Wells in Sloping Sand," *J. Geophys. Res.,* v. 69, n. 12, pp. 2551–2560, 1964b.

HARR, M. E., *Groundwater and Seepage,* McGraw-Hill Book Company, New York, 1962.

HEATH, R. C., *Basic Ground-Water Hydrology,* Water Supply Paper 2220, U.S. Geological Survey, Washington, D.C., 1983.

JACOB, C. E., "Flow of Ground Water," in *Engineering Hydraulics,* ed. H. Rouse, John Wiley & Sons, Inc., New York, 1950.

JENKINS, C. T., *Computation of Rate and Volume of Stream Depletion by Wells,* Techniques of Water Resources Investigations, Chap. D1, Book 4, U.S. Geological Survey, Washington, D.C., 1968.

LANG, S. M., *Methods for Determining the Proper Spacing of Wells in Artesian Aquifers,* Water-Supply Paper 1545-B, U.S. Geological Survey, Washington, D.C., 1961.

LI, W. -H., *Differential Equations of Hydraulic Transients, Dispersion and Groundwater Flow,* Prentice-Hall, Inc., Englewood Cliffs, N.J., 1972.

LINSLEY, R. K., KOHLER, M. A., and PAULHUS, J. L. H., *Hydrology for Engineers,* 3rd ed., McGraw-Hill Book Company, New York, 1982.

LOHMAN, S. W., *Ground-Water Hydraulics,* Professional Paper 708, U.S. Geological Survey, Washington, D.C., 1972.

MARIÑO, M. A., and LUTHIN, J. N., *Seepage and Groundwater*, Developments in Water Science Series No. 13, Elsevier Science Publishing Co., Inc., New York, 1982.

McWHORTER, D. B., and SUNADA, D. K., *Ground-Water Hydrology and Hydraulics,* Water Resources Publications, Fort Collins, Colo., 1977.

POLUBARINOVA-KOCHINA, P. YA., *Theory of Groundwater Movement* (translated from Russian by R. J. M. DeWiest), Princeton University Press, Princeton, N.J., 1962.

PRICKETT, T.A., and LONNGUIST, C. G., *Selected Digital Computer Techniques for Groundwater Resource Evaluation,* Bulletin 55, Illinois State Water Survey, Dept. of Registration and Education, Urbana, Ill., 1971.

RAMSAHOYE, L. E., and LANG, S. M., *A Simple Method for Determining Specific Yield from Pumping Tests,* Water-Supply Paper 1536-C, U.S. Geological Survey, Washington, D.C., 1961.

REED, J. E., *Type Curves for Selected Problems of Flow to Wells in Confined Aquifers,* Techniques of Water Resources Investigations, Chap. B3, Book 3, U.S. Geological Survey, Washington, D.C., 1980.

SPANGLER, M. G., and HANDY, R. L., *Soil Engineering,* Harper & Row, Publishers, Inc., New York, 1982.

STALLMAN, R. W., *Aquifer-Test Design Observation and Data Analysis,* Techniques of Water Resources Investigations, Chap. B1, Book 3, U.S. Geological Survey, Washington, D.C., 1971.

TODD, D. K., "Groundwater," in *Handbook of Applied Hydrology*, ed. V. T. Chow, McGraw-Hill Book Company, New York, 1964.

TODD, D. K., *Groundwater Hydrology*, 2nd ed., John Wiley & Sons, Inc., New York, 1980.

TRESCOTT, P. C., *Documentation of Finite-Difference Model for Simulation of Three-Dimensional Groundwater Flow*, Open File Report 75–438, U.S. Geological Survey, Washington, D.C., 1975 (reprint July 1976).

TRESCOTT, P. C., PINDER, G. F., and LARSON, S. P., *Finite-Difference Model for Aquifer Simulation in Two Dimensions with Results of Experiments*, Techniques of Water Resources Investigations, Chap. C1, Book 7, U.S. Geological Survey, Washington, D.C., 1976.

U.S. BUREAU OF RECLAMATION, *Studies of Groundwater Movement*, Technical Memorandum 657, U.S. Dept. of Interior, Denver, Colo., 1960.

U.S. DEPARTMENT OF INTERIOR, Water and Power Resources Service, *Ground Water Manual,* U.S. Government Printing Office, Washington, D.C., 1977 (reprint 1981).

VERRUIJT, A., *Theory of Groundwater Flow*, Gordon and Breach, Science Publishers, Inc., New York, 1970.

VIESSMAN, W., KNAPP, J. W., and LEWIS, G. L., *Introduction to Hydrology*, 2nd ed., IEP, A Dun-Donnelley Publisher, New York, 1977.

DE VRIES, J. J., "Some Calculation Methods for Determination of the Travel Time of Groundwater," *Aqua-Vu,* ser. A, no. 5, Communications of the Institute of Earth Sciences, Free

Reformed University, Amsterdam, 1972.

DE VRIES, J. J., "Groundwater Hydraulics," *Aqua-Vu,* ser. A, no. 6, Communications of the Institute of Earth Sciences, Free Reformed University, Amsterdam, 1975.

WALTON, W. C., *Groundwater Resources Evaluation,* McGraw-Hill Book Company, New York, 1970.

WANG, H. F., and ANDERSON, M. P., *Introduction to Groundwater Modeling: Finite Difference and Finite Element Methods,* W. H. Freeman and Company, Publishers, San Francisco, 1982.

WISLER, C. O., and BRATER, E. F., *Hydrology,* John Wiley & Sons, Inc., New York, 1959.

WRIGHT, C. E. (ed.), *Surface Water and Ground-Water Interaction,* UNESCO, Paris, 1980.

CHAPTER SIX

BOYER, M. C., "Streamflow Measurement," in *Handbook of Applied Hydrology,* ed. V. T. Chow, McGraw-Hill Book Company, New York, 1964.

BUCHANAN, T. J., and SOMERS, W. P., *Discharge Measurements at Gaging Stations,* Techniques of Water Resources Investigations, Chap. A8, Book 3, U.S. Geological Survey, Washington, D.C., 1969.

CARTER, R. W., and DAVIDIAN, J., *General Procedure for Gaging Streams,* Techniques of Water Resources Investigations, Chap. A6, Book 3, U.S. Geological Survey, Washington, D.C., 1968.

CHOW, V. T., *Open-Channel Hydraulics,* McGraw-Hill Book Company, New York, 1959.

CORBETT, D. M., et al., *Stream-Gaging Procedure,* Water-Supply Paper 888, U.S. Geological Survey, Washington, D.C., 1943.

DAILY, J. W., and HARLEMAN, D. R. F., *Fluid Dynamics,* Addison-Wesley Publishing Company, Inc., Reading, Mass., 1966.

DICKINSON, W. T., *Accuracy of Discharge Determinations,* Hydrology Paper 20, Colorado State University, Fort Collins, Colo., 1967.

FITZGERALD, M. G., and KARLINGER, M. R., *Daily Water and Sediment Discharges from Selected Rivers of the Eastern United States: A Time-Series Modeling Approach,* Water-Supply Paper 2216, U.S. Geological Survey, Washington, D.C., 1983.

GILS, H., "Discharge Measurement in Open Water by Means of Magnetic Induction," *Hydrometry,* vol. I, Proceedings of the Koblenz Symposium in Sept. 1970, p. 374–381, UNESCO, Paris, 1973.

GONZALEZ, D. D., SCOTT, C. H., and CULBERTSON, J. K., *Stage-Discharge Characteristics of a Weir in a Sand-Channel Stream,* Water-Supply Paper 1898-A, U.S. Geological Survey, Washington, D.C., 1969.

GUPTA, R. S., *Hydrology of Liberia,* Ministry of Lands and Mines, Government of Liberia, Dec. 1978.

GUPTA, R. S., *Hydrologic Study of the Saint John River Basin in Liberia, West Coast of Africa,* Design Project for the degree of Engineer, Polytechnic Institute of New York, 1980.

HAAN, C. T., *Statistical Methods in Hydrology,* Iowa State University Press, Ames, Iowa, 1977.

HAMMER, M. J., and MACKICHAN, K. A., *Hydrology and Quality of Water Resources,* John Wiley & Sons, Inc., New York, 1981.

HERSCHY, R. W., *Streamflow Measurement,* Elsevier Applied Science Publishers, Barking, Essex, England, 1985.

HIRANANDANI, M. G., and CHITALE, S. V., *Stream Gauging: A Manual,* Government of India, Ministry of Irrigation and Power, Central Water and Power Research Station, Poona, India, 1960.

HOLMES, H., WHIRLOW, D. K., and WRIGHT, L. G., "LE Flowmeter—A Unique Device for Open Channel Discharge Measurement," in *Hydrometry*, vol. I, *Studies and Report in Hydrology*, UNESCO, Paris, 1973.

KENNEDY, E. J., *Computation of Continuous Records of Streamflow*, Techniques of Water Resources Investigations, Chap. A13, Book 3, U.S. Geological Survey, Washington, D.C., 1983.

LINSLEY, R. K., and FRANZINI, J. B., *Water Resources Engineering*, 3rd ed., McGraw-Hill Book Company, New York, 1979.

LINSLEY, R. K., KOHLER, M. A., and PAULHUS, J. L. H., *Hydrology for Engineers*, 3rd ed., McGraw-Hill Book Company, New York, 1982.

MITCHELL, W. D., *Stage-Fall-Discharge Relations for Steady Flow in Prismatic Channels*, Water-Supply Paper 1164, U.S. Geological Survey, Washington, D.C., 1954.

MOSS, M. E., et al., *Design of Surface-Water Data Networks for Regional Information*, Water-Supply Paper 2178, U.S. Geological Survey, Washington, D.C., 1982.

RANTZ, S. E., "Characteristics of Logarithmic Rating Curves," in *Selected Techniques in Water Resources Investigations, 1966–67*, compiled by E. G. Chase and F. N. Payne, Water-Supply Paper 1982, U.S. Geological Survey, Washington, D.C., 1968.

RANTZ, S. E., et al., *Measurement and Computation of Streamflow*, vol. 1, *Measurement of Stage and Discharge*, Water-Supply Paper 2175, U.S. Geological Survey, Washington, D.C., 1982a.

RANTZ, S. E., et al., *Measurement and Computation of Streamflow*, vol. 2, *Computation of Discharge*, Water-Supply Paper 2175, U.S. Geological Survey, Washington, D.C., 1982b.

RIGGS, H. C., *Some Statistical Tools in Hydrology*, Techniques of Water Resources Investigations, Chap. A1, Book 4, U.S. Geological Survey, Washington, D.C., 1968.

RIGGS, H. C., *Regional Analyses of Streamflow Characteristics*, Techniques of Water Resources Investigations, Chap. B3, Book 4, U.S. Geological Survey, Washington, D.C., 1973.

ROBERTSON, J. A., and CROWE, C. T., *Engineering Fluid Mechanics*, 2nd ed., Houghton Mifflin Company, Boston, 1980.

SAVINI, J., and BODHAINE, G. L., *Analysis of Current-meter Data at Columbia River Gaging Stations*, Water Supply Paper 1869-F, U.S. Geological Survey, Washington, D.C., 1971.

SEARCY, J. K., "Graphical Correlation of Gaging-Station Records," in *Manual of Hydrology, Part 1, General Surface-Water Techniques*, Water-Supply Paper 1541-C, U.S. Geological Survey, Washington, D.C., 1960.

SEARCY, J. K., and HARDISON, C. H., "Double-Mass Curves," in *Manual of Hydrology, Part 1, General Surface-Water Techniques*, Water-Supply Paper 1541-B, U.S. Geological Survey, Washington, D.C., 1960.

SMITH, W., *Techniques and Equipment Required for Precise Stream Gaging in Tide-Affected Fresh-Water Reaches of the Sacramento River, California*, Water-Supply Paper 1869-G, U.S. Geological Survey, Washington, D.C., 1971.

SMOOT, G. F., and NOVAK, C. E., *Measurement of Discharge by the Moving-Boat Method*, Techniques of Water-Resources Investigations, Chapter A11, Book 3, U.S. Geological Survey, Washington, D.C., 1969.

SPEIGEL, M. R., *Theory and Problems of Statistics*, Schaum's Outline Series, McGraw-Hill Book Company, New York, 1961.

THOMAS, D. M., and BENSON, M. A., *Generalization of Streamflow Characteristics from Drainage-Basin Characteristics*, Water-Supply Paper 1975, U.S. Geological Survey, Washington, D.C., 1970.

UNESCO, *Hydrologic Information Systems*, ed. G. W. Whetstone and V. J. Grigoriev, Studies and Reports in Hydrology, UNESCO/WMO, Paris/Geneva, 1972.

UNESCO, *Three Centuries of Scientific Hydrology*, UNESCO/WMO/IAHS, Paris, 1974.

References

U.S. Bureau of Reclamation, *Water Measurement Manual,* revised reprint, U.S. Dept. of Interior, Washington, D.C., 1984.

U.S. Geological Survey, *National Handbook of Recommended Methods for Water-Data Acquisition,* Chap. 1, "Surface Water," Office of the Water Data Coordination, U.S. Geological Survey, Reston, Va., 1977.

Vanoni, V. A., "Velocity Distribution in Open Channels," *Civ. Eng.,* v. 11, n. 6, pp. 356–357, June 1941.

Vanoni, V. A., and Nomicos, G. A., "Resistance Properties of Sediment-Laden Streams," *Trans. ASCE,* Paper 3055, pp. 1140–1175, 1960.

Viessman, W., Knapp, J. W., and Lewis, G. L., *Introduction to Hydrology,* 2nd ed., IEP, A Dun-Donnelley Publisher, New York, 1977.

Villemonte, J. R., "Submerged-Weir Discharge Studies," *Eng. News Rec.,* v. 139, n. 26, Dec. 25, 1947.

Winter, T. C., "Uncertainties in Estimating the Water Balance of Lakes," *Water Res. Bull.,* v. 17, n. 1, pp. 82–115, Feb. 1981.

Wisler, C. O., and Brater, E. F., *Hydrology,* John Wiley & Sons, Inc., New York, 1959.

Yevdjevich, V. M., "Statistical and Probability Analysis of Hydrologic Data, Part II: Regression and Correlation Analysis," in *Handbook of Applied Hydrology,* ed. V. T. Chow, McGraw-Hill Book Company, New York, 1964.

CHAPTER SEVEN

Anderson, M. G., and Burt, T. P. (eds.), *Hydrological Forecasting,* John Wiley & Sons Ltd., Chichester, West Sussex, England, 1985.

Beard, L. R., *Simulation of Daily Streamflow,* Tech. Paper 6, Hydrologic Engineering Center, U.S. Army Corps of Engineers, Davis, Calif.

Beard, L. R., *Statistical Methods in Hydrology,* Hydrologic Engineering Center, U.S. Army Corps of Engineers, Davis, Calif., 1962.

Beard, L. R., *Use of Interrelated Records to Simulate Streamflow,* Tech. Paper 1, Hydrologic Engineering Center, U.S. Army Corps of Engineers, Davis, Calif., 1965.

Beard, L. R., *Streamflow Synthesis for Ungaged Rivers,* Technical Paper 5, Hydrologic Engineering Center, U.S. Army Corps of Engineers, Davis, Calif., 1967.

Beard, L. R., et al., *Estimating Monthly Streamflows within a Region,* Tech. Paper 18, Hydrologic Engineering Center, U.S. Army Corps of Engineers, Davis, Calif., 1970.

Bender, D. L., and Roberson, J. A., "The Use of Dimensionless Unit Hydrograph to Derive Unit Hydrograph for Some Pacific Northwest Basins," *J. Geophys. Res.,* v. 66, n. 2, pp. 521–527, 1961.

Bernard, M., "An Approach to Determinate Streamflow," *Trans. ASCE,* v. 100, 1935.

Betson, R. P., "What is Watershed Runoff?" *J. Geophys. Res.,* v. 69, n. 8, pp. 1541–1552, 1964.

Box, G. E. P., and Jenkins, G. W., *Time Series Analysis, Forecasting and Control,* Holden-Day, Inc., San Franscisco, 1976.

Bras, R. L., and Rodriguez-Iturbe, I., *Random Functions and Hydrology,* Addison-Wesley Publishing Company, Inc., Reading, Mass., 1985.

Carlson, R. E., MacCormick, A. J. A., and Watts, D. G., "Application of Linear Random Models to Four Annual Streamflow Series," *Water Resour. Res.,* v. 6, pp. 1070–1078, 1970.

Cheremisinoff, N. P., *Practical Statistics for Engineers and Scientists,* Technomic Publishing Co., Inc., Lancaster, Pa., 1987.

Chow, V. T., "Runoff," in *Handbook of Applied Hydrology,* ed. V. T. Chow, McGraw-Hill Book Company, New York, 1964.

CLARK, C. O., "Storage and Unit Hydrograph," *Trans. ASCE,* v. 110, 1943.

COMMONS, G., "Flood Hydrographs," *Civil Eng.,* v. 12, pp. 571–572, 1942.

CRAIG, G. S., JR., and RANKL, J. G., *Analysis of Runoff from Small Drainage Basins in Wyoming,* Water Supply Paper 2056, U.S. Geological Survey, Washington, D.C., 1978.

DAVIS, M. L., and CORNWELL, D. A., *Introduction to Environmental Engineering,* PWS Publishers, Boston, 1985.

DUNNE, T., "Models of Runoff Processes and Their Significance," in *Studies in Geophysics: Scientific Basis of Water Resources Management,* prepared by Geophysics Study Committee, Geophysics Research Board, Assembly of Math and Physics Sciences, National Research Council, National Academy Press, Washington, D.C., 1982.

DUNNE, T., and BLACK, R. D., "Partial Area Contributions to Storm Runoff in Small New England Watershed," *Water Resour. Res.,* v. 6, n. 5, pp. 1296–1311, 1970.

EICHERT, B. S., et al., *Methods of Hydrological Computations for Water Projects,* International Hydrological Program, UNESCO, Paris, 1982.

FIERING, M. B., and JACKSON, B. B., *Synthetic Streamflows,* Water Resources Monograph 1, American Geophysicists Union, Washington, D.C., 1971.

FITZGERALD, M. G., and KARLINGER, M. R., *Daily Water and Sediment Discharges from Selected Rivers of the Eastern United States: A Time-Series Modeling Approach,* Water-Supply Paper 2216, U.S. Geological Survey, Washington, D.C., 1983.

FLEMING, G., *Computer Simulation Techniques in Hydrology,* American Elsevier Publishing Company, Inc., New York, 1975.

FREEZE, R. A., "Role of Subsurface Flow in Generating Surface Runoff, 1: Baseflow Contribution to Channel Flow," *Water Resour. Res.,* v. 8, n. 3, pp. 609–623, 1972a.

FREEZE, R. A., "Role of Subsurface Flow in Generating Surface Runoff, 2: Upstream Source Areas," *Water Resour. Res.,* v. 8, n. 3, pp. 1272–1283, 1972b.

FREEZE, R. A., "Reply to Comments by Knisel," *Water Resour. Res.,* v. 9, n. 4, p. 1101, 1973.

GRAY, D. M. (ed.), *Handbook on the Principles of Hydrology,* Water Information Center, Inc., New York, 1973.

GRAY, D. M., "Synthetic Unit Hydrograph for Small Watersheds," *ASCE J. of Hydraul. Div.,* v. 87, n. HY4, pp. 33–53, 1961.

HAMMER, M. J., and MACKICHAN, K. A., *Hydrology and Quality of Water Resources,* John Wiley & Sons, Inc., New York, 1981.

HEDMAN, E. R., *Mean Annual Runoff as Related to Channel Geometry of Selected Streams in California,* Water Supply Paper 1999-E, U.S. Geological Survey, Washington, D.C., 1970.

HEDMAN, E. R., and OSTERKAMP, W. R., *Streamflow Characteristics Related to Channel Geometry of Streams in Western United States,* Water Supply Paper 2193, U.S. Geological Survey, Washington, D.C., 1982.

HEWLETT, J. D., "Comments on Letters Relating to Role of Subsurface Flow in Generating Surface Runoff, 2: Upstream Source Areas," by A. Freeze, *Water Resour. Res.,* v. 10, n. 3, June 1974.

HEWLETT, J. D., *Principles of Forest Hydrology,* University of Georgia Press, Athens, Ga., 1982.

HEWLETT, J. D., and HIBBERT, A. R., "Factors Affecting the Response of Small Watersheds to Precipitation in Humid Areas," in *International Symposium on Forest Hydrology,* ed. W. E. Sopper and H. W. Lull, Pergamon Press Ltd., Oxford, 1967.

HICKOK, R. B., KEPPEL, R. V., and RAFFERTY, B. R., "Hydrograph Synthesis for Small Arid-land Watersheds," *Agr. Eng.,* v. 40, pp. 608–615, 1959.

HORTON, R. E., "The Role of Infiltration in the Hydrologic Cycle," *Trans. Am. Geophys. Union,* v. 14, pp. 446–460, 1933.

HROMADKA, T. V., DURBIN, T. J., and DEVRIES, J. J., *Computer Methods in Water Resources,* Lighthouse Publications, Mission Viejo, Calif., 1985.

HROMADKA, T. V., BEECH, B. L., and CLEMENTS, J. M., *Computational Hydraulics for Civil Engineers,* Lighthouse Publications, Mission Viejo, Calif., 1986.

Hydrologic Engineering Center, *Hydrologic Engineering Methods for Water Resources Development: Hydrograph Analysis,* Vol. 4, U.S. Contribution to I.H.D., U.S. Army Corps of Engineers, Davis, Calif., 1973.

Hydrologic Engineering Center, *Hydrologic Engineering Methods for Water Resources Development: Hydrologic Frequency Analysis,* Vol. 3, U.S. Contribution to I.H.D., U.S. Army Corps of Engineers, Davis, Calif., 1975.

International Hydrological Decade, *Design of Water Resources Projects with Inadequate Data,* Vols. 1 and 2, Studies and Reports in Hydrology, UNESCO, Paris, 1974.

KIRBY, M. J. (ed.), *Hillslope Hydrology,* Wiley-Interscience, New York, 1978.

KLEMES, V., "Physically Based Stochastic Hydrologic Analysis," in *Advances in Hydrosciences,* Vol. 11, ed. V. T. Chow, Academic Press, Inc., New York, 1978.

KNAPP, B. J., *Elements of Geographical Hydrology,* George Allen & Unwin (Publisher) Ltd., London, 1979.

KNISEL, W. G., "Comments on Role of Subsurface Flow in Generating Surface Runoff, 2: Upstream Source Areas," by A. Freeze, *Water Resour. Res.,* v. 9, n. 4, pp. 1097–1100, Aug. 1973.

KOTTEGODA, N. T., *Stochastic Water Resources Technology,* Halsted Press, New York, 1980.

LANE, W. L., *Applied Stochastic Techniques, User's Manual,* U.S. Bureau of Reclamation, Denver, Colo., 1979.

LINSLEY, R. K., and FRANZINI, J. B., *Water Resources Engineering,* 3rd ed., McGraw-Hill Book Company, New York, 1979.

LINSLEY, R. K., KOHLER, M. A., and PAULHUS, J. L. H., *Hydrology for Engineers,* 3rd ed., McGraw-Hill Book Company, New York, 1982.

MAASS, A., et al., *The Design of Water Resources Systems,* Harvard University Press, Cambridge, Mass., 1962.

MATALAS, N. C., "Mathematical Assessment of Synthetic Hydrology," *Water Resour. Res.,* v. 3, n. 4, pp. 937–945, 1967.

MATALAS, N. C., and JACOBS, B., *A Correlation Procedure for Augmenting Hydrological Data,* Professional Paper 434-E, U.S. Geological Survey, Washington, D.C., 1964.

MATALAS, N. C., and WALLIS, J. R., "Generation of Synthetic Flow Sequence," in *Systems Approach to Water Management,* ed. A. K. Biswas, McGraw-Hill Book Company, New York, 1976.

MEINZER, O. E. (ed.), *Physics of the Earth: Hydrology,* McGraw-Hill Book Company, New York, 1942.

MEJIA, J. M., and ROUSSELLE, J., "Disaggregation Models in Hydrology Revisited," *Water Resour. Res.,* v. 12, n. 2, pp. 185–186, 1976.

MITCHELL, W. D., *Unit Hydrographs in Illinois,* Division of Waterways, State of Illinois, Springfield, Illinois, 1948.

MUSGRAVE, G. W., and HOLTON, H. N., "Infiltration," in *Handbook of Applied Hydrology,* ed. V. T. Chow, McGraw-Hill Book Company, New York, 1964.

PILGRIM, D. H., HUFF, D. D., and STEELE, T. D., "A Field Evaluation of Subsurface and Surface Runoff, II: Runoff Process," *J. Hydrol.,* v. 38, pp. 319–341, 1978.

RIGGS, H. C., *Some Statistical Tools in Hydrology,* Techniques of Water Resources Investigations, Chap. A1, Book 4, U.S. Geological Survey, Washington, D.C., 1968.

SABOL, G. V., "Clark Unit Hydrograph and R-Parameter Estimation," *ASCE J. Hydraul. Eng.,* v. 114, n. 1, pp. 103–111, 1988.

SALAS, J. D., DELLEUR, J. W., YEVJEVICH, V., and LANE, W. L., *Applied Modeling of Hydrologic Time Series,* Water Resources Publications, Littleton, Colo., 1980.

SEARCY, J. K., *Graphical Correlation of Gaging-Station,* Water Supply Paper 1541-C, U.S. Geological Survey, Washington, D.C., 1960.

SEARCY, J. K., and HARDISON, C. H., *Double-Mass Curves,* Water Supply Paper 1541-B, U.S. Geological Survey, Washington, D.C., 1960.

SEN, Z., "Wet and Dry Periods of Annual Flow Series," *ASCE, J. Hydraul. Div.,* v. 102, n. HY10, pp. 1503–1514, 1976.

SEN, Z., "Autorun Analysis of Hydrologic Times Series," *ASCE, J. Hydraul.,* v. 36, pp. 75–85, 1978.

SEN, Z., "Autorun Model for Synthetic Flow Generation," *ASCE, J. Hydraul.,* v. 81, pp. 157–170, 1985.

SHERMAN, L. K., "Streamflow from Rainfall by the Unit-graph Method," Eng. News Record, v. 108, pp. 501–502, 1932.

SHERMAN, L. K., "The Unit Hydrograph Method," Ch. X-IE, in *Physics of the Earth-Hydrology,* ed. O. E. Meinzer, McGraw-Hill Book Company, Inc., New York, 1942.

SINGH, V. P., *Hydrologic Systems: Rainfall–Runoff Modeling,* vol. 1, Prentice-Hall, Inc., Englewood Cliffs, N.J., 1988.

SNYDER, F. F., "Synthetic Unit Graphs," *Trans. Amer. Geophys. Union,* v. 19, pp. 447–454, 1938.

SRIKANTHAN, R., "Sequential Generation of Monthly Streamflows," *ASCE, J. Hydraul.,* v. 38, pp. 71–80, 1978.

STEDINGER, J. R., and VOGEL, R. M., "Disaggregation Procedures for Generating Serially Correlated Flow Vectors," *Water Resour. Res.,* v. 20, n. 1, pp. 47–56, 1984.

STEINGER, J. R., LETTENMAIER, D. P., and VOGEL, R. M., "Multisite ARMA (1, 1) and Disaggregation Models for Annual Streamflow Generation," *Water Resour. Res.,* v. 21, n. 4, pp. 497–509, 1985.

SVANIDIZE, G., *Mathematical Modeling of Hydrologic Series,* Water Resources Publications, Littleton, Colo., 1980.

TAYLOR, A. B., and SCHWARZ, H. E., "Unit Hydrograph Lag and Peak Flow Related to Basin Characteristics," *Trans. Amer. Geophys. Union,* v. 33, pp. 235–246, 1952.

THOMAS, D. M., and BENSON, M. A., *Generalization of Streamflow Characteristics from Drainage-Basin Characteristics,* Water Supply Paper 1975, U.S. Geological Survey, Washington, D.C., 1970.

THOMAS, H. A., *Improved Methods for National Assessment,* Water Resources Contract WR 15249270, Water Resources Group, Harvard University, Cambridge, Mass., 1981.

U. S. Soil Conservation Service, *Engineering Handbook, Section 4, Hydrology,* Supplement A, U.S. Dept. of Agriculture, Washington, D.C., 1957.

U.S. Soil Conservation Service, *National Engineering Handbook, Section 4, Hydrology,* U.S. Dept. of Agriculture, Washington, D.C., 1971.

U.S. Water Resources Council, *Guidelines for Determining Flood Flow Frequency,* Bull. 17B, Hydrology Committee, U.S. Water Resources Council, Washington, D.C., 1981.

VALENCIA, R. D., and SCHAAKE, J. C., "Disaggregation Process in Stochastic Hydrology," *Water Resour. Res.,* v. 9, n. 3, pp. 580–585, 1973.

VIESSMAN, W., KNAPP, J. W., and LEWIS, G. L., *Introduction to Hydrology,* 2nd ed., IEP, A Dun-Donnelley Publisher, New York, 1977.

WETHERILL, G. B., et al., *Regression Analysis with Applications,* Chapman & Hall Ltd., London, 1986.

WILLIAMS, H. M., "Discussion on Military Airfields: Design of Drainage Facilities," *Trans. ASCE,* v. 110, pp. 820–826, 1945.

World Meteorological Organization, *Hydrological Network Design and Information Transfer,* WMO No. 433, Report 8, WMO, Geneva, 1976.

YEVDJEVICH, V. M., "Regression and Correlation Analysis," in *Handbook of Applied Hydrology,* ed. V. T. Chow, McGraw-Hill Book Company, New York, 1964.

CHAPTER EIGHT

ADAMOWSKI, K., "Plotting Formula for Flood Frequency," *Water Resour. Bull.,* v. 17, n. 2, pp. 197–202, 1981.

ALTMAN, D. G., ESPEY, W. H., and FELDMAN, A. D., *Investigation of Soil Conservation Service Urban Hydrology Techniques,* Tech. Paper 77, Hydrologic Engineering Center, U.S. Army Corps of Engineers, Davis, Calif., 1980.

BARROWS, H. K., *Floods: Their Hydrology and Control,* McGraw-Hill Book Company, New York, 1948.

BEARD, L. R., *Statistical Methods in Hydrology,* Civil Works Investigation Project CW-151, U.S. Army Corps of Engineers, Sacramento, Calif., 1962.

BEARD, L. R., *Hypothetical Flood Computation for a Stream System,* Tech. Paper 12, Hydrologic Engineering Center, U.S. Army Corps of Engineers, Davis, Calif., 1968.

BENSON, M. A., "Characteristics of Frequency Curves Based on a Theoretical 1000-Year Record," in *Flood Frequency Analysis,* by T. Dalrymple, Water Supply Paper 1543-A U.S. Geological Survey, Washington, D.C., 1960.

BODHAINE, G. L., *Measurement of Peak Discharge at Culverts by Indirect Methods,* Techniques of Water Resources Investigations, Book 3, Chap. A3, U.S. Geological Survey, Washington, D.C., 1968 (reprint 1982).

BRATER, E. F., and KING, H. W., *Handbook of Hydraulics,* 6th ed., McGraw-Hill Book Company, New York, 1976.

BURNHAM, M. W., *Adoption of Flood Flow Frequency Estimates at Ungaged Locations,* Training Doc. 11, Hydrologic Engineering Center, U.S. Army Corps of Engineers, Davis, Calif., 1980.

CARTER, R. W., and GODFREY, R. G., *Storage and Flood Routing,* Water Supply Paper 1543-B, U.S. Geological Survey, Washington, D.C., 1960.

CHOW, V. T., "A General Formula for Hydrologic Frequency Analysis," *Trans. Am. Geophys. Union,* v. 32, pp. 231–237, 1951.

CHOW, V. T., "Frequency Analysis," in *Handbook of Applied Hydrology,* ed. V. T. Chow, McGraw-Hill Book Company, New York, 1964.

CREAGER, W. P., JUSTIN, J. D., and HINDS, J., *Engineering for Dams,* v. I, John Wiley and Sons, Inc., New York, 1945.

CUNNANE, C., "Unbiased Plotting Positions — A Review," *J. Hydrol.,* v. 37, pp. 205–222, 1978.

DALRYMPLE, T., *Flood-Frequency Analysis,* Water Supply Paper 1543-A, U.S. Geological Survey, Washington, D.C., 1960.

DALRYMPLE, T., "Flood Characteristics and Flow Determination," in *Handbook of Applied Hydrology,* ed. V. T. Chow, McGraw-Hill Book Company, New York, 1964.

DALRYMPLE, T., and BENSON, M. A., *Measurement of Peak Discharge by the Slope-Area Method,* Techniques of Water Resources Investigations, Book 3, Chap. A2, U.S. Geological Survey, Washington, D.C., 1967 (reprint 1984).

ELY, P. B., and PETERS, J. C., *Probable Maximum Flood Estimation—Eastern United States,* Tech. Paper 100, Hydrologic Engineering Center, U.S. Army Corps of Engineers, Davis, Calif., 1984.

FELDMAN, A. D., *Flood Hydrograph and Peak Flow Frequency Analysis,* Tech. Paper 62, Hydrologic Engineering Center, U.S. Army Corps of Engineers, Davis, Calif., 1979.

GRAY, D. M., "Statistical Methods—Fitting Frequency Curves, Regression Analysis," in *Handbook on the Principles of Hydrology,* ed. D. M. Gray, Water Information Center, Inc., Port Washington, N.Y., 1973 (published by the National Committee for the IHD in 1970).

GUNLACH, D. L., and THOMAS, W. A., *Guidelines for Calculating and Routing a Dam-Break Flood,* Research Note 5, Hydrologic Engineering Center, U.S. Army Corps of Engineers. Davis, Calif., 1977.

HAAN, C. T., *Statistical Methods in Hydrology,* Iowa State University Press, Ames, Iowa, 1977.

HERSHFIELD, D. M., "Estimating the Probable Maximum Precipitation," *ASCE J. Hydraul. Div.,* v. 87, n. HY5, pp. 99–116, Sept. 1961.

HINDS, J., CREAGER, W. P., and JUSTIN, J. D., *Engineering for Dams,* v. 2, John Wiley and Sons, Inc., New York, 1945.

HJELMFELT, A. T., and CASSIDY, J. J., *Hydrology for Engineers and Planners,* Iowa State University Press, Ames, Iowa, 1975.

HORN, D. R., "Graphic Estimation of Peak Flow Reduction in Reservoirs," *ASCE J. Hydraul. Eng,* v. 113, n. 11, 1987.

HUDSON, H. E., and HAZEN, R., "Droughts and Low Streamflows," in *Handbook of Applied Hydrology,* ed. V. T. Chow, McGraw-Hill Book Company, New York, 1964.

HULSING, H., *Measurement of Peak Discharge at Dams by Indirect Method,* Techniques of Water Resources Investigations, Book 3, Chap. A5, U.S. Geological Survey, Washington, D.C., 1967 (reprint 1984).

HYDROLOGIC ENGINEERING CENTER, *Hydrologic Frequency Analysis,* Hydrologic Engineering Methods for Water Resources Development, vol. 3, U.S. Army Corps of Engineers, Davis, Calif., 1975.

HYDROLOGIC ENGINEERING CENTER, *Feasibility Studies for Small Scale Hydropower Additions,* vol. 3, U.S. Army Corps of Engineers, Davis, Calif., 1979.

KILPATRICK, F. A., and SCHNEIDER, V. R., *Use of Flumes in Measuring Discharge,* Techniques of Water Resources Investigations, Book 3, Chap. A14, U.S. Geological Survey, Washington, D.C., 1983.

KOELZER, V. A., and BITOURI, M., "Hydrology of Spillway Design: Large Structures—Limited Data," *ASCE J. Hydraul. Div.,* v. 90, n. HY3, pp. 261–293, May 1964.

LAWLER, E. A., "Flood Routing," in *Handbook of Applied Hydrology,* ed. V. T. Chow, McGraw-Hill Book Company, New York, 1964.

MATTHAI, H. F., *Measurement of Peak Discharge at Width Contractions by Indirect Methods,* Techniques of Water Resources Investigations, Book 3, Chap. A4, U.S. Geological Survey, Washington, D.C., 1967 (reprint 1984).

MORRIS, E. C., *Mixed Population Frequency Analysis,* Training Document 17, Hydrologic Engineering Center, U.S. Army Corps of Engineers, Davis, Calif., 1982.

OGROSKY, H. O., "Hydrology of Spillway Design: Small Structures—Limited Data," *ASCE J. Hydraul. Div.,* v. 90, n. HY3, pp. 265–310, May 1964.

RAUDKIVI, A. J., *Hydrology: An Advanced Introduction to Hydrological Processes and Modeling,* Pergamon Press Ltd., Oxford, 1979.

References   713

RIGGS, H. C., *Frequency Curves,* Techniques of Water Resources Investigations, Book 4, Chap. A2, U.S. Geological Survey, Washington, D.C., 1968.

RIGGS, H. C., *Low-Flow Investigations,* Techniques of Water Resources Investigations, Book 4, Chap. B1, U.S. Geological Survey, Washington, D.C., 1972.

RIGGS, H. C., *Streamflow Characteristics,* Developments in Water Science Series No. 22, Elsevier Science Publishing Co., Inc., New York, 1985.

SAUER, V. B., et al., *Flood Characteristics of Urban Watersheds in the United States,* Water Supply Paper 2207, U.S. Geological Survey, Washington, D.C., 1983.

SINGH, V. P., and SINGH, K., "Parameter Estimation for Log-Pearson Type III Distribution by POME," *ASCE J. Hydraul. Eng.,* v. 114, n. 1, pp. 112–122, 1988.

SNYDER, F. F., "Hydrology of Spillway Design: Large Structures — Adequate Data," *ASCE J. Hydraul. Div.,* v. 90, n. HY3, pp. 239–259, May 1964.

SOKOLOV, A. A., RANTZ, S. E., and ROCHE, M., *Floodflow Computation: Methods Compiled from World Experience,* Studies and Reports in Hydrology, No. 22, UNESCO, Paris, 1976.

SPEIGEL, M. R., *Statistics,* Schaum's Outline Series, McGraw-Hill Book Company, New York, 1961.

SRIKANTHAN, R., and McMAHON, T. A., "Recurrence Interval of Long Hydrologic Events," *ASCE J. Hydraul. Eng.,* v. 112, n. 6, 1986.

STRELKOFF, T., et al., *Comparative Analysis of Flood Routing Methods,* Research Document 24, Hydrologic Engineering Center, U.S. Army Corps of Engineers, Davis, Calif., 1980.

TUNG, Y. K., and MAYS, L. W., "Generalized Skew Coefficients for Flood Frequency Analysis," *Water Resour. Bull.,* v. 17, n. 2, pp. 262–269, 1981.

UNESCO, *Flood Studies: An Internation Guide for Collection and Processing of Data,* Technical Papers in Hydrology 8, UNESCO, Paris, 1971.

U.S. ARMY CORPS OF ENGINEERS, *Flood Hydrograph Analyses and Computations,* Engineering Manual 1110-2-1405, U.S. Government Printing Office, Washington, D.C., 1959.

U.S. ARMY CORPS OF ENGINEERS, *Routing of Floods through River Channels,* Engineering Manual EM 1110-2-1408, U.S. Government Printing Office, Washington, D.C., 1960.

U.S. Bureau of Reclamation, *Design of Small Dams,* U.S. Dept. of Interior, Washington, D.C., 1965.

U.S. NATIONAL WEATHER SERVICE, *Probable Maximum Precipitation Estimates — U.S. East of the 105 Meridian,* Hydrometeorology Report 51, National Oceanic and Atmospheric Administration, U.S. Dept. of Commerce, Washington, D.C., 1978 (reprinted 1986).

U.S. NATIONAL WEATHER SERVICE, *Seasonal Variation of 10-Sq. Mi. Probable Maximum Precipitation Estimates — U.S. East of the 105 Meridian,* Hydrometeorology Report 53, National Oceanic and Atmospheric Administration, U.S. Dept. of Commerce, Washington, D.C., 1980.

U.S. NATIONAL WEATHER SERVICE, *Application of Probable Maximum Precipitation Estimates — U.S. East of the 105 Meridian,* Hydrometeorology Report 52, National Oceanic and Atmospheric Administration, U.S. Dept. of Commerce, Washington, D.C., 1982.

U.S. WATER RESOURCES COUNCIL, *Guidelines for Determining Flood Flow Frequency,* Bulletin 17B, Hydrology Committee of the Water Resources Council, Washington, D.C., 1981.

U.S. WEATHER BUREAU, *Seasonal Variation of the Probable Maximum Possible Precipitation, East of the 105 Meridian for Areas from 10 to 1000 Square Miles and Durations of 6, 12, 24, and 48 hours,* Hydrometeorology Report 35, Washington, D.C., 1956.

U.S. WEATHER BUREAU, *Generalized Estimates of Maximum Possible Precipitation for the United States West of the 105 Meridian,* Tech. Paper 38, U.S. Weather Bureau, Washington, D.C., 1960.

VIESSMAN, W., KNAPP, J. W., and LEWIS, G. L., *Introduction to Hydrology,* 2nd ed., IEP, A Dun-Donnelley Publisher, New York, 1977.

WANDLE, S. W., *Estimating Peak Discharges of Small, Rural Streams in Massachusetts*, Water Supply Paper 2214, U.S. Geological Survey, Washington, D.C., 1983.

WANG, B. H., and JAWED, K., "Transformation of PMP to PMF: Case Studies," *ASCE J. Hydraul. Eng.*, v. 112, n. 7, pp. 547–561, 1986.

WORLD METEOROLOGICAL ORGANIZATION, *Estimation of Maximum Floods*, Technical Note 98, Report of a Working Group of the Commission for Hydrometeorology, WMO, Geneva, 1969.

YEVDJEVICH, V., *Probability and Statistics in Hydrology*, Water Resources Publications, Fort Collins, Colo., 1972.

CHAPTER NINE

AMERICAN SOCIETY OF CIVIL ENGINEERS, *Report on Reevaluating Adequacy of Existing Dams*, Committee on Hydrometeorology of the Hydraulics Division, ASCE, New York, 1973.

BEARD, L. R., *Methods for Determination of Safe Yield and Compensation Water from Storage Reservoirs*, Technical Paper 3, Hydrologic Engineering Center, U.S. Army Corps of Engineers, Davis, Calif., 1965.

BRUNE, G. M., "Trap Efficiency of Reservoirs," *Trans. Am. Geophys. Union*, v. 34, n. 3, 1953.

BRUNE, G. M., and ALLEN, R. E., "A Consideration of Factors Influencing Reservoir Sedimentation in the Ohio Valley Region," *Trans. Am. Geophys. Union*, v. 22, pp. 649–655, 1941.

CASSAGRANDE, A., "Seepage through Dams," *J. New Engl. Water Works Assoc.*, June 1937.

CHADWICK, A. J., and MORFETT, J. C., *Hydraulics in Civil Engineering*, George Allen & Unwin (Publisher) Ltd., London, 1986.

CHOW, V. T., *Open-Channel Hydraulics*, McGraw-Hill Book Company, New York, 1959.

CHOW, V. T. (ed.), *Handbook of Applied Hydrology*, McGraw-Hill Book Company, New York, 1964.

CREAGER, W. P., JUSTIN, J. D., and HINDS, J., *Engineering for Dams*, Vol. I, John Wiley & Sons, Inc., New York, 1945.

DAVIS, C. V., and SORENSON, K. E. (eds.), *Handbook of Applied Hydraulics*, 3rd ed., McGraw-Hill Book Company, New York, 1969.

FREDRICH, A. J., *Techniques for Evaluating Long-Term Reservoir Yields*, Technical Paper 14, Hydrologic Engineering Center, U.S. Army Corps of Engineers, Davis, Calif., 1969.

GOLZE, A. R. (ed.), *Handbook of Dam Engineering*, Van Nostrand Rheinhold Co., New York, 1977.

GRZYWIENSKI, A., "Anti-vacuum Profiles for Spillways of Large Dams," *Trans. 4th Congress on Large Dams*, vol. 2, pp. 105–124, International Commission on Large Dams of the World-Power Conference, New Delhi, India, 1951.

HAGER, W. H., "Lateral Outflow over Side Weirs," *ASCE J. Hydraul. Eng.*, v. 113, n. 4, pp. 491–504, 1987a.

HAGER, W. H., "Continuous Crest Profile for Standard Spillway," *ASCE J. Hydraul. Eng.*, v. 113, n. 11, pp. 1453–1456, 1987b.

HINDS, J., CREAGER, W. P., and JUSTIN, J. D., *Engineering for Dams*, vol. 2, John Wiley and Sons, Inc., New York, 1945.

HWANG, N. H. D., and HITA, C. E., *Fundamentals of Hydraulic Engineering Systems*, 2nd ed., Prentice-Hall, Inc., Englewood Cliffs, N. J., 1987.

HYDROLOGIC ENGINEERING CENTER, *Reservoir Storage-Yield Procedures,* U.S. Army Corps of Engineers, Davis, Calif., 1967.

HYDROLOGIC ENGINEERING CENTER, *Hydrologic Engineering Methods for Water Resources Development,* Vol. 8, *Reservoir Yield,* U.S. Army Corps of Engineers, Davis, Calif., 1975.

HYDROLOGIC ENGINEERING CENTER, *Feasibility Studies for Small Scale Hydropower Additions,* vol. 3, U.S. Army Corps of Engineers, Davis, Calif., 1979.

JUSTIN, J. D., HINDS, J., and CREAGER, W. P., *Engineering for Dams,* vol. 3, John Wiley and Sons, Inc., New York, 1945.

MARTIN, R. O. R., and HANSON, R. L., *Reservoirs in the United States,* Water Supply Paper 1838, U.S. Geological Survey, Washington, D.C., 1966.

MCCARTHY, D. F., *Essentials of Soil Mechanics and Foundations,* 2nd ed., Reston Publishing Co., Inc., Reston, Va., 1982.

MORRIS, H. M., and WIGGERT, J. M., *Applied Hydraulics in Engineering,* 2nd ed., John Wiley and Sons, Inc., New York, 1972.

MURPHY, T. E., *Spillway Crest Design, Misc. Paper H-73-5,* Waterways Experimentation Station, U.S. Army Corps of Engineers, Vicksburg, Miss., 1973.

REESE, A. J., and MAYNORD, S. T., "Design of Spillway Crest," *ASCE J. Hydraul. Eng.,* v. 113, n. 4, pp. 476–490, 1987.

RIGGS, H. C., and HARDISON, C. H., *Storage Analyses for Water Supply,* Techniques of Water Resources Investigations, Book 4, Chap. B2, U.S. Geological Survey, Washington, D.C., 1983.

SIMON, A. L., *Hydraulics,* 3rd ed., John Wiley & Sons, Inc., New York, 1986.

TERZAGHI, K., and PECK, R. B., *Soil Mechanics in Engineering Practice,* 2nd ed., John Wiley & Sons, Inc., New York, 1967.

THOMAS, H. H., *The Engineering of Dams,* parts I and II, John Wiley and Sons Ltd., Chichester, West Sussex, England, 1976.

U.S. BUREAU OF RECLAMATION, *Dams and Control Works,* 3rd ed., U.S. Government Printing Office, Washington, D.C., 1954.

U.S. BUREAU OF RECLAMATION, *Design of Gravity Dams,* Water Resources Technical Publication, U.S. Dept. of Interior, Denver, Colo., 1976.

U.S. BUREAU OF RECLAMATION, *Design of Arch Dams,* Water Resources Technical Publication, U.S. Dept. of the Interior, Denver, Colo., 1977a.

U.S. BUREAU OF RECLAMATION, *Design Criteria for Concrete Arch and Gravity Dams,* Engineering Monograph 19, U.S. Dept. of Interior, Washington, D.C., 1977b.

U.S. BUREAU OF RECLAMATION, *Design of Small Dams,* 2nd ed., revised reprint, U.S. Dept. of Interior, Washington, D.C., 1977c.

U.S. BUREAU OF RECLAMATION, *Design of Small Canal Structures,* U.S. Dept. of Interior, Denver, Colo., 1978.

U.S. BUREAU OF RECLAMATION, *Water Measurement Manual,* revised reprint, U.S. Dept. of Interior, Denver, Colo., 1984.

U.S. DEPARTMENT OF THE ARMY, *Hydraulic Design of Spillways,* Engineering Manual 1110-2-1603, Office of Chief of Engineers, Washington, D.C., 1965.

U.S. DEPARTMENT OF THE ARMY, *Hydraulic Design of Spillways, Draft of Chapter 3 on Spillway Crest,* Engineering Manual 1110-2-1603, Office of the Chief of Engineers, Washington, D.C., 1986.

VIESSMAN, W., and WELTY, C., *Water Management: Technology and Institutions,* Harper & Row, Publishers, Inc., New York, 1985.

CHAPTER TEN

AMERICAN SOCIETY OF CIVIL ENGINEERS, Task Force on Friction Factors in Open Channels, "Friction Factors in Open Channels," *ASCE J. Hydraul. Div.,* v. 89, n. HY4, Mar. 1963.

AMERICAN SOCIETY OF CIVIL ENGINEERS AND WATER POLLUTION CONTROL FEDERATION, *Design and Construction of Sanitary and Storm Sewers*, 5th printing, ASCE, New York, 1982.

BAKHMETEFF, B. A., *Hydraulics of Open Channels*, McGraw-Hill Book Company, New York, 1932.

BENEFIELD, L. D., JUDKIN, J. F., and PARR, A. D., *Treatment Plant Hydraulics for Environmental Engineers*, Prentice-Hall, Inc., Englewood Cliffs, N.J., 1984.

BLENCH, T., "Regime Theory for Self-Formed Sediment Bearing Channels," *Trans. ASCE*, v. 117, pp. 383–400, 1952.

BRATER, E. F., and KING, H. W., *Handbook of Hydraulics*, 6th ed., McGraw-Hill Book Company, New York, 1976.

BREBBIA, C. A., and FERRANTE, A. J., *Computational Hydraulics*, Butterworth & Company (Publishers) Ltd., London, 1983.

CHEREMISINOFF, N. P., *Fluid Flow—Pumps, Pipes and Channels*, Ann Arbor Science Publishers, Ann Arbor, Mich., 1981.

CHOW, V. T., *Open-Channel Hydraulics*, McGraw-Hill Book Company, New York, 1959.

DAVIDIAN, J., *Computation of Water Surface Profiles in Open Channels*, Techniques of Water Resources Investigations, Chap. A15, Book 3, U.S. Geological Survey, Washington, D.C., 1984.

DAVIS, C. V., and SORENSON, K. E. (eds.), *Handbook of Applied Hydraulics*, Secs. 2, 6, 7, McGraw-Hill Book Company, New York, 1969.

FEATHERSTONE, R. E., and NALLURI, C., *Civil Engineering Hydraulics: Essential Theory with Worked Examples*, Granada Publishing Ltd., London, 1982.

HENDERSON, F. M., *Open Channel Flow*, Macmillan Publishing Company, New York, 1966.

HYDRAULIC RESEARCH STATION, *Charts for the Hydraulic Design of Channels and Pipes*, 5th ed., Hydraulic Research, Wallingford, England, 1983.

INGLIS, SIR CLAUDE, *Annual Report (Technical)*, Central Irrigation and Hydropower Research Station, Poona, India, 1940–41.

KENNEDY, R. G., "The Prevention of Silting in Irrigation Canals," *Proceedings of Institution of Civil Engineers*, London, v. 119, pp. 281–290, 1895.

KOUTITAS, C. G., *Elements of Computational Hydraulics*, Pentech Press, Plymouth, Devon, England, 1983.

LACEY, G., *Regime Flow in Incoherent Alluvium*, Central Board of Irrigation and Power, Publication 20, Government of India, Simla, India, 1940.

LACEY, G., "Stable Channels in Alluvium," *Proceedings of Institution of Civil Engineers*, London, v. 229, pp. 259–384, 1930.

LAL, J., *Hydraulics*, Metropolitan Book Co. Pvt. Ltd., New Delhi, 1958.

LANE, E. W., "Stable Channels in Erodible Material," *Trans. ASCE*, v. 102, pp. 123–142, 1937.

LANE, E. W., "Design of Stable Channels," *Trans. ASCE*, v. 120, pp. 1234–1260, 1955.

LIMERINOS, J. T., *Determination of the Manning Coefficient from Measured Bed Roughness in Natural Channels*, Water Supply Paper 1898-B, U.S. Geological Survey, Washington, D.C., 1970.

LINDLEY, E. S., "Regime Channels," *Minutes of Proceedings, Punjab Engineering Congress*, Lahore, India, v. 7, pp. 63–74, 1919.

LINSLEY, R. K., and FRANZINI, J. B., *Water Resources Engineering*, 3rd ed., McGraw-Hill Book Company, New York, 1979.

PICKARD, W. F., "Solving the Equations of Uniform Flow," *Proc. ASCE*, v. 89, n. HY4, Part I, July 1963.

POMEROY, R. D., "Flow Velocities in Small Sewers," *J. Water Poll. Control Fed.*, 39, 1525, 1967.

RAUDKIVI, A. J., *Loose Boundary Hydraulics*, 2nd ed., Pergamon Press, Inc., Elmsford, N.Y., 1976.

RICHARDS, K., *Rivers: Form and Process in Alluvial Channels*, Methuen Inc., New York, 1982.

ROBERSON, J. A., and CROWE, C. T., *Engineering Fluid Mechanics*, Houghton Mifflin Company, Boston, 1975.

ROUSE, H. (ed.), *Engineering Hydraulics*, Chap. 9, John Wiley & Sons, Inc., New York, 1950.

SIMON, A. L., *Practical Hydraulics*, John Wiley & Sons, Inc., New York, 1976.

STREETER, V. L., *Fluid Mechanics*, 5th ed., McGraw-Hill Book Company, New York, 1971.

TERRELL, P. W., and WHITNEY, M. B., "Design of Stable Canals and Channels in Erodible Material," *Trans. ASCE*, v. 123, pp. 101–115, 1958.

U.S. BUREAU OF RECLAMATION, *Design of Small Canal Structures*, U.S. Dept. of Interior, Denver, Colo., 1978 (reprint 1983).

U.S. SOIL CONSERVATION SERVICE, *Handbook of Channel Design for Soil and Water Conservation*, SCS-Technical Paper 61, U.S. Dept. of Agriculture, Washington, D.C., 1954.

WHITAKER, S., *Introduction to Fluid Mechanics*, Prentice-Hall, Inc., Englewood Cliffs, N.J., 1968.

WOODWARD, S. M., and POSEY, C., *Hydraulics of Steady Flow in Open Channels*, John Wiley & Sons, Inc., New York, 1941.

CHAPTER ELEVEN

ABBOTT, M. B., *An Introduction to the Method of Characteristics*, American Elsevier Publishing Company, Inc., New York, 1966.

ALLIEVI, L., *Theory of Water Hammer*, two vol., (trans. E. E. Halmos), sponsored by the ASCE and ASME, 1925.

AMERICAN SOCIETY OF CIVIL ENGINEERS, *Pipeline Design for Water and Wastewater*, Committee on Pipeline Planning, ASCE, New York, 1975.

AMERICAN SOCIETY OF CIVIL ENGINEERS AND WATER POLLUTION CONTROL BOARD, *Design and Construction of Sanitary and Storm Sewers*, Manual on Engineering Practice 37, ASCE, New York, 1982.

ASTHANA, K. C., "Transformation of Moody Diagram," *ASCE J. Hydraul. Div.*, v. 100, n. HY6, pp. 797–808, June 1974.

BARR, D. I. H., "Explicit Working for Turbulent Pipe Flow Problems," *ASCE J. Hydraul. Div.*, v. 102, n. HY5, pp. 667–673, May 1976.

BAUER, W. J., LOUIS, D. S., and VOORDUIN, W. L., "Basic Hydraulics, Part I: Closed Conduit," in *Handbook of Applied Hydraulics*, ed. C. V. Davis and K. E. Sorenson, McGraw-Hill Book Company, New York, 1969.

BENEFIELD, L. D., JUDKINS, J. F., and PARR, A. D., *Treatment Plant Hydraulics for Environmental Engineers*, Prentice-Hall, Inc., Englewood Cliffs, N.J., 1984.

BLASIUS, H., "Das Ähnlichkeitsgesetz bei Reibungsvorgängen in Flüssigkeiten" (The Law of Simulitude for Frictional Motions in Fluids), Forschungsheft Des Vereins Detscher Ingenieure, n. 131, Berlin, 1913.

BRATER, E. F., and KING, H. W., *Handbook of Hydraulics*, 6th ed., McGraw-Hill Book Company, New York, 1976.

BREBBIA, C. A., and FERRANTE, A. J., *Computational Hydraulics*, Butterworth & Company (Publishers) Ltd., London, 1983.

CHADWICK, A., and MORFETT, J., *Hydraulics in Civil Engineering*, George Allen & Unwin (Publisher) Ltd., London, 1986.

CHAUDHRY, M. H., *Applied Hydraulic Transients*, 2nd ed., Van Nostrand Reinhold Company, Inc., New York, 1987.

CHEREMISINOFF, N. P., *Fluid Flow: Pumps, Pipes, and Channels*, Ann Arbor Science Publishers, Ann Arbor, Mich., 1981.

DARCY, H., "Sur des recherches expérimentales relatives au mouvement des eaux dans les tuyaux" ("Experimental Researches on the Flow of Water in Pipes"), *Comptes Rendus Des Séances De L'Académie Des Sciences,* v. 38, pp. 1109–1121, 1854 (Name of Darcy is associated with the equation for his research on flow in pipes although Weisbach first formulated the equation).

DAUGHERTY, R. L., FRANZINI, J. B., and FINNEMORE, E. J., *Fluid Mechanics with Engineering Applications,* 8th ed., McGraw-Hill Book Company, New York, 1985.

FEATHERSTONE, R. E., and NALLURI, C., *Civil Engineering Hydraulics,* Granada Publishing Ltd., London, 1982.

FOX, J. A., *Hydraulic Analysis of Unsteady Flow in Pipe Networks,* John Wiley & Sons, Inc., New York, 1977.

HALLIWELL, A. R., "Velocity of a Waterhammer Wave in an Elastic Pipe," *ASCE J. Hydraul. Div.,* v. 89, n. HY4, pp. 1–21, 1963.

HWANG, N. H. C., and HITA, C. E., *Fundamentals of Hydraulic Engineering Systems,* 2nd ed., Prentice-Hall, Inc., Englewood Cliffs, N.J., 1987.

HYDRAULIC RESEARCH STATION, *Charts for the Hydraulic Design of Channels and Pipes,* 5th ed., Hydraulic Research Station Ltd., London, 1983.

JAIN, A. K., "Accurate Explicit Equation for Friction Factor," *ASCE J. Hydraul. Div.,* v. 102, n. HY5, pp. 674–677, May 1976.

JAIN, A. K., MOHAN, D. M., and KHANNA, P., "Modified Hazen-Williams Formula," *ASCE J. Environ. Eng. Div.,* v. 104, n. EE1, pp. 137–146, Feb. 1978.

KARASSIK, I. J., and CARTER, R., *Centrifugal Pumps: Selection, Operation and Maintenance,* McGraw-Hill Book Company, New York, 1960.

KARASSIK, I. J., KRUTZSCH, W. C., and FRASER, W. H., *Pump Handbook,* McGraw-Hill Book Company, New York, 1976.

KERR, S. L., "New Aspects of Maximum Pressure Rise in Closed Conduits," *Trans. Am. Soc. Mech. Eng.,* v. HYD-51-3, pp. 13–30, 1929.

LAI, R. Y. S., and LEE, K. K., "Moody Diagram for Direct Pipe Diameter Calculation," *ASCE J. Hydraul. Div.,* v. 101, n. HY10, pp. 1377–1379, Oct. 1975.

LAL, J., *Hydraulics,* Metropolitan Book Co. Pvt. Ltd., New Delhi, 1958.

LI, W. H., "Direct Determination of Pipe Size," *ASCE Civ. Eng.,* p. 74, June 1974.

LINSLEY, R. K., and FRANZINI, J. B., *Water Resources Engineering,* 3rd ed., McGraw-Hill Book Company, New York, 1979.

LYLE, N. H., and WEINBERG, G., "Pipeline Network Analysis by Electronic Digital Computer," *J. Am. Water Works Assoc.,* v. 49, pp. 517–529, 1957.

MCCLAIN, C. H., *Fluid Flow in Pipes,* 2nd ed., Industrial Press, Inc., New York, 1963.

MOODY, L. F., "Friction Factors for Pipe Flow," *Trans. ASME,* v. 66, p. 671, 1944.

MOTT, R. L., *Applied Fluid Mechanics,* 2nd ed., Charles E. Merrill Publishing Company, Columbus, Ohio, 1979.

POWELL, R. W., "Diagram Determines Pipe Sizes Directly," *ASCE Civ. Eng.,* pp. 45–46, Sept. 1950.

PRANDTL, L., "The Mechanics of Viscous Fluids," v. III, div. G, in *Aerodynamics Theory* by Durand, W. F., (editor-in-chief), Springer-Verlag, Berlin, 1935 (Prandtl modified the expression developed by von Kármán in 1930)

RICH, G. R., *Hydraulic Transients,* 2nd ed., Dover Publications, Inc., New York, 1963.

SIMON, A. L., *Practical Hydraulics,* John Wiley & Sons, Inc., New York, 1976.

SMITH, P. D., *Basic Hydraulics,* Butterworth & Company (Publishers) Ltd., London, 1982.

STEEL, E. W., and MCGHEE, T. J., *Water Supply and Sewage,* 5th ed., McGraw-Hill Book Company, New York, 1979.

STREETER, V. L., "Steady Flow in Pipes and Conduits," in *Engineering Hydraulics,* ed. H. Rouse, John Wiley & Sons, Inc., New York, 1950.

STREETER, V. L., *Fluid Mechanics,* 5th ed., McGraw-Hill Book Company, New York, 1971.

STREETER, V. L., and WYLIE, E. B., *Fluid Mechanics,* 8th ed., McGraw-Hill Book Company, New York, 1985.

SWAMEE, P. K., and JAIN, A. K., "Explicit Equations for Pipe-Flow Problems," *ASCE J. Hydraul. Div.,* v. 102, n. HY5, pp. 657–664, May 1976.

WATTERS, G. Z., *Analysis and Control of Unsteady Flow in Pipe Lines,* 2nd ed., Butterworth Publishers, Woburn, Mass., 1984.

WEISBACH, J., *Lehrbuch Der Ingenieur-und Maschinenmechanik (Textbook of Engineering Mechanics),* Brunswick, Germany, 1845.

WHITAKER, S., *Introduction to Fluid Mechanics,* Prentice-Hall, Inc., Englewood Cliffs, N.J., 1968.

WOOD, D. J., "An Explicit Friction Factor Relationship," *ASCE Civ. Eng.,* pp. 60–61, Dec. 1966.

WOOD, D. J., and RAYES, A. G., "Reliability of Algorithms for Pipe Network Analysis," *ASCE J. Hydraul. Div.,* v. 107, n. 10, 1981.

WYLIE, E. B., and STREETER, V. L., *Fluid Transients,* FEB Press, Ann Arbor, Mich., 1983.

CHAPTER TWELVE

AL-LAYLA, M. A., AHMAD, S., and MIDDLEBROOKS, E. J., *Handbook of Wastewater Collection and Treatment: Principles and Practice,* Garland Publishing, Inc., New York, 1980.

AMERICAN SOCIETY OF CIVIL ENGINEERS AND WATER POLLUTION CONTROL FEDERATION, *Design and Construction of Sanitary and Storm Sewers,* Manual on Engineering Practice No. 37, ASCE, New York, 1969.

AMERICAN SOCIETY OF CIVIL ENGINEERS, *Some Notes on the Rational Method of Storm Drain Design,* ASCE Urban Water Resources Research Program, TMN.6, ASCE, New York, 1969.

AMERICAN SOCIETY OF CIVIL ENGINEERS, *Pipeline Design for Water and Wastewater,* Committee on Pipeline Planning, Pipeline Division, ASCE, New York, 1975.

AMERICAN SOCIETY OF CIVIL ENGINEERS AND WATER POLLUTION CONTROL FEDERATION, *Gravity Sanitary Sewer Design and Construction,* Manual on Engineering Practice No. 60, ASCE, New York, 1982.

BARTLETT, R. E., *Surface Water Sewerage,* Halsted Press, New York, 1976.

BARTLETT, R. E., *Public Health Engineering: Sewerage,* 2nd ed., Applied Science Publishers Ltd., Barking, Essex, England, 1979.

BELL, F. C., "Generalized Rainfall–Duration–Frequency Relationships," *ASCE J. Hydraul. Div.,* v. 95, n. HY1, pp. 311–327, 1969.

BODHAINE, G. L., *Measurement of Peak Discharge at Culverts by Indirect Methods,* Techniques of Water Resources Investigations, Chap. A3, Book 3, U.S. Geological Survey, Washington, D.C., 1982.

BRAS, R. L., and PERKINS, F. E., "Effects of Urbanization on Catchment Response," *ASCE J. Hydraul. Div.,* v. 101, n. HY3, 1975.

CEDERGREN, H. R., *Drainage of Highway and Airfield Pavements,* John Wiley & Sons, Inc., New York, 1974.

CHOW, V. T., "Hydrologic Design of Culverts," *ASCE J. Hydraul. Div.,* v. 88, n. HY2, pp. 39–55, 1962.

CHOW, V. T. (ed.), *Handbook of Applied Hydrology,* Sec. 20, McGraw-Hill Book Company, New York, 1964.

ESCRITT, L. B., *Sewerage and Sewage Disposal,* 4th ed., Vol. II, Macdonald & Evans Ltd., Plymouth, Devon, England, 1972.

FAIR, G. M., GEYER, J. C., and OKUN, D. A., *Water and Wastewater Engineering,* Vol. 1, John Wiley & Sons, Inc., New York, 1966.

FEDERAL AVIATION AGENCY, *Airport Drainage,* Advisory Circular 150/5320-5A, U.S. Government Printing Office, Washington, D.C., 1965.

GEYER, J. C., and LENTZ, J. J., "An Evaluation of the Problems of Sanitary Sewer System Design," *J. Water Poll. Control Fed.,* v. 38, pp. 1138–1147, 1966.

GUPTA, B. R. N., *Design Aids for Public Health Engineers,* Halsted Press, New York, 1986.

HROMADKA, T. V., CLEMENTS, J. M., and SALUJA, H., *Computer Methods in Urban Watershed Hydraulics,* Lighthouse Publications, Mission Viejo, Calif., 1984.

HROMADKA, T. V., DURBIN, T. J., and DeVRIES, J. J., *Computer Methods in Water Resources,* Lighthouse Publications, Mission Viejo, Calif., 1985.

JOHNSON, R. E. L., "Development of Sanitary Sewer Design Criteria," *J. Water Poll. Control Fed.,* v. 37, pp. 1597–1606, 1965.

KINORI, B. Z., *Manual of Surface Engineering,* vol. I, Elsevier Science Publishing Co., Amsterdam, 1970.

LINSLEY, R. K., and FRANZINI, J. B., *Water Resources Engineering,* 3rd ed., McGraw-Hill Book Company, New York, 1979.

LUTHIN, J. N., Drainage Engineering, John Wiley & Sons, Inc., New York, 1966.

McCUEN, R. H., *A Guide to Hydrologic Analysis Using SCS Methods,* Prentice-Hall, Inc., Englewood Cliffs, N.J., 1982.

McCUEN, R. H., WONG, S. L., and WALTER, J. R., "Estimating Urban Time of Concentration," *ASCE J. Hydraul. Div.,* v. 110, n. 7, 1984.

McPHERSON, M. B., *Some Notes on the Rational Method of Storm Drain Design,* Tech. Memo. No. 6, ASCE, Water Resources Research Program, Harvard Univ., Cambridge, MA, 1969.

NATIONAL ASSOCIATION OF HOME BUILDERS, *Residential Wastewater Systems,* NAHB, Washington, D.C., 1980.

OKUN, D. A., and PONGHIS, G., *Community Wastewater Collection and Disposal,* World Health Organization, Geneva, 1975.

OVERTON, D. E., and MEADOWS, M. E., *Storm Water Modeling;* Academic Press, Inc., New York, 1976.

SAUER, V. B., et al., *Flood Characteristics of Urban Watersheds in the United States,* Water Supply Paper 2207, U.S. Geological Survey, Washington, D.C., 1983.

SCHAAKE, J. C., GEYER, J. C., and KNAPP, J. W., "Experimental Examination of the Rational Method," *ASCE J. Hydraul. Div.,* v. 93, n. HY6, pp. 353–363, 1967.

SCHILFGAARDE, J. V. (ed.), *Drainage for Agriculture,* American Society of Agronomy, Madison, Wis., 1974.

SINGH, V. P., *Hydrologic Systems: Rainfall-Runoff Modeling,* Vol. 1, Prentice-Hall, Inc., Englewood Cliffs, N.J., 1988.

SOKOLOV, A. A., RANTZ, S. E., and ROCHE, M., *Floodflow Computation: Methods Compiled from World Experience,* Studies and Report in Hydrology No. 22, UNESCO, Paris, 1976.

STANLEY, W. E., and KAUFMAN, W. J., "Sewer Capacity Design Practice," *J. Boston Soc. Civ. Eng.,* v. 317, 1953.

STEEL, E. W., and McGHEE, T. J., *Water Supply and Sewerage,* 5th ed., McGraw-Hill Book Company, New York, 1979.

STEPHENSON, D., *Stormwater Hydrology and Drainage,* Developments in Water Science Series No. 14, Elsevier Science Publishing Co., Amsterdam, 1981.

THOLIN, A. L., and KEIFER, C. J., "The Hydrology of Urban Runoff," *Trans. ASCE,* v. 125, p. 1308, 1960.

URBAN LAND INSTITUTE AMERICAN SOCIETY OF CIVIL ENGINEERS, National Association of Home Builders, *Residential Storm Water Management: Objective, Principles and Design Considerations,* NAHB, Washington, D.C., 1975.

U.S. BUREAU OF RECLAMATION, *Design of Small Dams,* 2nd ed., U.S. Dept. of Interior, U.S. Government Printing Office, Washington, D.C., 1977.

References

U.S. Bureau of Reclamation, *Design of Small Canal Structures,* U.S. Dept. of the Interior, Denver, Colo., 1978.

U.S. Bureau of Reclamation, *Drainage Manual,* U.S. Dept. of Interior, U.S. Government Printing Office, Denver, Colo., 1984.

U.S. Department of Transportation, *Hydraulic Design of Improved Inlets for Culverts,* Hydraulic Engineering Circular 13, Federal Highway Administration, Washington, D.C., 1972.

U.S. Department of Transportation, *Guidelines for the Design of Subsurface Drainage Systems for Highway Structural Sections,* Federal Highway Administration, Washington, D.C., 1973.

U.S. Department of Transportation, *Design of Urban Highway Drainage, The State-of-the-Art,* Federal Highway Administration, Washington, D.C., 1979.

U.S. Soil Conservation Service, *Engineering Field Manual for Soil Conservation Practices,* Chap. 14, U.S. Dept. of Agriculture, Washington, D.C., 1975a.

U.S. Soil Conservation Service, *Urban Hydrology for Small Watersheds,* Technical Release 55, U.S. Dept. of Agriculture, Washington, D.C., 1975b.

U.S. Soil Conservation Service, *Urban Hydrology for Small Watersheds,* revised Technical Release 55, U.S. Dept. of Agriculture, Washington, D.C., 1986.

Viessman, W., Knapp, J. W., and Lewis, G. L., *Introduction to Hydrology,* 2nd ed., IEP, A Dun-Donnelley Publisher, New York, 1977.

Wandle, S. W., Jr., *Estimating Peak Discharges of Small, Rural Streams in Massachusetts,* Water Supply Paper 2214, U.S. Geological Survey, Washington, D.C., 1983.

Whipple, W., et al., *Stormwater Management in Urbanizing Areas,* Prentice-Hall, Inc., Englewood Cliffs, N.J., 1983.

White, J. B., *Wastewater Engineering,* Edward Arnold (Publishers) Ltd., London, 1978.

GLOSSARY

Joint Editorial Board of the American Public Health Association, American Society of Civil Engineers, American Water Works Association, and Water Pollution Control Federation, *Glossary: Water and Wastewater Control Engineering,* 3rd ed., 1981.

Langbein, W. B., and Iseri, K. T., *General Introduction and Hydrologic Definitions,* Water Supply Paper 1541-A, U.S. Geological Survey, Washington, D.C., 1960.

Lohman, S. W., et al., *Definition of Selected Groundwater Terms,* Water Supply Paper 1988, U.S. Geological Survey, Washington, D.C., 1972.

Meinzer, O. E., *Outline of Ground-Water Hydrology With Definitions,* Water Supply Paper 494, U.S. Geological Survey, Washington, D.C., 1923 (reprint 1960)

UNESCO/WMO, *International Glossary of Hydrology,* Secretariat of World Meteorological Organization, Switzerland, 1974.

Answers to Even-Numbered Problems

Chapter 2

2.2 (a) 45.8 thousands (b) 45.9 thousands

2.4 39.75 yrs

2.6 112.5 yrs

2.8 (a) 118.69 thousands (b) 110.0 thousands

2.10 22.3 thousands, 24.0 thousands

2.12 472.5 gpcd

2.14 3.74 MG

2.16 (a) 15.75 mgd (b) 15.75 (c) 15.75 (d) low lift 21.0 (e) 25.40 (f) 32.77

2.18 29 in.

2.20 18.56 in.

2.22 $P = 4210$ Kw, $E = 25.7$ Gwh, CF $= 0.70$

2.24 61%

Chapter 3

3.2 7.15 acre-ft, 3.65 acre-ft

3.4 (−) 3 in.

3.6 4.25 in.

3.8 Breakpoint at 1970
Data from 1960 to 1970 adjusted by factor 0.78

3.10 (a) 4.29 in. (b) 4.36 in.

3.12 (a) 800 mm (b) 881 mm

3.14

	5 min	10 min	15 min	20 min	30 min	60 min
5 yr	4.24 in./hr	3.50	2.96	2.50	1.80	
3 yr	3.78	3.00	2.56	2.16	1.56	

3.16 (a) 0.42 in. (b) 0.42 in.

3.18 0.44 in./day

3.20

Mar	Apr	May	June	July	Aug.	Sept.	Oct.	Nov.
1.69 cm	3.84	7.04	12.90	14.92	12.98	4.91	3.06	1.12

Seasonal = 19.71 in.

3.22

Time, min	0	10	20	40	60	80
f_p, in./hr	9	4.85	2.91	1.57	1.28	1.22
F_p, in.		1.16	1.81	2.56	3.04	3.46
storm, f_p, in./hr				5.35	2.11	1.40

3.24

Period, min	0–20	20–40	40–60	60–80	80–100	100–120	120–140
Runoff, in.	0	0	0.75	0.78	0	0	0.16

3.26

Pd, min	0–20	20–40	40–45	45–60	60–80	80–100	100–120	120–140
RO, in.	0	0	0	0.19	0.69	0.10	0.08	0.58

3.28

Period, min	0–20	20–40	40–60	60–80	80–100	100–120	120–140
Runoff, in.	0	0	0.51	0.72	0.20	0.18	0.74

3.30 (a) 6.17 in. (b) 2.5 in./hr

3.32 (a) 1.24 in./hr (b) 10.25 in.

3.34

	Zone 1	Zone 2	Zone 3
First day	0.46 in.	0.31	0.15
Second day	0.46	0.29	0.12

Chapter 4

4.2 0.346, 0.53

4.4 10.5 cm^3 per unit area, 0.21 cm^3 per unit vol.

4.6 (a)

Head (–), cm	0	20	40	60	80	100
w, %	53.2	52.5	50.4	46.2	43.12	42.28

(b) 9.43 cm^3

4.8 11000 m^3

4.10 42%, 58%

4.12 2.31×10^{-4}

4.14 6.6×10^{-9} ft^2

4.16 78×10^{-4} cm^2/s

4.18 156.3 m^3/day

4.20 5.95 m/day

4.22 $R_e = 0.18$, no departure from Darcy's law

Chapter 5

5.2 (a) 109 m^3/d/m

(b)

Dist., m	0	1	6	9.5	...	28.6	30.0
u, KN/m^2	99.1	90.2	81.2	73.2		27.7	18.7

5.4 (a) 0.057 cft/min/ft (b) 749 psf, 686 psf

5.6 6.75 m^3/d/m

5.8 $0.05 \text{ m}^3/\text{s}$

5.10 $0.2 \text{ m}^3/\text{s}$

5.12   12.15 yrs

5.14   39.8 days

5.16 $0.091 \text{ m}^3/\text{s}$

5.18   5.32 m

5.20   1.19 ft

5.22 $T = 3.67 \text{ gpm/ft}, S = 6.7 \times 10^{-3}$

5.24 $Q = 2826 \text{ m}^3/\text{d}, S = 2.1 \times 10^{-4}$

5.26   18 ft

5.28 $Q = 2800 \text{ m}^3/\text{d}, S = 2.0 \times 10^{-4}$, valid after 2 min.

5.30 $T = 71.4 \text{ gpm/ft}, S = 18.6 \times 10^{-5}$

5.32 $T = 46.56 \text{ gpm/ft}, S = 0.0032$

5.34   6 wells, spacing = 350 ft

5.36 $226 \text{ m}^2/\text{d}$

5.38   5.6 ft

Chapter 6

6.2   1.99 ft/s, 6.97 cfs/ft

6.4   1.97 ft/s, 6.908 cfs/ft

6.6   3.33 ft/s

6.8   **(a)** 4.27 m   **(b)** 4.42 m, 6.98 m

6.10   111 cfs

6.12   109 cfs

6.14   110 cfs

6.16 $2 \text{ m}^3/\text{s}$

6.18 $8.52 \text{ m}^3/\text{s}$

6.20   159 cfs

6.22   47.11 cfs

6.24   1 : 4

6.26   845 cfs

6.28   11.86 cfs

6.30   **(a)** 89.56 cfs   **(b)** 89.51 cfs

6.32   2 min 30 sec

6.34   98.35 cfs

6.36   **(a)** 24 cm   **(b)** 25 cm

6.38 $Q = 2.65(\text{h} - 0.61)^{1.47}$

6.40 $Q = 92.5(\text{h} - 0.1)^{2.39}$

Chapter 7

7.2   0.73 cm

8.18

P_r, %	1	5	10	50	80	90
Q, m³/s	1155	803	661	334	231	168

8.20

P	1	10	50	90	99
$Q_{5\%}$, m³/s	683	401	210	117	74
$Q_{95\%}$, m³/s	427	289	166	84	46

8.22

P_c, %	5	6	20.7	35.8	56.0	75.4	89.7
Q, cfs	26500	25000	20000	17500	15000	12500	10000

$T = 500$ yrs

8.24 $Q_{peak} = 10790$ cfs, $Q_{50} = 5400$, $Q_{75} = 8090$, $W_{50} = 4.5$ hr, $W_{75} = 2.5$ hr

8.26 0.1%

8.28 1450 cfs

8.30 567 cfs

8.32 $Q_{10} = 103.0$ cfs

8.34

$P(X<)$	0.1	0.2	0.6	0.8	0.95	0.999
Q, cfs	106.9	114.0	140.1	168.0	226.4	507.8

8.36

Time, hr	0	1.0	2.0	3.0	4.0	5.0	6.0
I, cfs	0	70	280	140	40	0	0
O, cfs	0	15	105	185	140	66	29
El., ft	15.30	16.93	18.90	18.90	17.90	16.75	16.03

8.38 $x = 0.38$, $K = 26$ hrs

8.40

Time, hr	6 am	noon	6 pm	midnight	$\cdots$	4th, 6am	noon
O, cfs	150	155.4	192.0	338.9		108.3	85.6

Chapter 9

9.2 5600 MG

9.4 117 yrs

9.6 18.6 cft/d/ft

9.8 14.3 cft/d/ft

9.10 1.05 cft/d/ft

9.12 21.6 cft/d/ft

9.14 8 cft/d/ft

9.16 $V = 375.6$ K, $H = 232.4$ K, $d = 56.7$ ft

9.18 $SF = 5.4$

9.20 Middle-third law is satisfied, no tension
$p_{heel} = -2.1$ Ksf, $p_{toe} = -12.1$ Ksf

9.22 For proof, determine the distance to the resultant and equate to 2/3L.

9.24 $p_{max} = -12.1$ Ksf < -20, $\tau = -5.2$ Ksf, ok

9.26

Depth, ft	0	20	40	60	100	140	160
Thickness, ft	2.5	2.5	2.69	3.50	4.07	3.22	2.26

9.28

D/S quad. x, m	0.5	1.0	1.5	2.0
y, m	0.06	0.23	0.48	0.82
U/S quad. x, m	0.2	0.4	0.56	
y, m	0.02	0.08	0.19	

9.30 13100 cfs

9.32 15705 cfs

9.34 Crest height = 114.6 ft

D/S quad. x, ft	5	10	15	20	25
y, ft	1.21	4.35	9.20	15.67	23.68
U/S quad. x, ft	1	2	3		
y, ft	0.09	0.40	1.13		

9.36 $S_c = 0.0036$, $F_r = 1.32$, stable rapid flow

9.38 $\alpha = 11°$, transition $L = 266$ ft

9.40

x from D/S, ft	30	60	90	110
Δy, ft	0.27	0.19	0.11	0.03

9.42 $R_s = 8.5$ ft

H_a, ft	20	22	24	28	30
R, ft	5.72	5.59	5.47	5.26	5.17

9.44

H_a, ft	11	13	15	17	20
R, ft	1.59	1.52	1.47	1.42	1.37

Chapter 10

10.2 (a) $R = 1.35$ ft, $D = 5.31$, $Z_c = 34.43$, $Z_n = 18.26$
(b) 4.39 ft

10.4 $y_2 = 2.8$ ft, $v_2 = 13.86$ f/s

10.6 $y_1 = 7.53$ ft, $y_2 = 3.0$, $v_2 = 19.8$ f/s

10.8 (b) $y_c = 2.55$ ft (c) $y_2 = 1.7$ ft (d) supercritical

10.10 $y_c = 2.41$ m, $v_c = 3.44$ m/s

10.12 $y_c = 1.3$ m, $v_c = 3.0$ m/s

10.14 10.41 cfs

10.16 $y = 8.26$ ft, $Q = 857$ cfs

10.18 (a) $y_n = 2.1$ m (b) $S_c = 0.014$ (c) subcritical

10.20 $y_n = 3.8$ ft, freeboard = 2.5 ft

10.22 $y_n = 5.4$ ft, freeboard = 3.0 ft

10.24 (a) $d_0 = 1$ ft (b) $y = 0.3$ ft, $v = 3$ f/s (c) $y = 0.15$ ft

10.26 $d_0 = 130$ mm

10.28 $y = 1.45$ m, $b = 10$ m

10.30

Depth, ft	0.6	0.8	1.0	1.2	1.4		2.0	2.2
Dist., ft		20	41	63	85	⋯		146

10.32

Depth, ft	0.6	0.8	1.0	1.2	1.4		2.0	2.2
Dist., ft		18	38	59	80	⋯		140

10.34 $D_1 = 2.10$ m, $E = 0.03$ m

Chapter 11

11.2 5.72 m

11.4 Slope = 0.026, piezometric head = 72.05 ft

11.6 0.017

11.8 50.55 ft

11.10 $d = 8.5$ in.

11.12 2.0 ft

11.14 0.52 cfs

11.16 $0.66 \text{ m}^3/\text{s}$

11.18 (a) $0.21 \text{ m}^3/\text{s}$ (b) BHP = 125 Kw

11.20 2.05 cfs

11.22 $h_f = 4 \text{ m}$, $Q_1 = 0.48 \text{ m}^3/\text{s}$, $Q_2 = 0.17$, $Q_3 = 0.35$

11.24 3.19 cfs

11.26

Pipe	Q, gpm	Head loss, ft	Node	Pressure, ft.
AB	360	4.5	A	132.4
BC	618	12.1	B	177.9
CD	382	20.2	C	40.0
DE	382	20.2	D	169.8
EF	440	26.3	E	99.6
AF	440	6.5		
BE	978	28.4		

11.28 At the valve: 16.6 m, at 500 m: 11.4 m

11.30 410.5 m

11.32

Time, s	0	2.5	5.0	10.0	15.0	20.0
Head, m	10	16.03	28.87	55.60	48.1	43.0

11.34 1739, radial flow pump

11.36 Capacity reduction 35%, head 58%, power 72.5%

11.38

Q	H	P
2850	180.5	150
1900	199.5	123
950	206.7	92

11.40 1125 rpm, 95 ft

11.42 At the operating point, $Q = 1800$ gpm, $H = 223$ ft = 76%, pump ok, $HP = 134$

11.44 (a) At the operating point, $Q = 1040$ gpm, $H = 204$ ft
(b) Head = 102 ft each (c) $HP = 77.7$

11.46 (a) 21.1 (b) 0.21 (c) no cavitation problem

Chapter 12

12.2 29.4 mgd

12.4 $d_0 = 500$ mm at slope 0.0018

12.6

Block	A	B	C	D	E	F	G	H
d_0, in.	4	5	5	4	4	7	8	8
Grade	0.011	0.010	0.012	0.012	0.012	0.005	0.003	0.012

12.8 Invert levels
MH 7, 90.29 ⟶ MH 6, 86.79   MH 6, 86.69 ⟶ MH 8, 82.73
MH 5, 91.17 ⟶ MH 6, 86.88

12.10 $\overline{C} = 0.775$

12.12 12.7 min

12.14 $Q_5 = 95.5$ cfs, $Q_{10} = 121$ cfs, $Q_{20} = 143$ cfs

12.16

Time, hr	11.9	12.0	12.1	...	12.8	13.0	13.2
Q, cfs	28.7	54.8	88.9		27.5	18.0	13.6

12.18

MH	1	2	3
Q, cfs	18	46.9	58.9

12.20

Manhole from	Block A	B & C	D	E	F	G
Q, cfs	24.0	57.0	21.4	35.5	77.3	83.0

12.22

Location	Ave. D	Ave. D	Ave. D	St. 1	Ave. C	Ave. C	Ave. C	St. 1
Manhole	1–2	2–3	3–4	4–8	5–6	6–7	7–8	8–12
Q, m³/s	0.111	0.294	0.413	0.466	0.076	0.198	0.253	0.753
d_0, mm	400	580	670	495	295	420	420	565

Location	Ave. B	Ave. B	Ave. B	St. 1
Manhole	9–10	10–11	11–12	12–13
Q, m³/s	0.117	0.297	0.417	1.18
d_0, mm	345	445	490	705

12.24   15 in.

12.26   400 ft

12.28   1.24 cfs

12.30 $d = 4.5$ ft

12.32 $d = 10$ ft

12.34   94000 cft

12.36   83560 cft

Index